Procedures

Derivative for	Derivative Type, Procedure No.	Page
Alcohols	Arylurethanes, Procedure 1	156
Alcohols	3,5-Dinitrobenzoates, Procedure 2	156
Alcohols	Benzoates, *p*-Nitrobenzoates, Procedure 3	157
Alcohols	3-Nitrophthalates, Procedure 4	158
Alcohols, polyhydric	Acetates, Procedure 5	160
Aldehydes	Oxidation, Procedure 6	178
	Permanganate, Procedure 6a	178
	Hydrogen peroxide, Procedure 6b	178
Aldehydes	Dimedon, Procedure 10	180
Aldehydes, ketones	Semicarbazones, Procedure 7	179
Aldehydes, ketones	*p*-Nitrophenylhydrazones, Procedure 8	179
Aldehydes, ketones	2,4-Dinitrophenylhydrazones, Procedure 9	179
Aldehydes, ketones	Oximes, Procedure 11	181
Aldehydes, ketones	Reduction of diaryl ketones with sodium borohydride, Procedure 11a	182
Amides	9-Acylamidoxanthenes, Procedure 38	284
Amides	Hydrolysis, Procedure 37	282
Amines	Hinsberg test, Procedure 16	230
Amines (primary, secondary)	Amides (including Schotten-Baumann method), Procedure 17	232
Amines (primary, secondary)	Sulfonamides, Procedure 18	233
Amines (primary, secondary)	Phenylthioureas, Procedure 19	234
Amines (tertiary)	Quaternary ammonium salts, Procedure 20	235
Amines ($ArNR_2$)	Nitroso compounds, Procedure 21	236
Amines (tertiary)	Picrates, Procedure 22	236
Amines (tertiary)	Chloroplatinates, Procedure 23	237
Amines (primary, secondary)	Amides (via acid chloride), Procedure 35	271
Amino acids	*p*-Toluenesulfonates, Procedure 24	244
Amino acids	*N*-2,4-Dinitrophenyl derivatives, Procedure 25	244
Anilines	See amino derivatives	230–272
Anilines	Anilides, Procedure 36	272
Anilines (nitrated)	Hydrolysis, Procedure 37	282
Aromatics	Nitration, Procedure 26	249
Aromatics (ArR)	Side-chain oxidation, Procedure 27	253
Aromatics	Aroylbenzoic acids, Procedure 28	255
Aromatics	Friedel-Crafts acylation, Procedure 29	256
Aromatics	Chlorosulfonation, Procedure 30	259
Aromatics	Sulfonamides, Procedure 3 (via Procedure 30)	260
Aromatics	Picrates, Procedure 22	236
Carboxylic acids	Amides, Procedure 35	271
Carboxlylic acids	Anilides, *p*-Toluidides, and *p*-Bromoanilides, Procedure 36	272
Carboxylic acids	Neutralization equivalent, Procedure 32	268
Carboxylic acids	Phenacyl and substituted phenacyl esters, Procedure 33	270

Continued on next page

Procedures

Derivative for	Derivative Type, Procedure No.	Page
Carboxylic acids	Phenylhydrazides and phenylhydrazonium salts, Procedure 34	271
Carboxylic acids (and their salts)	Arylthiuronium salts, Procedure 39	287
Esters	Saponification and acid hydrolyses, Procedure 40	293
Esters	Saponification equivalent, Procedure 41	296
Esters	*N*-Benzylamides, Procedure 42	300
Esters	Acid hydrazides, Procedure 43	300
Esters	3,5-Dinitrobenzoates, Procedure 44	300
Ethers	3,5-Dinitrobenzoates, Procedure 45	306
Ethers (aromatic)	Bromination, Procedure 46	308
Ethers (aromatic)	Chlorosulfonation, Procedure 47	308
Ethers (aromatic)	Sulfonamides, Procedure 48 (via Procedure 47)	308
Ethers (phenolic)	Picrates, Procedure 49	309
Halides	Grignard reagents, organomercurials, Procedure 12	214
Halides	Aryl urethane, Procedure 13 (via Procedure 12)	214
Halides	Conversion to nitriles, Procedure 14a	215
Halides	α-Naphthyloxy ethers, Procedure 14	215
Halides	*S*-Alkyl isothiuronium picrates, Procedure 15	217
Ketones	See aldehydes, ketones	178–182
Nitriles	Hydrolysis, Procedure 37	282
Nitriles	Acid hydrolysis, Procedure 51	328
Nitriles	Controlled hydration to an amide, Procedure 51a	329
Nitriles	Reduction with sodium in alcohol, Procedure 52	329
Nitriles	Oxidation to the aldehyde, Procedure 52a	330
Nitriles	α-Iminoalkylmercaptoacetic acid hydrochlorides, Procedure 53	331
Nitro compounds	Tin/hydrochloric acid reduction to amines, Procedure 50	322
Nitro compounds	Base-induced condensation with benzaldehyde, Procedure 50b	324
Nitro compounds (poly)	Selective reduction to nitroamines, Procedure 50a	322
Olefins (alkenes)	Oxidative hydroboration, Procedure 11b	198
Phenols	Aryloxyacetic acids, Procedure 59	354
Phenols	Bromination, Procedure 60	354
Phenols	Phenylurethanes, Procedure 61	355
Phenolic ethers	Picrates, Procedure 49	309
Sulfonamides	Hydrolysis, Procedure 57	345
Sulfonamides	*N*-Xanthylsulfonamides, Procedure 58	346
Sulfonic acids	Neutralization equivalent, Procedure 32	268
Sulfonic acids (and salts)	Benzylthiuronium sulfonates, Procedure 53a	341
Sulfonic acids (and salts)	Sulfonyl chlorides and sulfonamides, Procedure 54	342
Sulfonic acids (and salts)	*p*-Toluidine sulfonates, Procedure 55	342
Sulfonic acids (amino substituted)	Chloroaryl sulfonamides and sulfonanilides, Procedure 56	343
Sulfonyl halides	Sulfonamides, Procedure 48	308

THE SYSTEMATIC IDENTIFICATION OF ORGANIC COMPOUNDS

Other books by the same authors

Robert M. Silverstein, G. Clayton Bassler, and Terence C. Morrill
Spectrometric Identification of Organic Compounds
Ralph L. Shriner and Rachel H. Shriner
Organic Syntheses. Cumulative Indices to Collective Volumes I, II, III, IV, and V.

THE SYSTEMATIC IDENTIFICATION OF ORGANIC COMPOUNDS

a laboratory manual

6th. ed.

Ralph L. Shriner

Visiting Professor of Chemistry
Southern Methodist University, Dallas, Texas 75275
Formerly Professor and Head of the
Department of Chemistry
University of Iowa, Iowa City, Iowa

The late Reynold C. Fuson

Formerly Distinguished Visiting Professor
University of Nevada, Reno, Nevada 89507
Formerly Professor of Organic Chemistry and
Member of the Center for Advanced Study
University of Illinois, Urbana, Illinois

David Y. Curtin

Professor of Chemistry
University of Illinois
Urbana, Illinois 61701

Terence C. Morrill

Professor of Chemistry
Rochester Institute of Technology
Rochester, N.Y. 14623

JOHN WILEY & SONS

New York Chichester Brisbane Toronto

Library of Congress Cataloging in Publication Data

Main entry under title:
The Systematic identification of organic compounds.

Fifth ed., published in 1964, by R. L. Shriner,
R. C. Fuson, and D. Y. Curtin.
Includes bibliographical references and index.
1. Chemistry, Organic—Laboratory manuals.
I. Shriner, Ralph Lloyd, 1899- II. Shriner, Ralph
Lloyd, 1899- The systematic identification of
organic compounds.

QD261.S5 1979 547′.34 79-13365
ISBN 0-471-78874-0

Printed in the United States of America

10 9 8 7 6 5 4 3

DEDICATION

Reynold C. Fuson (June 1, 1895–August 4, 1979)

We dedicate this edition to our coauthor, Reynold C. Fuson, who died in the summer of 1979. Bob Fuson, as he was called by his many friends, was a distinguished organic chemist known for original research and was the author of several books. Throughout his long career, he gave invaluable advice and guidance to hundreds of students, both undergraduate and graduate.

PREFACE

The 43 years since the publication of the first edition of this textbook have seen a proliferation of new techniques for the separation, purification, identification, and complete characterization of organic compounds. Our primary objective, teaching fundamental organic chemistry, remains unchanged. We also recognize that we have a responsibility to teach organic chemistry by using the modern techniques that, to a certain degree, are the same techniques practicing chemists will use on completion of their academic training. Finally, we believe that we have a responsibility to make chemists aware of the vast amount of published material that can be used both as a supplement to this book and as general references.

The major theme of this book continues to be the development of techniques for the characterization and identification of ''unknown'' organic compounds. Therefore, we provide students with systematic instructions for identification of compounds that have been previously described in the literature of organic chemistry. The same procedures can frequently be used to identify structures for compounds that have not hitherto been prepared or described.

The book is designed to be both a text for an organic qualitative analysis course and a stepping-stone to research. This means that many of the students who will use the book will be seniors or graduate students, and thus we must discuss the more sophisticated techniques that can be used to identify unknowns. The core of the learning experience associated with the identification of unknown compounds is still present, and we have supplemented this with more detailed discussions of a variety of the techniques available to the organic chemist. These techniques should be discussed even though they may not be used in the undergraduate laboratory. It is important that the student be made aware of how unknowns are identified in the laboratories of graduate research and industry. This awareness can be gained without adversely affecting the extent to which students learn fundamental organic chemistry. For example, it may be necessary to tell students that mass spectrometry, although a powerful tool, is not appropriate for routine operation in all undergraduate laboratories. It has also come to our attention that qualitative analysis has worked its way down into sophomore organic chemistry. We believe that the level of this book may also make it very useful for sophomore courses.

General directions for progress through Chapters 3 to 6 are outlined in Chapter 2. Especially useful are the report forms in Section 2.6. After a student has gained some knowledge of the type of unknown (for example, by the solubility class, elemental content, and ir and nmr spectra described in Chapters 3 to 5), he or she can determine the structure of the unknown from the detailed ir and nmr analysis and from supporting chemical tests (Chapter 6).

Chemical classification tests (''wet'' tests), detailed spectral analyses (ir and nmr), and derivative preparations have all been integrated into one large chapter (Chapter 6). This chapter is organized on the basis of functional groups and describes ir, nmr, chemical (''wet'') tests, and derivatization procedures for each major class of organic compounds (alcohols, ketones, etc.).

Thus, to summarize our organization to this point, a student may have very likely determined a number of possible candidates for the ''unknown'' compound on completion of the procedures in Chapters 3 through 5. The procedures and discussion contained in Chapter 6 should then allow the student to determine which of the candidates is the correct one and to prove this beyond a shadow of a doubt.

Chapter 7 describes procedures useful for separating organic compounds. The standard solvent extractions, using, for example, acids and bases as described in previous editions, have been supplemented by descriptions of chromatographic (lc, tlc, gc, etc.) procedures. Students can be directed toward sections of this chapter at any time that the instructor feels it is appropriate. Chapter 8 outlines special procedures (such as cmr, ord, etc.) that may be utilized for special unknowns or in those cases where exceptional students seek out extra challenges.

Chapter 9 includes all of the "roadmap" problems that were included in the last edition; in some instances, spectral data have been added. Chapter 10 discusses the literature that is useful to organic chemistry, both for characterization of unknowns and for research problems in general.

We have also retained the m.p. and b.p. tables and indexes that were a popular part of previous editions. These tables have been extensively supplemented by appendixes containing other tables, including practical data that can be used in organic laboratories (chromatographic solvents, etc.).

We have retained most of the preliminary characterization procedures described in previous editions (m.p., b.p., solubility classification, etc.) and supplemented them with related, modern classification techniques (tlc purity checks, etc.). We have also supplemented qualitative elemental analysis (by sodium fusion) with a description of how mass spectrometry, as well as other modern techniques, can be used for both qualitative and quantitative analysis. We recommend that molecular weight should be determined by procedures such as we describe using mass spectrometry or vapor phase osmometry, rather than by the freezing-point depression (Rast) method of earlier editions; all too frequently the Rast procedure yielded extremely poor results. By popular request, we have returned to the solubility classification categories decribed by letters (S_1, S_2, A_1, etc.) as used in the fourth edition, and we have supplemented these with solubility characterization involving organic solvents. The latter should yield results useful for spectral analysis, chromatographic analysis, and recrystallizations.

Infrared and nuclear magnetic resonance analyses have been integrated into the fundamental characterization scheme. Chapter 5 tells how to run ir and nmr spectra, and Chapter 6 tells how to relate these spectra to structure and to chemical tests. We have focused on the determination of ir and nmr spectra and the correlation of information from these spectra with structures. The student is directed toward introductory organic textbooks and instrumental analysis textbooks that provide the theory of ir and nmr spectrometry.

Of great concern to all chemists is safety in the laboratory. Chapter 1, as well as being a very brief introduction to this book, contains extensive discussion of and references to first aid procedures and to the toxic properties of reagents encountered in the laboratory. In numerous cases throughout the book we have pointed out the properties of both toxic compounds and compounds that are hazardous for other reasons. These include compounds that are irritating, carcinogenic, and explosive.

In recent years there has been much discussion about the toxic properties of benzene. In fact, many laboratories in which this book is used may forbid the use of benzene. It has been suggested that toluene can be used as a substitute for benzene. We must point out that the slightly different solvent properties of toluene mean that some procedures that go very well in benzene will go poorly (for example, slowly) in toluene or not at all. In addition, the toxicity of toluene should be of concern. In summary, care should be exercised when handling these arenes or any laboratory reagent. Hoods should be used at every opportunity, and skin contact should be scrupulously avoided.

In view of the complex situation regarding safety and hazardous materials, we have included the maximum allowable exposures prescribed by OSHA. We believe that every institution should be aware of the details so that it can adjust to the limitations that presently exist.

The bulk of the work required for this revision was carried out by Terence C. Morrill. The primary role of Ralph L. Shriner was that of providing well-tried and improved chemical tests. In addition, his advice, due to his many years of teaching and association with organic chemistry and qualitative analysis, was invaluable in determining the approach to this revision. A very substantial portion of the manuscript for this book was prepared while Terence C. Morrill was a visiting associate professor at the University of Rochester. The gracious hospitality and constructive suggestions of the chemists at the University of Rochester is greatly appreciated.

A number of industrial concerns contributed to this edition. We especially want to thank Ace Glass Co., Perkin-Elmer Corporation, Varian Associates, Sadtler Laboratories, and Waters Associates.

We are grateful to the following chemists for contributing their time and ideas to this edition: C. F. H. Allen (Rochester Institute of Technology), William Bigler (Rochester Institute of Technology), William Closson (SUNY-Albany), Louis Freidrich (University of Rochester), Robert E. Gilman, (Rochester Institute of Technology), Jack Kampmeier (University of Rochester), P. A. S. Smith (University of Michigan), David Strack (Waters Associates), and Thomas P. Wallace (Rochester Institute of Technology). Those who assisted in the reproduction and typing of the manuscript and in the preparation of the illustrations include Joyce Bennett, Becky Davis, Sue Hubregson, Cheryl Kane, Gaylene Morrill, and Sharon Valyear. Finally, T. C. M. owes a special debt of gratitude to his wife, Gaylene Morrill, and to his editor, Gary Carlson, for their patience during the preparation of this manuscript.

In conclusion, we hope that we have prepared a textbook that will be of use in organic qualitative analysis as well as a shelf reference for all chemists. We welcome comments and suggestions. The development of this book was facilitated by continual and extensive review, and any future revisions will be successful only if we continue to be made aware of the opinions of those people who use this book.

R. L. SHRINER
T. C. MORRILL

PREFACE TO THE FIRST EDITION

Laboratory courses designed to teach methods of identification of organic compounds have become increasingly popular during the past twenty-five years. Since the foundations in this field were laid by Mulliken, whose classic work, *The Identification of Pure Organic Compounds*, was published in 1904, several other excellent treatises of the subject have appeared, and the teaching of systematic identification has become widespread.

The importance of this type of course in the training of the chemist is now universally recognized. The ability to identify compounds—valuable as it is to organic chemists—is, however, not the primary reason for the great popularity of laboratory courses in the subject. The great difference between this and other types of laboratory courses usually included in chemical curricula is that as yet no scheme has been devised which reduces this work to the mere following of directions. At every step in the identification of compounds by present methods the student is required to exercise his own judgment. The student's faculty for careful observation, his ability to make correct deductions from his observations, and his originality in planning his work are at a premium in this type of course. From this point of view it is obvious that this sort of training is the best kind of experience for those preparing for research. In this work students not only become aware of the necessity for research but also are introduced to the methods which it involves.

A natural and important outgrowth of this interest in identification methods is the large amount of research which has been done recently in this field, particularly in connection with the preparation of derivatives suitable for characterization and identification work. A consequence of this is that the subject matter of these courses is in constant need of revision.

The present book is the outgrowth of several years of experience with the subject both on the pedagogical side and from the research point of view. Interest in this work at the University of Illinois was initiated by Professor C. G. Derick, who first gave a course of this sort in 1908. The course was subsequently developed by Professor Oliver Kamm, whose excellent textbook on the subject appeared in 1922. The laboratory exercises herein presented are those used at the present time at the University of Illinois in a one-semester course of two three-hour laboratory periods a week. The work is of such nature, however, that it can be readily adapted to longer or shorter terms by merely increasing or decreasing the number of unknown compounds assigned for identification. The course is designed for students who have had a year of organic chemistry.

In the preparation of the book, use has been made of many methods to be found in works of a similar nature. Chief among these are Mulliken's *The Identification of Pure Organic Compounds*, Kamm's *Qualitative Organic Analysis*, Clarke's *Handbook of Organic Analysis*, Staudinger's *Introduction to Qualitative Organic Analysis*, Porter, Stewart, and Branch's *Methods of Organic Chemistry*, and Bargellini's *Esercizi numerici di chimica organica*. To the authors of these, grateful acknowledgment is hereby made. For many of the innovations contained in this book the authors are indebted to other teachers of the subject here and elsewhere. Throughout the preparation of the manuscript Professor C.

S. Marvel has rendered constant and invaluable assistance. Professors John R. Johnson, A. W. Ingersoll, S. M. McElvain, G. H. Coleman, Wallace R. Brode, Ralph Connor, and C. F. H. Allen have all contributed helpful suggestions which the authors gladly acknowledge. Finally, to the hundreds of students who have used these directions at the University of Illinois and who have been the final judges of the worth of the new features which appear in this book—to these, especial acknowledgment is made for indispensable assistance.

Urbana, Illinois
September, 1935

R. L. S.
R. C. F.

CONTENTS

CHAPTER ONE

introduction

1.1 THE SYSTEMATIC IDENTIFICATION OF ORGANIC COMPOUNDS: COMPOUNDS PREVIOUSLY DESCRIBED IN THE CHEMICAL LITERATURE

The characterization of carbon compounds obtained from living organisms (i.e., from plants and animals), from fossils (coal, petroleum, gas, peat, lignite), and from laboratory syntheses has been reported in the chemical journals for more than 150 years. As of the mid-1970s, it is estimated that more than 5 million compounds have been characterized. Thousands of new compounds are added every year.

A chemist who has a compound in hand, and who needs to know what it is, cannot possibly be aware of all the reported data for comparison with the properties of the "unknown." Thus a *systematic* approach is essential. This approach must first exclude as many structural possibilities as possible; then reduce the number of possible structures to just a very few (say, three or four) possibilities; and finally, establish and confirm one structure.

It is quite clear that an instructor teaching an organic qualitative analysis course will not have 5 million compounds available for use. Chemical supply houses can provide only a limited number of compounds for use in chemical laboratories, in academic institutions, and in industrial laboratories. Eastman

Organic Chemicals, for example, lists about 4000 compounds, and Aldrich Chemical Company lists about 9000. Many other chemical companies list only 100 to 1000 compounds. Industrially produced organic compounds number about 6000 to 10,000. Thus the list of more common and more readily available chemicals is much smaller than 5 million.

In this text we have focused our attention on an even smaller list of compounds that can be used as "unknowns." The melting point–boiling point tables give a very accurate idea of the focus of this book. Instructors using this book may very well use other references (e.g., CRC reference volumes,[1] the Aldrich Company catalog, etc.) for a more extensive list of possibilities for "unknown" compounds.

Organic chemists are often confronted with either of the following extreme situations:

1. Determination of the identity of a compound that has no prior history. This is often the case for a natural products chemist who must study a very small amount of sample isolated from a plant or animal. A similar situation applies to the forensic chemist who analyzes very small samples related to a lawsuit or crime.

2. The industrial chemist or college laboratory chemist who must analyze a sample that contains a major *expected* product and minor products, all of which could be expected from a given set of reagents and conditions. It is entirely possible that such a sample with a well-documented history will allow one to have a properly preconceived notion as to how the analysis should be conducted.

1.2 RELATIONSHIP OF THE STUDY OF THE IDENTIFICATION OF ORGANIC COMPOUNDS TO ORGANIC RESEARCH

The theory and technique for identifying organic compounds constitute an essential introduction to research in organic chemistry. This study organizes the accumulated knowledge concerning physical properties, structures, and reactions of thousands of carbon compounds into a systematic, logical identification scheme. Although its initial aim is the characterization of previously known compounds, the scheme of attack constitutes the first stage in the elucidation of structure of newly prepared organic compounds.

---

[1] For example; *Handbook of Tables for Organic Compound Identification*, 3rd ed., edited by Z. Rappoport (CRC Press, Cleveland, Ohio, 1967).

If, for example, two known compounds A and B are dissolved in a solvent C, a catalyst D is added, and the whole subjected to proper reaction conditions of temperature and pressure, a mixture of new products plus unchanged starting materials results.

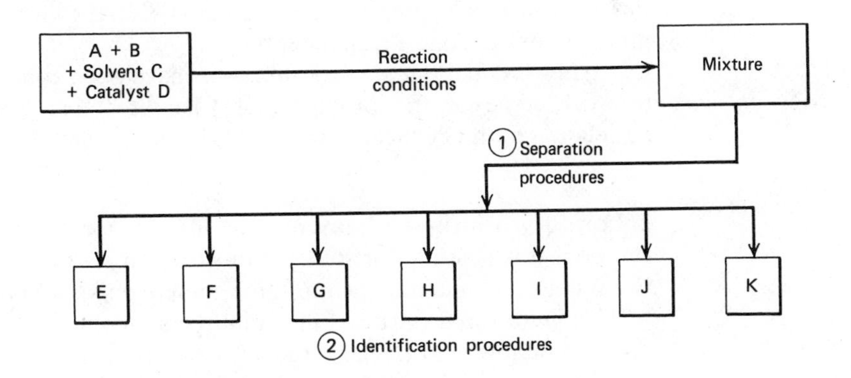

Immediately two problems arise:

1. What procedure shall be chosen to separate the mixture into its components?
2. How are the individual compounds (E through K) to be definitely characterized? Which ones are unchanged reactants? Which compounds have been described previously by other chemists? Finally, which products are new?

These two problems are intimately related. Separations of organic mixtures use both chemical and physical processes and are dependent on the structures of the constituents.

The present course of study centers attention on the systematic identification of individual compounds first. The specific steps are given in Chapter 2. Then the use of these principles for devising efficient procedures for the separation of mixtures is outlined in Chapter 7. The practical laboratory methods and discussion of principles for the steps in identification are given in Chapters 3 through 6.

1.3 SUGGESTIONS TO STUDENTS AND INSTRUCTORS

Schedule. An exact time schedule applicable to all schools cannot be set because of the varied use of semester, quarter, trimester, and summer session terms of instruction. However, for a semester of 15 weeks, two 3-hr laboratory periods per week, plus one "lab lecture" per week works well. Modification can be made to adapt the course to individual schools.

Lecture Material. The experiments and procedures (Chapter 6), and instructions for use of infrared and of nuclear magnetic resonance spectroscopy (Chapters 5 and 6) have been described such that students can study them as their work progresses. The first lecture should describe the course overview as outlined in Chapter 2. It is clearly not necessary to lecture on the specific "recipes" listed in this text (e.g., Chapter 6). Each student's unknown is an individual research problem and is done independently.

After the first one or two unknowns have been completed, it will be valuable to work some of the problems of Chapter 9 in class and discuss structure correlation with chemical reactions and spectral data.

Laboratory Work—Unknowns. By use of spectroscopic data and chemical reactions it is possible for students to work out six to eight single compounds and two mixtures (containing two to three components each) in a 15-week semester.

To get a rapid start and illustrate the systematic scheme, it may be useful to give a titratable acid to each student for a first unknown. The student is told that the substance is titratable and that he or she is to get the elemental analysis, melting or boiling point, and neutralization equivalent and to calculate the possible molecular weights.[2] Then, if the unknown contains halogen or nitrogen, the student is to select and try three to four (but no more) classification tests. Next a list of possible compounds with derivatives is prepared by consulting the table of acids (Table, p. 533). One derivative is made and turned in with the report (see pp. 20–30). This first unknown should be completed in two 3-hour laboratory periods.

The other unknowns should be selected so as to provide experience with compounds containing a wide variety of functional groups.

It is often desirable to check the student's progress after the preliminary tests, solubility classification, and elemental analyses have been completed. This checking procedure is highly recommended for the first one or two unknowns for each student.

Purity of Unknowns. Although every effort is made to provide samples of compounds with a high degree of purity, students and instructors should recognize the fact that many organic compounds decompose or react with oxygen, moisture, or carbon dioxide when stored for a considerable time. Such samples will have wide melting- or boiling-point ranges, frequently lower than the literature values. Hence, for each unknown the student should make a preliminary report of the observed values for melting or boiling point. The instructor should verify these data and if necessary tell the student to purify the sample by recrystallization or distillation and to repeat the determination of the physical constant in question. This avoids waste of time and frustration from conflicting data. (Read also p. 13).

---

[2] Alternatively, the student can be given a compound with mass spectral data.

Amounts of Unknowns. As a general guide, the following amounts are suggested:

Unknown No. 1, a titratable acid[2], 4 g of a solid or 10 ml of a liquid

Unknown No. 2, 3 g of a solid or 8 ml of a liquid

Unknown No. 3, 2 g of a solid or 5 ml of a liquid

Unknown No. 4, 1 g of a solid or 5 ml of a liquid

Mixtures should contain 4 to 5 g of each component. *Note*: If repurification of a sample is required, an additional amount should be furnished to the student.

The amounts listed above are essentially "macro-size" unknowns; use of analytical techniques and instrumentation such as thin-layer chromatography and gas chromatography may very well allow sample sizes of unknowns to be ca. 20% of that listed above. *In the case of these latter, that is, for micro-size samples, it is imperative that chemical test and derivatization procedures described in Chapter 6 be scaled down correspondingly.*

Toward the end of the term, when the student's laboratory technique has been perfected and facility in interpreting reactions has been obtained, it is possible to work with still smaller samples of compounds by using smaller amounts of reagents in the classification tests and making derivatives by means of semimicro techniques.[3] The directions can be modified and semimicro-scale apparatus used. Such apparatus for this course is available under the name Bantam Ware in kits from the Kontes Glass Co., Vineland, N.J.

Time-Saving Hints. It is important to plan laboratory work in advance. This can be done by getting the elemental analyses, physical constants, solubility behavior, and infrared and nmr spectra on several unknowns during one laboratory period. This information should be carefully recorded in the notebook and then reviewed (along with the discussion in each of these steps) the evening before the next laboratory period. A list of a few selected classification tests to be tried is made and carried out in the laboratory the next day. In some cases a preliminary list of possible compounds and desirable derivatives can be made. It is important to note that few of the 30 classification tests should be run on a given compound. It should not be necessary to make more than two derivatives; usually one derivative will prove to be unique. The object is to utilize the sequence of systematic steps outlined in Chapter 2 in the most efficient manner possible.

The instructor should guide the students so that the correct identification results by a process of logical deductive reasoning. Once the structure of the unknown is established, understanding of the test reactions and spectra becomes

---

[3] N. D. Cheronis and J. B. Entrikin, *Semimicro Qualitative Organic Analysis*, 2nd ed. (Interscience Publishers, New York, 1957); N. D. Cheronis, *Semimicro Experimental Organic Chemistry* (John DeGraff, New York, 1958). N. D. Cheronis and T. S. Ma, *Organic Functional Group Analysis by Micro and Semimicro Methods* (Wiley-Interscience, New York, 1964).

clear. Practice in this phase of reasoning from laboratory observations to structure is facilitated by early reading of Chapter 9. One method for developing this ability is for the instructor to write a structural formula on the blackboard and ask the students to predict the solubility behavior and to select the appropriate classification tests.

To tie together the identification work in this course with actual research, the instructor can select a few typical examples of naturally occurring compounds, such as nicotine, D-ribose, quinine, penicillin G, and vitamin B_1, and review the identifying reactions used to deduce these structures. The recent literature also furnishes examples of the value of infrared, ultraviolet, and nmr spectra in establishing structures. Knowledge of the mechanisms of the reactions used for classification tests and for preparing derivatives requires an understanding of the functional groups and their electronic structures.

Throughout this book, references to original articles, monographs, and reference works are given. Many of these will not be used during a one-semester course. However, the citations have been selected to furnish valuable starting sources for future work and are of great use in senior and graduate research.

The use of this manual will be greatly facilitated by the preparation of a set of index tabs for each chapter and parts of chapters. The time spent in preparing the index tabs is more than recovered in speeding up the location of experiments for functional groups, derivative procedures, and tables of derivatives.

At all times, the instructor and students should observe safety rules. They should always wear safety glasses in the laboratory and should become familiar with emergency treatment.

1.4 LABORATORY SAFETY AND FIRST AID

A chemistry laboratory can be one of the most dangerous work areas known, especially when the laboratory is not treated with respect. In this section we shall discuss those safety procedures that must be used when there is insufficient time to seek help from outside the laboratory.

It is important to be aware of the location of people and facilities needed when a laboratory accident results that demands expert assistance. The following phone numbers should be obtained by the laboratory instructor so that expert assistance can be quickly solicited:

Ambulance

Fire department

Institute (i.e., college) physician

Police

Poison control center

American Red Cross

Institute (i.e., college) health center

Table 1.1. Some Common Poisons and the Symptoms They Induce

Acids and alkalies	Will burn and corrode the tissues they contact and can literally burn a hole in a person's stomach.
Alcohol	Acts as a strong depressant on the central nervous system.
Cyanide	With all but very small doses, the victim collapses. Death follows quickly as a result of respiratory paralysis. May be taken orally, inhaled, through a wound, or absorbed through the skin. (Used in mole and ant killers.)
Cyanide and carbon monoxide	Produce death by asphyxiation by combining with oxygen-carrying system of the blood and thus preventing the transfer of oxygen to vital portions of the human system.
Hydrogen sulfide	Flammable, poisonous gas with the odor of rotten eggs; perceptible in a dilution of 0.002 mg/liter of air. Very hazardous; collapse, coma, and death may come within a few seconds after one to two inspirations. Insidious, as sense of smell may become fatigued upon continued exposure. Lower concentrations may cause irritation of mucous membranes, headache, nausea, and lassitude.
Lead	Acute lead poisoning may lead to anorexia, vomiting, malaise, convulsions, and permanent brain damage. Chronic cases may show weight loss, weakness, and anemia.
Mercury	Dangerous because it is fairly volatile (vapor pressure of 2×10^{-3} mm at 25°C) and readily absorbed by the respiratory tract, intact skin and gastrointestinal tract. Acute poisoning by metal or its salts can cause corrosion of skin and mucous membranes, severe nausea, vomiting, abdominal pain, bloody diarrhea, kidney damage, and death within 10 days. Chronic poisoning can result in inflammation of mouth and gums, excessive salivation, loosening of teeth, kidney damage, muscle tremors, spasms, depression, and other personality changes, irritability, and nervousness. Antidote: dimercaprol (BAL).
Methyl alcohol	Has a specific degenerating effect on the optic nerve, which may result in permanent damage and blindness even if only a small quantity has been consumed.
Phenylhydrazine	Causes hemolysis of erythrocytes.
Pyrethrin	(Found in insecticides.) Produces hyperexcitability, incoordination, and paralysis of teneral muscles and of respiratory actions.
Silver nitrate	Contact with skin or mucous membranes can be caustic and irritating. Swallowing can cause severe gastroenteritis and eventually death.

Most of the following items should be readily available in the chemistry laboratory; items on this list or their description may vary due to local safety regulations:

Fire blanket

Fire extinguisher

Eye-wash fountain

Shower

First aid kit

Washes for acid or base (alkali) burns

Table 1.2. Poisons for Which Vomiting Should Not Be Induced

DO NOT INDUCE VOMITING...

... if the person has swallowed anything listed below. Give milk or water; 1 to 2 cups from ages 1 to 5, and up to 1 quart if 5 years or older.

Ammonia[a]	Lime (calcium oxide)[a]
Benzene	Lye (sodium hydroxide)[a]
Bleach (sodium hypochlorite)[a]	Naphtha (petroleum ether)
Carbolic acid[b] disinfectants	Paint thinners and removers
Creosote (phenols)	Pine oil
Detergents[a]	Sodium carbonate[a]
Dry cleaning fluids	Strong acids
Gasoline	Strychnine
Kerosene	Washing soda (sodium carbonate)[a]

[a] Vinegar or any fruit juices may be given as well as milk or water, for these marked a are alkali corrosives.

[b] Phenolic.

It is imperative that all laboratory accidents be reported, whether or not they are treated. It is also important that someone accompany an injured person who is sent out of the laboratory for special care; if the injured person should faint, the injury could easily become compounded.

Most communities have agencies that provide instruction in first aid; these include the Red Cross, local (e.g., volunteer) ambulance corps, and institutions allied with the health professions.

Most laboratory reagents are poisonous; an understanding of the symptoms induced by poisons is important (Table 1.1).[4] If one can determine the type of poison that a person has swallowed, one should consult lists of poisons in order to decide whether vomiting should be induced. Table 1.2 lists poisons for which vomiting should *not* be induced; in most of these cases vomiting would result in passing the corrosive poison back through delicate bodily tissues. In these cases

Table 1.3. Poisons for Which Vomiting Should Be Induced[a]

Alcohol (rubbing, ethyl)
Alcohol (wood, methyl)
Antifreeze (ethylene glycol)
Borax
Camphor
Formaldehyde
Insect repellents

[a] To induce vomiting, tickle back of throat.

[4] Additional information on poisons can be found in *The Toxic Substances List*, edited by E. Christensen and T. Lugenbyhl (U.S. Department of Health, Education and Welfare, Rockville, Md., 1974).

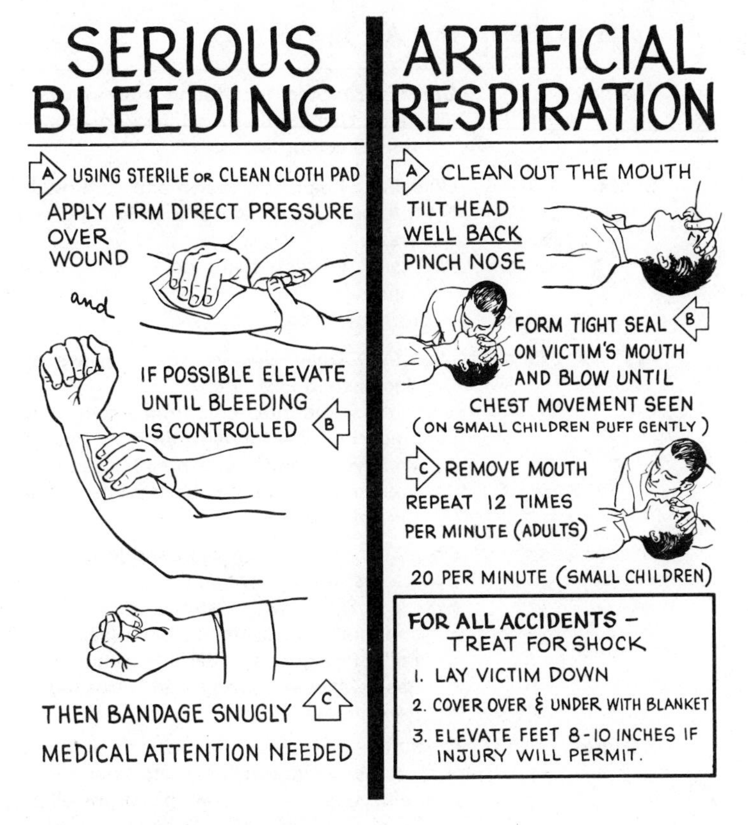

Figure 1.1. First aid instructions, American Red Cross.

liquids should be administered to dilute the poisons, as instructed in Table 1.2. Table 1.3 lists those poisons for which vomiting *should* be induced; instructions as to how to induce vomiting are listed in this table.

Figure 1.1 illustrates methods of treating bleeding and of artificial respiration. Table 1.4 briefly outlines a variety of standard first aid procedures.

Finally, a number of references recommended for the chemistry library are listed here; all of these are relevant to the areas of first aid and safety.

CRC Handbook of Laboratory Safety, 3rd ed., edited by N. V. Steere, (Chemical Rubber Co., Cleveland, Ohio, 1971).

Merck Index, 9th ed., edited by Martha Windholz (Merck Company, Rahway, N.J., 1976).

Dangerous Properties of Industrial Materials, 4th ed., edited by N. F. Sax (Van Nostrand Reinhold, New York, 1975).

The Toxic Substances List (U.S. Department of Health, Education and Welfare, Rockville, Md., 1974).

Safety in the Chemical Laboratory, edited by N. V. Steere (Division of Chemical Education of the American Chemical Society, Easton, Pa., Vol. 1, 1967; Vol. 2, 1971; Vol. 3, 1974).

Table 1.4. Handy First Aid Tips

Wounds	*Objective:* Protect the wound from infection and control bleeding. *First aid:* Use sterile dressings and apply direct pressure on the wound until bleeding stops.
Shock	*Objective:* Keep lying down and comfortable. *Symptoms:* Moist *pale* skin, breathing shallow, eyes lack luster, pulse weak. *First aid:* 1. Keep lying down and elevate feet if no head or chest injury. 2. Cover with blanket (do not cause sweating). 3. Give water to allay thirst.
Artificial respiration	*Objective:* Open airways, keep airways open, alternating increase and decrease in size of chest. *Symptoms:* Not breathing from electric shock, drowning, gas poisoning. *First aid:* Pull jaw forward, tilt head back. Seal victim's nose and mouth with your mouth and breathe directly into victim's mouth. Remove your mouth and turn head while victim exhales. Repeat 15–20 times per minute.
Poisons	*Objective:* Dilute the poison and induce vomiting, except as advised. *Symptoms:* Burns around mouth, empty bottle. *First aid:* Dilute with water or milk, induce vomiting by strong baking soda solution or finger. Universal antidote: 1 part strong tea, 1 part milk of magnesia, 2 parts burnt toast. Do not cause vomiting if victim has swallowed strong acid, kerosene, or strychnine. Check all labels for antidote.
Fractures	*Objective:* Keep broken bone ends and adjacent joints quiet. *Symptoms:* Pain, swelling, deformity *First aid:* Use stiff material, pillow, or blanket and splint as you find it.
Burns	*Objective:* Relieve pain and prevent infection. *Symptoms:* 1st degree = redness. 2nd degree = blisters. 3rd degree = deep tissue damage. *First aid:* Cover with thick layer of sterile dry dressing. Chemical burns: wash with water.
Transportation	*Objective:* If moving victim is necessary, DO NOT BEND, TWIST, OR SHAKE victim. *First aid:* Drag victim on coat, blanket, or rug; use chair, stretcher, or human carry—do not further injure victim.
Fainting	Have person lie flat if possible or have person lower head between knees and breathe deeply. Use ammonia ampule if available.
Heart attack	If person has medication, assist in giving; keep lying down; allow for comfortable breathing. Call doctor.

1.5 EXPLOSION HAZARDS OF COMMON ETHERS

A number of reports of violent explosions due to accidental detonation of peroxides, which can build up in common ether solvents, have been reported. These ethers include ethyl ether, isopropyl ether, dioxane, and tetrahydrofuran, as well as many others. Isopropyl ether seems to be especially prone to peroxide formation. Apparently the greatest hazard exists when ethers have been exposed to air, especially for extended periods of time. The danger is enhanced when the ethers are concentrated, for example, by distillation. *Any ether solvent that displays a precipitate or that seems to be more viscous than usual may very well contain peroxides; do not handle such samples and report this situation to your instructor IMMEDIATELY.*

The situation described just above involves ethers that are no longer acceptable for laboratory use. We shall now consider testing for small amounts of peroxides in ethers and the removal of such peroxides from the ethers.

There are a number of qualitative tests for the presence of peroxides in ethers; we describe two here.

Procedure A. Ferrous Thiocyanate Test for Peroxides

A fresh solution of 5 ml of 1% ferrous ammonium sulfate, 0.5 ml of 1 *N* sulfuric acid and 0.5 ml of 0.1 *N* ammonium thiocyanate are mixed (and if necessary, decolorized with a trace of zinc dust) and shaken with equal quantity of the solvent to be tested: if peroxides are present, a red color will develop.

Procedure B. Potassium Iodide Test for Peroxides

Add 1 ml of a freshly prepared 10% solution of potassium iodide to 10 ml of ethyl ether in a 25-ml glass-stoppered cylinder of colorless glass protected from light: when viewed transversely against a white background, no color is seen in either liquid.

If any yellow color appears when 9 ml of ethyl ether are shaken with 1 ml of a saturated solution of potassium iodide there is more than 0.005% peroxide and the ether should be purified or discarded.

Ferrous sulfate can be used to remove peroxides from ethers. Each liter of ether should be treated with 40 g of 30% aqueous ferrous sulfate. The reaction may well be vigorous and produce heat if the ethers contain appreciable amounts of peroxide; care should thus be exercised in using this procedure. The ether can then be dried (e.g., with magnesium sulfate) and distilled.

A simple method for removing peroxides from high-quality ether samples, without need for distillation apparatus or appreciable loss of ether, consists of percolating the solvent through a column of Dowex-1 ion exchange resin. A column of alumina can be used to remove peroxides and traces of water from ethyl ether, butyl ether, dioxane, and hydrocarbon solvents and for removing peroxides from tetrahydrofuran, decahydronaphthalene (decalin), 1,2,3,4-tetrahydronaphthalene (tetralin), cumene, and isopropyl ether.

Additional information regarding peroxides in ethers can be obtained from the *CRC Handbook of Laboratory Safety* (complete reference on p. 9) and from *The Chemistry of the Ether Linkage*, edited by S. Patai (Wiley-Interscience, New York, 1967).

CHAPTER TWO

identification of unknowns

2.1 INTRODUCTION

There are two uses to which the organization and approach outlined in this chapter can be applied. The first use, the more pedagogical, is that in which a student is being asked to identify a compound already described in the literature. The second use is the characterization of hitherto unknown compounds. This text is intended for both uses. This chapter will outline the former use as employed in a classroom setting.

The following set of directions is intended as a guide to the student in the process of identifying an unknown. Students will, of course, be expected to make a careful and systematic record of observations; the preparation of such a record will be greatly simplified by the use of the suggested sequence of operations.

We shall begin by assuming that the student has a sample in hand that is primarily one compound. As an aside, we should point out that if it is at all possible that the sample is comprised of more than one component, each of substantial proportion, then Chapter 7 on separation techniques should be consulted. Here in Chapter 2, we will, however, simply proceed assuming that the structure of a single, major component is of interest. The rest of this chapter is actually essentially a brief outline of Chapters 3–6 followed by sample report forms.

2.2 PRELIMINARY EXAMINATION

Refer to Chapter 3, pp. 31–33.

Note whether the substance is homogeneous or not, and record its physical state (solid or liquid), color, and odor. Perform the ignition test (p. 32), and record the results.

Thin-Layer and Gas Chromatography. Refer to Chapter 3, respectively pp. 33–37 and pp. 61–67. Simple thin-layer chromatography (tlc), which for solid samples can be complemented by melting-point (m.p.) analysis (see below), and gas chromatography (gc) are very easy and direct methods of purity determination. Observation of only a single developed spot on a thin-layer chromatogram (after using solvents of different polarity), a single peak on gas chromatographic columns of a variety of polarities, and a sharp melting point are all strong support for high sample purity. If the sample is a liquid or a solid, tlc should definitely be applied. If the sample is a liquid, gc should be tried. Gas chromatography of reasonably volatile solids is also possible.

2.3 PHYSICAL CONSTANTS

Refer to Chapter 3, pp. 37–61.

If the unknown is a solid, determine the melting point (pp. 37–45). If the melting-point range is more than 2.0°C, the compound should be recrystallized. Some compounds, however, may not have a sharper melting point no matter how pure, if they are of a type that undergoes decomposition or, in fact, any chemical change, at the temperatures used in the melting-point determination. If the unknown is a liquid or a very low-melting solid, determine its boiling point (pp. 46–52); the range of this constant should not exceed 5.0°C except for very high-boiling compounds. Distillation is recommended if the boiling-point range indicates extensive contamination, if the compound is inhomogeneous, or if it appears to be discolored. Distillation at reduced pressure may be necessary for those compounds that show evidence of decomposition in the boiling-point test.

As mentioned earlier, a sharp melting point is strong support for sample purity.[1] Narrow boiling-point (b.p.) ranges do not, however, strongly substantiate purity. Specific gravity (sp gr) can support sample purity. These sp gr values are, however, rarely used in the early stages of structure determination. An exception are the cases of very inert compounds (e.g., hydrocarbons); in such cases sp gr may be one of the very few early probes for structure determination. Refractive index values can easily be obtained and are a good indication of purity and, eventually, a support for identification.

---

[1] Sharp melting points of eutectic *mixtures* can be misleading; these do not, however, occur very frequently.

2.4 MOLECULAR WEIGHT DETERMINATION

Next to molecular formula, molecular weight can be the most useful factor in determining organic structure; an educated guess at the molecular formula can often be made from the molecular weight. The Rast method (freezing-point depression) has classically been used to determine the molecular weight; this method has been omitted in this edition because the Rast method does not give accurate results for a wide enough range of organic compounds. Mass spectrometry, discussed on pp. 70–77 of Chapter 3, gives molecular weights for a wide range of organic compounds; this technique may not, however, be available in all instructional laboratories.[2] Osmometry (pp. 67–70) is a simple technique that allows access to the molecular weight; one should, however, be wary of erroneously high osmometric weights due to aggregations of molecules. Neutralization equivalents (N.E.) and saponification equivalents (S.E.) may be determined and lead to molecular weights (or simple fractions of the molecular weight); because these, however, apply to specific functional groups (acids or amines and esters, respectively), their development is outlined in Chapter 6. For certain classes of compounds (for example, alcohols), mass spectral analysis is difficult and thus the other methods of molecular weight determination become more appropriate.

2.5 MOLECULAR FORMULA DETERMINATION

Consult Chapter 4. Simple "wet" or "test-tube" tests can be used to determine the presence of certain elements in the compound.

Test the compound for the presence of nitrogen, sulfur, chlorine, bromine, and iodine (pp. 78–83). If a residue was noted in the ignition test, identify the metal it contains.

Control Experiments. If the student is unfamiliar with the procedure for decomposing the compound or with the tests for the elements, he or she should carry out control experiments on a known compound at the same time that the unknown is tested. The compound to be used for the control experiment should, of course, contain nitrogen, sulfur, and a halogen.

In those laboratories where mass spectrometry is available, an attempt should be made to determine the molecular formula of the organic compound from the cluster of peaks in the area of the molecular ion in the mass spectrum; these peaks are due to isotopic contributions of elements in the molecular ion (see pp. 86–89). Mass spectral data can also be used to determine the presence and number of elements in the molecule that make unusually large or unusually small contributions to peaks in the molecular ion cluster (see pp. 87–89).

---

[2] Alternatively, instructors may feel compelled to provide the student with mass spectral data (or osmometric weights or % C, H, N, etc., data) in order to allow the student to have the experience of interpreting these data and applying them to structure determination.

Combustion analysis and other quantitative techniques are usually quite useful in determining the structure of organic compounds; these procedures are not usually carried out in organic qualitative analysis labs, but the data from such might be provided by the instructor.

The next stage in structure determination evolves in two steps. First, we shall be concerned with determining solubility and spectral properties (Chapter 5) to allow the placing of the "unknown" compounds in general structural classes. Second, we shall be interested in determining the exact structure of the compound by chemical tests, by detailed interpretation of spectra, and finally by chemical derivatization (Chapter 6).

2.6 SOLUBILITY TESTS

Refer to Chapter 5, pp. 90–133.

Determine the solubility of the unknown in those of the following reagents that may be expected to provide useful information: water, dilute hydrochloric acid, dilute sodium hydroxide, sodium bicarbonate solution, and cold concentrated sulfuric acid (pp. 90–95). If the classification is doubtful, treat an accurately weighed sample with an accurately measured volume of solvent.

We also recommend solubility studies (pp. 109–112) in organic solvents; results of these studies will be useful in choosing solvents for spectral analyses and for chromatographic analyses and for purification by recrystallization.

Practice Experiments. Accurate deduction on the basis of the solubility tests requires practice on known compounds, and for this purpose an experiment (pp. 93–95) has been provided.

When testing the solubility of the compound in water, the reaction of the solution or suspension to litmus and phenolphthalein should be determined.

When the solubility behavior of the unknown has been determined, draw up a list of the chemical classes to which the compound may belong.

Preliminary Report. To avoid loss of time through mistaken observation, it is recommended that at this point the student submit to the instructor his or her data on the physical constants, elementary composition, and solubility behavior.

2.7 INFRARED AND NUCLEAR MAGNETIC RESONANCE ANALYSES

Infrared and nuclear magnetic resonance analyses are usually crucial to organic structure determination. Infrared (ir) analysis (pp. 112–124) is an excellent functional group probe, which can be used hand in hand with chemical tests for functional groups; use of both ir and chemical tests *may* lead to structure diagnosis. This structure diagnosis is also frequently aided by nuclear magnetic resonance (nmr) analysis; nmr is essentially a method of determining the relative positions and numbers of spin-active nuclei (for example, protons).

Interim Report. After interpretation of solubility results and ir and nmr analyses (recalling all results in the preliminary report), one can usually propose one or a few reasonable structures and then proceed to the final characterization. Note that the instructor may well wish to review your interim report before you go on to final characterization.

The final characterization stage involves application of the "wet" classification tests, detailed scrutiny of the nmr, ir (and possibly other) spectra, culminated by derivatization of the compound; all of these steps are discussed below and outlined in detail in Chapter 6.

2.8 CLASSIFICATION TESTS

Refer to Chapter 6, pp. 134–355.

From the evidence that has been accumulated, deduce what functional group or groups are most likely to be present in the unknown, and test for them by means of suitable classification reagents. More than 30 of the most important of these are mentioned in Chapter 6 (p. 138), where directions for their use may be found. In Table 2.1 these tests are arranged according to the type of compounds they are suitable to detect.

The student is strongly advised against carrying out unnecessary tests, since they not only are a waste of time but also increase the possibility of error. Thus it is pointless to begin the functional group tests of a basic nitrogen-containing compound with tests for the keto or alcohol group. On the other hand, tests that can be expected to give information about the amino group, almost certainly present, are clearly indicated.

Several of the tests for ketones and aldehydes are, in general, easier to carry out and more reliable than tests for other oxygen functions. It is advisable, therefore, in the classification of a neutral compound suspected of containing oxygen, to begin with carbonyl tests, especially when ir analysis has implied a carbonyl group.

In this book we have provided directions on how to obtain ir and nmr spectra; we have also included sample spectra of most of the typical organic functional groups. We have not, however, attempted to develop the theoretical and instrumental foundations of ir and nmr analyses. Thus, as an aid to interpreting the ir and nmr spectra of these compounds, other texts should be consulted. Most lecture textbooks for basic organic chemistry (e.g., those by Morrison and Boyd and by Solomons) have introductory sections on this topic. Other textbooks that would be useful are discussed in Chapter 10.

After deducing the structure of an unknown compound (or perhaps deduction of a few possible structures), derivatization should be carried out to confirm the structure of the compound. Although the melting point of the derivative may be sufficient to allow correct choice of the identity of the unknown, it may also be useful to characterize the derivative by chemical and spectral means, in a fashion similar to the procedure used for the unknown.

Table 2.1. Chemical Tests for Functional Groups

Amines
- (*a*) Benzenesulfonyl chloride (p.230)
- (*b*) Nitrous acid (p. 220)
- (*c*) Acid chlorides (p.218)

Halogens
- (*a*) Ethanolic silver nitrate solution (p.202)
- (*b*) Sodium iodide in acetone (p.204)

Aldehydes and Ketones
- (*a*) Phenylhydrazine (p.165)
- (*b*) 2,4-Dinitrophenylhydrazine (p.162)
- (*c*) Sodium bisulfite solution (p.165)
- (*d*) Iodine and sodium hydroxide (p.167)
- (*e*) Hydroxylamine hydrochloride (p.163)

Aldehydes
- (*a*) Benedict's solution (p.172)
- (*b*) Tollens' reagent (p.170)
- (*c*) The fuchsin-aldehyde reagent (p.171)
- (*d*) Sodium bisulfite solution (p.165)

Phenols
- (*a*) Bromine water (p. 350)
- (*b*) Ferric chloride solution (p.348)
- (*c*) Acid chlorides (p.218)

Carboxylic Acids

No test; a strongly acidic substance, not a phenol or enol, is generally assumed to be a carboxylic acid. Neutralization equivalents should be determined (p. 268)

Unsaturation
- (*a*) Potassium permanganate solution (p. 192)
- (*b*) Bromine in carbon tetrachloride (p.190)

Nitro Compounds
- (*a*) Zinc and ammonium chloride (p. 320)
- (*b*) Ferrous hydroxide (p. 319)
- (*c*) Sodium hydroxide solution (p. 321)

Esters
- (*a*) Sodium hydroxide solution (p. 293), (also see saponification equivalents, p. 296)
- (*b*) Hydroxylamine hydrochloride (p. 163)

Alcohols
- (*a*) The hydrochloric acid-zinc chloride reagent (p. 150)
- (*b*) Acid chlorides (p. 218)
- (*c*) Ceric nitrate (p. 146)
- (*d*) Metallic sodium (p. 145)
- (*e*) Periodic acid (p. 159)
- (*f*) Chromic anhydride (p. 149)

Aromatic Hydrocarbons
- (*a*) Fuming sulfuric acid (p. 245)
- (*b*) Anhydrous aluminum chloride and azoxybenzene or chloroform (p. 247)

Ethers
- (*a*) Hydriodic acid (p. 305)
- (*b*) Bromine water (p. 350)

2.9 THE PREPARATION OF DERIVATIVES

Refer to Chapter 6; pp. 134–355.

The list of possible compounds that results from the preceding steps in the examination of an unknown may contain a number of structural possibilities. The next step in identification is the confirmation of the identity of one of these possibilities with the unknown and the simultaneous demonstration that each of the remaining possibilities differs in some way from the unknown. This final proof can be done by the preparation of derivatives.

In eliminating compounds from the list of possibilities, one is not restricted to the use of derivatives. Any sufficiently characteristic property, such as specific gravity, refractive index, optical rotation, neutralization equivalent, or spectra may be employed.

The Properties of a Satisfactory Derivative. (1) A satisfactory derivative should be easily and quickly made and readily purified. This generally means that the derivative must be a solid, because, in the isolation and purification of small amounts of material, solids afford greater ease of manipulation, and melting points are more accurately and more easily determined than boiling points. The most suitable derivatives melt above 50°C but below 250°C. Most compounds that melt below 50°C are difficult to crystallize, and a melting point above 250°C is undesirable on account of possible decomposition and because the stem correction of the thermometer often amounts to several degrees.

(2) The derivative must be prepared by a reaction that occurs in good yield. Processes accompanied by rearrangements and side reactions are to be avoided.

(3) The derivative should possess properties distinctly different from those of the original compound. Generally, this means that there should be a marked difference between its melting point and that of the parent substance.

(4) The derivative chosen should be one that will single out uniquely one compound from among all the possibilities. Hence the melting points of the derivatives to be compared should differ from each other by at least 5°C.

Consult Chapter 6 and select a suitable derivative from those suggested. It will be noted that derivatives that are satisfactory for purposes of identification are numerous but often of limited scope.

In deciding whether a compound actually possesses the physical constants observed, considerable latitude must be allowed for experimental error. Thus, if the boiling point is very high or the melting point very low, the range must be extended somewhat beyond 5°C. Other constants such as specific gravity (p. 52), refractive index (p. 58), and neutralization equivalents (p. 268) may be used, with proper allowance for experimental error, to exclude compounds from the list of possibilities. A complete list of possible compounds with derivatives for each should always be made even though a product obtained in the classification tests appears to be a suitable derivative.

Examination of the list of possibilities often suggests further functional group tests to be attempted. For example, if a list of possible nitro compounds contains a nitro ketone, carbonyl tests may be valuable, especially if the ir spectrum is consistent with the presence of carbonyl.

If this text does not describe a useful derivatization procedure, a literature search can be made for derivatives of the substance. The most direct way to make a thorough search for a particular compound is to look for the molecular formula in the formula indices of each of the following works in order:

Beilstein's *Handbuch der organischen Chemie,* 4th ed., Index to the 2nd to 4th Supplements. This covers the literature to 1959. The use of Beilstein will be discussed further below.[3]

Chemisches Zentralblatt, Collective Indices: 1922–1924, 1925–1929, 1930–1934.

Chemisches Zentralblatt, Annual Formula Indices, 1935–1939.

Consult Chapter 10 for references that are useful for after 1939. The most reliable indexes are the *Chemical Abstracts* Subject, Author, and Formula Indexes. Decennial Indexes cover the years 1937–1946, 1947–1956; after this the cumulative indexes cover 5-year spans only. Cumulative indexes for 1957–1961 and 1962–1966, 1967–1971, and 1972–1976 have been published. The annual indices must be consulted for later years. Although there is a Collective Formula Index to *Chemical Abstracts* for the years 1920–1946 as well as annual Formula Indices for later years, these are not complete and should not be depended on when a thorough search is desired.

The importance of Beilstein's *Handbuch* is such that further discussion of its use is warranted. The main work or *Hauptwerk* covers the literature through 1909 in 27 volumes. The organization is based on structure in such a way that it is possible to find a desired compound rather easily without using the index once one is familiar with the work. Thus, acyclic hydrocarbons, alcohols, aldehydes, and ketones are in Vol. 1; acyclic acids in Vol. 2; acyclic hydroxy, aldehydo, and keto acids in Vol. 3; sulfonic acids, amines, and phosphines in Vol. 4. In Vol. 5 cyclic (including aromatic) hydrocarbons are treated, and the presentation of cyclic compounds continues along these lines until Vol. 17, where the discussion of heterocyclic compounds begins.

Once a particular compound is located in the main work it is easily found in the supplements. The First Supplement (*Erstes Ergänzungswerk*), which covers the literature from 1910 to 1919, has, as do the later supplements, the same arrangement as the main work. Thus a compound found in Vol. 1 of the main work is in Vol. 1 of the First Supplement. Furthermore, an auxiliary set of page numbers at the top center of each page of each of the supplements relates the material on that page to the corresponding page in the main work. The 2nd, 3rd, and 4th Supplements cover 1920–29, 1930–49, and 1950–59 in the same way. Newer supplements are anticipated.

Elsevier's *Encyclopedia of Organic Chemistry* was initially intended to cover the literature of organic chemistry with the same thoroughness as Beilstein's *Handbuch.* Unfortunately, publication was discontinued after the appearance of a

---

[3] Consult Chapter 10 for references that are guides to the use of Beilstein.

few volumes. The published volumes, however, provide a valuable supplement to Beilstein's *Handbuch*, for they are devoted to subjects such as steroids that were omitted from the original organization of Beilstein.

2.10 MIXTURES

Refer to Chapter 7, pp. 356–411.

At some time during the course, one or more mixtures may be assigned. After obtaining the mixture from the instructor, proceed with the separation according to the methods outlined in Chapter 7. Many of the mixtures contain a volatile component which may be removed by heating the mixture on a steam bath. In dealing with a mixture of unknown composition, it is inadvisable to attempt distillation at higher temperatures.

When the components of the mixture have been separated, identify each according to the procedure followed for simple unknowns.

2.11 REPORTS OF UNKNOWNS

After the identification of an unknown has been completed, the results should be reported on special forms supplied by the instructor. The following specimen reports illustrate the correct procedure.

REPORT FORM 1

Compound *n-Butyl alcohol* Name *John Smith*
Unknown No. *1* Date *June 1, 1976*

1. Physical Examination:
   (*a*) Physical state *Liquid* (*b*) Color *None* (*c*) Odor *Choking*
   (*d*) Ignition test *Burns with bluish flame, no residue.* (*e*) tlc (*f*) gc
   (*e*) tlc (*f*) gc
2. Physical Constants:
   (*a*) m.p.: observed ; corr. (*b*) Sp gr *0.812*$_4^{20°C}$.
   b.p.: observed *114–117*°C; corr. *115–118*°C. (*c*) n_D^{20} *1.3988.*
3. Elemental Analysis:
   F – , Cl – , Br – , I – , N – , S – , Metals *None.*
4. Solubility Tests:

H_2O	NaOH	$NaHCO_3$	HCl	H_2SO_4
+				

Reaction to litmus *None*; to phenolphthalein *None.*

5. Molecular Weight Determination: *None*

6. Preliminary Classification Tests:

Reagent	Results	Inferences
2,4-Dinitrophenylhydrazine	*No ppt.*	*No carbonyl group*
Acetyl chloride	*Reaction—heat—fruity odor*	*Presence of hydroxyl group*
Ceric nitrate	*Red color*	
Lucas reagent	*Dissolved in reagent—no separation of oily layer*	*Probably a primary alcohol*

Functional group indicated by these tests: *Alcohol, probably primary.*

7. Spectroscopic Results:

Type of spectrum (and solvent)	Significant frequencies	Inferences
(in CCl_4)	*3600, 3300 cm^{-1}*	*—O—H*
	1025 cm^{-1} (very broad)	*—C—O—*
(in $CDCl_3$)	*δ0.95, 3H, t*	*CH_3CH_2—*
	δ1.25–1.90, 4H, m	*aliphatic*
	δ2.15,[a] 1H, s	*OH*
	δ3.65, 2H, t	*—CH_2CH_2OH*

[a]This chemical shift was observed to be concentration dependent.

8. Preliminary Examination of the Literature:

Possible compounds	m.p. or b.p. (°C)	Suggestions for further tests
Isobutyl alcohol	*108*	
Methylisopropylcarbinol	*113*	*A sec-methyl carbinol, should give iodoform test*
3-Pentanol	*116*	
n-Butyl alcohol	*117*	
2-Pentanol	*119*	*A sec-methyl carbinol, should give iodoform test*
Methyl-t-butylcarbinol	*120*	

9. Further Classification and Special Tests:

Reagent	Results	Inferences
Iodoform test	*No ppt.*	*Not a sec-methyl carbinol*

10. Probable Compounds:

Name	Useful derivatives and their m.p.'s, neut equiv, etc.			
	3,5-Dinitro-benzoate (°C)	**α-Naphthyl-urethan (°C)**	**Phenylurethan (°C)**	**Sp. gr.**
n-Butyl alcohol	*64*	*71*	*61*	*0.810*
Isobutyl alcohol	*86*	*104*	*86*	*0.805*
3-Pentanol	*97*	*71*	*49*	*0.820*

11. Preparation of Derivatives:

Name of derivative	Observed m.p. (°C)	Reported m.p. (°C)
3,5-Dinitrobenzoate	*62–63*	*64*
α-Naphthylurethan	*68–69*	*71*
Phenylurethan	*57–59*	*61*

12. Special Comments:
Lanthanide shift reagents [specifically Eu(dpm)$_3$] could be used to simplify the nmr spectrum (see Chapter 10 of Shriner et al.).

13. Literature Used:
Tables in this text; Huntress and Mulliken, Identification of Pure Organic Compounds, Order I; The Aldrich Library of Nmr Spectra.

REPORT FORM 2

Compound *2-Amino-4-nitrotoluene** Name *John Smith*
Unknown No. *2* Date *July 13, 1976*

1. Physical Examination:
   (*a*) Physical state *Solid.* (*b*) Color *Yellow.* (*c*) Odor
   (*d*) Ignition test *Yellow flame, no residue.*
   (*e*) tlc (*f*) gc
2. Physical Constants:
   (*a*) m.p.: observed *107–108*°C; corr. *109–110*°C. (*b*) Sp. gr.
   b.p.: observed ; corr. (*c*) n_D

* Also called 2-methyl-5-nitroaniline or 2-methyl-5-nitrobenzenamine.

3. Elemental Analysis:
F –, Cl –, Br –, I –, N +, S –, Metals –.
4. Solubility Tests:

H_2O	NaOH	$NaHCO_3$	HCl	H_2SO_4
–	–		+	

Reaction to litmus ; to phenolphthalein

5. Molecular Weight Determination: *150±4(osmometric, in methanol).*
6. Preliminary Classification Tests:

Reagent	Results	Inferences
Hinsberg	*NaOH: clear soln. HCl: ppt.*	*Primary amine*
Nitrous acid	*Orange ppt. with 2-naphthol*	*Primary aromatic amine*

Functional group indicated by these tests: *Primary aromatic amine.*

7. Spectroscopic Results:

Type of spectrum (and solvent)	Significant frequencies	Inferences
NMR ($CDCl_3$/DMSO-d_6)	*δ2.20, 3H, t*	*Ar**CH_3***
	δ4.70, 2H, bs	***—NH_2***
	δ6.9-7.6, 3H, m	*aromatic protons*

8. Preliminary Examination of the Literature:

Possible compounds	m.p. or b.p. (°C)	Suggestions for further tests
p-Aminoacetophenone	*106*	*Tests for methyl ketone indicated*
2-Amino-4-nitrotoluene	*107*	*Tests for nitro group indicated*
β-Naphthylamine	*112*	
m-Nitroaniline	*114*	*Test for nitro group*
4-Amino-3-nitrotoluene	*116*	*Test for nitro group*
		run uv spectrum

9. Further Classification and Special Tests:

Reagent	Results	Inferences
2,4-Dinitrophenylhydrazine	*No ppt.*	*Not p-aminoacetophenone*
Iodine and sodium hydroxide	*No iodoform*	*Not a methyl ketone*
Zinc and ammonium chloride on benzoyl derivative of unknown	*Silver mirror*	*Nitro group present*

10. Probable Compounds:

Name	Useful derivatives and their m.p.'s, neut equiv., etc.			
	m.p. (°C)	*Benzene-sulfonamide* (°C)	*Acetamide* (°C)	*Phenol* (°C)
2-Amino-4-nitrotoluene	*107*	*172*	*150*	*118*
m-Nitroaniline	*114*	*136*	*155*	*97*
4-Amino-3-nitrotoluene	*116*	*102*	*96*	*32*

11. Preparation of Derivatives:

Name of derivative	Observed m.p. (°C)	Reported m.p. (°C)
Benzenesulfonamide	*170–171*	*172*
2-Hydroxy-4-nitrotoluene	*116–117*	*118*

12. Special Comments:

4-Amino-3-nitrotoluene has been reported to be hydrolyzed to 4-hydroxy-3-nitrotoluene with sodium hydroxide solution [Neville and Winther, Ber., **15,** *2893 (1882)]. The unknown gave only starting material under these conditions.*

The unknown was converted to the phenol by the method reported by Ullmann and Fitzenkam, Ber., **38,** *3790 (1905).*

13. Literature Used:

Tables in this text; special references cited above; Aldrich Library of Nmr Spectra

REPORT FORM 3

Compound *β-Naphthol* Name *John Smith*
Unknown No. *3* Date *September 13, 1976*

1. Physical Examination:
   (*a*) Physical state *Solid.* (*b*) Color *White.* (*c*) Odor *Suggests moth balls.*
   (*d*) Ignition test *Smoky flame—no residue—suggests aromatic compound.* (*e*) tlc (*f*) gc
2. Physical Constants:
   (*a*) m.p.: observed *121–122.5°C*; corr. . (*b*) Sp. gr.
   b.p.: observed *284–286°C*; corr. . (*c*) n_D
3. Elementary Analysis:
   F –, Cl –, Br –, I –, S –, Metals *None.*

4. Solubility Tests:

H_2O	NaOH	$NaHCO_3$	HCl	H_2SO_4
–	+	–		

Reaction to litmus ; to phenolphthalein

5. Molecular weight: *144 (mass spectrometry; unit mass resolution). Molecular formula from M+1, M+2 peaks:* $C_{10}H_8O$.
6. Preliminary Classification Tests:

Reagent	Results	Inferences
Bromine water	*Precipitate*	*Probably a phenol*
Ferric chloride	*Green solution*	

Functional group indicated by these tests: *Phenol*

7. Spectral Results:

Type of spectrum (and solvent)	Significant frequencies	Inferences
ir (in $CHCl_3$*)*	*3300* cm^{-1} *(broad)*	*Hydroxyl group*
	3600 (sharp)	
	1632 cm^{-1} *(medium)*	*Carbonyl group? C=C?*
	1605 cm^{-1}	*Aromatic*
	1200 cm^{-1} *(very broad)*	*C—O*
nmr	*δ7.0–7.8, 7H, m*	*aromatic protons*
(acetone)	*δ8.35,*[a] *1H, bs*	*—OH*

[a] Increasing the concentration of substrate in acetone shifted this signal to lower field.

8. Preliminary Examination of the Literature:

Possible compounds	**m.p. or b.p. (°C)**	**Suggestions for further tests**
p-Hydroxybenzaldehyde	*115*	*Should give carbonyl tests*
Hydroquinone monobenzyl ether	*122*	*Should be cleaved by acid*
Picric acid	*122*	*Contains nitrogen—strong acid, unlikely*
β-Naphthol	*122*	
Toluhydroquinone	*124*	*Should be readily oxidized to yellow quinone*

9. Further Classification and Special Tests:

Reagent	**Results**	**Inferences**
2,4-Dinitrophenylhydrazine	*No ppt.*	*p-Hydroxybenzaldehyde unlikely*
$Ag(NH_3)_2^+$	*Silver mirror*	*Toluhydroquinone not ruled out*

10. Probable Compounds:

Name	Useful derivatives and their m.p.'s, neut equiv. etc.			
	***Bromide* (°C)**	***Acetate* (°C)**		
β-Naphthol	*(1-) 84*	*70*		
Toluhydroquinone	*(tri-) 204*	*(di-) 52*		

11. Preparation of Derivatives:

Name of derivative	Observed m.p. (°C)	Reported m.p. (°C)
Monobromo	*84–86*	*84*
Acetate	*67–68*	*70*

12. Special Comments:

The bromo derivative was prepared by treatment with bromine in acetic acid [*Smith,* J. Am. Chem. Soc., **35,** *789* (*1913*)].

Toluhydroquinone unlikely on several grounds. Reported to be readily soluble in water. Ferric chloride in concentrated aqueous solution gives a brownish red color and in dilute solution a yellow color (*Beil., VI, 874*), *while the unknown gave a green color* (*as does β-naphthol: Huntress and Mulliken, p. 234*).

β-Naphthol has been reported to give a positive Tollens' test (Beil., VI, 628). α- and β-Substituted naphthalenes can have an aromatic ring absorption as high as 1650 cm^{-1}

13. Literature Used:

Tables in this text; *references cited above*; *Bellamy*, The Infra-red Spectra of Complex Molecules; *Sadtler Standard Spectra* (*nmr*).

CHAPTER THREE

preliminary examination, purity determination, and physical properties of organic compounds

The investigator begins at this point when he or she has in hand a sample that is believed to be primarily one compound; if the investigator believes that the sample contains more than one component, Chapter 7 on separations should be consulted. The assumption that predominantly one component is present may be based on (1) instructor's guidance, (2) the method of synthesis, and/or (3) the method of isolation.

3.1 PRELIMINARY EXAMINATION

3.1.1 Physical State

Note whether the unknown substance is a liquid or solid; tables of compounds (Appendix III) are subdivided on the basis of phase. In addition, insofar as the phase relates to solubility and volatility, an aid to choice or purification method is provided: liquids are usually purified by distillation or by gas chromatography (Chapter 7), and solids are purified by recrystallization (or sublimation).

3.1.2 Color

The color of the original sample is noted, as well as any change in color that may occur during the determination of the boiling point (Section 3.3.2), distillation, or after chromatographic separation.

The color of some compounds is due to impurities; frequently these are produced by slow oxidation of the compound by oxygen in the air. Aniline, for example, is usually reddish brown, but a freshly distilled sample is nearly colorless.

Many liquids and solids are definitely colored because of the presence of chromophoric groups in the molecule. Many nitro compounds, quinones, azo compounds, stable carbocations, and carbanions, and compounds with extended conjugated systems, are colored. If an unknown compound is a stable, colorless liquid or a white crystalline solid, this information is valuable because it excludes chromophoric functional groups as well as many groups that by oxidation would become chromophores.

3.1.3 Odor

Many types of organic compounds have characteristic odors. It is not possible to describe odors in a precise manner, but the student should become familiar with the odors of common compounds. Alcohols have odors different from those of esters; phenols from amines; aldehydes from ketones. Mercaptans, isonitriles, and pentamethylenediamine usually are described as possessing disagreeable odors; however, they differ from each other. Moreover, the odor is most pronounced in the lower-molecular-weight members of a class, because these are more volatile. Benzaldehyde, nitrobenzene, and benzonitrile all have the odor of bitter almond oil. Eugenol, coumarin, vanillin, methyl salicylate, and isoamyl acetate have characteristic odors that are easily remembered. Hydrocarbons also differ in their odors—toluene, hexane, isoprene, indene, pinene, and naphthalene possess unique odors.

The student should cautiously note the odors of common organic compounds and compare them with the odors of the unknowns. Safety precautions are usually noted on the labels for such known compounds; if one is in doubt, references such as the *Merck Index* (see Chapter 10) should be consulted.

3.1.4 Ignition Test

Procedure. A sample of 0.1 g of the substance is placed in a porcelain crucible cover (or any piece of porcelain) and brought to the edge of a flame to determine flammability. It is then heated gently over a low flame and behind a safety shield and eventually heated strongly to accomplish thorough ignition. A note is made of (1) flammability and nature of the flame (is the compound explosive?); (2) if the compound is a solid, whether it melts and the manner of its melting; (3) the odor of gases or vapors evolved (*caution*!); (4) residue left after ignition. Will it fuse? If a residue is left, the lid is allowed to cool, a drop of distilled water is added, and the solution tested with litmus. A drop of dilute hydrochloric acid is added. Is a gas evolved? A flame test with a platinum wire is made on the hydrochloric acid solution to determine the metal present.

When using this test, as well as any other test, control runs should be made on standard compounds of known composition. When possible, a known compound of similar composition to that suspected for the unknown should be chosen. Control compounds, which can be used to indicate the range of results possible for this ignition test, are ethanol, toluene, barium benzoate, copper acetate, sodium potassium tartrate, and sucrose.

Discussion. Many liquids burn with a characteristic flame that assists in determining the nature of the compound. Thus, an aromatic hydrocarbon (which has a relatively high carbon content) burns with a yellow, sooty flame. Aliphatic hydrocarbons burn with flames that are yellow but much less sooty. As the oxygen content of the compound increases, the flame becomes more and more clear (blue). If the substance is flammable, the usual precautions must be taken in subsequent manipulation of the compound. This test also shows whether a melting point of a solid should be taken and indicates whether the solid is explosive.

If an inorganic residue is left after ignition, it should be examined for metallic elements. A few simple tests will often determine the nature of the metal present.[1] If the flame test indicates sodium, a sample of the compound should be ignited on a platinum foil instead of a porcelain crucible cover. (Why?)

3.1.5 Summary and Applications

The tests of this Section are extremely useful for decisions as to whether further purification is necessary and as to what type of purification procedures should be used. If the various tests in this section indicate that the compound is very impure, recrystallization or adsorption chromatography are almost certainly required (Chapter 7). Although liquids are very often easily analyzed by gas chromatography (Section 3.4), those that leave residues upon ignition should *not* be injected into the gas chromatograph.

3.2 THIN-LAYER CHROMATOGRAPHY

Thin-layer chromatography (tlc) is perhaps the most rapid, easiest, and most often applied method for assessing the purity (complexity, and often the nature) of organic compounds. Chromatography may be defined as a separation by differential partitioning of chemical components by distribution between two phases, one phase being immobile (e.g., a surface) and the other being the transport medium (e.g., the solvent or eluant). The general subject of chromatography will be discussed thoroughly in Chapter 7. In tlc, the immobile phase is a

[1] One should consult a standard book on inorganic qualitative analysis, for example, T. Moeller, *Qualitative Analysis* (McGraw-Hill, New York, 1958).

thin layer of adsorbent spread over a sheet of glass or plastic. Calcium sulfate or an organic polymer serves to bind the adsorbent to the sheet. A small amount of sample is placed at the bottom of the slide and this spotted slide is placed on end in a container with a shallow layer of solvent (see Fig. 3.2); the distance to which the solvent moves the compound up the chromatogram sheet is dependent on the ability of the compound to adhere to this adsorbent system, as well as many other factors. More often than not, adsorbent–solvent systems can be found to separate most components of a given mixture. This procedure is especially useful for compounds that are heat-sensitive or nonvolatile, that is, those compounds that are not amenable to boiling point or gas chromatographic determination.

Procedure. Commercial sheets, precoated with alumina (Al_2O_3) or silica gel ($SiO_2 \cdot xH_2O$), often containing fluorescent material, are available.[2] A preliminary aid to choice of solvent is the following: prepare fine capillary tubes (e.g., drawn m.p. tubes) containing solutions of the sample in each of a series of solvents by capillary action. Table 3.1 lists those solvents that can be screened for possible use. Merely touch a chromatographic sheet with the tube (Fig. 3.1). When the solvent circle has reached its final position, a useful solvent will have moved the sample ring to a position about 0.3 to 0.5 that of the solvent circle radius. If no other solvent information (e.g., from recrystallization attempts) is available, chloroform and solvents of similar polarity may be tried.

Table 3.1. Chromatographic solvents[a]

Solvent	
Petroleum ether	Increasing polarity[b] ↓
Cyclohexane	
Carbon tetrachloride	
Benzene	
Methylene chloride	
Chloroform (alcohol free)	
Ethyl ether	
Ethyl acetate	
Pyridine	
Acetone	
n-Propyl Alcohol	
Ethanol	
Methanol	
Water	
Acetic acid	

[a] Mixtures of two or more solvents can be used as developing solvents in chromatographic separations.

[b] Polarity, in this context, is meaningful only for chromatography and is not necessarily the same as polarity as measured by, for example, dielectric constant.

[2] Eastman Organic Chemistry Department, 343 State St., Rochester, N.Y. 14650. One could also consult the annual Instrument (or Equipment) Guides of *Science* or of the journal *Analytical Chemistry* for guides to sources of chromatography equipment.

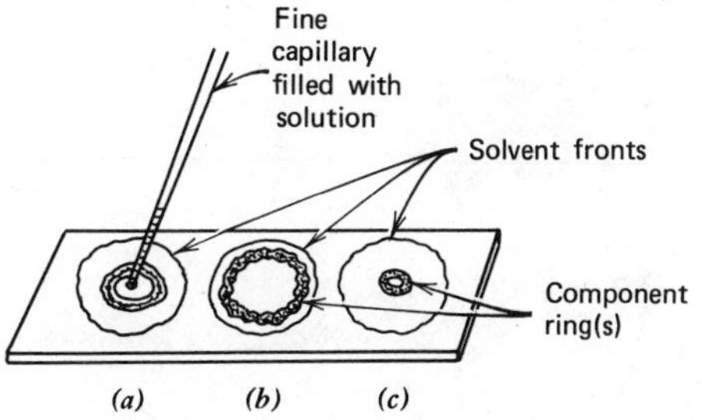

Figure 3.1. Thin-layer chromatography: determination of solvent choice. (*a*) Good development. (*b*) Overdeveloped. (*c*) Underdeveloped.

A development chamber (see Fig. 3.2) that can be capped is prepared by filling it to a depth of ca. 3 mm with the solvent of choice. A solution (1–10%) of the sample of interest is prepared in any volatile solvent. Fine capillary tubing is prepared by, for example, drawing out m.p. capillary tubing; touching the fine tubing to the surface of the solution causes some of the solution to rise up into the tubing (capillary action). The chromatographic sheet is spotted with sample by touching the solution-charged tubing to a position ca. 0.5 cm from one end of the plate (Fig. 3.3). The sheet *must* be placed such as to position the spot above the solvent at the beginning of the analysis. This sheet is now placed, spotted-end down, upright in the development chamber. Solvent is allowed to climb up the plate until the solvent front is ca. 1 cm from the upper end of the adsorbent-covered region (Fig. 3.2); during this development, the system should be enclosed to ensure that the atmosphere is saturated with solvent vapor. Saturation may be further ensured by fitting a filter paper (wick) to the inside wall of the chamber. After development (ca. 5 min for sheets of microscope-slide size, and 0.5–1.0 hr for large plates), the position to which the solvent front has moved is marked by scratching a line along this solvent front. Any obvious sample spots are scratch-marked (circled) also. Allow the chromatogram to dry (usually for 5–30 min) before visualization.

Special procedures must be used to visualize colorless compounds. If the chromatogram originally contained fluorescent material, the developed chromatogram may be observed under ultraviolet light (*caution*!); dark spots are observed where sample spots block out the fluorescence. These spots (and any detected by alternative methods below) should be marked by circling. Alternatively, spots may be brought out by placing the developed chromatogram in a(n)

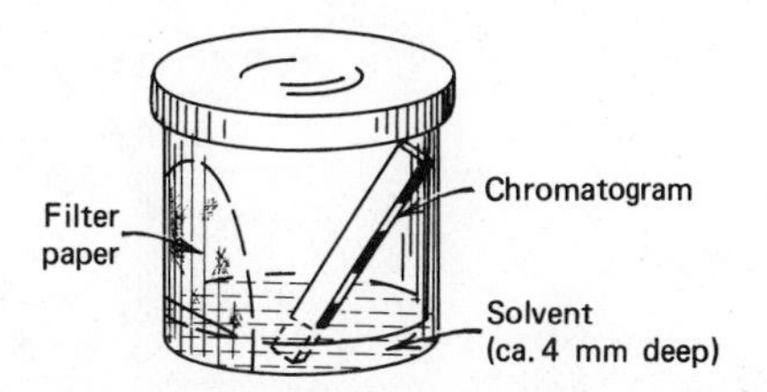

Figure 3.2. Chromatographic development unit.

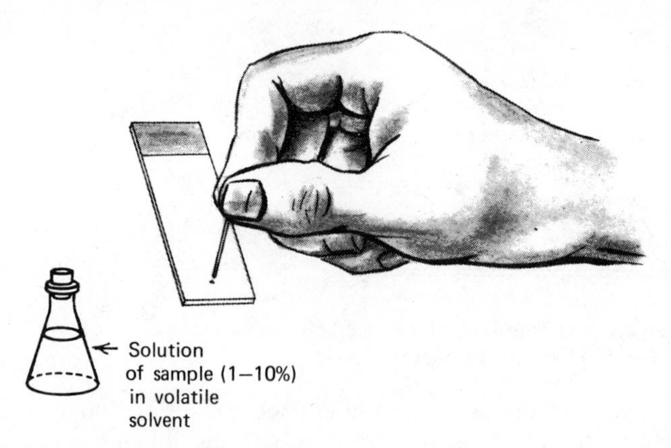

Figure 3.3. Preparation for tlc.

(encloseable) chamber containing iodine crystals (Fig. 3.4). Nearly all compounds (except alkanes and aliphatic halides) form iodine charge-transfer complexes; the formation of these complexes is (often rapidly) reversible, so the position of the dark spot should be marked quickly. Another visualization technique involves spraying the chromatogram with sulfuric acid to cause the colorless compounds to darken by charring; this is, of course, a sample-destructive analytical technique.

If commercial chromatographic sheets are not available, glass plates are easily prepared. Prepare a slurry of adsorbent (35 g of silica gel G in 100 ml of chloroform or 60 g of alumina in 100 ml of 67 vol%/33 vol% chloroform/methanol). Stir the slurry thoroughly; dip a pair of back-to-back plates (e.g., standard microscope slides) into the slurry and slowly withdraw them. Wipe excess adsorbent from the edges and separate the plates. Wipe adsorbent from the back of the plates and allow them to dry (ca. 5 min). Such plates may be used immediately or stored in a desiccator.

Discussion. Development of different chromatograms (alumina, silica gel, silica acid) of the same sample that clearly yield only a single spot in a variety of solvents is a very good indication of purity. In addition, the identity of an unknown can be supported (but not completely proved; why not?) by comparison of spot positions of known and unknown materials (Fig. 3.5). Figure 3.5 also implies how the progress of a reaction can be monitored. The most common

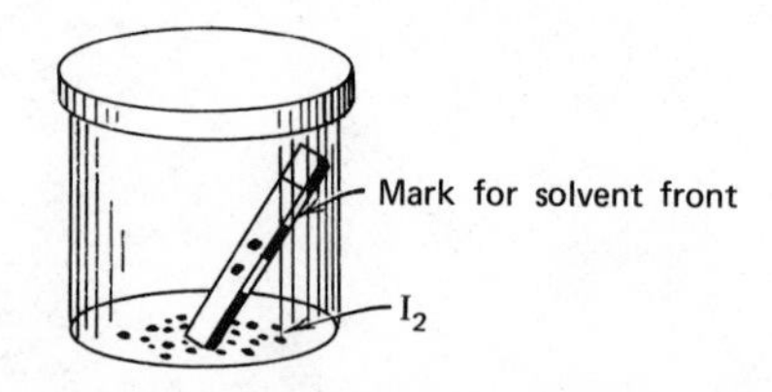

Figure 3.4. Iodine tlc spot-marking chamber.

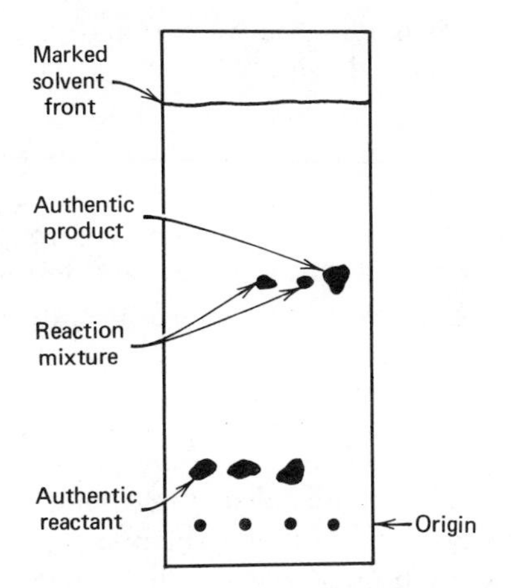

Figure 3.5. Thin-layer chromatogram; reaction mixture composition analysis.

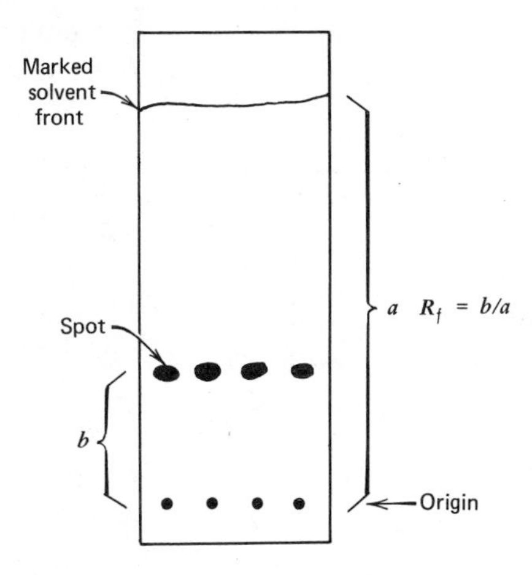

Figure 3.6. Determination of R_f on tlc chromatogram.

method of reporting tlc results is by the R_f value (see also Fig. 3.6) in a specific solvent on a specific type of chromatographic sheet:

$$R_f = \frac{\text{distance that a spot of interest has moved from origin } (b)}{\text{distance the solvent front has moved from origin } (a)}$$

Alternatively, the results may be reported as R_x, where

$$R_x = \frac{\text{distance spot of interest has moved from the origin}}{\text{distance spot of reference compound has moved from the origin}}$$

Information obtained from tlc is useful for solvent and adsorbent choices in (preparative-scale) column chromatography (see Chapter 7).

3.3 THE DETERMINATION OF PHYSICAL PROPERTIES

3.3.1 Melting Points[3]

The melting point of a compound is the range of temperature at which the solid phase changes to liquid. Since this process is frequently accompanied by decomposition, the value may be not an equilibrium temperature but a temperature of transition from solid to liquid only. If the ignition test indicates that a solid melts easily (25 to 300°C), the melting point should be determined by Procedure A. For higher melting-point ranges (300 to 500°C), use special equipment (see, e.g., Fig. 3.10). If a melting-point determination by Procedure A indicates definite decomposition or transition from one crystalline state to another, Procedure B is

[3] E. L. Skau, J. C. Arthur, jr. and H. Wakeham, Chapter 7 in *Physical Methods of Organic Chemistry*, 3rd ed., Vol. I, Part I, edited by A. Weissberger (Interscience, New York, 1959), chap. 7.

recommended. Compounds melting between 0 and 25°C may be analyzed by the freezing-point method described on p. 45.

Melting points for a large number of compounds and their derivatives are listed in this book. Frequently a small amount of impurity will cause a depression (and broadening) of the observed melting point. Thus the procedure of determining melting points of mixtures described below is strongly recommended. If, for any of a number of reasons, one has a compound that is contaminated by minor amounts of impurities, the section on recrystallization should be consulted (see Chapter 7).

Mixture Melting Points. The "mixed melting-point" method provides a means of testing for the identity of two solids (which should, of course, have identical melting points) by examination of the melting-point behavior of a mechanical mixture of the two. In general, a mixture of samples of nonidentical compounds shows a melting-point depression. Although the use of mixed melting points is valuable at certain points of the identification procedure, a mixed melting point of an unknown with a known sample from the side shelf will not be accepted in this course as proof of the structure.

A few pairs of substances when mixed show no melting-point depression, but more frequently the failure to depress may be observed only at certain compositions. It requires little additional effort to measure the melting points of mixtures of several compositions if the following method is used.

Make small piles of approximately equal sizes of the two components (A and B) being examined. Mix one-half of pile A with one-half of pile B. Now separate the mixture of A and B into three equal parts. To the first add the remainder of component A, and to the third, the remainder of component B. It is seen that three mixtures with the compositions 80% A, 20% B; 50% A, 50% B; and 20% A, 80% B are obtained. The melting points of all three mixtures may be measured at the same time by any of the following procedures.

Procedure A. For many melting-point determinations, 6-cm-long "melting-point tubes" of capillary size (ca. 1 mm width) are used; if such capillary tubes are not available, they may be made by melting and drawing out (clean) soft glass tubing (e.g., 15-mm tubing) or drawing soft-glass test tubes. A modest amount of the compound (powdered, if necessary) is placed on a hard, clean surface; a small amount is picked up by tapping the mouth of the capillary tube on the compound (Fig. 3.7*a*). The capillary tube is then held vertically (Fig. 3.7*b*, mouth up) and rubbed with a file or a coin with a milled edge, or tapped on the table top or dropped through a glass tube (Fig. 3.7*c*) to pack the compound at the bottom. This charged tube can now be used in any of the devices shown in Figs. 3.8 through 3.10 (consult these figures for further directions). Use of the Fisher-Johns stage apparatus (Fig. 3.12) requires microscope cover glasses rather than capillary tubes. The capillary tube should contain a 2–3 mm column of sample.

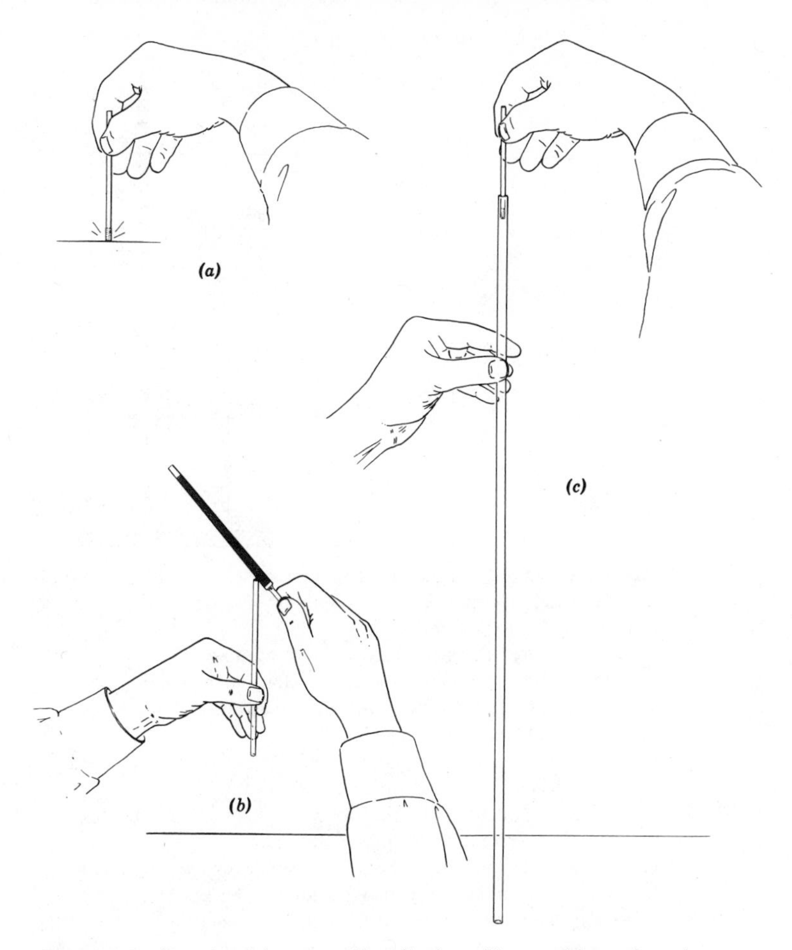

Figure 3.7. Charging (a) and packing (b, c) capillary melting-point tubes.

It is often time-saving to run a preliminary melting-point determination, raising the temperature of the bath very rapidly. After the approximate melting point is known, a second determination is carried out by raising the temperature rapidly until within 5°C of the approximate value and then proceed slowly as described above. A fresh sample of the compound should always be used for each determination.

Failure to use clean capillary melting tubes is one of the causes of low melting points and wide melting-point ranges. These are due to alkali on the surface of the glass, which catalyzes the aldol condensation of aldehydes or ketones, or the mutarotation of sugars and their derivatives, and so on. For example, in an uncleaned tube α-D-glucose softened at 133°C and melted at 143–146°C, whereas in a dealkalinized tube the softening occurs at 142°C and the melting at 146–147°C. The use of clean tubes made from Pyrex glass avoids these problems.

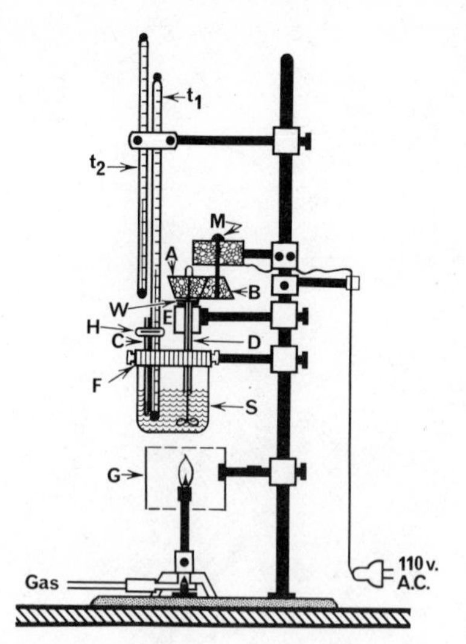

Figure 3.8. Simple melting-point apparatus. Mineral oil (e.g., Nujol), silicone oil, or cottonseed oil is used in vessel. Thermometer t_2 is optional (for stem corrections).

M Motor (induction, sparkless)
A No. 9 rubber stopper; fitted to shaft of stirrer
B No. 1 rubber stopper; attached to shaft at the stirring motor
W Metal washer; fitted between stopper A and bearing D
S Vessel containing heating oil
D Bearing (e.g., use a brass cork borer; the bearing is fitted to the shaft that extends from the stirring motor to the propeller blade)
E Clamp-held rubber stopper (no. 6), which holds the bearing
G Cylindrical metal protector around flame
H Metal or Tygon ring to hold m.p. capillary to thermometer t_2
C Standard m.p. capillary loaded with sample (see Fig. 3.7)
F Ring stand clamp attached to large hose clamp fitted around the vessel S

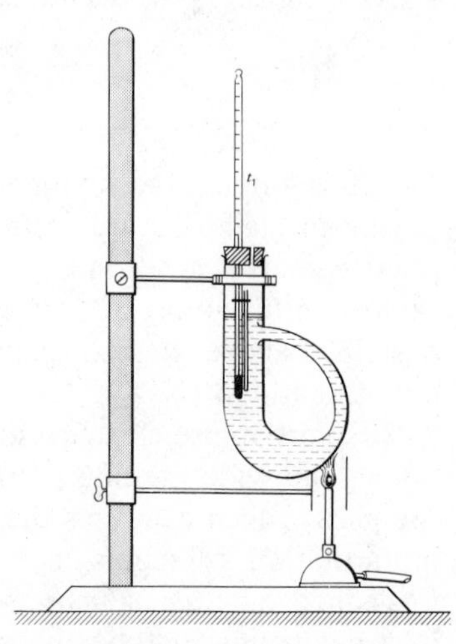

Figure 3.9. Thiele tube melting-point apparatus. See auxiliary information in Fig. 3.8. A stem correction (see thermometer t_2 on Fig. 3.8) can be made.

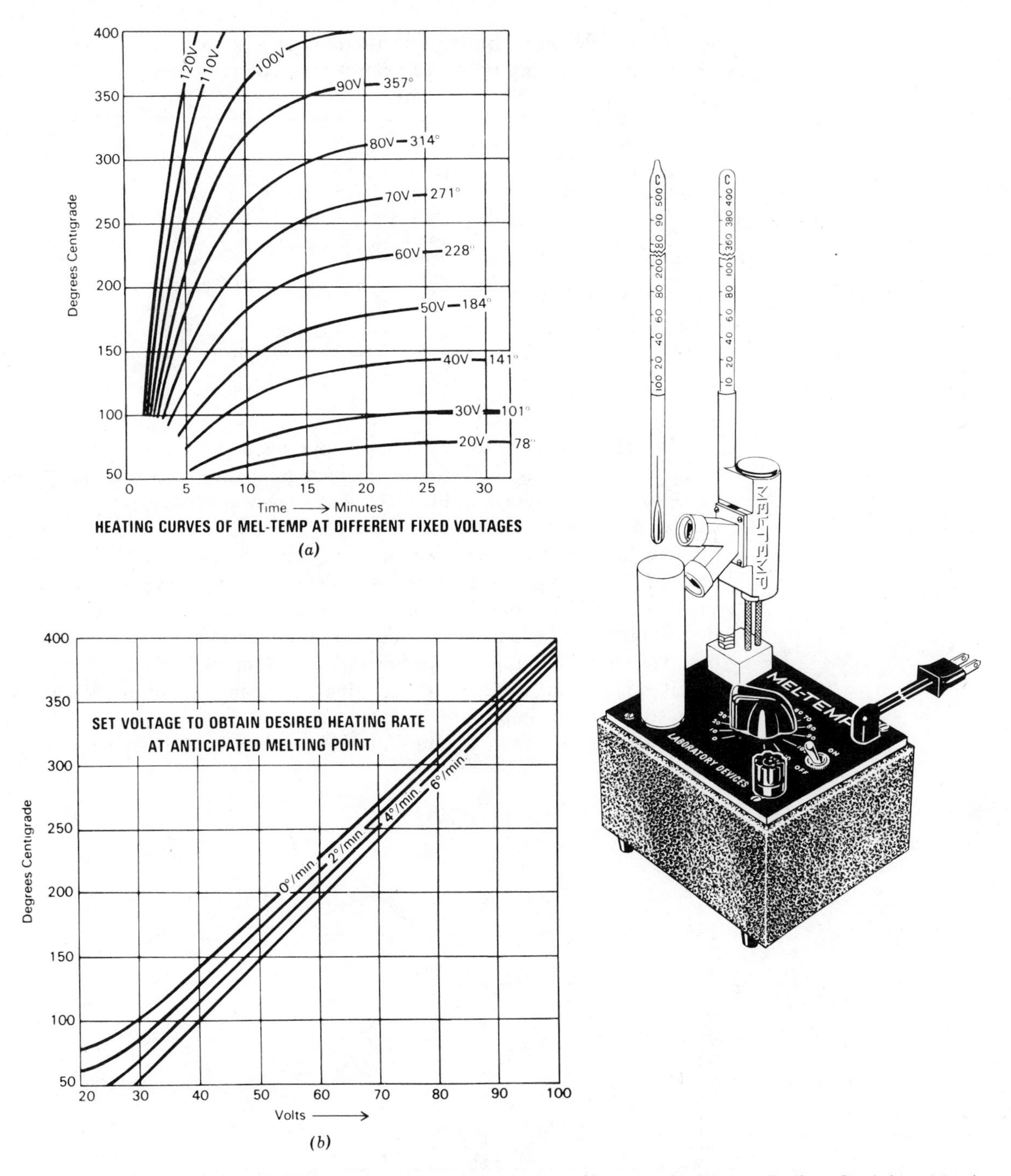

Figure 3.10. Mel-Temp melting-point apparatus. (Courtesy of Laboratory Devices, Cambridge, Mass.)

Corrected Melting Points. The thermometer should always be calibrated by observing the melting points of several pure compounds such as the following:

m.p. (corr.) (°C)		m.p. (corr.) (°C)	
0	Ice	187	Hippuric acid
53	*p*-Dichlorobenzene	200	Isatin
90	*m*-Dinitrobenzene	216	Anthracene
114	Acetanilide	238	Carbanilide
121	Benzoic acid	257	Oxanilide
132	Urea	286	Anthraquinone
157	Salicylic acid	332	N,N′-Diacetylbenzidine

If care is taken to use the same apparatus and thermometer in all melting-point determinations, it is convenient and time-saving to prepare a calibration curve such as that shown in Fig. 3.11. The observed melting point of the standard compound is plotted against the corrected value, and a curve, *DA*, is drawn through these points. In subsequent determinations the observed value, *B*, is projected horizontally to the curve and then vertically down to give the corrected value, *C*. Such a calibration curve includes corrections for inaccuracies in the thermometer and stem correction. The thermometer should be calibrated by observing the melting points of several pure compounds.

It is important to record the melting-point range of an unknown compound, because this is an important index of purity. A large majority of pure organic compounds melt within a range of 0.5°C or melt with decomposition over a

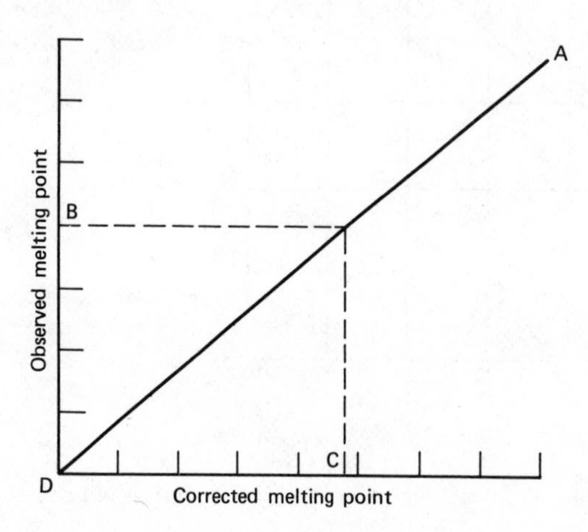

Figure 3.11. Melting-point calibration curve.

narrow range of temperature (about 1°C). If the melting-point range or decomposition range is wide, the compound should be recrystallized from a suitable solvent and the melting or decomposition point determined again. Some organic compounds, such as amino acids, salts of acids or amines, and carbohydrates, melt with decomposition over a considerable range of temperature.

In lieu of the calibration curve above, a stem correction (see Fig. 3.8) can be made using the formula

$$\text{Correction} = +N(t_1 - t_2)(1.54 \times 10^{-3})$$

where N = number of degrees of mercury above the liquid level at t_1 = observed melting point and t_2 = average temperature of the surfaced mercury thread. Since the last figure (t_2) is measured only approximately (see Fig. 3.8, t_2 temperature) and since calibration curves encompass stem corrections, this procedure is seldom used.

If a compound melts above 250°C, its melting point may be determined by means of the Maquenne block.[4]

The Fisher-Johns apparatus, shown in Fig. 3.12, has an electrically heated aluminum block fitted with a thermometer reading to 300°C. The sample is placed between two 18-mm microscope cover glasses which are put in the depression of the aluminum block. The temperature is regulated by means of a variable transformer, and the melting point is observed with the aid of the illuminator and magnifying glass. A calibration curve must be prepared for the instrument by reference to known compounds as described above.

Procedure B. The Nalge-Axelrod instrument, shown in Fig. 3.13, consists of a 25-power microscope with Polaroid inserts, a light source, an electrically heated aluminum block controlled by a variable transformer, and a thermometer reading to 400°C. The sample is placed between two 18-mm microscope cover glasses in the depression of the heating block. The cover, which has a small aperture, is put in place, the light turned on, and the microscope focused. The tube is rotated so as almost to cross the Polaroids, and the crystals are observed as the temperature is raised. The melting points of anisotropic crystals are easily noted, because the polarization colors disappear when melting occurs. Also, the transition from one allotropic crystal modification to another is easily observed. A calibration curve must be prepared (above).

In the usual m.p. experiment (Procedure A, p. 38), the m.p. is defined as the temperature range between the first appearance of a liquid phase and the disappearance of the last crystals of the solid phase. With the microscope slide techniques, there are difficulties with the application of the definitions. Since the crystals are largely separated in the microscope slide procedure, the melted crystals are normally not in contact with the unmelted crystals. Thus, there is

---

[4] Consult the books on laboratory techniques listed in Chapter 10.

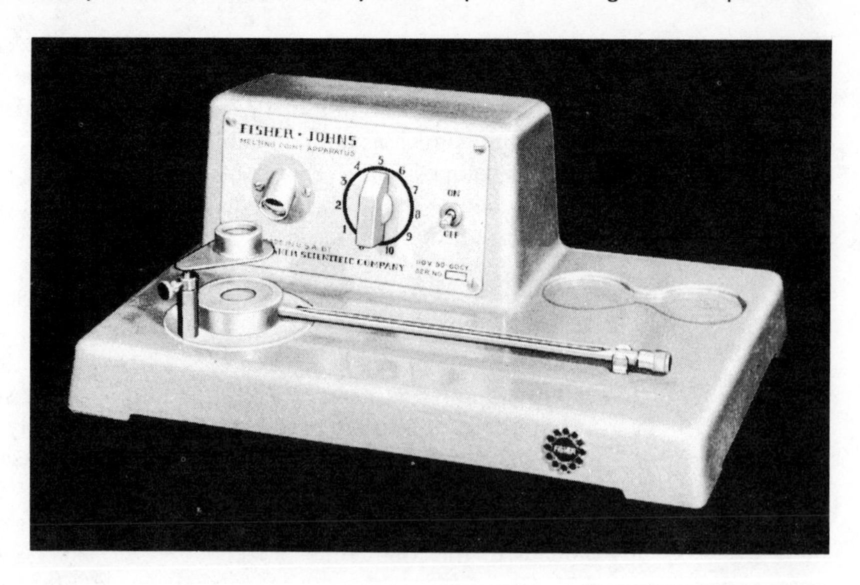

Figure 3.12. Fisher-Johns melting-point apparatus. (Courtesy of Fisher Scientific Co., Pittsburgh, Pa.)

Figure 3.13. Nalge-Axelrod melting-point apparatus. (Courtesy of the Nalge Co., Rochester, N.Y.)

often a very wide range of temperature over which the melting can occur. This procedure is therefore to be avoided; the m.p. determined in a capillary tube is a more accurate and more meaningful criterion of purity.

Freezing Points

A few milliliters of the liquid is placed in an ordinary test tube fitted with a thermometer and a wire stirrer (made of copper, nickel, or platinum). The tube is fastened in a slightly larger test tube by means of a cork and cooled in an ice or ice-salt bath or an acetone-Dry Ice mixture, and the liquid is stirred vigorously (Fig. 3.14). As soon as crystals start to form, the tube is removed from the bath, and vigorous stirring is continued while the temperature on the thermometer is being read. The freezing point is the temperature reached after the initial supercooling effect has disappeared. The temperature of the cooling bath should not be too far below the freezing point of the compound. Freezing points of most organic liquids as ordinarily determined are only approximate (note that the determination requires a relatively large sample).

A more elaborate apparatus for determining freezing points (down to −65°C) has been described.[5]

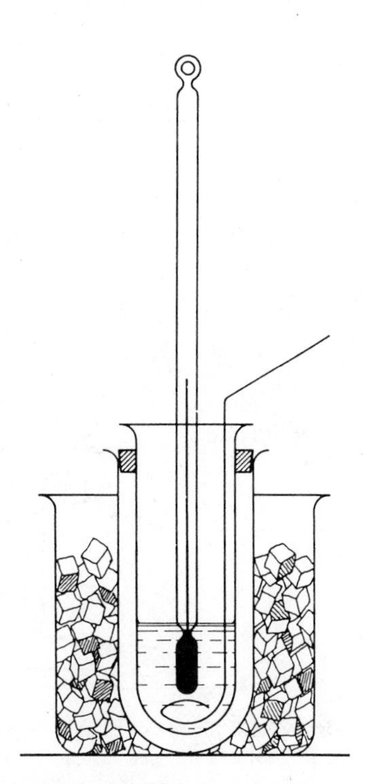

Figure 3.14. Simple freezing-point apparatus.

[5] R. J. Curtis and A. Dolenga, *J. Chem. Educ.*, **52**, 201 (1975).

3.3.2 Boiling Points

The use of boiling points (b.p.) for compound identification has been introduced in Chapter 2 (p. 13).

Procedure A. A small-scale distillation apparatus similar to that shown in Fig. 3.15 is set up. Alternatively, an apparatus using ground-glass joints (e.g., standard taper) as shown in Fig. 3.16 may be used. Test tubes immersed in a beaker of ice are used to condense the vapors and act as receivers. A few boiling chips and 10 ml of the liquid whose boiling point is to be determined are added. The thermometer is inserted so that the top of the mercury bulb is just *below* the side arm. The liquid is heated to boiling by means of a low, shielded burner.[6] The liquid is distilled at as uniform a rate as possible. After the first 2 to 3 ml of distillate has collected, the receiver is changed without interruption of the distillation, and the next 5 to 6 ml is collected in a clean, dry test tube. There will be a considerable lag of the thermometer reading, but usually the boiling-point range can be determined during the collection of the second portion of distillate. This boiling-point range should be recorded.

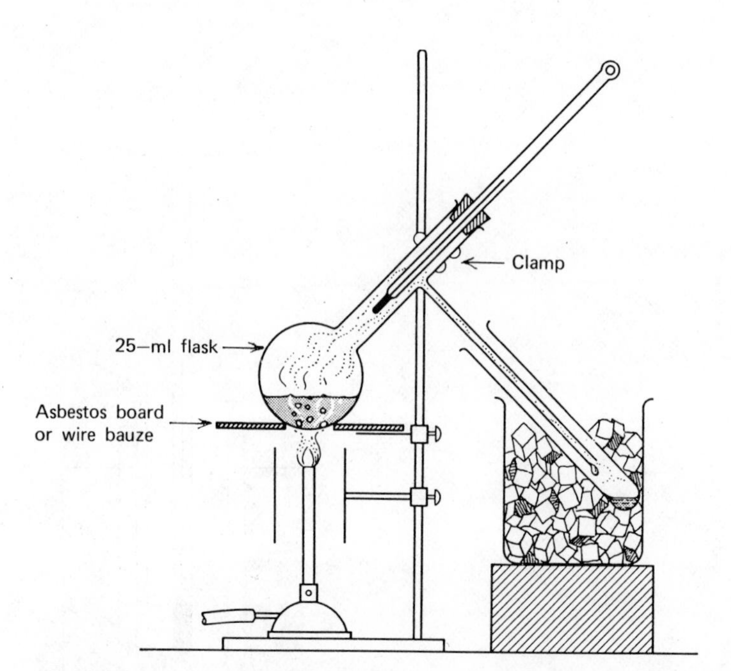

Figure 3.15. Simple distillation apparatus. Use boiling chips in main flask. A heating mantle (Chapter 7, p. 366) can be used instead of the burner. Heating mantles are usually safer, as they are less likely to ignite any flammable liquids.

[6] If available, use electric heating mantles (see, e.g., Chapter 7, p. 366) to induce boiling. This avoids the danger of fires that could be started by use of the burner.

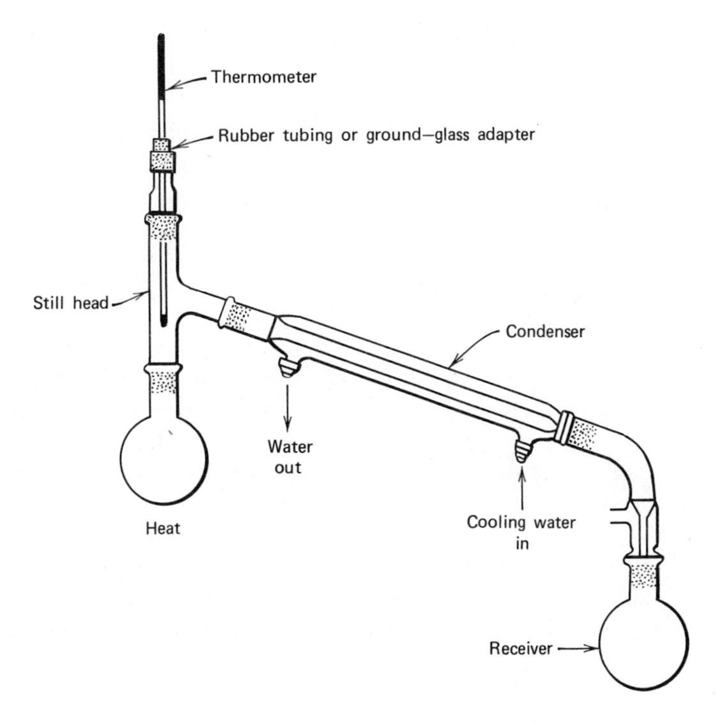

Figure 3.16. Standard taper apparatus for simple distillation. Heating and cooling devices and clamps are not shown.

The boiling-point range is a useful index of purity of the sample. Many organic liquids are hygroscopic, and some decompose on standing. Generally the first few milliliters of distillate will contain any water or more volatile impurities, and the second fraction will consist of the pure substance. If the boiling-point range is large, the liquid should be refractionated through a suitable column (see Chapter 7).

The boiling point determined by the distillation of a small amount of liquid as described above is frequently in error. Unless special care is taken, the vapor may be superheated; also, the boiling points observed for high-boiling liquids may be too low because of the time required for the mercury in the thermometer bulb to reach the temperature of the vapor. The second fraction collected above should be used for a more accurate boiling-point determination by Procedure B below. *Portions of the main fraction should also be used for the determination, as far as possible, of all subsequent chemical, spectral, and physical tests.*

Procedure B. A *micro* boiling-point tube is made as shown in Fig. 3.17. The outer tube is a 5-mm micro test tube (or is made from glass tubing); a capillary m.p. tube is inverted and placed in the larger tube (Fig. 3.17). Two drops of the liquid whose boiling point is to be determined is added, and the tube is fastened to the thermometer in the apparatus used for the determination of melting points (Figs. 3.8 and 3.9). The temperature is raised until a rapid and continuous stream

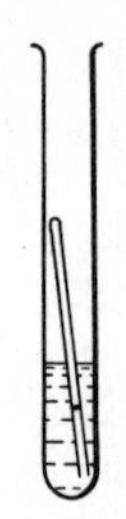

Figure 3.17. Micro boiling-point tube (ca. 5 cm tall).

of bubbles comes out of the small capillary and passes through the liquid. The flame is then removed and the bath allowed to cool while being stirred continuously. The temperature is noted at the instant bubbles cease to come out of the capillary and just before the liquid enters it. This temperature is the boiling point; such a result is usually much more accurate than that determined by Procedure A.

Effect of Pressure on Boiling Point. At the time the boiling point is being determined, the atmospheric pressure should be recorded. Table 3.2 illustrates the magnitude of such barometric "corrections" of boiling point for pressures that do not differ from 760 mm by more than about 30 mm.

It is evident that small deviations in pressure from 760 mm, such as 5 mm, may be neglected in ordinary work.

Investigators working in laboratories at high altitudes[7] and low barometric pressures have found it convenient to determine a set of empirical corrections to be added to observed boiling points in order to get boiling points at 760 mm. The corrections are obtained by distilling a number of different types of compounds with different boiling points. The difference between the boiling point recorded in the literature and the observed boiling point gives the correction.

Nomographs for b.p. versus pressure data of organic compounds have been devised; these charts are useful for vacuum distillations and are provided in Appendix I to this text.

In order to give an idea of the change in boiling point with pressure, the data on three pairs of nonassociated and associated compounds are given in Table 3.3. The temperatures are given to the nearest whole degree. The data indicate that, as the pressure is reduced, the boiling point of an associated compound does not fall off as much as the boiling point of a nonassociated liquid.

Correlations of Boiling Point with Structure. The boiling points of the members of a given homologous series increase as the series is ascended. The boiling points rise in a uniform manner as shown in Fig. 3.18, but the increment per CH_2 group is not constant, being greater at the beginning of the series than for the higher members (Table 3.4).

---

[7] At the top of Mt. Evans in Colorado, water boils at 81°C (average pressure: 460–470 mm; altitude 14,200 ft). Water boiling at the University of Colorado (ca. 5000 ft) will have a temperature of ca. 90°C.

Table 3.2. Boiling-Point Changes per Slight Pressure Change

b.p (°C)	b.p. (°K)	Correction in °C for 10-mm Difference in Pressure: Nonassociated[a] liquids	Associated[a] liquids
50	323	0.38	0.32
100	373	0.44	0.37
150	423	0.50	0.42
200	473	0.56	0.46
300	573	0.68	0.56
400	673	0.79	0.66
500	773	0.91	0.76

[a] Associated liquids are those liquids that have substantial intermolecular associations due to hydrogen bonding; an example is methanol.

Table 3.3. Boiling Points (°C) at Reduced Pressures

Compound	Pressure in millimeters of mercury (torr): 760	700	650	600	550	ΔT[a]
n-Heptane	98	96	94	91	88	10
n-Propyl alcohol	97	95	93	91	89	8
Iodobenzene	188	185	182	179	175	13
n-Valeric acid	186	183	180	178	175	11
Flourene	298	294	290	286	282	16
β-Naphthol	295	292	288	284	280	15

[a] $\Delta T = (\text{b.p.})_{760} - (\text{b.p.})_{550}$.

If a hydrogen atom of a saturated hydrocarbon (alkane) is replaced by another atom or group, an elevation of the boiling point results. Thus alkyl halides, alcohols, aldehydes, ketones, acids, and so on, boil higher than the hydrocarbons with the same carbon skeleton.

If the group introduced is of such a nature that it promotes association, a very marked rise in boiling point occurs. This effect is especially pronounced in the alcohols (Fig. 3.18) and acids, because hydrogen bonding can occur. For example, the difference in boiling point between propane (nonassociated) and *n*-propyl alcohol (associated) is 142°C—a difference far greater than the change in molecular weight would indicate. As more hydroxyl groups are introduced the boiling point rises, but the change is not as great as that caused by the first hydroxyl group. The increment per hydroxyl group is much greater than the increment per methylene group (Table 3.5).

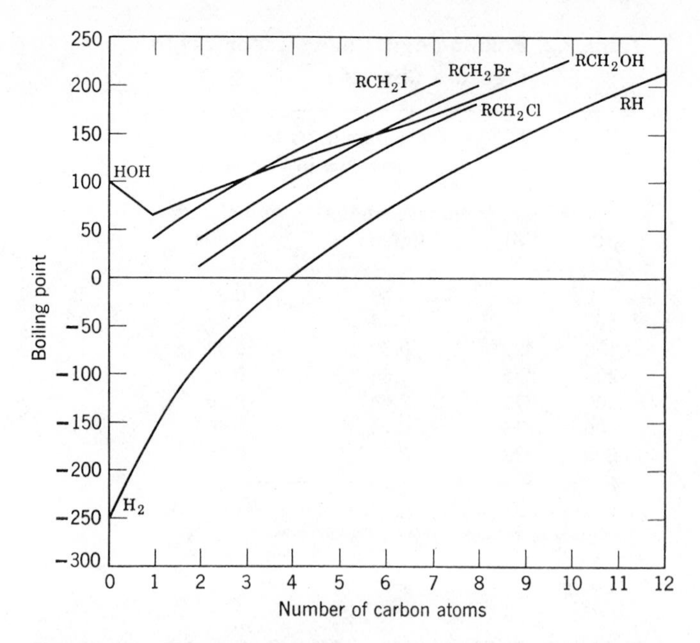

Figure 3.18. Relationship between boiling point and molecular weight.

If the hydroxyl groups are converted to ether linkages, the association due to hydrogen bonds is prevented and the boiling point drops. The following series illustrates this effect:

CH_2OH	$CH_2OC_2H_5$	$CH_2OC_2H_5$	$CH_2OC_2H_5$
$CHOH$	$CHOH$	$CHOH$	$CHOC_2H_5$
CH_2OH	CH_2OH	$CH_2OC_2H_5$	$CH_2OC_2H_5$
+290°C	+230°C	+191°C	+185°C

A comparison of oxygen derivatives with their sulfur analogs also shows that association is a more potent factor than molecular weight. The thiol (RSH) compounds are associated only slightly and hence boil lower than their oxygen analogs even though the former have higher molecular weights than the latter:

	b.p. (°C)		b.p. (°C)
HOH	100	HSH	−62
CH_3OH	66	CH_3SH	+6
CH_3COOH	119	CH_3COSH	93

Ethers and thio ethers are not associated, and hence the alkyl sulfides boil higher than the ethers because they have higher molecular weights:

	b.p. (°C)		b.p. (°C)
$(CH_3)_2O$	−24	$(CH_3)_2S$	+38
$(C_2H_5)_2O$	+35	$(C_2H_5)_2S$	+92

Table 3.4. Boiling Point and Chain Length for Straight-Chain Alkanes

	b.p. (°C)	Δ[a]
Pentane	36	
		32
Hexane	68	
		30
Heptane	98	
		27
Octane	125	
		24
Nonane	149	
		24
Decane	173	
		21
Undecane	194	
		21
Dodecane	215	

[a] Δ = change in b.p. for addition of one methylene group.

These data on sulfur and oxygen compounds, and on hydrocarbons, alkyl chlorides, bromides, and iodides, illustrate the general rule that replacement of an atom by an atom of higher atomic weight causes a rise in the boiling point, provided that no increase or decrease in the extent of association takes place as a result of this substitution.

Just as with solubility relationships (Chapter 5), branching of the chain and position of the functional group influence the boiling point. The saturated aliphatic alcohols (Table 3.6) serve to illustrate the following generalizations:

1. Among isomeric alcohols, the straight-chain isomer has the highest boiling point.
2. If comparisons are made of alcohols of the same type, the greater the branching of the chain the lower the boiling point.
3. A comparison of the boiling points of isomeric primary, secondary, and tertiary alcohols shows that primary alcohols boil higher than secondary alcohols, which, in turn, boil higher than tertiary alcohols provided that isomeric alcohols with the same maximum chain length are compared.

A knowledge of the boiling points of some simple compounds is frequently of value in excluding certain types of compounds. The following simple generalizations are helpful.

Table 3.5. Boiling Point and Hydroxyl Group Substitution

	CH_3	CH_3	CH_2OH	CH_2OH
	CH_2	CH_2	CH_2	$CHOH$
	CH_3	CH_2OH	CH_2OH	CH_2OH
	−45°C	+97°C	+216°C	+290°C
Δ/OH =	142°C	119°C	74°C	

1. An organic chloro compound that boils below 132°C must be aliphatic. If it boils above 132°C, it may be either aliphatic or aromatic. This follows from the fact that the simplest aryl halide, chlorobenzene, boils at 132°C.
2. Similarly, an organic bromo compound that boils below 157°C or an iodo compound that boils below 188°C must be aliphatic. Other bromo and iodo compounds may be either aliphatic or aromatic.

3.3.3 Specific Gravity

The use of specific gravity in compound identification has been outlined in Chapter 2, p. 13). Recall that specific gravity (sp gr), for substance 2, is defined as

$$\text{sp gr}\,\frac{T_2}{T_1} = \frac{w_2}{w_1}$$

where w_2 = weight of a precise volume of substance 2 (the unknown), and w_1 = weight of precisely the same volume of substance 1 (usually water) and T_2, T_1 are the temperatures of these substances. The density (d) of substance 2 can be obtained from

$$d_2 = \left(\text{sp gr}\,\frac{T_2}{T_1}\right)_2 (d_1)_{T_1}$$

where $(d_1)_{T_1}$ = the density of water (or other reference substance) at temperature T_1, and d_2 is the density of substance 2 at temperature T_2. Such densities are available from standard chemistry handbooks.

Specific gravity may be determined by means of a small pycnometer.

Procedures. If a small pycnometer with a capacity of 1 to 2 ml is not available, either of the two forms shown in Figs. 3.19 and 3.20 may be used.

The pycnometer in Fig. 3.19 is made from a piece of capillary tubing (1–2 mm). It is bent into the shape shown, a small bulb blown in the middle, and one end drawn out to a fine capillary. A scratch is made on the other arm at the same height as the tip of the capillary. The pycnometer is suspended by means of

Table 3.6. Alcohol Boiling Point and Branching

Primary alcohols		Secondary alcohols		Tertiary alcohols	
Formula	b.p.(°C)	Formula	b.p.(°C)	Formula	b.p.(°C)
CH_3OH	66				
CH_3CH_2OH	78				
$CH_3CH_2CH_2OH$	97	$CH_3CH(OH)CH_3$	82		
$CH_3CH_2CH_2CH_2OH$	116	$CH_3CH_2CH(OH)CH_3$	99	$CH_3C(CH_3)(OH)CH_3$	83
$CH_3CH(CH_3)CH_2OH$	108				
$CH_3CH_2CH_2CH_2CH_2OH$	138	$CH_3CH_2CH_2CH(OH)CH_3$	119		
$CH_3CH(CH_3)CH_2CH_2OH$	131				
		$CH_3CH_2CH(OH)CH_2CH_3$	115		
$CH_3CH_2CH(CH_3)CH_2OH$	129				
$CH_3—C(CH_3)_2—CH_2OH$	114	$CH_3CH(CH_3)—CH(OH)CH_3$	111	$CH_3CH_2—C(CH_3)(OH)—CH_3$	102

a fine Nichrome, aluminum, or platinum wire, and its weight is determined.[8] The pycnometer is then filled with distilled water to a point beyond the mark and suspended in a constant-temperature bath (for example, at 20°C). After about 10 min, the amount of liquid in the tube is adjusted by holding a piece of filter paper to the capillary tip until the meniscus in the open arm coincides with the mark. The pycnometer is then removed from the beaker, dried, and weighed.

Figure 3.20 shows another specific gravity device. Commercial 1.00-ml (or somewhat larger) volumetric flasks may be used. The weight of the empty bulb is determined. It is then filled with distilled water and suspended by a wire in a constant-temperature bath (for example, at 20°C). The level of the water in the bulb is adjusted to the mark by means of a disposable pipet. The bulb is then removed from the beaker, dried, and weighed.

[8] Analytical balance (±0.5 mg).

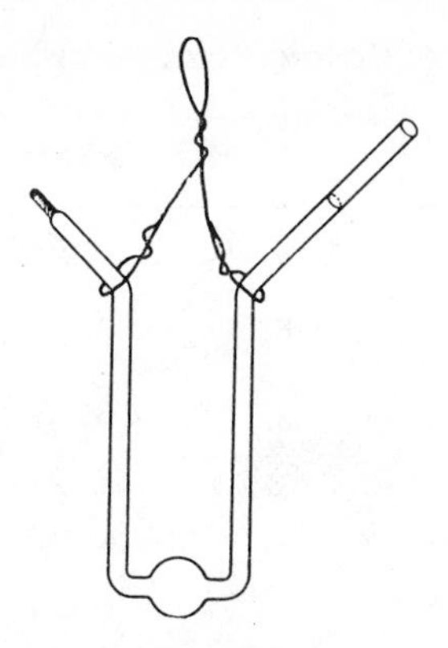

Figure 3.19. Micropycnometer.

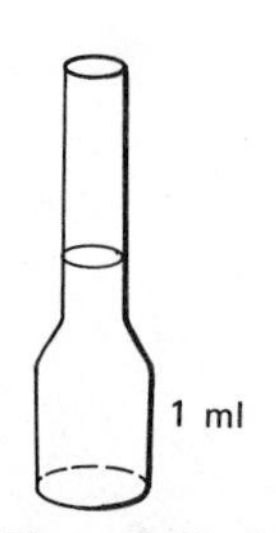

Figure 3.20. Specific gravity bulb (small volumetric flask).

The weight of the empty pycnometer and its weight when filled with distilled water are recorded and kept permanently. To determine the specific gravity of a liquid the bubb or pycnometer is filled with the liquid at 20°C and its weight determined.

$$\text{sp gr}_{20}^{20} = \frac{\text{weight of sample}}{\text{weight of water}}$$

The apparatus must be carefully cleaned and dried immediately after use.

Care must be taken that the sample used for this determination is pure. It is best to use a portion of the center fraction collected from distillation or a gas chromatographic collection corresponding to a single peak (Chapter 7). Sometimes it is necessary to determine the density with reference to that of water at 4°C. This may be done by means of the factor 0.99823:

$$\text{sp gr}_{4}^{20} = \frac{\text{weight of sample}}{\text{weight of water}} \times 0.99823$$

Discussion. The specific gravity of a liquid may often be used to exclude certain compounds from the list of possibilities. It varies with the composition as well as the structure of the compound.

Hydrocarbons are usually lighter than water. As a given homologous series of hydrocarbons is ascended, the specific gravity of the members increases, but the increment per methylene radical gradually diminishes. Curves I, II, and III in Fig. 3.21 show the change in density for the alkanes, 1-alkenes, and 1-alkynes. It will be noted that the specific gravity of the acetylenic hydrocarbon is greater than that of the corresponding olefin, which in turn is more dense than the alkane hydrocarbon with the same number of carbon atoms. The position the unsaturated linkage occupies also influences the density. Moving the double bond nearer the middle of the molecule causes an increase in the specific gravity. The data in Table 3.7 illustrate this change.

The replacement of one atom by another of higher atomic weight usually increases the density. Thus curve IV, Fig. 3.21, which represents the specific

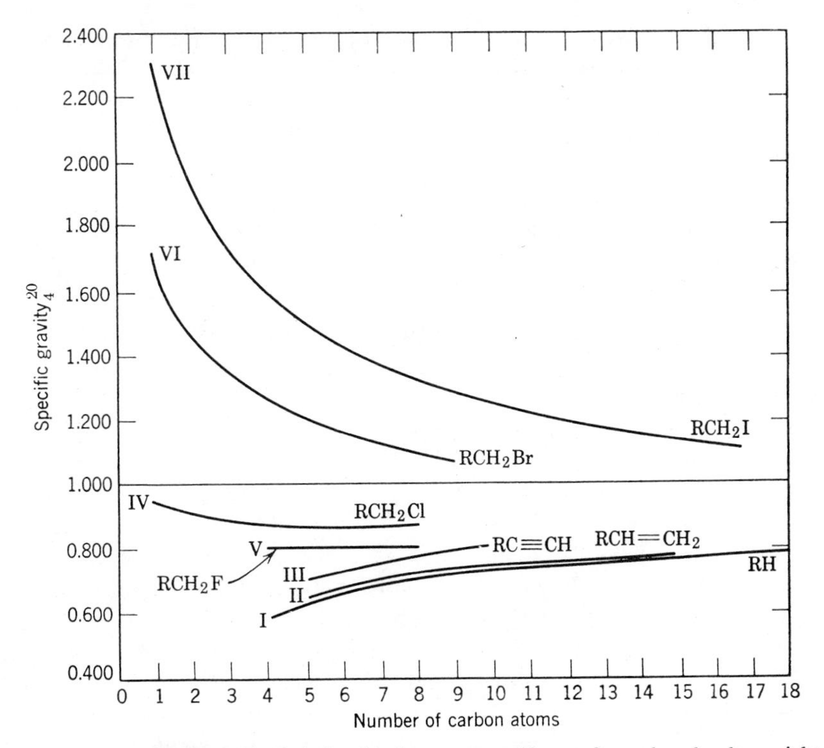

Figure 3.21. Relationship between specific gravity and molecular weight.

gravities of the normal alkyl chlorides, lies above the curves of the hydrocarbons. It will be noted that the alkyl chlorides are lighter than water and that the specific gravities *decrease* as the number of carbon atoms is increased.

The rather limited data on the alkyl fluorides are shown by curve V, Fig. 3.21. The graph is interesting because it reveals only a very slight change in density as the number of carbon atoms is increased.

Curves VI and VII in Fig. 3.21 show that the specific gravity of the primary alkyl bromides and iodides is greater than 1.0 and that in these homologous series the specific gravity decreases as the number of carbon atoms is increased. The slopes of curves IV, VI, and VII are decreasing due to the fact that the halogen atom constitutes a smaller and smaller percentage of the molecule as the molecular weight is increased by increments of methylene units. The relative position of the curves in Fig. 3.21 shows that the specific gravity increases in the order

$$RH < RF < RCl < RBr < RI$$

provided that comparisons are made on alkyl halides with the same carbon skeleton and of the same class. Similar relationships are exhibited by secondary and tertiary chlorides, bromides, and iodides.

The specific gravities of aryl halides also arrange themselves in order of increasing weight of the substituent (Table 3.8).

Table 3.7. Specific Gravity and Double Bond Position

Name	Compound	sp gr
1-Pentene	$CH_2{=}CHCH_2CH_2CH_3$	0.645_4^{25}
2-Pentene	$CH_3CH{=}CHCH_2CH_3$	0.651_4^{25}
1,4-Pentadiene	$CH_2{=}CHCH_2CH{=}CH_2$	0.659_4^{20}
1,3-Pentadiene	$CH_2{=}CHCH{=}CHCH_3$	0.696_4^{20}
2,3-Pentadiene	$CH_3CH{=}C{=}CHCH_3$	0.702_4^{20}
1-Hexene	$CH_2{=}CHCH_2CH_2CH_2CH_3$	0.673_4^{20}
2-Hexene	$CH_3CH{=}CHCH_2CH_2CH_3$	0.681_4^{20}
3-Hexene	$CH_3CH_2CH{=}CHCH_2CH_3$	0.722_4^{20}

An increase in the number of halogen atoms present in the molecule increases the specific gravity. Compounds containing two or more chlorine atoms or one chlorine atom together with an oxygen atom or an aryl group will generally have a specific gravity greater than 1.000 (Table 3.9).

The introduction of functional groups containing oxygen causes an increase in the specific gravity. The curves in Fig. 3.22 represent the change in specific gravity of some of the common types of compounds. The ethers (curve VIII) are the lightest of all the organic oxygen compounds. The aliphatic alcohols (curve IX) are heavier than the ethers but lighter than water. The specific gravity of the alcohols becomes greater than 1.0 if a chlorine atom (ethylene chlorohydrin), a second hydroxyl (ethylene glycol), or an aromatic nucleus (benzyl alcohol) is introduced. The dip in curve IX is due to the fact that methanol is more highly associated than ethanol. The amines (curve X) are not as dense as the alcohols and are less associated. Association also causes the specific gravity of formic acid and acetic acid to be greater than 1.000; the higher liquid fatty acids are lighter than water (curve XI).

The simple esters (curve XII) and aldehydes (RCHO) are lighter than water, whereas esters of polybasic acids (curve XIII), halogenated, keto, or hydroxy esters are heavier than water. Introduction of the aromatic ring also may cause esters to be heavier than water. Examples of esters of these types that are heavier than water are phenyl acetate, methyl benzoate, benzyl acetate, ethyl salicylate,

Table 3.8. Boiling Point and Specific Gravity of Aryl Halides

Compound	b.p. (°C)	sp gr_4^{20}
Benzene	79.6	0.878
Fluorobenzene	86	1.024
Chlorobenzene	132	1.107
Bromobenzene	156	1.497
Iodobenzene	188	1.832

Table 3.9. Specific Gravity Change per Number of Chlorine or Oxygen Atoms

Compound	sp gr	Compound	sp gr
Benzyl chloride	1.1026^{18}_{4}	Carbon tetrachloride	1.595^{20}_{4}
Benzal chloride[a]	1.2557^{14}_{4}	Ethylene chlorohydrin	1.213^{20}_{4}
Benzotrichloride	1.3800^{20}_{20}	Chloroacetone	1.162^{16}_{4}
Methylene chloride	1.336^{20}_{4}	Methyl chloroacetate	1.235^{20}_{20}
Chloroform	1.4984^{15}_{4}		

[a] Benzylidene chloride, $C_6H_5CHCl_2$.

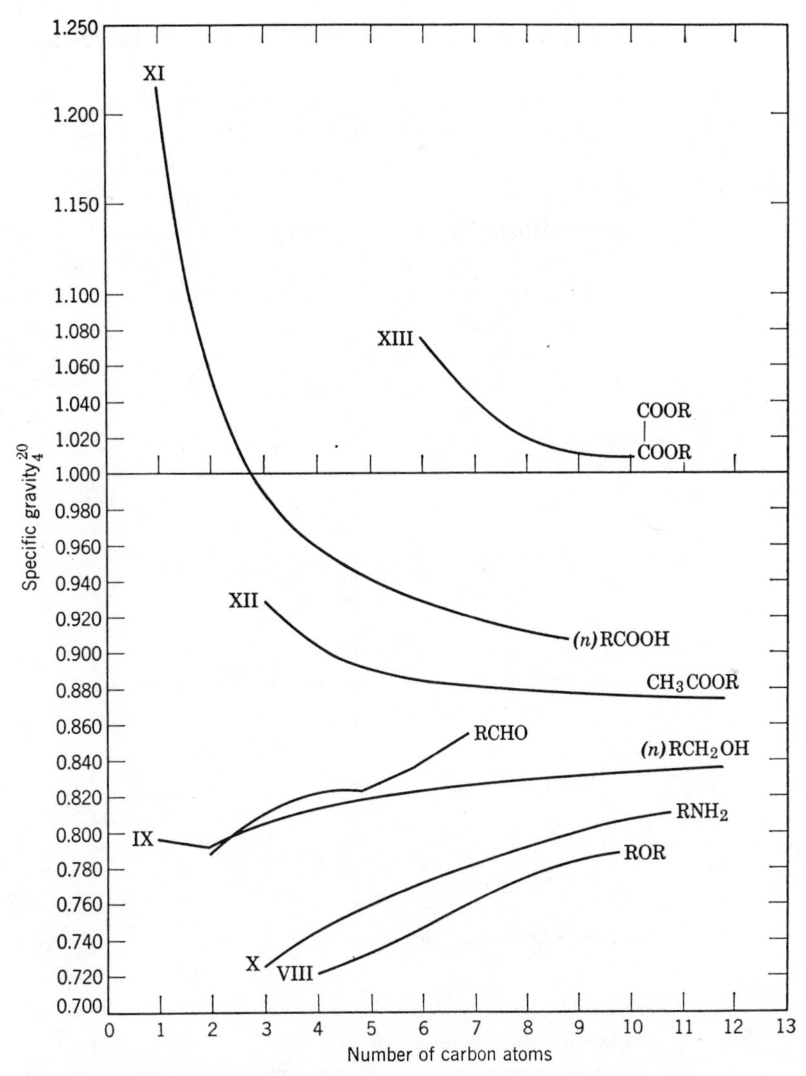

Figure 3.22. Relationship between specific gravity and molecular weight.

n-butyl oxalate, triacetin, isopropyl tartrate, and ethyl citrate. Since the hydrocarbons are lighter than water, it is to be expected that esters containing long hydrocarbon chains will show a correspondingly diminished specific gravity.

In general, compounds containing several functional groups—especially those groups that promote association—will have a specific gravity greater than 1.0. Merely noting whether a compound is lighter or heavier than water gives some idea of its complexity. This is of considerable value in the case of neutral liquids. If the compound contains no halogen and has a specific gravity less than 1.0, it probably does not contain more than a single functional group in addition to the hydrocarbon or ether portion. If the compound is heavier than water, it is probably polyfunctional.

3.3.4 The Index of Refraction of Liquids

The refractive index of a liquid is equal to the ratio of the sine of the angle of incidence of a ray of light in air to the sine of the angle of refraction in the liquid (Fig. 3.23). The ray of light undergoes changes in wave velocity ($V_{air} \rightarrow V_{liquid}$) and in direction at the boundary interface, and these changes are dependent on temperature (T) and wavelength (λ) of light. Direct measurements of the angles of incidence and refraction are not feasible; hence optical systems have been devised that are dependent on the critical angle of reflection at the boundary of the liquid with a glass prism of known refractive index. The Abbe type of refractometer operates on this principle, and a number of instruments are available commercially.[9] The advantages of the Abbe type of refractometer are

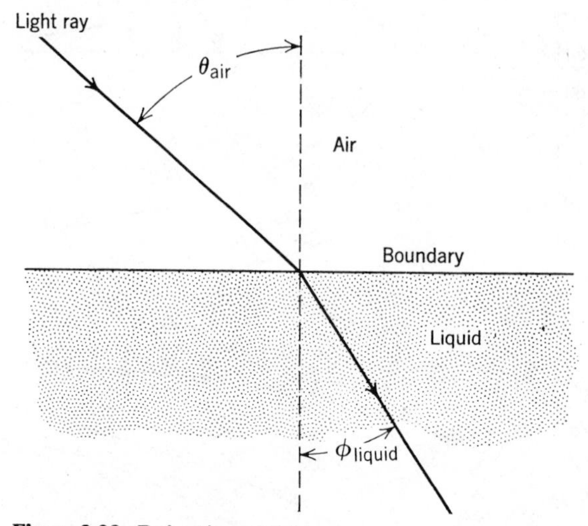

Figure 3.23. Refraction of light.

$$n_{\lambda}^{T} = \frac{\sin \theta_{air}}{\sin \phi_{liquid}} = \frac{V_{air}}{V_{liquid}}$$

where n = index of refraction.

[9] Bausch & Lomb Optical Co., Rochester, N.Y.; consult the annual instrument guides in the journals *Science* or *Analytical Chemistry* for additional sources.

that (1) a white-light source of illumination may be used, but the prism system gives indices of refraction for the sodium D line; (2) only a few drops of liquid are needed; (3) provision for temperature control of prisms and sample is incorporated; and (4) the compensating Amici prisms permit the determination of specific dispersion.

A schematic drawing of the optical system of the Bausch & Lomb Abbe 3L refractometer is shown in Fig. 3.24, and the instrument in Fig. 3.25. Water at 20°C is allowed to flow through the jackets (*J*) surrounding the prisms (P_1, P_2). If the liquid sample is free-flowing, it is introduced by means of a pipet through one of the channels beside the prisms (*D*). If the sample is viscous, the upper prism is lifted and a few drops spread on prism P_2 with a wooden applicator. The prisms are then closed slowly, any excess liquid being squeezed out. The lamp (*L*) is turned on. While one looks in the eyepiece (*E*), the coarse adjusting wheel (*A*) is turned and the position of the lamp (*L*) is also adjusted so as to obtain a uniformly lighted field. Then the eyepiece (*E*) is focused on the cross hairs (*H*)

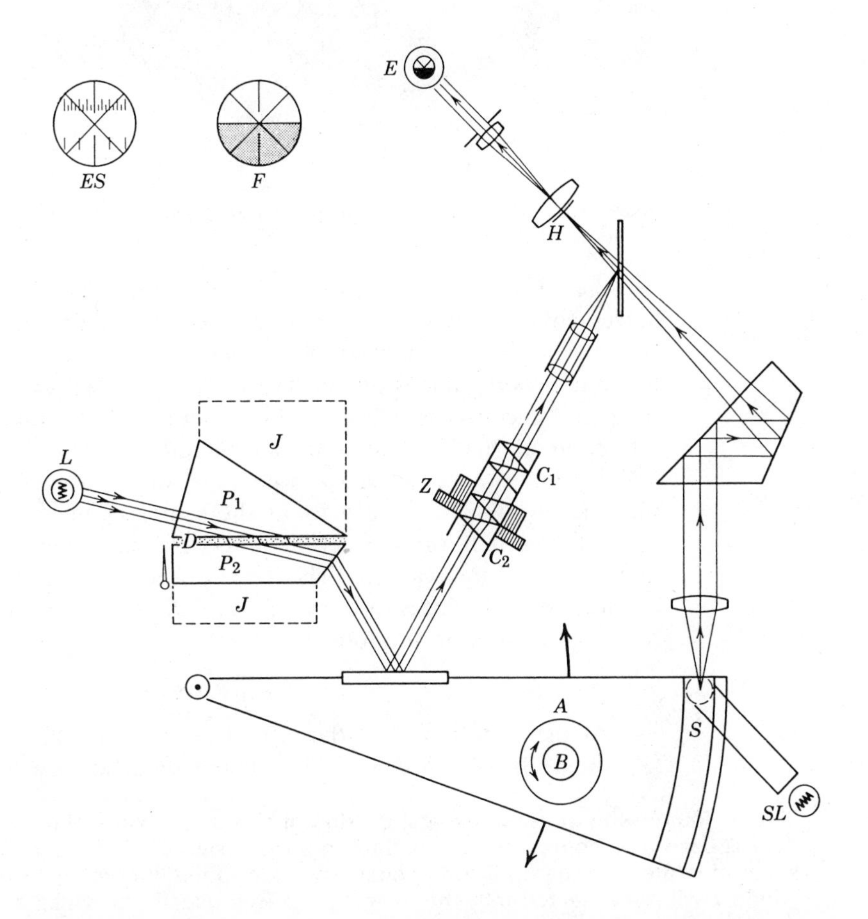

Figure 3.24. Schematic diagram of the optical system of the Abbe 3L refractometer. (Redrawn from the operating manual by courtesy of Bausch & Lomb Optical Co., Rochester, N.Y.)

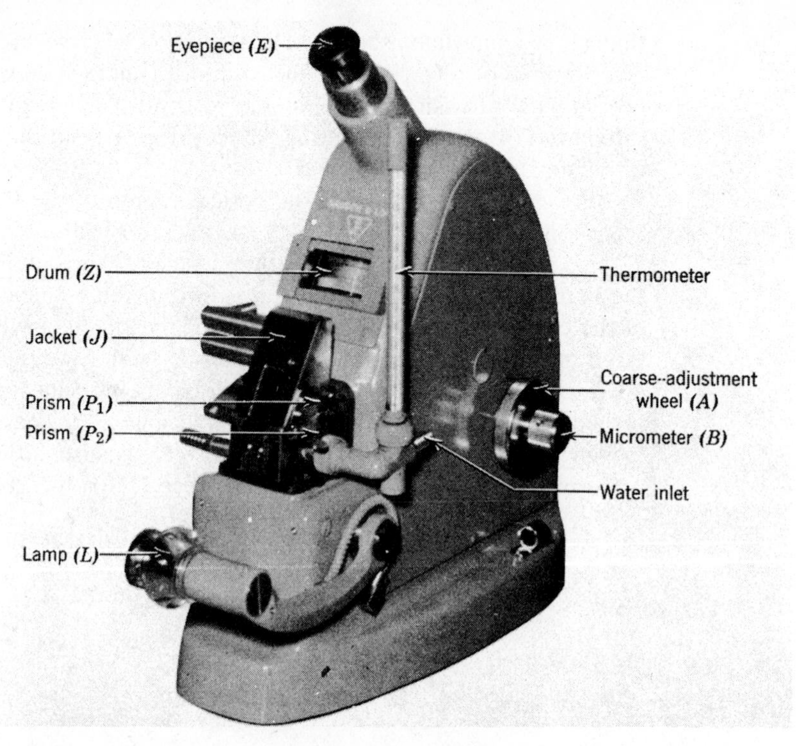

Figure 3.25. Bausch & Lomb Abbe 3L refractometer. (Courtesy Bausch & Lomb Optical Co., Rochester, N.Y.)

and the wheel (A) is turned so that the dividing line between the light and dark halves coincides with the center of the cross hairs. Usually the borderline is colored and is achromatized by turning the milled screw (Z) until a sharp black-to-white dividing line is obtained at the center.[10] This adjusting screw (Z) rotates the two Amici prisms (C_1, C_2), which compensate for differences in the degree of refraction of light of different wavelengths present in white light. After the dividing line is made as sharp as possible, the micrometer screw (B) is turned so that the dividing line is exactly at the center of the cross hairs, as shown at F. Then the switch on the left side of the instrument is depressed to cause illumination of the scale (S). The index of refraction for the sodium D line is read to three decimal places in the eyepiece (ES) and the fourth place estimated. The result is recorded in the following form:

$$n_{\mathrm{D}}^{20} = 1.4357$$

At the same time that the refractive index is read, the reading on the drum (Z) should be noted. The prisms should then be cleaned by means of a cotton

[10] Certain liquids with unusual dispersions do not give a sharp line of demarcation when an electric light is used as a source of illumination. In such cases, the lamp should be swung down and a sheet of white cardboard placed at such an angle to the illuminating prism that daylight from a window will be reflected into the instrument. This usually provides a sharp division line with careful adjustment of the dispersion screw (Z).

swab dipped in toluene or petroleum ether for water-insoluble compounds.[11] Distilled water is used to remove water-soluble compounds. Extreme care must be taken not to scratch the prisms. Metal or glass applicators should be avoided, and only clean absorbent cotton (free from dust) used to clean the prisms. Manuals for operating various instruments should be consulted for variations in operating procedures. A complete discussion of the principles of refractometry and refractometers is available.[12]

Discussion. The values for density and refractive index are useful in excluding certain compounds from consideration in the identification of an unknown. Care must be taken, however, that the sample is pure. It is best to determine these physical constants on a center cut from distillation (b.p. determination) or a gas chromatography collection.

These two constants can also serve as a final check on an unknown after its identity and structure have been established. They are of value in research work for checking structures. This checking is accomplished by comparing the observed index to the literature value (see tables in Appendix III or consult references in Chapter 10).

3.4 GAS CHROMATOGRAPHY

Gas chromatography (gc) has also been referred to as vapor phase chromatography (vpc). Gas chromatography requires a stationary phase that is usually a heat-stable and nonvolatile liquid; this phase is coated uniformly over an inert solid support (e.g., over crushed firebrick that is finely and evenly divided). The mixture of stationary phase and solid support is packed into a tube (metal or, less often, glass). The sample to be analyzed is injected at a temperature sufficient to vaporize it; this gaseous sample is caught up by the carrier gas (i.e., the mobile phase). This movement allows contact with, and thus adsorption of the sample by, the stationary liquid phase. Components of the sample are separated by their relative abilities to be adsorbed by the supported stationary liquid;[13] separation is thus a function of both the polarity and volatility of the components of the sample.

Gas chromatography is useful for purity determinations and component analysis of sufficiently volatile organic compounds. Observation of a single, large peak in a variety of gc determinations (various columns, temperatures, etc.) for a given sample is a strong indication of its purity. Samples must thus be thermally stable to volatilization conditions for gc analysis.

---

[11] Do *not* use acetone to clean prisms.

[12] N. Bauer, K. Fajans, and S. Z. Lewin, "Refractometry," chap. 18, pp. 1140–1281, in *Physical Methods of Organic Chemistry*, 3rd ed., Vol. I, Part II, edited by A. Weissberger (Interscience, New York, 1960).

[13] The detailed mechanism for the separation of components by gc is quite complex; for more discussion one should consult one or more of the specialized monographs listed in Chapter 10.

A simplified schematic diagram of a gas chromatograph is shown in Fig. 3.26. The instrument consists of the following:

1. A tank of carrier gas (see point 1, Fig. 3.26), which is usually helium (or, less frequently, nitrogen). Helium is used because it is inert and has a high thermal conductivity.
2. A method of controlling the gas flow. This usually involves a valve on the tank (points 1 and 2, Fig. 3.26) and a regulator on the instrument (point 3, Fig. 3.26).
3. An injection port (see point 4, Fig. 3.26, and *A* and *B* on Fig. 3.28). This is a metal cap, with a hole over a piece of rubber or plastic material that can be pierced with a syringe.
4. A heated column (see point 5, Fig. 3.26). This is a metal (or glass) tube that contains the solid support and stationary phase (Table 3.10); this "column" is coiled to fit in the heated compartment and is connected from the injection port to the detector.
5. A detector. The two most common detectors are of the thermal conductivity and flame ionization types. The thermal conductivity detector measures changes, at a filament, in the conductivity of the carrier gas as a function of sample content; this change is electrically passed on to the recorder (Fig. 3.28). Flame ionization measures sample content by burning the eluted sample in a small hydrogen flame, and counting the ions so produced.

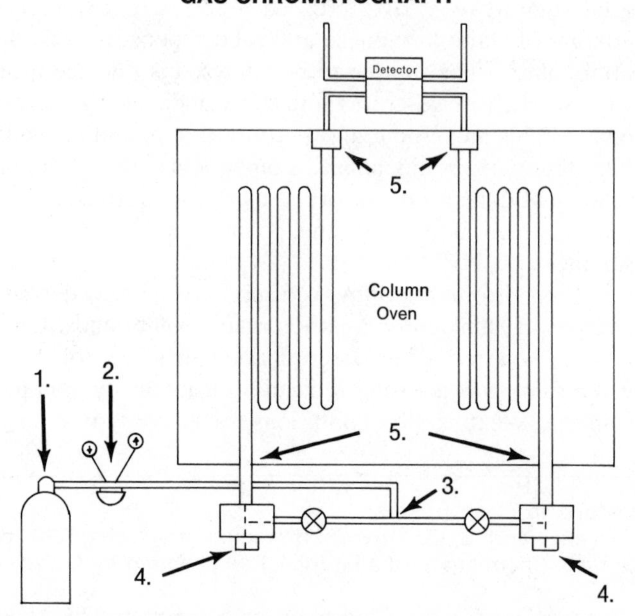

Figure 3.26. Schematic diagram of a gas chromatograph.

Table 3.10. Abbreviated List of gc Stationary Phases[a]

Stationary phase	Use[b]	Solvent[c]	Max. temp.[d]
Apiezon L	B, S, P, II–V	B, T	300
Squalene	B, N, P, III–V; gases	T	140
DEGA	III	A	200
DEGS	P, S; II–IV	A, C	200
Tricresyl phosphate (TCP)	S; II–V; gases	A, M	125
Carbowax 20M	P, S, I–V	C	250
Versamid 900	N, II–V	[a]	350
DC 710 (silicone oil)	N, P, (0); IV, V	A, C	300
SE 30	N, P, S, (0); II–V; gases	C, T	350
FFAP	S, Si, (0); I–IV	CH_2Cl_2	275
Porapak (porous polymers)[a]			
Silver nitrate	Olefins	M	50

[a] Consult A. Gordon and R. Ford, *The Chemist's Companion* (Wiley, New York, 1972), for more details.

[b] Code: Specific elements in compound (B, Si, P, etc.); for inorganic and organic compounds: (0); I, extensive hydrogen bonders (glycols, aminoalcohols, polyacids, etc.); II, polar hydrogen bonders and active hydrogen compounds (alcohols, phenols, primary and secondary amines, nitriles (α-H), RNO_2(α-H); III, polar hydrogen bonds, but no exchangeable hydrogen (e.g., no α-H); IV, modestly exchangeable hydrogen; V, neutral compounds alkanes, mercaptans, sulfides).[a]

[c] Solvents for column preparation (heat purge often necessary): A, acetone; B, benzene; C, chloroform; M, methanol; T, toluene.

[d] Upper temperature limit (°C); manufacturer's recommendation.

Procedure. Most instruments have a single temperature meter and a knob that can be set to choose the portion of the instrument that needs to have the temperature monitored (Fig. 3.27) by the meter. The detector and injector blocks should be heated continuously; these points come to temperature very slowly (overnight) and are usually several tens of degrees warmer than the column. The column temperature is usually set such that it is 10 to 20°C above the boiling point of the sample of interest;[14] it should, however, *always* be below the upper limit recommended for the stationary phase (Table 3.10). Typical column temperatures are 50 to 200°C, and the corresponding detector and injector temperatures are 80 to 250°C; clearly, sample stability should influence temperatures chosen. To begin instrument operation, one should be *certain* that there is helium flow[15] before turning on the filament (bridge) current in a thermal conductivity instrument; flow is assured by checking the tank and instrument gas gauges *and* by checking the detector exit port with a bubble meter. Allow about 30 min for the flow rate and column temperatures to reach a steady state and measure the flow rate with the instrument gauge or with a bubble meter. The current is then turned on and set typically between 100 and 200 mA (Fig. 3.27). The chart pen should be zeroed and the recorder started. A sample is injected (e.g., 1 μl if neat, or 10–100 μl if a 1–10% solution in a volatile solvent) by means of a microsyringe

[14] If the sample is a solid and is being introduced in solution, an educated guess as to these temperature settings is in order.

[15] Failure to have an inert gas around the detector filament will very likely result in destruction of the filament.

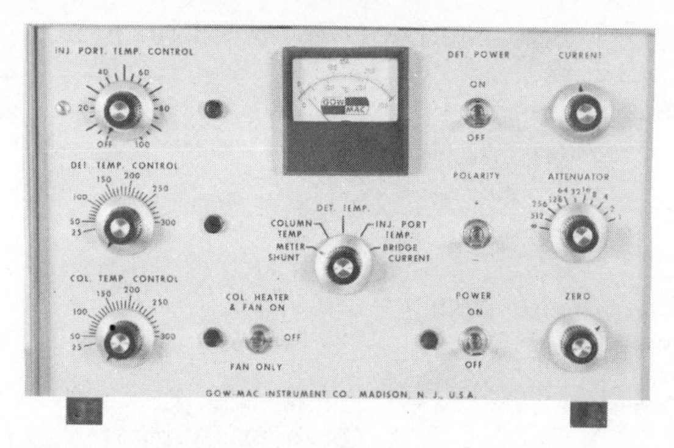

Figure 3.27. Instrument panel of typical gas chromatograph: GOW-MAC Instrument Co., thermal conductivity detector.

into port *A* or *B*. The chart should be marked at the time of injection; a record should also be made of the column size and type used, the column temperature, gas flow rate, and other details (Fig. 3.29). As the peaks are traced on the recorder, one should be ready to make attenuation (detector response) changes (see the Attenuator knob in Fig. 3.27 and see Fig. 3.29) such that the top of the peak can be observed. Retention time is taken as the time from injection until the time this peak maximum is obtained. After all of the components have been eluted, the chromatograph is shut down by turning off the filament current and only then reducing the carrier flow to a trickle.

Discussion. Retention times (or retention volumes = flow rate × retention time) are characteristic of the compound of interest. One should, however, avoid precise, quantitative comparisons to literature retention values; too many parameters have to be reproduced to justify such correspondence. However, injection of an unknown mixed with an authentic sample of the suspected compound resulting in a single, sharp peak is strong support for identity of the two components of this mixture. This mixture should be analyzed on both polar and nonpolar gc columns. Reaction progress (reactant, product, side product quantities, etc.), as well as purity, can be monitored by gas chromatography.

If a peak is due to only one compound, the area under a peak is proportional to the number of moles of compound causing that peak. In Fig. 3.29, for example, areas could be determined by measuring the heights of each of the water and amine peaks, and multiplying these heights by the widths of the corresponding peaks at half-height.[16] The area for the water peak should be multiplied by 8 (it was measured at one-eighth of the recorder sensitivity at which the amine peak was measured).

A concept that is important for quantitative gas chromatography is the molar response. Molar response (m.r.) is simply the area of a peak per mole of

[16] This triangulation method for measuring peak areas should be used only if the peaks are quite symmetrical.

compound that is producing that peak. The molar response can be measured by use of the following simple formula:

$$\text{m.r.} = \frac{\text{area of peak}}{\text{moles of compound producing the peak}} \qquad (\text{cm}^2/\text{mole} = \text{common units})$$

Where possible, the molar response of all components in a gas chromatogram should be determined. This is done by chromatographing known volumes of

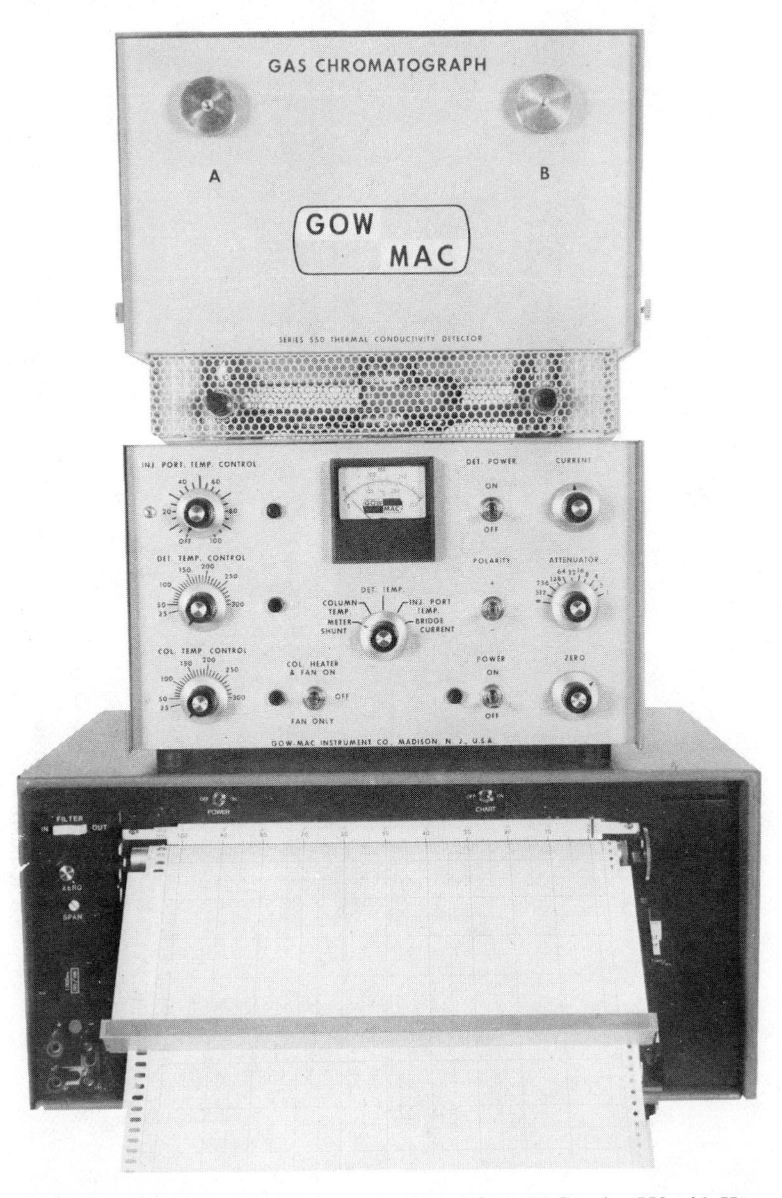

Figure 3.28. Gas chromatograph and recorder: GOW-MAC series 550 with Honeywell recorder.

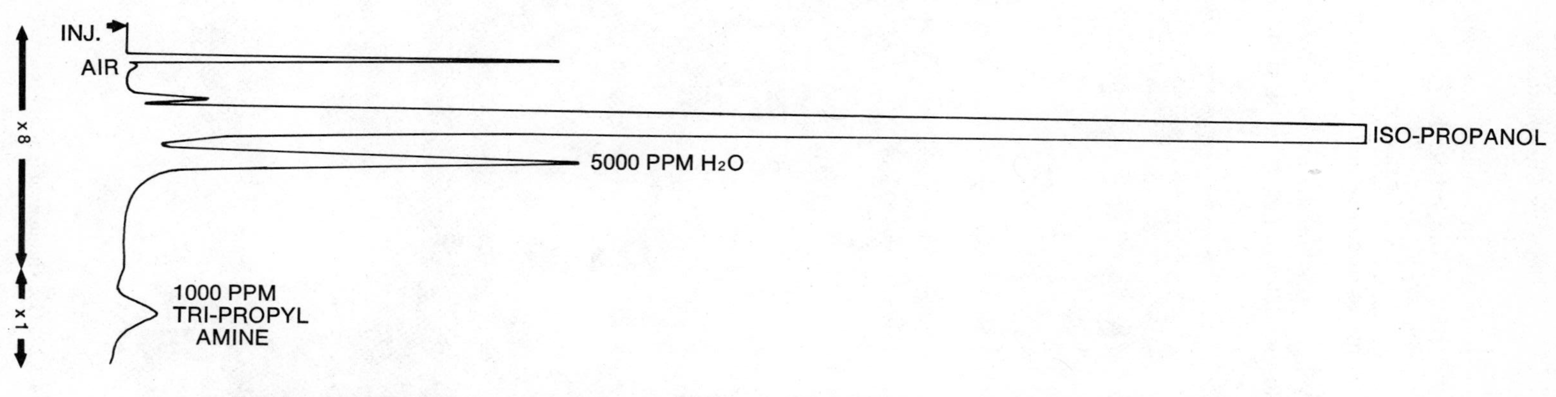

Figure 3.29. Gas chromatograph trace recording. Note that the water and amino samples are measured at different sensitivities.

standard solutions of each component. These molar responses are then used to convert peak areas of all components in a mixture to moles of compound and thus mole percentages of all components in the mixture can be calculated.

3.5 MOLECULAR WEIGHT DETERMINATION

We shall mention here five methods of molecular weight determination: the Rast method (freezing-point depression), vapor phase osmometry, mass spectrometry, neutralization equivalents, and saponification equivalents. The Rast method requires very simple equipment; in addition, this method is often useful for compounds that cannot be measured mass spectrometrically. The results are usually only approximations, and we shall not describe the procedure.[17] Vapor phase osmometry and mass spectrometry require fairly sophisticated instruments. Mass spectrometry provides the most accurate molecular weights (and often provides the molecular formula and structure). Compounds that are thermally sensitive or that do not have significant vapor pressures or do not give stable molecular ions should, however, be measured by other than mass spectral techniques. Neutralization equivalents (for acids and amines) and saponification equivalents (for esters) are titration techniques; they do, however, require information about the number and type of functional groups in the unknown compound, so these methods are not discussed until the appropriate part of Chapter 6. Vapor phase osmometry does not require the high thermal stability and volatility that mass spectrometry does. Vapor phase osmometry and mass spectrometry are discussed just below.

3.5.1 Vapor Phase Osmometry

A technique that is based on a colligative property that can be used to determine the molecular weight of organic compounds is vapor phase (or pressure) osmometry. The vapor phase osmometer, or more correctly, the thermoelectric differential vapor pressure instrument, takes advantage of the fact that the vapor pressure of a solution depends on the molal concentration of the solute in that solution. A schematic of a commercial instrument is provided in Fig. 3.30; it consists of two thermistor beads in a container holding an atmosphere saturated in solvent vapor. At the bottom of the container there is a solvent cup and wick that maintains this atmosphere (e.g., methanol vapor). The thermistor beads are connected to a Wheatstone bridge circuit that is initially balanced with a drop of solvent on each bead. One bead is then covered with a drop of solution of known concentration. The time between drop placement and measurement (500 sec in Fig. 3.31) should be the same for every determination. The solvent begins to condense on the bead bearing the drop of solution because of the lower vapor

[17] See the reference in footnote 18, p. 70 for a discussion of the Rast procedure.

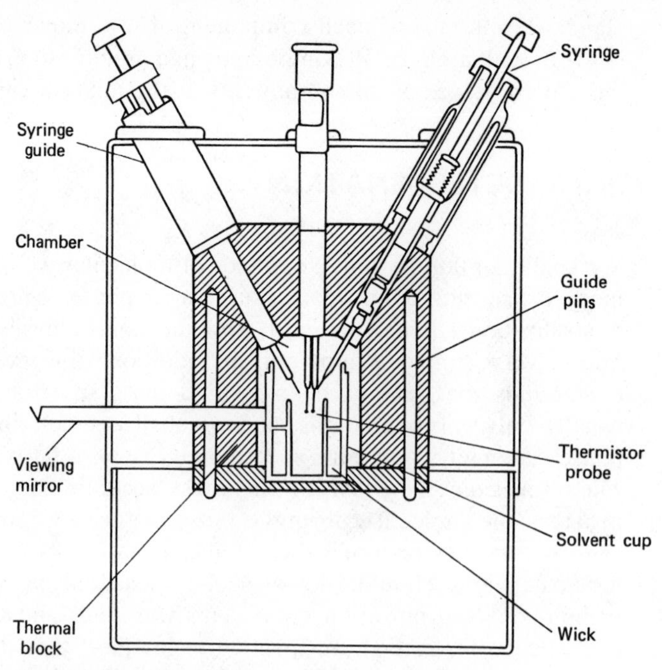

Figure 3.30. Schematic of vapor phase osmometer.

pressure of its solution. The resulting increase in temperature (due to condensation) is measured in potential by the Wheatstone bridge circuit. The instrument and solvent are calibrated with an organic solute of known composition. For example, benzil, $(C_6H_5\overset{O}{\overset{\|}{C}}—)_2$, can be obtained in a pure form, is easily dried, is nonvolatile, and is soluble in organic solvents and thus is a good calibration standard. The solvent used for calibration and for analysis of the unknown must be the same.

The typical commercial instrument is capable of temperature change measurements of about 10°C (i.e., potential changes in a microvolt range of 10 to 10^5 μV) with a precision of a few percent. The temperature range of the system is selected between room temperature and ca. 130°C. Separate probes are required for aqueous and nonaqueous systems. The volume of solution required for a single measurement need be only 10 μl to 0.1 ml.

Procedure. Prepare standard solutions (molalities of Table 3.11) of benzil (molecular weight = 210) in methanol. Balance the osmometer at 37.5°C containing saturated methanol vapor and drops of methanol solvent on both thermistor beads. Cover one of the beads with a calibrating solution (e.g., with the first solution of Table 3.11, which was made from 30 mg of benzil per 1000 g of methanol); measure and record the change in voltage (μV). Do the same for the remaining calibration solutions. Plot $\Delta V/m$ versus m (where m = molality) and

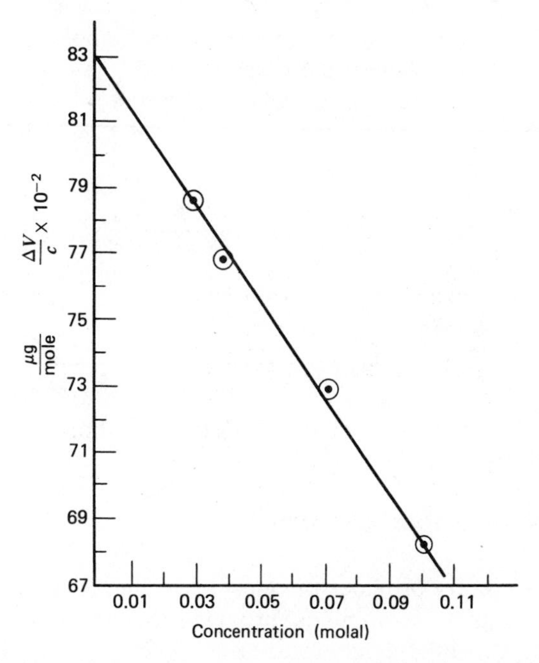

Figure 3.31. Vapor phase osmometry calibration curve. Benzil in methanol at 37.5°C, 500 sec.

extrapolate to $m = 0$ (see Fig. 3.31); the value of $\Delta V/m$ at this point, $(\Delta V/m)_{m=0}$, is 83.1×10^2 μV kg/g for this example (Fig. 3.31). The calibration constant for the instrument–solvent system is then determined from the equation

$$K_{\text{solvent}} = (M)\left(\frac{\Delta V}{m}\right)_{m=0} = (210)(83.1 \times 10^2)$$

where M = the molecular weight of calibrating solute (g/mole).

$$K_{\text{MeOH}} = (210 \text{ g/mole})(83.1 \times 10^2) \frac{\mu\text{V} \cdot \text{kg solvent}}{\text{g solute}}$$

$$= 1.75 \times 10^6 \frac{\mu\text{V} \cdot \text{kg solvent}}{\text{mole solute}}$$

Table 3.11. Calibration Data—Vapor Phase Osmometry

Concentration[a]	$\Delta V(\mu V^b)$	$\Delta V/m \times 10^{-2}$
0.03	236	78.6
0.04	308	77.0
0.07	510	72.9
0.10	680	68.0
(0.00)	—	(83.1)

[a] m = molality = grams of solute/kilograms of solvent; benzil in methanol.
[b] μV = microvolts.

Now prepare solutions of the unknown in methanol of roughly the same molalities as for benzil above. Measure ΔV versus m for the unknown; make a plot of $\Delta V/m$ versus m for these data. Extrapolate the graph to $m = 0$ to obtain $(\Delta V/m)_{m=0}^{\text{unknown}}$. The molecular weight of the unknown (M_u) can then be obtained from the equation

$$M_u = \frac{K_{\text{MeOH}}}{(\Delta V/m)_{m=0}^{\text{unknown}}}$$

If other than benzil is used as a calibrating solute, the following formula is used:

$$M_u = \frac{K}{(\Delta V/m)_{m=0}^{u}} = \frac{(M_k)(\Delta V/M)_{m=0}^{k}}{(\Delta V/m)_{m=0}^{u}}$$

where M_k = molecular weight of (known) calibrating solute and $(V/m)_{m=0}^{k}$ = intercept for the calibrating solute.

3.5.2 Mass Spectrometry

Mass spectrometry provides the most accurate method for determining molecular weights of organic compounds, provided that the sample is of sufficient thermal stability at the inlet temperature and is structurally amenable to providing a substantial molecular (or parent) ion peak. If evidence of decomposition is shown in boiling point or gas chromatographic analyses (above), alternative methods of molecular weight determination (vapor phase osmometry, Rast method[18]) should be considered.

A schematic of a typical mass spectrometer is given in Fig. 3.32. In electron-impact mass spectrometry, a sample is volatilized and the gas is led directly into a beam of ionizing electrons; less frequently the sample (especially if it is non-volatile) is thrust directly into the ionizing beam. The most important process is the displacement of one electron from the molecule (M) to form the molecular ion ($M^{\cdot+}$, a radical cation):

$$M \xrightarrow{e^-} M^{\cdot+} + 2e^-$$

This ionization does not significantly change the mass of the particle and allows particle acceleration by means of an applied potential. The accelerated ions are then passed into the field of a magnet; the magnetic field induces a circular path, the radius of curvature of which is proportional to the mass-to-charge ratio (m/e) of the particle. Since the magnitude of the charge (e) is normally 1, the magnitude of m/e is a measure of mass. Although the mass spectrum is complicated by peaks due to fragmentation of the molecular ion, for example,

$$M^{+} \rightarrow F_1^{+} + F_2^{\cdot}$$

---

[18] E. L. Skau and H. Wakeham, *Physical Methods of Organic Chemistry*, 2nd ed., Vol. I, Part I, edited by A. Weissberger (Wiley-Interscience, New York, 1949), pp. 90ff.

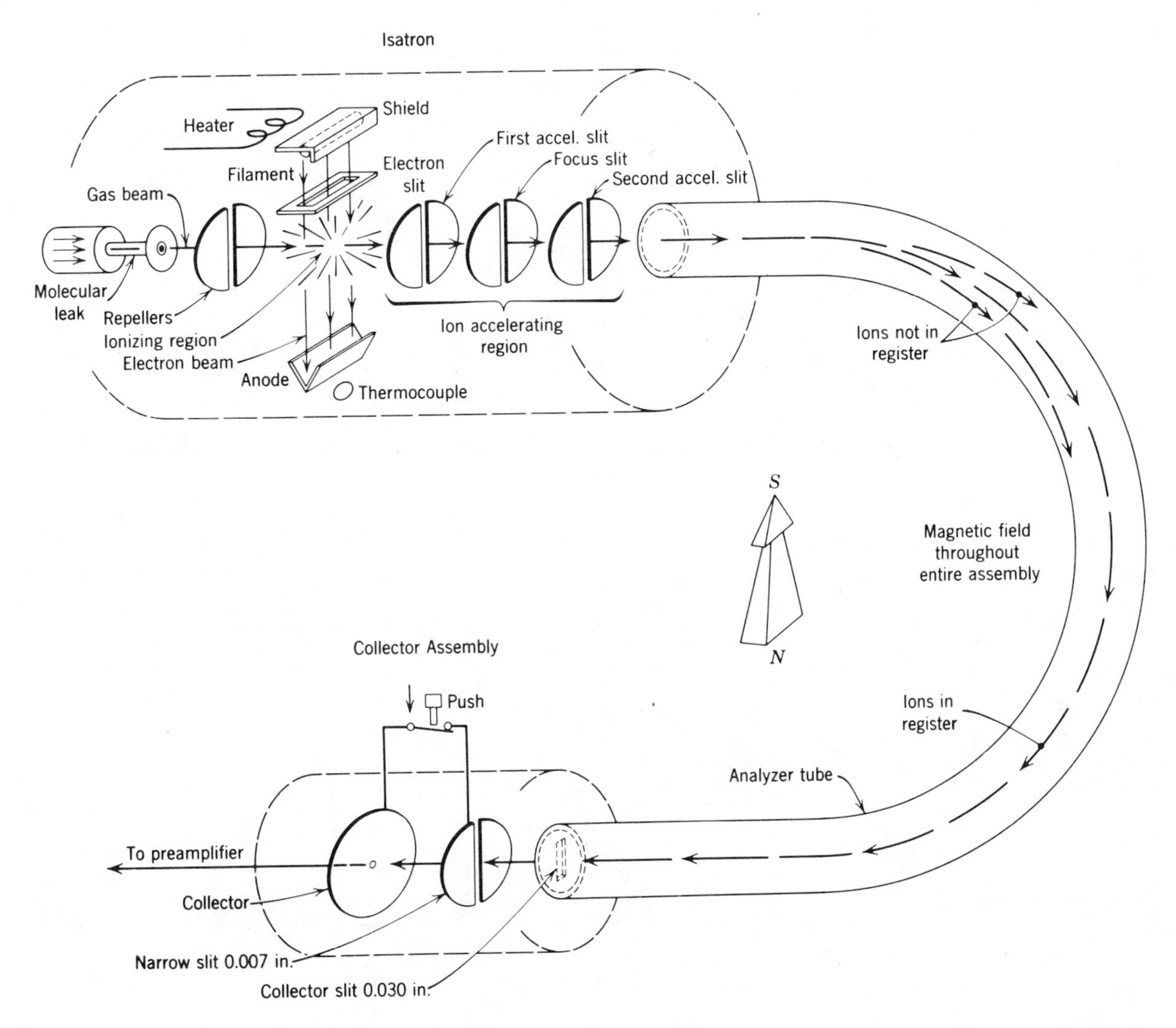

Figure 3.32. Schematic diagram of CEC model 21-103 mass spectrometer. The magnetic field is perpendicular to the page.

the molecular ion of a pure sample can often be identified as the largest peak in the cluster of peaks of highest *m/e* values (Fig. 3.33; often the molecular ion is not accompanied by a cluster).

Procedure (see Figs. 3.34–3.40). Close the heated inlet's pump-out value after pumping out the previous sample (Fig. 3.35). Place sample (Fig. 3.36) in the inlet piston; push this sample into the heated inlet. Dial a leak rate (Fig. 3.37) from the inlet to the source that results in a measurable sample pressure (determined by trial and error). Set the energy (Fig. 3.38) of the ionizing beam (e.g., 70 eV), the scan starting point (e.g., 50 amu, Fig. 3.39), and the scan rate

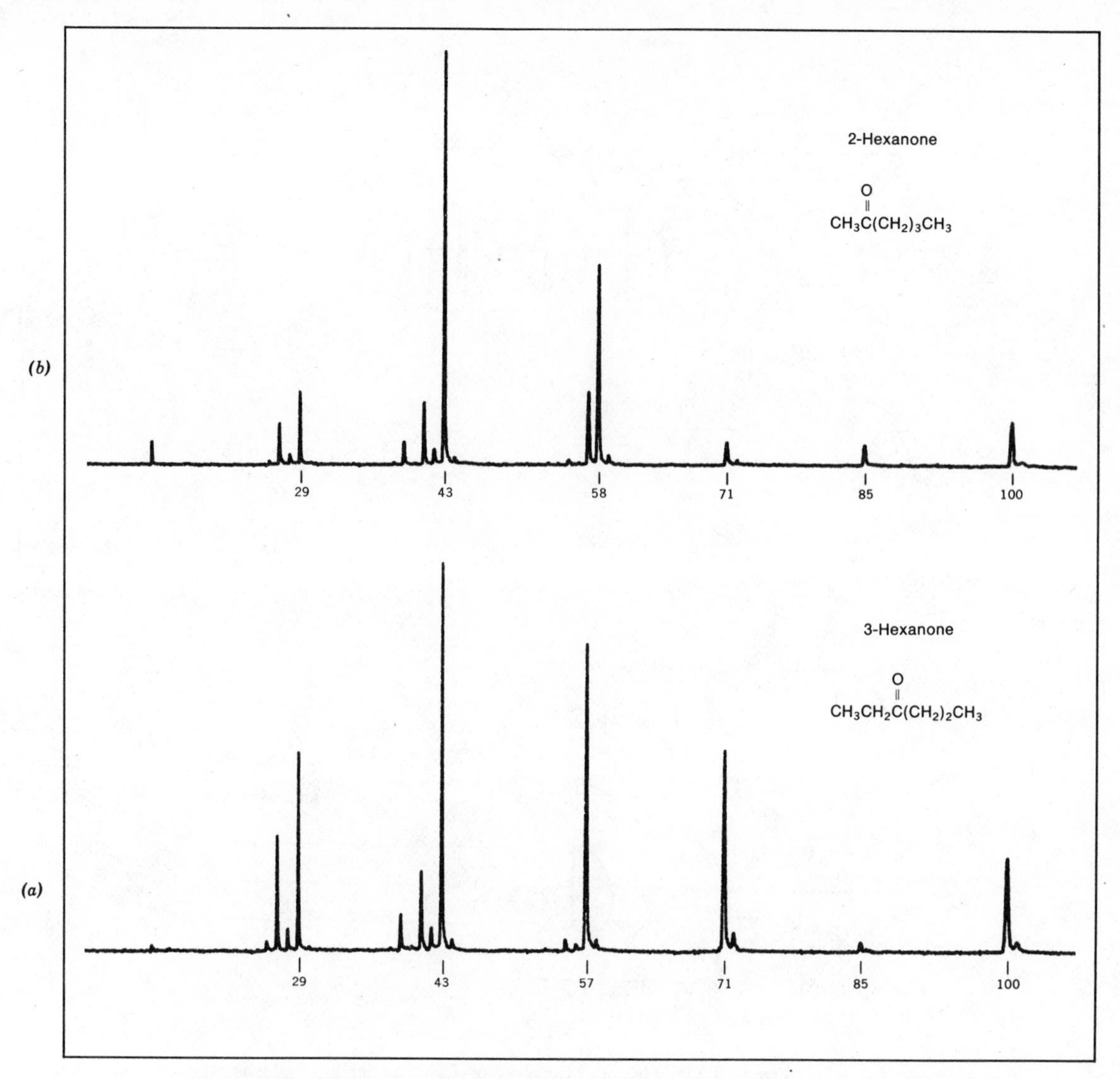

Figure 3.33. (*a*) Mass spectrum of 3-hexanone. Molecular ion (M) at *m/e* 100. The M+1 peak is at *m/e* 101, and the base (most intense) peak is at *m/e* 43 = CH_3CO^+. (*b*) Mass spectrum of 2-hexanone. (Courtesy of Varian Instruments Division, Palo Alto, Calif.)

(e.g., 10 amu/min). Start the recorder and push the scan-start button (Fig. 3.40). Set the spectrum amplitude so as to have the most intense peak on scale. The scan rate can be used to mark off the chart paper on linear response instruments; thus the mass of the molecular ion peak is determined by a simple count.

Discussion. The cluster of peaks near the molecular ion peak, if observable, is useful for molecular formula determination and will be discussed in Chapter 4. Masses of fragmentation peaks are also useful and will be discussed in Chapter 6.

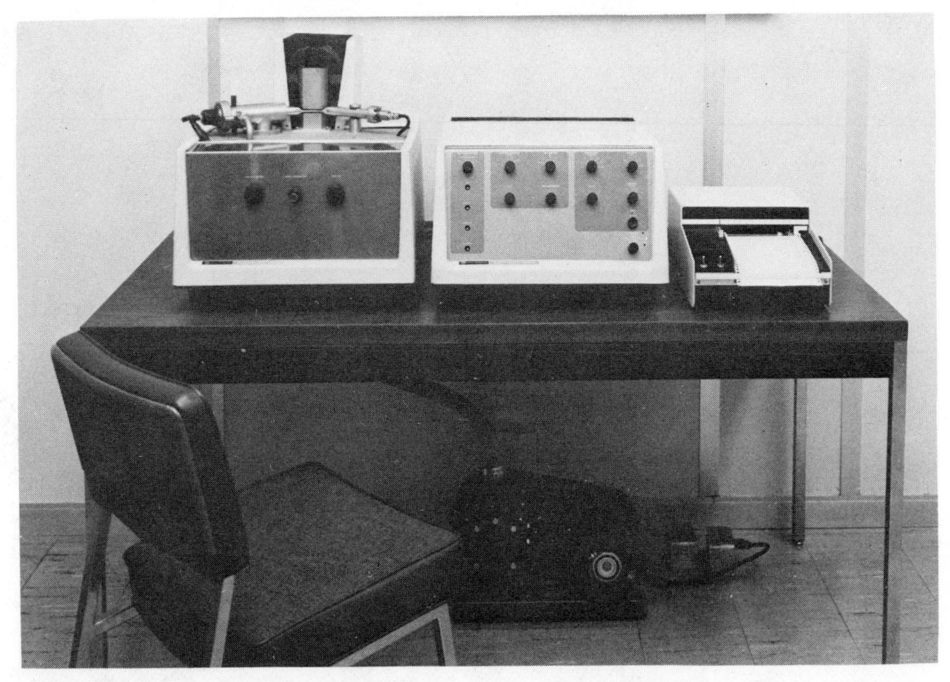

Figure 3.34. Components of the EM-600 mass spectrometer. (Courtesy of Varian Instruments Division, Palo Alto, Calif.)

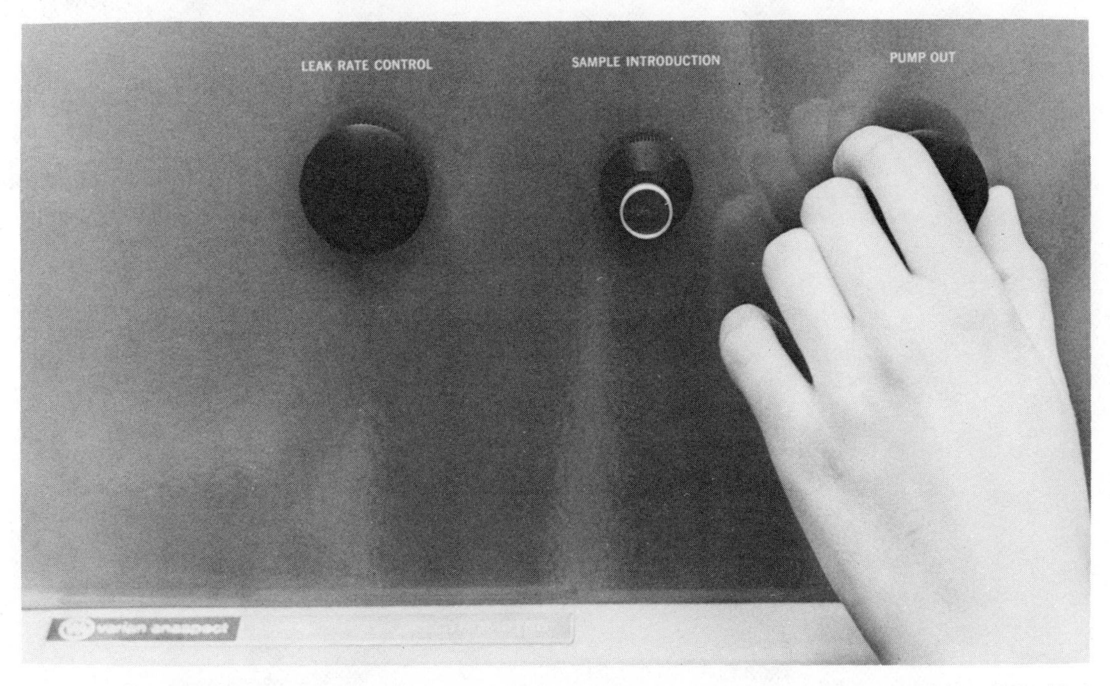

Figure 3.35. Closing the pump-out valve. (Courtesy of Varian Instruments Division, Palo Alto, Calif.)

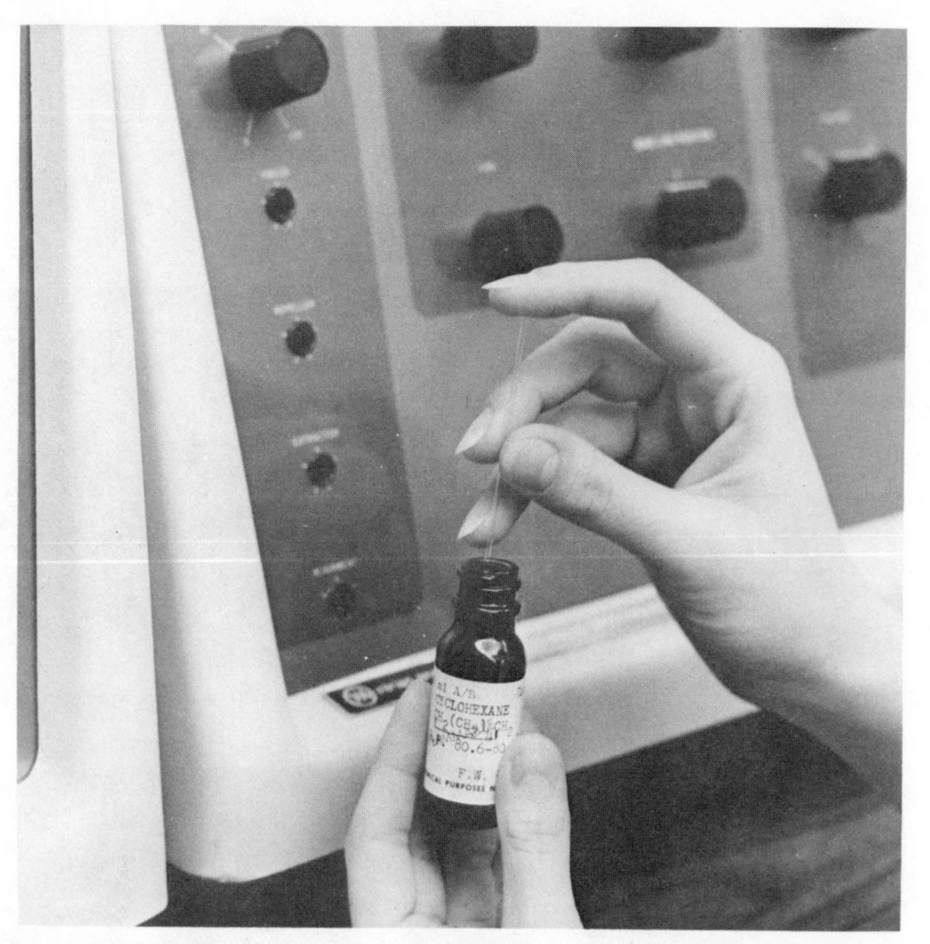

Figure 3.36. Placing sample in the inlet piston. (Courtesy of Varian Instruments Division, Palo Alto, Calif.)

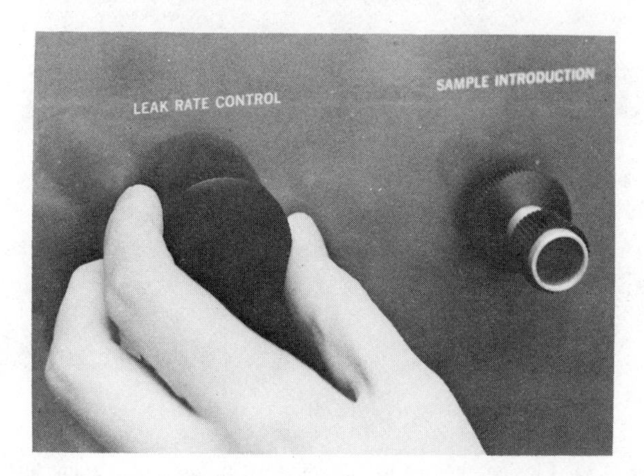

Figure 3.37. Dialing a leak rate. (Courtesy of Varian Instruments Division, Palo Alto, Calif.)

Figure 3.38. Setting the energy of the ionizing beam. (Courtesy of Varian Instruments Division, Palo Alto, Calif.)

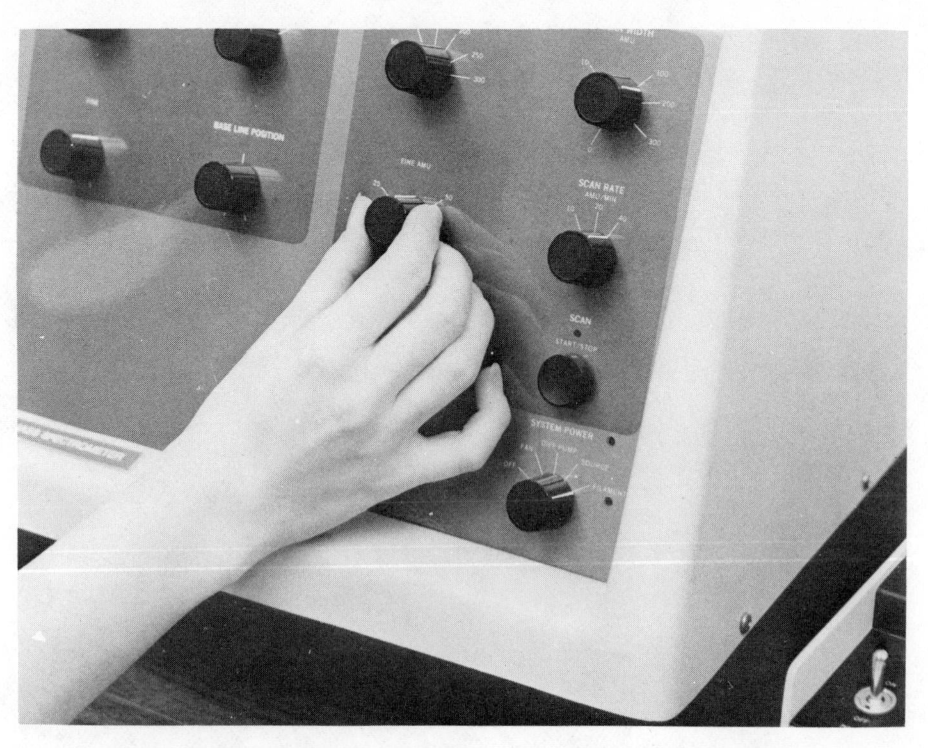

Figure 3.39. Setting the scan starting point. (Courtesy of Varian Instrument Division, Palo Alto, Calif.)

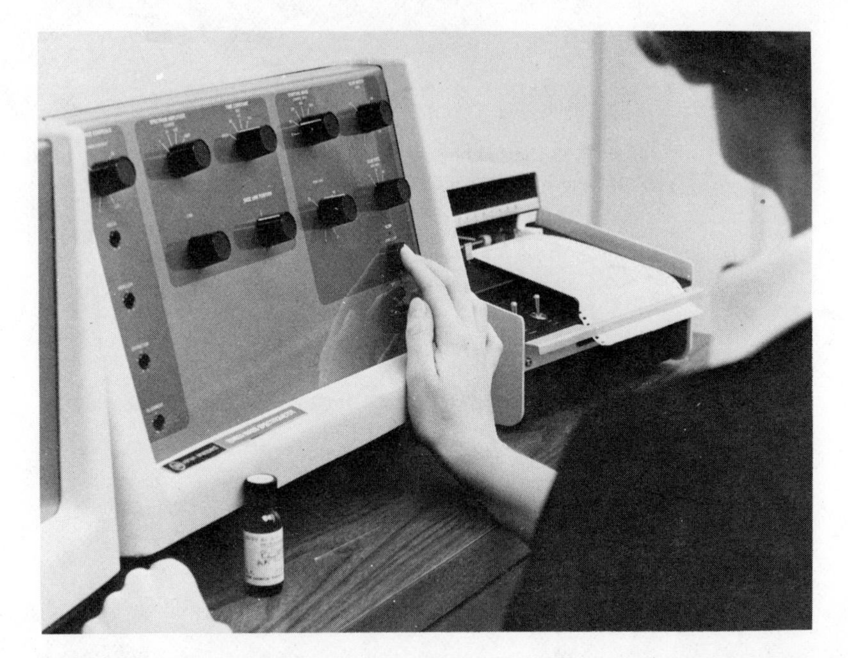

Figure 3.40. Starting the recorder and pushing the scan-start button. (Courtesy of Varian Instrument Division, Palo Alto, Calif.)

CHAPTER FOUR

determination of molecular formula

Once the chemist has established the purity and molecular weight of an organic compound (Chapter 3), the next fact to be determined is the molecular formula. Even if only the empirical formula (simplest formula) is determined, the relationship of this to the molecular weight gives rise to the molecular formula (see below). We shall first describe methods that simply tell what elements are present in the organic compound (qualitative analysis), and follow this by discussions of methods that describe quantitatively the amounts of each element (quantitative analysis) present in a compound.

4.1 QUALITATIVE ELEMENTAL ANALYSIS

Organic chemists do not usually employ chemical tests for carbon, hydrogen, and oxygen. It is often valuable, however, to determine the existence of common companion elements: nitrogen, sulfur, fluorine, chlorine, bromine, and iodine. The detection of these companion elements by means of chemical ("wet") tests (after these elements have been leached from the compound by reaction with molten sodium) is usually straightforward. Many of these chemical tests are very sensitive, so all aqueous solutions should be *carefully* prepared with distilled (or better, deionized) water. Samples that show indications of explosive character in the ignition test should either (1) not be analyzed by the sodium fusion procedure or (2) be analyzed by a smaller-scale procedure than described below. Compounds that are known to react explosively with molten sodium are nitroalkanes, organic azides, diazo esters, diazonium salts, and some aliphatic polyhalides (chloroform, carbon tetrachloride). *Safety glasses (with side shields) should be in place when this procedure is being conducted; the safety of neighbors should be preserved by proper orientation of the fusion reaction vessels (i.e., do not point vessels at your neighbors).*

Controls should be run for all tests for which there is the least doubt about the decisiveness of the results; for example, if one is unsure about the validity of the observations associated with a positive nitrogen test run on an unknown, the same test should be carried out on a compound that is known to contain nitrogen. It may even be necessary to run a control on a compound that is known *not* to contain the element of interest; observations associated with this control allow one to draw conclusions about tests yielding negative results or about the purity of the reagents involved.

Knowledge of the elemental composition of an organic compound being studied is essential for the following.

1. Selection of the appropriate experiments that serve as tests for functional groups and selection of procedures for the preparation of derivatives; that is, knowledge of the chemical composition of a compound serves as one of the guides for use of Chapter 6.
2. Efficient use and intrepretation of mass, ir, nmr, and uv spectra for deduction of molecular formula (Chapter 4), functional groups (Chapters 4 and 5), and structure (Chapter 6).

Almost all elements listed in the periodic table can be a part of an organic compound. In this text we shall, however, be concerned with detection of only a few of the elements more commonly found in organic compounds; detection of the other elements is a subject that should be dealt with in, for example, a course on instrumental analysis that covers atomic absorption, emission, flame photometric, and other instrumental procedures. For an introduction to the identification of "unknown" organic compounds, it is recommended that the possible elements be limited to C, H, P, N, O, S, F, Cl, Br, and I. A few common salts, such as those containing Na, K, and Ca, could be used (recall the ignition test on p. 32).

4.1.1 Fusion of Organic Compounds with Sodium

$$\text{C, H, O, N, S, X} \xrightarrow{\text{Na}} \begin{matrix} \text{NaX} \\ \text{NaCN} \\ \text{Na}_2\text{S} \\ \text{NaSCN} \end{matrix} \qquad \text{X} = \text{halogen}$$

Procedure A.[1] *Sodium.* A small test tube (100 × 13 mm) made of *soft* glass is fastened in a vertical position in a Bunsen clamp from which the rubber has been removed. A piece of *clean* sodium metal about 4 mm on an edge is placed in the test tube. The lower part of the tube is heated until the sodium melts and sodium vapors rise in the tube. Then one-third of a mixture of 0.1 g of the compound, mixed with *ca.* 50 mg of powdered[2] sucrose is added and heat is again applied. The addition and heating are repeated a second and third time, and then the bottom of the tube is heated to a dull red. The tube is allowed to cool, and 1 ml of ethanol is added to dissolve any unchanged sodium. The tube is heated to a dull red again and while still hot is dropped into a small beaker containing 20 ml of distilled water (*caution*!). During the heating, alcohol vapors may ignite at the mouth of the tube; this should not affect the analysis. The tube is broken up with a stirring rod, and the solution is heated to boiling and filtered. The filtrate, which should be colorless, is used for the specific tests described below.

Procedure B. A small (100 × 13 mm), *pyrex* glass test tube is charged with *ca.* 10 mg (*ca.* 0.010 ml = 10 μl for liquids) of *unknown* and a freshly cut, pea-size (*ca.* 50 mg) piece of sodium metal (*caution*!). Heating is carried out as described in Procedure A; the glowing and charred residue is allowed to cool to room temperature. A few drops of methanol are added, with stirring, to ensure decomposition of all sodium metal; this is repeated until no further hydrogen gas (bubbles) escape. To this solution is added *ca.* 2 ml of distilled water; this solution is boiled (add boiling chips or glass rod) and filtered. This filtrate, which should be colorless, is used for the specific tests outlined below; any indication of incomplete fusion (e.g., color) at this point requires that the entire procedure be repeated on fresh unknown.

If a sharp report or explosion occurs when the first portion of the unknown is heated with the sodium, the procedure is interrupted and about 0.5 g of the unknown is reduced by boiling gently with 5 ml of glacial acetic acid and 0.5 g of zinc dust. *Caution*: Do not heat too strongly, for compounds such as amine acetates of low molecular weight may be lost by evaporation. After most of the zinc has dissolved, the mixture is evaporated to dryness and the entire residue is then decomposed by Procedure A or B above.

Alternatively, procedures involving fusion of magnesium and potassium carbonate[3] and of zinc/calcium oxide mixtures[4] have been used.

---

[1] The Lassaigne sodium decomposition test; see K. N. Campbell and B. K. Campbell, *J. Chem. Educ.*, **27**, 261 (1950). The technique of gentle heating of the mixture, addition of the sample mixture in portions, and final ignition provides opportunity for the reactions to take place without violence and without premature loss by volatilization.

[2] Powdered sucrose (sugar) is sold in supermarkets as "confectioner's sugar." It contains 97% sucrose and 3% starch. The mixture of the unknown and powdered sugar provides a charring and reducing action so that compounds containing nitrogen as amide, nitroso, nitro, azo, hydrazo, and heterocyclic rings produce sodium cyanide. Sulfur compounds such as sulfides, sulfoxides, sulfones, sulfonamides, and heterocyclic sulfur compounds produce sodium sulfide.

[3] R. H. Baker and C. Barkenbus, *Ind. Eng. Chem. Anal. Ed.*, **9**, 135 (1937).

[4] E. L. Bennet, C. W. Gould, E. H. Swift, and C. Niemann, *Ind. Eng. Chem. Anal. Ed.*, **19**, 1035 (1947).

Specific Tests for Elements

Sulfur. (a) A few milliliters of the above solution are acidified with acetic acid, and a few drops of lead (II) acetate solution are added. A black precipitate of lead sulfide indicates sulfur.

$$Na_2S + Pb(O_2CCH_3)_2 \rightarrow \underset{\text{(black ppt.)}}{PbS} + 2CH_3CO_2Na$$

(b) To another 1 ml sample of the filtrate from the Na decomposition is added 2 drops of a dilute solution of sodium nitroprusside; a deep blue-violet color indicates the presence of sulfur:

$$Na_2S + \underset{\text{sodium nitroprusside}}{Na_2Fe(CN)_5NO} \rightarrow \underset{\text{blue-violet "thio-nitro complex"}}{Na_4[Fe(CN)_5NOS]} + 2NaOH$$

Nitrogen. The reagents needed are a 1.5% solution of *p*-nitrobenzaldehyde in 2-methoxyethanol (methyl cellosolve); a 1.7% solution of *o*-dinitrobenzene in 2-methoxyethanol, and a 2% solution of sodium hydroxide in distilled water.

Procedure.[5] In a small test tube, 1 ml of the *p*-nitrobenzaldehyde solution, 1 ml of the *o*-dinitrobenzene solution, and 2 drops of the sodium hydroxide solution are mixed. Then 2 drops of the filtrate from the sodium decomposition of the unknown are added. A deep blue-purple compound is produced by the sodium cyanide formed from the nitrogen-containing functional group in the original. A yellow or tan coloration is a negative test. This test for nitrogen is much more sensitive than the Prussian blue (or Turnbull's blue) test described in our earlier editions.

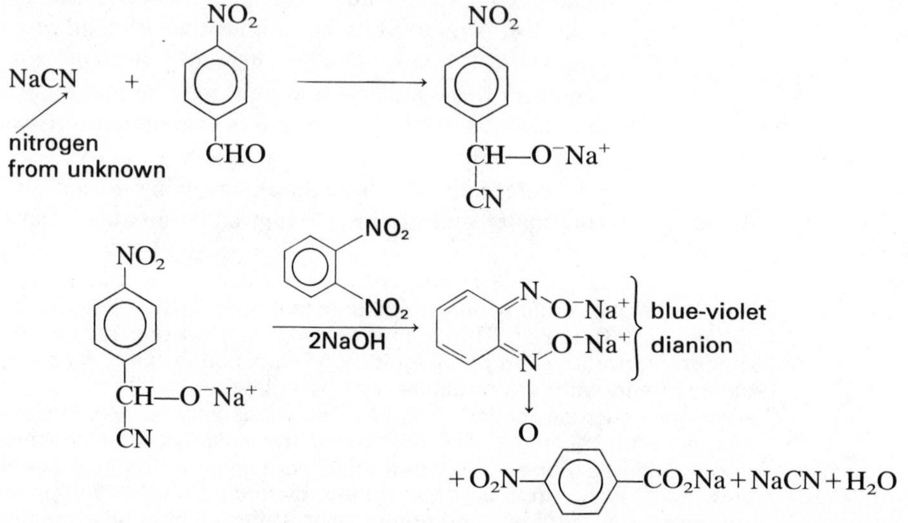

$$NaCN + p\text{-}O_2NC_6H_4CHO \longrightarrow p\text{-}O_2NC_6H_4CH(CN)\text{—}O^-Na^+$$

NaCN: nitrogen from unknown

$$p\text{-}O_2NC_6H_4CH(CN)\text{—}O^-Na^+ + o\text{-}C_6H_4(NO_2)_2 \xrightarrow{2NaOH} C_6H_4(=N\text{—}O^-Na^+)(=N(\rightarrow O)\text{—}O^-Na^+) + O_2N\text{—}C_6H_4\text{—}CO_2Na + NaCN + H_2O$$

blue-violet dianion

[5] G. G. Guilbault and D. N. Kramer, *Anal. Chem.*, **38**, 834 (1966); *J. Org. Chem.*, **31**, 1103 (1966).

This test is valid in the presence of NaX (where X = halide) or Na_2S (which will be present if the original unknown contained halogens or sulfur).

The products of the above reactions provide the explanation of why this test is so sensitive. Acidification of the solution of the purple dianion results in a yellow solution of *o*-nitrophenylhydroxylamine (an acid–base indicator):

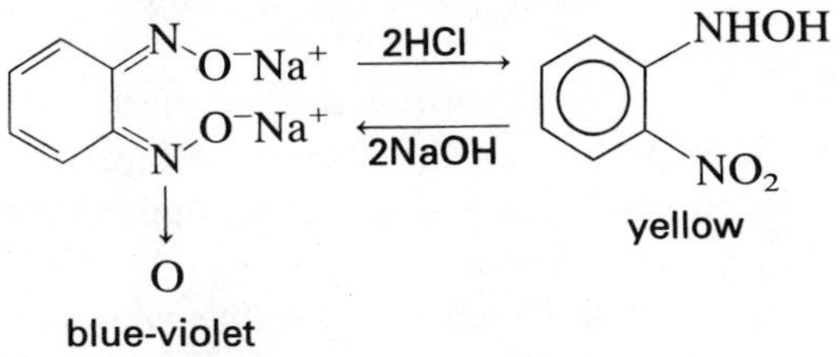

$$\text{blue-violet dianion } (N{-}O^-Na^+,\ N(\rightarrow O){-}O^-Na^+) \underset{2NaOH}{\overset{2HCl}{\rightleftharpoons}} C_6H_4(NHOH)(NO_2) \text{ (yellow)}$$

blue-violet

yellow

Alternative Nitrogen Procedure. About 2 drops of ammonium polysulfide solution are added to 2 ml of the stock solution, and the mixture is evaporated to dryness on a steam bath. Dilute hydrochloric acid (5 ml) is added, and the solution is warmed and filtered. A few drops of ferric chloride solution are added to the filtrate. A red coloration indicates nitrogen:

$$NaCN + (NH_4)_2S_x \rightarrow NaSCN + (NH_4)_2S_{x-1}$$

$$6\,NaSCN + FeCl_3 \rightarrow \underset{\text{(Red)}}{Na_3Fe(SCN)_6} + 3NaCl$$

A sample of *p*-bromobenzenesulfonamide (m.p. 166°C) is recommended as a control. This sample should be decomposed by Na and analyzed for N, Br, and S by the procedures described in this chapter (pp. 78–83).

Certain nitro compounds are converted to orange or red complexes when treated with sodium hydroxide (see p. 321). Amides and imides may liberate ammonia or amines when treated with sodium hydroxide (see p. 282). These tests essentially specify the type of functional group that could contain nitrogen and thus should be kept in mind when a positive nitrogen is obtained here.

The Halogens. (*a*)[6] About 2 ml of the Na-fusion filtrate is acidified with dilute nitric acid and boiled gently for a few minutes (to expel any hydrogen cyanide or hydrogen sulfite that may be present).[7] A few drops of (0.1 *M*) silver nitrate solution are added. A heavy precipitate indicates the presence of chlorine, bromine, or iodine. Silver chloride is white, silver bromide is pale yellow, and silver iodide is yellow. If only a faint turbidity or opalescence is produced, it is probably due to the presence of impurities in the reagents or in the glass of the test tube used in the original sodium decomposition.

(*b*) *Beilstein's Test.* A small loop is made in the end of a copper wire and is heated in the Bunsen flame until the flame is no longer green. The wire is cooled;

[6] If this test gives ambiguous results, one or more of the succeeding tests should be employed.

[7] If a flocculent white precipitate of silicic acid separates, the solution should be filtered or centrifuged and the clear filtrate or decantate should be used. Silicic acid could arise from sodium silicate which, in turn, could have arisen from the reaction of the glass with the molten sodium in the fusion procedure.

the loop is dipped in a little of the original compound and heated in the edge of the Bunsen flame. A green flame indicates halogen.

This test is extremely sensitive and should always be cross-checked by the silver nitrate test because minute traces of impurities containing halogen suffice to produce a green flame. Very volatile liquids may evaporate completely before the wire can be heated sufficiently to cause decomposition, thus causing failure of the test.

Certain nitrogen compounds not containing halogen cause a green color to be imparted to the flame; among them are quinoline and pyridine derivatives, organic acids, urea, and copper cyanide. Some inorganic compounds also give green flames.

(*c*) *Bromine and Iodine.* About 3 ml of the solution from the sodium decomposition is acidified with 10% sulfuric acid and boiled for a few minutes. The solution is cooled, and 1 ml of carbon tetrachloride is introduced; then a drop of freshly prepared chlorine water[8] is added. The production of a purple color in the carbon tetrachloride indicates iodine.

$$2NaI + Cl_2(H_2O) \rightarrow 2NaCl + I_2(CCl_4)$$

The addition of chlorine water is continued drop by drop, the solution being shaken after each addition. The purple will gradually disappear and will be replaced by a reddish brown color if bromine is present.

$$I_2(CCl_4) + Cl_2(H_2O) \rightarrow 2ICl$$

$$2NaBr + Cl_2(H_2O) \rightarrow 2NaCl + Br_2(CCl_4)$$

(*d*) *Bromine.* To 3 ml of the stock solution in a test tube are added 3 ml of glacial acetic acid and 0.1 g of lead dioxide. A piece of filter paper, moistened with a 1% solution of fluorescein, is placed over the mouth of the test tube, and the contents of the tube are heated to boiling. If bromide is present in the solution, bromine vapors cause the yellow fluorescein to turn pink owing to the formation of eosin. Chlorides and cyanides do not interfere with this test. Iodides give a brown color.

(*e*) *Chlorine.* If the above tests for bromine and iodine are negative and a good precipitate was produced by silver nitrate, the presence of chlorine is indicated. If bromine and iodine have been found to be present, one of the following procedures should be used to detect the presence of chlorine.

(*f*) *Chlorine, Bromine, and Iodine.* About 10 ml of the original filtrate is acidified with 10% sulfuric acid and boiled for a few minutes. The solution is cooled and tested for iodine by adding 0.5 ml of carbon tetrachloride to 1 ml of the solution and adding a few drops of sodium nitrite solution. A purple color indicates iodine. If iodine is present the remainder of the solution is treated with sodium nitrite and the iodine is extracted with carbon tetrachloride. The solution is finally boiled for a minute and then cooled. To 1 ml of this solution, 0.5 ml of carbon tetrachloride and 2 drops of chlorine water are added. A brown color

[8] Since chlorine water does not keep well, a stabilized sodium hypochlorite such as Chlorox may be used, but the test solution must be kept acid to litmus.

indicates bromine. The remaining solution is diluted to 60 ml, 2 ml of concentrated sulfuric acid, and then 0.5 g of potassium persulfate ($K_2S_2O_8$) are added, and the solution is boiled for 5 min. After the mixture has been cooled, silver nitrate solution is added; a white precipitate indicates chlorine.

A similar procedure for the detection of chlorine, in which lead dioxide (PbO_2) and acetic acid replace potassium persulfate and sulfuric acid as the oxidizing agent, may also be used.

(*g*) *Chlorine in the Presence of Nitrogen, Sulfur, Bromine, and Iodine.* About 10 ml of the original filtrate is acidified with dilute nitric acid and boiled to expel hydrogen cyanide and hydrogen sulfide. Sufficient 0.1 *M* silver nitrate is added to precipitate completely all the halogens as silver halides, and the precipitate is removed. If both nitrogen and sulfur are present, the precipitate is boiled for 10 min with 30 ml of concentrated nitric acid to destroy any silver thiocyanate that may be present. The mixture is then diluted with 30 ml of distilled water and filtered. The precipitate of silver halides is then boiled with 20 ml of 0.1% sodium hydroxide for 2 min. The solution is filtered, the filtrate is acidified with nitric acid, and silver nitrate solution is added. A white precipitate indicates chlorine.

(*h*) *Fluorine.* About 2 ml of the stock solution is acidified with acetic acid, and the solution is boiled and cooled. One drop of the solution is placed on a piece of zirconium-alizarin test paper. A yellow color on the red paper indicates the presence of fluoride. The test paper is prepared by dipping a piece of filter paper into a solution composed of 3 ml of 1% ethanolic alizarin solution and 2 ml of a 0.4% solution of zirconium chloride (or nitrate). The red filter paper is dried and, just before use, is moistened with a drop of 50% acetic acid.

4.1.2 Detection of Metals and Other Inorganic Elements

Metallic and other inorganic elements may be determined by standard qualitative tests.[9] A procedure has been outlined for the preparation of organic compounds for metallic elemental analysis.[10] Spectral tests for inorganic elements may also be used.[11] Spectral tests are usually carried out more readily and normally give much more reliable results than do "wet" tests.

4.2 QUANTITATIVE ELEMENTAL ANALYSIS

4.2.1 Combustion and Related Analyses

Quantitative analytical data are routinely reported for confirmation of structure of new organic compounds; these data are also extremely useful for structure

---

[9] F. Feigl and V. Anger, *Spot Tests in Inorganic Analysis*, 6th ed. (Elsevier, New York, 1972).

[10] D. J. Pasto and C. Johnson, *Organic Structure Determination* (Prentice-Hall, Englewood Cliffs, N.J., 1969), p. 320.

[11] H. H. Willard, L. L. Merritt, and J. A. Dean, *Instrumental Methods of Analysis*, 5th ed. (Van Nostrand, New York, 1974).

determinations of unknown compounds. Such microanalyses are usually determined by commercial firms[12] equipped with combustion or other appropriate analytical equipment. Samples are prepared (after having passed purity standards, Chapter 3) by drying (usually in an Abderhalden drying pistol, see Fig. 4.1). A small amount of the material is spread thinly in a porcelain boat (or on glazed weighing paper); the boat is then placed in the horizontal portion of the drying pistol. The bulb of the drying pistol is charged with fresh, anhydrous drying agent, for example, phosphorus pentoxide (or, less efficiently, calcium sulfate or calcium chloride). The entire system is evacuated with a vacuum pump. The speed at which the sample loses water may be increased by allowing toluene (or xylene) to reflux up from the lower flask; the sample must, of course, be stable to the boiling point of the solvent. If it is believed that hydrocarbon solvents should be removed from the sample, wax shavings are substituted for the desiccating agent. Samples can be examined under a magnifying lens for filter paper fibers or other

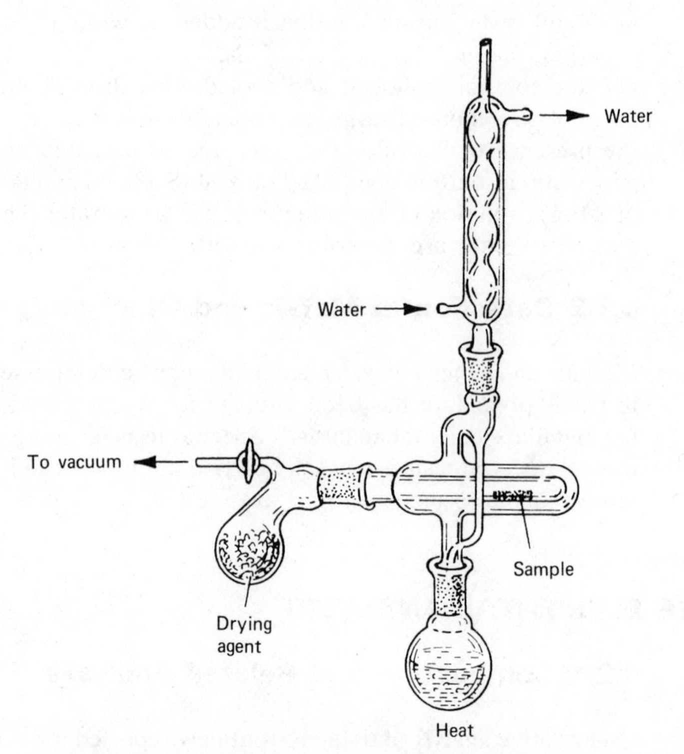

Figure 4.1. Abderhalden drying pistol. (Courtesy of John A. Landgrebe, author of *Theory and Practice of Organic Chemistry*, and of D. C. Heath, Publishers.)

---

[12] Quantitative analyses of C, H, N, S, and X are determined by Galbraith Analytical Laboratories, Knoxville, Tenn.; Baron Consulting, Orange, Conn.; and Huffman Microanalytical Laboratories, Wheatridge, Colo.

Table 4.1. Quantitative Elemental Composition and Molecular Formula

Element	Observed Percent[a]	Percent/at. wt.	Adjusted ratios[b]	Simplest no. of atoms	Formula[c]
C	49.90	49.90/12.011 = 4.154	8.99	9	
H	4.19	4.19/1.008 = 4.157	9.00	9	$(C_9H_9O_4Cl_1)_x$
Cl	16.37	16.37/35.453 = 0.462	1.000	1	
O	(29.54)[d]	29.54/16.000 = 1.846	3.996	4	

[a] Analytical data obtained from a consulting firm.
[b] Obtained by dividing each datum in third column by lowest result in third column.
[c] x = 1, 2, or other integer.
[d] Obtained by difference; assumes no other elements in compound. 100 − (49.90 + 4.19 + 16.37) = 29.54.

extraneous materials. Normally 5 mg of sample are needed for C and H analysis and another 5 mg for each additional analysis (e.g., sulfur, halogen, deuterium).[13] Oxygen analyses are not usually obtained; percent of oxygen is normally obtained by difference (see below). Confirmation of a molecular formula is satisfactory when the calculated and experimentally determined percentages agree within ±0.3%; for example,

Anal. Calcd. for $C_{13}H_{16}O$: C, 82.93; H, 8.57
Found: C, 82.87; H, 8.67

When the molecular formula is unknown prior to analysis, the empirical (simplest) formula may be obtained as outlined on Table 4.1. There are a number of essentially equivalent ways of manipulating these data; most methods start by

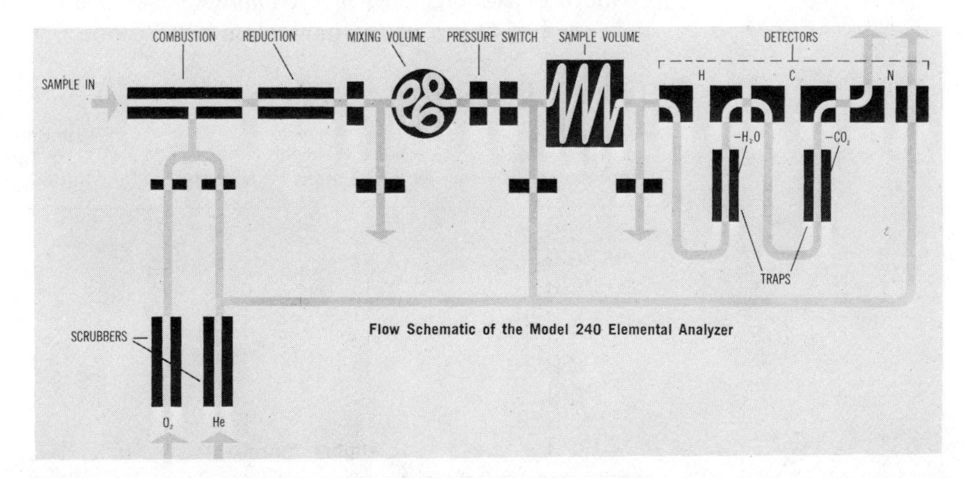

Figure 4.2. Model 240, Elemental Analyser. This provides combined determinations of C, H, and N with a single sample of about 1–3 mg (or smaller in some cases). It is readily convertible for oxygen analysis. (Courtesy of Perkin Elmer Corporation, Norwalk, Conn.)

[13] Figure 4.2 is a schematic representation of a typical combustion analysis instrument. Hydrogen and carbon are determined in terms of their combustion products: H_2O and CO_2.

dividing the percent composition by the atomic weight (taken from the periodic table). These ratios are then adjusted, as shown in Table 4.1, to obtain the lowest whole number of atoms. In Table 4.1, the formula $C_9H_9O_4Cl_1$ is the simplest formula; it would be the molecular formula only if $x = 1$. The value of x is ascertained by an independent determination of the molecular weight; for example, if vapor phase osmometry yields a molecular weight that is within 10% of 216.6 amu (Table 4.1, $x = 1$), the *molecular* formula is indeed C_9H_9OCl. Mass spectrometry ideally and frequently yields a molecular weight that *exactly* corresponds (for example, 216 here, using mass of most abundant isotope of Cl, ^{35}Cl) to that for the molecule of interest. It is, of course, possible to have molecular weights that are multiples of 216.6; for example, if the compound of Table 4.1 yields a molecular weight of 435 via another method, a formula of $C_{18}H_{18}O_4Cl_2$ can be assumed.

4.2.2 Formula Determination by Mass Spectrometry

As described in Chapter 3, a sample of sufficient thermal stability and of significant volatility that results in a measurable substantial molecular ion is amenable to molecular weight determination by mass spectrometry. In addition, use of the isotope-abundance ratios (M+1, M+2 intensity data) of unit-resolution mass spectrometry, or the exact masses of high-resolution mass spectrometry, often yield the molecular formula of the unknown.

In unit-resolution spectrometry (mass to nearest whole number), the cluster of peaks of successive mass numbers usually observed in the region of the molecular ion is due to the significant abundance of more than one isotope of one or more of the elements in a compound (see Table 4.2). The molecular ion is composed of atoms of the most abundant isotope (^{12}C, ^{1}H for methane); these,

Table 4.2. Isotope Abundance Aspects of Methane (CH_4)

Possible molecular species	Nominal mass	Nuclide	Abundance relative to the most abundant isotope
$^{12}C^{1}H_4$	16	^{12}C	(100)[a]
$^{13}C^{1}H_4$	17	^{13}C	1.08
$^{14}C^{1}H_4$	18	^{14}C	[b]
$^{12}C^{1}H_3{}^{2}H_1$	17	^{2}H	0.016
$^{12}C^{1}H_2{}^{2}H_2$	18	—	—
$^{12}C^{1}H_1{}^{2}H_3$	19	—	—
$^{12}C^{2}H_4$	20	—	—
$^{12}C^{1}H_3{}^{3}H_1$	18	^{3}H	[b]
$^{13}C^{2}H_4$	Highly improbable species		

[a] Arbitrary standard; set at 100%.
[b] Radioactive isotope; relative abundance varies with time and is negligible for mass spectral considerations.

[14] Tables for species of mass 12–250 are found in R. M. Silverstein, G. C. Bassler, and T. C. Morrill, *Spectrometric Identification of Organic Compounds*, 3rd ed. (Wiley, New York, 1974), p. 41ff.

virtually without exception, also correspond to the isotope of lowest mass. Contributions are made to the smaller peaks (of higher mass) in this cluster by molecules composed of less abundant isotopes (e.g., ^{13}C, ^{2}H). Peaks from species containing minor isotopes of more than two elements are usually too weak to make significant contributions to the mass spectrum. Data arising from the M+1 and M+2 intensities are often very useful for determination of the molecular formula. Since there usually is a large number of formulas corresponding to a given molecular weight, it is indeed fortunate that computer generation of possible formulas has been carried out.[14] A portion of one of these tables, relevant to methane, is presented in Table 4.3. It is clear that experimental measurement of an M+1 (at *m/e* 17) peak intensity of ca. 1.0% is strong support for the molecular formula of methane.

Clearly, tables analogous to Table 4.3 for heavier organic compounds would be quite lengthy. Fortunately, simple rules allow dispensing with many candidate formulas. First, the nitrogen rule states that standard organic compounds of even-valued mass may not have odd numbers of nitrogen and compounds of odd-numbered masses must contain an odd number of nitrogen atoms. Obviously, the second formula of Table 4.3 could be eliminated by the nitrogen rule. A less quantitative rule is the "no-nonsense" rule. The molecular formula must correspond to that of a stable organic molecule. On this basis, both of the first two candidates in Table 4.3 can be discarded. The nitrogen rule and the "no-nonsense" rule are not really needed to determine the molecular formula of methane (Table 4.3). These rules are, however, very useful in the process of selecting the molecular formula of compounds of higher molecular weight.

Figure 3.33 includes the mass spectrum of 3-hexanone; since the peaks are well defined, the intensity is measured as the height of the peak. The intensity of the M+1 peak (*m/e* 101) would be of modest value in ascertaining the molecular formula of this keto compound, since this M+1 peak is of such low intensity.

Determination of the nature and number of atoms, other than C, H, O, and N, in a molecular formula is clearly often simplified greatly by the use of mass spectrometry. Table 4.4 allows separation of atoms into those (for example, F, I,

Table 4.3. Masses and Isotope Abundance Ratios for All Combinations of C, H, N, and O Corresponding to Mass 16

Mass 16	M+1[a]	M+2[a]
O	0.04	0.20
NH_2	0.41	—
CH_4	1.15	—

[a] Based on molecular ion (M, *m/e* 16) = 100% intensity. See reference in footnote 14 on p. 86.

Table 4.4. Isotope Abundances Based on the Common Isotope Set 100%

Element	Abundance (%)		
Carbon	^{12}C 100	^{13}C 1.08	
Hydrogen	^{1}H 100	^{2}H 0.016	
Nitrogen	^{14}N 100	^{15}N 0.38	
Oxygen	^{16}O 100	^{17}O 0.04	^{18}O 0.20
Fluorine	^{19}F 100		
Silicon	^{28}Si 100	^{29}Si 5.10	^{30}Si 3.35
Phosphorus	^{31}P 100		
Sulfur	^{32}S 100	^{33}S 0.78	^{34}S 4.40
Chlorine	^{35}Cl 100		^{37}Cl 32.5
Bromine	^{79}Br 100		^{81}Br 98.0
Iodine	^{127}I 100		

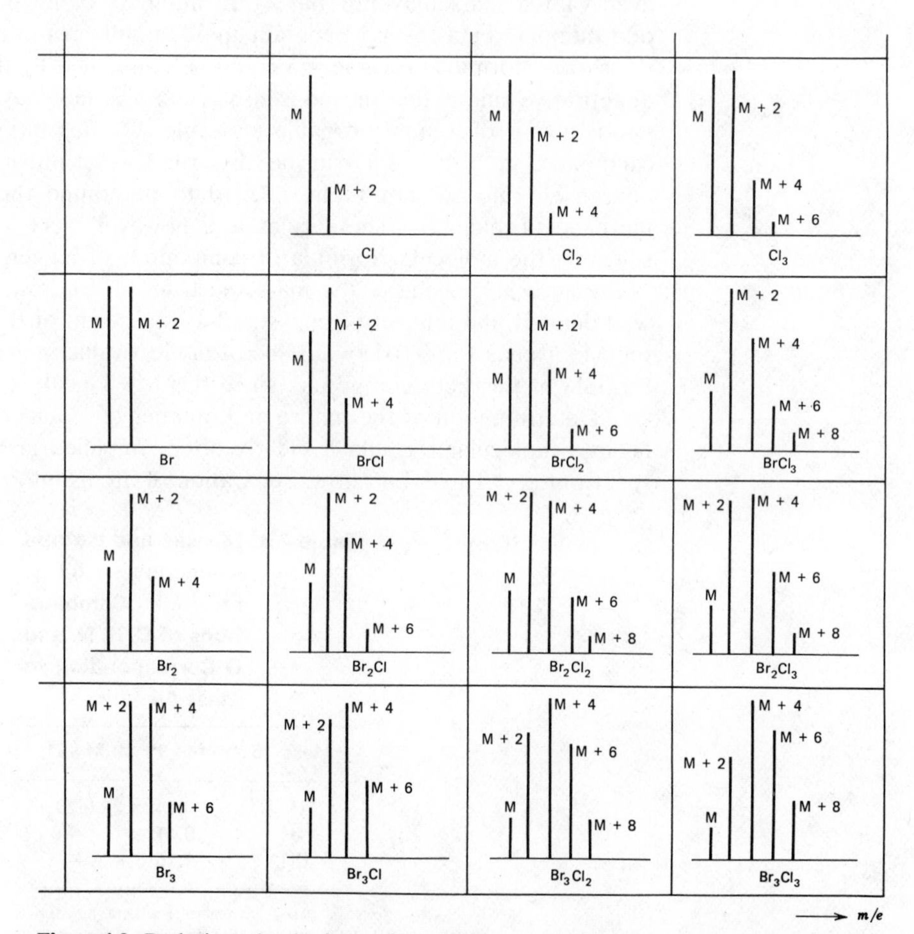

Figure 4.3. Peaks in molecular ion region of bromo and chloro compounds. Contributions due to C, H, N, and O are usually small compared to those for Br and Cl.

P) that will result in low M+1 and M+2 peak intensities and those (e.g., Br, Cl, S, Si) that will result in higher intensities for these peaks than the intensities expected from the C, H, N, O tables. Atoms in the first category are not at all discernible from initial evaluation of the molecular ion cluster; atoms in the second category are usually identified quickly from the mass spectra.

An example of the second category is worthy of discussion. An alkyl chloride would give rise to a mass spectrum in which the only substantial contribution to the M+2 peak would be due to ^{37}Cl. This is true because the abundance of ^{37}Cl, the heavy isotope of chlorine, is so much greater than the abundance of the heavy isotopes of carbon and hydrogen (see Table 4.4). Thus, the molecular ion region of any alkyl chloride (containing one chlorine atom per molecule) is expected to look like the diagram in the second box of the first row of Fig. 4.3.

High-resolution mass spectrometry is also very useful for determining molecular formulas; this subject is discussed in many of the spectroscopy references listed in Chapter 10.

CHAPTER FIVE

classification of organic compounds by solubility and nuclear magnetic resonance (nmr) and infrared (ir) spectra

In this chapter we begin the process of determining the structural composition of organic compounds. Both solubility and spectral analyses can lead to the same kinds of information. For example, water-insoluble carboxylic acids are soluble in aqueous sodium hydroxide. In addition, these acids yield ir spectral bands that are very characteristic of their carboxyl groups. Thus, it is clear that deductions based on an integrated interpretation of solubility and spectral analyses are powerful in organic structure determination.

5.1 SOLUBILITY

The solubility of organic compounds can be divided into two major categories: solubility in which a chemical reaction is the driving force, for example, the acid–base reaction

$$O_2N-C_6H_4-CO_2H \xrightarrow[H_2O]{OH^-} O_2N-C_6H_4-CO_2^-Na^+$$

p-nitrobenzoic acid
water-insoluble

sodium *p*-nitrobenzoate
water-soluble

and solubility in which only simple miscibility is involved (e.g., dissolving ethyl ether in carbon tetrachloride for nmr analysis). Although the two solubility sections below interrelate, the first section is used mainly to identify functional groups and the second to determine solvents for recrystallizations, spectral analyses, and chemical reactions.

5.1.1 Solubility in Water, Aqueous Acids and Bases, and Ether

Three kinds of information can often be obtained about an unknown substance by a study of its solubility behavior in water, 5% sodium hydroxide solution, 5% sodium bicarbonate solution, 5% hydrochloric acid, and cold concentrated sulfuric acid. First, the presence of a functional group is often indicated. For instance, because hydrocarbons are insoluble in water, the mere fact that an unknown such as ethyl ether is partially soluble in water indicates that a polar functional group is present. Second, solubility in certain solvents often leads to more specific information about the functional group. For example, benzoic acid is insoluble in the polar solvent, water, but is converted by dilute sodium hydroxide to a salt, sodium benzoate, which is readily water-soluble. In this case, then, the solubility in 5% sodium hydroxide solution of a water-insoluble unknown is a strong indication of an acidic functional group. Finally, certain deductions about molecular weight may sometimes be made. For example, in many homologous series of monofunctional compounds, the members with fewer than about five carbon atoms are water-soluble, whereas the higher homologs are insoluble.

Compounds are first tested for solubility in water. In considering solubility in water, a substance is arbitrarily said to be "soluble" if it dissolves to the extent of 3 g/100 ml of solvent. This standard is dictated by the limitations inherent in the method employed, which depends on rough semiquantitative visual observations, as will be seen. Care is needed in interpreting classifications of "soluble" and "insoluble" in other references, because different definitions of these words are often used.

When solubility in dilute acid or base is being considered, the significant observation to be made is not whether the unknown is soluble to the extent of 3% or to any arbitrary extent, but, rather, whether it is significantly more soluble in aqueous acid or base than it is in water. Such increased solubility is the desired positive test for an acidic or basic functional group.

Acidic compounds are discovered by their solubility in 5% sodium hydroxide. Strong and weak acids (classes A_1 and A_2, respectively; see Table 5.1 and Fig. 5.1) are differentiated by the solubility of the former but not the latter in the weakly basic solvent, 5% sodium bicarbonate. Compounds that behave as bases in aqueous solution are detected by their solubility in 5% hydrochloric acid; no attempt is made here to differentiate among the various strengths of bases in this class (class B).

Many compounds that are neutral toward 5% hydrochloric acid behave as bases in more acidic solvents such as concentrated sulfuric acid or syrupy phosphoric acid. In general, compounds containing sulfur or nitrogen have an

Table 5.1. Organic Compounds Comprising the Solubility Classes of Fig. 5.1[a]

Class	Compounds
S_2	Salts of organic acids (RCO_2Na, RSO_3Na), amine hydrochlorides (RNH_3Cl); amino acids ($R{-}CH(NH_3^+){-}CO_2^-$); polyfunctional compounds (functional groups are hydrophilic), i.e., carbohydrates (sugars), polyhydroxy compounds, polybasic acids, etc.
S_A	Monofunctional carboxylic acids with five carbons or fewer; arenesulfonic acids.
S_B	Monofunctional amines with six carbons or fewer.
S_1	Monofunctional alcohols, aldehydes, ketones, esters, nitriles, and amides with five carbons or fewer.
A_1	Strong organic acids: carboxylic acids with more than six carbons; phenols with electron-withdrawing groups in the ortho and para positions, β-diketones.
A_2	Weak organic acids: phenols, enols, oximes, imides, sulfonamides, thiophenols, all with more than five carbons. β-diketones, nitro compounds with α-hydrogens, sulfonamides.
B	Aliphatic amines with eight or more carbons, anilines (only one phenyl group attached to nitrogen), some oxy ethers.
MN	Miscellaneous neutral compounds containing nitrogen or sulfur and having more than five carbon atoms.
N_1	Alcohols, aldehydes, methyl ketones, cyclic ketones, and esters with one functional group and more than five but fewer than nine carbons; ethers with fewer than eight carbon atoms, epoxides.
N_2	Alkenes, alkynes, ethers, some aromatic compounds (especially those with activating groups), ketones (other than those cited in class N_1).
I	Saturated hydrocarbons, haloalkanes, aryl halides, diaryl ethers, deactivated aromatic compounds.

[a] Carboxylic acid halides and anhydrides have not been classified because of their high reactivity.

atom with a nonbonded pair of electrons and would be expected to dissolve in such strongly acidic media. No additional information would be gained, therefore, by determining such solubility, and for this reason, when the elemental analysis has shown the presence of sulfur or nitrogen, no solubility tests beyond those for acidity and basicity in aqueous solution are carried out. Compounds that contain nitrogen or sulfur and are neutral in aqueous acid or base are placed in solubility class MN.

Most compounds neutral in water and containing oxygen in any form are reasonably strong bases in concentrated sulfuric acid. Solubility in, or any other evidence of a reaction with, this reagent is indicative of an oxygen atom or else of a reactive hydrocarbon function such as an olefinic bond or easily sulfonated aromatic ring; such compounds are said to be in class N. A further distinction is made with syrupy phosphoric acid, class N_1 compounds being soluble in this reagent as well as in sulfuric acid and class N_2 compounds being soluble only in the latter. Compounds that are too weakly basic to dissolve in sulfuric acid are placed in class I (inert compounds).

Since the solubility behavior of water-soluble compounds gives no information about the presence of acidic or basic functional groups, this information must

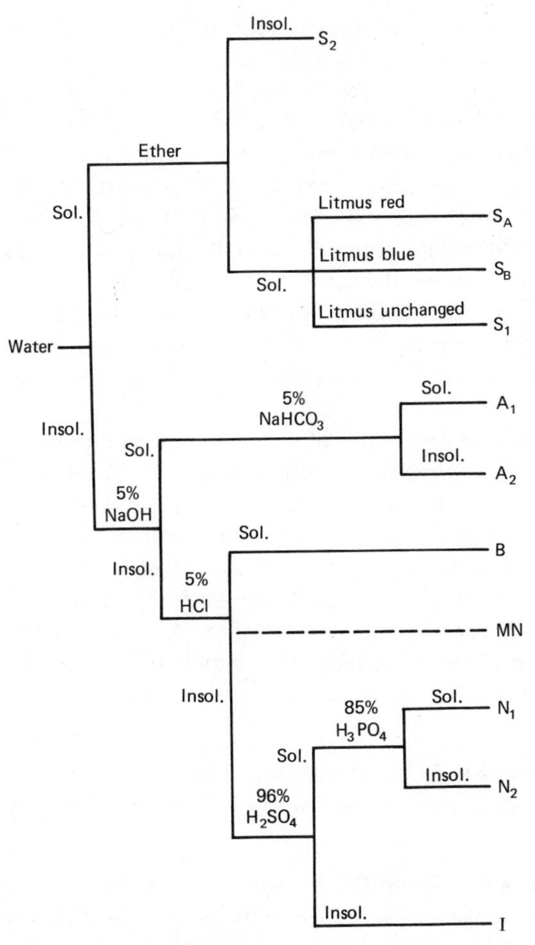

Figure 5.1. Classification of organic compounds by solubility: determinations in water, acids, bases, and ethers (see Table 5.1 for compounds comprising each class). sol. = soluble, insol. = insoluble.

be obtained by testing their aqueous solutions with litmus or pH paper. Note that the behavior of an acid in 5% hydrochloric acid and of a base in sodium hydroxide solution should be examined routinely, since the molecule may have both acidic *and* basic functional groups (class A_1-B or A_2-B).

A more detailed survey of the solubility classes is given below.

Determination of Solubilities

Procedure for Water Solubility. Place 0.2 ml (0.1 g of a solid) of the compound in a small test tube, and add in portions 3 ml of water. Shake vigorously after the addition of each portion of solvent, being careful to keep the mixture at room temperature. If the compound dissolves completely, record it as soluble.

Solids should be finely powdered to increase the rate of solution. If the solid appears to be insoluble in water or ether, it is sometimes advisable to heat the mixture gently. If solution is effected in this way, the liquid again is cooled to room temperature and is shaken to prevent supersaturation. It is always well in such cases to "seed" the cooled solution by adding a crystal of the solid. Care should be taken in weighing the sample; it should weigh 0.10 g within 0.01 g.

Liquids are handled most conveniently by means of a graduated pipet that permits accurate measurement of the amount added. When two colorless liquid phases lie one above the other, it is often possible to overlook the boundary between them and thus to see only one phase. This mistake can generally be avoided by shaking the test tube vigorously when a liquid unknown seems to have dissolved in the solvent. If two phases are present the solution will become cloudy. In the rare cases where two colorless phases have the same refractive index, the presence of a second phase will escape detection even if this precaution is taken.

Acid-base properties of water-soluble compounds should be determined with litmus paper; compounds with a $pK_a < 8$ will fall in class S_A (see Table 5.1 and Fig. 5.1). Compounds with a $pK_b < 9$ will fall in class S_B. Consequently, phenols with pK_a of about 10 give aqueous solutions too weakly acidic (pH about 5) to turn litmus paper red. Litmus is red at pHs below 4.5 and blue above 8.3. For similar reasons, an aromatic amine such as aniline is too weak a base (pK_b 9.4) to turn litmus blue in aqueous solution. Although more refined procedures can be developed using a pH-indicating paper, it is preferable to rely more on the tests discussed in Chapter 6.

Procedure for Ether Solubility. Use the amounts of solute and solvent (here ether) as described above for the water-solubility procedure.

Procedure for Solubility in Aqueous Acid or Aqueous Base. (Use amounts of solute and solvent as described in the water-solubility procedure above.)

To test for solubility in sodium hydroxide, sodium bicarbonate, or hydrochloric acid solution, shake the mixture thoroughly, separate (filter if necessary) the aqueous solution from any undissolved unknown, and *neutralize* with acid or base. Examine the solution very carefully for any sign of the separation of the original unknown. Even a cloudy appearance of the neutralized filtrate is a positive test.

When solubility in acid or alkali is being determined, heat should *not* be applied because it might cause hydrolysis to occur. If the mixture is shaken thoroughly, the time required for solution to take place should not be more than 1 or 2 min.

Often it is possible to utilize a single portion of solute for tests with several different solvents. Thus if the compound is found to be insoluble in water, a fairly accurate measure of its solubility in dilute sodium hydroxide solution can be obtained by adding about 1 ml of a 20% solution of sodium hydroxide. The resulting 4 ml of solvent will contain about 5% sodium hydroxide. If the substance is very insoluble, it often may be recovered and used subsequently for the hydrochloric acid test.

Procedure for Solubility in Concentrated Acids. Use the procedure described above for water solubility; employ sulfuric or phosphoric acids in the amount described for water above.

When sulfuric acid is used, it is more convenient to place the 3 ml of solvent in the test tube and then add the solute. With this reagent significant reactions sometimes take place, and it is important to look for such manifestations as the production of heat, a change of color, the formation of a precipitate, or the evolution of a gas. Careful notes should be made of all such observations, because they may be very useful at a later stage of the identification.

Theory of Solubility

Polarity and Solubility. When a solute dissolves, its molecules or ions become distributed more or less randomly among those of the solvent. In crystalline sodium chloride, for example, the average distance between sodium and chloride ions is 2.8 Å. In a 1 *M* solution the solvent has interspersed itself in such a way that sodium and chloride ions are about 10 Å apart. The difficulty of separating such ions is indicated by the high melting point (800°C) and boiling point (1413°C) of pure sodium chloride.

The work required to separate two oppositely charged plates is lowered by the introduction of matter between them by a factor called the *dielectric constant* of the medium. It is not suprising, then, that water, with the high dielectric constant of 80, facilitates the separation of sodium and chloride ions and, in fact, dissolves sodium chloride readily, whereas ether (dielectric constant 4.4) or hexane (dielectric constant 1.9) is an extremely poor solvent for salts of this type. Water molecules between two ions (or the charged plates of a condenser) are small dipoles, which orient themselves end to end in such a way as to neutralize partially the ionic charges and thus stabilize the system. It might be expected, therefore, that solvating ability and dielectric constant should be parallel. This is not entirely true, however. A high dielectric constant is necessary but not sufficient for an effective ion solvent. For example, hydrogen cyanide, with a dielectric constant of 116, is a very poor solvent for salts such as sodium chloride. Although the situation is complex, one major factor responsible for the efficiency of water and other hydroxylic solvents is their ability to form hydrogen bonds.

The high dielectric constant and hydrogen bonding ability of water, which make it a good solvent for salts, also make it a poor solvent for nonpolar substances. In pure water, molecules are oriented in such a way that positive and negative centers are adjacent. To attempt to dissolve a nonpolar substance such as benzene in water is to try to separate unlike charges in a medium of low dielectric constant. In general, then, a polar solvent may be expected to dissolve readily only polar solutes, and a nonpolar solvent, only nonpolar solutes. This generalization has been summarized briefly as "like dissolves like."

Table 5.1, related to Fig. 5.1, describes various compounds, the structures of which cause them to fall in a particular solubility class. A discussion of these trends and a rationalization of this behavior is given below.

Since water is a polar compound, it is a poor solvent for hydrocarbons.

Olefinic, acetylenic, or benzenoid character do not affect the polarity greatly. Hence, unsaturated or aromatic hydrocarbons are not very different from alkanes in their water solubility. The introduction of halogen atoms does not alter the water solubility by polarity change but rather by simple molecular weight change. Thus, as a halogen is substituted for a hydrogen, the water solubility decreases. On the other hand, salts are extremely polar, the ones encountered in this work generally being water-soluble (class S_2).

Other compounds lie between these two extremes. Here are found the alcohols, esters, ethers, acids, amines, nitriles, amides, ketones, and aldehydes, to mention a few of the classes of frequent occurrence.

As might be expected, acids and amines generally are more soluble than neutral compounds. The amines probably owe their abnormally high solubility to their tendency to form hydrogen-bonded complexes with water molecules. This theory is in harmony with the fact that the solubility of amines diminishes as the basicity decreases. It also explains the observation that many tertiary amines are more soluble in cold than in hot water. Apparently, at lower temperatures the solubility of the hydrate is involved, whereas at higher temperatures the hydrate is unstable and the solubility measured is that of the free amine.

Monofunctional ethers, esters, ketones, aldehydes, alcohols, nitriles, amides, acids, and amines may be considered together with respect to water solubility. *In most homologous series of this type, the upper limit of water solubility will be found in the neighborhood of the member containing five carbon atoms.*

Since most organic molecules have both a polar and a nonpolar part, it might be expected that the solubility would depend on the balance between the two parts. As the hydrocarbon part of the molecule increases, the properties of the compounds approach those of the hydrocarbons from which the compounds may be considered to be derived. This means that water solubility decreases and ether solubility increases. A similar change in solubility occurs as the number of aromatic hydrocarbon residues in the molecule increases. Thus α-naphthol and p-hydroxybiphenyl are less soluble than phenol:

Water Solubility

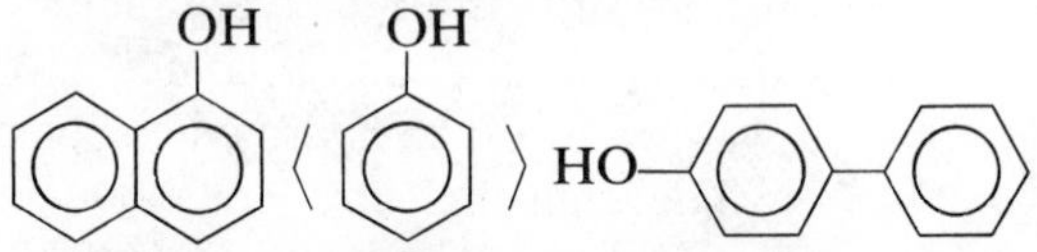

α-naphthol < phenol > *p*-hydroxybiphenyl

The phenyl group, when present as a substituent in aliphatic acids, alcohols, aldehydes, and similar compounds, has an effect on solubility approximately equivalent to a four-carbon-atom aliphatic unit. Benzyl alcohol, for example, is about as soluble as n-amyl (or pentyl) alcohol, and hydrocinnamic acid exhibits a solubility similar to that of heptanoic acid:

Water Solubility

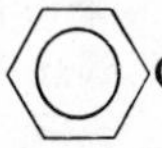

$C_6H_5CH_2OH \cong CH_3(CH_2)_3CH_2OH$ $\qquad C_6H_5CH_2CH_2CO_2H \cong CH_3(CH_2)_4CO_2H$

benzyl alcohol — *n*-pentyl alcohol — heptanoic

The solubility of a substance is a measure of an equilibrium between the pure substance and its solution. It is seen, then, that such an equilibrium is affected not only by the solvent–solute interactions already discussed but also by the intermolecular forces in the pure solute. These forces are independent of the polarity or other properties of the solvent, and their relative strengths may be estimated by a comparison of melting and boiling points, since the processes of the melting of a solid or the boiling of a liquid involve a separation of molecules that is somewhat related to the separation which occurs on solution.

The dicarboxylic acids illustrate the inverse relationship of melting point and solubility. The data in Table 5.2 show that each member with an even number of carbon atoms melts higher than either the immediately preceding or following acid (with an odd number of carbon atoms). The intracrystalline forces in the members with an even number of carbon atoms evidently are greater than in those with an odd number. Since the solubility limit for solids is set at 3.3 g/100 ml of water, it is evident that adipic acid (six carbon atoms) is water-insoluble but pimelic acid (seven carbon atoms) is water-soluble.

Table 5.2. Water Solubility of Dicarboxylic Acids

$HO_2C—(CH_2)_{x-2}CO_2H$

Even number (x) of carbon atoms	m.p. (°C)	Solubility (g/100 g water at 20°C)	Odd number (x) of carbon atoms	m.p. (°C)	Solubility (g/100 g of water at 20°C)
Oxalic (2)	189	9.5	Malonic (3)	135	73.5
Succinic (4)	185	6.8	Glutaric (5)	97	64
Adipic (6)	153	2	Pimelic (7)	103	5
Suberic (8)	140	0.16	Azelaic (9)	106	0.24
Sebacic (10)	133	0.10			

The concomitance of high melting point and low solubility is further illustrated by the isomers maleic and fumaric acids. Fumaric acid sublimes at 200°C and is insoluble in water. Maleic acid melts at 130°C and is soluble in water. Among cis-trans isomers, the cis form generally is the more soluble. Similarly, with polymorphous substances such as benzophenone,[1] the lower-melting forms possess the higher solubilities.

HCCOOH ‖ HCCOOH
maleic acid (*cis*)
Z-isomer

HOOCCH ‖ HCCOOH
fumaric acid (*trans*)
E-isomer

The diamides of dicarboxylic acids constitute another group of compounds in which the melting point is a valuable index of the forces present in the crystals.

[1] Benzophenone occurs in at least four forms.

Urea (m.p. 132°C) is water-soluble. On the other hand, oxamide has the very high melting point of 420°C, which correlates with its low solubility in water. Substitution of methyl groups for the hydrogen atoms in the amide group lowers the melting point by reducing intermolecular hydrogen bonding and increases the solubility in water; *N,N′*-dimethyl- and *N,N,N′,N′*-tetramethyloxamide are water-soluble. Adipamide is water-insoluble, whereas its *N,N,N′,N′*-tetraethyl derivative is water-soluble.

urea	oxamide		
$(H_2N)_2C{=}O$	$CONH_2$–$CONH_2$	$CONHCH_3$–$CONHCH_3$	$CON(CH_3)_2$–$CON(CH_3)_2$
m.p. 132°C	m.p. 420°C	m.p. 217°C	m.p. 80°C

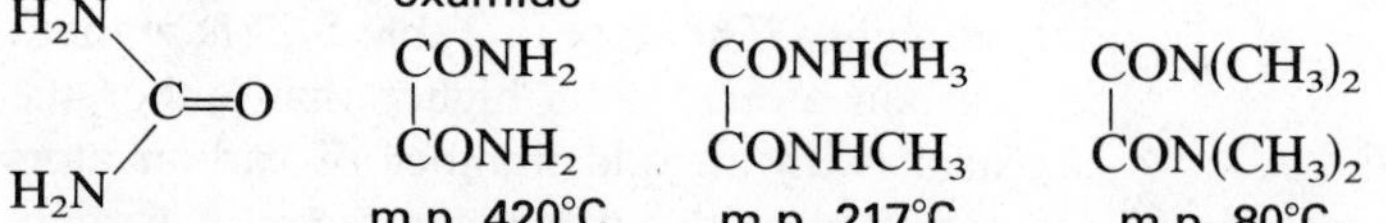

Amides of the type $RCONH_2$ and RCONHR obey the general rule that the boderline compounds contain about five carbon atoms (above). However, *N,N*-dialkylamides ($RCONR_2$) melt lower than the corresponding unsubstituted amides and are much more soluble in water, the solubility limit being in the neighborhood of nine to ten carbon atoms. It has been found that amides having the group $—CONH_2$ are associated owing to the fact that they may act both as acceptors and as donors in forming hydrogen bonds. Such an association is not possible for the *N,N*-disubstituted amides ($RCONR_2$), and hence their state of molecular aggregation is low, as indicated by their low melting points and higher solubilities.

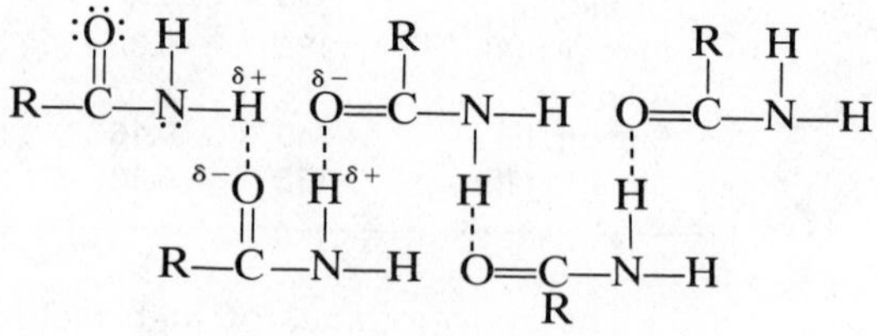

```
   :Ö: H                 R              R   H
    ‖  |      δ+    δ-   |              |   |
R—C—N—H  O=C—N—H  O=C—N—H
              :     :                  |          :
          δ-O  H δ+        H          H
            ‖  |           :          |
      R—C—N—H  O=C—N—H
                     R
```

In general, an increase in *molecular weight* leads to an increase in intermolecular forces in a solid. Polymers and other compounds of high molecular weight generally exhibit low solubilities in water and ether. Thus formaldehyde is readily soluble in water, whereas paraformaldehyde is insoluble.

$$CH_2O \qquad HO(CH_2O—)_xH \quad \text{paraformaldehyde}$$

water-soluble water-insoluble

Glucose is soluble in water, but its polymers—starch, glycogen, and cellulose—are insoluble. Many amino acids are soluble in water, but their condensation polymers, the proteins, are usually insoluble. The tendency of some types such as proteins, dextrins, and starches to form colloidal dispersions may be deceptive.

Another method of increasing the molecular weight of a molecule is by the introduction of halogens. Usually this introduction merely results in a decreased water solubility with the result that some water-soluble compounds when substituted by halogens then fall in the water-insoluble class.

The five-carbon upper limit for water solubility follows from a very general principle, that increased structural similarity between the solute and the solvent is accompanied by increased solubility. Because of the polar nature of water, compounds owe their solubility in it almost entirely to the polar groups they may contain. As a homologous series is ascended, the hydrocarbon (nonpolar) part of the molecule continually increases while the polar function remains essentially unchanged. There follows, then, a trend toward a decrease in the solubility in polar solvents such as water.

The fact that the upper limits of water solubility for many series lie in the same neighborhood is due to the similarity of the polarities of many functional groups. The particular region (that of the member containing five carbon atoms) in these several series at which the upper limit of water solubility is reached is determined wholly by the altogether arbitrary proportions of solvent and solute chosen for use in this scheme of separation. It would have been equally easy and perhaps as satisfactory to employ a ratio of solute to solvent that would place the limit elsewhere.

The tendency of certain oxygen-containing compounds to form hydrates also contributes to water solubility. The stability of these hydrates is, therefore, a factor in determining water and ether solubility. Such compounds as chloral probably owe their great solubility in water to hydrate formation.

$Cl_3C{-}CHO$	$Cl_3C{-}CH(OH)_2$	$HCO_2CH_2CH_3$	$CH_3C({=}O)CO_2CH_3$
chloral	chloral hydrate	ethyl formate	methyl pyruvate

Low-molecular-weight esters of formic and pyruvic acids are hydrolyzed by water at room temperature, as indicated by the fact that the aqueous solution becomes distinctly acid to litmus.

Effect of Chain Branching on Solubility. It might be anticipated from a consideration of the effect of branching of the hydrocarbon chain on boiling points of the lower homologous series, such as the hydrocarbons and alcohols, that branching lowers intermolecular forces and decreases intermolecular attraction. It is not surprising, then, that a compound having a branched chain is more soluble than the corresponding straight-chain compound. This is a very general rule and is particularly useful in connection with simple aliphatic compounds. For example, the solubility of an iso compound differs widely from that of its normal isomer and is close to that of the next lower normal member of the homologous series in question. Effects of chain branching are shown in Table 5.3. In general, the more highly branched of two isomeric compounds possesses the greater solubility.

The position of the functional group in the carbon chain also affects solubility. For example, 3-pentanol is more soluble than 2-pentanol, which in turn is more soluble than 1-pentanol. When the branching effect is combined with moving the functional group toward the center of the molecule, as illustrated by 2-methyl-2-butanol, a very marked increase in solubility is noted. *Normally, the*

Table 5.3. Water Solubility of Various Organic Compounds

Types of compounds	General formula	Soluble	Borderline	Insoluble
Acids	RCO_2H	Pivalic (C_5)	Isovaleric (C_5)	*n*-Valeric (C_5)
Acid chlorides	RCOCl	Isobutyryl (C_4)	*n*-Butyryl (C_4)	
Alcohols	ROH	Neopentyl (C_5)	2-Methyl-3-butanol (C_5)	*n*-Amyl (C_5)
Amides	$RCONH_2$	Isobutyramide (C_4)	*n*-Butyramide (C_4)	
Esters	RCO_2R'	Isopropyl acetate (C_5)	*n*-Propyl acetate (C_5)	
Ketones	RCOR′	Isopropyl methyl (C_5)	Methyl *n*-propyl (C_5)	
Nitriles	RCN		Isobutyronitrile (C_5)	*n*-Butyronitrile (C_5)

more compact the structure, the greater the solubility, provided that comparisons are made on compounds of the same type.

Theory of Acid–Base Solubility

Effect of Structure on Acidity and Basicity In general, the problem of deciding whether a water-insoluble unknown should dissolve in dilute acid or base is primarily a matter of estimating its approximate acid or base strength. We are thus concerned with structural features that will stabilize the organic anion, A^-, and position the following equilibrium[2] farther to the right:

$$HA \rightleftharpoons H^+ + A^-$$

that is, we increase the magnitude of K_A, where

$$K_A = \frac{[H^+][A^-]}{[HA]}$$

or decrease the magnitude of $pK_a = \log(1/K_a) = -\log K_a$.

As in the case for most organic mechanistic phenomena, the two principal effects influencing structural control of acid–base strength are electronic and steric; these are now discussed in turn.

Electronic Effects on Acidity and Basicity Extensive studies have been made on the quantitative correlation of structure with acid or base strength of substituted organic compounds. These effects have been rationalized[3] on an electronic basis; a variety of evidence indicates that the para and ortho positions of benzenoid compounds are more sensitive to electronic alteration than are the

---

[2] H_3O^+ is often written H^+ for brevity.

[3] Consult any of the books on physical organic chemistry listed in Chapter 10. Results of ion cyclotron resonance studies have complicated these rationalizations.

meta positions:

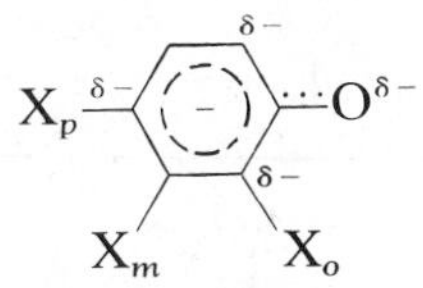

This is illustrated schematically for the phenoxide ion just above; resonance theory dictates that ortho and para positions are centers of partial negative charge and thus these centers respond dramatically to substitution by polar groups; ortho-substitution considerations are, however, always clouded by steric factors (see below).

Most organic carboxylic acids have dissociation constants in water at 25°C of 10^{-6} or greater and are therefore readily soluble in 5% sodium hydroxide. Phenols, on the other hand, because they are generally less acidic (the dissociation constant of phenol is about 10^{-10}) although soluble in strongly basic sodium hydroxide solution, are insoluble in dilute sodium bicarbonate (carbonic acid has a first dissociation constant[4] of 4×10^{-7}). The introduction of substituent groups, however, may have a profound effect on acidity. Thus *o*- and *p*-nitrophenol have dissociation constants of about 6×10^{-8}; in other words, the introduction of an ortho or para nitro group increases the acidity of phenol by a factor of about 600. As might be anticipated, the addition of two nitro groups, as in 2,4-dinitrophenol, increases the acidity to such an extent that the compound is soluble in dilute sodium bicarbonate solution. The acidity-increasing effect of the nitro group is due to stabilization of the phenoxide anion by additional distribution of the negative charge on the nitro group.

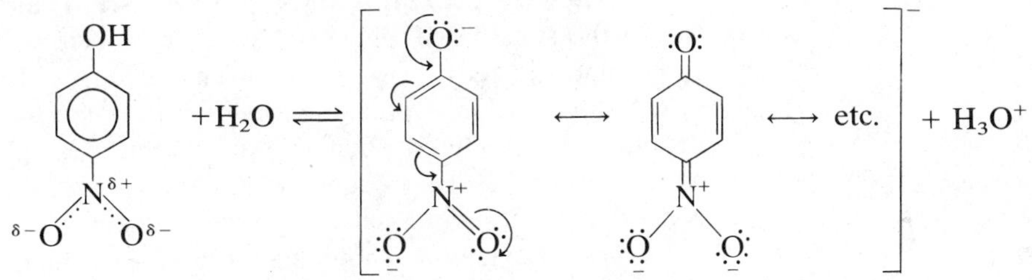

A similar acid-strengthening effect is observed when halogen is introduced into phenol. Thus an ortho bromine atom increases the acidity of phenol by a factor of about 30 and a para bromine atom by a factor of about 5. It is not suprising, then, that 2,4,6-tribromophenol is a sufficiently strong acid to dissolve in sodium bicarbonate solution. These changes due to bromine have been rationalized by inductive effects.

Similar electronic influences affect the basicity of amines. Aliphatic amines in aqueous solution have basicity constants of about 10^{-3} or 10^{-4}, nearly that of

---

[4] After correction for the amount of dissolved CO_2, the first ionization constant for carbonic acid is estimated to be ca. 10^{-4}.

ammonia (10^{-5}). Introduction of a conjugated phenyl group,[5] however, lowers the basicity by some 6 powers of 10. Thus aniline has a K_b of 5×10^{-10}. The effect of the phenyl ring is to stabilize the free amine on the left side of the equilibrium by resonance (see below); the phenyl group also decreases the basicity of nitrogen inductively.

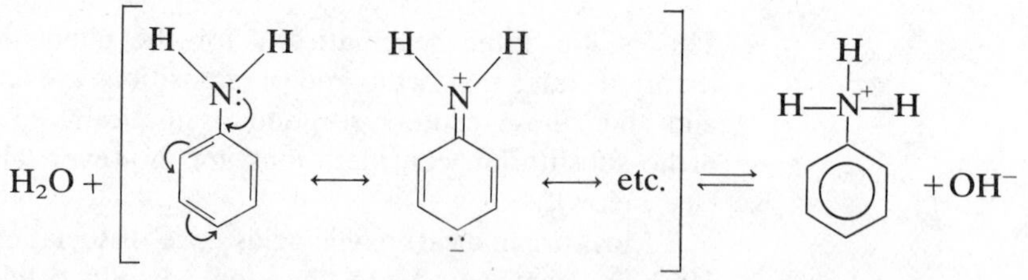

It is not suprising that a second phenyl substituent decreases the basicity to such an extent that the amine is no longer measurably basic in water. Thus, diphenylamine is insoluble in dilute hydrochloric acid. Substitution of a nitro group on the phenyl ring of aniline lowers the base strength, because this electron-withdrawing group destabilizes the anilinum ion (conjugate acid) and stabilizes the free base.

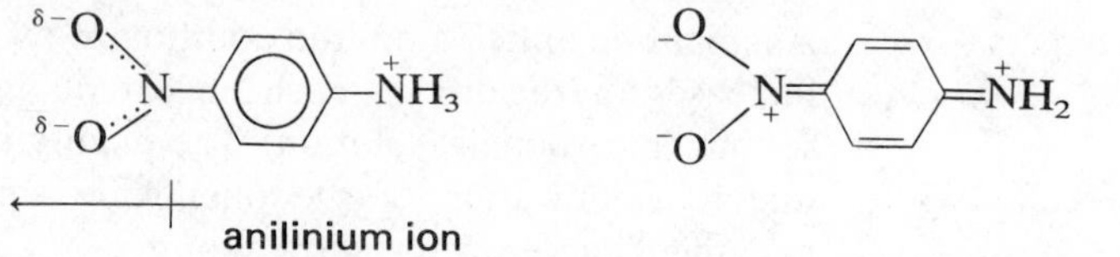

anilinium ion

Steric Effects on Acidity and Basicity It has been known for some time that ortho-substituted phenols have very much reduced solubility in aqueous alkali, and the term "cryptophenol" has been used to emphasize this behavior. Claisen's alkali (35% potassium hydroxide in methanol-water) has been used to dissolve such hindered phenols. An extreme example is 2,4,6-tri-*t*-butylphenol, which fails to dissolve in aqueous sodium hydroxide or Claisen's alkali. It can be converted to a sodium salt only by treatment with sodium in liquid ammonia. 2,4,6-Tri-*t*-butylaniline shows similarly unusual behavior. It is such a weak base that the pK_a of the conjugate acid is too low to measure in aqueous solution. 2,6-Di-*t*-butylpyridine is also significantly weaker as a base than the corresponding dimethylpyridine. It has been suggested that the weakening of the base strengths of the amines is due to steric strain introduced when a proton is added to the nitrogen atom; it seems likely that the instability of 2,6-di-*t*-butylphenoxide ion and of the hindered ammonium ions mentioned is due largely to steric interference with solvation of ions.

Steric strain may either increase or decrease the acidity of carboxylic acids. For example, substitution of alkyl groups on the α-carbon atom of acetic acid

[5] The nonbonded electron pair on nitrogen is said to be conjugated with formal "double" bonds of the benzene ring; i.e., it possesses a $-\ddot{N}-C{=}C$ unit.

tends to decrease acidity via destabilization of the conjugate base by steric inhibition of solvation. Ortho-substituted benzoic acids, on the other hand, are very appreciably stronger than the corresponding para isomers. The ortho substituents cause the carboxyl group to be out of the plane of the ring, with consequent greater destabilization of the acid than of the anion.[6] This latter case exemplifies a second kind of steric effect, which is frequently important and is referred to as *steric inhibition of resonance.* As a second example, although *p*-nitrophenol is some 2.8 pK_a units stronger than phenol, 3,5-dimethyl-4-nitrophenol is only about 1.6 pK_a units stronger.

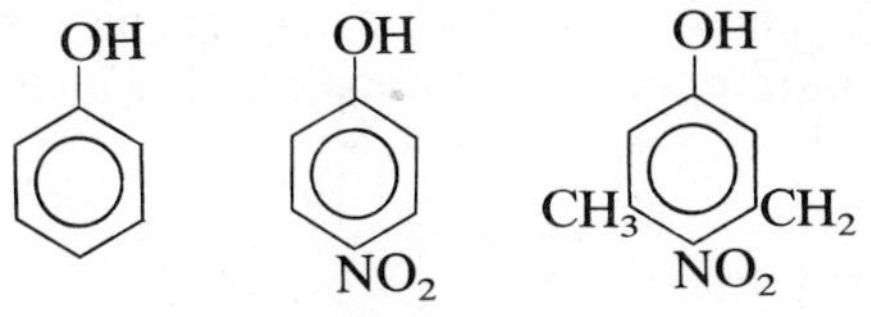

pK_a at 25°C: 9.99 (phenol), 7.21 (p-nitrophenol), 8.24 (3,5-dimethyl-4-nitrophenol)

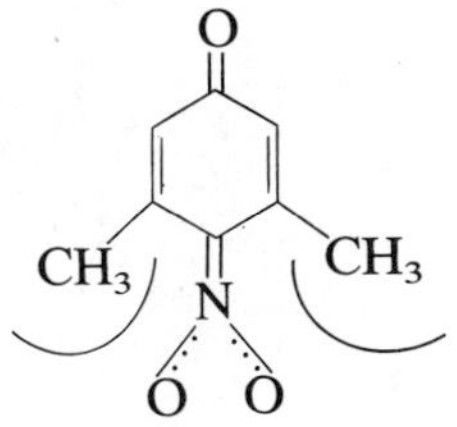

A

A part of the effect of the two methyl groups in reducing the acidity of the nitrophenol seems to be due to steric inhibition of resonance in the anion. Thus, structure A, above, requires coplanarity or near coplanarity of all the nitro group's atoms and the aromatic ring. Such coplanarity is inhibited by the presence of the methyl groups in the ion.

Solubility in Dilute Hydrochloric Acid Aliphatic amines, primary, secondary, and tertiary, form polar, ionic salts with hydrochloric acid. Hence, aliphatic amines are readily soluble in dilute hydrochloric acid (class B, if water-insoluble).

Conjugated aryl groups diminish the basicity of the nitrogen atom; primary aromatic amines, although more weakly basic than primary aliphatic amines, are soluble in dilute hydrochloric acid; diarylamines and triarylamines are not soluble. Diphenylamine, triphenylamine, and carbazole, for example, are insoluble. Arylalkylamines containing not more than one aryl group are soluble.

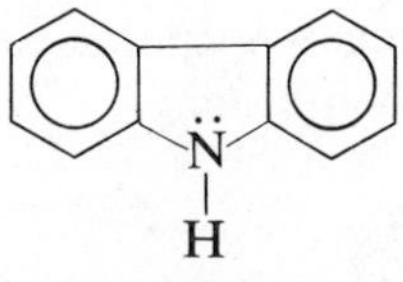

carbazole 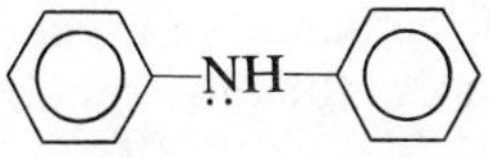 diphenylamine 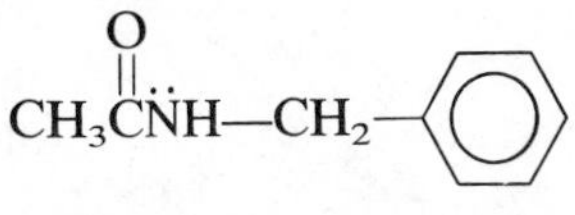 *N*-benzylacetamide

Disubstituted amides ($RCONR_2$) that are of sufficiently high molecular weight to be water insoluble are soluble in dilute hydrochloric acid. This behavior

[6] More details on this can be found in the books listed in Chapter 10 for the areas of either reaction mechanisms or physical organic chemistry.

contrasts with that of the simple amides ($RCONH_2$), which are neutral compounds.[7] Most monosubstituted amides (RCONHR) also are neutral. *N*-Benzylacetamide, however, is basic.

It should be noted that amines may react with 5% hydrochloric acid to form *insoluble* hydrochlorides. Compounds of this type may be incorrectly classed. For example, certain arylamines, such as α-naphthylamine, form hydrochlorides that are sparingly soluble in dilute hydrochloric acid. By warming the mixture slightly and diluting it with water, solution sometimes may be effected. The appearance of the solid usually will show if the amine has undergone a change. In order to decide doubtful cases, the solid should be separated and its melting point compared with that of the original compound. A halogen test with alcoholic silver nitrate indicates formation of a hydrochloride.

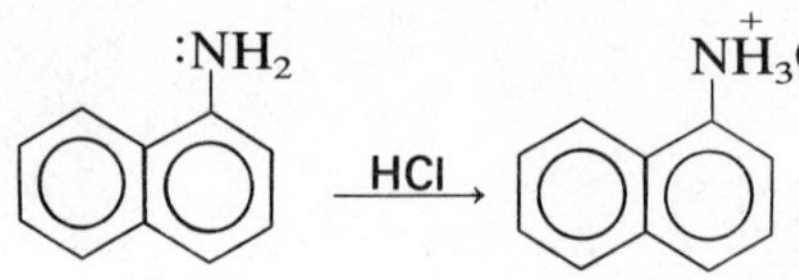

$:NH_2$ —HCl→ $\overset{+}{N}H_3\overset{-}{Cl}$

α-naphthylamine hydrochloride

Another possibility is to dissolve the suspected base in ether, and treat that with 5% HCl. Formation of solid at the interface indicates a basic amine.

A few types of oxygen-containing compounds that form oxonium salts upon treatment with hydrochloric acid also are basic.

Solubility in 5% Sodium Hydroxide and 5% Sodium Carbonate Solutions A list of various organic acids, related to Table 5.1, is given in Table 5.4. Most of the reasons for classifications such as these can be understood in terms of control of the stability of the conjugate base (anion) by structural factors.

Aldehydes and ketones are sufficiently acidic to react with aqueous alkali to yield anions that serve as reaction intermediates in such processes as the aldol condensation; they are far too weakly acidic, however, to dissolve to any measurable extent in sodium hydroxide solution. When two carbonyl groups are attached to the same carbon atom, as they are in acetoacetic and malonic esters and in 1,3-diketones, the acidity increases sharply because of the added stabilization of the anion, in which the negative charge can be distributed on two oxygen atoms as well as the central carbon atom.

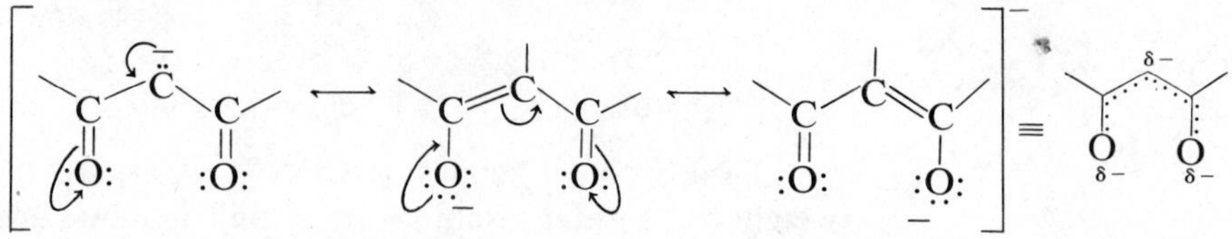

[resonance structures of the 1,3-dicarbonyl anion ↔ ↔]$^-$ ≡ δ− delocalized structure with O δ− and O δ−

[7] Amides are generally comparable to water in basicity and small structural changes, such as alkylation, need change their K_b by only $\sim 10^2$ to move them into the "basic" category.

Table 5.4. Solubility Classes of Various Organic Acids

Compounds		
Name	**General structure**	**Solubility class**[a]
Carboxylic acids	RCO_2H	A_1
Sulfonic acids	RSO_3H	A_1
Sulfinic acids	RSO_2H	A_1
Enols	$-\overset{\vert}{C}=\overset{\vert}{C}-OH$	A_2
Imides	$-\overset{O}{\overset{\Vert}{C}}-NH-\overset{O}{\overset{\Vert}{C}}-$	A_2
Nitro[b]	$>CH-NO_2$	A_2
Arenesulfonamides[c]	$ArSO_2NHR$	A_2
β-Dicarbonyl compounds[d]	$-\overset{O}{\overset{\Vert}{C}}-\underset{\vert}{CH}-\overset{O}{\overset{\Vert}{C}}-$	A_2[d]
Oximes	$>C=N-OH$	A_2

[a] Borderline cases are named in Table 5.5.
[b] Primary (RCH_2NO_2) and secondary (R_2CHNO_2) nitroalkanes only.
[c] The acidity of the N—H proton is utilized in the Hinsberg test (Procedure 19). This category also includes sulfonamides of ammonia and other sulfonamides of primary amines.
[d] Highly electronegative groups, e.g., trifluoromethyl, on the carbonyl group can move these compounds into class A_1.

It should be noted that, although β-dicarbonyl compounds are approximately as acidic as the phenols, the rate of proton removal from carbon may be a relatively slow reaction and the rate of solution of such substances may be so slow that they appear to be insoluble in base.

It is of interest that nitro compounds have a tautomeric form, the *aci* form, which is approximately as strong an acid as the carboxylic acids. The aci form of nitroethane has a K_a of 3.6×10^{-5}.

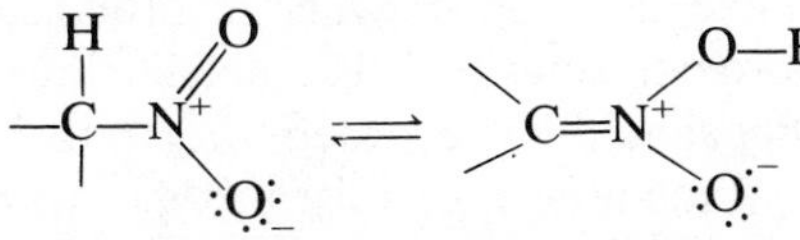

$$-\overset{H}{\overset{\vert}{\underset{\vert}{C}}}-N^{+}(=O)(-\ddot{O}:^{-}) \rightleftharpoons\ >C=N^{+}(-O-H)(-\ddot{O}:^{-}) \quad \textit{aci form}$$

Even one nitro group confers sufficient acidity on a substance to make it soluble in dilute sodium hydroxide. Thus nitroethane has a K_a of about 3.5×10^{-9}. This should be compared to the K_a values for the following β-dicarbonyl compounds:

$CH_3COCH_2COOC_2H_5$	2×10^{-11}
$CH_3COCH(C_2H_5)COOC_2H_5$	2×10^{-13}
$CH_2(COOC_2H_5)_2$	5×10^{-14}
$CH_3COCH_2COCH_3$	1×10^{-9}

Just as the grouping

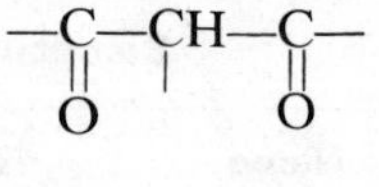

$$-\underset{\underset{O}{\|}}{C}-\underset{|}{CH}-\underset{\underset{O}{\|}}{C}-$$

is acidic, so is the imide grouping

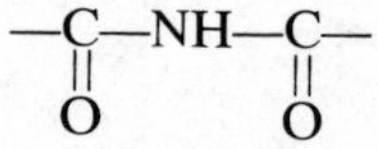

$$-\underset{\underset{O}{\|}}{C}-NH-\underset{\underset{O}{\|}}{C}-$$

and imides are soluble in dilute sodium hydroxide solution but not in sodium bicarbonate. A *p*-nitrophenyl group makes the —CONH— function weakly acidic in aqueous solution. Thus *p*-nitroacetanilide[8] dissolves in sodium hydroxide solution but not sodium bicarbonate. Sulfonamides show the same base solubility as *p*-nitroacetanilide. Oximes, which have a hydroxyl group attached to a nitrogen atom, show similar solubility behavior. The structures

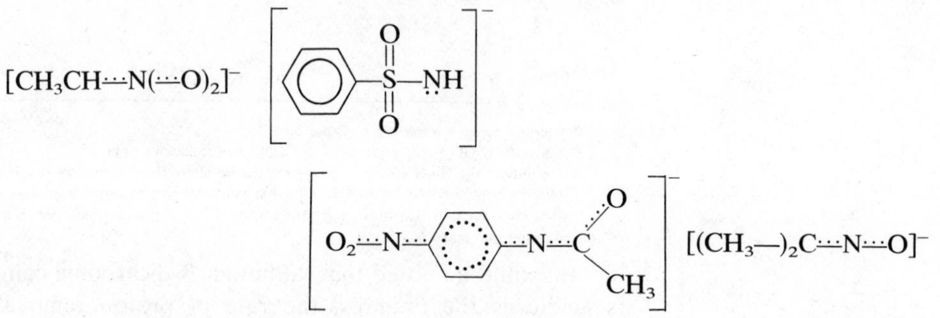

$$[CH_3CH\text{---}N(\text{---}O)_2]^- \qquad \left[C_6H_5-\underset{\underset{O}{\|}}{\overset{\overset{O}{\|}}{S}}-\ddot{N}H\right]^-$$

$$\left[O_2\text{---}N\text{---}C_6H_4\text{---}N\text{---}C(\text{---}O)CH_3\right]^- \qquad [(CH_3-)_2C\text{---}N\text{---}O]^-$$

are the conjugate bases of

nitroethane benzenesulfonamide *p*-nitroacetanilide acetone oxime

Esters (containing five or six carbon atoms) that are almost completely soluble in water may be hydrolyzed by continued shaking with dilute sodium hydroxide solution.[9] The alkali should not be heated, and the solubility or insolubility should be recorded after 1 to 2 min.

Monoesters of dicarboxylic acids are bicarbonate soluble; these esters are also rapidly hydrolyzed even with aqueous bases as weak as sodium bicarbonate.

---

[8] Compounds of this type may form adducts (Meisenheimer complexes) by bonding hydroxide to the carbon bearing the amide group:

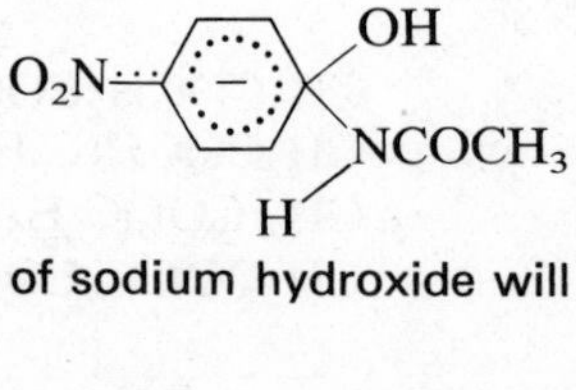

$$O_2N\text{---}C_6H_4^{(-)}\langle^{OH}_{N(H)COCH_3}$$

[9] Use of lithium hydroxide in place of sodium hydroxide will often yield water-soluble salts.

Fatty acids containing 12 or more carbon atoms react with the alkali slowly, forming salts that are soaps. The mixture is not clear but consists of an opalescent colloidal dispersion that foams when shaken. Once this behavior has been observed it is easily recognized.

Certain of the sodium salts of highly substituted phenols are insoluble in sodium hydroxide. This property may be detected by trying the solubility of any residue in water. Certain phenols that are very insoluble in water may precipitate due to hydrolysis and hence appear to be insoluble in alkali.

Solubility of Amphoteric Compounds Compounds containing both an acidic and a basic group are amphoteric. Low-molecular-weight amino acids exist largely as dipolar salts. They are soluble in water and may give solutions neutral to litmus (thus are class S_2).

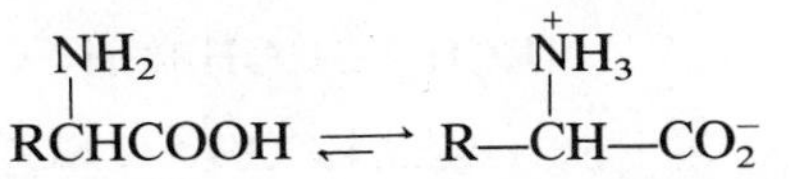

$$RCH(NH_2)COOH \rightleftharpoons R{-}CH(\overset{+}{N}H_3){-}CO_2^-$$

The water-insoluble amphoteric compounds act both as bases and as strong or weak acids, depending on the relative basicity of the amino group, since the basicity determines the extent to which the acidic group will be neutralized by inner salt formation. If the α-amino group carries only aliphatic substituents, the compounds will dissolve in hydrochloric acid and sodium hydroxide but not in sodium bicarbonate, class $A_2(B)$:

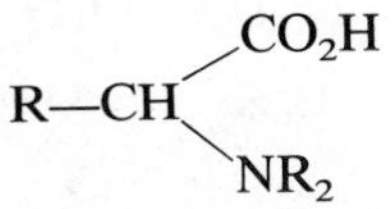

$$R{-}CH(CO_2H)(NR_2)$$

The presence of an aryl group on the nitrogen atom, however, diminishes its basicity so that such compounds are soluble even in aqueous bicarbonate solution. This is illustrated by the following compounds (class A_1):

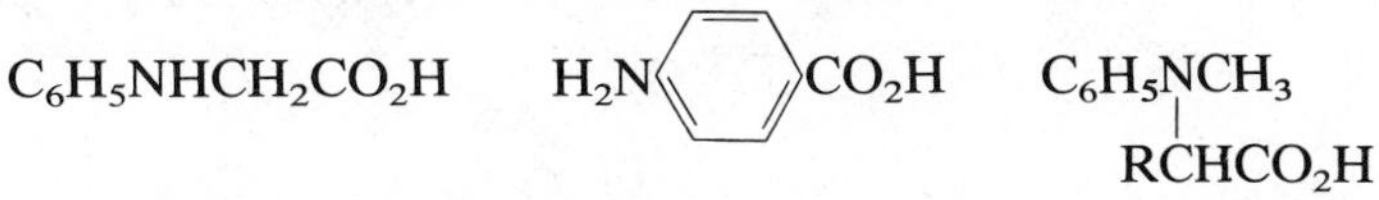

$$C_6H_5NHCH_2CO_2H \qquad H_2N{-}C_6H_4{-}CO_2H \qquad C_6H_5N(CH_3){-}RCHCO_2H$$

If two aryl groups are attached to the nitrogen atom the compound is not basic but behaves simply as a strong acid (class A_1):

$$(C_6H_5)_2NCH_2CO_2H$$

Solubility in Cold Concentrated Sulfuric Acid Cold concentrated sulfuric acid is used with neutral, water-insoluble compounds containing no elements other than carbon, hydrogen, and oxygen. If the compound is unsaturated, is readily sulfonated, or possesses a functional group containing oxygen, it will dissolve in cold concentrated sulfuric acid. Solution in sulfuric acid frequently is accompanied by a reaction such as sulfonation, polymerization, dehydration, or addition of the sulfuric acid to olefinic or acetylenic linkages; but in many cases ions are produced from which the solute may be recovered by dilution with ice water. The following illustrate some of the more common reactions:

$$RCH{=}CHR + H_2SO_4 \longrightarrow [RCH_2{-}\overset{+}{C}HR] \xrightarrow{HSO_4^-} RCH_2\overset{OSO_3H}{\overset{|}{C}}HR$$

$$RCH_2OH + 2H_2SO_4 \longrightarrow RCH_2OSO_3H + H_3O^+ + HSO_4^-$$

(The water arising from sulfate ester formation is converted to hydronium ion by concentrated sulfuric acid.)

$$(RCH_2)_2CHOH + 2H_2SO_4 \rightarrow (RCH_2)_2CHOSO_3H + H_3O^+ + HSO_4^-$$

$$(RCH_2)_3COH + H_2SO_4 \rightarrow (RCH_2)_2C{=}CHR + H_3O^+ + HSO_4^-$$

$$CH_3O{-}C_6H_5 + 2H_2SO_4 \rightarrow CH_3O{-}C_6H_4{-}SO_3H + H_3O^+ + HSO_4^-$$

$$(CH_3)_3C_6H_3 + 2H_2SO_4 \rightarrow (CH_3)_3C_6H_2{-}SO_3H + H_3O^+ + HSO_4^-$$

$$(C_6H_5)_3COH + 2H_2SO_4 \rightarrow (C_6H_5)_3C^+ + H_3O^+ + 2HSO_4^-$$

$$\left.\begin{array}{l}(C_6H_5)_2C{=}O + H_2SO_4 \rightarrow (C_6H_5)_2\overset{+}{C}OH + HSO_4^- \\ C_6H_5\overset{O}{\overset{\|}{C}}OH + H_2SO_4 \rightarrow C_6H_5\overset{+}{C}(OH)_2 + HSO_4^-\end{array}\right\} \xrightarrow{H_2O} \text{Substrate regenerated}$$

$$(CH_3)_3C_6H_2{-}CO_2H + 2H_2SO_4 \rightarrow (CH_3)_3C_6H_2{-}\overset{+}{C}{=}O + H_3O^+ + 2HSO_4^-$$

$$CH_3\overset{O}{\overset{\|}{C}}O\overset{O}{\overset{\|}{C}}CH_3 + 3H_2SO_4 \rightarrow 2CH_3\overset{+}{C}(OH)_2 + HSO_4^- + HS_2O_7^-$$

Alkanes, cycloalkanes and their halogen derivatives are insoluble in sulfuric acid. Simple aromatic hydrocarbons and their halogen derivatives do not undergo sulfonation under these conditions and are insoluble. However, the insertion of two or more alkyl groups[10] in the benzene nucleus permits the compound to be sulfonated easily; hence, polyalkylbenzenes dissolve rather readily in sulfuric acid.

---

[10] Other activating groups will also (usually) facilitate sulfonation.

For this reason isodurene and mesitylene are soluble. Occasionally the solute may react in such a manner as to yield an insoluble product. A few high-molecular-weight ethers such as phenyl ether undergo sulfonation so slowly at room temperature that they may not dissolve.

Many secondary and tertiary alcohols are dehydrated readily by concentrated sulfuric acid to give olefins which then undergo polymerization. The resulting polymers are insoluble in cold concentrated sulfuric acid and hence form a distinct layer on top of the acid. Benzyl alcohol (and similar alcohols) react with concentrated sulfuric acid; colored precipitates result.

Borderlines Between Solubility Classes. In Table 5.5 are listed a number of compounds selected in such a way as to show the position of the most important of the various borderlines between solubility classes. These compounds have been grouped as far as possible according to chemical nature. In each group an attempt has been made to include the borderline members together with one or more members at either side of the borderlines. The solubility class of a compound not listed will be evident by considering its relation to the borderline members of the series to which it belongs. Thus the table shows *n*-butyl alcohol to be in class S_1; it follows that the other butyl alcohols and all lower homologs are also in this class. Similarly, since isopentyl alcohol is in class N_1, it follows that *n*-pentyl alcohol and all higher alcohols are in N_1 or N_2.

Table of Solubilities. Although it often is possible to predict the solubility class of a compound by reference to its structural formula, there are many exceptions. Moreover, it occasionally is difficult to classify a compound even by reference to actual experiment; many compounds, as shown in Table 5.5, occupy borderline positions. In Appendix I are listed a number of the more common compounds the solubility classes of which are difficult to predict.

New combinations of functional groups are continually being discovered that provide further exceptions to the rules. It is recommended therefore that, if a structure is of a type that is unfamiliar, the original literature should be consulted.

5.1.2 Solubility in Organic Solvents

The solubility of organic compounds in organic solvents should be determined in order to plan a variety of laboratory operations. A range of solvents, all useful to the organic chemist, is tabulated in Table 5.6. These solvents are useful for running organic reactions, dissolving substrates for spectral and analysis, and standard laboratory maintenance (cleaning glassware, etc.); solvent uses, cross-referenced to Table 5.6, are listed in Table 5.7. Compositions listed in Table 5.7 are only approximate guidelines; variations in these data depend on, for example, available instrument sophistication, detail of absorption information needed, and so on. Ultraviolet determinations (Chapter 8) are usually carried out much later and are thus not discussed here. Virtually all of the solvents listed, as well as mixtures of these and of other solvents, are useful for column and thin-layer chromatography, for recrystallizations (of solids, Chapter 7), and for extractions

Table 5.5. Borderlines Between Solubility Classes

Compound	Solubility class	Compound	Solubility class
Acids		Ethyl benzoate	N_2
Chloroacetic	S_1	Methyl carbonate (C_3)	S_1–N_1
n-Butyric	S_1	Ethyl oxalate	S_1–N_1
α-Chloropropionic	S_1	Methyl malonate	S_1–N_1
Crotonic	S_1	Ethyl carbonate (C_5)	S_1–N_1
Isovaleric (C_5)	S_1–A_1	Ethyl succinate	N_1
Valeric (C_5)	A_1	Ethyl phthalate	N_1
Alcohols		Ethyl malonate	N_2
n-Butyl	S_1	*n*-Butyl carbonate	N_1–N_2
t-Pentyl	S_1	*n*-Butyl oxalate	N_2
Isopropyl-		Ethers	
methylcarbinol	S_1–N_1	Ethyl methyl	S_1
Isopentyl	S_1–N_1	Ethyl	S_1–N_1
Benzyl	N_1	Ethyl isopropyl	S_1–N_1
Cyclopentyl	N_1	Isopropyl	N_1
Aldehydes		*n*-Butyl	N_2
Isobutyraldehyde	S_1	Hydrocarbons (aromatic)	
n-Butyraldehyde	S_1–N_1	Mesitylene	N_2
Isovaleraldehyde (C_5)	N_1	Isodurene	N_2
Amides		Cymene	I
Formamide	S_1–S_2	*p*-Xylene	N_2–I
Acetamide	S_1–S_2	Diphenylmethane	N_2–I
Propionamide	S_1–S_2	*m*-Xylene	N_2–I
Isobutyramide	S_1–S_2	*o*-Xylene	N_2–I
n-Butyramide	S_1–M	Naphthalene	I
Formanilide	S_1–M	Ketones	
Acetanilide	M	Ethyl methyl	S_1
Amines		Isopropyl methyl	S_1
Diethyl	S_1	Methyl *n*-propyl	S_1–N_1
Isopentyl	S_1	Pinacolone	S_1–N_1
n-Pentyl	S_1	Diethyl	S_1–N_1
Benzyl	S_1	Cyclopentanone	S_1
Piperidine	S_1	Cyclohexanone	S_1
Cyclohexyl	S_1	Acetophenone	N_1
Di-*n*-propyl	S_1–B	Di-*n*-butyl	N_1–N_2
Di-*n*-butyl	B	Benzil	N_2
Aniline	B	Benzophenone	N_2
Tri-*n*-propyl	B	Nitriles	
Esters		Propionitrile	S_1
Ethyl acetate	S_1–N_1	Isobutyronitrile	S_1–MN
Methyl propionate	S_1	Succinonitrile	S_1–S_2–MN
n-Propyl formate	S_1	Glutaronitrile[a]	S_2–MN
Isopropyl acetate	S_1	*n*-Butyronitrile	MN
n-Propyl acetate	S_1–N_1	Nitro compounds	
Methyl isobutyrate	S_1–N_1	Nitromethane	S_1–A_2
n-Butyl formate	S_1–N_1	Nitroethane	A_2
Methyl isovalerate		Nitrobenzene	MN
(C_6)	N_1	Phenols	
sec-Butyl acetate	N_1	Hydroquinone	S_1
n-Butyl acetate	N_1	Chlorohydroquinone	S_1–A_2
Benzyl acetate	N_1	Phloroglucinol	S_2–A_2
Ethyl caprylate (C_{10})	N_2	Phenol	S_1–A_2

[a] Trimethylene cyanide.

Table 5.6. Common Organic Solvents[a]

Solvents			
Name	Formula	Common use (Code)[b]	Dielectric constant (25°C)
Acetone[a]	$CH_3(C{=}O)CH_3$	C, R, nmr[d]	21
Acetonitrile	CH_3CN	R, uv, nmr[d]	36
Benzene[e]	C_6H_6	R, nmr[d]	4.2
Carbon disulfide	CS_2	R, ir, nmr	2.6
Carbon tetrachloride[c]	CCl_4	R, ir, uv, nmr	2.2
Chloroform[c]	$CHCl_3$	All five[d]	4.7
N,*N*-Dimethylformamide	$HC({=}O)N(CH_3)_2$	R, nmr[d], R	37
Dimethyl sulfoxide (DMSO)	$CH_3(S{=}O)CH_3$	nmr,[d] R	49
Ethanol[c]	CH_3CH_2OH	R, uv, C, nmr[d]	24
(Ethyl) ether[c]	$(CH_3CH_2)_2O$	R, C	4
Hexane	$CH_3(CH_2)_4CH_3$	uv, C, R	2.0
Methanol	CH_3OH	R, uv, nmr,[d] C	33
Methylene chloride	CH_2Cl_2	R, C, ir, nmr[d]	9
Pyridine	(pyridine ring, N)	R, nmr[d]	12
Tetrahydrofuran	(tetrahydrofuran ring, O)	R, nmr[d]	7.3
(Water)	(H_2O)	(All five)[d]	(78.5)

[a] The ir and nmr (proton) spectra of many of these compounds may be found in the Sadtler collection and in R. M. Silverstein, G. C. Bassler, and T. C. Morrill, *Spectrometic identification of Organic Compounds*, 3rd ed. (Wiley, New York, 1974).

[b] C = glassware cleaning; R = reaction medium; solvents to dissolve samples for spectral analysis are denoted ir, nmr, or uv.

[c] Preliminary solubility analysis should employ these solvents.

[d] Deuterated solvents, e.g., acetone-d_6 = CD_3COCD_3, are available for determination of proton magnetic resonance spectra.

[e] Toluene can be substituted to reduce toxicity problems.

during work-up of reaction products. To conserve sample, only those solvents marked need be used for preliminary characterization of unknowns; sample can, of course, be recovered (by evaporation) from virtually all of these solvents. Somewhat speculative structural conclusions can be drawn from solubility in such solvents; for example, dinitrophenyl benzyl sulfide (a thioether) is not water-soluble, but it is sufficiently soluble in polar organic solvents (e.g., DMSO, Table

Table 5.7. Solubility Standards for Use of Organic Solvents

Uses[a]	Solubility (%)[b]
R	10
ir	5
nmr[c]	25

[a] See Table 5.6.

[b] Weight/volume; only an approximate guide.

[c] Research instruments, computer averaging, and Fourier Transform techniques demand much smaller minimum concentration levels.

5.6, dielectric constant[11] = 49) to allow determination of nmr spectra. Thus, the necessity of using DMSO-d_6, rather than $CDCl_3$, suggests the presence of one or more polar groups in the molecule. Finally, since all solvents listed in Table 5.6 are often used for a variety of purposes, they are often encountered as impurities in samples of interest.

Procedure. Carry out solubility tests in organic solvents using the simple procedure described for water solubilities on p. 91. Use 3 ml of organic solvent and the following weights (or volumes) for the organic substance (see Table 5.7): R, 0.3 g (0.20 ml); ir, 0.15 g (150 μl); nmr, 0.75 g (750 μl).[12]

5.2 INFRARED (ir) ANALYSIS OF ORGANIC COMPOUNDS

A detailed discussion of the theory of infrared spectra and of the instruments involved will not be presented here. Introductory and advanced theory (and instrumentation) are described in a gamut of books (see Chapter 10) running from that by Morrison and Boyd (an introductory organic text) to those of Bellamy and Conley (specialized ir books) and branching off to instrumental analysis textbooks. Note that ir absorption spectrometry is very sensitive to the bond multiplicity and to the atomic composition of functional groups. Thus ir analysis is largely a functional group probe; some general structural conclusions can also often be determined.

One can deduce the essentials of how an ir instrument functions from the somewhat superficial schematic diagram of Fig. 5.2. We shall now consider a very brief, qualitative discussion of the optics and electronics of an infrared spectrometer related to Fig. 5.2. A source, for example, a heated filament, provides radiation that is directed, via mirrors, through the sample and reference cells. These two beams of radiation are directed, via a series of additional mirrors, to a chopper. These mirrors and the chopper effectively produce a single beam that possesses alternate pulses of each of the sample and the reference beams. Mirrors focus this "chopped" beam into the slit entrance to the monochromator. Prior to reaching the monochromator, this beam is composed of the various energies emitted by the source. The beam's energy is dispersed by gratings (or prisms) and slits within the monochromator area such that at any one instant (grating setting) a *specific* magnitude of radiational energy is being directed toward the exit slits and on the detector. When a beam contains radiation that has been partly absorbed by the sample, this absorption is perceived by the detector as an off-null perturbation. Such perturbations are transmitted by the detector, as an electrical impulse, to the servo motor. The servo motor causes an attenuator to be pushed into the

---

[11] Dielectric constant is one of the few numerical estimates of solvent polarity; see the discussion of dielectric in the section on water solubility (p. 95). The correlation of solvent polarity with dielectric constant is, at best, crude.

[12] Note: 1 ml = 1000 μl.

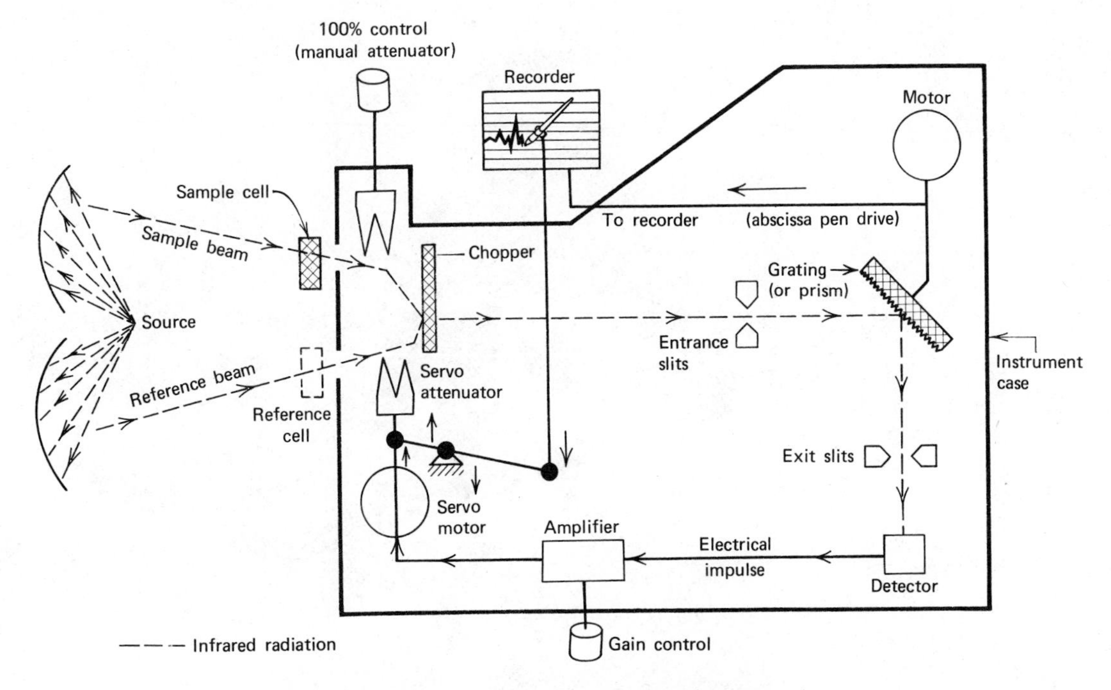

Figure 5.2. Schematic diagram of a double-beam infrared spectrometer.

reference beam to cause the sample and reference beams to be rebalanced. The detector thus, by detecting infrared absorptions and electrically creating a reference beam compensation, maintains the combined beam at an optical null. The recorder receives two dimensions of input. One is the monitoring of changes in position of the grating (or prism) which is connected electrically to changes in position on the abscissa (i.e., it records the energy or wavelength under consideration). The other is a recording of the insertion of the servo attenuator; that is, this is a measure of the degree of absorption (measured on the ordinate).

We shall be most concerned here with the sampling area of the ir instrument (Fig. 5.3). As in any double-beam absorption technique, we are interested in the absorption due to the sample cell relative to a reference cell (solvent or air).

Compounds that are to be subjected to ir (or other spectral analysis) should be pure. Solids should be samples from recrystallization (resulting in sharp melting points) or collected from chromatography. Liquids should be samples from distillation (or from gas chromatographic collections, Chapter 7).

Samples can be prepared for analysis by any one of the following methods. Liquids can be examined directly as a thin film ("neat") between plates (Fig. 5.4); in principle, however, the spectrum of a compound dissolved in a nonpolar organic solvent, for example, carbon tetrachloride, provides a spectrum that is least distorted due to associations caused by solvent–solute or sample–sample aggregates. All solution techniques (Fig. 5.5) require reference cells; a cell containing pure solvent (of path length identical to that of the sample cell) should

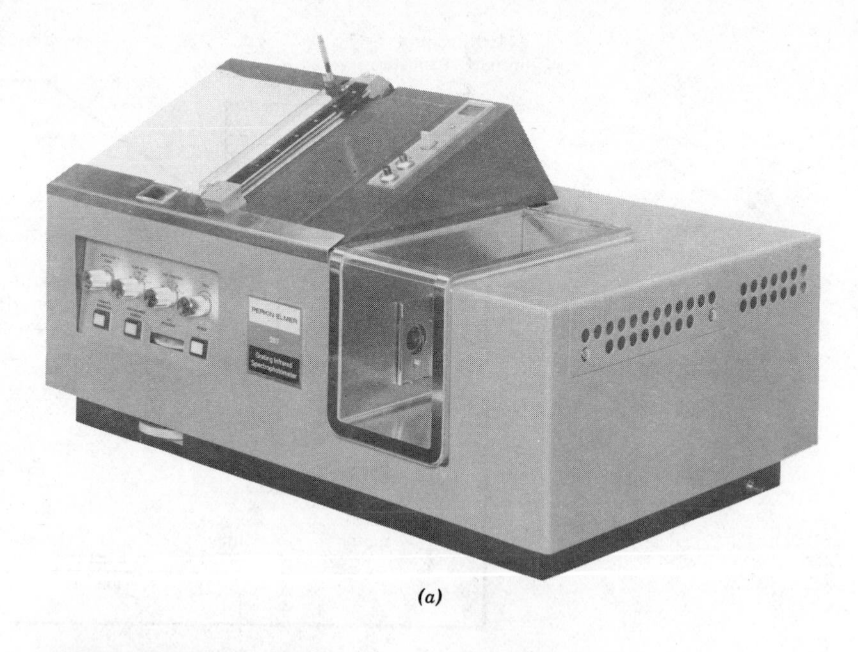

(a)

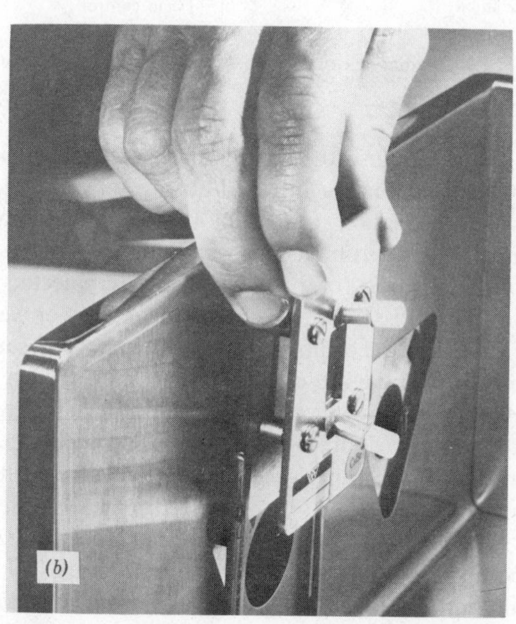

Figure 5.3. (*a*) Grating infrared spectrometer; sample cell holder is visible. (*b*) Sampling area of an ir spectrometer (double-beam); placing the sample cell in its holder. (*c*) Matched cells for liquids are placed in the spectrometer (p. 115). (*d*) Gas cells are shown in the sampling area. Note the typically greater size of the gas cell compared to the solution cells of (*a*)–(*c*). Also gases yield more detailed, more highly resolved spectra (as shown) compared to solutions (p. 115).

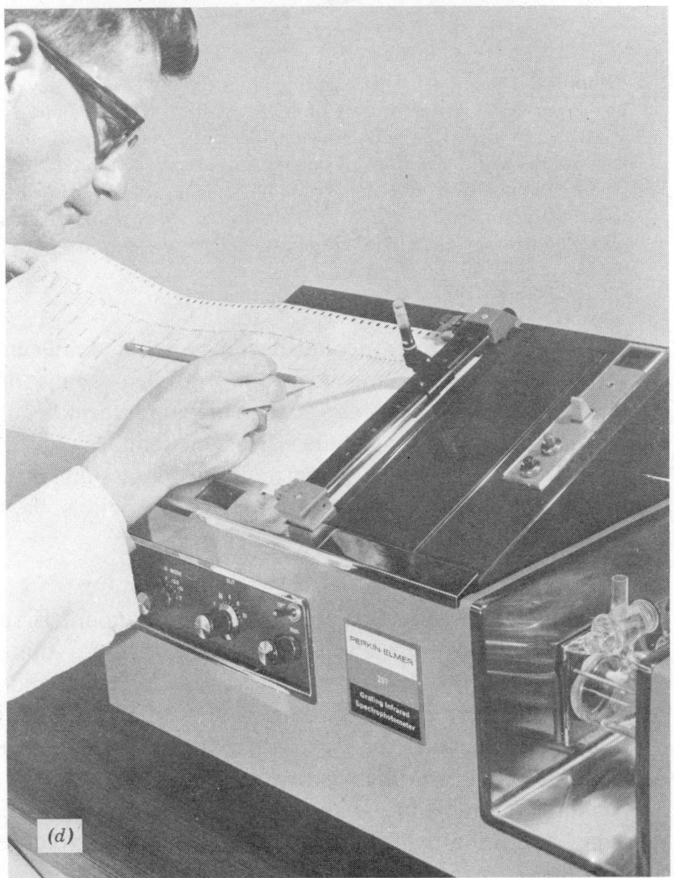

Fig. 5.3, (continued)

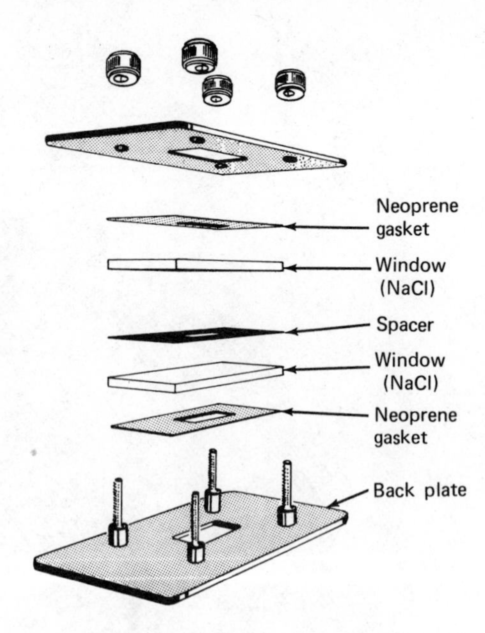

Figure 5.4. Demountable cells for ir analysis of liquids. Assembly procedure: (1) Place bottom gasket and lower NaCl plates in nest of bolts on back (lower) plate. Note: NaCl plates should not be scratched or wetted on surfaces; handle them carefully by their edges. (2) Place spacer on lower window and add a few drops of sample; spacer may be omitted for sufficiently nonvolatile or viscous liquids. Extremely volatile liquids should be analyzed in enclosed solution cells (Fig. 5.5). (3) Place upper plate, top gasket, and front plate on the cell; carefully tighten nuts until the sample is evenly dispersed between the plates. *Do not overtighten the nuts, as this may break the salt plates.*

be placed in the reference beam. Positions of significant absorption of standard ir solvents and mulls are given in Fig. 5.6. Mulls do not involve substantial miscibility of the substrate in the solvent; intermolecular associations (for example, hydrogen bonding) persist in such samples. From the standpoint of lack of absorption by the supporting medium, potassium bromide mulls ("pellets") are the most attractive; preparation of such pellets, however, demands the most careful technique of any method (Fig. 5.7).

Procedure. Be sure that the infrared instrument is running to ensure sufficient preanalysis warm-up. Prepare the sample for a thin-film cell (Fig. 5.4), for a solution (ca. 10% for 0.1-mm cells; see Table 5.6 and Fig. 5.5) cell or in a mull. Mulls are prepared by blending ca. 1% of the sample in the mulling medium by thoroughly mashing the pair with a pestle in a smooth, agate mortar. Potassium bromide mulls are analyzed as in Fig. 5.7; nujol, Fluorolube®, and hexachlorobutadiene mulls are analyzed by the cells in Fig. 5.4. Make sure that the chart paper and initial wavelength setting are correctly zeroed and interrelated on the instrument; exact operation guidelines should be obtained from the instructor owing to variations in instrument characteristics. Gain and balance control

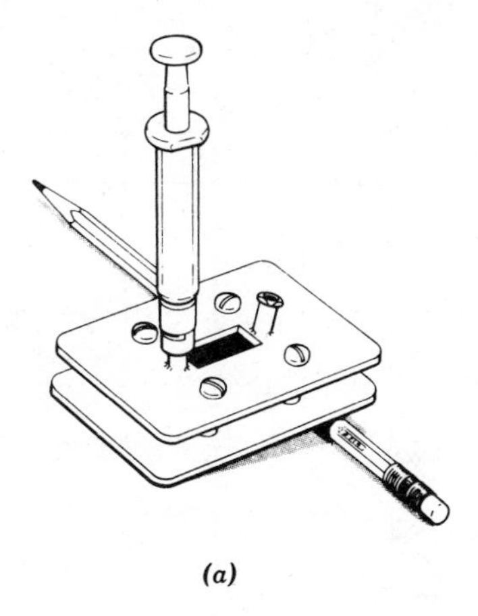

(a)

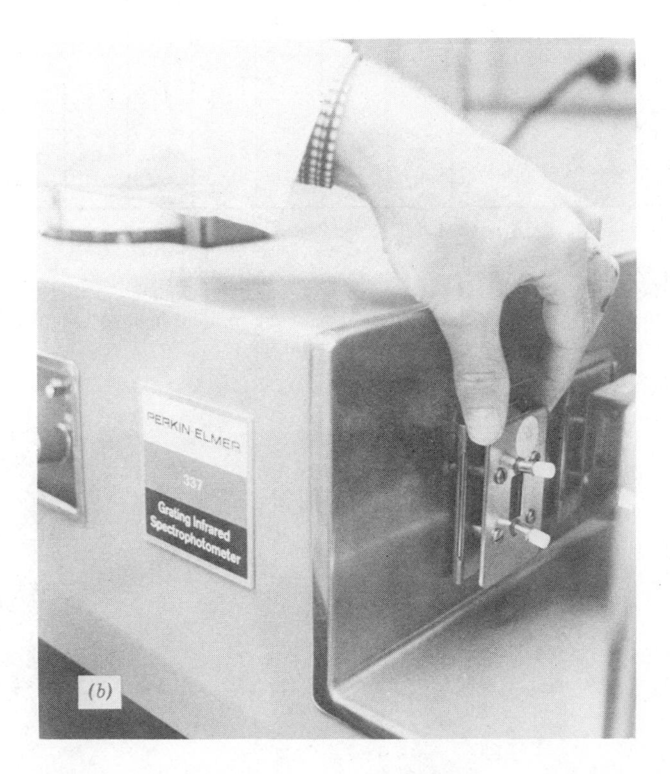

Figure 5.5. (*a*) Correct way to fill a sealed cell. (*b*) Enclosed solution cell. (1) Remove both plugs. (2) Add, with needle-less syringe as shown in (*a*), sufficient solution (ca. 0.050 ml of a 5% solution if cell has 0.10-mm path length) to fill cell; hold cell at an angle during filling to allow bubbles to escape. (3) Replace upper plug while syringe is still in place; then replace syringe with plug. (4) Clean the cell by washing with pure solvent (introduced via syringe); dry the cell's inner chamber by connecting a charged drying tube to one port and drawing air (with syringe or vacuum) through the cell. The cells should be stored in a dessicator when not in use.

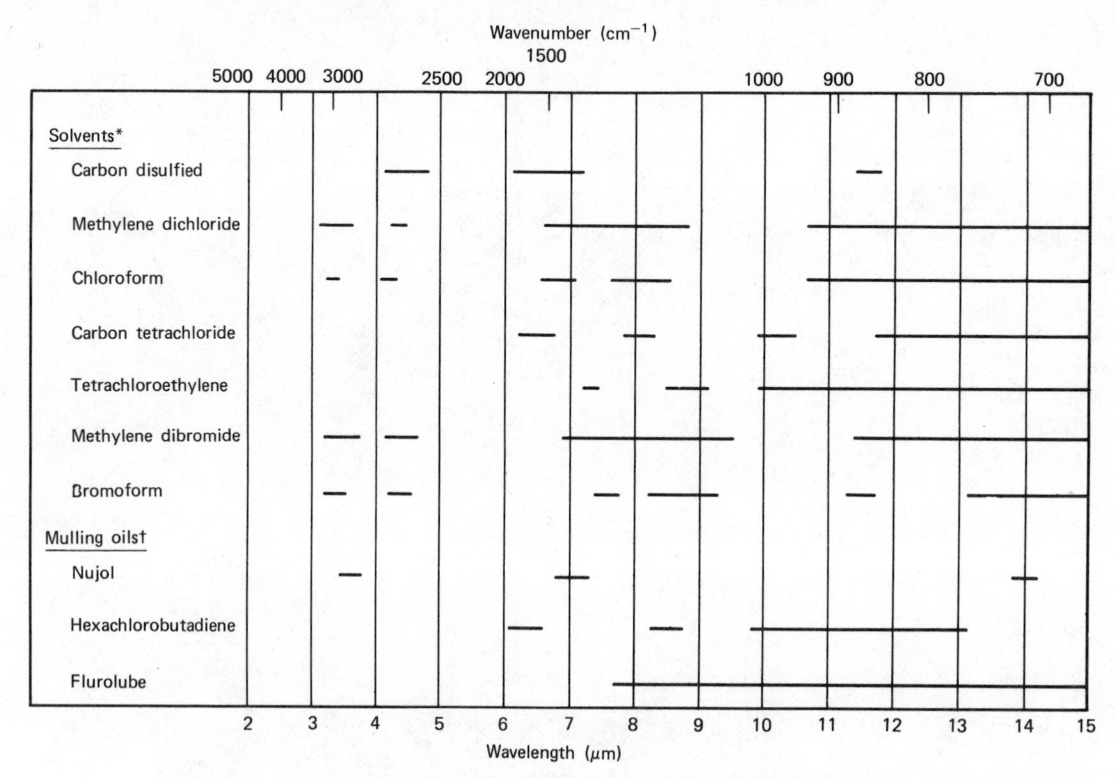

Figure 5.6. Transparent regions of ir solvents and mulling oils. *The open regions are those in which the solvent transmits more than 25% of the incident light at 1 mm thickness. †The open regions for mulling oils indicate transparency of thin films.

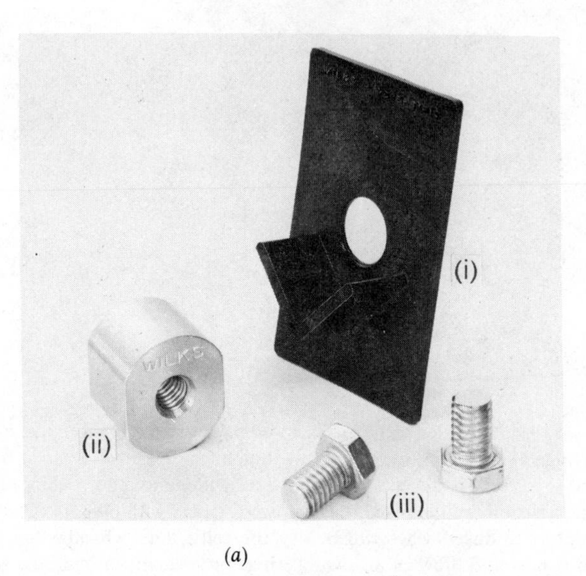

(*a*)

(*b*)

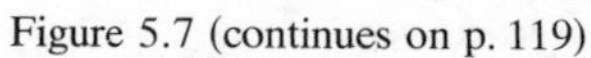

Figure 5.7 (continues on p. 119)

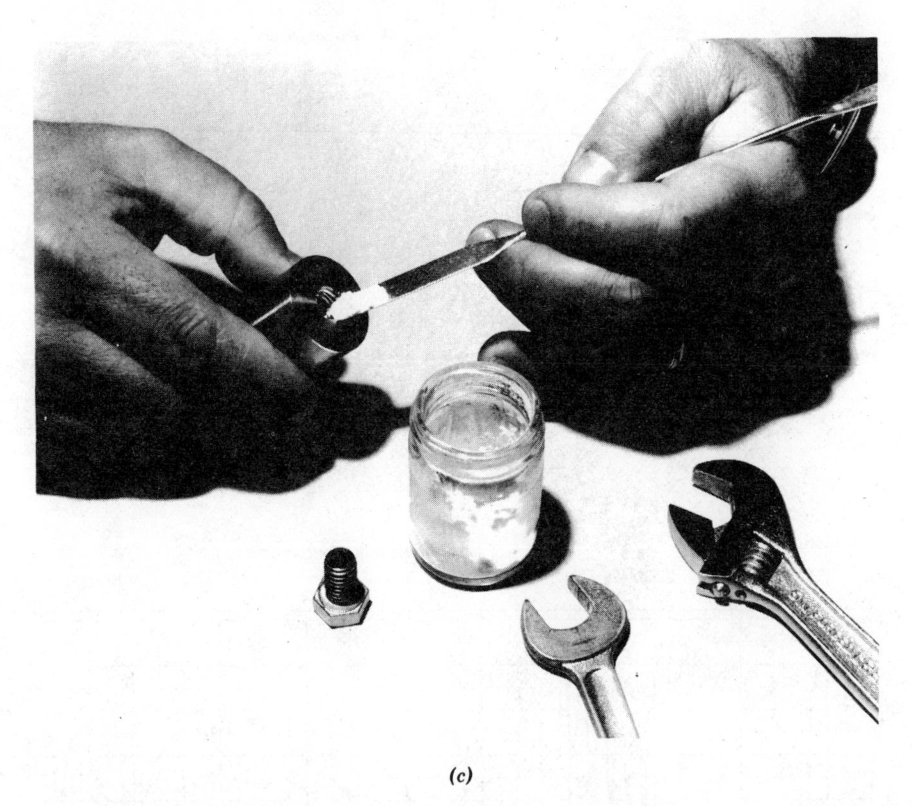

(c)

Figure 5.7 (continued). A Mini-Press (Wilks Scientific, South Norwalk, Conn.) for preparation of potassium bromide pellets Procedure:

1. Prepare a mull of ca. 1.0 mg of sample* in 100 mg of oven-dried, anhydrous KBr by thoroughly grinding in an agate mortar and pestle.
2. Place one of the bolts in the cell (part ii in Fig. 5.7*a*) and insert the mixture into the cavity of the cell (see Fig. 5.7*c*).
3. Assemble the press apparatus by inserting the other bolt and tightening the bolts as shown in Fig. 5.7*b*.
4. The KBr disk so-formed can be scanned in the holder (see part i of Fig. 5.7*a*). Note that only the most careful techniques (e.g., use of evacuated die presses, oven-dried equipment, etc.) will minimize the appearance of water absorption (O—H stretch at 3800–3000 cm^{-1} and O—H bend at 1750–1520 cm^{-1}).

These special techniques are described in R. T. Conley, *Infrared Spectroscopy*, 2nd ed. (Allyn & Bacon, Boston, Copyright © 1972), chap. 4, p. 47.

* These amounts serve only as rough guidelines; actual amounts are determined only after trial and error.

variations require instructor guidance, as does any arbitrary manipulation of the reference beam area. Set the "100% adjust" (attenuator) dial to minimize loss of resolution (% transmittance, *T*, too low) or off-scale baseline (% *T* too high); frequently an initial value (at 4000 cm^{-1}) of % T = ca. 85–90 is appropriate. Choose a slit setting (usually normal), set the pen on the paper, and start the scan (slower chart speeds are less demanding on the pen and recorder).

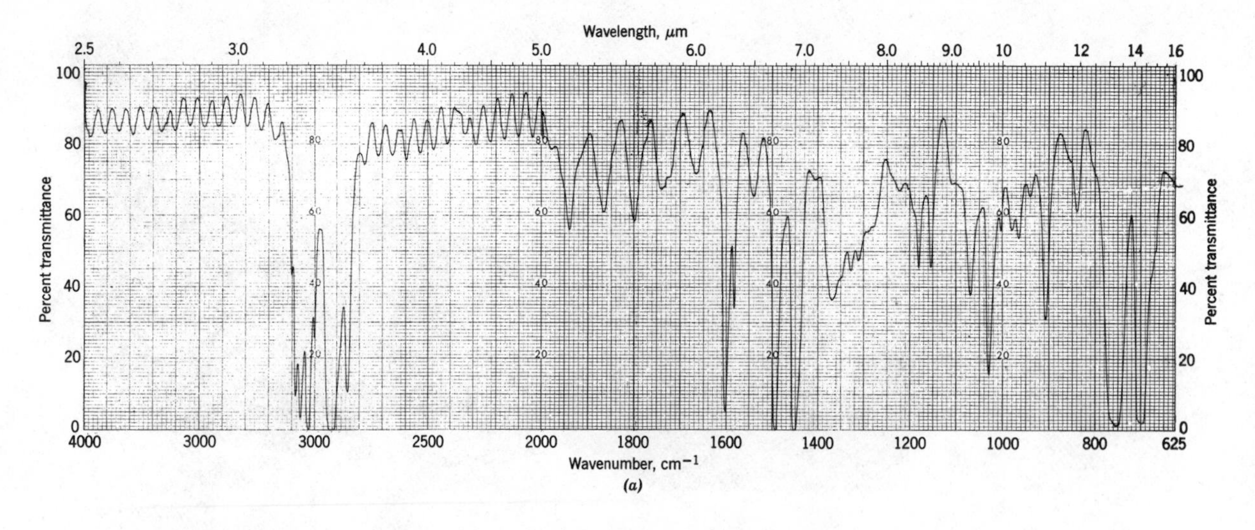

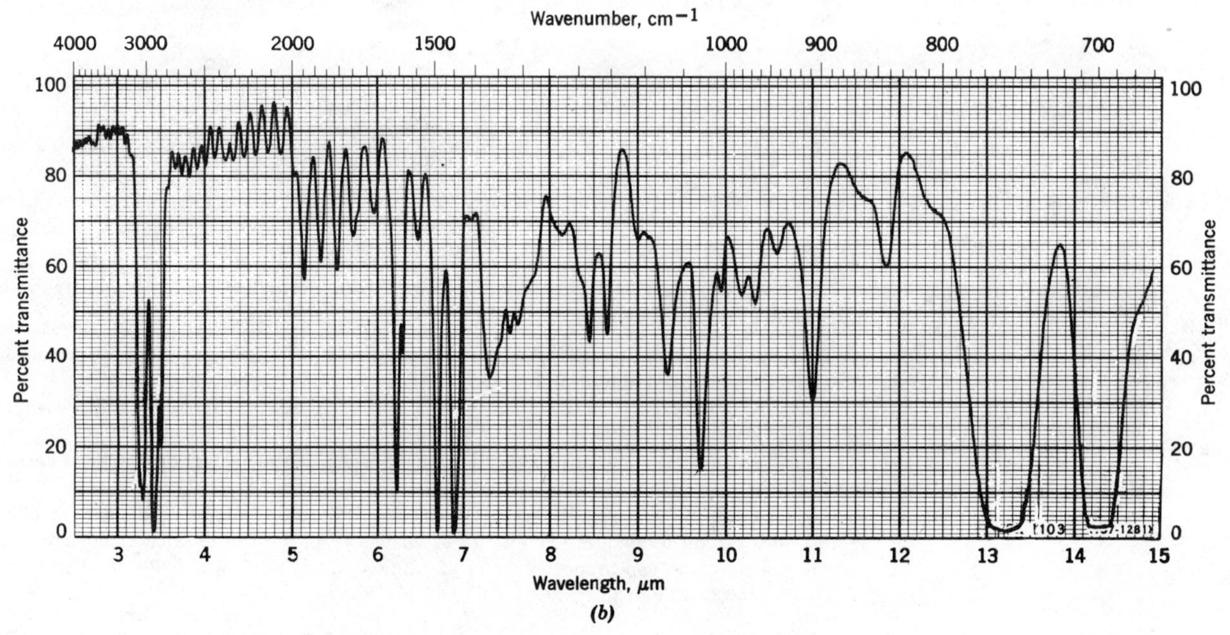

Figure 5.8. Infrared spectra of polystyrene. (*a*) Linear wavenumber (cm^{-1}) scale. (*b*) Linear wavelength (μm) scale.

Discussion Most ir analyses require trial and error, coupled with examination of the prepared sample (e.g., opacity of mull or color intensity of solution) to achieve success. Precise band positions are obtainable only after calibrating with a reference sample (commonly, the polystyrene window); for example, if the 6.24 μm (1603 cm^{-1}) band of polystyrene is out of position by a given increment (e.g., 0.05 μm), the sample bands should be adjusted by that increment. Infrared spectra of polystyrene are given in Fig. 5.8; note the drastic difference in spectra as

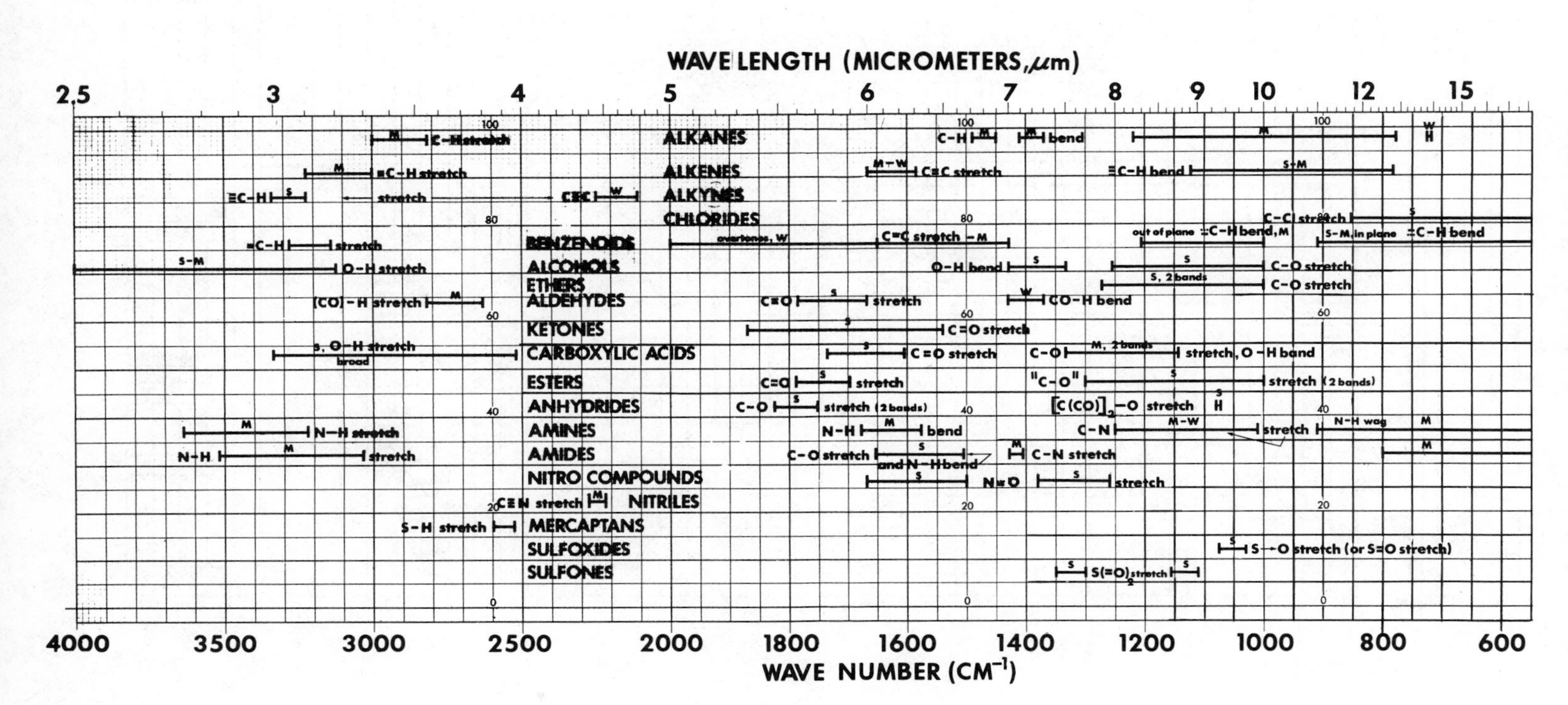

Figure 5.9. Correlation of infrared absorption with organic functional groups. Rows below the first row show only unique bands for the new functional group (e.g., the alkene row shows *only* bands due to the double bond and does not show C—H absorption bands for *saturated* side chains). Additional ir details are organized by functional group in Chapter 6.

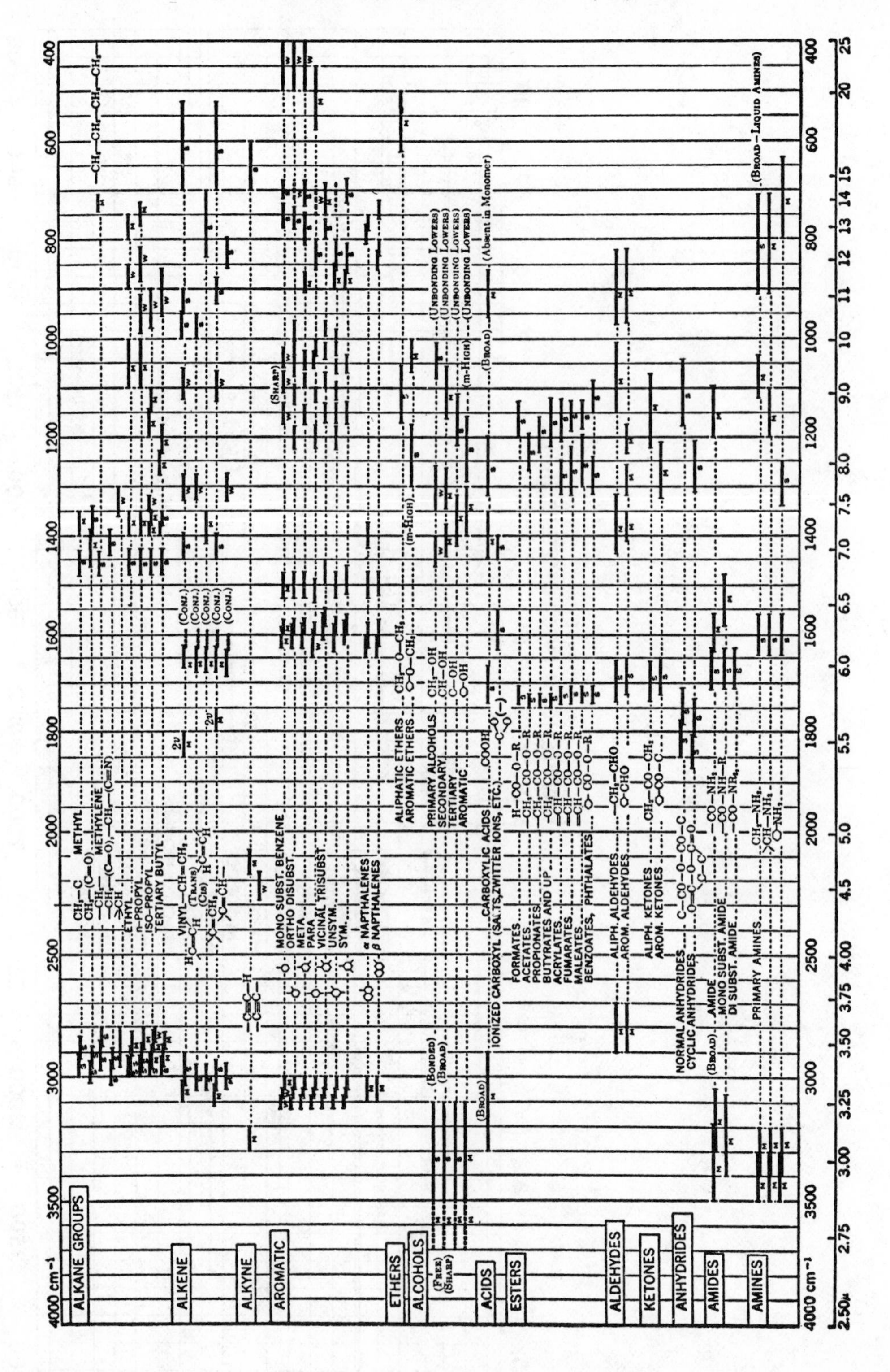
4000 cm⁻¹
3500
3000
2500
2000
1800
1600
1400
1200
1000
800
600
400
ALKANE GROUPS
ALKENE
ALKYNE
AROMATIC
ETHERS
ALCOHOLS
ACIDS
ESTERS
ALDEHYDES
KETONES
ANHYDRIDES
AMIDES
AMINES
METHYL
METHYLENE
ETHYL
n-PROPYL
ISO-PROPYL
TERTIARY BUTYL
VINYL
MONO SUBST. BENZENE
ORTHO DISUBST
META
PARA
VICINAL TRISUBST.
UNSYM.
SYM.
α NAPTHALENES
β NAPTHALENES
ALIPHATIC ETHERS
AROMATIC ETHERS
PRIMARY ALCOHOLS
SECONDARY
TERTIARY
AROMATIC
CARBOXYLIC ACIDS
IONIZED CARBOXYL (SALTS, ZWITTER IONS, ETC.)
FORMATES
ACETATES
PROPIONATES
BUTYRATES AND UP
ACRYLATES
FUMARATES
MALEATES
BENZOATES, PHTHALATES
ALIPH. ALDEHYDES
AROM. ALDEHYDES
ALIPH. KETONES
AROM. KETONES
NORMAL ANHYDRIDES
CYCLIC ANHYDRIDES
AMIDE
MONO SUBST. AMIDE
DI SUBST. AMIDE
PRIMARY AMINES
(Absent in Monomer)
(Broad – Liquid Amines)
2.50μ
2.75
3.00
3.25
3.50
3.75
4.00
4.5
5.0
5.5
6.0
6.5
7.0
7.5
8.0
9.0
10
11
12
13
14
15
20
25

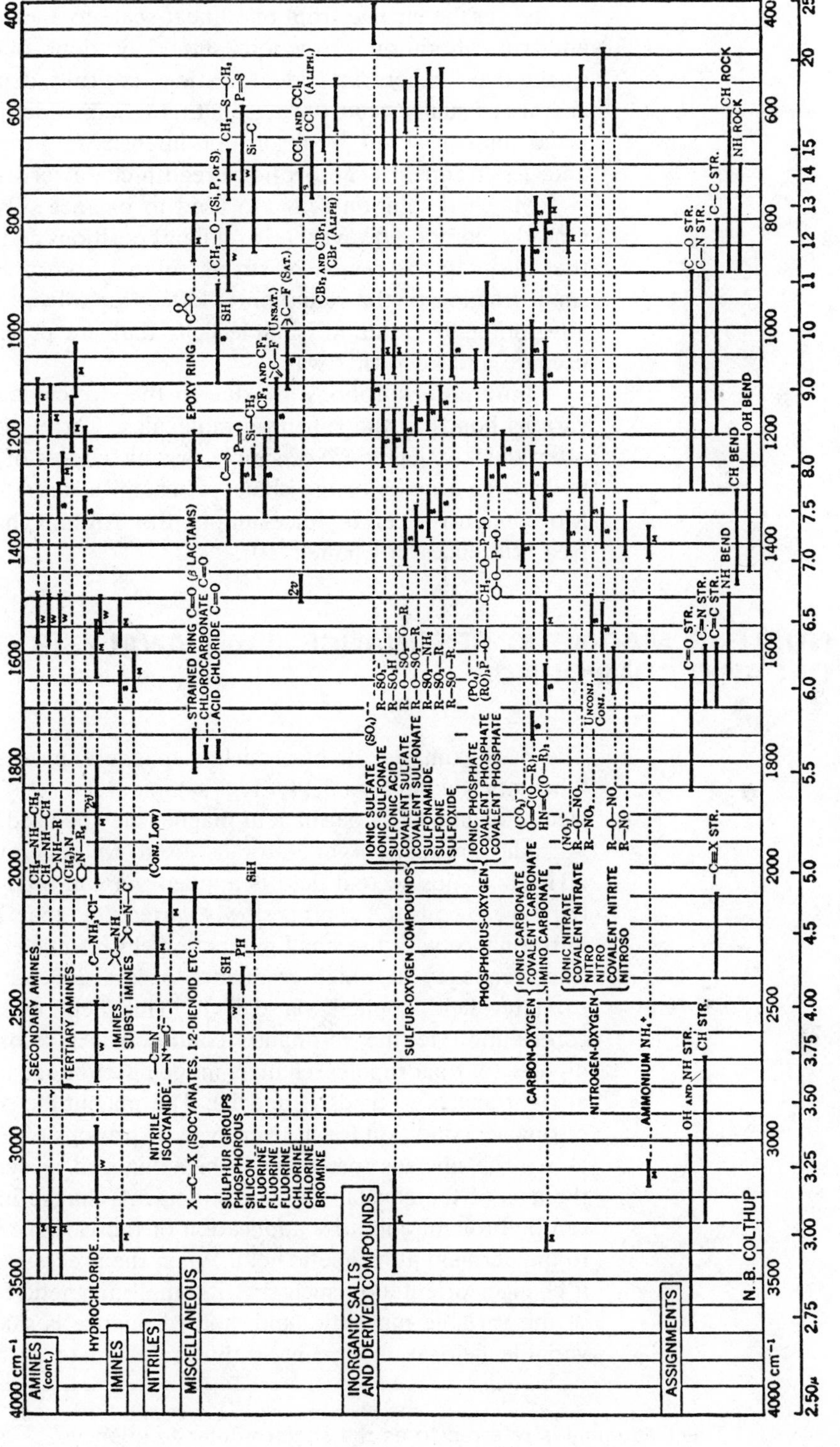

Figure 5.10. Colthup chart correlating infrared absorption with organic functional groups.

a result in the change from one linear scale to another. This is important because additive calibration corrections should be done only with the linear scale, and qualitative "fingerprint" identifications critically depend on appearance. Organic chemists usually report wavelength (λ, μm, micrometers) or more frequently, wave number (ν, cm^{-1}); absorption intensity (s, strong; m, medium; and w, weak) and, less frequently, absorption breadth characteristics are sometimes mentioned.

Matched solution cells are used to balance solvent absorptions (see above). One should not, however, rely on band positions assumedly due to sample (solute) that occur in the region of strong solvent absorption (Fig. 5.6), because only a small fraction of the radiation is left (after solvent absorption) to detect sample absorption. In addition, sample cells that are precisely matched are difficult to prepare and keep that way.

Infrared absorptions are due to the stretching and/or bending vibrations of various bonds in the substrate molecules. Correlations tabulated in Fig. 5.9 are often quickly diagnostic of simple organic classes of molecules; this figure should be used as a guide to the choice of chemical tests to be employed in Chapter 6. More extensive charts, for example, the Colthup chart of Fig. 5.10, are available (see references in Chapter 10).

5.3 NUCLEAR MAGNETIC RESONANCE (nmr) ANALYSIS OF ORGANIC COMPOUNDS

The basic components of a nuclear magnetic resonance spectrometer are shown schematically in Fig. 5.11. The sample tube (A) is placed in the field of an electromagnet (E; a permanent magnet may be used instead of an electromagnet). A radio frequency generator is used to generate a second field, often of 60 MHz;[13] this second field is applied via coil B at right angles to the first field (E). Sweep coils (C) apply a slowly varied field that, at various field values, will be in resonance with the small magnetic fields of nuclei (e.g., protons) in the sample. The frequency at which resonance occurs depends on the magnitude of spectrometer field applied via coil B. This field makes by far the predominant contribution to the resonance condition. At resonance, the nuclear magnetic dipoles (or quadrupoles) change, inducing a voltage in the detector coil D (which is at right angles to fields B and E). The magnitude of the sweep coil fields at the resonance condition for a given type of protons of a particular sample is denoted as ν_s. The nucleus (proton) can be visualized as symmetrically precessing about the axis of the constant magnetic field. A change in the angle of this precession can be brought about by application of the variable magnetic field at right angles to the permanent magnetic field. When the precessing nucleus changes spin states, it changes orientation such that its small magnetic field corresponds to the value of the variable magnetic field, and resonance is said to occur. The value of the variable field at that point is thus referred to as the resonance value for that

[13] This frequency is referred to as the spectrometer frequency.

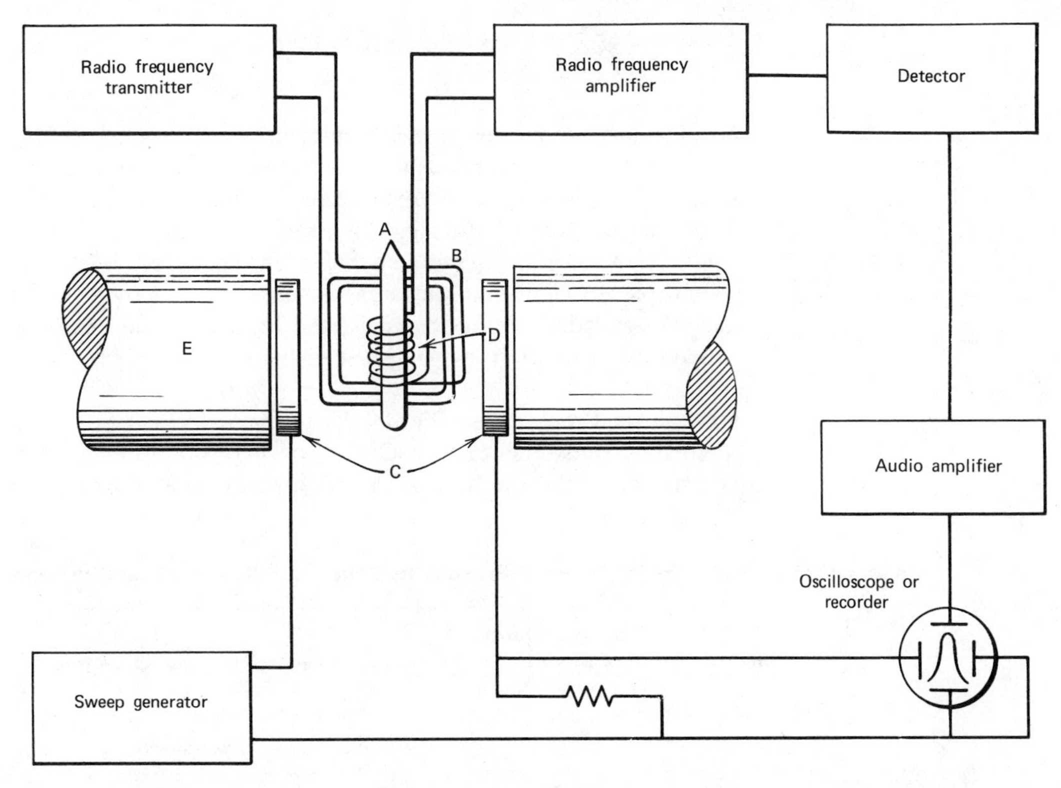

Figure 5.11. Schematic diagram of nmr spectrometer. A, sample tube; B, coils for application of permanent field; C, sweep coils; D, detector coil; E, electromagnet (or permanent magnet).

nucleus. Thus, the receiver coil (D) must be placed at right angles to both the sweep coils (C) and to the permanent field (B). Resonance frequencies for protons of standard organic compounds are found in a range of 0–600 Hz (hertz, identical to cycles per second) lower than the resonance, ν_r, of a reference compound; the reference compound is usually tetramethylsilane [TMS, $(CH_3)_4Si$]. Although it is the field that is normally the parameter value that is varied (swept), it is conventional to report the signals in terms of frequency (hertz). It is very useful to convert the frequency of the midpoint of such signals to a dimensionless chemical shift position called δ (delta).

$$\delta = \frac{|\nu_s - \nu_r|}{\nu_{\text{appl.}}} \times 10^6$$

For example, if the sample frequency is 60 Hz lower than the reference signal, we calculate the chemical shift as

$$\delta = \frac{60\text{ Hz}}{60\text{ MHz}} \times 10^6 = 1.0\text{ parts per million (ppm)}$$

on the common 60-MHz (megahertz = 1 million cycles per second) instrument.

Another less frequently used scale is the τ scale:

$$\tau = 10.00 - \delta$$

The instrument thus provides information. Resonance of the variable magnetic field with the magnetic field of nuclei that are changing spin states in the sample gives rise to the chemical shift (δ), which is plotted on the abscissa of the recorder; the area of the signal is proportional to the number of nuclei in the sample that resonate at this frequency. The first piece of information (δ) and the second piece of information (area) is supplemented by a third piece of information, the splitting of these signals; the latter may be translated into additional structural information, as will be seen below.

Procedure. Prepare a solution of 0.10 g (ca. 0.15 ml) of pure sample in ca. 0.5 ml of solvent (most commonly $CDCl_3$ or carbon tetrachloride; see Table 5.8). Transfer all of the resulting solution with a disposable pipette (dropper) to a nmr

Table 5.8. Shift Positions of Residual Protons in Commercially Available Deuterated nmr Solvents[a]

Solvent	Isotopic purity of atom (%D)	Positions of residual protons (δ values)					
		Group	δ	Group	δ	Group	δ
Acetic acid-d_4	99.5	Methyl	2.05	Hydroxyl	11.53[b]		
Acetone-d_6	99.5	Methyl	2.05				
Acetonitrile-d_3	98	Methyl	1.95				
Benzene-d_6	99.5	Methine	7.20				
Chloroform-d	99.8	Methine	7.25				
Cyclohexane-d_{12}	99	Methylene	1.40				
Deuterium oxide	99.8	Hydroxyl	4.75[b]				
1,2-Dichloroethane-d_4	99	Methylene	3.69				
Diethyl-d_{10} ether	98	Methyl	1.16	Methylene	3.36		
Dimethylformamide-d_7	98	Methyl	2.76	Methyl	2.94	Formyl	8.05
Dimethyl sulfoxide-d_6	99.5	Methyl	2.50				
p-Dioxane-d_8	98	Methylene	3.55				
Ethyl alcohol-d_6 (anh.)	98	Methyl	1.17	Methylene	3.59	Hydroxyl	2.60[b]
Hexafluoroacetone deuterate	99.5	Hydroxyl	9.00[b]				
Methyl alcohol-d_4	99	Methyl	3.35	Hydroxyl	4.84[b]		
Methylcyclohexane-d_{14}	99	Methyl	1.92	Methylene	1.54	Methine	1.65
Methylene chloride-d_2	99	Methylene	5.35				
Pyridine-d_5	99	Alpha	8.70	Beta	7.20	Gamma	7.58
Silanar-C[c] ($CDCl_3$ + 1% TMS)	99.8	Methyl (TMS)	0.00[d]	Methine	7.25		
Tetrahydrofuran-d_8	98	α-Methylene	3.60	β-Methylene	1.75		
Tetramethylene-d_8 sulfone	98	α-Methylene	2.92	β-Methylene	2.16		

[a] Data furnished by Merck Sharp and Dohme of Canada, Ltd.
[b] This value may vary considerably, depending on the solute.
[c] Trademark.
[d] By definition.

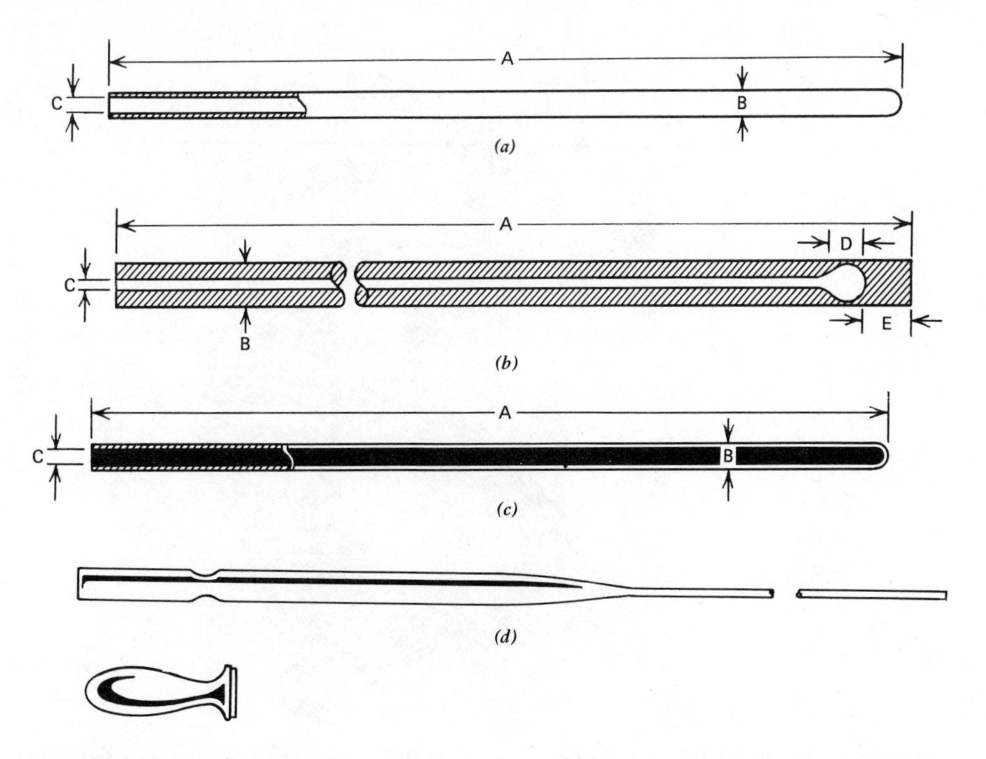

Figure 5.12. Accessories for nmr. (*a*) Standard nmr tube, takes ca. 5–30% solution of sample in 0.4–0.5 ml of solutions. Dimensions: A, 6–8 in.; B, 4 mm; C, ca. 3.2 mm. (*b*) nmr Microtube for nmr (requires 25 μl minimally). Dimensions: A, 7 in.; B, 5 mm; C, 1.5 or 2.0 mm; D, 25 μl; E, 10 mm. Tubes with cavity depths that can be adjusted are superior. (It is also suggested that rectangular, rather than spherical, cavities offer a better chance for regularity of shape and, therefore, better spinning properties.) (*c*) Tube of amber glass (to protect light-sensitive samples); dimensions similar to (*a*). (*d*) Disposable pipette (70–100 mm length) and bulb used for nmr analysis; the constriction near the top can be used for cotton plugs (allowing sample to be filtered as introduced). (Illustrations courtesy of Wilmad Glass Co., Buena, N.J.)

tube (Fig. 5.12). Add the internal standard; this usually consists of a drop or two of tetramethylsilane.[14] The tube should be inserted to the proper length in the spinner *before* placing it in the probe. Figure 5.13 shows a tube, properly equipped with spinner (which is just below the index finger), being inserted into the probe of the nmr spectrometer. The tube should be set spinning rapidly (ca. 10 revolutions per second); the spinning rate should be increased to remove spinning side bands.[15] Set the sweep width dial (usually 100 Hz or 10 ppm, Fig. 5.14) and the sweep time dial (the slowest scan will allow the best pen response).

---

[14] Often commercial solvents for nmr already contain internal standard; check the label of the solvent bottle.

[15] Spinning side bands are identifiable as one or more pairs of small peaks flanking a large peak. Each peak of a pair is equally displaced to the right or left of the large peak. If the increase in spinning rate does not move or obliterate the side bands, the peaks are not spinning side bands.

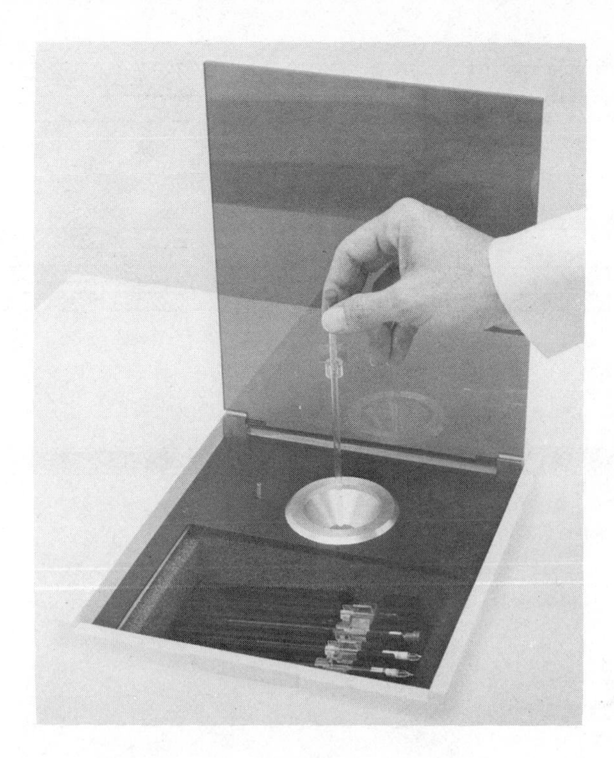

Figure 5.13. Sample tube for nmr, equipped with a spinner, being inserted into the probe of a nmr spectrometer.

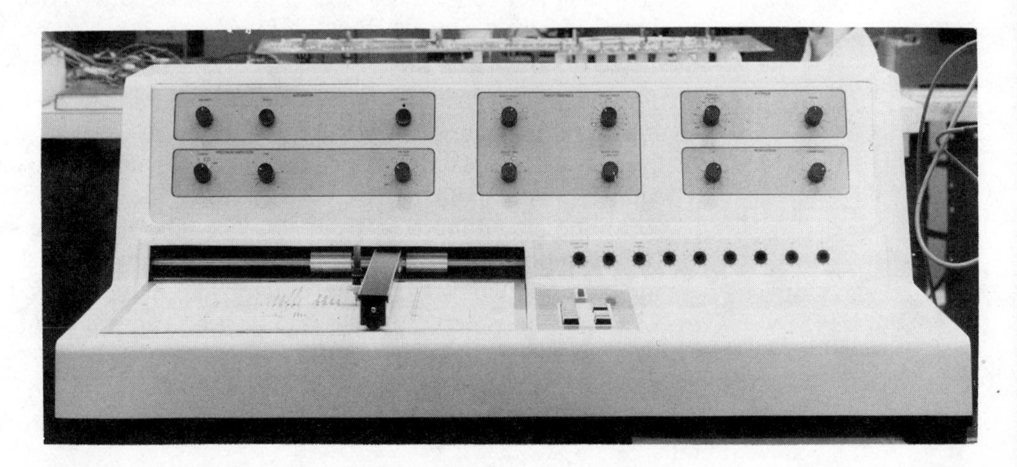

Figure 5.14. Recorder and control panel for a Varian EM-360 nmr spectrometer. (Courtesy of Varian Associates, Palo Alto, Calif.)

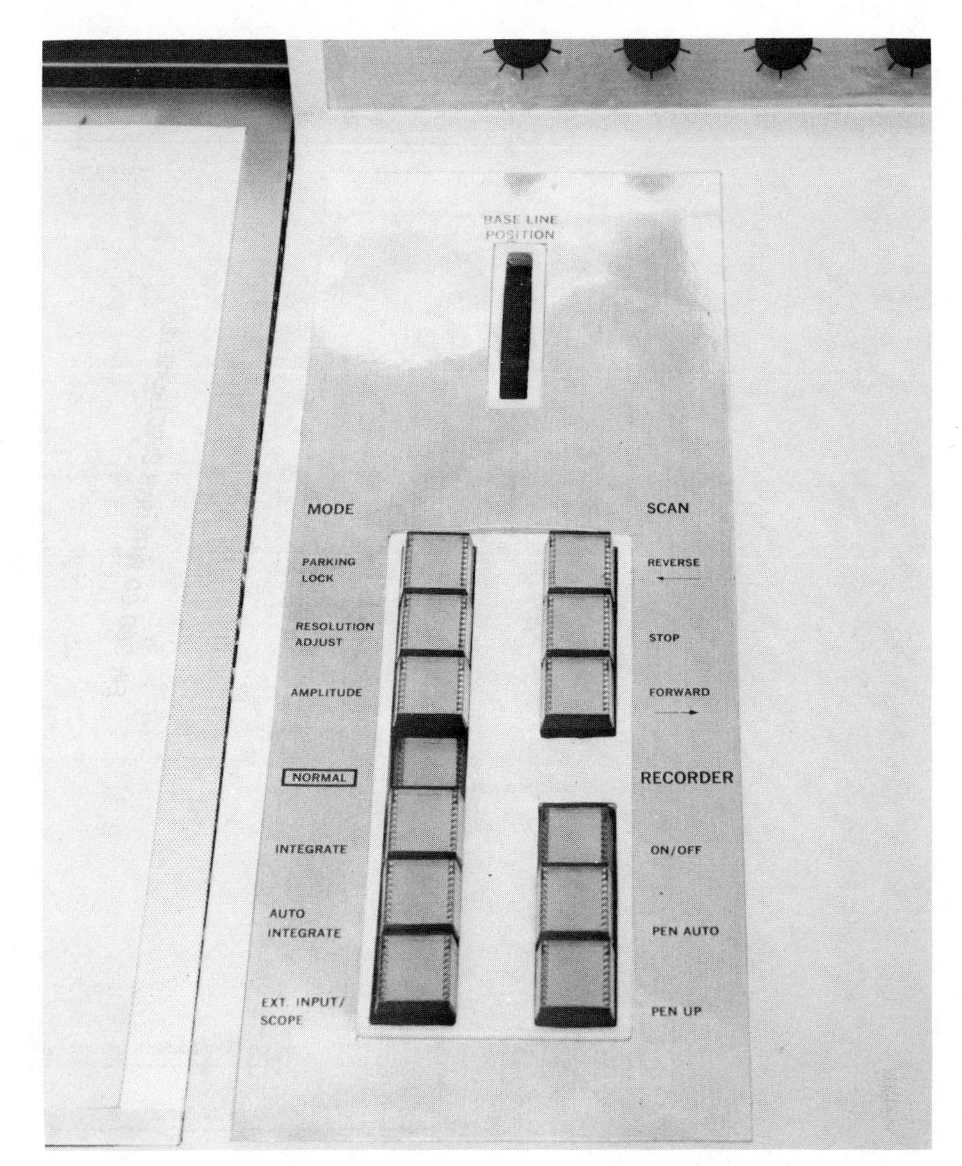

Figure 5.15. Control panel for the recorder of a Varian EM-360 nmr spectrometer. (Courtesy of Varian Associates, Palo Alto, Calif.)

Choose a spectrum amplitude setting that will give rise to peak intensities such that the most intense peak does not go off scale; this may be done by viewing the entire spectrum on an oscilloscope display (if available) or by readjusting this setting after one complete scan has been recorded on paper. Radio frequency and filter settings may be adjusted to improve resolution; these and any more sophisticated adjustments demand instructor guidance. *This instrument is delicate and expensive!* Switch on the recorder, making sure the pen is in contact with the

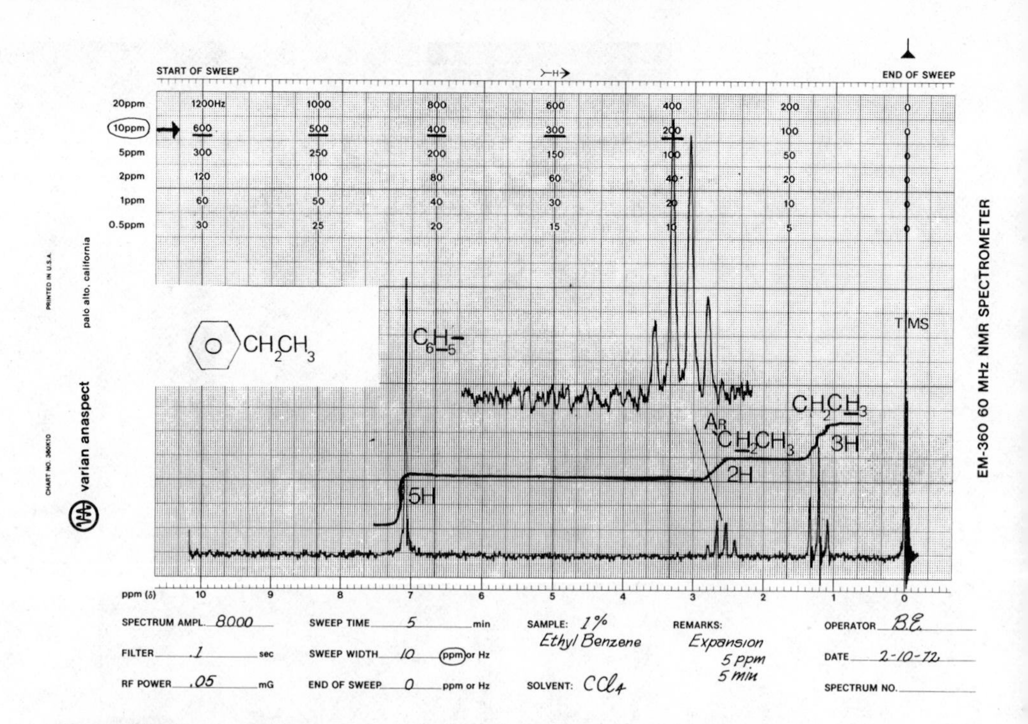

Figure 5.16. Proton nmr spectrum of ethylbenzene (Varian EM-360 nmr spectrometer; courtesy of Varian Associates, Palo Alto, Calif.) Proton assignments are aided by the ranges cited in Fig. 5.17 and by the discussions of nmr in Chapter 6.

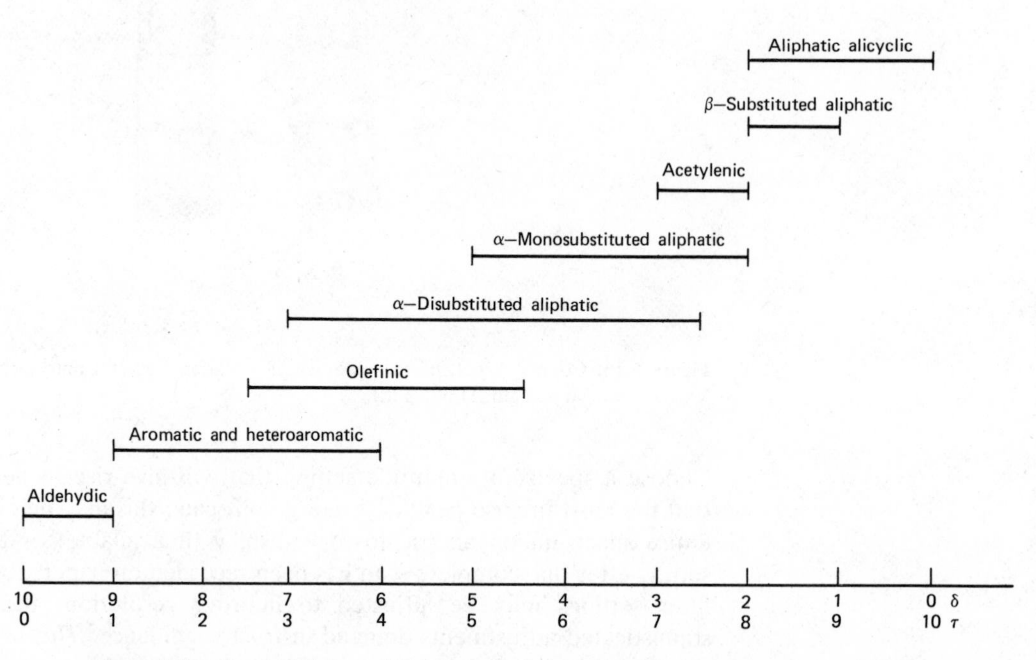

Figure 5.17. Chemical shift positions for protons in various classes of organic compounds.

paper, and start the scan (Fig. 5.15). Electronic integration techniques should be carried out only after consulting with the instructor.

Discussion The nmr spectrum of a typical organic compound is shown in Fig. 5.16. Three types of information can be taken from this spectrum.

1. The midpoints (δ) of the signals fall into regions (see Fig. 5.17) typical of their position in the structure.
2. The area under the signals (measured by triangulation, planimeter, or taking the heights of the separately recorded integration curve) is proportional to the number of protons causing the signal.
3. The multiplicity (number of peaks in a given signal) is often an indication of the number of adjacent protons in a structure; that is, if $n+1$ lines are observed, the protons have n equivalent neighbors. Consult other textbooks for more details on this subject (see Chapter 10).

5.4 SUMMARY

The purpose of Chapter 3 was to describe the determination of simple properties, purity, and molecular weight of an unknown. Chapter 4 described how to determine both the nature of the elements in the sample and the molecular formula of the sample. This chapter has described how to determine solubility and spectral (ir, nmr) properties of a sample, the structure of which is (very likely) unknown as yet. Note that almost all of the preceding tests will be routinely carried out on all samples; we shall, however, be very selective as to the additional tests chosen (i.e., those tests in Chapter 6) to deduce completely and prove the structure of the compound. It is thus crucial to know how to plan the remainder of the investigation (using Chapter 6 and further information); this will be illustrated with an example:

Compound, Unknown No.: *1* Investigator: *John Smith*
Compound Name: Date: *June 1, 1975*
Structure:
1. Physical Examination
   (a) State: *Liquid.* (b) Color: *None.* (c) Odor: *Choking.*
   (d) Ignition test: *Burns with bluish flame, no residue.*
2. Thin Layer or (Gas Chromatography)
   (a) Adsorbent: (b) Eluant:
   (c) Observations: (d) Conditions:
3. Physical Constants
   b.p. 114–117°C (uncorr.), $n_D^{20} = 1.3988$
4. Qualitative Elemental Analysis
   (a) Halides: *None.* (b) Heteroatoms: *None (other than possibly O).*
   (c) Metals: *None.*

5. Molecular Weight Determinations
   (a) Method: *Osmometric.*
   (b) Results: *87, 82 (second run on smaller amount of sample).*
6. Solubility Class: S_1.
7. Reaction to phenolphthalein: *None.*
8. Organic Solvent Solubility

Solvent:	Acetone	CCl_4	$CHCl_3$	Hexane	Ethanol	Ether
Solubility:	*very soluble*	*v.s.*	*v.s.*	*v.s.*	*v.s.*	*v.s.*

9. Infrared Analysis — Instrument: *Perkin Elmer 237*
   Phase: *Thin film (neat).*
   Comments: *Spacer was used for salt plates.* — Linear scale: (cm^{-1}) or μm

Band (cm^{-1})	Intensity	Appearance	Preliminary assignment	Final assignment
3600	*Weak*	*Sharp*	*O—H stretch*	
3300	*Strong*	*Broad*	*O—H stretch*	
3000–2850	*Strong*	—	*C—H stretch*	
1025	*Medium*	*Broad*	*C—O stretch*	

10. nmr Analysis — Nucleus: *proton.* — Field strength: *60 MHz.*
    Solvent: *$CDCl_3$.* Concentration: *ca. 10%.* Instrument: *Varian A60.*

Resonance	Integration	Multiplicity	Preliminary assignment	Final assignment
δ 0.91	*3H*	*Triplet (distorted)*	*$C\underline{H}_3$—CH_2*	
ca. 1.20–1.95	*4H*	*Multiplet*	*$C\underline{H}_2$—$C\underline{H}_2$—CH_2*	
3.47	*2H*	*Triplet*	*—CH_2—$C\underline{H}_2OH$*	
4.58	*1H*	*Singlet*	*—OH*	
0.00	—	*Singlet*	*TMS*	*TMS*

Even the student who is not yet trained in organic chemistry can interpret the preceding report as follows: Preliminary properties imply a very simple compound. Solubility results indicate that the compound is oleophilic as well as hydrophilic. Spectral analysis clearly indicates an aliphatic alcohol. The ir spectrum identifies the —OH functional group and the nmr spectrum identifies the straightness of the chain. Solubility data also indicate that there are four carbons in the chain; one thus proposes *n*-butyl alcohol as the compound: $CH_3CH_2CH_2CH_2OH$.

Although the mystery is virtually eliminated, it is *imperative* that a more thorough characterization be carried out. A modest indication of this need for complete characterization is the poor value reported above for the molecular weight; this is likely due to molecular aggregations caused by hydrogen bonding in the osmometric determination. The trained chemist would quickly turn to Chapter 6 to review chemical tests (e.g., Na metal treatment) for alcohols, as well as additional spectral tests (e.g., concentration dependence of the OH absorption in both the ir and nmr spectra). Chemical treatment to produce a derivative (Chapter 6) of an alcohol is usually straightforward; those derivatives for which the physical properties (e.g., melting point) are not reported in the literature should also be completely characterized as above.

CHAPTER SIX

the detection and confirmation of functional groups: complete structure determination

As a result of earlier purity (Chapter 3), physical (Chapter 3), spectral (Chapter 4), and chemical (Chapters 3, 4, and 5) determinations, the student probably has a reasonable idea regarding the identity of the unknown compound; it is virtually certain that additional, thorough characterization is in order. A very large proportion of organic compounds lend themselves to final characterization by the chemical and spectral tests described in this chapter. As a result of the characterizations described in earlier chapters, most unknowns should be understood well enough to allow the student to choose the section (or sections) below that describe testing procedures for that particular class of compound. These tests are indexed below by functional group as well as by reagent.

Each section corresponding to a functional group (or groups) is composed of the following parts:

1. Chemical tests
2. Illustrations and descriptions of ir and nmr spectral characteristics followed by a brief statement of mass spectral characteristics
3. Derivatization procedures
4. Reference to unusual physical or chemical characterization procedures

Despite the tremendous importance and ease of spectral analyses, chemical tests are indispensable to complete characterization; thus these "wet" tests are nearly always used for compound identification. For example, the double bond of 1,2-dimethylcyclohexene is difficult to detect spectrally; chemical addition reactions (e.g., of bromine or of potassium permanganate) to such double bonds are, however, usually rapid. Thus, any student enchanted with spectral analysis to the exclusion of chemical analysis may be severely hampered in his or her attempt to identify a compound.

In earlier sections we have merely described how to determine ir and nmr spectra; detailed discussion of the theory necessary to interpret such spectra is left to other textbooks. We shall, however, describe many of the spectral absorptions (for example, nmr bands) of standard molecular classes without discussing the theory (for example, of spin–spin coupling of nmr spectrometry).

Derivatization procedures have somewhat diminished in stature with the advent of organic spectrometry. Although it certainly is no longer as important to make, for example, *both* a semicarbazone and a dinitrophenylhydrazone of an aldehyde, derivatization procedures still provide both physical data (for example, melting point) and an insight to the chemistry of the new substrate. Chemists should also remember that certain "derivatizations" are really "conversions" of one common organic compound into another. Conversion (e.g., oxidizing a secondary alcohol to a ketone) may yield a compound that should also be thoroughly characterized. Indeed, many derivatizations are really syntheses or "preps."

The tables of data of derivatives in various references should be consulted to see if m.p. data are available; if the m.p. is not known, the chemist may have a sample that is essentially another unknown to characterize. Tabulated physical constants such as refractive indices are useful for confirming chemical constitution. One should supplement these comparisons with references to ir, nmr, and less frequently mass spectra that should be compared to the spectra of the unknown for "fingerprint" identification.

As far as spectral analysis is concerned, we shall be concerned largely with an interpretive rather than a "fingerprint" approach. We shall be concerned with identifying spectral bands that can be used to detect organic functional groups and structural units in the molecule. We shall then combine these results with results from chemical reactions to characterize completely the sample at hand.

Finally, it is intuitively reasonable that special characterizations should be employed for certain classes of organic compounds. One expects, for example, that the specific rotation, $[\alpha]$, of amino acids (in view of the fact that they often have chiral centers) would be useful.

It cannot be overemphasized that the thorough organic chemist must consult additional references (see Chapter 10) to seek out additional procedures and data to characterize many compounds.

Since no textbook intended for student use can hope to include all chemical tests, tables of reference to many chemical tests described in *Organic Structure Determination*, by D. J. Pasto and C. R. Johnson (Prentice-Hall, Englewood Cliffs, N.J., 1969), and in *Semimicro Qualitative Organic Analyses*, 3rd ed., by N. D. Cheronis, J. B. Entrikin, and E. M. Hodnett (Wiley-Interscience, New York,

1965), have been provided throughout this chapter. We again urge students to consult other texts, such as those mentioned above and in Chapter 10, and research journals for additional procedures.

INDEX TO CHARACTERIZATION PROCEDURES FOR FUNCTIONAL GROUP CLASSES

[a] Includes anilines.

INDEX TO REAGENTS USED FOR CHEMICAL CLASSIFICATION TESTS AND DERIVATIZATION PROCEDURES

Experiments

Reagent, Experiment No.	Class Tested for	Page
Acid chloride, Experiment 18	Alcohols, amines, enols	218
Azoxybenzene, Experiment 22	Aromatics	247
Benedict's solution, Experiment 13	Aldehydes, ketones	172
Benedict's solution, Experiment 29b	Mercaptans, thiophenols	334
Bromine in carbon tetrachloride, Experiment 14	Alkenes, alkynes	190
Bromine water, Experiment 30	Phenols, enols	350
Cerium(IV) oxidation, Experiment 2	Alcohols	146
Chloroform-aluminum trichloride, Experiment 23	Aromatics	248
Chromium trioxide, Experiment 3	Alcohols (primary, secondary)	149
Dinitrophenylhydrazine, Experiment 6	Aldehydes, ketones	162
Ferric chloride-pyridine, Experiment 29c	Phenols, enols	348
Ferrous hydroxide, Experiment 27	Nitro compounds	319
Fuchsin reagent, Experiment 12	Aldehydes	171
Fuming sulfuric acid, Experiment 21	Aromatics	245
Hydriodic acid, (Zeisel's test) Experiment 26	Ethers	305
Hydroxylamine hydrochloride, Experiment 7	Aldehydes, ketones	163
Iodoform test, Experiment 10	Methyl ketones	167

Experiments

Reagent, Experiment No.	Class Tested for	Page
Lucas test (Zn/HCl), Experiment 4	Alcohols	150
Ninhydrin test, Experiment 20	Amino acids	242
Nitrous acid, Experiment 19	Amines, anilines	220
Nitrous acid, Experiment 29a	Mercaptans, thiophenols	332
Periodic acid, Experiment 5	Vicinal diols, polyols	159
Phenylhydrazine (and *p*-Nitrophenylhydrazine) Experiments 8, 24	Aldehydes and ketones (Experiment 8), carbohydrates (Experiment 24)	165 261
Potassium permanganate, Experiment 15	Alkenes, alkynes, vicinal diols, mercaptans, sulfides	192
Silver nitrate (ethanolic), Experiment 16	Halides (labile)	202
Sodium (metal), Experiment 1	Alcohols (and other active hydrogen compounds)	145
Sodium bisulfite, Experiment 9	Aldehydes, ketones	165
Sodium hydroxide, Experiment 25	Ammonium and amine salts	288
Sodium hydroxide, Experiment 29	Polynitro compounds	320
Sodium iodide (in acetone), Experiment 17	Halides	204
Tollens Test, Experiment 11	Aldehydes	170
Zinc/ammonium chloride, Experiment 28	Nitro compounds	320

Procedures

Derivative for	Derivative Type, Procedure No.	Page
Alcohols	Arylurethanes, Procedure 1	156
Alcohols	3,5-Dinitrobenzoates, Procedure 2	156
Alcohols	Benzoates, *p*-Nitrobenzoates, Procedure 3	157
Alcohols	3-Nitrophthalates, Procedure 4	158
Alcohols, polyhydric	Acetates, Procedure 5	160
Aldehydes	Oxidation, Procedure 6	178
	Permanganate, Procedure 6a	178
	Hydrogen peroxide, Procedure 6b	178
Aldehydes	Dimedon, Procedure 10	180
Aldehydes, ketones	Semicarbazones, Procedure 7	179
Aldehydes, ketones	*p*-Nitrophenylhydrazones, Procedure 8	179
Aldehydes, ketones	2,4-Dinitrophenylhydrazones, Procedure 9	179
Aldehydes, ketones	Oximes, Procedure 11	181
Aldehydes, ketones	Reduction of diaryl ketones with sodium borohydride, Procedure 11a	182
Amides	9-Acylamidoxanthenes, Procedure 38	284
Amides	Hydrolysis, Procedure 37	282
Amines	Hinsberg test, Procedure 16	230
Amines (primary, secondary)	Amides (including Schotten-Baumann method), Procedure 17	232
Amines (primary, secondary)	Sulfonamides, Procedure 18	233
Amines (primary, secondary)	Phenylthioureas, Procedure 19	234

Procedures

Derivative for	Derivative Type, Procedure No.	Page
Amines (tertiary)	Quaternary ammonium salts, Procedure 20	235
Amines ($ArNR_2$)	Nitroso compounds, Procedure 21	236
Amines (tertiary)	Picrates, Procedure 22	236
Amines (tertiary)	Chloroplatinates, Procedure 23	237
Amines (primary, secondary)	Amides (via acid chloride), Procedure 35	271
Amino acids	*p*-Toluenesulfonates, Procedure 24	244
Amino acids	*N*-2,4-Dinitrophenyl derivatives, Procedure 25	244
Anilines	See amine derivatives	230–272
Anilines	Anilides, Procedure 36	272
Anilines (nitrated)	Hydrolysis, Procedure 37	282
Aromatics	Nitration, Procedure 26	249
Aromatics (ArR)	Side-chain oxidation, Procedure 27	253
Aromatics	Aroylbenzoic acids, Procedure 28	255
Aromatics	Friedel-Crafts acylation, Procedure 29	256
Aromatics	Chlorosulfonation, Procedure 30	259
Aromatics	Sulfonamides, Procedure 31 (via Procedure 30)	260
Aromatics	Picrates, Procedure 22	236
Carboxylic acids	Amides, Procedure 35	271
Carboxylic acids	Anilides, *p*-Toluidides, and *p*-Bromoanilides, Procedure 36	272
Carboxylic acids	Neutralization equivalent, Procedure 32	268
Carboxylic acids	Phenacyl and substituted phenacyl esters, Procedure 33	270

Procedures

Derivative for	Derivative Type, Procedure No.	Page
Carboxylic acids	Phenylhydrazides and phenylhydrazonium salts, Procedure 34	271
Carboxylic acids (and their salts)	Arylthiuronium salts, Procedure 39	287
Esters	Saponification and acid hydrolyses, Procedure 40	293
Esters	Saponification equivalent, Procedure 41	296
Esters	*N*-Benzylamides, Procedure 42	300
Esters	Acid hydrazides, Procedure 43	300
Esters	3,5-Dinitrobenzoates, Procedure 44	300
Ethers	3,5-Dinitrobenzoates, Procedure 45	306
Ethers (aromatic)	Bromination, Procedure 46	308
Ethers (aromatic)	Chlorosulfonation, Procedure 47	308
Ethers (aromatic)	Sulfonamides, Procedure 48 (via Procedure 47)	308
Ethers (phenolic)	Picrates, Procedure 49	309
Halides	Grignard reagent, organomercurials, Procedure 12	214
Halides	Arylurethane, Procedure 13 (via Procedure 12)	214
Halides	Conversion to nitriles, Procedure 14a	215
Halides	α-Naphthyloxy ethers, Procedure 14	215
Halides	*S*-Alkylthiuronium picrates, Procedure 15	217
Ketones	See aldehydes, ketones	178–182
Nitriles	Hydrolysis, Procedure 37	282

Procedures

Derivative for	Derivative Type, Procedure No.	Page
Nitriles	Acid hydrolysis, Procedure 51	328
Nitriles	Controlled hydration to an amide, Procedure 51a	329
Nitriles	Reduction with sodium in alcohol, Procedure 52	329
Nitriles	Oxidation to the aldehyde, Procedure 52a	330
Nitriles	α-(Imidioylthio) acetic acid hydrochlorides Procedure 53	331
Nitro compounds	Tin/hydrochloric acid reduction to amines, Procedure 50	322
Nitro compounds	Base-induced condensation with benzaldehyde, Procedure 50b	324
Nitro compounds (poly)	Selective reduction to nitroamines, Procedure 50a	322
Olefins (alkenes)	Oxidative hydroboration Procedure 11b	198
Phenols	Aryloxyacetic acids, Procedure 59	354
Phenols	Bromination, Procedure 60	354
Phenols	Phenylurethans, Procedure 61	355
Phenolic ethers	Picrates, Procedure 49	309
Sulfonamides	Hydrolysis, Procedure 57	345
Sulfonamides	*N*-Xanthylsulfonamides, Procedure 58	346
Sulfonic acids	Neutralization equivalent, Procedure 32	268
Sulfonic acids (and salts)	Benzylthiuronium sulfonates, Procedure 53a	341
Sulfonic acids (and salts)	Sulfonyl chlorides and sulfonamides Procedure 54	342

Procedures

Derivative for	Derivative Type, Procedure No.	Page
Sulfonic acids (and salts)	*p*-Toluidine sulfonates, Procedure 55	342
Sulfonic acids (amino substituted)	Chloroaryl sulfonamides and sulfonanilides, Procedure 56	343
Sulfonyl halides	Sulfonamides, Procedure 48	308

6.1 ALCOHOLS

Chemical and spectral analysis of alcohols are essentially controlled by the very polar hydroxyl group in these molecules; hydrogen bonding and chemical exchange of the hydrogen are other fundamental properties of the OH (hydroxyl) group that should be kept in mind.

EXPERIMENT 1. SODIUM DETECTION OF ACTIVE HYDROGEN

Sodium metal, especially samples that have been freshly cut, frequently react readily with hydroxyl groups to liberate hydrogen:

$$2ROH + 2Na \rightarrow 2RO^{-}Na^{+} + H_2(g)$$

The order of reactivity of alcohols with sodium is known to decrease with increasing size of the alkyl portion of the molecule. This test is subject to many limitations, and the results should be interpreted with caution.

Procedure. To 1 ml of *n*-butyl alcohol, add thin slices of metallic sodium until no more will dissolve. Cool the solution, and observe. Add an equal volume of ether. What is the precipitate? Apply the test to acetone, *n*-butyl ether, and toluene.

This test may be applied to solid compounds or very viscous liquids by dissolving them in an inert solvent such as anhydrous ligroin or benzene.

Discussion. This reagent is of most value in testing *neutral* compounds for the presence of groups that contain easily replaceable hydrogen atoms. Functional groups containing a hydrogen atom attached to oxygen, nitrogen, or sulfur may react with sodium to liberate hydrogen.

$$2ROH + 2Na \rightarrow 2RONa + H_2$$

$$2R_2NH + 2Na \rightarrow 2R_2NNa + H_2$$

$$2RSH + 2Na \rightarrow 2RSNa + H_2$$

This test is most useful in connection with alcohols of intermediate molecular weight, that is, those containing from three to about eight carbon atoms. Lower alcohols are difficult to obtain in anhydrous condition. The presence of traces of moisture causes the test to be positive. Alcohols of high molecular weight react slowly with sodium, and the evolution of gas is often so slow as to make the test of little value. Metallic sodium when cut in moist air adsorbs water on its surface so that, when placed in a perfectly dry solvent such as benzene, it gives off hydrogen produced by the interaction of the metal with the adsorbed moisture.

Hydrogen atoms attached to carbon are not displaced by metals unless there are adjacent functional groups that activate the hydrogen atoms. Compounds with active methine groups, such as acetylene or monosubstituted acetylenes, react with sodium.

$$HC{\equiv}CH + 2Na \rightarrow NaC{\equiv}CNa + H_2$$

$$2RC{\equiv}CH + 2Na \rightarrow 2RC{\equiv}CNa + H_2$$

Frequently the hydrogen produced is not observed, as this hydrogen undergoes reaction with unsaturated functional groups as rapidly as it is produced.

A methylene group adjacent to an activating group, especially one between two such groups, possesses hydrogen atoms that may be displaced by sodium. This hydrogen may also be difficult to observe due to its subsequent reaction with unsaturation in the original organic compound.

$$2CH_3COCH_2COOC_2H_5 + 2Na \rightarrow 2[CH_3COCHCOOC_2H_5]^-Na^+ + H_2$$

Reactive methyl groups are present in certain compounds, especially methyl ketones such as acetone and acetophenone. These react with sodium to give the sodium derivative of the ketone and a mixture of products formed by reduction and condensation. For example, acetone yields sodium acetonide, sodium isopropoxide, sodium pinacolate, mesityl oxide, and phorone.

Metallic sodium is thus a useful reagent for detecting the types of reactive hydrogen compounds that are not sufficiently active to produce hydrogen ions in an ionizing solvent. It is obviously unnecessary to try the action of sodium on compounds known to be acids.

Questions

1. Predict the action of sodium on phenol, benzoic acid, oximes, nitromethane, and benzenesulfonamide. Why is this test never used with these compounds? What effect would the presence of moisture have on this test?
2. What is the principal defect of metallic sodium as a classification reagent?

The "active" hydrogen of the hydroxyl group can often be detected by another procedure, involving the use of acid halides, described in Experiment 18.

EXPERIMENT 2. AMMONIUM HEXANITRATOCERIUM(IV) OXIDATION

$$-\overset{|}{C}H-OH + 2Ce(IV) \rightarrow -\overset{|}{C}=O + 2Ce(III) + 2H^+$$

Reagent. Add 13 ml of concentrated nitric acid to 400 ml of distilled water and then dissolve 109.6 g of yellow $(NH_4)_2Ce(NO_3)_6$ in this dilute nitric acid solution. After solution is complete, dilute to 500 ml. The test is carried out at room temperature (20 to 25°C). Hot solutions (50 to 100°C) of Ce(IV) oxidize many types of organic compounds. The above reagent is usable for about 1 month.

Procedure

(a) ***For Water-Soluble Compounds.*** To 1 ml of the ammonium hexanitratocerate reagent add 4 to 5 drops of a liquid unknown or 0.1 to 0.2 g of a solid. Mix thoroughly and note if the yellow color of the reagent changes to red. If a red color develops, watch the solution carefully and note the time for the mixture to become colorless. If no change is noted in 15 min, the test tube may be stoppered and allowed to stand several hours or overnight. Also note if bubbles of carbon dioxide are liberated.

Controls. Try this test on (1) methanol; (2) isopropyl alcohol; (3) glycerol; (4) glucose; (5) lactic acid (85%).

(b) For Water-Insoluble Compounds. Add 4 ml of pure dioxane[1] to 2 ml of the reagent. If a red color develops or if the solution becomes colorless, the dioxane must be purified. If the mixture remains yellow or is only a light orange-yellow, it may be used to test water-insoluble compounds. Divide the 6 ml of the solution in half, reserving 3 ml for observation as a control. To the other 3 ml of the dioxane containing reagent, add 4 to 5 drops of a liquid unknown or 0.1 to 0.2 g of a solid. Mix thoroughly and make the same observations as in part (a) above.

Controls. Try this test on (1) 1-heptanol; (2) benzyl alcohol; (3) *dl*-mandelic acid.

Discussion. This reagent, which consists of a yellow solution of ammonium hexanitratocerate in dilute nitric acid, forms red complexes with compounds that contain alcoholic hydroxyl groups. Positive tests are given by primary, secondary, and tertiary alcohols containing up to 10 carbon atoms. Also, all types of glycols, polyols, carbohydrates, hydroxy acids, hydroxy aldehydes, and hydroxy ketones give red solutions.

The red complex has been shown to be the intermediate for the oxidation of alcohols by Ce(IV) solutions. Hence, *a second phase* of this test involves disappearance of the red color due to oxidation of the coordinated alcohol and reduction of the colored Ce(IV) complex to the colorless Ce(III) anion.

The overall sequence of reactions for a primary alcohol is

(a) $\underset{\text{yellow}}{(NH_4)_2Ce(NO_3)_6} + RCH_2OH \rightarrow \underset{\text{red complex}}{[\text{alcohol} + \text{reagent}]}$

(b) $[\text{complex}] \rightarrow RCH_2O\cdot + \underset{\text{colorless}}{(NH_4)_2Ce(NO_3)_5} + HNO_3$

(c) $RCH_2O\cdot + \underset{\text{yellow}}{(NH_4)_2Ce(NO_3)_6} \rightarrow RCHO + \underset{\text{colorless}}{(NH_4)_2Ce(NO_3)_5} + HNO_3$

The rate of the oxidation steps (b and c) depends on the structure of the hydroxy compound. Table 6.1 gives times for the reduction of the red Ce(IV) complex to the colorless Ce(III) complex at 20°C using the above test conditions.

The products from other hydroxylic compounds are

$$R_2CHOH \rightarrow R_2CO + H_2O$$

$$RCHOHCHOHR \rightarrow 2RCHO + H_2O$$

$$\underset{\text{OH}}{R_2C}\text{—}\underset{\text{OH}}{CR_2} \rightarrow 2R_2CO + H_2O$$

$$RCHOHCO_2H \rightarrow RCHO + CO_2 + H_2O$$

---

[1] The dioxane should be checked with the ceric nitrate solution to be sure that it does *not* give a positive test. The dioxane sold as "histological grade" is usually pure enough so that it may be used. Pure dioxane does not give a red complex. Commercial dioxane sometimes contains glycols or antioxidants as preservatives and must be purified.

Table 6.1. Approximate Times for Reduction of Red Ce(IV) complexes at 20°C to Colorless Ce(III) Nitrato Anion with Oxidation of Alcohols to Aldehydes or Ketones

	Time[a]		Time[a]
Primary alcohols		**Diols, Triols, . . . , Polyols**	
Allyl alcohol	6 min	Pinacol	5 sec
Methyl cellosolve	1.2 hr	Mannitol	38 sec
1-Propanol	3.6 hr	2,3-Butanediol	1 min
Benzyl alcohol	4.0 hr	Glycerol	10 min
1-Butanol	4.1 hr	Propylene glycol	15 min
2-Methyl-1-propanol	4.1 hr	Diethylene glycol	3 hr
1-Heptanol	5.0 hr	Ethylene glycol	5 hr
Ethanol	5.5 hr	⋮	
Methanol	7.0 hr	1,4-Butanediol	1 hr
2-Methyl-1-butanol	7.0 hr	1,4-Butynediol	36 min
1-Decanol	12.0 hr	1,4-Butenediol (mostly cis)	3 min
Secondary alcohols		**Carbohydrates**	
Cyclohexanol	3.7 hr	Glucose	1 min
2-Propanol	6.0 hr	Fructose	30 sec
2-Butanol	9.0 hr	Galactose	1 min
2-Pentanol	17.0 hr	Lactose	5 min
2-Octanol	16.0 hr	Maltose	8 min
Diphenylcarbinol	12.0 hr	Sucrose	12 min
		Cellulose—insoluble—no red	
		Starch—insoluble—no red	
Tertiary alcohols		**Hydroxy acids**	
tert-Butyl alcohol	>48 hr	Lactic acid	15 sec + CO_2
tert-Pentyl alcohol	>48 hr	Malic acid	30 sec + CO_2
3-Methyl-3-hydroxy-		Tartaric acid	1 min + CO_2
-1-butyne	36 hr	Mandelic acid	1 min + CO_2
		Citric acid	1 min + CO_2
		Hydroxy ketones	
		3-Hydroxy-2-butanone	15 sec
		3-Methyl-3-hydroxy-	
		2-butanone	10 sec

[a] Variations in time of oxidation can be expected due to variable size of reagent drops and to the age of the reagent.

Among simple monohydroxy compounds, methanol gives the deepest red color. As the molecular weight of the alcohols increases, the color becomes less intense and somewhat brownish-red.

A red color is produced by aqueous 40% formaldehyde (formalin). This is due to methanol present in the solution. Acetaldehyde frequently gives a red color due to the presence of acetaldol: $CH_3CHOHCH_2CHO$. Alternatively, these aldehydes may hydrate in aqueous solution to form gem-diols [e.g., $RCH(OH)_2$], which may be the species that are oxidized.

Negative tests (no red complex with retention of the yellow color of the reagent) are given by all pure aldehydes, ketones, saturated and unsaturated acids, ethers, esters, dibasic and tribasic acids. The dibasic acids, oxalic and malonic, do *not* give a red color (negative test), but do reduce the yellow Ce(IV) to colorless Ce(III) solutions.

Phenols do not give the characteristic red color. When tested in dioxane solution, phenols are oxidized to brown or black products.

Basic aliphatic amines cause precipitation of white ceric hydroxide. If the amines are dissolved in dilute nitric acid (forming the amine nitrate) and this solution is treated with the ceric reagent, no red colors develop provided that there are no alcoholic hydroxyl groups present in addition to the amino groups. If alcoholic groups are present, then dilute nitric acid solutions of such compounds do give red colors. For example, dilute nitric acid solutions of

$$HOCH_2CH_2NH_2 \qquad (HOCH_2CH_2)_2NH \qquad (HOCH_2CH_2)_3N$$

all give positive tests.

Alcohols containing halogens give positive tests: For example, $ClCH_2CH_2OH$, $BrCH_2CH_2OH$, $ClCH_2CH_2CH_2OH$, and $CH_3CHOHCH_2Cl$ form red complexes.

Very insoluble alcohols of high molecular weight such as 1-hexadecanol, triphenylcarbinol, or benzpinacol fail to react even in the dioxane solutions and do not give a red color.

Long-chain alcohols, C_{12} through C_{18}, *will* give a positive test when added to an acetonitrile solution of ammonium hexanitratocerate at the boiling point, 82°C. Procedure C, below, is especially useful for such long-chain alcohols.

Test Procedure C. The reagent consists of 21.5 g of ammonium hexanitratocerate (G. F. Smith Co.) dissolved in 100 ml of acetonitrile. Add about 0.1 g of the unknown compound to 2 ml of this reagent in a test tube. Stir the mixture with a glass rod and heat just to boiling. In 1 to 6 min the color will change from yellow to red. Even cholesterol, $C_{27}H_{45}OH$, gives an orange to red color. The red color disappears as oxidation of the alcohol group takes place.

All the lower alcohols and glycols also give a red color with this acetonitrile solution at room temperature, but the rates of oxidation as indicated by change in color are quite different from those cited above.

EXPERIMENT 3. CHROMIC ANHYDRIDE (CHROMIUM TRIOXIDE, JONES OXIDATION)

This test detects the presence of a hydroxyl substituent that is on a carbon bearing at least one hydrogen, and therefore oxidizable.

Procedure

$$3RCH_2OH + 4CrO_3 + 6H_2SO_4 \rightarrow 3RCO_2H + 9H_2O + 2Cr_2(SO_4)_3$$

$$3R_2CHOH + 2CrO_3 + 3H_2SO_4 \rightarrow 3R_2CO + 6H_2O + Cr(SO_4)_3$$

$$3RCHO + 2CrO_3 + 3H_2SO_4 \rightarrow 3RCO_2H + 3H_2O + Cr_2(SO_4)_3$$

To 1 ml of acetone in a small test tube, add 1 drop of a liquid or about 10 mg of a solid compound. Then add 1 drop of the chromic acid/sulfuric acid reagent and note the result *within 2 sec.* Run a control test on the acetone and compare the result. A positive test for primary or secondary alcohols consists in the production of an opaque suspension with a green to blue color. Tertiary alcohols give no visible reaction in 2 sec, the solution remaining orange in color. *Disregard* any changes after 2 sec.

Reagent. A suspension of 25 g of chromic anhydride in 25 ml of concentrated sulfuric acid is poured slowly with stirring into 75 ml of water. The deep orange-red solution is cooled to room temperature before use.

A good grade of commercial acetone may be used. Some samples of acetone may become cloudy in appearance in 20 sec, but this does not interfere, providing the test solution remains yellow. If the acetone gives a positive test it should be purified by adding a small amount of potassium permanganate and distilling.

Discussion. This test is a rapid method for distinguishing primary and secondary alcohols from tertiary alcohols. Positive tests are given by primary and secondary alcohols without restriction as to molecular weight. Even cholesterol ($C_{27}H_{46}O$) gives a positive test. Aldehydes give a positive test but would be detected by other classification experiments. Olefins, acetylenes, amines, ethers, and ketones give negative tests in 2 sec provided that they are not contaminated with small amounts of alcohols. Enols may give a positive test, and phenols produce a dark-colored solution entirely unlike the characteristic green-blue color of a positive test. (Preparative scale: see Pasto and Johnson, p. 363.)

EXPERIMENT 4. HYDROCHLORIC ACID/ZINC CHLORIDE (LUCAS TEST)

This test often provides classification information for alcohols, as well as a probe for the existence of the hydroxyl group. Substrates that easily give rise to cationic character at the carbon bearing the hydroxyl group undergo this test readily; primary alcohols (e.g., CH_3CH_2OH) do not give a positive result.

Procedure

$$R_2CHOH + HCl \xrightarrow{ZnCl_2} R_2CHCl + H_2O$$

$$R_3COH + HCl \xrightarrow{ZnCl_2} R_3CCl + H_2O$$

(a) To 1 ml of the alcohol in a test tube add 10 ml of the hydrochloric acid/zinc chloride reagent at 26–27°C. Stopper the tube and shake; then allow the mixture to stand. Note the time required for the formation of the alkyl chloride, which appears as an insoluble layer or emulsion. Carry out the test on each of the following alcohols, and note by means of a watch the time required for the reaction to take place: (1) 1-butanol; (2) 2-pentanol; (3) 1-propanol; (4) *tert*-butyl alcohol; (5) pentyl alcohol; (6) allyl alcohol; (7) benzyl alcohol.

Reagent. The hydrochloric acid/zinc chloride reagent is made by dissolving 136 g (1 mole) of anhydrous zinc chloride in 105 g (1 mole) of concentrated hydrochloric acid, with cooling.

(*b*) To 1 ml of the alcohol in a test tube add 6 ml of concentrated hydrochloric acid. Shake the mixture, and allow it to stand. Observe carefully during the first 2 min. Test the following alcohols, and record your results: (1) *n*-propyl alcohol; (2) 2-pentanol; (3) *tert*-butyl alcohol; (4) benzyl alcohol.

Discussion. Since the Lucas test depends on the appearance of the alkyl chloride as a second liquid phase, it is normally applicable only to alcohols that are soluble in the reagent. This limits the test in general to monofunctional alcohols lower than hexyl and certain polyfunctional molecules.

The reaction of alcohols with halogen acids is a displacement reaction in which the reactive species is the conjugate acid of the alcohol $R—OH_2^+$, and, as might be expected, is analogous to the replacement reactions of organic halides and related compounds with silver nitrate and iodide ion (Experiments 16 and 17). The effects of structure on reactivity in these reactions are closely related. Thus primary alcohols do not react perceptibly with hydrochloric acid even in the presence of zinc chloride at ordinary temperatures; chloride ion is too poor a nucleophilic agent to effect a concerted displacement reaction, on the one hand, and the primary carbonium ion is too unstable to serve as an intermediate in the carbonium ion mechanism, on the other. Hydrogen bromide and hydrogen iodide, which have anions with nucleophilic reactivity increasing in that order, are increasingly reactive toward primary alcohols. These are nucleophilicity orders to be expected in hydroxylic solvents.

Tertiary alcohols react with concentrated hydrochloric acid so rapidly that the alkyl halide is visible within a few minutes at room temperature, at first as a milky suspension and then as an oily layer. The acidity of the medium is increased by the addition of the anhydrous zinc chloride (a strong Lewis acid), and the reaction rate is increased still further. This reaction is not a nucleophilic displacement comparable to that undergone by primary alcohols but rather proceeds by way of a carbonium ion intermediate. The high reactivity of tertiary alcohols is a consequence of the relatively great stability of the intermediate carbonium ion. Allyl alcohol, although a primary alcohol, yields a carbonium ion that is relatively stable because its charge is distributed equally on the two terminal carbon atoms.

$$CH_2{=}CH{-}CH_2OH \xrightarrow{H^+} [CH_2{=}CHCH_2^+ \leftrightarrow {}^+CH_2CH{=}CH_2]$$

As might be expected, it reacts rapidly with Lucas reagent with the evolution of heat. Allyl chloride may be caused to separate by dilution of the mixture with ice water.

Secondary alcohols are intermediate in reactivity between primary and tertiary alcohols. Although they are not appreciably affected by concentrated hydrochloric acid alone, they react with it fairly rapidly in the presence of anhydrous zinc chloride; a cloudy appearance of the mixture is observed within 5 min, and in 10 min a distinct layer is usually visible.

For more extended discussion of the effect of structure on reactivity in replacement reactions of this type see the discussion of the silver nitrate test (Experiment 16).

Questions

1. Write the structural formulas and names of the isomeric five-carbon saturated alcohols that were not used in this experiment. How would they react with this reagent?
2. How would you account for the difference in the behavior of allyl alcohol and *n*-propyl alcohol? Benzyl alcohol and *n*-pentyl alcohol?

Infrared analysis of alcohols is useful not only for recognizing functional groups, but also for additional structural information. Although a number of assignments are made for the infrared spectrum of Fig. 6.1*a*, by far the most diagnostic bands are the O—H and C—O stretching bands. When infrared analysis of the O—H band is carried out on the compound diluted with solvent (Fig. 6.1*b*), the broad band due to the hydrogen-bonded hydroxyl group (Fig. 6.1*a*) is replaced (or accompanied) by the sharper O—H band at higher wave number (shorter wavelength); this new band is due to the O—H stretch of a hydroxyl group free of hydrogen-bonded association.

$$\underset{\underset{\displaystyle H}{|}}{R\overset{\delta-}{O}}\text{---}\overset{\delta+}{H}\text{—}\overset{\delta-}{O}\text{—}R$$

Analysis by nmr of the protons in alcohol structures is also a valuable source of information about structure and functional groups. Figure 6.2*a* illustrates the nmr spectrum of a simple alcohol and points out typical interproton coupling; the magnitude of vicinal coupling $\left(\begin{matrix} C\text{—}C \\ | \quad | \\ H \quad H \end{matrix}\right)$ of protons on adjacent sp^3-hydridized carbons is ca. 7 Hz. Because of the rapid exchange of the hydroxylic protons with each other, the nmr signal of the OH group is often observed as a singlet. When care is taken to exclude traces of acid from the solvent, or when the nmr analysis is carried out in DMSO-d_6 (or DMSO), the O—H peak often displays multiplicity (Fig. 6.2*b*) corresponding to the class of alcohol:

Class	Primary	Secondary	Tertiary
Structural unit	RCH_2OH	R_2CHOH	R_3COH
Multiplicity of OH resonance	triplet (*t*)[a]	doublet (*d*)	singlet (*s*)

[a] This signal will be a quartet (*q*) if R = H.

The molecular ion of alcohols, measured mass spectrometrically, is often

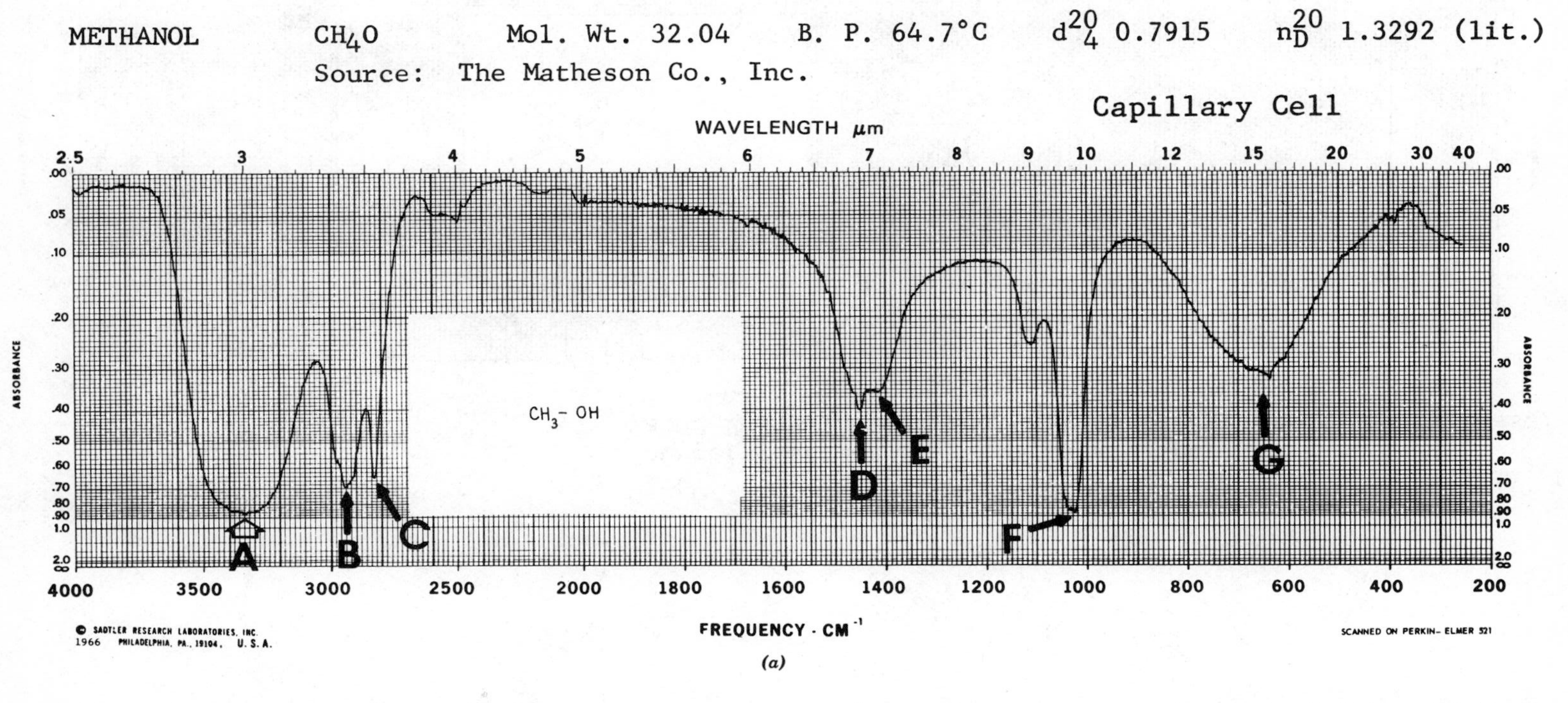

Figure 6.1. (*a*) Infrared spectrum of methanol: thin film. O—H stretch: **A**, 3340 cm^{-1} (2.99 μm), $\nu_{O—H}$, associated. C—H stretch: **B**, 2940 cm^{-1} (3.40 μm), $\nu_{asym\,CH_3}$; **C**, 2830 cm^{-1} (3.53 μm), $\nu_{sym\,CH_3}$. Bending vibrations: **D**, 1450 cm^{-1} (6.90 μm), $\delta_{asym\,CH_3}$ of CH_3O; **E**, 1420 cm^{-1} (7.04 μm), in-plane O—H bend. C—O stretch: **F**, 1060 cm^{-1} (9.43 μm), $\nu_{C—O}$. Bending vibrations: **G**, 650 cm^{-1} (15.39 μm), out-of-plane O—H bend.

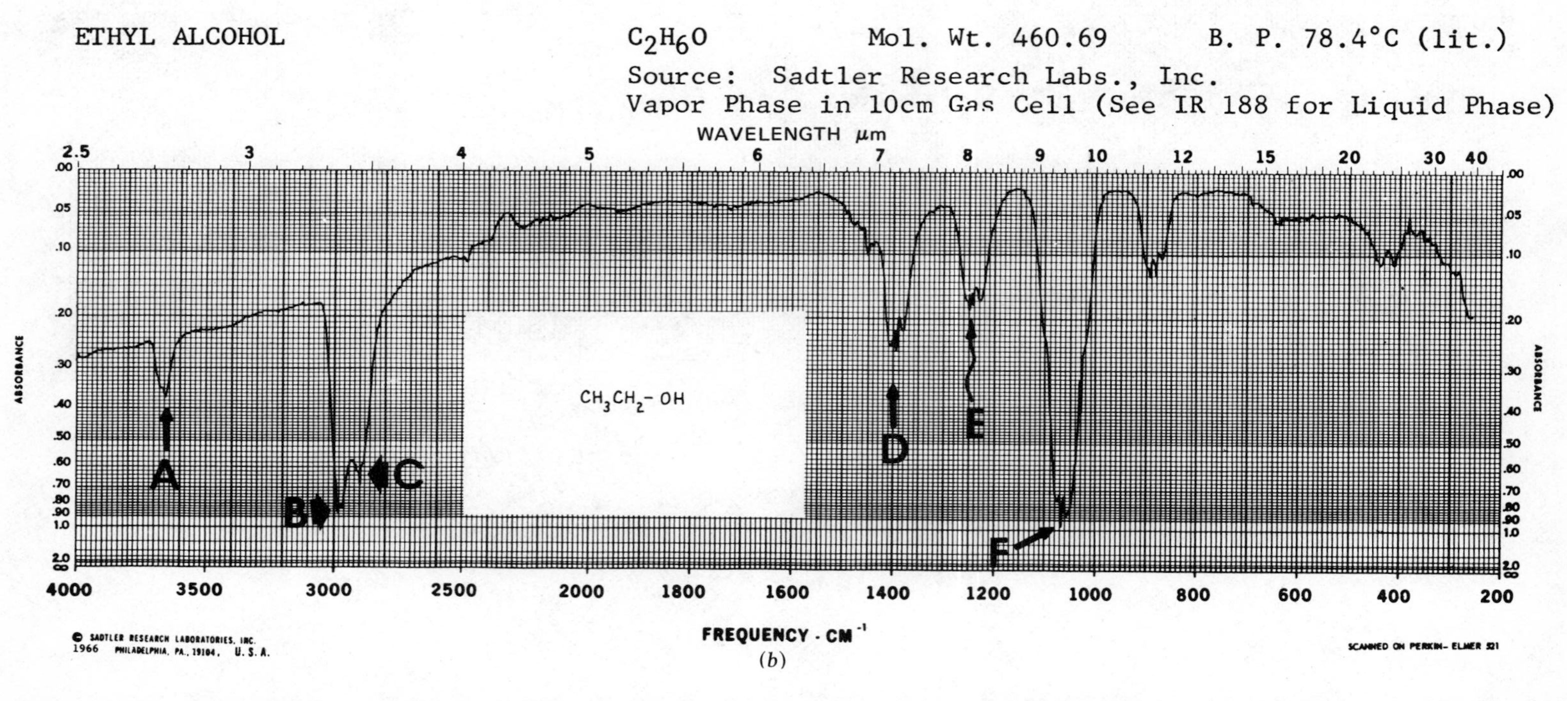

(*b*)

Figure 6.1. (*Continued*) (*b*) Infrared spectrum of ethanol, gas phase. O—H stretching band has a comparatively weak appearance here; alcohols when analyzed *in solution* (see part *a*) normally give intense O—H stretch bands. O—H stretch: **A**, 3660 cm^{-1} (2.73 μm), $\nu_{O—H}$, free. C—H stretch: **B**, 2970 cm^{-1} (3.37 μm), $\nu_{asym\,CH_3}$; **C**, 2900 cm^{-1} (3.45 μm), $\nu_{sym}\,CH_3$. C—H bend: **D**, 1400 cm^{-1} (7.15 μm), $\delta_{sym}\,CH_2$; **E**, 1250 cm^{-1} (8.00 μm), CH_2 (wag). C—O stretch: **F**, 1060 cm^{-1} (9.43 μm), $\nu_{C—O}$.

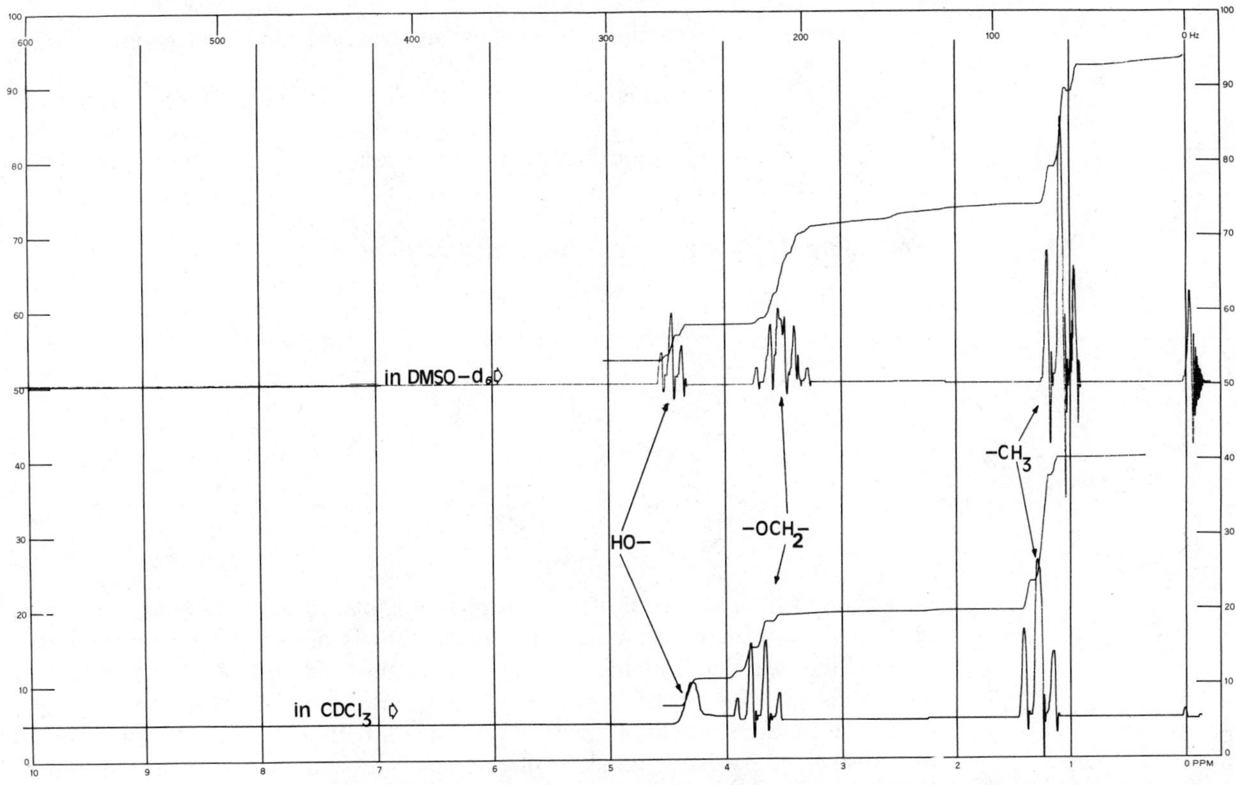

Figure 6.2. Nuclear magnetic resonance (proton) spectrum of ethanol: 60 MHz, 600-Hz sweep width. (*a*) (lower scan), $CDCl_3$ solvent. (*b*) (upper scan), DMSO-d_6 solvent.

difficult to detect; the following generalizations, based on structure, can be made:

Alcohol class	Intensity of Molecular Ion Peak
Primary	Low
Secondary	Low
Tertiary	Usually nonexistent

The molecular weight of alcohols can be determined by mass spectrometry by first converting the alcohol to a silyl ether (see Pasto and Johnson, p. 368). A number of the following fragment ion peaks are often useful for deducing alcohol ($R_1R_2R_3COH$) structures:

Fragment structure	*m/e*	Comment
$[CH_2OH]^+$	31	Usually strongest for primary alcohols
$[R_1R_2COH]^+$	$[29+R_1+R_2]$	$M-R_3$, loss of largest R favored
$[M-H_2O]$	$[M-18]$	Favored by longer chains; 1,4-elimination

Common derivatization of alcohols involves either esterification, —OH→ —O(C=O)R, (Procedures 1–3), or oxidation, $-\overset{|}{C}HOH \rightarrow\ >C{=}O$ (references described below).

Oxidation, under other than the most vigorous conditions, will not occur with tertiary alcohols, for they have no α-hydrogen.

Procedure 1. Phenyl- and α-Naphthylurethans

$$C_6H_5{-}N{=}C{=}O + ROH \longrightarrow C_6H_5{-}NH{-}\overset{O}{\overset{\|}{C}}{-}OR$$

$$C_{10}H_7{-}N{=}C{=}O + ROH \longrightarrow C_{10}H_7{-}NH{-}\overset{O}{\overset{\|}{C}}OR$$

One gram of the anhydrous alcohol or phenol is placed in a test tube, and 0.5 ml of phenyl isocyanate[2] or α-naphthyl isocyanate is added. If the reactant is a phenol, the reaction should be catalyzed by the addition of 2 to 3 drops of anhydrous pyridine or triethylamine. If a spontaneous reaction does not take place, the solution should be warmed on a steam bath for 5 min. It is then cooled in a beaker of ice, and the sides of the tube are scratched with a glass rod to induce crystallization. The urethan is purified by dissolving it in 5 ml of petroleum ether or carbon tetrachloride, filtering the hot solution, and cooling the filtrate in an ice bath. The crystals are collected on a filter and dried on a clay plate; the melting point is then determined.

Procedure 2. 3,5-Dinitrobenzoates

$$ROH + Cl{-}\overset{O}{\overset{\|}{C}}{-}C_6H_3(NO_2)_2 \longrightarrow R{-}O{-}\overset{O}{\overset{\|}{C}}{-}C_6H_3(NO_2)_2 + HCl$$

About 0.5 g of 3,5-dinitrobenzoyl chloride is mixed with 2 ml of the alcohol in a test tube and the mixture boiled gently for 5 min. Then 10 ml of distilled water is added and the solution cooled in an ice bath until the product solidifies. The precipitate is collected on a filter, washed with 10 ml of 2% sodium carbonate solution, and recrystallized from 5 to 10 ml of a mixture of ethyl alcohol and water of such composition that the ester will dissolve in the hot solution but will

---

[2] *Caution!* The isocyanates are lachrymatory.

separate when the solution is cooled. After the crystals have been removed by filtration and dried on a porous plate, the melting point is determined.

If 3,5-dinitrobenzoyl chloride is not available, it may be made by mixing 0.5 g of 3,5-dinitrobenzoic acid with 1 g of phosphorus pentachloride in a test tube. The mixture is warmed gently to start the reaction. After the initial rapid reaction has subsided, the mixture is heated for about 4 min at such a rate as to cause vigorous bubbling. While still liquid the mixture is poured on a watch glass, and the mass is allowed to solidify. The material is transferred to a clean clay plate and rubbed with a spatula in order to remove phosphorus oxychloride. The residual acid chloride is used immediately for the preparation of the derivative as described above.

$$3,5\text{-}(O_2N)_2C_6H_3\text{—}CO_2H \xrightarrow{PCl_5} 3,5\text{-}(O_2N)_2C_6H_3\text{—}\overset{O}{\overset{\|}{C}}\text{—}Cl + POCl_3 + HCl$$

3,5-Dinitrobenzoates may also be prepared by the pyridine method described in Procedure 3a.

Procedure 3. Benzoates and p-Nitrobenzoates

$$ROH + Ar\overset{O}{\overset{\|}{C}}Cl \xrightarrow{(C_5H_5N)} RO\overset{O}{\overset{\|}{C}}Ar + HCl^{+}\left(C_5H_5NH^{+}Cl^{-}\right)$$

$$Ar = C_6H_5\text{—} \text{ or } p\text{-}NO_2C_6H_4\text{—}$$

(*a*) One milliliter of the alcohol is dissolved in 3 ml of anhydrous pyridine, and 0.5 g of benzoyl or *p*-nitrobenzoyl chloride is added. After the initial reaction has subsided, the mixture is warmed over a low flame for a minute and poured, with vigorous stirring, into 10 ml of water. The precipitate is allowed to settle, and the supernatant liquid is decanted. The residue is stirred thoroughly with 5 ml of 5% sodium carbonate solution, removed by filtration, and purified by recrystallization from alcohol.

(*b*) One milliliter of the alcohol, mixed with 0.5 g of benzoyl or *p*-nitrobenzoyl chloride, is boiled over a low flame for a few minutes. The mixture is poured into water and purified as in (a).

References to a number of important reactions of alcohols are listed here:

t-Butyl hypochlorite oxidation:

$$H\text{—}\overset{|}{\underset{|}{C}}\text{—}OH \xrightarrow{Me_3COCl} \rangle C{=}O$$

Pasto and Johnson, p. 363, preparative scale

$$(Me_3COH + NaOCl \rightarrow Me_3COCl)$$

Sarett reagent oxidation:

$$H\text{—}\overset{|}{\underset{|}{C}}\text{—}OH \xrightarrow[C_5H_5N]{CrO_3} \rangle C{=}O$$

Pasto and Johnson, p. 362; *be careful of order of mixing reagents*

The use of lanthanide shift reagents (Chapter 8) to simplify the nmr spectra of alcohols is well documented; alcohols are one of the most useful substrates for application of this technique. Ultraviolet spectra (Chapter 8) of alcohols usually display weak bands and thus are only infrequently useful for structure diagnoses.

Procedure 4. 3-Nitrophthalates

O, O, O, NO_2 $\xrightarrow{ROH}$ CO_2H, CO_2R, NO_2

Not all alcohols give solid derivatives by this procedure (see below); the carboxyl group in the product does, however, provide advantageous characteristics. The product can be manipulated by acid-base extractions (Chapter 7) and esters arising from optically active alcohols can often be resolved by treatment with chiral amines (Chapter 7).

An extremely useful application of this derivative would be the determination of the neutralization equivalent (Procedure 32) of such an acid ester; this could very possibly lead to an estimate of the molecular weight of this compound.

(a) ***From Alcohols Boiling Below 150°C.*** A mixture of 0.4 g of 3-nitrophthalic anhydride and 0.5 ml of the alcohol is boiled gently in a test tube fitted with a glass tube for a condenser. The heating is continued for 5 to 10 min after the mixture liquefies. The mixture is cooled, diluted with 5 ml of water, and heated to boiling. If solution is not complete, an additional 5 to 10 ml of hot water is added. The solution is cooled and the ester allowed to crystallize. Sometimes the derivative separates as an oil and must be allowed to stand overnight to crystallize. The product is recrystallized once or twice from hot water.

(b) ***From Alcohols Boiling Above 150°C.*** A mixture of 0.4 g of 3-nitrophthalic anhydride, 0.5 g of the alcohol, and 5 ml of dry toluene is boiled until all the anhydride has dissolved and then for 15 min more. The toluene is then removed by suction, using a water pump. The residue is extracted twice with 5 ml of hot water, the residual oil is dissolved in 10 ml of 95% alcohol, and the solution is heated to boiling. If the hot solution is not clear, it should be filtered. Water is added to the hot solution until a turbidity is produced that is cleared up by the addition of a drop or two of alcohol. The solution is allowed to cool slowly and finally permitted to stand. Many of the higher alkyl 3-nitrophthalates derived from the monoalkyl ethers of ethylene glycol and diethylene glycol separate as oils and must be allowed to stand several days to solidify. Occasionally toluene may be substituted for the alcohol-water mixture for recrystallization. It is sometimes desirable to determine the neutralization equivalent of the alkyl acid phthalate as well as the melting point.

Additional tests for alcohols have been described in Cheronis, Entrikin, and

Hodnett:

Test	Page
Vanadium oxine	368
Ferric hydroxamate	368
Xanthate	368
N-Bromosuccinimide	369

6.1.1 Polyhydroxy Alcohols

The chemistry of certain members of this class of alcohols is differentiated from the simple alcohol class by the presence of vicinal pairs of hydroxyl groups,

$$\left(\begin{array}{cc} OH & OH \\ | & | \\ -C- & -C- \\ | & | \end{array} \right)_x$$

Chemical identification and derivatization procedures are thus concerned with detecting this arrangement. Derivatization of the hydroxyl functions very often enhances the manageability of these compounds. These procedures are especially useful in identifying and in characterizing the poly-ol portions of carbohydrates and other related compounds.

EXPERIMENT 5. PERIODIC ACID DETECTION OF VICINAL DIOLS AND RELATED COMPOUNDS

$$\begin{array}{l} RCHOH \\ \;\;| \\ RCHOH \end{array} + HIO_4 \rightarrow 2RCHO + H_2O + HIO_3$$

$$\begin{array}{l} RCHOH \\ \;\;| \\ R'C{=}O \end{array} + HIO_4 \rightarrow RCHO + R'COOH + HIO_3$$

$$\begin{array}{l} RC{=}O \\ \;\;| \\ RC{=}O \end{array} + HIO_4 + H_3O \rightarrow 2RCOOH + HIO_3$$

$$RCHOHCHNH_2R' + HIO_4 \rightarrow RCHO + R'CHO + NH_3 + HIO_3$$

Place 2 ml of the periodic acid reagent in a small test tube, add 1 drop (no more) of concentrated nitric acid, and shake thoroughly. Then add 1 drop or a small crystal of the compound to be tested. Shake the mixture for 10 to 15 sec, and add 1 to 2 drops of aqueous silver nitrate solution (5%). The instantaneous formation of a *white* precipitate (silver iodate) indicates that the organic compound has been oxidized by the periodate, which is thereby reduced to iodate. This constitutes a positive test. Failure to form a precipitate, or the appearance of a brown precipitate that redissolves on shaking, constitutes a negative test.

Dioxane may be added to facilitate the reaction of water-insoluble poly-ols.

$$HIO_3 + AgNO_3 \rightarrow HNO_3 + \underline{AgIO_3(s)}$$

Apply the test to the following substances: isopropyl alcohol, acetone, ethylene glycol, glycerol, glucose, formalin, tartaric acid, lactic acid.

Reagent. The periodic acid reagent is made by dissolving 0.5 g of paraperiodic acid (H_5IO_6) in 100 ml of distilled water.

Discussion. Periodic acid has a very selective oxidizing action on 1,2-glycols, α-hydroxy aldehydes, α-hydroxy ketones, 1,2-diketones, α-hydroxy acids, and α-amino alcohols. The rate of the reaction decreases in the order mentioned. Under the conditions specified above, α-hydroxy acids sometimes give a negative test. β-Dicarbonyl compounds and other active methylene compounds also react.

It is important that the exact amounts of reagent and nitric acid be used. The test depends on the fact that silver iodate is only very slightly soluble in dilute nitric acid whereas silver periodate is very soluble. If too much nitric acid is present, however, the silver iodate will fail to precipitate.

Olefins, secondary alcohols, 1,3-glycols, ketones, and aldehydes are not affected by periodic acid under the above conditions. The periodic acid test is best suited for water-soluble compounds.

The following mechanism has been proposed to account for the oxidation of vicinal diols:

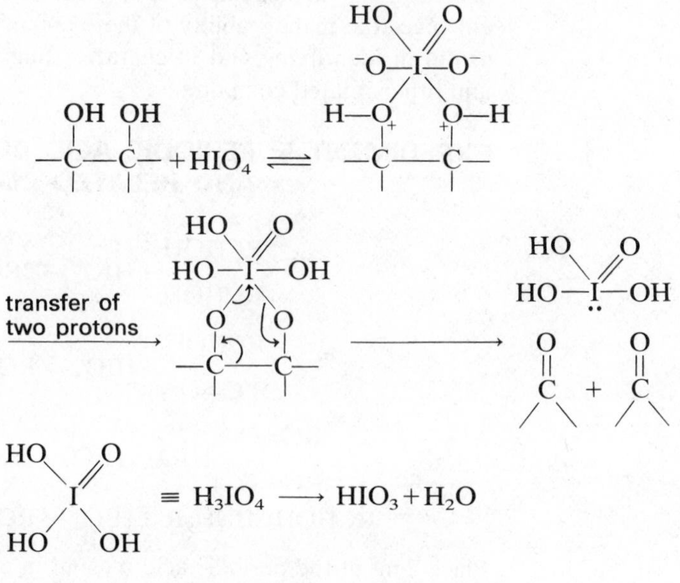

$$\mathrm{-\overset{OH}{\underset{|}{C}}-\overset{OH}{\underset{|}{C}}- + HIO_4 \rightleftharpoons -\overset{H-\overset{+}{O}}{\underset{|}{C}}-\overset{\overset{+}{O}-H}{\underset{|}{C}}-}\quad \text{(O's bridged by } {}^{-}O-I(OH)(=O)-O^{-})$$

transfer of two protons $\longrightarrow$

$$\mathrm{-C(-O-)-C(-O-)-}\ \text{bridged by HO–I(HO)(=O)–OH} \longrightarrow \mathrm{HO-\ddot{I}(HO)(=O)-OH} + \mathrm{>C{=}O} + \mathrm{O{=}C<}$$

$$\mathrm{(HO)_2I(=O)OH \equiv H_3IO_4 \longrightarrow HIO_3 + H_2O}$$

Procedure 5. Acetates of Polyhydroxy Compounds

This procedure, as well as some others described below (e.g., trimethylsilylation), is useful for enhancing the volatility of carbohydrates and related compounds.

(a) Three grams of the anhydrous polyhydroxy compound is mixed with 1.5 g of powdered fused sodium acetate and 15 ml of acetic anhydride. The mixture is heated on the steam bath, with occasional shaking, for 2 hr. At the end of this time the warm solution is poured, with vigorous stirring, into 100 ml of ice water. The mixture is allowed to stand, with occasional stirring, until the excess of acetic anhydride has been hydrolyzed. The crystals are removed by filtration, washed thoroughly with water, and purified by recrystallization from alcohol.

(*b*) Two grams of the polyhydroxy compound is added to 20 ml of anhydrous pyridine. Eight grams of acetic anhydride is added, with shaking, and after any initial reaction has subsided the solution is boiled for 3 to 5 min under a reflux condenser. The mixture is cooled and poured into 50 to 75 ml of ice water. The acetyl derivative is removed by filtration, washed with cold 2% hydrochloric acid, and then washed with water. It is purified by recrystallization from alcohol.

Procedures for identifying and characterizing polyhydroxy compounds, as described by Pasto and Johnson, are outlined here:

Borate test

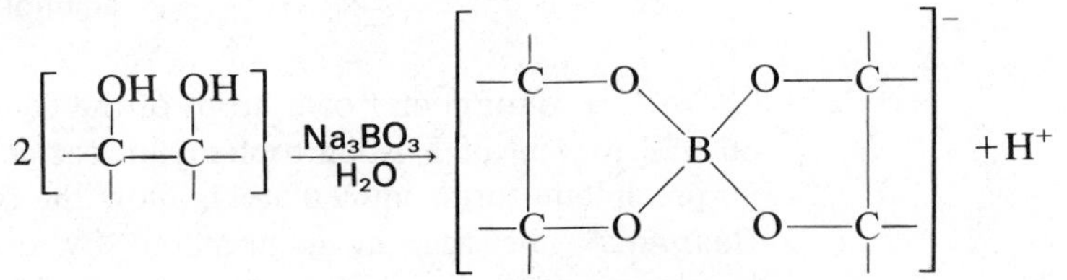

$$2\left[-\overset{\text{OH}}{\underset{|}{\overset{|}{C}}}-\overset{\text{OH}}{\underset{|}{\overset{|}{C}}}-\right] \xrightarrow[H_2O]{Na_3BO_3} \text{[borate ester anion]}^- + H^+$$

borate ester anion
or
tetraalkoxyborate ion

Qualitative test for acidity of borate, Pasto and Johnson, p. 365.

Acetonides:

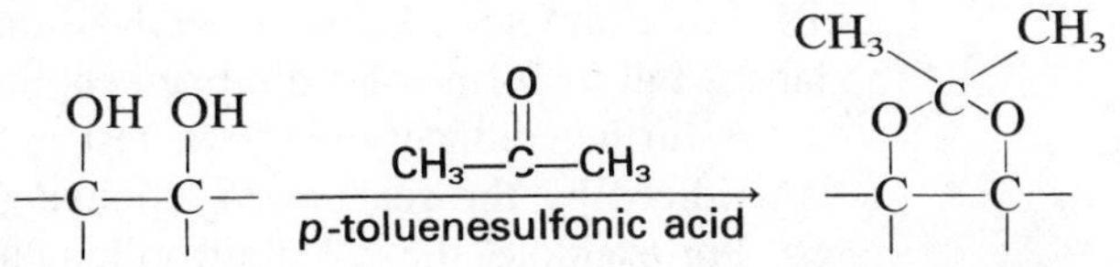

$$-\overset{\text{OH}}{\underset{|}{\overset{|}{C}}}-\overset{\text{OH}}{\underset{|}{\overset{|}{C}}}- \xrightarrow[p\text{-toluenesulfonic acid}]{CH_3-\overset{O}{\overset{\|}{C}}-CH_3} \text{acetone ketal}$$

acetone ketal
(1,3-dioxolane)

Preparative-scale procedure, Pasto and Johnson, p. 367.

Trimethylsilylation (Pasto and Johnson, preparative scale, p. 368):

$$2ROH + (CH_3)_3SiNHSi(CH_3)_3 \rightarrow 2ROSi(CH_3)_3 + NH_3$$

hexamethyldisilazane (TRI-SIL®) — trimethylsilyl ether

6.2 ALDEHYDES (RCHO) AND KETONES (RCOR)

Both chemical and spectral tests are profoundly influenced by the carbonyl (C=O) group of aldehydes and ketones. Aldehydes are normally more reactive than ketones; in addition, the C—H bond of the CHO (aldehyde) group allows spectral differentiation of this carbonyl group from the simple carbonyl group of ketones.

First, we shall discuss chemical procedures that detect the presence of nearly all ketonic and aldehydic carbonyl groups. This will be followed by those tests that are more selective, including those that work only with aldehydes.

EXPERIMENT 6. 2,4-DINITROPHENYLHYDRAZINE

This reaction probably represents the most studied and most successful of all qualitative tests and derivatizing procedures. In addition, the general details of the reaction serve as a model for a number of other chemical reactions (osazone, semicarbazone, oxime, and other arylhydrazone preparations).

$$\mathrm{>C{=}O} + \mathrm{O_2N{-}C_6H_3(NO_2){-}NHNH_2} \xrightarrow[\text{alcohol}]{H_2SO_4} \mathrm{>C{=}NNH{-}C_6H_3(NO_2)_2}$$

Add a solution of 1 or 2 drops of the compound to be tested in 2 ml of 95% ethanol to 3 ml of 2,4-dinitrophenylhydrazine reagent. Shake vigorously, and, if no precipitate forms immediately, allow the solution to stand for 15 min.

Reagent.[3] The reagent is prepared by dissolving 3 g of 2,4-dinitrophenylhydrazine in 15 ml of concentrated sulfuric acid. This solution is then added, with stirring, to 20 ml of water and 70 ml of 95% ethanol. The solution is mixed thoroughly and filtered.

Discussion. Most aldehydes and ketones yield dinitrophenylhydrazones that are insoluble solids. The precipitate may be oily at first and become crystalline on standing. A number of ketones, however, give dinitrophenylhydrazones that are oils. For example, methyl *n*-octyl ketone, di-*n*-pentyl ketone, and similar substances fail to form solid dinitrophenylhydrazones.

A further difficulty with the test is that certain allyl alcohol derivatives may be oxidized by the reagent to aldehydes or ketones, which then give a positive test. For example, the 2,4-dinitrophenylhydrazones of the corresponding carbonyl compounds have been obtained in yields of 10 to 25% from cinnamyl alcohol, 4-phenyl-3-buten-2-ol, and vitamin A_1. Benzhydryl alcohol (diphenylcarbinol) also was found to be converted to benzophenone dinitrophenylhydrazone in low yield. Needless to say, there is always the further danger that an alcohol sample may be contaminated by enough of its aldehyde or ketone, formed by air oxidation, to give a positive test. If the dinitrophenylhydrazone appears to be formed in very small amount, it may be desirable to carry out the reaction on the scale employed for the preparation of a derivative (Procedure 9, p. 179) and to make an estimate of the yield (based on the molecular weight of carbonyl compound, which can be guessed closely enough by consulting the list of possibilities gathered from the tables). The melting point of the solid should be checked to be sure it is different from that of 2,4-dinitrophenylhydrazine (m.p. 198°C, d).

---

[3] This reagent is sometimes called "Brady's Reagent."

If necessary, this hydrazone derivative can be recrystallized from a solvent such as ethanol. Solvents containing reactive carbonyl groups (for example, acetone) should not be used, as they may result in formation of another hydrazone.

The color of a 2,4-dinitrophenylhydrazone may give an indication as to the structure of the aldehyde or ketone from which it is derived. Dinitrophenylhydrazones of aldehydes or ketones in which the carbonyl group is not conjugated with another functional group are yellow. Conjugation with a carbon-carbon double bond or with a benzene ring shifts the absorption maximum toward the visible and is easily detected by an examination of the ultraviolet spectrum.[4] However, this shift is also responsible for a change in color from yellow to orange-red. In general, then, a yellow dinitrophenylhydrazone may be assumed to be unconjugated. However, an orange or red color should be interpreted with caution, since it may be due to contamination by an impurity (for example, 2,4-dinitrophenylhydrazine is orange-red).

In difficult cases it may be desirable to try the preparation of a dinitrophenylhydrazone in diethylene glycol dimethyl ether (diglyme), ethylene glycol monomethyl ether (glyme), DMF, or DMSO; difficulty in the work-up due to removal of non-volatile solvent can be encountered in these cases. Methanol can be used as an alternative to ethyl alcohol; the more volatile alcohol may, however, result in mixtures that are more difficult to purify.

EXPERIMENT 7. HYDROXYLAMINE HYDROCHLORIDE

This reaction has a mechanism similar to that for the preceding experiment involving arylhydrazone formation; formation of oximes results in the liberation of hydrogen chloride, which can be detected by an indicator.

$$RCHO + H_2NOH{\cdot}HCl \rightarrow RCH{=}NOH + HCl + H_2O$$

$$R_2CO + H_2NOH{\cdot}HCl \rightarrow R_2C{=}NOH + HCl + H_2O$$

(a) ***For Neutral Compounds.*** To 1 ml of the reagent add a drop or a few crystals of the compound, and note the color change. If no pronounced change occurs at room temperature, heat the mixture to boiling. A change in color from orange to red constitutes a positive test. Try the test on (1) *n*-butyraldehyde; (2) acetone; (3) benzophenone; (4) glucose.

(b) ***For Acidic or Basic Compounds.*** To 1 ml of the indicator solution add about 0.2 g of the compound, and adjust the color of the mixture so that it matches 1 ml of the reagent in a separate test tube of the same size. This is done by adding a few drops of dilute (1%) sodium hydroxide or hydrochloric acid solution. Then add the resulting solution to 1 ml of the reagent and note if a red color is produced. Try this test on (1) salicylaldehyde; (2) *p*-dimethylaminobenzaldehyde. Try the tests on tartaric acid by Procedures (a) and (b).

---

[4] Z. Rappoport and T. Sheradsky, *J. Chem. Soc., B*, 277 (1968); L. A. Jones, J. C. Holmes, and R. B. Seligman, *Anal. Chem.*, **28**, 191 (1956).

Reagents. To a solution of 5 g of hydroxylamine hydrochloride in 1 liter of 95% ethanol is added 3 ml of Bogen or Grammercy Universal Indicator. The color of the solution is adjusted to a bright orange shade (pH 3.7 to 3.9) by adding dilute (5%) ethanolic sodium hydroxide solution dropwise. The reagent is stable for several months.

Indicator Solution. A solution of the indicator is made by adding 3 ml of either of the above indicators to 1 liter of 95% ethanol.

Discussion. The change in color of the indicator is due to the hydrochloric acid liberated in the reaction of the carbonyl compound with hydroxylamine hydrochloride, the oxime not being sufficiently basic to form a hydrochloride. All aldehydes and most ketones give an immediate change in color. Some higher-molecular-weight ketones such as benzophenone, benzil, benzoin, and camphor require heating. Sugars, quinones, and hindered ketones (such as *o*-benzoylbenzoic acid) give a negative test.

Many aldehydes undergo autoxidation in the air and contain appreciable amounts of acids; hence the action of an aqueous solution or suspension on litmus must always be determined. If the solution is acidic, Procedure (b) must be used; this is also true of compounds whose solubility behavior shows them to be acids or bases.

$$\mathrm{RCHO} \xrightarrow{\text{air}} \mathrm{RCO_2H}$$

The mechanism of the reaction of carbonyl compounds with hydroxylamine is apparently very closely related to the mechanisms of the reactions with phenylhydrazine, dinitrophenylhydrazine, and semicarbazide. Semicarbazone formation is the reaction of this group that has been most thoroughly studied. The mechanism in aqueous solution, as catalyzed by the acid HA, is shown below.

The mechanism shows that the reaction is strongly retarded by solutions that are either too acidic or too basic. Since the reaction requires both free semicarbazide and a proton source (HA), a too strongly acidic solution depresses the reaction rate by converting virtually all of the semicarbazide to its inactive conjugate acid, whereas a strongly basic solution depresses the rate by converting the acid HA to its inactive conjugate base. The most favorable conditions are those in which the product of the concentrations of HA and $NH_2NHCONH_2$ is at a maximum.

$$R_2C{=}O + HA + H_2NNHCONH_2 \rightleftharpoons R_2C{=}O{\cdot}HA + H_2NNHCONH_2$$

$$\downarrow\uparrow$$

$$R_2C\begin{matrix}\diagup OH \\ \diagdown NHNHCONH_2\end{matrix} + HA \rightleftharpoons R_2C\begin{matrix}\diagup OH \\ \diagdown \overset{+}{N}H_2NHCONH_2\end{matrix} + A^-$$

$$\downarrow\uparrow$$

$$R_2C\begin{matrix}\diagup OH{\cdot}HA \\ \diagdown NHNHCONH_2\end{matrix} \rightleftharpoons R_2C{=}\overset{+}{N}HNHCONH_2 + H_2O + A^-$$

$$\downarrow\uparrow$$

$$R_2C{=}NNHCONH_2 + H_2O + HA$$

The reaction is reversible. Although in relatively weakly acidic solutions the equilibrium may be made to lie well toward the right, strong acid removes semicarbazide by converting it to the conjugate acid and shifts the equilibrium toward the free ketone.

Preparation of semicarbazones for derivatization purposes is described below in Procedure 7, p. 179.

EXPERIMENT 8. PHENYLHYDRAZINE AND p-NITROPHENYLHYDRAZINE[5]

$$>C{=}O + H_2NNHC_6H_5 \rightarrow >C{=}NNHC_6H_5 + H_2O$$

(a) To 5 ml of water add 0.5 ml of phenylhydrazine or 0.5 g of *p*-nitrophenylhydrazine and then acetic acid drop by drop until the phenylhydrazine just dissolves. Then add 0.5 ml of acetone and shake. If no precipitation occurs, warm gently for a minute in a low flame, add 2 ml of water, and cool.

(b) Dissolve 0.5 ml of acetophenone in 2 ml of ethanol, and add water dropwise until the cloudiness just disappears on shaking. If too much water is added, a little alcohol must be introduced to clarify the solution. To this clear solution add 0.2 ml of pure phenylhydrazine or 0.2 g of *p*-nitrophenylhydrazine. If the solution remains clear for several minutes, catalyze the reaction by the addition of a drop of acetic acid; warm gently for a few minutes and then cool. What advantage has this procedure over that given for acetone?

Discussion. Although most aldehydes and ketones may be depended upon to form phenylhydrazones under the conditions employed here, the use of phenylhydrazine has the serious disadvantage that many phenylhydrazones are oils and are therefore difficult to detect visually. *p*-Nitrophenylhydrazine obviates this difficulty to a large extent, and its use is recommended. The mechanisms of these reactions are probably similar to that of semicarbazone formation, which is discussed above.

At least one source of difficulty in the use of phenylhydrazine has been demonstrated. The phenylhydrazones that are first produced not only can exist in two different stereochemical forms, but also may tautomerize in solution to the isomeric azo compound. Such different forms may make the separation process very complex. Nitro groups on the ring help stabilize the hydrazone form and so prevent tautomerization.

EXPERIMENT 9. SODIUM BISULFITE ADDITION COMPLEXES

The precipitation of a bisulfite addition complex is indicative of a variety of carbonyl compounds. Since this reaction is apparently very greatly influenced by

---

[5] *Caution:* Arylhydrazines tend to be carcinogenic and thus should be handled carefully.

the steric environment of the carbonyl group, not all ketones (see below) form such a derivative.

$$RCHO + NaHSO_3 \rightarrow R\text{—}\underset{\displaystyle H}{\overset{}{C}}\begin{matrix} \diagup OH \\ \diagdown \overset{-}{SO_3}\overset{+}{Na} \end{matrix}$$

Prepare an alcoholic solution of sodium bisulfite by adding 3 ml of ethanol to 12 ml of a 40% aqueous solution of the salt. A small amount of salt will be precipitated by the alcohol and must be separated by decantation or filtration before the reagent is ready for use.

Place 1 ml of the reagent in a test tube and add 0.3 ml of benzaldehyde. Stopper the test tube and shake vigorously. Repeat the test with (1) heptanal; (2) acetophenone.

Discussion. The formation of bisulfite addition compounds, which have been shown to be α-hydroxyalkanesulfonates, is a general reaction of aldehydes. Most methyl ketones, low-molecular-weight cyclic ketones (up to cyclooctanone), and certain other compounds having very active carbonyl groups behave similarly. Some methyl ketones, however, form the addition compounds only slowly or not at all. Examples are aryl methyl ketones, pinacolone, and mesityl oxide. On the other hand, cinnamaldehyde forms an addition compound containing two molecules of bisulfite.

The bisulfite is in equilibrium with the carbonyl compound, and, since sodium bisulfite is decomposed by either acids or alkalies, the addition compounds are stable only in neutral solutions. Compounds derived from low-molecular-weight carbonyl compounds are soluble in water. They are useful because they are solids—easily purified—and because they are readily decomposed by acids and by alkalies to regenerate the original compounds.

The nitrogen analogs of aldehydes, imines (or Schiff bases) also add sodium bisulfite; the product is identical with that formed by the action of a primary amine on the aldehyde bisulfite compound. The carbon-sulfur bond in these

$$RCH\text{=}N\text{—}R' + NaHSO_3 \rightarrow R\text{—}\underset{\displaystyle SO_3Na}{\underset{|}{CH}}\text{—}NHR'$$

$$\underset{\displaystyle SO_3Na}{\underset{|}{RCHOH}} + R'NH_2 \nearrow$$

compounds is reactive, the sulfonate grouping being displaced by reactions with anions such as CN^-.

Questions

1. Suggest an explanation of the fact that cyclohexanone reacts with sodium bisulfite readily whereas diethyl ketone does not.
2. What is the explanation of the failure of pinacolone to react? Compare this case with that of acetophenone.
3. Explain the behavior of cinnamaldehyde.

4. Why is an alcoholic solution of sodium bisulfite used? Try the test on acetone, using an aqueous solution.

EXPERIMENT 10. THE IODOFORM TEST

The iodoform test is a test for methyl ketones; it will also give positive results for compounds that can be oxidized to methyl ketones under the test conditions.

$$\mathrm{RCHOHCH_3 + I_2 + 2NaOH \rightarrow R{-}\overset{\overset{\displaystyle O}{\|}}{C}{-}CH_3 + 2NaI + 2H_2O}$$

$$\mathrm{RCOCH_3 + 3I_2 + 3NaOH \rightarrow R{-}\overset{\overset{\displaystyle O}{\|}}{C}{-}CI_3 + 3NaI + 3H_2O}$$

$$\big\downarrow \mathrm{NaOH}$$

$$\mathrm{RCO_2Na + CHI_3}$$

Apply the following test to (1) isopropyl alcohol; (2) acetone; (3) ethyl acetate; (4) ethyl acetoacetate; (5) acetophenone; (6) pure methanol. (Note that some lower commercial grades of methanol give misleading positive results due to impurities.)

Place 4 drops of the liquid (0.1 g of the solid) to be tested in a test tube (15 mm wide). Add 5 ml of dioxane,[6] and shake until all the sample has gone into solution. Add 1 ml of 10% sodium hydroxide solution, and then iodine-potassium iodide solution (see below) with shaking, until a slight excess yields a definite dark color of iodine. If less than 2 ml of the iodine solution is decolorized, place the test tube in a water bath maintained at a temperature of 60°C. If the slight excess of iodine already present is decolorized, continue the addition of iodine solution (keeping the dioxane solution at 60°C), with shaking, until a slight excess of iodine solution again yields a definite dark color. The addition of iodine is continued until the dark color is not discharged by 2 min of heating at 60°C. The excess of iodine is removed by the addition of a few drops of 10% sodium hydroxide solution, with shaking. Now fill the test tube with water and allow to stand for 15 min. A positive test is indicated by the formation of a foul-smelling yellow precipitate (iodoform). The precipitate should be collected (filtration) and dried and its melting point checked; iodoform (CHI_3) melts at 119–121°C (d) and has a distinctive odor. If the iodoform is reddish, dissolve in 3 to 4 ml of dioxane, add 1 ml of 10% sodium hydroxide solution, and shake until only a light lemon color remains. Dilute with water and filter.

Reagent. The iodine-potassium iodide solution is made by adding 200 g of potassium iodide and 100 g of iodine to 800 ml of distilled water and stirring until solution is complete.

$$\mathrm{I_2 + KI \xrightarrow{H_2O} KI_3}$$

The solution is deep brown due to the triiodide anion (I_3^-).

---

[6] Dioxane is appropriate for water-*insoluble* compounds; water-soluble compounds may be treated by substituting 2 ml of water for the dioxane solvent.

Discussion. This test is positive for compounds that contain the grouping CH_3CO—, CH_2ICO—, or CHI_2CO— when joined to a hydrogen atom or to a carbon atom that does not have highly active hydrogens or groups that provide an excessive amount of steric hindrance. The test will, of course, be positive also for any compound that reacts with the reagent to give a derivative containing one of the requisite groupings. Conversely, compounds that contain one of the requisite groupings will not give iodoform if that grouping is destroyed by the hydrolytic action of the reagent before iodination is complete.

Following are the principal types of compounds that give the test:

$$CH_3CHO \qquad CH_3CH_2OH \qquad CH_3COR \qquad CH_3CHOHR$$

$$RCOCH_2COR \qquad R\overset{OH}{\overset{|}{C}}HCH_2\overset{OH}{\overset{|}{C}}HR$$

(R = any alkyl or aryl radical except a di-ortho-substituted aryl radical.) R, however, if large, will sterically inhibit this reaction. The test is negative for compounds of the following types:

$$CH_3COCH_2CO_2R \qquad CH_3COCH_2CN \qquad CH_3COCH_2NO_2 \text{ etc.}$$

In such compounds the reagent removes the acetyl group and converts it to acetic acid, which resists iodination.[7]

A modified reagent[8] has been suggested for distinguishing methyl ketones from methyl carbinols. It consists of a solution of 1 g of potassium cyanide,[8a] 4 g of iodine, and 6 ml of concentrated ammonium hydroxide solution in 50 ml of water.

$$CH_3-\overset{O}{\overset{\|}{C}}-R \xrightarrow[KCN/NH_4OH]{I_2} CHI_3$$

$$CH_3CHOHR \xrightarrow[KCN/NH_4OH]{I_2} \text{N.R.}$$

This reagent produces iodoform from methyl ketones but not from methyl carbinols.

The cleavage of trihalo ketones with base, exemplified by the second step of the iodoform test, is related to the reversal of the Claisen condensation. In each case the reaction can proceed because of the stability of the final anionic

---

[7] For a general discussion of this test see R. C. Fuson and B. A. Bull, *Chem. Rev.*, **15**, 275 (1934).

[8] E. Rothlin, *Arch. Escuela Farm. Fac. ci. Med. Córdoba* [R.A.] *Secc. ci.*, **1939**, No. 10, p. 1; *C.A.*, **35**, 5091 (1941).

[8a] Potassium cyanide is *extremely* dangerous; use only with instructor's permission. Do not mix with acid.

fragment:

$$\mathrm{R\overset{O}{\overset{\|}{C}}{-}CI_3 + OH^- \rightarrow \left[R\overset{O^-}{\overset{|}{C}}{-}CI_3 \atop \quad | \atop \quad OH \right] \rightarrow RC\begin{matrix} {}^{\diagup\!\!\diagup}O \\ {}_{\diagdown}OH \end{matrix} + {}^-CI_3}$$

$$\mathrm{R\overset{O}{\overset{\|}{C}}{-}CH_2\overset{O}{\overset{\|}{C}}R + OH^- \rightarrow \left[R\overset{O^-}{\overset{|}{C}}{-}CH_2\overset{O}{\overset{\|}{C}}R \atop \quad | \atop \quad OH \right] \rightarrow RC\begin{matrix} {}^{\diagup\!\!\diagup}O \\ {}_{\diagdown}OH \end{matrix} + {}^-CH_2\overset{O}{\overset{\|}{C}}R}$$

Secondary alcohols, and ketones of the structures

$$\mathrm{RCH_2CHOHCH_2R} \quad \text{and} \quad \mathrm{RCH_2COCH_2R}$$

do not produce iodoform although they may undergo some halogenation on the methylene group adjacent to the carbonyl group. Occasionally commerical samples of diethyl ketone give a weak iodoform test. This is due to the presence of impurities such as 2-pentanone.

Bifunctional alcohols and ketones of the following types give positive iodoform tests:

$$\mathrm{CH_3CHOHCHOHCH_3} \qquad \mathrm{CH_3COCHOHCH_3}$$

$$\mathrm{CH_3COCOCH_3} \qquad \mathrm{CH_3COCH_2CH_2COCH_3}$$

β-Keto esters do not produce iodoform by the test method, but their alkaline solutions do react with sodium hypoiodite.

Acetoacetic acid is unstable; acidic aqueous solutions decompose to give CO_2 and acetone.

$$\mathrm{CH_3COCH_2CO_2H \xrightarrow{H^+} CH_3COCH_3 + CO_2}$$

The acetone will give a positive iodoform test. This behavior is generally useful if a β-keto ester is one of the possibilities being considered, since these esters are hydrolyzed by boiling with 5% sulfuric acid (acid-induced reverse condensation):

$$\mathrm{RCOCH_2CO_2R' + H_2O \xrightarrow{H^+} RCOCH_3 + R'OH + CO_2}$$

Procedure. A 0.5-g sample of the β-keto ester is placed in a 100-ml distilling flask and 50 ml of 5% sulfuric acid is added. A few boiling stones are added and the mixture is gently heated to boiling so that distillation takes place. About 6 to 8 ml of distillate is collected and divided into three portions. One portion is treated with sodium hypoiodite (Experiment 10); a second portion is treated with the 2,4-dinitrophenylhydrazine reagent (p. 162); and the third portion is tested for the alcohol (R'OH) with ammonium hexanitratocerate reagent (p. 146).

Easily hydrolyzable ethyl or isopropyl esters give positive iodoform tests.

Often, after finding chemical and spectral data showing the presence of a carbonyl group in a substrate, the chemist becomes concerned with determining whether this carbonyl group is that of an aldehyde or of a ketone. We shall now concern ourselves with tests that give positive results with aldehydes but negative results with ketones.

A simple chemical test for aldehydes involves the use of CrO_3; the procedure for alcohols described in Experiment 3 (Section 6.1) should be followed. Aldehydes are normally oxidized easily:

$$3RCHO + 2CrO_3 + 3H_2SO_4 \rightarrow 3RCO_2H + 3H_2O + 2Cr_2(SO_4)_3$$

Ketones normally do not react under these conditions, nor with potassium permanganate (Section 6.4), because of the absence of a C—H bond connected directly to the carbonyl group.

EXPERIMENT 11. TOLLENS REAGENT

$$RCHO + 2Ag(NH_3)_2OH \rightarrow 2Ag(s) + RCOONH_4 + H_2O + 3NH_3$$

Prepare Tollens reagent according to the directions below, and test its reaction with the following: (1) formalin; (2) acetone; (3) benzaldehyde; (4) glucose; (5) hydroquinone; (6) *p*-aminophenol. The test is conducted by adding a small amount of the reagent prepared as described below. If no reaction takes place in the cold, the solution should be warmed slightly.

Reagent. Into a thoroughly clean test tube place 2 ml of a 5% solution of silver nitrate, and add a drop of dilute sodium hydroxide solution (10%). Add a very dilute solution of ammonia (about 2%) drop by drop, with constant shaking, until the precipitate of silver oxide just dissolves. In order to obtain a sensitive reagent it is necessary to avoid a large excess of ammonia.

This reagent should be prepared just before use and should not be stored, because the solution decomposes on standing and deposits a highly explosive precipitate.

Question

1. Would the presence of a reactive halogen atom interfere with this test?

Discussion. It should be noted that acyloins, diphenylamine, and other aromatic amines, as well as α-naphthol and certain other phenols, give a positive Tollens test. α-Alkoxy and α-dialkylamino ketones have also been found to reduce ammoniacal silver nitrate. In addition, the stable hydrate of trifluoroacetaldehyde gives a positive test.

This test often results in a smooth deposit of silver metal on the inner surface of the test tube, hence the name the "silver mirror" test. In some cases, however, the metal forms merely as a granular gray or black precipitate, especially if the glass is not scrupulously clean.

The reaction is autocatalyzed by the silver metal and often involves an induction period of a few minutes.

EXPERIMENT 12. FUCHSIN-ALDEHYDE REAGENT

Place 2 ml of fuchsin-aldehyde reagent in a test tube, and add 1 drop of butyraldehyde. Shake the tube gently, and observe the color developed in 3 to 4 min. Repeat the test with (1) benzaldehyde; (2) formaldehyde solution; (3) acetophenone; (4) acetone.

In this test the reagent should not be heated, and the solution tested should not be alkaline. When the test is used on an unknown, a simultaneous test on a known aldehyde should be performed for comparison.

Reagent. Dissolve 0.5 g of pure fuchsin (*p*-rosaniline hydrochloride) in 500 ml of distilled water, and filter the solution. Saturate 500 ml of distilled water with sulfur dioxide, mix thoroughly with the filtered fuchsin solution, and allow to stand overnight. This produces a practically colorless and very sensitive reagent.

Discussion. Fuchsin is a pink triphenylmethane dye that is converted to the colorless leucosulfonic acid by sulfurous acid. Apparently the reaction involves 1,6-addition of sulfurous acid to the quinoid nucleus of the dye.

$$(H_2N\text{—}C_6H_4)_2C{=}C_6H_4{=}\overset{+}{N}H_2Cl^- + 3H_2SO_3 \longrightarrow$$

(pink solution)

$$H_3\overset{+}{N}\text{—}C_6H_4\text{—}\underset{SO_3H}{C}(\text{—}C_6H_4\text{—}NHSO_2H)_2\,Cl^- + 2H_2O$$

(Schiff's reagent—colorless)

This leucosulfonic acid is unstable and loses sulfurous acid when treated with an aldehyde to produce a violet-purple quinoid dye.

$$H_2\overset{+}{N}{=}C_6H_4{=}C\left[\text{—}C_6H_4\text{—}NH\text{—}\underset{O}{\overset{O}{S}}\text{—}C(OH)(H)R\right]_2 Cl^-$$

(violet-purple solution)

It is important to note that the color of this dye is different from that of the original fuchsin. It is not a light pink but has a blue cast bordering on a violet or purple. Some ketones and unsaturated compounds react with sulfurous acid to regenerate the pink color of the fuchsin. The development of a light pink color in the reagent is not, therefore, a positive test for aldehydes.

The fact that certain compounds cause the regeneration of the pink color of the original fuchsin has been made the basis of a test. It is reported that, when a specially prepared reagent is used and the reaction time is 1 hr, aldoses produce a pink color whereas ketoses and disaccharides (except maltose) do not. This modification of the Schiff test must be employed with caution, because many organic compounds produce a pink color with the reagent when shaken in the air; other compounds such as α,β-unsaturated ketones combine with sulfurous acid and thus reverse the first reaction given above.

EXPERIMENT 13. BENEDICT'S SOLUTION

Compounds Containing No Sulfur

$$
\begin{array}{c} 2Cu^{2+} \\ \text{Citrate complex} \end{array} + \left\{ \begin{array}{l} RCHO \rightarrow RCO_2H + Cu_2O \\ \begin{array}{c} RCHOH \\ | \\ RCO \end{array} \rightarrow \begin{array}{c} R\text{—}CO \\ | \\ R\text{—}CO \end{array} + Cu_2O \\ ArNHNHAr \rightarrow ArN{=}NAr + Cu_2O \\ ArNHNH_2 \rightarrow ArH + ArOH + N_2 + Cu_2O \\ ArCHO \rightarrow \text{No reaction} \end{array} \right.
$$

To a solution or suspension of 0.2 g of the compound in 5 ml of water add 5 ml of Benedict's solution. Note whether a yellow or yellowish green precipitate is formed; then heat the mixture to boiling and note whether a precipitate is formed and if so, its color. Try the test on (1) *n*-butyraldehyde; (2) acetoin; (3) benzoin; (4) glucose; (5) sucrose; (6) glycerol; (7) 2-butanol; (8) acetone; (9) phenylhydrazine.

To a solution of 0.2 g of sucrose in 5 ml of water add 2 drops of concentrated hydrochloric acid and boil the solution for a minute. Cool the solution, neutralize the acid with dilute sodium hydroxide solution, and try the action of Benedict's reagent. Explain the result.

Reagent. Benedict's solution is made by dissolving the following salts in distilled water.

Hydrated copper sulfate (17.3 g)
Sodium citrate (173.0 g)
Anhydrous sodium carbonate (100 g)

The citrate and carbonate are dissolved by heating with 800 ml of water. Additional water is added to bring the volume of solution to 850 ml. The copper sulfate is dissolved in 100 ml of water, and the resulting solution is poured slowly, with stirring, into the solution or citrate and carbonate. The final solution is made up to 1 liter by addition of water.

Discussion. Benedict's solution, which contains the copper bound in the complex anion, functions as a selective oxidizing agent. It was introduced as a reagent for reducing sugars to replace Fehling's solution, which is very strongly alkaline. Benedict's reagent will detect 0.01% of glucose in water. The color of the precipitate may be red, yellow, or yellowish green, depending on the nature and amount of the reducing agent present.

Benedict's reagent is reduced by α-hydroxy aldehydes, α-hydroxy ketones, and α-keto aldehydes. It does not oxidize simple aromatic aldehydes. Molecules containing only the alcohol functional group (primary, secondary, or tertiary alcohols or glycols) or only the keto grouping are not oxidized by Benedict's solution.

Hydrazine derivatives, as exemplified by phenylhydrazine and hydrazobenzene, are oxidized by this reagent. Other easily oxidizable systems, such as phenylhydroxylamine, aminophenol, and related photographic developers, also reduce Benedict's solution.

Additional tests for carbonyl compounds have been outlined in Cheronis, Entrikin, and Hodnett:

Test	Page
3,5-Dinitrobenzoic acid	393
N-Hydroxybenzenesulfonamide[9]	394
2-Hydrazinobenzothiazole	396

Mercaptans (thiols) interfere with this test; the use of this test to detect the SH group is described below (Experiment 28a).

The infrared spectrum of a simple aldehyde is shown in Fig. 6.3*a*. The strong C═O stretch at 1725 cm^{-1} (5.80 μm) and the pair of peaks at 2800 and 2700 cm^{-1} (3.57 and 3.70 μm) for the aldehydic C—H stretch are strong support for the —CH═O group. The spectrum of Fig. 6.3*a* should be compared to that of Fig. 6.3*b*; the doubly conjugated carbonyl group in Fig. 6.3*b* results in a carbonyl stretch at 1660 cm^{-1} (6.02 μm). This decrease in wave number (increase in wavelength) is a typical result of conjugation of a carbonyl group with a π-system.

Figure 6.3*c* is the ir spectrum of a simple ketone. Conjugating the ketone's carbonyl group with a π system would result in nearly the same change in the carbonyl band position as was observed above (Figs. 6.3*a*, 6.3*b*).

Cyclic ketones display absorption patterns very dependent on ring size; smaller rings have more strained carbonyl groups and thus their absorptions are at higher wave numbers: cyclohexanone, 1715 cm^{-1} (5.83 μm); cyclopentanone, 1740 cm^{-1} (5.75 μm); and cyclobutanone 1780 cm^{-1} (5.62 μm).

The nmr spectrum of a simple aldehyde is given in Fig. 6.4*a*; compare the low-field position of the highly deshielded aldehydic proton in Fig. 6.4*a* to the higher-field signals of the nmr spectrum of a simple ketone, Fig. 6.4*b*. Also, note the smaller magnitude of coupling (J = 2–3 Hz unresolved in Fig. 6.4*a*) between vicinal protons where one proton is attached to an sp^2-hybridized carbon atom and the other proton is attached to an sp^3-hybridized carbon; this should be contrasted to the typical (ca. 7 Hz) coupling involving vicinal protons that are both attached to adjacent sp^3-hybridized carbon atoms.

Mass spectral cleavage of aldehydes and ketones often results in the following useful fragments:

Fragments	Mass	Source
RCO^+	R+28	R—COR′ or R—CHO bond cleavage
$ArCO^+$	Ar+28	Ar—COR′ or Ar—CHO bond cleavage
R—C($\overset{+}{\ddot{O}}$H)═CR_1R_2	29+R+R_1+R_2	McLafferty cleavage
$(CHO)^{+}$[a]	29	R—CHO bond cleavage

[a] From aldehydes only.

[9] Also called benzenesulfonhydroxamic acid or Piloty's Acid.

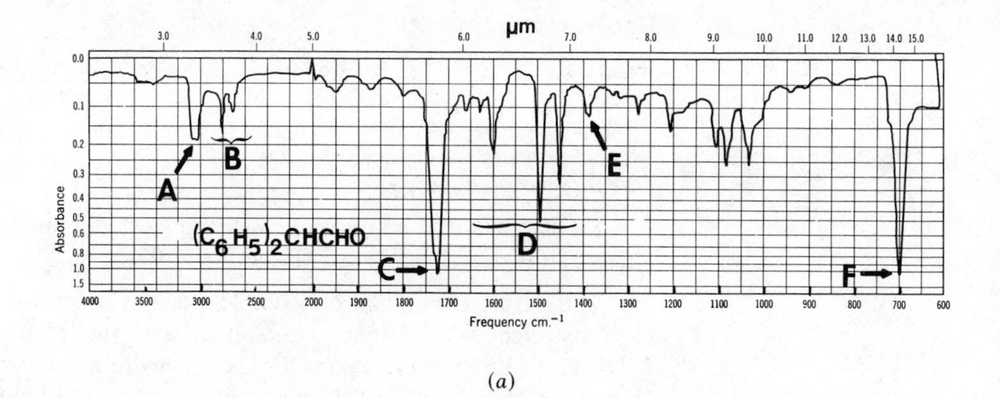

(*a*)

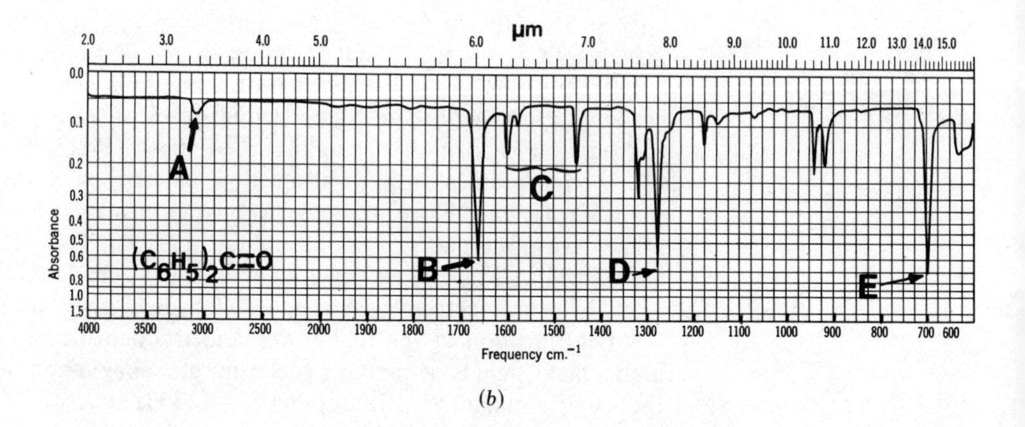

(*b*)

Figure 6.3. (*a*) Infrared spectrum of diphenylacetaldehyde: carbon tetrachloride solution, matched cells. C—H stretch: **A**, 3050 cm^{-1} (3.28 μm), $\nu_{C—H}$ aromatic; **B**, 2800 cm^{-1} (3.57 μm), 2700 cm^{-1} (3.70 μm), $\nu_{C—H}$ aldehydic doublet due to Fermi resonance with overtone of the 1380 cm^{-1} band. C=O stretch: **C**, 1725 cm^{-1} (5.80 μm), $\nu_{C=O}$. C$\cdots$C stretch: **D**, 1600 cm^{-1} (6.25 μm), 1490 cm^{-1} (7.15 μm), 1450 cm^{-1} (6.90 μm), $\nu_{C\cdots C}$. (C=O)—H bend: **E**, 1380 cm^{-1} (7.25 μm), $\delta_{C—H}$ (aldehydic). C$\cdots$C bend (out of plane): **F**, 700 cm^{-1} (14.3 μm), C$\cdots$C out-of-plane bend.

(*b*) Infrared spectrum of benzophenone (2% in carbon tetrachloride). C—H stretch: **A**, 3050 cm^{-1} (3.28 μm), $\nu_{C—H}$ (aromatic). C=O stretch: **B**, 1660 cm^{-1} (6.02 μm), $\nu_{C=O}$ (conjugated). C$\cdots$C stretch: **C**, 1600 cm^{-1} (6.25 μm), 1450 cm^{-1} (6.90 μm), $\nu_{C\cdots C}$. C—(C=O)—C stretch and bend: **D**, 1275 cm^{-1} (7.85 μm). C$\cdots$C bend: **E**, 700 cm^{-1} (14.3 μm).

With the exception of the fragment from McLafferty cleavage, all of the fragments arise from cleavage of a group (e.g., G_1) attached to the carbonyl group;

$$G_1 \mid \overset{\overset{:O^{+}}{\parallel}}{C} — G_2$$

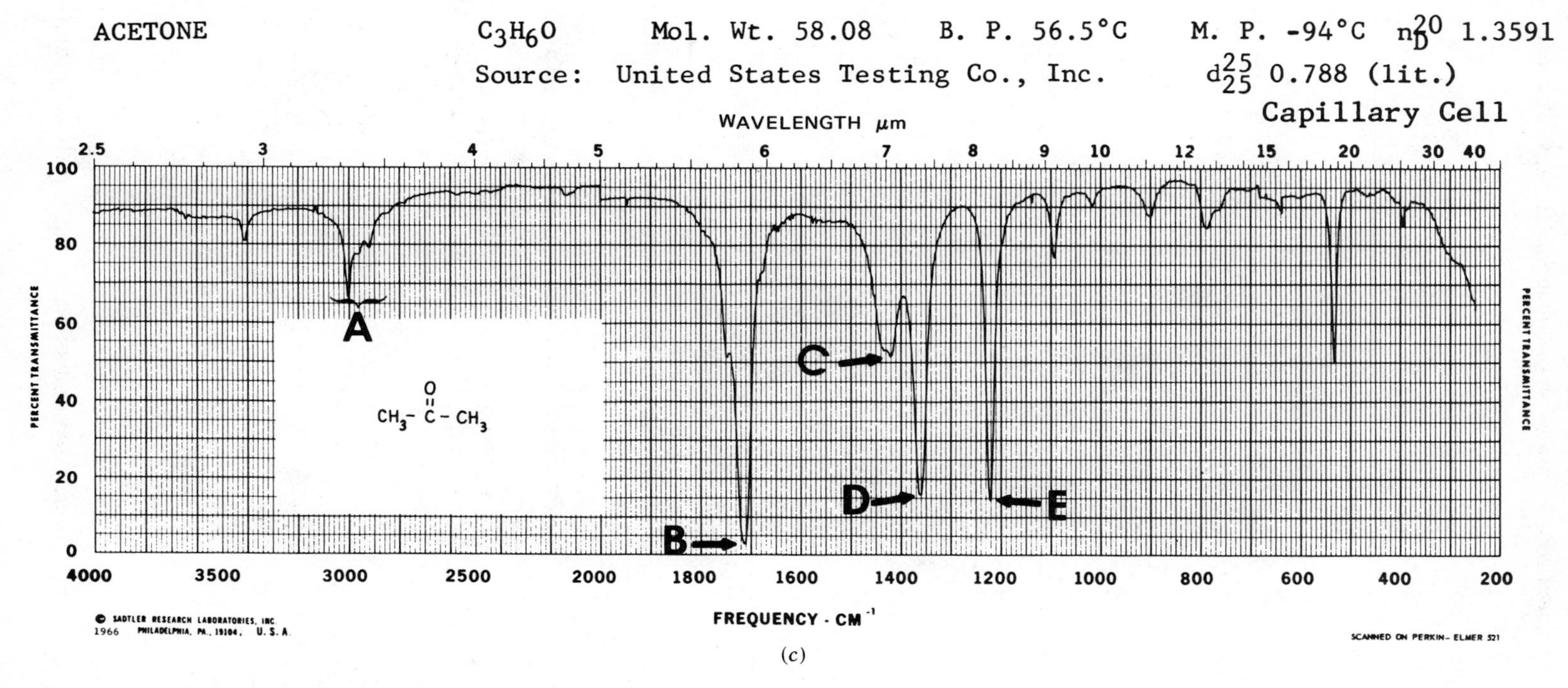

(*c*)

Figure 6.3. (*Continued*) (*c*) Infrared spectrum of acetone: neat. C—H stretch: **A**, 3005 cm^{-1} (3.33 μm), 2960 cm^{-1} (3.38 μm), 2920 cm^{-1} (3.42 μm), ν_{asym}, ν_{sym} CH_3, overtone of δ_{sym} (in order). C=O stretch: **B**, 1715 cm^{-1} (5.83 μm), $\nu_{C=O}$. C—H bend: **C**, 1420 cm^{-1} (7.04 μm), δ_{asym} CH_3 of CH_3CO; **D**, 1363 cm^{-1} (7.33 μm), δ_{sym} CH_3 of CH_3CO; **E**, 1220 cm^{-1} (8.20 μm), C—C(=O)—C stretch and bend.

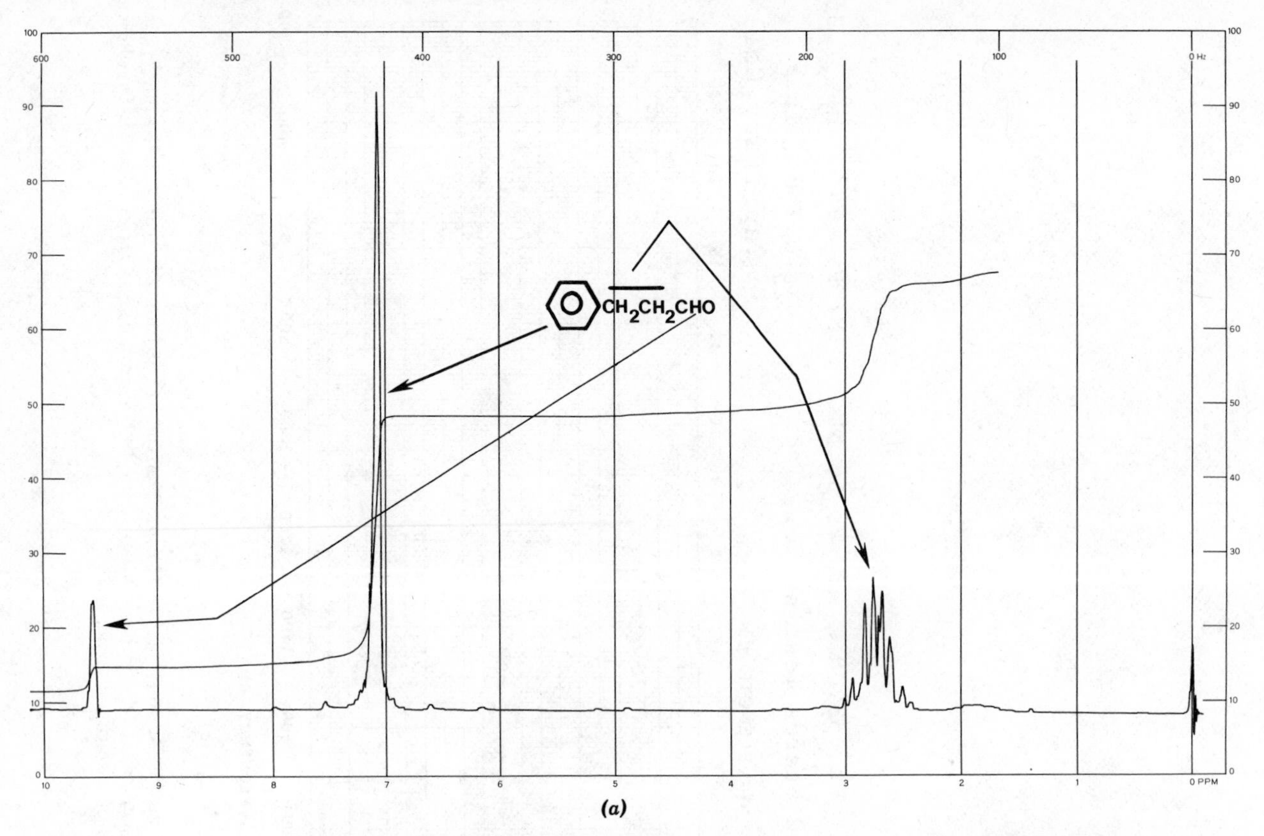

Figure 6.4. (*a*) Nuclear magnetic resonance (proton) spectrum of dihydrocinnamaldehyde (also called α-phenylpropionaldehyde or 3-phenylpropanal): 60 MHz, 600-Hz sweep width, $CDCl_3$ solvent.

gives rise to a charged fragment (here, G_2CO^+). The ability of groups to stabilize charge and thus induce large fragment peaks is in the order Ar>R>H; thus acetophenone gives rise to a very large *m/e* 105 peak ($C_6H_5CO^+$) and a very small *m/e* 43 peak (CH_3CO^+). McLafferty cleavage involves rearrangement and hydrogen transfer; a γ-hydrogen is required:

$$\underset{M^{+\cdot}}{R-C(=\ddot{O}^+\cdots H)-C(R_1)(R_2)-C-C} \xrightarrow{-\,[C=C]} R-C(-\ddot{O}^{+\cdot}H)=C(R_1)(R_2)$$

In fact, any system possessing a carbonyl group, a γ-hydrogen, and the geometric

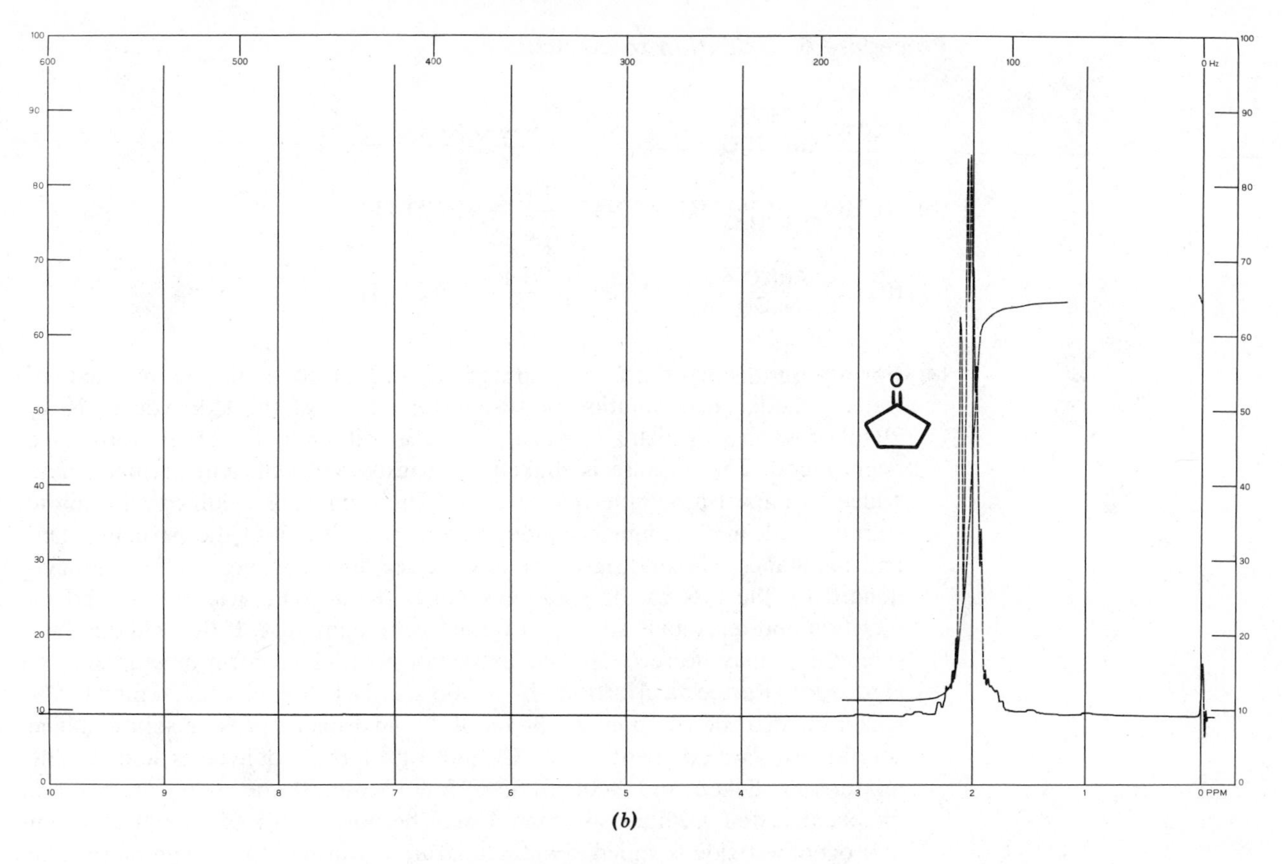

Figure 6.4. (*Continued*) (*b*) Nuclear magnetic resonance (proton) spectrum of cyclopentanone: 60 MHz, 600-Hz sweep width, $CDCl_3$ solvent.

flexibility necessary to achieve a six-membered ring arrangement can undergo the McLafferty rearrangement.

Although the peak due to CHO^+ (m/e 29) is usually small, such a peak might be very useful in supporting an aldehyde's identity.

A tool that is often very useful for characterizing carbonyl compounds is ultraviolet (uv) spectrometry. Information obtained from uv spectra is useful in relation to optical rotatory dispersion (ord), to the choice of wavelengths for carbonyl photochemistry, and to involved structural details of α,β-unsaturated carbonyl compounds (the Woodward-Fieser rules). Ultraviolet spectra are especially useful when the carbonyl group is conjugated to a double bond or to an aromatic ring. Discussions of these subjects may be found in Chapter 8.

Derivatives of aldehydes are usually produced by converting the carbonyl group to another group, such as $>C{=}N{-}NHR$, or by oxidizing the CHO group to a carboxyl group. Ketones can likewise be converted to substituted hydrazones or oximes. Both aldehydes and ketones can be reduced, to primary and secondary alcohols, respectively. These alcohols could then be derivatized as described in Section 6.1.

Procedure 6. Oxidation to an Acid

$$\text{(a) } RCHO \xrightarrow[OH^-/H_2O]{KMnO_4} RCO_2^-Na^+ \xrightarrow{H^+} RCO_2H$$

$$\text{(b) } RCHO \xrightarrow[OH^-/H_2O]{H_2O_2} RCO_2^-Na^+ \xrightarrow{H^+} RCO_2H$$

$$\text{(c) } RCHO \xrightarrow[NaOH]{AgNO_3} [RCO_2^-Na^+] \xrightarrow{HNO_3} RCO_2H$$

(a) ***Permanganate Method.*** A saturated solution of potassium permaganate in water is added to a solution or suspension of 1 g of the aldehyde in 10 to 20 ml of water to which a few drops of 10% sodium hydroxide solution have been added. The mixture is shaken vigorously, and sufficient permanganate added to impart a definite purple color. The mixture is acidified with dilute sulfuric acid, and sodium bisulfite solution is added until the permanganate and manganese dioxide have been converted to manganese sulfate, as evidenced by the loss of the solution's purple color. The acid is removed by filtration and recrystallized from a water/acetone mixture. If the acid does not separate, it may be recovered by extraction with chloroform or ether.

(b) ***Hydrogen Peroxide Method.*** In a 500-ml flask are placed 20 ml of 5% sodium hydroxide solution and 30 ml of 3% hydrogen peroxide solution. The solution is warmed to 65 to 70°C, and 1 g of the aldehyde is added. The mixture is shaken and kept at 65°C for 15 min. If the aldehyde has not dissolved, a few milliters of ethanol may be added. An additional 10 ml of hydrogen peroxide is added, and the mixture is warmed for 10 min more. The solution is made acid to Congo red, and the acid that separates is removed by filtration.

If the acid is water soluble it is best to make the solution neutral to phenolphthalein and evaporate to dryness. The sodium salt of the acid is then converted to a suitable derivative by Procedure 33, 34, or 39.

(c) ***Silver(I) Oxide Method.***[10] Dissolve 3.4 g of silver nitrate in 10 ml of distilled water in a 50-ml beaker. Add 10% sodium hydroxide solution dropwise with vigorous stirring until no further precipitation of silver oxide occurs (ca. 4 ml). Then add 1 g of the aldehyde with vigorous stirring and an additional 5 ml of 10% sodium hydroxide solution. The reaction mixture usually warms up as the oxidation proceeds and the silver oxide is converted to metallic silver. After stirring for 10 to 15 min, the silver and unchanged silver oxide are removed by filtration. The filtrate is acidified with dilute nitric acid (20%) and the solution cooled in an ice bath. The precipitated solid acid is collected on a filter, dried, and its melting point taken. If necessary, it may be recrystallized from a small amount of hot water or a 1:1 water/isopropyl

---

[10] S. C. Thomason and D. G. Kubler, *J. Chem. Educ.*, **45,** 546 (1968). These authors describe use of silver(I) oxide and silver(II) oxide.

alcohol mixture. If the acid is so soluble in water that it does not precipitate at the acidification step, it may be recovered by removal of the water (evaporation) followed by extraction with chloroform or ether.

Procedure 7. Semicarbazones

$$-\overset{|}{C}{=}O + H_2NNHCONH_2 \rightarrow -\overset{|}{C}{=}NNHCONH_2 + H_2O$$

(a) ***For Water-Soluble Compounds.*** One milliliter of the aldehyde or ketone, 1 g of semicarbazide hydrochloride, and 1.5 g of sodium acetate are dissolved in 10 ml of water in a test tube. The mixture is shaken vigorously, and the test tube is placed in a beaker of boiling water and allowed to cool. It is then placed in a beaker of ice, and the sides of the tube are scratched with a glass rod. The crystals of the semicarbazone are removed by filtration and recrystallized from water or 25 to 50% ethanol.

(b) ***For Water-Insoluble Compounds.*** One milliliter of the aldehyde or ketone is dissolved in 10 ml of ethanol. Water is added until the solution is faintly turbid; the turbidity is removed with a few drops of ethanol. Then 1 g of semicarbazide hydrochloride and 1.5 g of sodium acetate are added, and from this point Procedure (a) is followed.

Procedure 8. p-Nitrophenylhydrazones

$$-\overset{|}{C}{=}O + ArNHNH_2 \rightarrow -\overset{|}{C}{=}NNHAr + H_2O$$

A mixture of 0.5 g of *p*-nitrophenylhydrazine, 0.5 g of the aldehyde or ketone, and 10 to 15 ml of ethanol is heated to boiling, and a drop of glacial acetic acid is added. The mixture is kept hot for a few minutes, and more ethanol is added if necessary to obtain a clear solution. The solution is cooled, and the *p*-nitrophenylhydrazone is collected on a filter. It may be recrystallized from a small amount of ethanol.

If the derivative does not separate from the solution on cooling, the mixture is heated to the boiling point, water is added until the solution is cloudy, and then a drop or two of ethanol is added to clarify it. The hydrazone that separates on cooling is recrystallized from a water/ethanol mixture.

Procedure 9. 2,4-Dinitrophenylhydrazones

$$-\overset{|}{C}{=}O + ArNHNH_2 \rightarrow -\overset{|}{C}{=}NNHAr + H_2O$$

A solution of 2,4-dinitrophenylhydrazine is prepared in the following fashion. To 0.4 g of 2,4-dinitrophenylhydrazine in a 25-ml Erlenmeyer flask is added 2 ml of concentrated sulfuric acid. Water (3 ml) is added dropwise, with swirling or stirring until solution is complete. To this warm solution is added 10 ml of 95% ethanol.

A solution of the carbonyl compound in ethanol is prepared by dissolving 0.5 g of the compound in 20 ml of 95% ethanol. The freshly prepared 2,4-dinitrophenylhydrazine solution is added, and the resulting mixture is allowed to stand at room temperature. Crystallization of the 2,4-dinitrophenylhydrazone usually occurs within 5 to 10 min. If no precipitate is formed the mixture is allowed to stand overnight.

Recrystallization can usually be effected in the following way. The 2,4-dinitrophenylhydrazone is heated on a steam cone with 30 ml of ethanol (95%). If solution occurs immediately, water is added slowly until the cloud point is reached or until a maximum of 5 ml of water has been added. If the dinitrophenylhydrazone does not dissolve, ethyl acetate is added slowly to the hot mixture until solution is attained. The hot solution is filtered through a fluted filter and allowed to stand at room temperature until crystallization is complete (about 12 hr).

The reaction in diethylene glycol dimethyl ether is carried out with a solution of 4 g of 2,4-dinitrophenylhydrazine dissolved in 120 ml of the ether by warming and allowing to stand at room temperature for several days. To 5 ml of this solution at room temperature is added 0.1 g of the carbonyl compound in 1 ml of 95% ethanol or in 1 ml of diethylene glycol dimethyl ether. Three drops of concentrated hydrochloric acid is then added. If there is not an immediate precipitate of the hydrazone, the solution should be diluted with water and allowed to stand.

Procedure 10. Dimedon Derivative and Octahydroxanthene Preparations

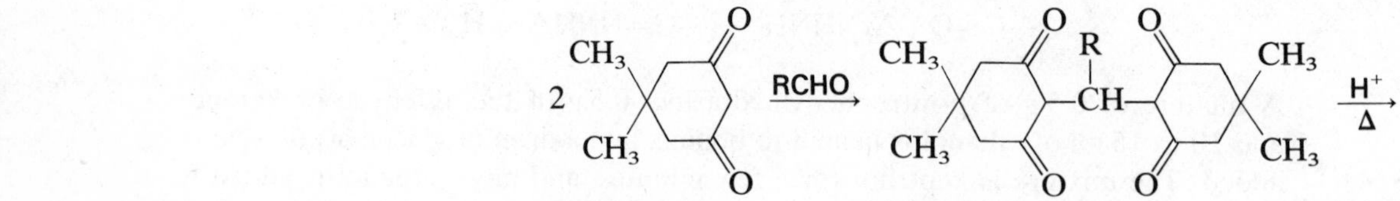

2 (methone) $\xrightarrow{RCHO}$ (methone derivative) $\xrightarrow[\Delta]{H^+}$ (octahydroxanthene)

methone (or "dimedon")[11]

"methone derivative" or "bis-methone"

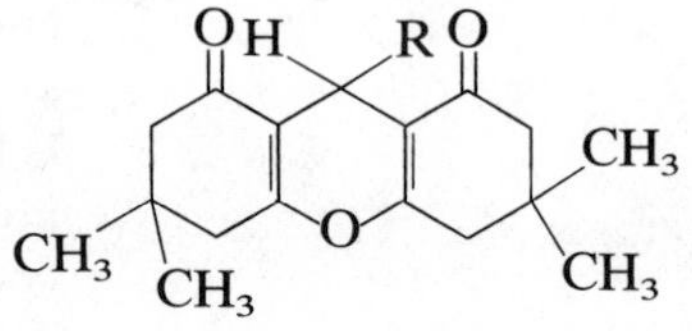

an octahydroxanthene

(a) ***Aldehyde Bismethone Condensation Products.*** To a solution of 0.1 g of the aldehyde in 4 ml of 50% ethanol is added 0.4 g of methone, if the aldehyde is aliphatic. If an aromatic aldehyde is a possibility, only 0.3 g of methone is used. One drop of piperidine is added, and the mixture is boiled gently for 5 min. If the hot solution is clear at the end of this time, water is

[11] "Dimedon"=***dime***thylcyclohexane***dione***.

added dropwise to the cloud point. The mixture is then chilled, and the derivative, after being separated by filtration, is washed with 2 ml of cold 50% ethanol. The derivative is recrystallized from mixtures of methanol and water.

(b) ***Cyclization to Substituted Octahydroxanthenes.*** If it is desirable to prepare an additional derivative, the above methone condensation product may be converted to a substituted octahydroxanthene by boiling a solution of 0.1 g of the methone derivative in 80% ethanol (3 to 6 ml) to which 1 drop of concentrated hydrochloric acid has been added. The cyclization is complete in 5 min, after which time water is added to the cloud point. The mixture is cooled and the substituted xanthene removed by filtration. This product is usually pure, but it may be recrystallized from aqueous methanol.

Ketones. The carbonyl reagents described under aldehydes (pp. 178–181) may be used to obtain solid derivatives of ketones. Low-molecular-weight ketones may be characterized by means of 2,4-dinitrophenylhydrazones (Procedure 9), *p*-nitrophenylhydrazones (Procedure 8), or semicarbazones (Procedure 7). For the higher-molecular-weight ketones, hydrazones, phenylhydrazones (Experiment 8, p. 165), and oximes (Procedure 11) are suitable.

Methyl ketones may be oxidized to acids selectively by means of sodium hypochlorite.[12] This reagent is particularly useful for unsaturated methyl ketones, because ordinary oxidizing agents attack the double bond.

$$RCH{=}CHCOCH_3 + 3NaOCl \rightarrow RCH{=}CHCOONa + CHCl_3 + 2NaOH$$

Hydantoins, formed by warming the carbonyl compound with a mixture of potassium (or sodium) cyanide and ammonium carbonate, have been recommended as derivatives, especially for keto ethers.[13]

$$RCOR \xrightarrow[(NH_4)_2CO_3]{KCN} \text{5,5-dialkylhydantoin (R}_2\text{C ring with C=O, N–H, C=O, N–H)}$$

Procedure 11. Oximes

$$\text{>C=O} \xrightarrow[\text{base}]{NH_2OH{\cdot}HCl} \text{>C=N–OH} + H_2O + \text{a salt}$$

(a) ***Pyridine Method.*** A mixture of 1 g of the aldehyde or ketone, 1 g of hydroxylamine hydrochloride, 5 ml of pyridine, and 5 ml of absolute ethanol

---

[12] A. M. Van Arendonk and M. E. Cupery, *J. Amer. Chem. Soc.*, **53**, 3184 (1931); C. D. Hurd and C. L. Thomas, *ibid.*, **55**, 1646 (1933).

[13] H. R. Henze and R. J. Speer, *J. Amer. Chem. Soc.*, **64**, 522 (1942).

is heated under reflux for 2 hr on a steam bath. The solvents are removed by evaporation in a current of air under a hood. The residue is triturated thoroughly with 5 ml of cold water, and the mixture is filtered. The oxime is recrystallized from methanol, ethanol, or an ethanol-water mixture.

(b) About 0.5 g of hydroxylamine hydrochloride is dissolved in 3 ml of water; 2 ml of 10% sodium hydroxide solution and 0.2 g of the aldehyde or ketone are then added. If the carbonyl compound is water-insoluble, just sufficient ethanol is added to the mixture to give a clear solution. The mixture is warmed on the steam bath for 10 min and cooled in an ice bath. In order to hasten crystallization, the sides of the flask are scratched with a glass rod. Occasionally the addition of a few milliliters of distilled water will assist in causing the oxime to separate. The product may be recrystallized from water or dilute ethanol.

Certain cyclic ketones, such as camphor, require an excess of alkali and a longer time of heating. If a ketone fails to yield an oxime by either Procedure (a) or (b), 1 g of it should be treated with 1 g of hydroxylamine hydrochloride, 4 g of potassium hydroxide, and 20 ml of 95% ethanol. The mixture is heated under reflux for 2 hr and poured into 150 ml of water. The suspension is stirred and allowed to stand to permit unchanged ketone to separate. The solution is filtered, acidified with hydrochloric acid, and allowed to stand to permit the oxime to crystallize. The product is recrystallized from ethanol or an ethanol/water mixture.

Diaryl ketones can be reduced with sodium borohydride to diarylcarbinols; these conditions are much milder than lithium aluminum hydride or sodium/alcohol conditions.

Procedure 11a. Borohydride Reduction of Diaryl Ketones to Diarylcarbinols

$$4Ar_2CO \xrightarrow[NaOH/(CH_3)_2CHOH]{NaBH_4} [(Ar_2CHO)_4B^-\overset{+}{Na}] \xrightarrow[H_2O]{NaOH} 4Ar_2CHOH$$

To a slurry of 0.4 g of sodium borohydride in 15 ml of isopropyl alcohol in a 50-ml flask is added 3 g of the diaryl ketone. The components are mixed thoroughly, a few boiling chips are added, and after inserting a reflux condenser the mixture is allowed to reflux for 30 min on a steam bath. The mixture is cooled to room temperature and 20 ml of 10% sodium hydroxide solution is added with vigorous stirring. The mixture is cooled to room temperature and extracted with three 10-ml portions of methylene chloride. The methylene chloride extract is dried with 1 g of anhydrous magnesium sulfate and the methylene chloride is removed by distillation (water bath). The residue is recrystallized from a mixture of methanol or isopropyl alcohol with water. Crystals are collected on a filter, dried, and the melting point is determined.

Sodium borohydride in isopropyl alcohol solution is a selective reducing agent for the carbonyl group in aldehydes and ketones. This method is valuable for those ketones that yield a solid carbinol. Aldehydes are more readily characterized by their 2,4-dinitrophenylhydrazones.

This procedure does not reduce olefinic groups (C=C), even when conjugated, hence is useful for special cases:

$$C_6H_5CH{=}CH{-}CHO \xrightarrow[NaOH+H_2O]{NaBH_4} C_6H_5CH{=}CH{-}CH_2OH$$

$$C_6H_5CH{=}CH{-}COR \xrightarrow[NaOH+H_2O]{NaBH_4} C_6H_5CH{=}CH\underset{\displaystyle OH}{\underset{|}{C}}HR$$

Under the above conditions, nitro groups, cyano groups, amides, and esters are not reduced. However, it must be remembered that ester and amide groups are hydrolyzed by the refluxing alcoholic sodium hydroxide solution. Then the alkaline solution must be *acidified* before extraction with methylene chloride. In many cases, the carbinol acid (or the corresponding lactone[14]) will precipitate and can be removed by filtration.

$$C_6H_5COC_6H_4COOCH_3 \xrightarrow[NaOH,\ H_2O]{NaBH_4} C_6H_5\underset{\displaystyle OH}{\underset{|}{C}}HC_6H_4COONa \xrightarrow{HCl} C_6H_5\underset{\displaystyle OH}{\underset{|}{C}}HC_6H_4CO_2H$$

Reference. H. C. Brown, *Boranes in Organic Chemistry* (Cornell University Press, Ithaca, N.Y., 1972), chaps. XII and XIII.

A number of procedures describing the reactions of simple carbonyl compounds are given in Pasto and Johnson. When water-soluble derivatives of carbonyl compounds are desired, the following procedures can be applied (p. 393):

$$>C{=}O \xrightarrow[\text{Girard's reagent T}]{(CH_3)_3\overset{+}{N}CH_2CONHNH_2\ Cl^-} >C{=}NNHCO\overset{+}{N}(CH_3)_3\ Cl^-$$

$$>C{=}O \xrightarrow[\text{Girard's reagent P}]{C_5H_5\overset{+}{N}{-}CH_2CONHNH_2\ Cl^-} >C{=}NNHCOCH_2{-}\overset{+}{N}C_5H_5\ Cl^-$$

$$>C{=}O \xrightarrow[\text{(p-toluenesulfonate salt)}^{15}]{\text{1-methyl-3-(CONHNH}_2\text{)pyridinium}} >C{=}NNHCO{-}(\text{N-methylpyridinium-3-yl})$$

---

[14] (lactone structure: 3-phenylphthalide, C_6H_5-substituted isobenzofuranone with C=O and ring O)

[15] C. F. H. Allen and J. W. Gates, Jr., *J. Org. Chem.*, **6**, 596 (1941).

Most reactions of the type described above can be carried out in the reverse direction.

6.2.1 Polyfunctional Carbonyl Compounds[16]

The procedures described in the preceding as well as the following sections should be utilized for characterizing saccharides and other classes of carbohydrates. The periodic acid test (Experiment 5) usually gives positive results for substrates containing vicinal carbonyl groups (e.g., α-diketones) as well as vicinal diols and is very useful for characterizing carbohydrates.

6.2.2 Acetals and Ketals

Acetals are identified by reference to the alcohol and the aldehyde or ketone that they yield when hydrolyzed. The hydrolysis is accomplished by heating the acetals with dilute acids; under these conditions the following reaction occurs:

$$\text{R(R)H}\,\text{C(OR}')_2 + \text{H}_2\text{O} \rightleftharpoons 2\text{R}'\text{OH} + \text{R(R)H}\,\text{C}{=}\text{O}$$

Low-molecular-weight acetals are readily hydrolyzed by heating with 1 to 2% hydrochloric acid for 3 to 5 min under reflux. Higher-molecular-weight acetals require a longer time (30 to 60 min). The hydrolysis of acetals of water-insoluble alcohols and aldehydes or ketones may be facilitated by using a 50% solution of dioxane as the solvent.

The exact procedure to be followed is dependent on the hydrolysis products. If the alcohol and carbonyl compound are both water-soluble, the hydrolysis mixture is made neutral to litmus and divided in half. One portion is used for the characterization of the aldehyde or ketone by the preparation of a suitable derivative. Semicarbazones (Procedure 7), phenylhydrazones (Experiment 8), and *p*-nitrophenylhydrazones (Procedure 8) will be found useful. The second portion of the hydrolysis mixture is treated with benzoyl chloride according to the Schotten-Baumann method (Procedure 18, p. 233). This converts the alcohol to the corresponding alkyl benzoate, which is separated and hydrolyzed by Method *A* of Procedure 25, p. 293. The alcohol is then characterized by one of the methods described in Section 6.1.

If one of the hydrolysis products is insoluble in the reaction mixture, it is separated and characterized first. The product in the aqueous solution is then treated by either of the methods mentioned above.

If both the alcohol and carbonyl compound are insoluble, they are separated from the aqueous layer. In the event that one of the products is an aldehyde or an

[16] Carbohydrates represent a special class of the title compounds, and will be dealt with elsewhere (Section 6.11).

aliphatic methyl ketone, it may be separated from the alcohol by means of its bisulfite compound (Experiment 9, p. 165). Sometimes fractional distillation may be used to separate the alcohol and carbonyl compound.

Acetals and ketals may contain dangerous peroxides just as ethers often do; the discussion of these peroxides in Section 6.14 should be consulted.

It has been recommended that 2,4-dinitrophenylhydrazones be formed directly from the acetal (or ketal) by treatment with the acidic hydrazine reagent (Pasto and Johnson, p. 378 and Procedure 9 above):

$$\mathrm{R{-}\underset{\displaystyle OR''}{\overset{\displaystyle OR''}{C}}{-}R'} \xrightarrow[\mathrm{H^+/C_2H_5OH}]{\mathrm{ArNHNH_2}} \left[\mathrm{R{-}\overset{\displaystyle O}{\overset{\|}{C}}{-}R'}\right] \longrightarrow \mathrm{\begin{matrix} R \\ \quad\diagdown \\ \quad\; C{=}NNHAr \\ \quad\diagup \\ R' \end{matrix}}$$

Figure 6.5 shows the ir spectrum of a typical acetal. The most important bands are those strong ones associated with various coupled C—O stretching modes:

cm^{-1}	μm
1190–1160	8.40–8.62
1195–1125	8.37–8.89
1098–1063	9.11–9.41
1055–1035	9.48–9.66
1116–1103	8.96–9.02 (acetals only)

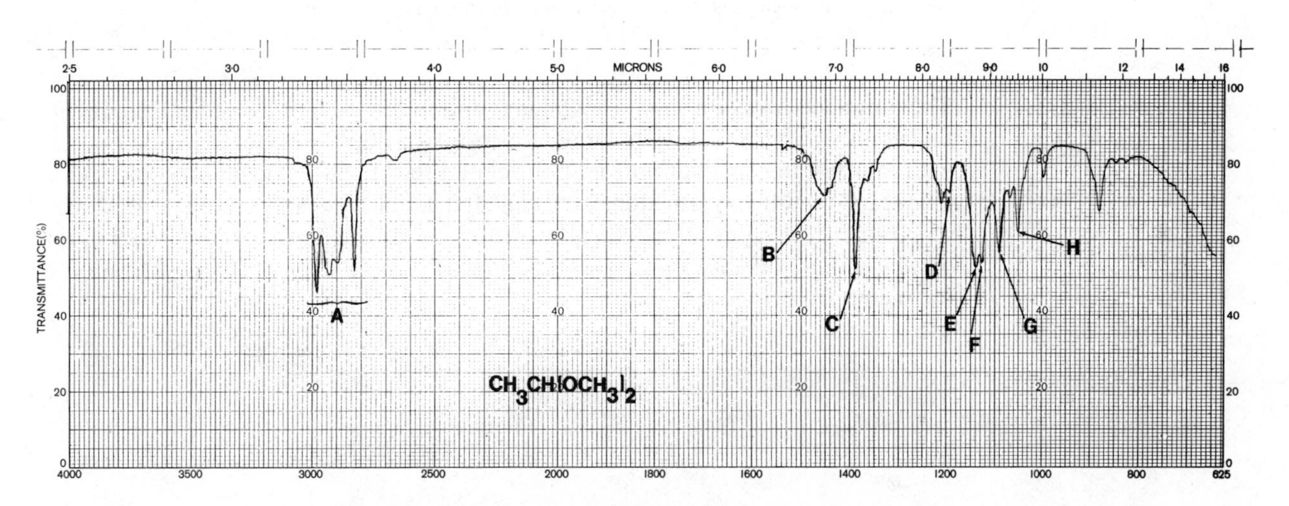

Figure 6.5. Infrared spectrum of 1,1-dimethoxyethane (dimethyl acetal of acetaldehyde): neat. C—H stretch: **A**, 3000–2800 cm^{-1} (3.33–3.57 μm); $\nu_{\text{C—H}}$ (aliphatic). C—H bend: **B**, 1450 cm^{-1} (6.90 μm); δ_{asym} CH_3O; **C**, 1375 cm^{-1} (7.27 μm); δ_{sym} CH_3C. C—O—C—O—C stretch: **D**, 1185 cm^{-1} (8.44 μm); **E**, 1140 cm^{-1} (8.77 μm); **F**, 1128 cm^{-1} (8.87 μm) (acetals *only*); **G**, 1092 cm^{-1} (9.16 μm); **H**, 1050 cm^{-1} (9.52 μm).

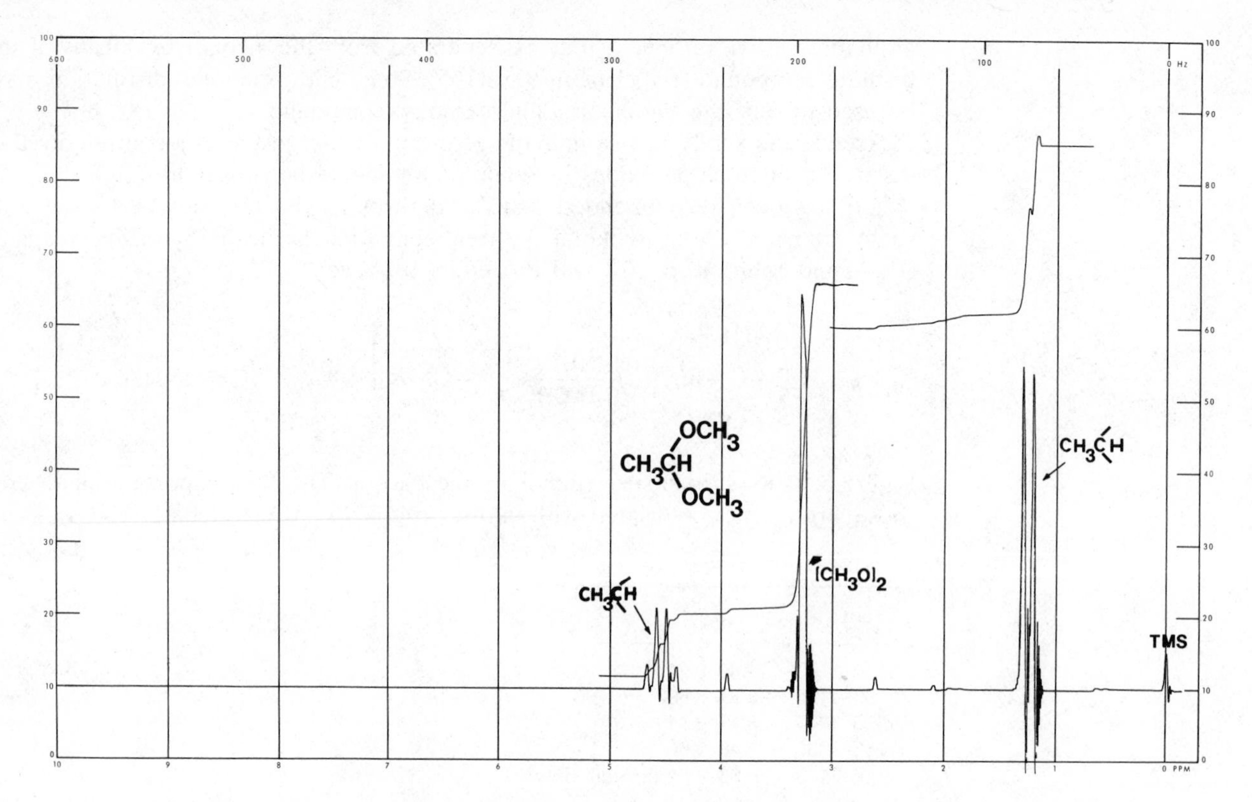

Figure 6.6. Nuclear magnetic resonance (proton) spectrum of 1,1-dimethoxyethane (the dimethyl acetal of acetaldehyde): 60 MHz, 600-Hz-sweep width, $CDCl_3$ solvent.

The nmr spectrum of a simple acetal is found in Figure 6.6; as we shall frequently find for this and a variety of other functional groups, we are concerned largely with interpreting spectral information related to the carbons and protons in close proximity to the oxygen atoms. Acetal and ketal cleavages, as measured mass spectrometrically, are dominated by patterns that are facilitated by neighboring oxygen atoms; priority rules can be deduced for both classes by consideration of fragments from a general acetal:

Fragment from $RCH(OR')_2$	Mass	Comment
$R\overset{+}{C}(OR')_2$	$44+R+2R'$	Larger R lost more readily if ketal
$[HC(OR')_2]^+$	$45+2R'$	Less important than process just above

The chemistry described here for acetals and ketals can often be applied to carbohydrates, because the latter often contain acetal- and/or ketal-like units:

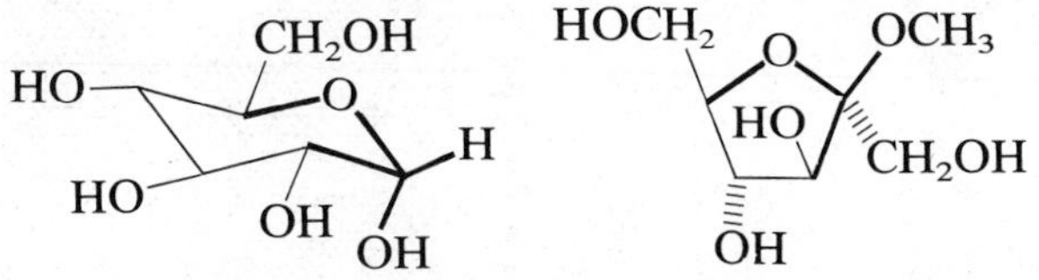

α-D-(+)-glucose

*hemi*acetal (C(OR)(OR)) unit highlighted

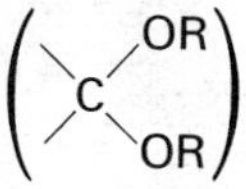

methyl β-D-fructofuranoside
ketal unit highlighted

These compounds can be characterized by alcohol and polyhydroxyl tests described above; in addition, carbonyl tests (Section 6.2) can be used because, in solution, the cyclic acetal and cyclic ketal forms of these molecules can equilibrate with acyclic forms that possess ketonic or aldehydic carbonyl groups. Treatment with acid will facilitate this equilibration.

6.3 ALKANES (PARAFFINS)

Alkanes are not usually characterized chemically, because they are quite inert to simple reaction conditions; one exception is the reaction with sulfuryl chloride mentioned below. Chemists usually rely heavily on physical (see p. 37) and spectral (see Figs. 6.7 and 6.8) characterization. In addition, one should be prepared to prove that the compound is *not* a similar, more reactive class (e.g., alkenes might have to be ruled out).

Derivatization procedures for saturated hydrocarbons are difficult to find because of the inertness of these compound. A few smooth chemical reactions do occur; highly strained, cyclic alkanes can be hydrogenated,[17]

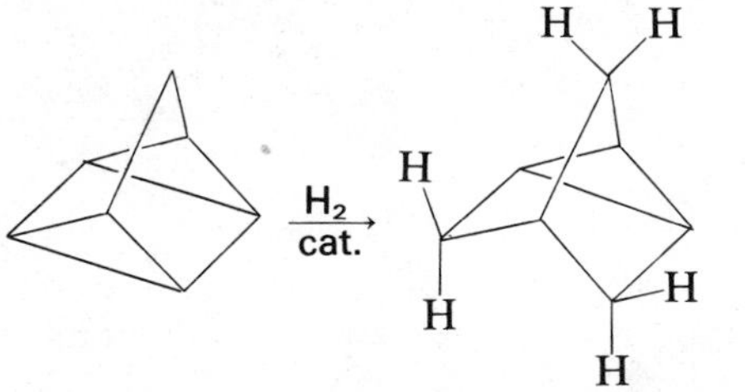

$\xrightarrow[\text{cat.}]{H_2}$

and alkyl chlorides can be prepared by treatment with sulfuryl chloride,[18]

$$RH \xrightarrow{SO_2Cl_2} RCl$$

[17] H. K. Hall, Jr., C. D. Smith, and J. H. Baldt, *J. Amer. Chem. Soc.*, **95**, 3197 (1973).
[18] H. Markgraf, *J. Chem. Educ.*, **46**, 610 (1969); J. M. Adduci, J. H. Dayton and D. C. Eaton, *J. Chem. Educ.*, **48**, 313 (1971).

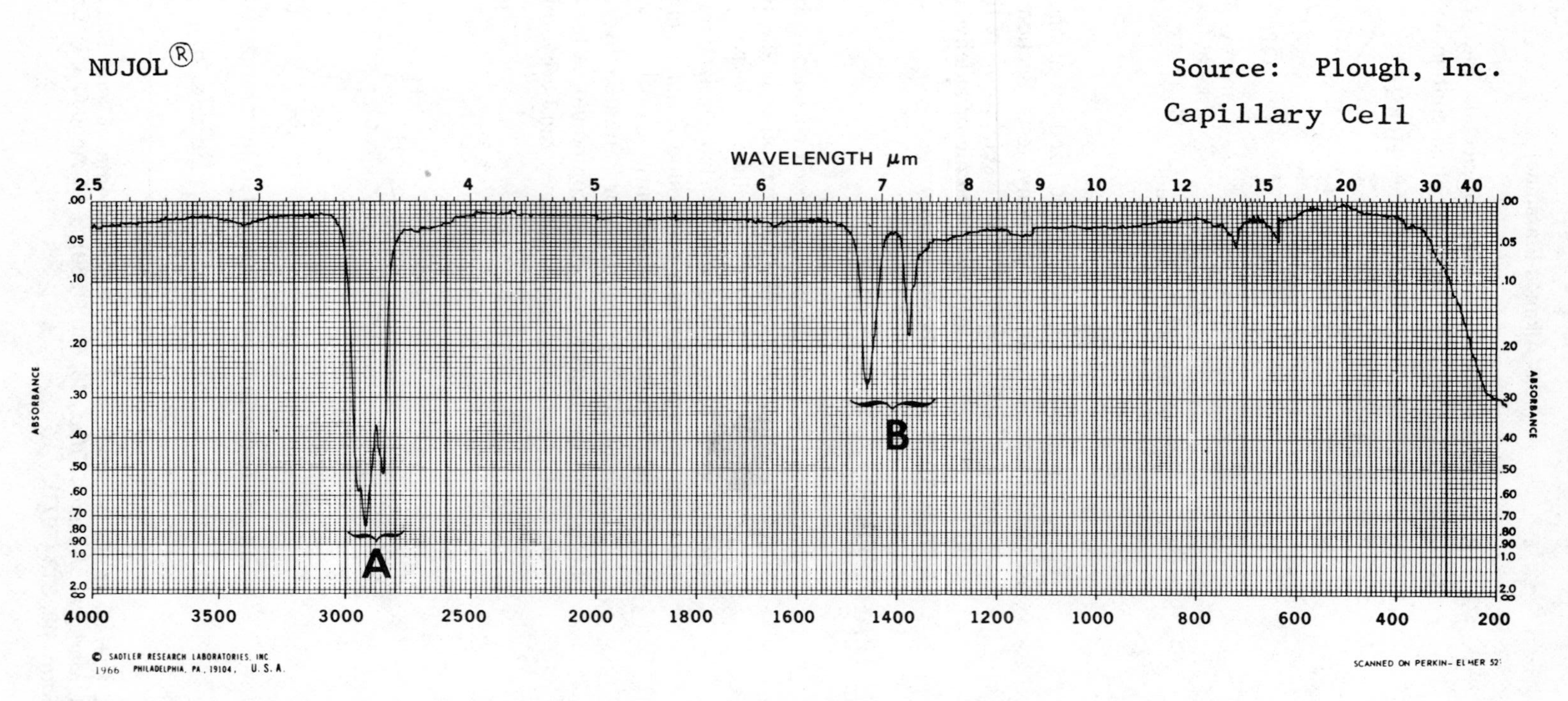

Figure 6.7. Infrared spectrum of Nujol: neat. Nujol is composed of saturated hydrocarbons. C—H stretch: **A**, 2950, 2920, 2850 cm^{-1} (3.39, 3.42, 3.51 μm), $\nu_{C—H}$ (alkane). C—H bend: **B**, 1460, 1375 cm^{-1} (6.85, 7.28) μm), $\delta_{C—H}$ (alkane).

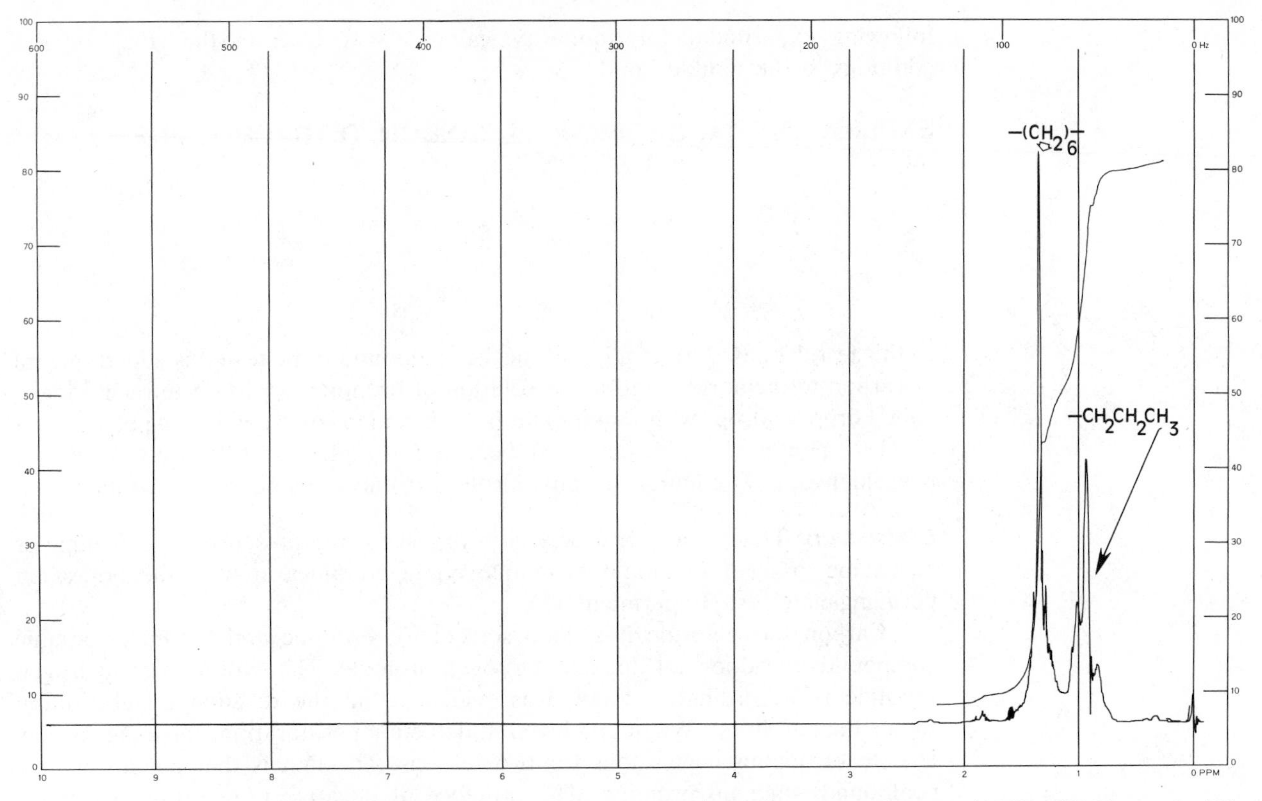

Figure 6.8. Nuclear magnetic resonance (proton) spectrum of octane: 60 MHz, 600-Hz sweep width, $CDCl_3$ solvent.

These procedures are, however, of severely limited use here; the former requires sophisticated conditions and characterization of a complex product. The latter reaction results in substantial yields of more than one product (e.g., polysubstituted products can be obtained); such reactions are usually avoided for simple derivatizations. Tables listing physical properties (b.p., m.p., $[\alpha]$, n_D, etc.) and spectra (ir and nmr) are usually much more useful for characterizing alkanes.

Mass spectral cleavage of alkanes is complex, and the student is referred to specialized tests (Chapter 10). Definite patterns are, however, easily discernible, and cleavage is facilitated by branching; thus information diagnostic of structure can be obtained.

6.4 ALKENES (OLEFINS)

The C=C unit of alkenes is usually easily detected by ir (C=C and =C—H stretch, see Chapter 5) and nmr (chemical shift of olefinic protons, Chapter 5) spectral methods. It can also be detected very easily by chemical tests. The following experiments are quite typical of C═C tests in that they involve additions to the double bond.

EXPERIMENT 14. BROMINE IN CARBON TETRACHLORIDE SOLUTION

$$\rangle C{=}C\langle \xrightarrow{Br_2} \rangle C(Br){-}C(Br)\langle$$

In this test 0.1 g (0.2 ml of a liquid) of the compound to be tested is added to 2 ml of carbon tetrachloride, and a 5% solution of bromine in carbon tetrachloride is added drop by drop (with shaking) until the bromine color persists. Apply the test to (1) 2-pentene; (2) hexane; (3) benzene; (4) phenol; (5) formic acid; (6) benzaldehyde; (7) ethanol; (8) allyl alcohol; (9) acetophenone; (10) aniline.

Discussion. This reagent is widely used to test for the presence of an olefinic or acetylenic linkage. It should be employed in conjunction with the potassium permanganate test (Experiment 15).

Carbon tetrachloride is a good solvent for bromine and for many organic compounds but does not dissolve hydrogen bromide. The evolution of hydrogen bromide is, accordingly, accepted as evidence that the reaction is substitution rather than addition. When employed in detecting unsaturation, this reagent may lead to erroneous conclusions for two reasons. The first is that not all olefinic compounds take up bromine. The presence of negative groups on the carbon atoms of an ethylenic bond causes the addition to be slow and in extreme cases prevents the reaction. The following illustrates this point:

$$C_6H_5CH{=}CH_2 + Br_2 \xrightarrow{\text{rapid}} C_6H_5CHBrCH_2Br$$

$$C_6H_5CH{=}CHCO_2H + Br_2 \xrightarrow{\text{slow}} C_6H_5CHBrCHBrCO_2H$$

A positive test for unsaturation is one in which the bromine color is discharged *without the evolution* of hydrogen bromide.

When bromination is employed as a qualitative test, carbon tetrachloride is the preferred solvent; it is, however, an unfortunate choice for a study of the mechanism of the reaction, because in nonpolar solvents the reaction is complicated. There is a heterogeneous reaction that occurs on the walls of the vessel. Furthermore, the reaction is powerfully catalyzed by traces of water or acids and may be inhibited by oxygen. It could be anticipated that such effects would make it difficult to obtain reproducible results and so render uncertain the interpretation of any negative qualitative test.

In a polar solvent such as water, methanol, or acetic acid, the reaction has been shown to be a two-stage process. The first is the reaction of olefin with bromine to give a bromonium ion. The second step is the reaction of the bromonium ion with bromide ion to give the product. Since the bromonium ion opens with inversion of configuration, the overall steric course of the reaction, leads to net addition of the two bromine atoms to opposite sides of the double bond.[19]

$$R_2C{=}CR_2 + Br{-}Br \rightarrow R_2C\overset{\overset{+}{Br}}{\frown}CR_2 \;\; Br^- \rightarrow R_2C(Br){-}CR_2(Br)$$

Substituents that help to stabilize the positively charged bromonium ion increase the rate; those that destabilize the positive ion retard the reaction. For example, substitution of an alkyl group on either of the doubly bound carbon atoms in olefins of the type $RCH{=}CH_2$ increases the rate of addition in acetic acid solution by a factor of 30 to 40.

Discharge of the bromine color *accompanied by the evolution of hydrogen bromide,* indicating that substitution has occurred, is characteristic of many compounds. In this category are enols, many phenols, and enolizable compounds. Methyl ketones appear to be more reactive than other ketones but, like other carbonyl compounds, may exhibit an induction period because the hydrobromic acid liberated acts as a catalyst for the enolization step of the bromination. Simple esters do not give this test. Ethyl acetoacetate decolorizes the solution immediately, whereas ethyl malonate may require as much as a minute. A number of active methylene compounds that do not discharge the color at room temperature readily give a test at 70°C. Among these substances are propionaldehyde and cyclopentanone. Aryl ethers behave similarly. Benzyl cyanide, even at 70°C, may require several minutes.

Aromatic amines are exceptional in that the first mole of hydrogen bromide formed is not evolved but reacts to convert the amine to its salt. For this reason the reaction is often mistaken for simple addition.

$$C_6H_5NH_2 + Br_2 \rightarrow p\text{-}BrC_6H_4\overset{+}{N}H_3Br^-$$

Benzylamine represents an unusual type that reacts readily with bromine. Substitution of the hydrogen atoms on the nitrogen atom appears to take place, followed by decomposition to benzonitrile:

$$3C_6H_5CH_2NH_2 + 2Br_2 \rightarrow [C_6H_5CH_2NBr_2] + 2C_6H_5CH_2\overset{+}{N}H_3Br^-$$
$$[C_6H_5CH_2NBr_2] \rightarrow C_6H_5CN + 2HBr$$

Certain tertiary amines such as pyridine form perbromides upon treatment

[19] See P. B. D. De La Mare and R. Bolton, *Electrophilic Additions to Unsaturated Systems* (Elsevier, New York, 1966).

with bromine:

$$C_5H_5N + Br_2 \rightarrow C_5H_5N^{+}\text{–}Br \quad Br^{-}$$

The bromine color is likewise discharged by aliphatic amines of all types.

EXPERIMENT 15. POTASSIUM PERMANGANATE SOLUTION[20]

$$2KMnO_4 + H_2O \rightarrow 2KOH + 2MnO_2 + 3(O)$$

$$>C{=}C< + (O) + H_2O \rightarrow >C(OH){-}C(OH)<$$

$$\xrightarrow{\text{further oxidation}} >C{=}O \quad O{=}C<$$

Method A. Baeyer Test—Aqueous Solutions

To 2 ml of water or ethanol add 0.1 g (or 0.2 ml) of the compound to be examined. Then add a 2% potassium permanganate solution drop by drop with shaking until the purple color of the permanganate persists. Apply this test to (1) 2-pentene; (2) toluene; (3) phenol; (4) benzaldehyde; (5) aniline; (6) formic acid; (7) cinnamic acid; (8) glucose. If the permanganate color is not changed in 0.5 to 1 min, allow the tubes to stand for 5 min with occasional vigorous shaking. Do not be deceived by a slight reaction, which may be due to the presence of impurities.

Although acetone rather than ethanol has sometimes been employed as a solvent for water-insoluble compounds, it has been found that certain carefully purified olefins show a negative test in acetone but a positive one in ethanol. Ethanol does not react with neutral, dilute potassium permanganate at room temperature within 5 min.

Potassium permanganate in aqueous acetic acid has been used to distinguish among simple primary, secondary, and tertiary alcohols. Under the conditions of the test, primary and secondary alcohols react but tertiary alcohols do not. In fact, *tert*-butyl alcohol can be a co-solvent for use in the Baeyer test. It is frequently used in place of ethanol or acetone above.

Method B. Phase Transfer Method Using Quaternary Ammonium Salts: "Purple Benzene" Reagent

Dissolve 10 g of sodium chloride and 0.05 g of potassium permanganate in 50 ml of distilled water. Place the solution in a 125-ml separatory funnel, add

---

[20] F. O. Ritter, *J. Chem. Educ.*, **30**, 395 (1953).

50 ml of benzene, agitate the mixture gently, and allow the layers to separate. Note that the benzene layer is colorless; only the lower aqueous layer is purple. Next add 0.1 g of tetrabutyl ammonium bromide to the mixture, agitate and allow the layers to separate. The benzene layer should become a deep purple due to the phase transfer of the permanganate anion along with the tetrabutyl ammonium cation from the aqueous layer into the benzene layer. The transfer is not complete; however, an equilibrium is established. Separate the layers and note the appearance of the purple benzene over 5 min. If the purple color changes to brown, the presence of impurities in the quaternary ammonium salt is indicated; the reagent should be discarded and the quaternary salt purified. Benzene is not oxidized by dilute permanganate at 20°C, since it is not olefinic but is aromatic.

To test a compound for unsaturation, place 5 ml of the above purple benzene solution in a test tube, add 1 drop of distilled water and add about 0.1 g of the unknown. Stopper the tube, shake, and then rock the tube back and forth holding it over a sheet of white paper to follow the change in color from purple to brown. The time required ranges from 30 sec. to 5–15 min. Carry out this test on:

(a) Cyclohexene
(b) Stilbene
(c) 1,1-Diphenylethylene
(d) Triphenylethylene
(e) Glyceryl trioleate (corn or cottonseed oil)
(f) Diphenylacetylene

Compare the times required for these reactions against those for Method A for the same compounds.

In addition to tetra-n-butyl ammonium bromide, tetrapentyl- or tetrahexyl ammonium chloride or bromide may be used and are available in sufficient purity. Other long chain quaternary ammonium halides have been used.

Discussion. Procedure A. A solution of potassium permanganate is decolorized by compounds having ethylenic or acetylenic linkages. This is known as Baeyer's test for unsaturation. In cold dilute aqueous solutions the chief product of the action of potassium permanganate on an olefin is the glycol. If the reaction mixture is heated, further oxidation takes place, leading ultimately to cleavage of the carbon chain:

$$\mathrm{RCH{=}CHR} \rightarrow \mathrm{R\underset{\displaystyle OH}{\underset{|}{C}}H{-}\underset{\displaystyle OH}{\underset{|}{C}}HR} \rightarrow \mathrm{R{-}\underset{\displaystyle OH}{\underset{|}{C}}{=}O} + \mathrm{O{=}\underset{\displaystyle OH}{\underset{|}{C}}{-}R}$$

Procedure B. This phase transfer technique, using either long chain quaternary ammonium salts or crown ethers, has become a valuable method for carrying out many types of reactions. For this qualitative Baeyer test, the use of the quaternary ammonium salts has given good results, whereas tests using samples of available crown ethers were erratic. During the past 10 years, the literature in this field has become quite extensive. Good discussions and many references are given by G. W. Gokel and W. P. Weber in *J. Chem. Ed.*, **55,** 350 (1978); and by C. M. Starks and C. Liotta in a monograph; "Phase Transfer Catalysis," Academic Press, New York, 1978.

Acetylenic linkages are usually cleaved by oxidation and yield acids:

$$\mathrm{R{-}C{\equiv}C{-}R'} + \mathrm{H_2O} + 3[\mathrm{O}] \rightarrow \mathrm{RCOOH} + \mathrm{R'COOH}$$

The speed with which unsaturated compounds decolorize permanganate depends on the solubility of the organic compound. If the compound is very insoluble it is necessary to powder the compound and shake the mixture vigorously for several minutes or to dissolve the substance in a solvent unaffected by permanganate. A few tetrasubstituted olefins such as

$$C_6H_5CBr{=}CBrC_6H_5 \quad \text{and} \quad (C_6H_5)_2C{=}C(C_6H_5)_2$$

fail to give positive tests with bromine in CCl_4 or the permanganate solutions described above.

Inspection of the following equation shows that, as the reaction proceeds, the solution becomes alkaline:

$$3\,{>}C{=}C{<} + 2KMnO_4 + 4H_2O \rightarrow 3\,{>}\underset{OH}{\underset{|}{C}}{-}\underset{OH}{\underset{|}{C}}{<} + 2MnO_2(s) + 2KOH$$

It is necessary, however, to avoid using a solution that is strongly alkaline, as this changes the nature of the test. In sodium carbonate solution, for example, even acetone gives a positive test. Frequently no actual precipitate of manganese dioxide is observed; the purple color gradually changes to a reddish brown.

However, the use of permanganate in neutral media is feasible. Thus, with zinc permanganate, the zinc hydroxide produced is only slightly soluble and the solution remains practically neutral. Also, it is possible to use potassium permanganate in the presence of magnesium sulfate to accomplish this objective. In this case, the hydroxyl ion is precipitated in the form of insoluble magnesium hydroxide.

The Baeyer test, though superior to the bromine test for unsaturated compounds, offers certain complications in its turn. All easily oxidizable substances give this test. Carbonyl compounds that decolorize bromine solutions generally give a negative Baeyer test. Acetone is a good example; although it decolorizes bromine solutions rapidly, it can be used as a solvent in the Baeyer test. Aldehydes give a positive Baeyer test; however, many of them, such as benzaldehyde and formaldehyde, do not decolorize bromine solutions. Formic acid and its esters, which have the O=CH— group, also reduce permanganate.

Alcohols form another important class of compounds that decolorize permanganate solutions but not bromine solutions. Pure alcohols do not give the test readily; however, they often contain impurities that are easily oxidized. Other types of compounds also are likely to contain slight amounts of impurities that may decolorize permanganate solutions. For this reason, the decolorization of only a drop or two of the permanganate solution cannot always be accepted as a positive test.

Phenols and arylamines also reduce permanganate solution and undergo oxidation to quinones; these may be further oxidized with an excess of the reagent to yield a series of oxidation products, among which are maleic acid, oxalic acid, and carbon dioxide.

$$\text{C}_6\text{H}_5\text{OH} \xrightarrow{(O)} \text{O=C}_6\text{H}_4\text{=O} \xrightarrow{(O)} \begin{matrix}\text{CHCO}_2\text{H}\\ \|\\ \text{CHCO}_2\text{H}\end{matrix} \xrightarrow{(O)} \begin{matrix}\text{CO}_2\text{H}\\ |\\ \text{CO}_2\text{H}\end{matrix} \xrightarrow{(O)} \text{CO}_2$$

Organic sulfur compounds in which the sulfur is present in the reduced state also reduce permanganate and undergo oxidation. Mercaptans are oxidized to disulfides, and thio ethers and thio ketals to sulfones.

$$2\text{RSH} + (\text{O}) \longrightarrow \text{RSSR} + \text{H}_2\text{O}$$

$$\text{RSR} \xrightarrow{(O)} \text{R}_2\text{SO} \xrightarrow{(O)} \text{RSO}_2\text{R}$$

$$\text{R}_2\text{C(SR)}_2 \xrightarrow{(O)} \text{R}_2\text{C(SO}_2\text{R)}_2$$

This oxidation of thio compounds takes place more readily in an acid medium. Hence, if the elementary analysis shows the presence of sulfur, it is desirable to dissolve 0.1 g of the compound in 2 ml of glacial acetic acid and add the dilute potassium permanganate solution drop by drop. If the purple color is discharged, the presence of an oxidizable sulfur-containing functional group is a possibility. When the sulfur is present in the form of sulfones, alkyl sulfates, or unsubstituted sulfonic acids, the permanganate solution is not reduced. However, certain substituted sulfonates, such as aldehyde and ketone bisulfite compounds, and phenolic sulfonic acids do reduce permanganate.

The fact that glycols formed from olefins that exist as geometrical isomers are the products of addition of the two hydroxyl functions from the same side of the double bond (called *cis* or *syn* addition) has suggested a cyclic manganese ester intermediate, analogous to that described above for oxidations by periodic acid (Experiment 5).

Questions

1. What functional groups respond to both the bromine and the permanganate tests?
2. Which of these tests is better for detecting the presence of multiple bonds?
3. In what instances is it helpful to use both reagents?

A number of tests for olefins have been described by Pasto and Johnson; the first of these involves formation of complexes with the double bonds:

$$\text{>C=C<} + \text{I}_2 \longrightarrow \text{>C=C<} \rightarrow \text{I}_2$$

liquid

colored charge-transfer complex (Pasto and Johnson, p. 328)

(similar color from ethers and arenes treated with iodine)

$$\text{>C=C<} + \text{C(NO}_2)_4 \xrightarrow{\text{CHCl}_3} \text{>C=C<} \rightarrow \text{C(NO}_2)_4$$

(Pasto and Johnson, p. 332)

Ozonolysis can be used for degradation studies:

$$R'CH{=}CR_2 \xrightarrow{O_3} R'CH\underset{O}{\overset{O-O}{\quad}}CR_2 \quad \begin{cases} \xrightarrow[H_2O]{Zn} R'CHO + R_2CO \\ \xrightarrow[H^+]{H_2O_2} R'CO_2H + R_2CO \end{cases}$$

Alkenes can also be oxidatively degraded by basic permanganate (Procedure 6, described above for aldehydes) or by permanganate oxidation aided by periodate ion (Pasto and Johnson, p. 333).

$$MnO_4^- + \;\rangle C{=}C\langle\; \xrightarrow[OH^-]{H_2O} \left[-\underset{OH}{\overset{|}{C}}-\underset{OH}{\overset{|}{C}}- \right] \longrightarrow RCO_2H + R_2CO + MnO_3^- \quad (MnO_3^- \xrightarrow{IO_4^-} MnO_4^-)$$

Additions to alkenes can be carried out via ionic pathways using reagents other than bromine; arenesulfenyl halides add to give products that are normally solid and are usually easily characterized:

$$C_6H_5CH{=}CH_2 + 2,4\text{-}(O_2N)_2C_6H_3SCl \longrightarrow C_6H_5\underset{Cl}{\underset{|}{C}H}-CH_2S-C_6H_3(NO_2)_2$$

2,4-dinitrobenzenesulfenyl chloride (Pasto and Johnson, p. 335).

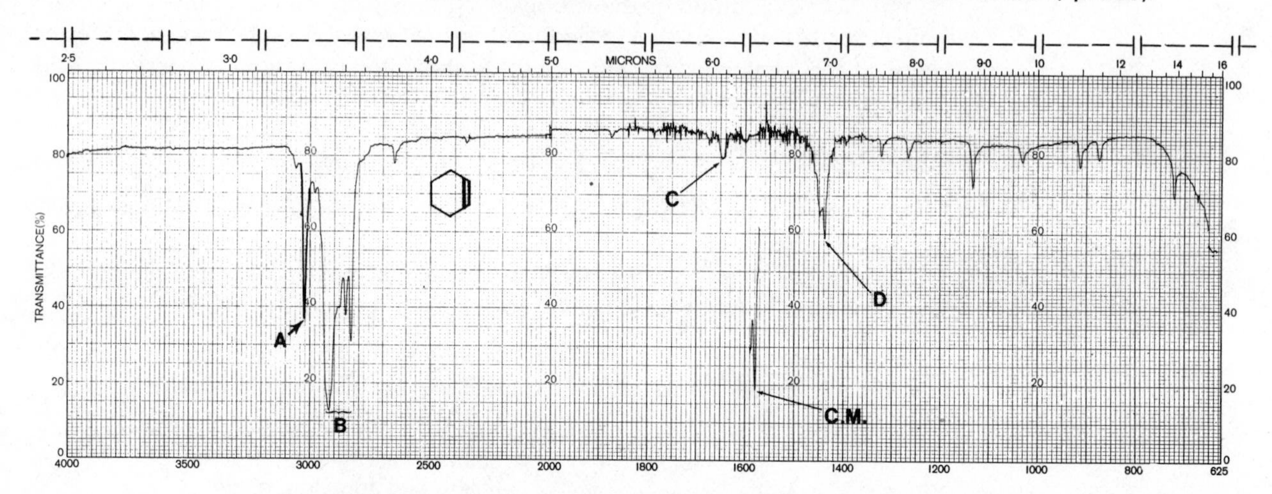

Figure 6.9. Infrared spectrum of cyclohexene: neat. C—H stretch: **A**, 3020 cm^{-1} (3.31 μm), =C—H stretch; **B**, 2960–2820 cm^{-1} (3.38–3.55 μm), aliphatic C—H stretch. C=C stretch: **C**, 1647 cm^{-1} (6.07 μm), $\nu_{C=C}$ (unconjugated). C—H bend: **D**, 1441 cm^{-1} (6.94 μm), olefinic C—H bend (cis double bond). C.M. = calibration marker (polystyrene window); all bands below 2000 cm^{-1} (above 5.00 μm) have been corrected by +3 μm (before tabulation), because this C.M. band should occur at 1583 cm^{-1} (6.32 μm).

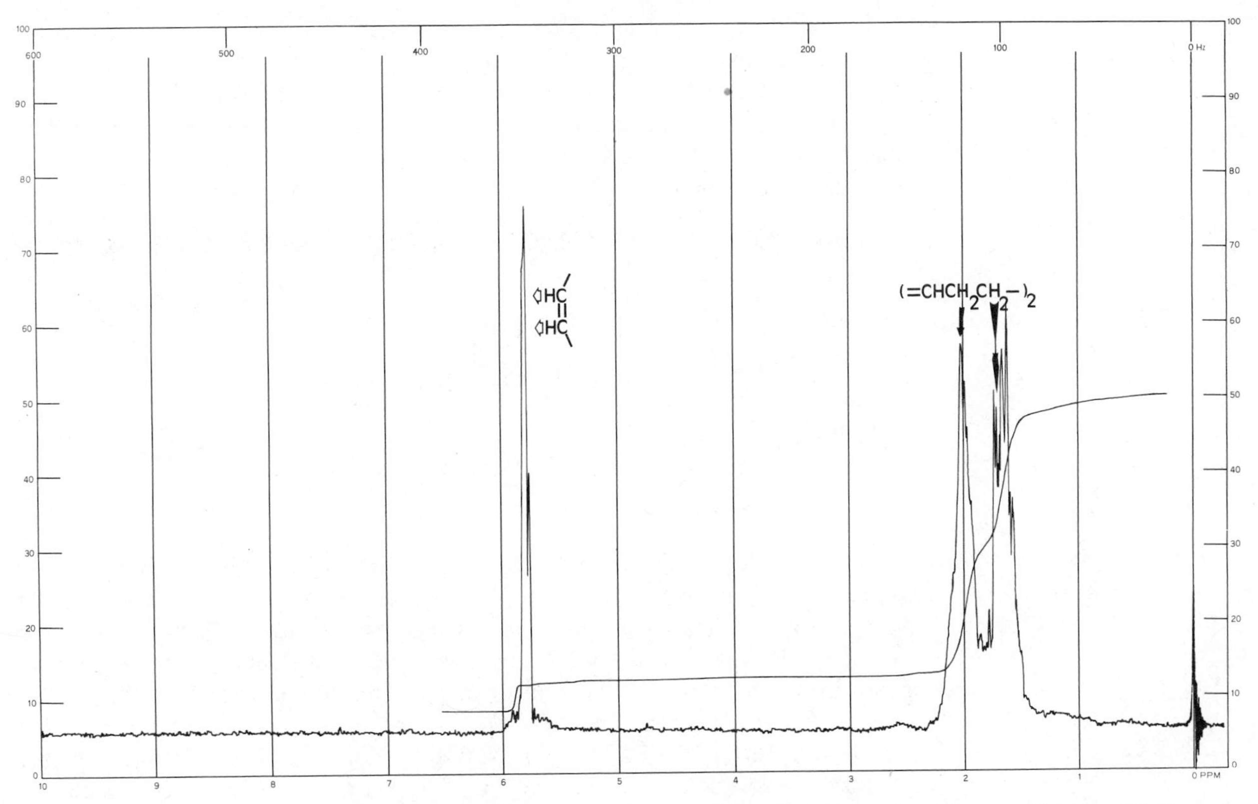

Figure 6.10. Nuclear magnetic resonance (proton) spectrum of cyclohexene: 60 MHz, 600-Hz sweep width, $CDCl_3$ solvent.

The ir spectrum of a simple olefin is shown in Fig. 6.9; a nmr spectrum of the same olefin is shown in Fig. 6.10. As has been discussed, the introduction of this double bond into the molecule usually results in easily discernible ir and nmr absorptions. Consultation of books (Chapter 10) on these subjects will quickly demonstrate that these absorptions often allow conclusions regarding (1) the number of substituents on the double bond, (2) geometric isomer character of the double bond, and (3) aromatic or electronegative substituents near the double bond.

The mass spectral patterns of alkenes are complex, and this topic is left largely to other books (see Chapter 10). It should be pointed out here, however, that cleavage takes place readily at allylic bonds:

$$\left[\rangle C{=}C{-}C{-}X \right]^{+} \longrightarrow \left[\rangle C{=}C{-}C\langle \right]^{+} + X\cdot$$

When choosing among proposed structures for conjugated (C═C—C═C) dienes, the Woodward-Fieser rules (Chapter 8) applied to uv spectrometry are often essential. Ultraviolet spectrometry is very useful in a variety of situations where π systems are conjugated.

A valuable way to prepare derivatives of alkenes involves their conversion, via hydroboration, to alcohols (Procedure 11b).

If the alcohol product is a solid, it may be recrystallized and will serve as a derivative. However, most common alcohols are liquids at 25°C and hence a third step is necessary to produce a solid, such as the conversion of the alcohol to either the 3,5-dinitrobenzoate or α-naphthylurethan (Procedures 2 and 13).

Procedure 11b. Oxidative Hydroboration for Conversion of Alkenes to Alcohols

$$\mathrm{>C{=}C<} \xrightarrow[\mathrm{THF, N_2, 25°C}]{\mathrm{NaBH_4/BF_3{\cdot}O(C_2H_5)_2}} \left[\mathrm{-\overset{H}{\underset{|}{\overset{|}{C}}}-\overset{B}{\underset{|}{\overset{|}{C}}}-} \right] \xrightarrow[\mathrm{NaOH,\ H_2O}]{\mathrm{H_2O_2}} \mathrm{-\overset{H}{\underset{|}{\overset{|}{C}}}-\overset{OH}{\underset{|}{\overset{|}{C}}}-} \quad (\rightarrow \text{alcohol derivatives})$$

Moisture and oxygen must be excluded during the first steps, hence the glassware must be cleaned and dried in an oven. A convenient assembly (shown in Fig. 6.10a, below) consists of a 100-ml three-necked flask fitted with a reflux condenser, a small dropping funnel, a thermometer, and a small tube reaching to the bottom for introducing a stream of *dry*[21] nitrogen gas. This stream serves to displace the oxygen and to mix the reactants. The flow of dry nitrogen is started, and 20 ml of dry tetrahydrofuran (THF), 0.25 g of $NaBH_4$, and 1.2 ml of the olefin are placed in the flask. A solution of 1.2 g of $BF_3{\cdot}(C_2H_5)_2O$[22] in 3 ml of dry THF is placed in the small dropping funnel. The temperature is kept at 25°C by means of the water bath, and the THF solution is added dropwise from the funnel at such a rate that the temperature is kept at ca. 25°C. This addition takes about 3 to 5 min. The mixture is allowed to stand for 1 hr with the slow stream of dry nitrogen gas passing through. At the end of this time, the excess hydride is decomposed by

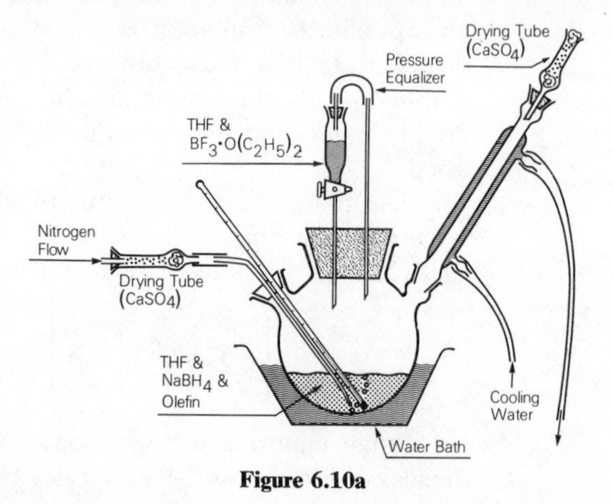

Figure 6.10a

[21] The gas should not be passed so rapidly that the liquids are driven off during the reaction.
[22] Vacuum distillation (aspirator) is the best way to purify BF_3 etherate.

dropwise addition of 10 ml of water. Then 3.2 ml of 3 M sodium hydroxide is added, followed by 3.2 ml of 30% hydrogen peroxide solution using the nitrogen gas stream to effect mixing. The mixture is then warmed to 30 to 40°C with warm water in the water bath. After 15 min, 4 g of sodium chloride is added to saturate the water phase and cause the THF solution of the alcohol to separate. The mixture is cooled to 20°C, the stream of N_2 gas is shut off, and 5 ml of absolute ether is added to facilitate formation of two phases. The THF-ether layer is separated, washed with 5 ml of saturated sodium chloride solution, dried with 1 g of anhydrous magnesium sulfate, and filtered through a small dry filter paper into a test tube. The solvents, ether and THF, are evaporated by directing a stream of dry nitrogen gas into the tube. The residual liquid alcohol can be converted either to the α-naphthylurethan (Procedure 1) or to the 3,5-dinitrobenzoate (Procedure 2). Sufficient amounts of the recrystallization solvents should be used in each case so as to recover only about one-half of the product.

Discussion. The hydroboration step can proceed to different degrees to produce mono-, di-, and trialkylboranes:

$$\mathrm{RCH{=}CH_2} \xrightarrow[\mathrm{BF_3\cdot(C_2H_5)_2O}]{\mathrm{NaBH_4;THF}} \left\{\begin{array}{l}\mathrm{(RCH_2CH_2{-})BH_2}\\ \mathrm{(RCH_2CH_2{-})_2BH}\\ \mathrm{(RCH_2CH_2)_3B}\end{array}\right.$$

The degree to which the successive B—H bonds add to the double bonds is largely controlled by the steric requirements of the R group; simple alkenes, with R groups of modest size, usually form trialkyl boranes ($x = 3$, $y = 0$ below).

Oxidative cleavage of the alkyl boranes by alkaline hydrogen peroxide to alcohols is one of the most extensively studied reactions of organoboranes:

$$\underset{x+y=3}{\mathrm{(RCH_2CH_2)_x BH_y}} \xrightarrow[\mathrm{H_2O}]{\mathrm{H_2O_2/NaOH}} x\mathrm{RCH_2CH_2OH} + \mathrm{Na_3(BO_3)_3}$$

Organoboranes can be converted to compounds other than alcohols (e.g., alkanes, amines, halides, etc.); the references by H. C. Brown cited below should be consulted.

The hydroboration of olefins is anti-Markovnikov and cis. For example, 1-methylcyclopentene yields almost pure *trans*-2-methylcyclopentanol by Procedure 11a:

CH$_3$ $\xrightarrow{\mathrm{BH_3}}$ B—, H, H, CH$_3$ $\xrightarrow[\mathrm{NaOH}]{\mathrm{H_2O_2}}$ OH, CH$_3$

If the olefin is either a symmetrical 1,2-disubstituted alkene or a cycloolefin, the secondary alcohol produced may be further oxidized to a ketone (Experiment 15), which is in turn characterized by its 2,4-dinitrophenylhydrazone (Procedure 9).

$\xrightarrow[\mathrm{(2)H_2O_2}]{\mathrm{(1)BH_3}}$ OH $\xrightarrow{\mathrm{[O]}}$ O $\longrightarrow$ NNHAr

Unsymmetrical internal olefins (RCH=CHR′) yield substantial proportions of two secondary alcohols and hence hydroboration is not useful. Such olefins are more easily characterized by other properties (b.p., sp gr, n_D, ir, nmr):

$$CH_3CH{=}CHCH_2CH_3 \xrightarrow[(2)H_2O_2]{(1)BH_3} CH_3CHOHCH_2CH_2CH_3 \text{ and } CH_3CH_2CHOHCH_2CH_3$$

In the hydroboration of terminal olefins, the boron atom adds to the terminal carbon atom to the extent of 90 to 99% to yield anti-Markovnikov adduct. Between 1 and 10% of the boron addition is to the internal carbon atom that would yield a secondary alcohol by oxidative cleavage; this secondary alcohol

$$\left[\begin{array}{c} CH_3 \\ | \\ RCH{-} \end{array}\right]_x BH_y \xrightarrow[H_2O_2]{H_2O;NaOH} \begin{array}{c} CH_3 \\ | \\ RCHOH \end{array}$$

would be present in the sample and would also be converted to the α-naphthylurethan or 3,5-dinitrobenzoate and would lower the m.p. of the desired derivative. However, by using larger amounts of the solvents for crystallization of the derivatives (to keep the smallest amount of the impurity in solution), it is possible to obtain a pure product for identification.

By means of this hydroboration procedure, conjugated diolefins yield 1,4-glycols, which can be converted to solid derivatives:[23]

$$CH_2{=}\underset{}{\overset{CH_3}{\overset{|}{C}}}{-}\overset{CH_3}{\overset{|}{C}}{=}CH_2 \longrightarrow \longrightarrow \underset{OH}{\underset{|}{CH_2}}{-}\overset{CH_3}{\overset{|}{CH}}{-}\overset{CH_3}{\overset{|}{CH}}{-}\underset{OH}{\underset{|}{CH_2}}$$

References. G. Zweifel and H. C. Brown, *Org. Reactions,* **13,** 1 (1963). This chapter has an excellent review of *Hydration of Olefins via Hydroboration.* See also H. C. Brown, *Hydroboration* (Benjamin, New York, 1962). See also H. C. Brown, *Boranes in Organic Chemistry* (Cornell University Press, Ithaca, N.Y., 1972); and H. C. Brown, *Organic Syntheses via Boranes* (Wiley, New York, 1975).

6.5 ALKYNES (ACETYLENES)

Since we are now dealing with multiple bonds (C≡C) that are somewhat similar to the double bond in alkenes, it should not be surprising that many of the tests described in the preceding section on alkenes are also useful here.

The following tests have been described (Pasto and Johnson) as being quite useful for alkynes; both mercury[24] and nitroaromatic compounds can be explosive, and thus extreme care in handling should be employed. Large reaction scale and high reflux temperatures should be avoided.

---

[23] These diols are often polar enough to require extra extractions with organic solvent in the work-up procedure.

[24] *Caution:* Copper and silver acetylides, as well as mercuric acetylide, are potential explosives.

$$R{-}C{\equiv}C{-}R' + O_2N{-}C_6H_3(NO_2){-}SCl \xrightarrow{ClCH_2CH_2Cl}$$

$$\begin{matrix} R & & SAr \\ & C{=}C & \\ Cl & & R' \end{matrix}$$ **(Pasto and Johnson p. 339)**

$$RC{\equiv}C{-}H + HgCl_2/KI \xrightarrow{KOH} (RC{\equiv}C)_2Hg$$ **(Pasto and Johnson, p. 339)**

Compare the latter to the sodium metal test (Experiment 1).

Examples of ir (Fig. 6.11) and nmr (Fig. 6.12) spectra of alkynes are provided; symmetrically substituted C≡C units are difficult to detect by ir. Usually ir spectra reveal the C≡C unit only when the alkyne is unsymmetrical about the unsaturation e.g., in terminal alkynes). Since the nonterminal C≡C units cannot have proton substituents, one must rely on the protons α to the multiple bond $\left(\begin{matrix} C{-}C{\equiv}C{-} \\ | \\ H \end{matrix}\right)$ to interpret the nmr spectrum; this is less dependable than, for example, using ≡C—H absorptions for analysis.

Mass spectral cleavage patterns of alkynes are similar to those of alkenes.

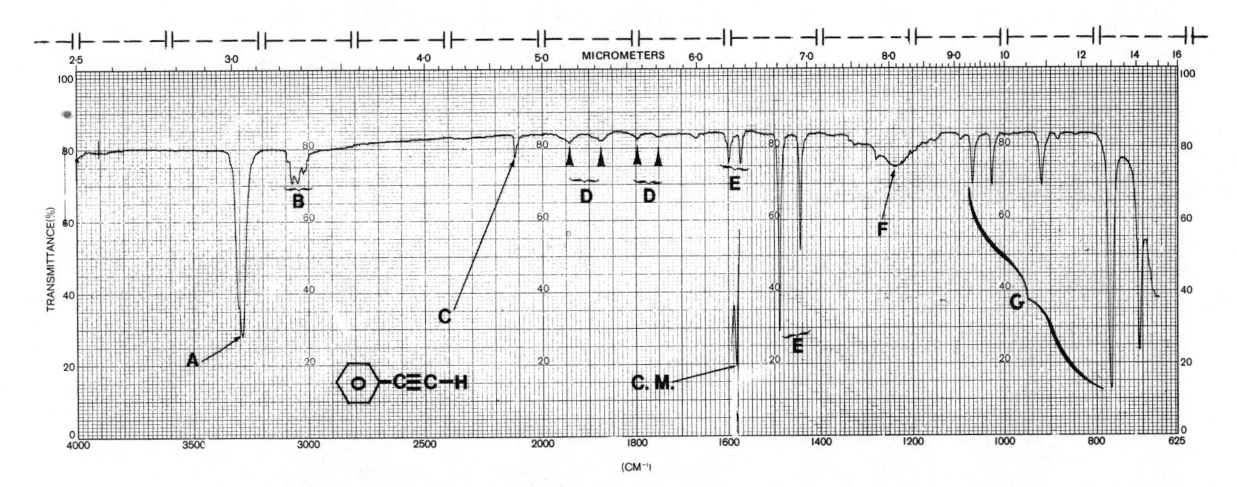

Figure 6.11. Infrared spectrum of phenylacetylene: neat. C—H stretch: **A**, 3290 cm^{-1} (3.04 μm), ≡C—H stretch; **B**, 3100–3000 cm^{-1} (3.23–3.33 μm), aromatic C—H stretch. C≡C stretch: **C**, 2115 cm^{-1} (4.73 μm), $\nu_{C\equiv C}$. Aromatic: **D**, 2000–1667 cm^{-1} (5.00–6.00 μm), aromatic overtones (see Fig. 6.21); **E**, 1667–1429 cm^{-1} (6.00–7.00 μm), C┄C stretch. Acetylenic: **F**, 1237 cm^{-1} (8.08 μm), overtone of ≡C—H bend. Aromatic: **G**, 800–625 cm^{-1} (12.50–16.00 μm), ┄C—H out-of-plane bend and/or C┄C ring breathing. C.M. = calibration band (1583 cm^{-1} band of a polystyrene window). Since the calibration marker, within the experimental error of observation of its position on the chart paper, is in the correct position, no corrections were necessary for tabulations of the band positions.

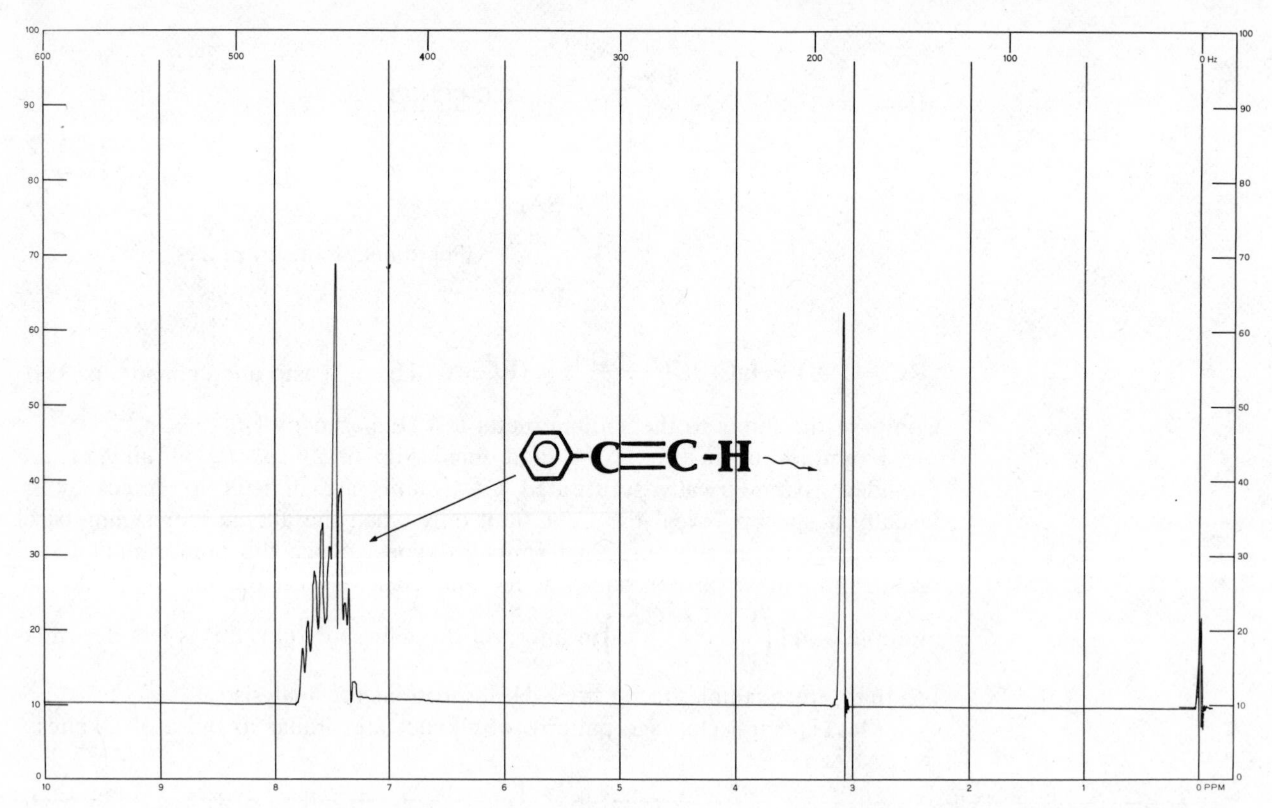

Figure 6.12. Nuclear magnetic resonance (proton) spectrum of phenylacetylene: 60 MHz, 600-Hz sweep width, $CDCl_3$ solvent.

6.6 ALIPHATIC HALIDES

Since aliphatic halides are often detected initially by elemental analysis (Chapter 4) for the halogens, it should not be surprising that further characterization takes advantage of the halo substituent and its ability to be displaced. The two tests for displaceable halogen (X) that are discussed below are complementary to some extent and are thus often very useful for classifying the structures of alkyl halides.

NaI/acetone test: *increasing* reactivity →

R_3CCl R_2CHCl RCH_2Cl

$AgNO_3$/ethanol test: *decreasing* reactivity →

The theory behind these halide displacement reactions is outlined after the discussion of the experimental procedures and results.

EXPERIMENT 16. SILVER NITRATE SOLUTION (ETHANOLIC)

$$RX + AgNO_3 \rightarrow \underline{AgX(s)} + RONO_2$$

This reagent is useful for classifying compounds known to contain halogen. Add 1 drop of the halogen compound to 2 ml of a 2% ethanolic silver nitrate solution. If no reaction is observed after 5 min standing at room temperature, heat the solution to boiling and note if a precipitate is formed. If there is a precipitate, note its color. Add 2 drops of dilute (5%) nitric acid, and note if the precipitate dissolves. Silver halides are insoluble in dilute nitric acid; silver salts of organic acids are soluble.

Since alkyl halides often contain small amounts of isomeric impurities, it may be advisable in certain borderline cases to collect and weigh the dry silver halide obtained from a weighed sample of unknown. Generally an approximate value for the molecular weight can be arrived at from a consideration of its physical constants and from an inspection of the list of possibilities; the estimated theoretical yield of the silver halide can thus be compared with the amount obtained. If the observed yield amounts to only a few percent, the test is negative. An alkyl halide that gives only a small amount of silver halide because it reacts slowly may be distinguished from a mixture of an inert halide with a small amount of reactive impurity by collecting the halide initially precipitated and then testing the filtrate with more silver nitrate.

Apply the test to (1) benzoyl chloride; (2) benzyl chloride; (3) ethyl bromide; (4) bromobenzene; (5) chloroform; (6) chloroacetic acid.

Discussion. Many halogen-containing substances react with silver nitrate to give an insoluble silver halide, and the rate of this reaction is an index of the degree of reactivity of the halogen atom in question. This information is valuable because it permits certain deductions to be drawn concerning the structure of the molecule.

The most reactive halides are those that are ionic. Among organic compounds, the amine salts of the halogen acids constitute the most common examples.

$$[RNH_3]^+X^-$$

Less frequently encountered are oxonium salts and carbonium salts that contain ionic halogen.

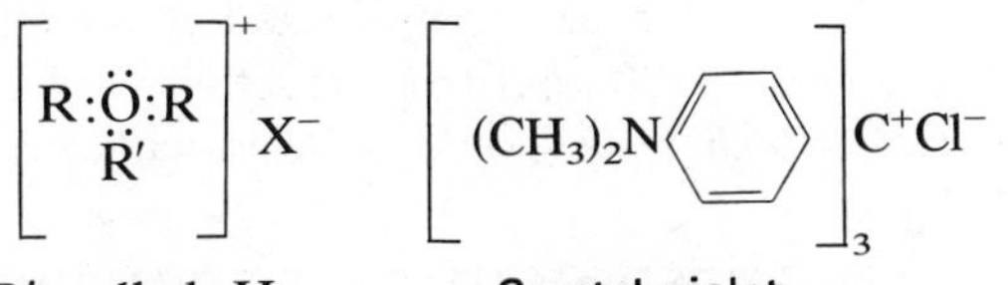

$$\left[\mathrm{R}:\ddot{\underset{\cdot\cdot}{\mathrm{O}}}:\mathrm{R} \atop \mathrm{R'}\right]^+ X^- \qquad \left[(CH_3)_2N\text{—}C_6H_4\text{—}\right]_3 C^+Cl^-$$

R′ = alkyl, H   **Crystal violet**

Aqueous solutions of these salts give an immediate precipitate of the silver halide with aqueous silver nitrate solution.

A summary of the results to be expected in the alcoholic silver nitrate test is given below.

I. The following water-soluble compounds give an immediate precipitate with aqueous silver nitrate.

1. Amine salts of halogen acids.

$$(RNH_3)^+X^- + Ag^+NO_3^- \rightarrow \underline{AgX(s)} + (RNH_3)^+NO_3^-$$

2. Oxonium salts.
3. Carbonium halides.
4. Low-molecular-weight acid chlorides. Many of these are hydrolyzed by water and so furnish the halide ion.

$$RCOCl + H_2O \rightarrow RCOOH + HCl$$

II. Water-insoluble compounds fall roughly into three groups with respect to their behavior toward alcoholic silver nitrate solutions.

1. Compounds in the first group give an immediate precipitation at room temperature.

$$RCOCl \qquad RCHClOR \qquad R_3CCl$$
$$RCH{=}CHCH_2X \qquad RCHBrCH_2Br \qquad RI$$

2. The second group includes compounds that react slowly or not at all at room temperature but that give a precipitate readily at higher temperatures.

$$RCH_2Cl \qquad R_2CHCl \qquad RCHBr_2 \qquad O_2N\text{-}C_6H_3(NO_2)\text{-}Cl$$

3. A final group is made up of compounds that are usually inert toward hot alcoholic silver nitrate solutions.

$$ArX \qquad RCH{=}CHX \qquad HCCl_3$$

In the reaction with silver nitrate, cyclohexyl halides exhibit a decreased reactivity when compared with the corresponding open-chain secondary halides. Cyclohexyl chloride is inactive, and cyclohexyl bromide is less reactive than 2-bromohexane, although it will give a precipitate with alcoholic silver nitrate. Similarly, 1-methylcyclohexyl chloride is considerably less reactive than acyclic tertiary chlorides. However, both 1-methylcyclopentyl and 1-methylcycloheptyl chloride are more reactive than the open-chain analogs.

Since, as emphasized above, reactivity toward alcoholic silver nitrate is often very different from reactivity toward sodium iodide in acetone (Experiment 17), *both* tests should be used with any halogen compound.

EXPERIMENT 17. SODIUM IODIDE IN ACETONE

$$RCl + NaI \rightarrow RI + \underline{NaCl}(s)$$

$$RBr + NaI \rightarrow RI + \underline{NaBr}(s)$$

To 1 ml of the acetone solution of sodium iodide in a test tube add 2 drops of the compound whose elemental analysis showed the presence of chlorine or bromine. If the compound is a solid, dissolve about 0.1 g in the smallest possible volume of acetone, and add the solution to the reagent. Shake the test tube, and allow the solution to stand at room temperature for 3 min. Note if a precipitate is formed and also if the solution turns reddish brown (because of the liberation of free iodine). If no change occurs at room temperature, place the test tube in a beaker of water at 50°C. At the end of 6 min, cool to room temperature and note if a reaction has occurred. Try this test on (1) *n*-butyl bromide; (2) *sec*-butyl bromide; (3) *tert*-butyl chloride; (4) ethylene bromide; (5) benzyl chloride; (6) benzoyl chloride; (7) benzenesulfonyl chloride; (8) α-chloroacetophenone.

Reagent. Fifteen grams of sodium iodide is dissolved in 100 ml of pure acetone. The solution, colorless at first, becomes a pale lemon yellow. It should be kept in a dark bottle and discarded as soon as a definite red-brown color develops.

Discussion. This test depends on the fact that sodium chloride and bromide are only very slightly soluble in acetone. As might be anticipated, the order of reactivity of simple halides is primary > secondary > tertiary. With sodium iodide, primary bromides give a precipitate of sodium bromide within 3 min at 25°C, whereas the chlorides give no precipitate and must be heated to 50°C in order to effect a reaction. Secondary and tertiary bromides react at 50°C; but the tertiary chlorides fail to react within the time specified. Tertiary chlorides will react if the test solutions are allowed to stand for a day or two.

In the reaction with sodium iodide in acetone, cyclopentyl chloride reacts at a rate comparable with that of acyclic secondary chlorides, whereas the reaction with cyclohexyl chloride is considerably slower. Thus, cyclohexyl chloride and bromide, bornyl chloride, and similar compounds fail to react appreciably with sodium iodide at 50°C within 6 min. This retardation of rate in the cyclohexyl system is due to the special geometry imposed upon the transition state by the cyclohexane ring.

Benzyl ($ArCH_2X$) and allyl ($>C{=}\overset{|}{C}{-}\underset{|}{\overset{|}{C}}{-}X$) halides are extremely reactive toward NaI/acetone and give a precipitate of sodium halide within 3 min at 25°C; the reason for this extreme reactivity is discussed below.

Although triphenylmethyl chloride would be expected to be too hindered to undergo a displacement reaction with iodide ion, it has been found to react very much faster than benzyl chloride with the sodium iodide/acetone reagent. However, the reaction is not a simple replacement to form trityl iodide, as is shown by the fact that the color of *iodine* is observed. This example emphasizes the care that needs to be taken in the interpretation of qualitative tests of this type. It must be remembered that members of a class of compounds under consideration may react with the same reagent in several ways and that for each kind of reaction the effect of structure on reactivity may be different.

Although picryl chloride has been reported to give picryl iodide on treatment with potassium iodide in ethanol, treatment with potassium iodide in an acidic medium such as boiling acetic acid or even acetone containing some acetic acid at room temperature gives trinitrobenzene and iodine.

Polybromo compounds such as bromoform and 1,1,2,2-tetrabromoethane react with sodium iodide at 50°C to give a precipitate and liberate iodine. Carbon tetrabromide reacts at 25°C.

Sulfonyl chlorides give an immediate precipitate and also liberate free iodine. Presumably the iodine is formed by the action of sodium iodide on the sulfonyl iodide

$$ArSO_2Cl + NaI \rightarrow ArSO_2I + \underline{NaCl}(s)$$

$$\downarrow NaI$$

$$ArSO_2Na + I_2$$

Benzenesulfonyl chloride gave a 60% yield of sodium benzenesulfinate; other products obtained were diphenyl disulfone (27%) and phenyl thiosulfinate (10%).

Alkyl sulfonates also react, producing the corresponding sodium sulfonates as precipitates.

$$ArSO_2OR + NaI \rightarrow ArSO_2ONa + RI$$

This reaction must be kept in mind in the event that one of the groups in the sulfonic ester contains halogen.

1,2-Dichloro and 1,2-dibromo compounds not only give a precipitate of sodium chloride or bromide but also liberate free iodine.

$$RCH(Br)-CH(Br)R + 2NaI \longrightarrow RCH{=}CHR + 2NaBr + I_2$$

Stepwise:

$$I^- \curvearrowright Br;\quad R-CH(Br)-CH(Br)-R \longrightarrow RCH{=}CHR + IBr + Br^-$$

$$IBr + I^- \longrightarrow I_2 + Br^-$$

A comparison of ethylene halides gave the following results:

	Ppt. at 25°C	
$BrCH_2CH_2Br$	1.5 min	
$BrCH_2CH_2Cl$	3 min	
$ClCH_2CH_2Cl$	None	(Ppt. at 50°C in 2.5 min)

THEORY OF SUBSTITUTION REACTIONS

Because the test with alcoholic silver nitrate and the replacement of halide ion with sodium iodide are so closely related, they will be discussed together. It is convenient to define two extreme mechanisms by which halides and related substances undergo nucleophilic substitution. The first of these has been called by various investigators "S_N2" (substitution nucleophilic bimolecular), "displacement," or "*N*" (nucleophilic); and the second "S_N1," "solvolysis," or "Lim." (limiting). Since the two mechanisms have widely differing structural requirements, any consideration of the effect of structure on the reactivity of halides must take account of the mechanism involved.

Detailed discussion of this subject is beyond the scope of this book; excellent reviews are available.[25] A brief mention of the two mechanisms concerned,

[25] A. Streitwieser, *Chem. Rev.*, **56**, 571 (1956), and *Solvolytic Displacement Reactions* (McGraw-Hill, New York, 1962); C. Ingold, *Structure and Mechanism in Organic Chemistry*, 2nd ed. (Cornell University Press, Ithaca, N.Y., 1969).

however, will be of considerable help in understanding the alcoholic silver nitrate test and also the test employing sodium iodide in acetone.

The S_N2 or direct displacement mechanism is a concerted reaction whereby a nucleophilic reagent such as iodide or hydroxide ion collides with the halide undergoing reaction, collision occurring at the rear of the carbon atom from which the halide ion is being displaced.

$$I^- + H{-}C(R)(H){-}X \rightarrow \overset{\delta-}{I} \cdots C(R)(H)(H) \cdots \overset{\delta-}{X} \rightarrow I{-}C(R)(H){-}H + X^-$$

(transition state)

It is seen that the transition state has the negative charge distributed largely on the entering and leaving groups. Since the reaction is initiated by formation of the covalent bond to I^-, the rear of the carbon at which displacement is occurring must be unhindered in order to permit facile entrance of the attacking atom or group. Thus this kind of reaction might be expected to be sensitive to steric blocking at the site of reaction but relatively insensitive to electronic effects. The S_N1 or carbocation mechanism involves two stages, of which the first is ionization of the reacting halide.

$$R{-}X \rightarrow R^+ + X^-$$
$$R^+ + Y^- \rightarrow R{-}Y$$

It seems clear that this mechanism should not be retarded by steric blocking at the rear by R and, on the other hand, should be helped very markedly by a change of R that aids in stabilization of positive charge. *In general, halides that react with sodium iodide in acetone must do so by the S_N2 mechanism, and those that react with alcoholic silver nitrate must do so by the S_N1 type.* These conclusions are dictated by the nature of these reagents. Sodium iodide in acetone has an ion, iodide ion, which is a very good displacing agent (said to be strongly nucleophilic), but acetone is a very poor solvent for ionization. This reagent, therefore, strongly favors the *N* mechanism. On the other hand, nitrate ion is a poor nucleophilic agent but ethanol is a moderately good ionizing solvent; and silver, by its power of coordinating with the leaving halogen, is excellent at assisting ionization. This reagent, then, favors the S_N1 mechanism.

Saturated Halides. Tertiary carbonium ions are more stable than secondary, which in turn are more stable than primary. One reason for these differences is that the tertiary ion has a method of distributing its positive charge, which is less efficient with secondary ions and still less with primary.

$$R{-}C^+(R)(R) > R{-}C^+(R)(H) > R{-}C^+(H)(H)$$

tertiary (3°) secondary (2°) primary (1°)

The second factor favoring the formation of tertiary carbonium ions from their halides, as compared with the corresponding formation of primary or secondary ions, is that steric strain which may be present in a bulky tertiary halide (with angles around the central carbon atom of 109°) is relieved to some extent in going to the planar ion (with 120° angles between groups attached to the central carbon atom). It should be expected, then, that the order of reactivity of alkyl halides with silver nitrate would be that actually found:

tertiary > secondary > primary

It is of interest that solvolytic reactivity is very great in such highly branched molecules as tri-*t*-butylmethyl chloride, which may react as much as 40,000 times as fast as *t*-butyl chloride.

As might be anticipated, the order of

reactivity of simple halides is reversed in the displacement reaction with alkali iodides because the order of steric accessibility to iodide ion is primary > secondary > tertiary.

Bicyclic Halides. Halogen at the bridgehead carbon atom of certain bicyclic systems is very unreactive in both the solvolysis and the displacement reactions. For example, apocamphyl chloride(I) is unaffected by treatment either with hot 30% potassium hydroxide in 80% ethanol for 24 hr or with hot silver nitrate in aqueous ethanol for 48 hr. Nucleophlic attack via an S_N2 process is blocked because the nucleophile cannot fit inside the ring skeleton; carbocation (Ia) formation is inhibited because the ring skeleton resists the coplanarity required at C_1, C_2, C_6, and C_7 (the geometry required by an sp^2 hybrid at the cationic center, C_1).

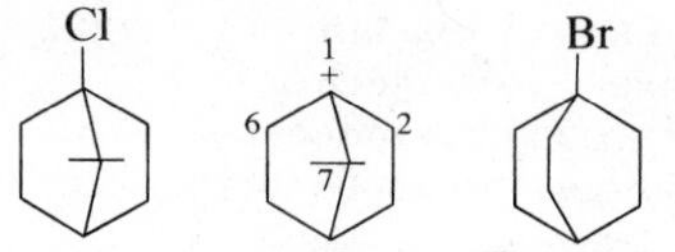

Cl　　　1 +　　　Br
　　6　　2
　　　7

I　　　Ia　　　II

1-Bromobicyclo[2.2.1]heptane is changed to the carbinol when treated for 2 days with aqueous silver nitrate at 150°C. When a 2.2.2-bicyclooctane cage is employed as in the bromide (II), aqueous silver nitrate gives the corresponding alcohol after 4 hr at room temperature. The greater reactivity of the 2.2.2 as compared to the 2.2.1 system is probably due to the superiority of bromine as a leaving group (compared to chlorine) and the greater flexibility of the ring skeleton of II, which can better provide the coplanarity at position 1 necessary for carbocation formation.

Aryl and Vinyl Halides. Vinyl halides and those aryl halides in which halogen is attached directly to the aromatic ring are usually unreactive toward either direct displacement or the solvolysis mechanism.

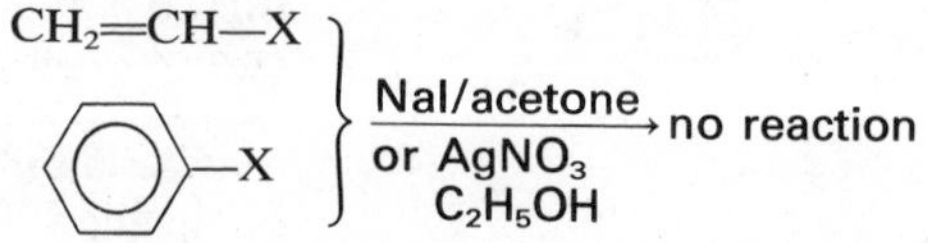

$$\left.\begin{array}{l} CH_2{=}CH{-}X \\ C_6H_5{-}X \end{array}\right\} \xrightarrow[\text{or } AgNO_3,\ C_2H_5OH]{\text{NaI/acetone}} \text{no reaction}$$

The high strength of the C—X bond and the instability of potential carbocation intermediates inhibit these reactions.[26]

Benzyl and Allyl Halides. Since the benzyl and the allyl carbocations have resonance stabilization as shown below, it is not surprising that benzyl and allyl halides react with silver nitrate more rapidly than do the corresponding saturated halides. Benzyl and allyl halides

Benzyl: $C_6H_5CH_2X \longrightarrow \left[C_6H_5CH_2^+ \longleftrightarrow {}^+C_6H_5{=}CH_2 \longleftrightarrow \text{Other structures} \right] \xrightarrow{Y^-} C_6H_5CH_2Y$

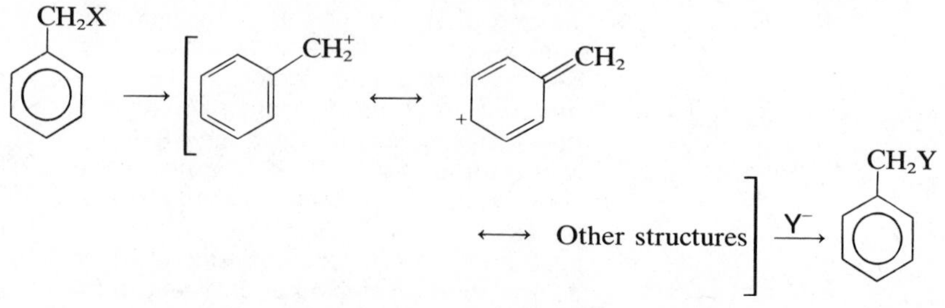

Allyl: $CH_2{=}CH{-}CH_2{-}X \longrightarrow \left[\begin{array}{c} CH_2{=}CH{-}CH_2^+ \\ \updownarrow \\ {}^+CH_2{-}CH{=}CH_2 \end{array} \right] \xrightarrow{Y^-} CH_2{=}CH{-}CH_2{-}Y$

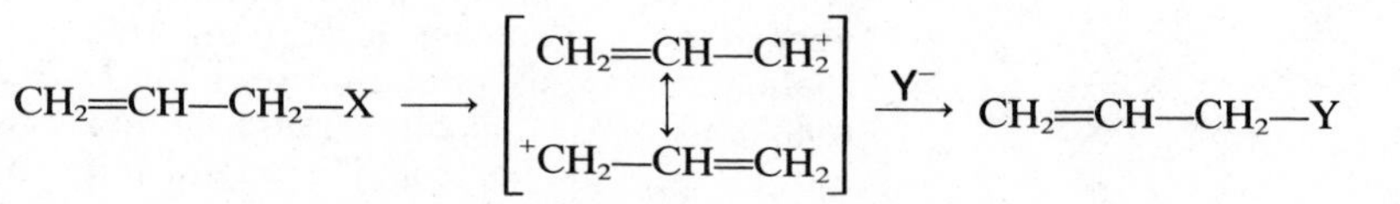

[26] For exceptions see Z. Rappoport, *Accts. Chem. Res.*, **9**, 265 (1976).

are thus reactive toward sodium iodide in acetone.

The fact that these systems are primary alkyl halides and also give stable carbocation intermediates partially explains the reactivity of these systems toward reactions involving each of the two different mechanisms.

The effect of additional aryl substitution is in general to increase the rate of the S_N1 reaction. Thus, benzhydryl chloride reacts readily with warm ethanol even without added silver nitrate, and triphenylmethyl chloride in liquid sulfur dioxide (a solvent chosen because it is effective in solvating ions but does not react) ionizes extensively to triphenylcarbonium chloride, $(C_6H_5)_3C^+Cl^-$.

Neighboring Group Participation. The following section should be considered only after the reader has become familiar with most of the fundamental functional groups covered elsewhere in this book; the material of the following nine paragraphs is more appropriate for students who have completed at least one semester of organic chemistry.

It might be expected that the effect of any electron-withdrawing atom or group adjacent to a carbon atom at which a solvolysis reaction is to occur would be to depress the rate, because the presence of a partial positive charge adjacent to the site at which a positive charge is to be induced is unfavorable. The lack of reactivity of β-alkoxy halides, α-halo ketones (such as α-chloroacetophenone), β-chloroalkyl toluenesulfonates, and vicinal dichlorides illustrates this effect. It should be noted, however, that α-halo ketones, esters, amides, and nitriles are highly reactive toward iodide ion and, when treated with sodium iodide in acetone at 25°C, give a precipitate of sodium halide within 3 min.

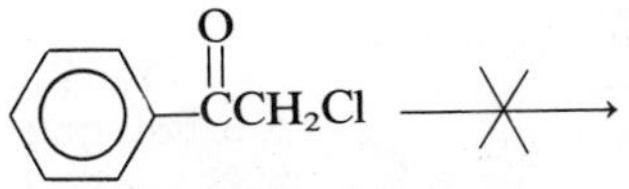

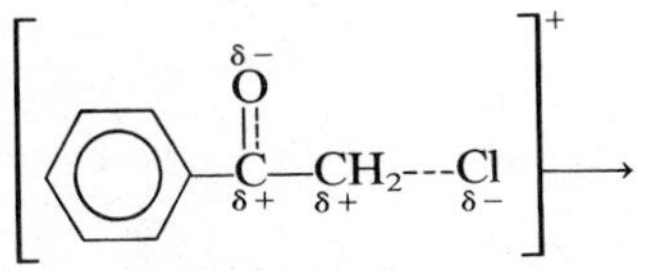

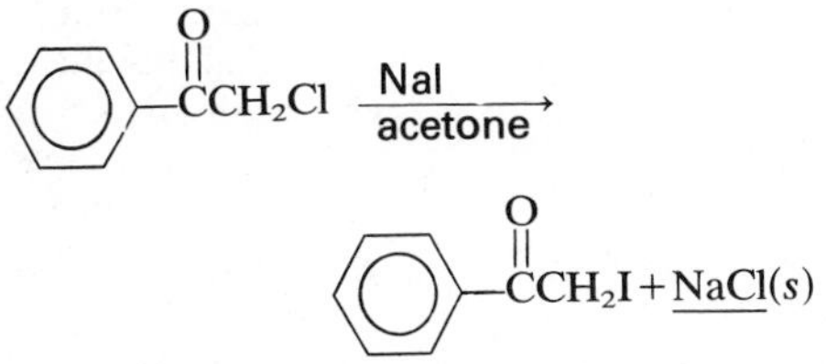

$$C_6H_5COCH_2Cl \not\longrightarrow$$

α-chloroacetophenone

$$\left[C_6H_5\overset{\delta-}{O}=\underset{\delta+}{C}-\underset{\delta+}{CH_2}\cdots\underset{\delta-}{Cl}\right]^+ \longrightarrow$$

$$C_6H_5COCH_2Cl \xrightarrow[\text{acetone}]{\text{NaI}} C_6H_5COCH_2I + \underline{NaCl(s)}$$

Evidence has been advanced that the acceleration of the displacement reactions by an α-carbonyl function is due to stabilization of the transition state by partial distribution of negative charge on the carbonyl oxygen atom:

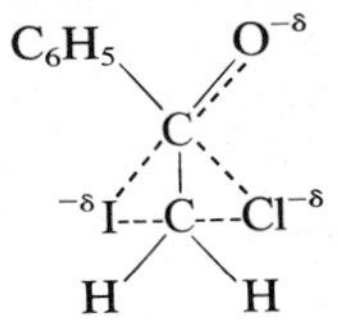

$$^{-\delta}I\cdots CH_2(C(=O^{-\delta})C_6H_5)\cdots Cl^{-\delta}$$

The groups $R_2\ddot{N}$—, $R\ddot{S}$—, $:\ddot{I}$—, *beta* to halogen, unlike alkoxyl or chlorine, increase the rate of solvolysis. Halides containing these groups react by an internal displacement as shown below.

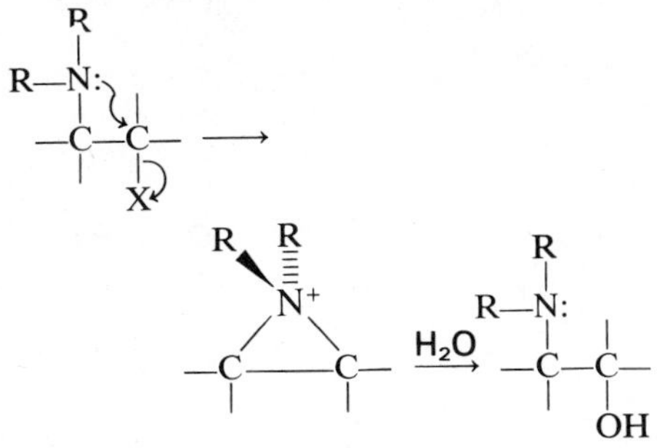

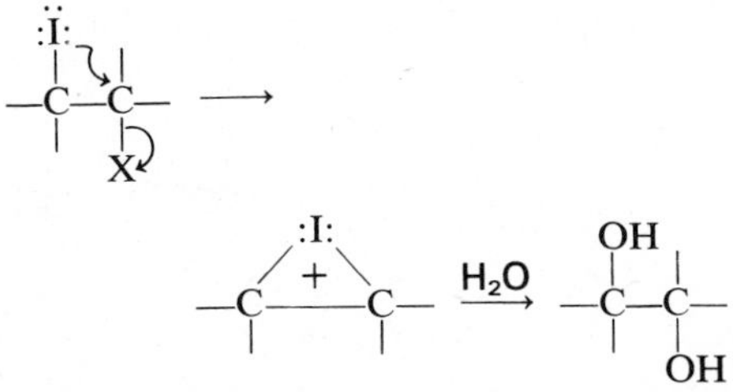

$$R_2\ddot{N}-C-C-X \longrightarrow \text{aziridinium ion } (R_2N^+) \xrightarrow{H_2O} R_2N-C-C-OH$$

$$:\ddot{I}-C-C-X \longrightarrow \text{iodonium ion } (:I^+:) \xrightarrow{H_2O} HO-C-C-OH$$

In the case of other neighboring groups such as bromine and acetoxyl, the depression due to the inductive effect in rate of solvolysis (analogous to the depression in rate by neighboring chlorine) is compensated for, to a large extent but not completely, by the internal displacing action, so that the rates of solvolysis of substances of the types below are only slightly lower than those of the unsubstituted analog.

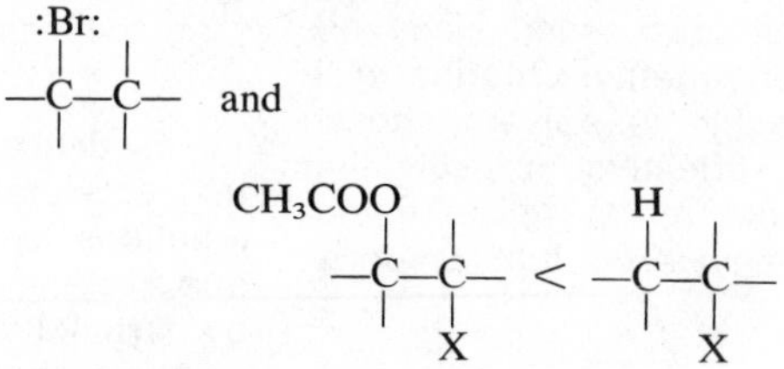

$$\overset{:\mathrm{Br}:}{-\mathrm{C}}-\mathrm{C}- \quad \text{and}$$

$$\overset{CH_3COO}{-\mathrm{C}}-\underset{X}{\mathrm{C}}- \; < \; \overset{H}{-\mathrm{C}}-\underset{X}{\mathrm{C}}-$$

Since the neighboring group, in order to displace a leaving group, X, must be able to participate in a Walden inversion with attack from the rear, cis groups in cyclic systems are ineffective. Thus, when cyclohexyl *p*-bromobenzenesulfonate is substituted with a *cis*-acetoxyl in the 2-position, its solvolysis rate in glacial acetic acid is lowered by a factor of nearly 5000 (inductive effect), but a *trans*-2-acetoxyl group lowers it by a factor of only 5, because the inductive effect is largely counterbalanced by the fact that acetoxyl is a good internal displacing agent and is in the proper steric position for trans displacement.

Other Halogen Compounds. Since most acid chlorides and bromides react rapidly with ethanol, even without silver nitrate, they give an immediate precipitate with alcoholic silver nitrate. They also react rapidly with sodium iodide in acetone, probably by a carbonyl addition mechanism.

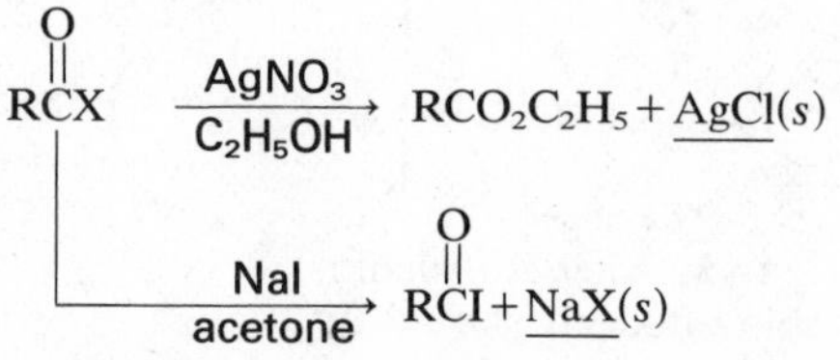

$$\mathrm{R\overset{O}{\overset{\|}{C}}X} \xrightarrow[C_2H_5OH]{AgNO_3} \mathrm{RCO_2C_2H_5} + \underline{\mathrm{AgCl}(s)}$$

$$\mathrm{R\overset{O}{\overset{\|}{C}}X} \xrightarrow[\text{acetone}]{NaI} \mathrm{R\overset{O}{\overset{\|}{C}}I} + \underline{\mathrm{NaX}(s)}$$

An α-halogen ether is solvolyzed rapidly; the resultant carbocation is relatively stable (the ion may be re-

$$\mathrm{R{-}O{-}\underset{Cl}{\underset{|}{\overset{|}{C}}}{-}} \rightarrow [\mathrm{R{-}O{-}\overset{+}{\underset{|}{C}}{-}} \leftrightarrow \mathrm{R{-}\overset{+}{O}{=}\underset{|}{C}{-}}]$$

garded as an ether of the conjugate acid of a ketone). Thus compounds of this type often give an immediate precipitate with alcoholic silver nitrate.

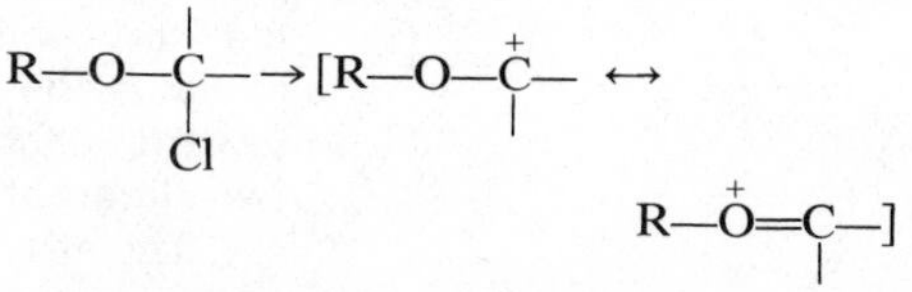

The accumulation of chlorine atoms on the same carbon atom in simple aliphatic compounds results in a remarkable degree of inertness toward silver nitrate, such as is seen in chloroform, carbon tetrachloride, 1,1,2,2-tetrachloroethane, and trichloroacetic acid. This is not true of bromo compounds; carbon tetrabromide reacts with silver nitrate at 25°C, and bromoform and tetrabromoethane give a precipitate with boiling alcoholic silver nitrate.

In compounds of the allyl chloride type, accumulation of chlorine atoms on a single carbon atom does not decrease their reactivity but actually seems to increase it. Benzotrichloride (α,α,α-trichlorotoluene) is more easily hydrolyzed than benzal chloride (α,α-dichlorotoluene, or benzylidene chloride), which, in turn, is more reactive than benzyl chloride:

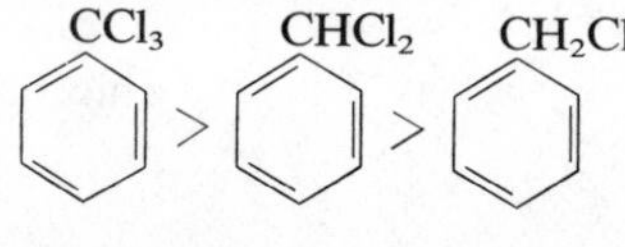

$$C_6H_5CCl_3 > C_6H_5CHCl_2 > C_6H_5CH_2Cl$$

Simple polychloro compounds with the halogens on a single carbon atom, such as chloroform, carbon tetrachloride, and trichloroacetic acid, fail to react with sodium iodide in acetone However, benzal chloride and benzotrichloride are reactive because of the ability to form a stable, benzylic cation:

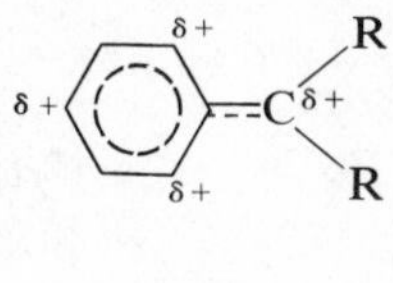

$$\text{(benzylic cation with } \delta+ \text{ on ring positions, } {=}C^{\delta+}R_2)$$

Halides (RX) can be characterized by preparing *N*-substituted phthalimide derivatives:

$$\text{Cl}_4\text{C}_6(\text{CO})_2\text{NH} \xrightarrow[\text{(or DMSO)}]{\text{(1) KOH/DMF} \quad \text{(2) RX/DMF}} \text{Cl}_4\text{C}_6(\text{CO})_2\text{N—R}$$

$$2\,\text{C}_6\text{H}_4(\text{CO})_2\overset{-}{\text{N}}\overset{+}{\text{K}} + \text{X(CH}_2)_n\text{X} \longrightarrow \text{C}_6\text{H}_4(\text{CO})_2\text{N—(CH}_2)_n\text{—N(CO)}_2\text{C}_6\text{H}_4 + 2\text{KX}$$

The tetrachlorophthalimide salt appears to be best for monohalo compounds[27] and the unsubstituted phthalimide for dihalides.[28] Procedures may be obtained from the literature.

Figures 6.13 and 6.14 show, respectively, the ir and nmr spectra of typical aliphatic halides.

Halogens give rise to reasonably intense ir absorptions at the following positions:

Bond	Absorption range (C—X stretch)
C—F	1250–960 cm^{-1} (8.00–10.42 μm)
C—Cl	830–500 cm^{-1} (12.04–20.0 μm)
C—Br	667–290 cm^{-1} (14.99–34.5 μm)
C—I	500–200 cm^{-1} (20.2–50.0 μm)

Since many commercial instruments do not scan below 625 cm^{-1}, the C—Br and C—I stretching absorptions are normally not useful. If a molecule contains a single carbon-halogen bond, a single absorption will be observed; increasing the number of carbon-halogen bonds in a molecule increases the intensity and complexity of absorption in that region.

The chemical shifts of protons in an nmr spectrum of an aliphatic halide depend on the nature of the halogen and the degree of alkyl branching at the

---

[27] C. F. H. Allen, W. R. Adams, and C. L. Myers, *Anal. Chem.*, **37**, 158 (1965).
[28] C. F. H. Allen and J. P. Glauser, *Anal. Chem.*, **44**, 1694 (1972).

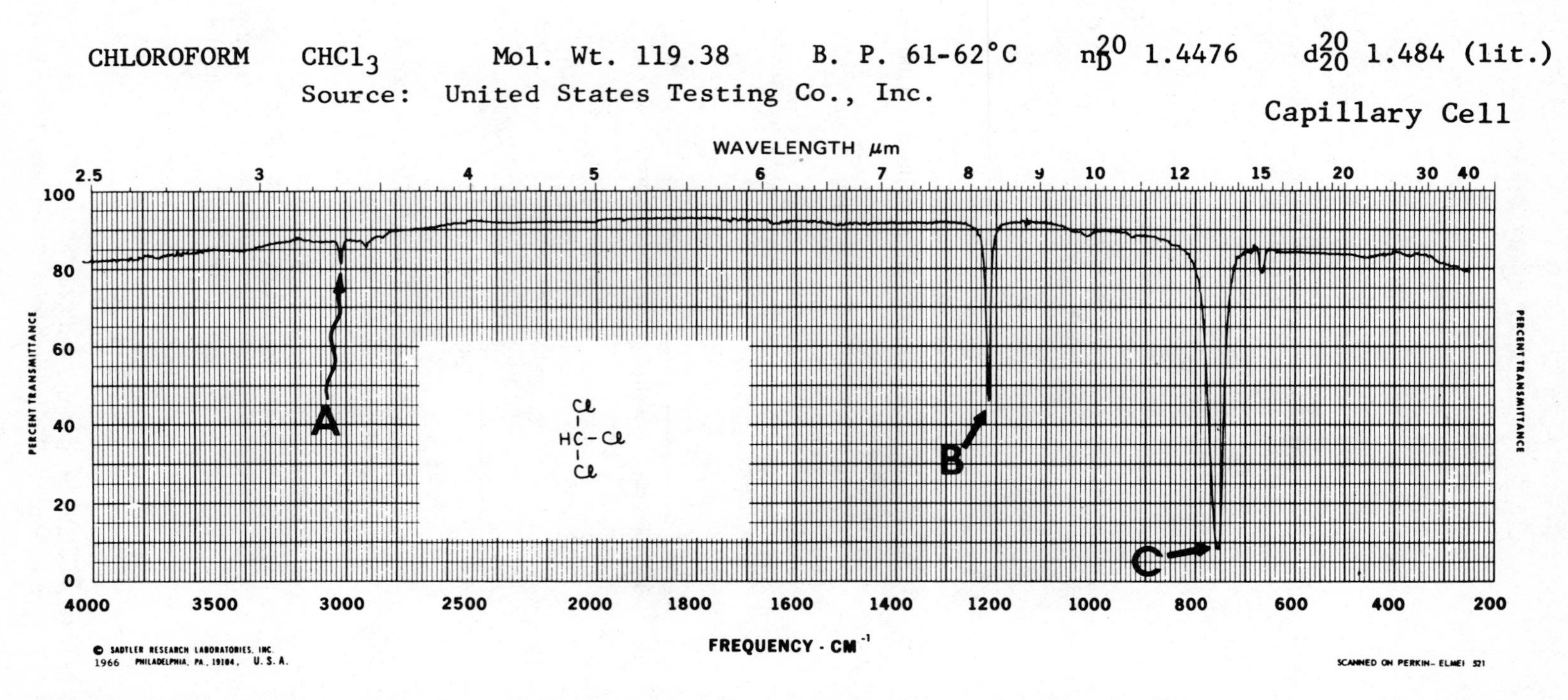

Figure 6.13. Infrared spectrum of chloroform: neat. C—H stretch: **A**, 3010 cm^{-1} (3.32 μm), $\nu_{C—H}$ (methine C—H stretch is characteristically weak). C—H bend: **B**, 1215 cm^{-1} (8.23 μm), $\delta_{C—H}$. C—Cl stretch: **C**, 755 cm^{-1} (13.25 μm), $\nu_{C—Cl}$.

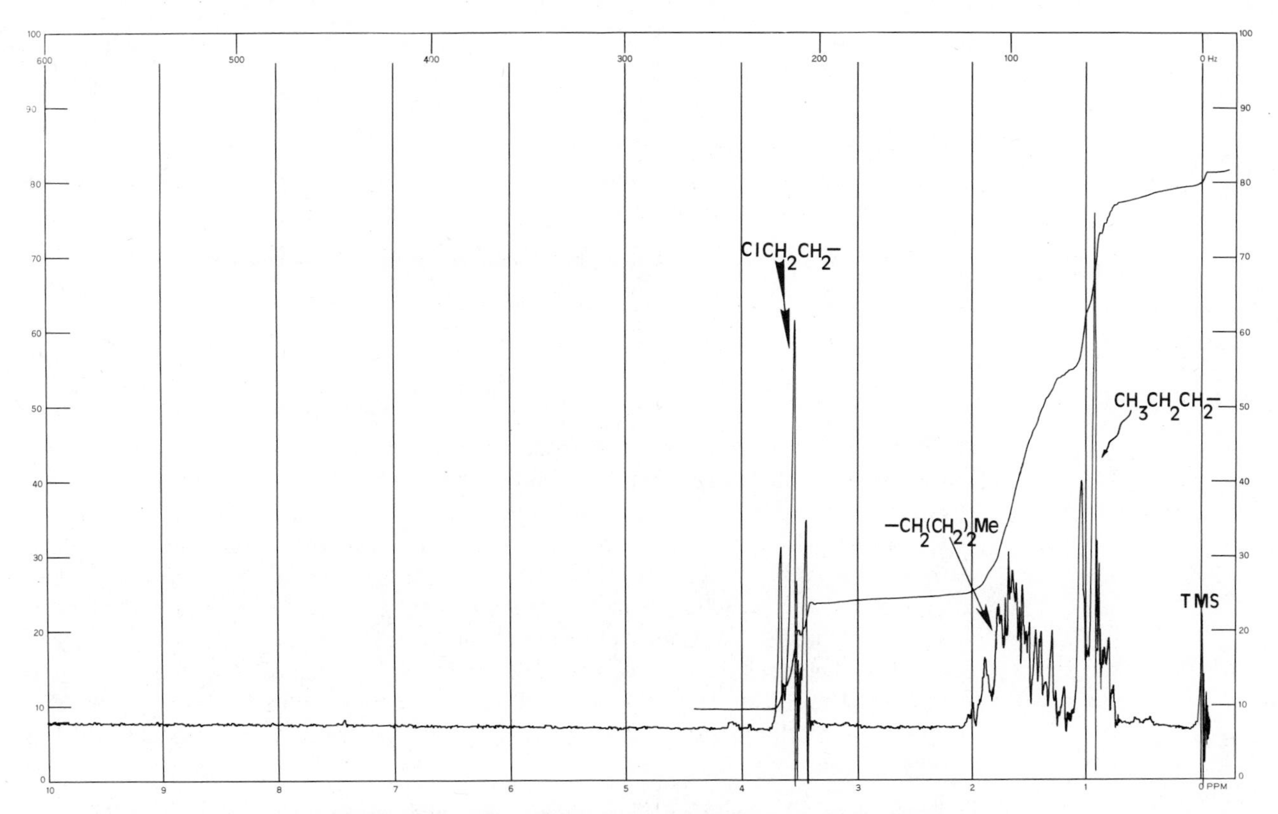

Figure 6.14. Nuclear magnetic resonance (proton) spectrum of 1-chlorobutane (*n*-butyl chloride): 60 MHz, 600-Hz sweep width, $CDCl_3$ solvent.

carbon bearing this proton:

	δ(TMS = δ0.00 ppm)[a]			
X	CH_3X	RCH_2X	R_2CHX	R_3CX
F	4.25	4.50	4.05	[b]
Cl	3.05	3.45	4.05	[b]
Br	2.70	3.40	4.10	[b]
I	2.15	3.15	4.25	[b]

[a] ±0.05 ppm, CCl_4 solvent.
[b] β-proton signals usually occur at δ2.2 or higher fields.

The mass spectra of organic chlorides and bromides are quite diagnostic of the number and type of halogenations; this has been discussed in Chapter 4.

Aliphatic halides, other than fluorides, can be converted to Grignard reagents; for example, cyclohexylmagnesium bromide can be made from cyclohexyl bromide by treating the alkyl halide with magnesium in dry ether (see Procedure

12):

$$C_6H_{11}Br \xrightarrow[(CH_3CH_2)_2O]{Mg} C_6H_{11}MgBr$$

Such Grignard reagents should be treated immediately with appropriate reagents to prepare isolable derivatives.

Procedure 12. Grignard Reagents and Alkylmercuric Halides

$$RX \xrightarrow[\text{ether}]{Mg} RMgX$$

$$RMgX \xrightarrow{HgX_2} RHgX$$

The Grignard reagent is made from the alkyl halide by treating 0.3 g of magnesium in 15 ml of dry ether with 1 ml of the alkyl halide in a clean, dry test tube. A crystal of iodine is added to start the reaction. When the reaction is complete, the solution is filtered through a little glass wool and the filtrate allowed to flow into a test tube containing 4 to 5 g of mercuric chloride, bromide, or iodide, depending on the halogen in the original alkyl halide. The reaction mixture is shaken vigorously, warmed on a steam cone for a few minutes, and evaporated to dryness. The residue is boiled with 20 ml of 95% ethanol, and the solution is filtered. The filtrate is diluted with 10 ml of water and cooled in an ice bath. The alkylmercuric halide that separates is collected on a filter and recrystallized from 60% ethanol.

Procedure 13. Anilides, Toluidides, and N-Naphthylamides from Alkyl Halides

$$RX + Mg \longrightarrow RMgX$$

$$RMgX \xrightarrow[(2)\ H_2O]{(1)\ C_6H_5NCO} C_6H_5NHCOR$$

$$RMgX \xrightarrow[(2)\ H_2O]{(1)\ CH_3C_6H_4NCO} CH_3C_6H_4NHCOR$$

$$RMgX \xrightarrow[(2)\ H_2O]{(1)\ C_{10}H_7NCO} C_{10}H_7NHCOR$$

The Grignard reagent is prepared as in Procedure 12 and is treated with 0.5 ml of phenyl, *p*-tolyl, or α-naphthyl isocyanate dissolved in 10 ml of absolute ether. The mixture is shaken and allowed to stand 10 min. About 25 ml of 2% hydrochloric acid is added, with very vigorous shaking. The ether layer is separated and dried with magnesium sulfate, and the ether is distilled. The residue is recrystallized from methanol, ether, or petroleum ether.

Procedure 14. Alkyl β-Naphthyl Ethers

$$\beta\text{-}C_{10}H_7OH \xrightarrow[C_2H_5OH]{NaOH} \beta\text{-}C_{10}H_7O^-Na^+ \xrightarrow{RX} \beta\text{-}C_{10}H_7OR$$

To a solution of 0.6 g of sodium hydroxide in 25 ml of ethanol are added 2 g of β-naphthol and 2 g of the alkyl halide. If the halide is a chloride, 0.5 g of potassium iodide is added also. The mixture is heated under reflux for 30 min and poured into 75 ml of cold water. This mixture is made distinctly alkaline to phenolphthalein and stirred vigorously. The alkyl β-naphthyl ether is removed by filtration and recrystallized from ethanol or an ethanol-water mixture.

This procedure is useful occasionally for making derivatives of dihalides of the type $X(CH_2)_nX$. Potassium iodide must not be added if the compound is a 1,2-dihalide ($n = 2$). Why?

Procedure 14a. Displacement of Chlorine from Primary and Secondary Alkyl Chlorides with Cyanide[29]

$$R\text{—}Cl + NaCN \xrightarrow{DMSO} R\text{—}CN + NaCl$$

Method for Preparing Nitriles from Primary Chlorides. A three-necked flask (fitted with a mechanical stirrer, reflux condenser, and thermometer) is charged with 40 ml of dry dimethyl sulfoxide (DMSO) and 7.5 g of dry powdered sodium cyanide.[29a] The slurry is heated to 90°C on a water bath, which is then removed, and with good stirring 12 g of the alkyl chloride (or 6 g of dichloride) is added slowly through the top of the reflux condenser. The temperature of the reaction mixture usually rises to 150 to 160°C. After all of the halide has been added (ca. 10 min), the mixture is stirred for an additional 10 min, and then cooled to room temperature.

If a *mononitrile* is being prepared, the mixture is poured into 100 ml of cold water and the nitrile is extracted with two 25-ml portions of methylene chloride.

---

[29] R. A. Smiley and C. A. Arnold, *J. Org. Chem.*, **25**, 257 (1960); a procedure one-tenth the scale described is adequate.

[29a] Cyanide salts are highly poisonous and thus extremely dangerous. Do **not** combine them with acid. Dispose of the residue only as directed by your laboratory instructor.

The methylene chloride extracts are washed three times with 25 ml of saturated sodium chloride solution, dried with 3 g of anhydrous calcium chloride, and the methylene chloride is removed by distillation from a water bath. The residual oil of the nitrile can then be converted to the amide (Procedure 51a, p. 329), or to the aldehyde 2,4-dinitrophenylhydrazone (Procedure 52a, p. 330).

If a *dinitrile* is the expected product, after cooling the reaction mixture, 50 ml of chloroform is added and this mixture is poured into 25 ml of saturated sodium chloride solution. About 50 ml of water is added, the chloroform layer is separated, washed with three portions of 25 ml each of saturated salt solution, dried with 3 g of anhydrous calcium chloride, and the chloroform removed by distillation. The residual, oily dinitrile is then converted to either the di-amide or dinitrophenylhydrazone (Procedures 51a or 52a, respectively).

Method for Preparing Nitriles from Secondary Chlorides.[30] These are prepared by heating the sodium cyanide slurry with an electric heating mantle to 90°C and then, with good stirring, adding the secondary halide slowly through the top of the condenser over a period of 30 min. The temperature is gradually raised to 150°C and the mixture is stirred for 3 hr. The mixture is then cooled and worked up in the same fashion as for primary nitriles above.

Discussion. Alkyl chlorides (and alkylene dichlorides) are very common compounds. The conversion of these chlorides into other useful compounds has been hampered by the fact that displacement of the chlorine atoms by nucleophilic reagents (S_N2 reactions) takes place very slowly (long reaction times, low yields, use of pressure autoclaves) in hydroxylic solvents. Use of dimethyl sulfoxide, however, makes it possible to effect displacement reactions in high yield in reasonably short reaction times:

$$RCH_2Cl \xrightarrow[\text{DMSO}]{\text{NaCN}} RCH_2CN + NaCl$$

$$R_2CHCl \xrightarrow[\text{DMSO}]{\text{NaCN}} R_2CHCN + NaCl$$

$$ClCH_2CH_2Cl \xrightarrow[\text{DMSO}]{\text{2NaCN}} NC{-}CH_2CH_2{-}CN + 2NaCl$$

Yields of nitriles prepared by this method are in the 70–90% range. Nitriles are liquids (or low-melting solids) at 25°C and thus must be converted to amides (Procedure 51a) or to arylhydrazones (Procedure 52a) if solids are desired:

Procedure 14a ↓ ; Procedure 51a ↓

$$RCl \xrightarrow[\text{DMSO}]{\text{NaCN}} RCN \xrightarrow[\text{b) NaOH/H}_2\text{O}]{\text{a) BF}_3\cdot 2\text{CH}_3\text{CO}_2\text{H}} RCONH_2$$

Procedure 52a: $RCN \xrightarrow[\text{75\% HCO}_2\text{H}]{\text{Raney Ni}} RCHO \xrightarrow{\text{ArNHNH}_2} RCH{=}NNHAr$

[30] The procedure is that followed for the primary chlorides, with the exceptions noted in this section.

The dimethyl sulfoxide must be dry.[31] DMSO boils at 189°C (760 mm) or 85–87°C at 20 mm; other DMSO properties include: d_4^{20} 1.1014; n_D 1.477; dipole moment 3.96 D; and dielectric constant 49.

Primary and secondary bromides also react with sodium cyanide in dimethyl sulfoxide to form nitriles. However, it is usually time-saving to characterize the bromides by conversion to the substituted urethane (via Grignard reagent, Procedures 12 and 13).

Since it is a highly polar solvent, DMSO dissolves many types of compounds and can be used for other S_N2 displacement reactions. The saponification of esters by alkali in aqueous dimethyl sulfoxide is reported to be rapid.[32] Commercial DMSO often has a characteristic odor. Care should be taken to avoid skin contact with DMSO. The pharmacological activity of pure dimethyl sulfoxide is under study.

Procedure 15. S-Alkylthiouronium Picrates

$$H_2N{-}\underset{\underset{S}{\|}}{C}{-}NH_2 \xrightarrow[C_2H_5OH]{RX} \left[(H_2N)_2C \overset{\cdots}{=} S{-}R \right]^+ X^- \xrightarrow{\text{picric acid } (O_2N)_2C_6H_2(OH)NO_2} \left[(H_2N)_2C \overset{\cdots}{=} SR \right]^+ ArO^-$$

A mixture of 0.5 g of powdered thiourea, 0.5 g of the alkyl bromide or iodide, and 5 ml of 95% ethanol is placed in a test tube and boiled for 2 min. In another test tube 0.4 g of picric acid is dissolved in the minimum amount of boiling ethanol. The two solutions are mixed and allowed to cool. The *S*-alkylthiouronium picrate is removed by filtration and recrystallized from ethanol.

Alkyl chlorides may sometimes be induced to react by adding 1 g of potassium iodide to the original reaction mixture. Sufficient ethanol or water must be added to produce a clear solution at the boiling point. The subsequent procedure is the same as outlined above.

Two additional procedures for derivatizing alkyl halides have been described by Pasto and Johnson:

$$RI + 3,5\text{-}(O_2N)_2C_6H_3{-}CO_2^-\overset{+}{Ag} \longrightarrow 3,5\text{-}(O_2N)_2C_6H_3{-}\overset{\overset{O}{\|}}{C}{-}OR + \underline{AgI(s)}$$

(Pasto and Johnson, p. 351)

---

[31] This very hygroscopic solvent can be dried by distilling from calcium hydride (b.p. 64°C/4 mm).

[32] J. A. Vison and E. K. Hocker, *J. Chem. Educ.*, **46**, 245 (1969).

Grignard reagents:

$$\mathrm{RX} \xrightarrow{\mathrm{Mg}} \mathrm{RMgX} \xrightarrow[(2)\ \mathrm{H_3O^+}]{(1)\ \mathrm{CO_2}} \mathrm{RCO_2H} \qquad \text{(Pasto and Johnson, p. 352)}$$

CO_2 may be introduced as a solid (Dry Ice) or as a gas.

The unusual properties of fluorocarbons have also been discussed by Pasto and Johnson (p. 353); ^{19}F nmr, when available, is clearly useful for characterizing these compounds.[33] Even when an instrument for detecting ^{19}F signals is not available, the fact that ^{19}F couples strongly with adjacent protons may allow detection of fluorine via proton magnetic resonance spectrometry.

6.7 AMINES

The nucleophilic nitrogen of amines (R_3N:) usually undergoes facile reactions, via its nonbonded electrons, with electrophilic reagents, such as Brønsted acids, H^+, or with acid chlorides,

$$\mathrm{R\overset{O}{\overset{\|}{C}}{-}Cl}$$

EXPERIMENT 18. ACID CHLORIDE TEST FOR ACTIVE HYDROGEN

(a)

$$\mathrm{R\overset{O}{\overset{\|}{C}}{-}Cl} + \mathrm{H_2O} \longrightarrow \mathrm{R\overset{O}{\overset{\|}{C}}{-}OH} + \mathrm{HCl}$$

Cautiously add a few drops of acetyl chloride to 1 ml of water and touch the test tube to see if heat is evolved. In a similar manner test the behavior of benzoyl chloride toward water. What difference do you observe?

Acid chlorides are usually strong lachrymators and should be destroyed with dilute ammonium hydroxide before disposal.

(b)

$$\mathrm{R\overset{O}{\overset{\|}{C}}{-}Cl} + 2\,\mathrm{C_6H_5\ddot{N}H_2} \longrightarrow \mathrm{C_6H_5NH\overset{O}{\overset{\|}{C}}{-}R} + \mathrm{C_6H_5\overset{+}{N}H_3\overset{-}{Cl}}$$

Repeat each of the foregoing tests, using 0.5 ml of aniline in place of water. What are the products of this reaction? Pour the mixture into 5 ml of water.

---

[33] E. F. Mooney, *An Introduction to ^{19}F NMR Spectroscopy,* (Heyden, London–Sadtler, Philadelphia, 1970); C. H. Dungan and U. R. van Wazer, *Compilation of ^{19}F NMR Chemical Shifts* (Wiley-Interscience, New York, 1951 to mid 1967).

What is the precipitate? What is in solution?

(c)

$$RC(=O)\text{—}Cl + C_5H_5N{:} \longrightarrow C_5H_5\overset{+}{N}\text{—}C(=O)R\ Cl^-$$

Treat 0.5 ml of pyridine or quinoline with a few drops of acetyl chloride. Account for the generation of heat in this reaction, not withstanding the fact that the amine is recovered on dilution of the mixture with water followed by neutralization.

(d)

$$R\overset{O}{\overset{\|}{C}}\text{—}Cl + R'OH \rightarrow R\overset{O}{\overset{\|}{C}}\text{—}OR' + HCl$$

Add drop by drop 1 ml of acetyl chloride to (1) 1 ml of ethanol; (2) 0.5 g of phenol. In each case allow the reaction mixture to stand for a minute or two and then pour it cautiously into 5 ml of water. In (2), remove the liquid and test its solubility in cold dilute sodium hydroxide solution.

(e) *The Schotten-Baumann Reaction*

$$R\overset{O}{\overset{\|}{C}}\text{—}Cl + R'OH + NaOH \rightarrow R\overset{O}{\overset{\|}{C}}\text{—}OR' + NaCl + H_2O$$

$$R\overset{O}{\overset{\|}{C}}\text{—}Cl + R'NH_2 + NaOH \rightarrow R\overset{O}{\overset{\|}{C}}\text{—}NHR' + NaCl + H_2O$$

In a small glass-stoppered bottle place 5 ml of ethanol, 10 ml of water, and 2 ml of benzoyl chloride. To this solution add in portions, with vigorous shaking, 10 ml of 20% sodium hydroxide solution. Shake the mixture for several minutes, and then test the solution with litmus paper to make sure that it is still alkaline. What are the products of the reaction? Repeat the above experiment, using 2 ml of aniline instead of the ethanol. What advantage does this procedure possess over benzoylation without the use of alkali as in (b)? On what factors does the success of the Schotten-Baumann reaction depend?

Discussion. Nucleophiles, such as alcohols and amines, can attack the electrophilic carbonyl group of acid halides, forming product by an addition-elimination mechanism:

$$R\text{—}\underset{\delta+}{\overset{\overset{\delta-}{O}}{\overset{\|}{C}}}\text{—}X + R'\ddot{N}H_2 \longrightarrow R\text{—}\overset{O^-}{\overset{|}{\underset{\underset{R'}{|}}{\underset{{}^+NH_2}{\underset{|}{C}}}}}\text{—}X \xrightarrow[-HX]{B} R\text{—}\overset{O}{\overset{\|}{C}}NHR' + X^- + BH^+$$

B = base

$$\overset{\delta-}{\underset{\delta+}{\mathrm{RC(=O)}}}\mathrm{-X + R'OH} \quad \mathrm{R-C(O^-)(X)-O^+(R')H} \xrightarrow[-HX]{B} \mathrm{R-C(=O)OR' + X^- + BH^+}$$

The Schotten-Baumann reaction is of particular interest because it might be expected that the water and hydroxyl ion present could compete with the alcohol or amine to be acylated and reduce seriously the yield of the product desired. The success of the reaction is probably due to a combination of circumstances. It generally occurs in a heterogeneous medium with the organic reagent and benzoyl chloride in the same phase. One could speculate that the organic reagents combine in the organic phase and that only acid-base neutralization reactions take place in the aqueous phase.

EXPERIMENT 19. NITROUS ACID

(*a*)

$$\mathrm{C_6H_5NH_2 + HONO + HCl} \xrightarrow{0°C} \mathrm{C_6H_5\overset{+}{N}_2\overset{-}{Cl} + 2H_2O}$$

$$\mathrm{C_6H_5\overset{+}{N}_2\overset{-}{Cl}} \xrightarrow{H_2O + heat} \mathrm{C_6H_5OH + N_2 + HCl}$$

Diazotization. Dissolve 1 ml of aniline in 3 ml of concentrated hydrochloric acid diluted with 5 ml of water, and cool the solution to 0°C in a beaker containing cracked ice. Dissolve 1 g of sodium nitrite in 5 ml of water, and add the solution slowly, with shaking, to the cold solution of aniline hydrochloride. Continue the addition until the mixture gives a positive test for nitrous acid. The test is carried out by placing a drop of the solution on starch-iodide paper; a blue color indicates the presence of nitrous acid. Remove 2 or 3 ml of the solution to another test tube, warm gently, and observe the evolution of gas. What is formed? What is the term applied to the last reaction?

(*b*)

$$\mathrm{C_6H_5\overset{+}{N}_2\overset{-}{Cl} + C_{10}H_7ONa + NaOH} \longrightarrow \mathrm{C_6H_5N{=}N{-}C_{10}H_6\overset{-}{O}\overset{+}{Na} + NaCl + H_2O}$$

red–orange azo dye

Coupling. Add a second portion of the cold diazonium solution—about 2 ml—to a solution of 0.1 g of β-naphthol in 2 ml of 10% sodium hydroxide solution and 5 ml of water. Note the formation of the orange-red dye.

(*c*)

$$\mathrm{R_2NH + HONO \rightarrow R_2NN{=}O + H_2O}$$

Dissolve 2 ml of methylaniline in 5 ml of concentrated hydrochloric acid, dilute with 5 ml of water, and cool in an ice bath. Add slowly and with shaking 1.5 g of sodium nitrite dissolved in 5 ml of water, and note the result.[34] What is the product? Is there any difference between aromatic and aliphatic secondary amines as far as this reaction is concerned?

(d)

$$C_6H_5NR_2 + HONO + HCl \rightarrow p\text{-}ON\text{-}C_6H_4\text{-}H\overset{+}{N}R_2Cl^- + H_2O$$

Repeat (c), using dimethylaniline. Note the color and character of the reaction product. Do tertiary aliphatic amines behave in the same manner?

(e)

$$C_6H_5OH + HONO \longrightarrow p\text{-}HO\text{-}C_6H_4\text{-}NO \rightleftharpoons O{=}C_6H_4{=}NOH + H_2O$$

yellow

C_6H_5OH | H_2SO_4 ↓

$$\left[HO\text{-}C_6H_4\text{-}N{=}C_6H_4{=}\overset{+}{O}H\right](HSO_4)^- \xrightarrow{H_2O} O{=}C_6H_4{=}N\text{-}C_6H_4\text{-}OH$$

blue → red

NaOH ↓

$$O{=}C_6H_4{=}N\text{-}C_6H_4\text{-}\bar{O}\ \overset{+}{Na}$$

blue

Add a crystal of sodium nitrite to 2 ml of concentrated sulfuric acid, and shake until dissolved. Add 0.1 g of phenol, and note the changes in color. Pour the solution into 20 ml of ice water, and note the color. Add sodium hydroxide solution until the mixture is alkaline, and again note the color.

The blue color observed in this reaction is due to phenolindophenol formed from the reaction of initially produced *p*-nitrosophenol (quinone monoxime) with excess phenol. This reaction, known as "Liebermann's nitroso" reaction, is characteristic of phenols in which an ortho or para position is unsubstituted. It may be used to test for the nitroso grouping by mixing equal amounts of the nitroso compound and phenol, adding the mixture to sulfuric acid, and proceeding as in the above test.

---

[34] *Caution:* Nitrosamines are in some instances powerful carcinogens; avoid inhalation or contact.

Discussion. Reaction of Primary Amines. Both aliphatic and aromatic primary amines react with nitrous acid to give initially the corresponding diazonium ion. As might be anticipated, the aliphatic diazonium compounds are very much less stable than the aromatic, so much so that their existence has not been directly detected. Nitrogen gas and the alcohol, olefin, and products of other displacement and carbonium ion reactions are formed. R'^+ represents the carbonium ion formed by 1,2-rearrangement.

$$RNH_2 \xrightarrow[HCl]{HONO} R\text{—}N_2^+ \begin{cases} R^+ \longrightarrow ROH + \text{olefin} + RCl + ROR \\ \qquad\qquad\qquad + N_2 \\ R'^+ \longrightarrow R'OH + \text{olefin} + R'Cl + R'OR' \end{cases}$$

Aromatic diazonium salts, on the other hand, are generally stable in solution at 0°C. When heated in aqueous solution, they lose nitrogen to give the aryl cation, Ar^+. This ion reacts rapidly with water to give phenol.

The coupling reaction between certain diazonium salts and phenols has been shown to involve reaction between diazonium ion and phenoxide ion.[35] If the solution is too acidic the phenoxide ion is converted to phenol, and thus reaction is retarded; if the solution is too basic the diazonium ion reacts with hydroxide ion to give diazotate, ArN_2O^-, which does not couple. The solution must, therefore, be properly buffered for a satisfactory coupling reaction.

The diazotization of anthranilic acid produces a zwitterion or a cyclic acyl diazotate. This compound is unstable and loses carbon dioxide and nitrogen to form a highly reactive intermediate called "benzyne," formula A. The reactive benzyne combines with ethanol to form phenetole, with diethyl maleate to form a substituted benzocyclobutene, and with anthracene to form triptycene. The best yields of products are obtained by carrying out the diazotization with amyl nitrite and the displacement reaction in an aprotic solvent such as methylene chloride, tetrahydrofuran, or acetonitrile.[36]

$$\underset{\text{anthranilic acid}}{C_6H_4(CO_2^-)(\overset{+}{N}H_3)} \xrightarrow[HCl]{NaNO_2} \underset{(B)}{C_6H_4(CO_2^-)(N_2^+)} \rightleftharpoons \text{cyclic acyl diazotate } (C=O,\ O,\ N{=}N) \longrightarrow \underset{(A)}{C_6H_4} + CO_2 + N_2$$

---

[35] C. K. Ingold, *Structure and Mechanism in Organic Chemistry*, 2nd ed. (Cornell University Press, Ithaca, N.Y., 1967), p. 387. See also N. A. Frigero, *J. Chem. Educ.*, **43**, 142 (1966).

[36] *Note:* Danger; *o*-diazoniobenzoate (B) is explosive in the solid state. Avoid solvent removal.

Reaction with Secondary Amines. Both aliphatic and aromatic secondary amines react with nitrous acid to form *N*-nitroso compounds, commonly called nitrosamines.[37] They are pale yellow oils or solids:

$$R_2NH \xrightarrow{\text{HONO}} \underset{\text{nitrosamine}}{R_2N\text{—}NO}$$

Reaction with Tertiary Amines. The chemistry of the reaction of tertiary amines is quite complex.[38] Under certain conditions, it may appear that tertiary amines undergo no reaction; this is actually true only at low pH, low temperature, and dilute conditions. The amine is simply protonated to form salts under these mild conditions:

$$R_3N: \underset{\text{base}}{\overset{H^+}{\rightleftharpoons}} R_3\overset{+}{N}H$$

These salts can be recognized by their reaction with base to regenerate the original amine.

Under higher temperatures, less acidic conditions and other conditions, a variety of reactions occur when tertiary amines are treated with nitrous acid. For aliphatic amines:

$$R_3N: + HONO \rightarrow R_2N\text{—}NO + ROH$$

and for aromatic amines:

$$C_6H_5\ddot{N}Ar_2 + HONO \longrightarrow ON\text{—}C_6H_4\text{—}\ddot{N}Ar_2$$

and for mixed aryl-alkyl amines:

$$C_6H_5\ddot{N}R_2 + HONO \longrightarrow C_6H_5\text{—}\underset{}{\overset{NO}{\overset{|}{N}}}\text{—}R + ON\text{—}C_6H_4\text{—}\ddot{N}R_2$$

Although nitrous acid is most useful for characterizing amines, other functional groups also react. A methylene group adjacent to a keto group is converted to an oximino group,[39] and alkyl mercaptans yield red *S*-alkyl thionitrites.

$$CH_3COCH_2CH_3 \xrightarrow{\text{HONO}} CH_3CO\text{—}\underset{\underset{N\text{—}OH}{\|}}{C}\text{—}CH_3 + H_2O$$

$$RCH_2SH \xrightarrow{\text{HONO}} \underset{\text{red}}{RCH_2SNO} + H_2O$$

A very simple test for amines involves spot-test treatment with aqueous copper(II) sulfate; amines often produce intense blue to blue-green solutions due

---

[37] *Caution:* Many of these compounds are carcinogenic and should be handled carefully.

[38] G. E. Hein, *J. Chem. Educ.*, **40,** 181 (1963).

[39] W. L. Semon and V. R. Damrell, *Org. Syntheses*, **Coll. Vol. II,** 204 (1943).

to complexation of the Cu(II):

$$x\text{RNH}_2 + \text{CuSO}_4 \xrightarrow{\text{H}_2\text{O}} \text{Cu(NH}_2\text{R)}_x^{2+} + \text{SO}_4^{2-}$$

The procedure is described by Pasto and Johnson (p. 415).

The ir and nmr spectra of typical amines are shown in Figs. 6.15 and 6.16, respectively. In principle, and very often in practice, the ir spectra of amines lead to conclusions as to the number of hydrogen atoms on the nitrogen atom of the amino group. The N—H stretch of amino groups that are not involved in hydrogen bonding occurs in the range 3550–3320 cm^{-1} (2.82–3.01 μm); hydrogen bonding moves the N—H stretch to the 3300–3000 cm^{-1} (3.03–3.33 μm) region. In view of the coupled interaction of the two N—H bonds of a primary amine, it is often possible to see two peaks due to N—H stretch:

—N(H)(H) symmetric N—H stretch, 3400 cm^{-1} (2.94 μm)

—N(H)(H) asymmetric N—H stretch, 3490 cm^{-1} (2.87 μm)

Secondary amino groups, since they have only one N—H bond, usually display only a single N—H stretching band; pure tertiary amines (R_3N) do not display N—H stretching bands.

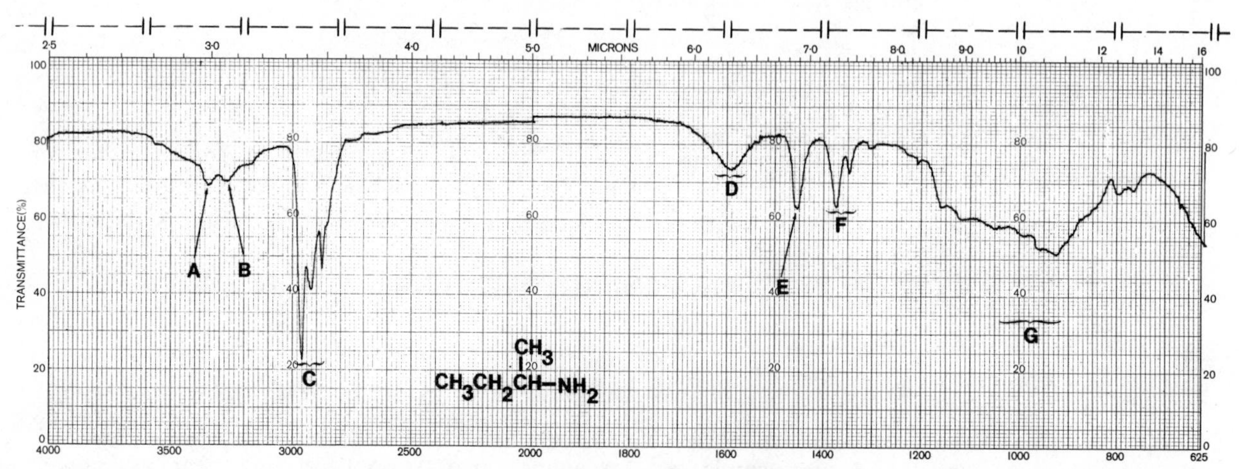

Figure 6.15. Infrared spectrum of *sec*-butylamine: neat. N—H stretch: **A**, 3347 cm^{-1} (2.99 μm), ν_{asym}. NH_2; **B**, 3285 cm^{-1} (3.04 μm), ν_{sym}. NH_2 (**A** and **B** coupled). C—H stretch: **C**, 3000–2850 cm^{-1} (3.33–3.51 μm), ν_{CH_3}, ν_{CH_2}. N—H bend: **D**, 1590 cm^{-1} (6.29 μm), δ_{NH_2} (scissoring). C—H bend: **E**, 1457 cm^{-1} (6.86 μm), δ_{CH_3} (scissoring); **F**, 1375, 1348 cm^{-1} (7.27, 7.42 μm), δ_{CH_3}, δ_{CH_2}. Nitrogen: **G**, 1200–800 cm^{-1} (8.33–12.50 μm), C—N stretch and N—H wag (neat sample), mostly N—H wag.

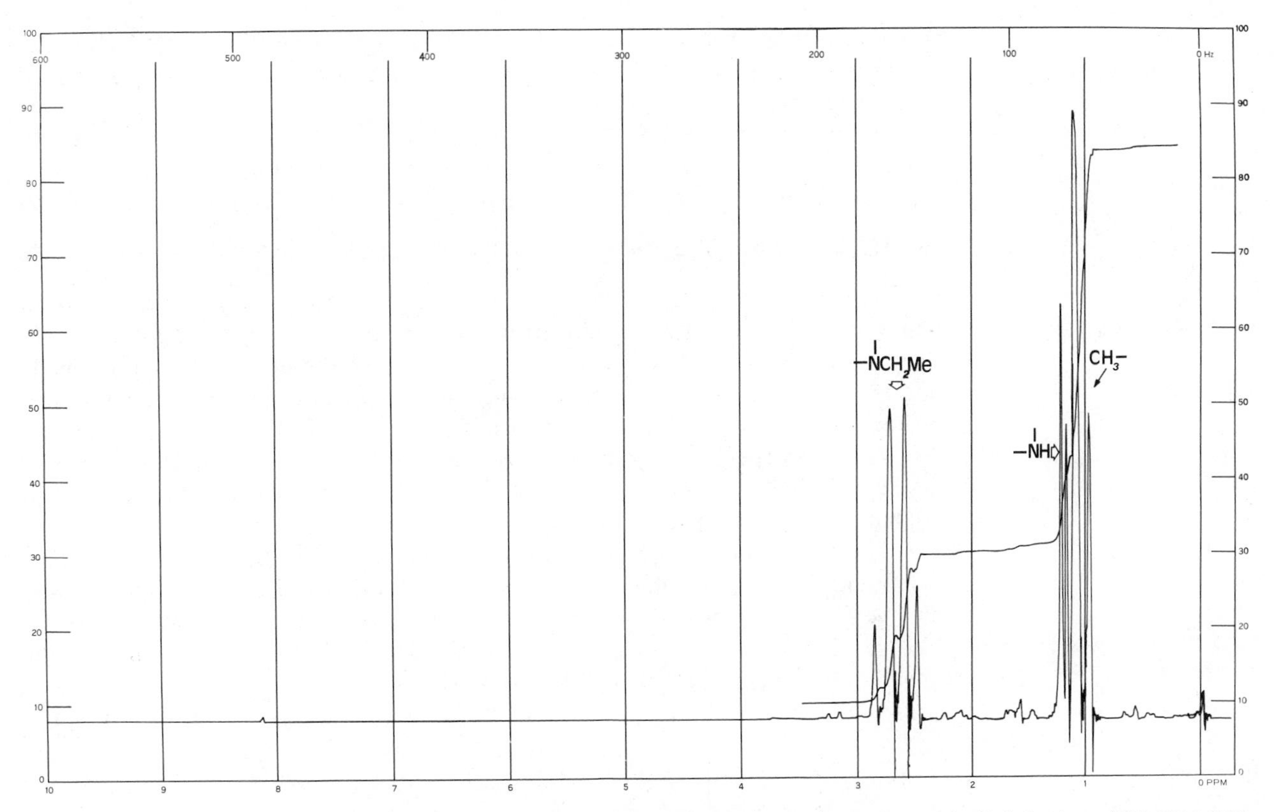

Figure 6.16. Nuclear magnetic resonance (proton) spectrum of diethylamine, $(CH_3CH_2)_2NH$: 60 MHz, 600-Hz sweep width, $CDCl_3$ solvent.

Unfortunately, there are a few exceptions to these correlations. The ir spectra of primary amines, when determined under conditions leading to inadequate resolution, can give rise to only a single N—H stretching band. In addition, the ir spectra of secondary amines under special conditions can give rise to more than one N—H stretching absorption. Thus it is extremely important that one ascertain the class of amine by both spectral *and* chemical tests to ensure the validity of the final structural assignment.

The ir spectra of the hydrochlorides of amines show broad bands due to N—H stretch in the 3175–2381 cm^{-1} (3.15–4.20 μm) region; multiple bands in this region are quite possible.

The nmr spectra of primary and secondary amines, determined under usual conditions, show the N—H absorptions as broad bands (with no discernible splitting) in the region $\delta 0.5$–3.0 ($\tau 9.5$–7.0). The broad character of these absorptions is due to rapid exchange of the N—H protons and to the quadrupole moment of nitrogen. Under these conditions, the N—H protons will not split the signals due to protons on carbon attached to these nitrogens (HC—NH). If the exchange is stopped, such as by carefully drying the amines with a Na—K alloy or by converting the amino group to an acetamido group $\left(CH_3\overset{\overset{\displaystyle O}{\|}}{C}NH\right)$, the *H*C—N*H* coupling will be observable in the C—H reasonance absorption (J = ~5 Hz). The N—H resonance will continue to be a broad singlet due to the aforementioned quadrupole moment of nitrogen.

It has been reported[40] that the classification of amines can be aided by determining the nmr spectrum of the amine in trifluoroacetic acid solvent. Under these conditions, the amine is completely converted to its trifluoroacetate salt ($RNH_3^{+}\,^{-}O_2CCF_3$). The multiplicity of the H_α signal in the H_α—C—$\overset{+}{N}$— unit will often indicate the number of protons on the nitrogen atom. Treatment of benzylamine, $C_6H_5CH_2NH_2$, with trifluoroacetic acid gives rise to a CH_2 absorption at $\delta 4.4$ ($\tau 5.6$) that is a quartet ($J = 7$ Hz); since the benzene ring protons do not couple with the methylene group, the multiplicity must be due to three protons on the nitrogen in the salt form ($C_6H_5CH_2NH_3^{+}$).

The mass spectra of amines can be very useful; molecular ion intensities of acyclic aliphatic amines are frequently weak and those of aromatic and cyclic aliphatic amines are usually strong.

The magnitude of *m/e* of the molecular ion of amines and all other nitrogen compounds is very useful: As discussed in Chapter 4, a nitrogen compound with an even number for the molecular ion mass *cannot* have an odd number of nitrogen atoms. A nitrogen compound with a odd-numbered molecular weight must contain an odd number of nitrogen atoms.

An important cleavage of amines is that facilitated by the nitrogen atom:

$$\left[-\overset{|}{\underset{|}{C}}-\overset{|}{\underset{|}{C}}-\overset{+}{N}\!\!< \right] \longrightarrow -\overset{|}{\underset{|}{C}}\cdot + \;>\!C{=}\overset{+}{N}\!<$$

Such fragmentation gives rise to the following peaks that are usually of substantial intensity and of significant use in structure determination:

Fragment	Mass (*m/e*)	Comment
$CH_2NH_2^+$	30	If strong, implies primary amine that is not α-branched
$RCHNH_2^+$	29+R	Will arise predominately via cleavage of largest (R′) α-substituent; mass of R′ and R allow conclusions regarding structure
$C_2H_4N^+$	42	Implies cyclic amine
$[M-HCN]^+$	66	Implies an aniline type of structure
$[M-HCN-H]^+$	65	Implies an aniline type of structure
$[M'-HCN]^+$	66+R	Implies aniline with R substituent on benzene ring
$[M'-HCN-H]^+$	65+R	Implies aniline with R substituent on benzene ring

[40] W. R. Anderson jr., and R. M. Silverstein, *Anal. Chem.*, **37**, 1417 (1965).

Because of the conjugation of the nonbonded pair of amine electrons with the aromatic ring of anilines, these compounds are usually much less basic than aliphatic amines (see Chapter 5); anilines are, however, still soluble in mineral acids:

$$Ar\ddot{N}H_2 + HCl \rightarrow ArNH_3^+Cl^-$$

Most of the chemical tests used to identify aliphatic amines described above are usable for anilines. Spectral analyses, as far as the amino group itself is concerned, are very similar to those analyses that were described for aliphatic amines. The aromatic ring does, however, induce certain unique spectral features; ir and nmr spectra of typical anilines are shown in Figures 6.17 and 6.18, respectively. The derivatization procedures described below for amines can virtually all be used for anilines.

Ultraviolet spectra can be very useful for the characterization of anilines. The aromatic nucleus means that the bands will be intense and reasonably well defined. The amino functional group attached to the aromatic nucleus is strongly supported when the following typical bathochromic shifts (shifts to longer wavelengths) are observed:

	E_2 band $\lambda_{max}(\varepsilon_{max})$	*B* band $\lambda_{max}(\varepsilon_{max})$
C_6H_5—$\ddot{N}H_2$ (water)	230 nm (8600)	280 nm (1430)
C_6H_5—NH_3^+ (aq. acid)	203 nm (7500)	254 nm (160)

The most useful derivatives of primary and secondary amines take advantage of their reactive N—H bond. Of these the benzenesulfonamides are frequently used because the preparation of these derivatives is involved in the Hinsberg method for classifying amines (Procedure 16 below). When the benzensulfonamides prove unsuitable, recourse may be had to numerous similar derivatives each of which presents advantages in the identification of certain types of amines. *p*-Toluene- (Procedure 18a), *p*-bromobenzene- (Procedure 18b), *m*-nitrobenzene- (Procedure 18b), methane- (Procedure 18b), and α-naphthalenesulfonamides (Procedure 18b) are recommended.

The phenylthioureas are especially valuable for characterizing low-molecular-weight, water-soluble amines. They are formed by reaction with phenyl isothiocyanate (Procedure 19):

$$RNH_2 + C_6H_5NCS \rightarrow C_6H_5NHCSNHR$$

$$R_2NH + C_6H_5NCS \rightarrow C_6H_5NHCSNR_2$$

The reagent phenyl isothiocyanate is not sensitive to water; this reaction may be carried out with dilute aqueous solutions of the amines.

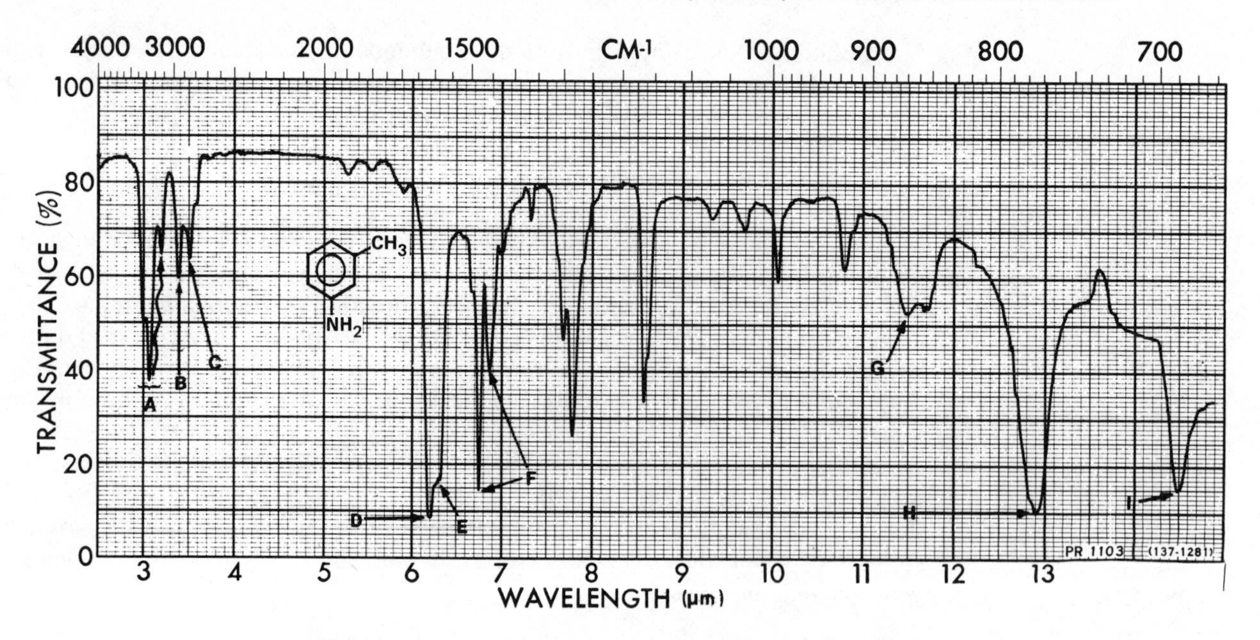

Figure 6.17. Infrared spectrum of *m*-toluidine: neat. N—H stretch (associated): **A**, 3257, 3145 cm^{-1} (3.07, 3.18 μm), νNH_2 (coupled pair). C—H stretch: **B**, 2959 cm^{-1} (3.38 μm), $\nu_{C—H}$ (aromatic); **C**, 2849 cm^{-1} (3.51 μm), $\nu_{C—H}$ (aliphatic). C⋯C stretch: **D**, 1613 cm^{-1} (6.20 μm), $\nu_{C⋯C}$ (aromatic). N—H bend: **E**, 1587 cm^{-1} (6.30 μm), $\nu_{N—H}$ (scissoring). C⋯C stretch: **F**, 1481, 1458 cm^{-1} (6.75, 6.86 μm), $\nu_{C⋯C}$ (aromatic). N—H wag: **G**, 870 cm^{-1} (11.5 μm). ⋯C—H bend (out of plane): **H**, 775 cm^{-1} (12.9 μm). C⋯C bend (out of plane): **I**, 690 cm^{-1} (14.5 μm).

The amides of acetic (Procedure 17a), benzoic (Procedure 17b), and *p*-nitrobenzoic acids (Procedure 17c) are conveniently prepared by treatment of the amine, respectively, with acetic anhydride, benzoyl chloride, or *p*-nitrobenzoyl chloride.

$$RNH_2 + (CH_3CO)_2O \rightarrow RNHCOCH_3 + CH_3COOH$$

$$2RNH_2 + C_6H_5COCl \rightarrow RNHCOC_6H_5 + RNH_3Cl$$

$$2RNH_2 + O_2N\text{-}C_6H_4\text{-}COCl \rightarrow O_2N\text{-}C_6H_4\text{-}CONHR + RNH_3Cl$$

Acetyl and benzoyl derivatives are known for nearly all primary and secondary amines, and for this reason these derivatives are very useful.

3-Nitrophthalic anhydride reacts with primary and secondary amines to produce phthalamic acids. The phthalamic acid from the primary amine undergoes dehydration when heated to 145°C and forms the alkyl 3-nitrophthalimide. That from the secondary amine is stable to heat.

As has been mentioned before, preparation of such phthalamic acid derivatives (Procedure 4) can provide an easily manipulated derivative of a chiral compound, for example, of an amino acid that is optically active. Manipulation of these compounds can be carried out by acid-base extraction via reactions of the

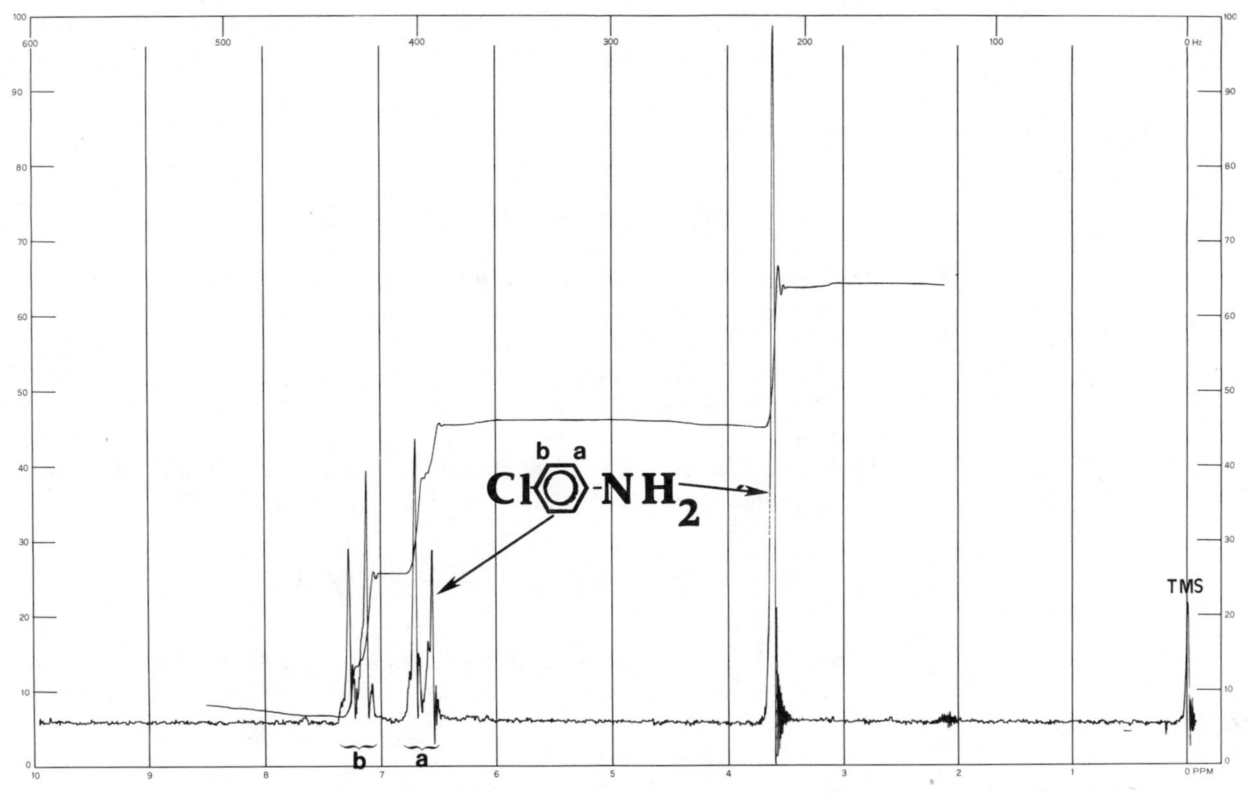

Figure 6.18. Nuclear magnetic resonance (proton) spectrum of *p*-chloroaniline: 60 MHz, 600-Hz sweep width, $CDCl_3$ solvent.

3-Nitrophthalic anhydride $\xrightarrow{RNH_2}$ 2-(CO_2H)-6-($CONHR$)-nitrobenzene $\xrightarrow[-H_2O]{\Delta}$ *N*-R-3-nitrophthalimide

3-Nitrophthalic anhydride $\xrightarrow{R_2NH}$ 2-(CO_2H)-6-($CONR_2$)-nitrobenzene $\xrightarrow{\Delta}$ no reaction

carboxyl group. This reagent therefore offers an additional means of distinguishing among primary, secondary, and tertiary amines and furnishes derivatives for the primary and secondary amines.

Other types of derivatives that have been found valuable in the characterization of amines are 2,4-dinitrobenzylidene derivatives, azo compounds (Experiment 19, p. 220), salts of *p*-toluenesulfonic acid, molecular compounds with phenol, compounds with *p*-nitrobenzyl halides, and picrates (Procedure 22).

Primary and secondary amines combine with aryl isocyanates or acyl azides to produce substituted ureas.

$$ArN{=}C{=}O + RNH_2 \longrightarrow ArNHCONHR$$

$$ArN{=}C{=}O + R_2NH \longrightarrow ArNHCONR_2$$

$$ArCON_3 \xrightarrow{\Delta} ArN{=}C{=}O \xrightarrow{RNH_2} ArNHCONHR$$

Phenyl-, α- and β-naphthyl-, *m*- and *p*-bromophenyl-, and *p*-chlorophenyl-, *m*- and *p*-nitrophenyl-, and 3,5-dinitrophenylureas have been described.

A number of amine hydrochlorides may be obtained by passing dry hydrogen chloride into an ether or benzene solution of the amine. Some of the hydrochlorides have satisfactory melting or decomposition points and hence may serve as derivatives.

Hydrochlorides of amino alcohols are best made by neutralizing *n*-propyl alcohol solutions of the alkanolamines with dry hydrogen chloride.

$$RNH_2 \xrightarrow{\text{HCl gas}} RNH_3^+Cl$$

Procedure 16. Benzenesulfonyl Chloride (Hinsberg's Method for Characterizing Primary, Secondary, and Tertiary Amines)

$$C_6H_5SO_2Cl + RNH_2 + 2NaOH \rightarrow Na^+[C_6H_5SO_2NR]^- + NaCl + 2H_2O$$

$$C_6H_5SO_2Cl + R_2NH + NaOH \rightarrow C_6H_5SO_2NR_2 + NaCl + H_2O$$

The amounts described below correspond to those for a qualitative test; a preparative scale, leading to isolable quantities of product, should involve a 5- to 50-fold increase in all reagents.

To 0.3 ml of aniline in a test tube add 5 ml of 10% sodium hydroxide solution and 0.4 ml of benzenesulfonyl chloride. Stopper the test tube, and shake the mixture very vigorously. Test the solution to make sure that it is alkaline. After all the benzenesulfonyl chloride has reacted, cool the solution and filter or decant from any residue (A). Note whether residue A is a solid or liquid and whether it is lighter or heavier than the alkaline solution. What information can you deduce from these observations? Test the solubility of residue A in water and in dilute hydrochloric acid. Note that solubility of A in hydrochloric acid indicates that the original was a tertiary amine. The sodium salts of certain sulfonamides of high molecular weight may be insoluble in the alkaline solution.[41] Usually they are soluble in water.

If the reaction mixture heats up considerably, it should be cooled. Certain *N,N*-dialkylanilines produce a purple dye if the mixture becomes too hot. This may be prevented by carrying out the reaction at 15 to 20°C.

Acidify the clear filtrate. Scratch the test tube to hasten crystallization of the product (B).

Repeat this test, using methylaniline and dimethylaniline instead of aniline. Some secondary amines react slowly, and it is occasionally necessary to warm the reaction mixture.

Show by means of a diagram how the above procedure distinguishes among primary, secondary, and tertiary amines.

When the Hinsberg method is used to separate mixtures of amines, it is necessary to recover the pure individual amines. The benzenesulfonamides may be hydrolyzed as follows.

$$C_6H_5SO_2NHR + H_2O + HCl \rightarrow C_6H_5SO_3H + RNH_3Cl$$

$$C_6H_5SO_2NR_2 + H_2O + HCl \rightarrow C_6H_5SO_3H + R_2NH_2Cl$$

The sulfonamide is hydrolyzed by heating 10 g of it with 100 ml of 25% hydrochloric acid under reflux. Sulfonamides of primary amines require 24 to 36 hr refluxing, whereas sulfonamides of secondary amines may be hydrolyzed in 10 to 12 hr. After solution is complete, the mixture is cooled, made alkaline with 20% sodium hydroxide solution, and extracted with three 50-ml portions of ether. The ether solution is dried, and, after the ether has been driven off, the amine is distilled. With certain very low- or very high-boiling amines it is often more convenient to recover them as hydrochlorides by passing dry hydrogen chloride gas into the dry ether solution.

Many sulfonamides are hydrolyzed only with great difficulty; a more satisfactory procedure involves 48% hydrobromic acid and phenol.[42] This reaction is not simple hydrolysis but is a reductive cleavage in which the hydrogen bromide is oxidized to bromine and the sulfonamide reduced to the disulfide. The primary purpose of the phenol is to remove the bromine by the formation of *p*-bromophenol.

$$2ArSO_2NR_2 + 5HBr + 5C_6H_5OH \rightarrow ArSSAr + 2R_2NH + 5BrC_6H_4OH + 4H_2O$$

---

[41] The benzenesulfonamides of cyclohexyl- through cyclodecylamine and certain high-molecular-weight amines are insoluble in 10% sodium hydroxide solution. See P. E. Fanta and C. S. Wang, *J. Chem. Educ.*, **41,** 280 (1964). Certain primary amines may yield alkali-insoluble disulfonyl derivatives. These may be hydrolyzed by boiling for 30 min with 5% sodium ethoxide in absolute ethanol.

[42] H. R. Snyder and R. E. Heckert, *J. Amer. Chem. Soc.*, **74,** 2006 (1952); H. R. Snyder and H. C. Geller, *ibid.*, **74,** 4864 (1952).

Discussion. Arenesulfonyl chlorides can be useful in characterizing primary and secondary amines. The Hinsberg method for separating amines is based on the fact that the sulfonamides of primary amines are soluble in alkali whereas those of secondary amines are not. Since tertiary amines do not give amides, the method provides a means of classifying and separating the three types of amines. However, the results of the Hinsberg test must not be used alone in classifying amines; it is necessary to consider also the solubility of the original compound. If that compound is amphoteric, that is, soluble in both acids and alkalies, the Hinsberg method fails to distinguish among the classes. For example, *p*-(*N*-methylamino)-benzoic acid reacts with benzenesulfonyl chloride and alkali to give a *solution* of the sodium salt of the *N*-benzenesulfonyl derivative.

$$p\text{-}(CH_3NH)C_6H_4COOH + C_6H_5SO_2Cl + 2NaOH \longrightarrow p\text{-}(CH_3NSO_2C_6H_5)C_6H_4COONa + NaCl + 2H_2O$$

Acidification precipitates the free acid; this fact taken by itself would indicate that the original compound was a primary rather than a secondary amine.

Procedure 17. Substituted Amides from Amines

$$\text{(a) } RNH_2 \text{ (or } R_2NH) \xrightarrow{HCl,\ NaOH} \xrightarrow[CH_3CO_2Na]{(CH_3CO)_2O} RNHCCH_3 \text{ (or } R_2NCCH_3) \quad (C{=}O)$$

$$\text{(b) } RNH_2 \text{ (or } R_2NH) \xrightarrow[C_5H_5N,\ C_6H_6]{C_6H_5COCl} C_6H_5CNHR \text{ (or } C_6H_5CONR_2) \quad (C{=}O)$$

$$\text{(c) } RNH_2 \text{ (or } R_2NH) \xrightarrow[\text{base}]{ArCOCl} ArCNHR \text{ (or } ArCONR_2) \quad (C{=}O)$$

$$Ar = C_6H_5,\ p\text{-}O_2NC_6H_4—$$

(a) ***Substituted Acetamides from Water-Insoluble Amines.*** A solution of the amine is prepared by dissolving about 0.5 g of the compound in 25 ml of 5% hydrochloric acid. Small portions of a 5% sodium hydroxide solution are added until the mixture becomes cloudy; the turbidity is then removed by adding 2 to 3 ml of 5% hydrochloric acid. A few chips of ice are added, followed by 5 ml of acetic anhydride. The mixture is stirred or swirled vigorously, and a previously prepared solution of 5 g of sodium acetate (trihydrate) in 5 ml of water is added in one portion. If the product does not crystallize, the mixture is chilled overnight.

Recrystallization may be effected from cyclohexane or from a mixture of cyclohexane and benzene. The acetamide must be thoroughly dry before recrystallization is attempted from these solvents. An ethanol-water mixture may also be used for recrystallization.

(b) *Substituted Benzamides*: *Pyridine Method.* To a solution of 0.5 g of the compound in 5 ml of dry pyridine and 10 ml of dry benzene is added, dropwise, 0.50 ml of benzoyl chloride. The resulting mixture is heated in a water bath at 60 to 70°C for 30 min and is then poured into 100 ml of water. The benzene layer is separated, and the aqueous solution is washed once with 10 ml of benzene. The combined benzene solutions are washed with water and with 5% sodium carbonate solution and dried with a little anhydrous magnesium sulfate. The drying agent is removed by filtration through a fluted filter, and the benzene is evaporated to a small volume (3 to 4 ml). Hexane (about 20 ml) is stirred into the mixture, and the crystalline benzoyl derivative is removed by filtration and washed with hexane. Recrystallization may usually be effected from a mixture of cyclohexane and hexane or from a mixture of cyclohexane and ethyl acetate. Ethanol or aqueous ethanol may also be used with many compounds.

(c) *Substituted Benz- and p-Nitrobenzamides.* The regular procedure for the Schotten-Baumann reaction described under Experiment 18(e), p. 219, may be used. Two modified procedures are the following.

1. A mixture of 20 ml of 5% sodium hydroxide solution, 5 ml of chloroform, 0.5 g of the compound, and 0.5 ml of benzoyl chloride is shaken or stirred for about 20 min and then allowed to stand for 12 hr. The chloroform layer is separated, and the aqueous layer is washed with 10 ml of chloroform. The combined chloroform solutions are washed with water, dried with anhydrous magnesium sulfate, and evaporated to a small volume (2 to 3 ml). Hexane (20 to 25 ml) is stirred into the solution, and the derivative is removed by filtration and washed with hexane.

2. About 1 ml of the amine is added to a solution of 1 g of benzoyl chloride or *p*-nitrobenzoyl chloride in 20 ml of dry benzene. The resulting solution is boiled for 15 min under a reflux condenser and is then allowed to cool. The solution is filtered, and the precipitate is washed with 10 ml of warm benzene, the washings being added to the original filtrate. The benzene solution is next washed with 10 ml of 2% sodium carbonate solution, then with 10 ml of 2% hydrochloric acid, and finally with 10 ml of distilled water. The benzene is evaporated, and the residue is recrystallized from dilute ethanol.

Procedure 18. Sulfonamides from Amines

$$R_2NH \text{ (or } RNH_2) \xrightarrow[\text{(or } CH_3SO_2Cl)]{ArSO_2Cl} R_2N{-}\overset{\overset{O}{\uparrow}}{\underset{\underset{O}{\downarrow}}{S}}{-}Ar \text{ (or } RNH{-}\overset{\overset{O}{\uparrow}}{\underset{\underset{O}{\downarrow}}{S}}{-}Ar)$$

(a) *Benzene- and p-Toluenesulfonamides.* These derivatives are prepared by the procedure outlined in Procedure 17, sufficient amounts of material being used to permit recrystallization of the final product from 95% ethanol.

(b) *Benzyl-, p-Bromobenzene-, m-Nitrobenzene-, α-Naphthyl-, and Methanesulfonamides.* A solution of 1 g of the sulfonyl chloride in 25 ml of dry benzene is prepared, and 2 ml of the amine is added. The solution is shaken and allowed to stand for 10 min. The amine hydrochloride is removed by filtration, and the filtrate is evaporated. The residue is recrystallized once or twice from dilute ethanol.

Procedure 19. Phenylthioureas from Amines

$$C_6H_5\text{—}NCS + R_2NH \text{ (or } RNH_2) \longrightarrow C_6H_5\text{—}NH\text{—}\overset{S}{\overset{\|}{C}}\text{—}\overset{S}{\overset{\|}{C}}NR_2$$

Equal amounts of the amine and phenyl isothiocyanate are mixed in a test tube and shaken for 2 min. If no reaction occurs spontaneously, the mixture is heated for 3 min over a low flame. The aliphatic amines usually react immediately, whereas the aromatic amines require heating. The mixture is then kept in a beaker of ice until the mass solidifies. The solid is powdered and washed with ligroin and 50% ethanol in order to remove any excess of either reactant. The residue is then recrystallized from 95% ethanol.

Other tests have been described that utilize the N—H bond of primary and secondary amines. The periodic acid test (Experiment 5 above) can be used as a test for a NH_2 group vicinal to a hydroxyl group $\left(\text{—}\overset{OH}{\overset{|}{\underset{|}{C}}}\text{—}\overset{NH_2}{\overset{|}{\underset{|}{C}}}\text{—}\right)$; the periodic acid reagent can also be used to degrade such an amino alcohol as a method of derivative preparation.

Treatment of primary and secondary amines with 2,4-dinitrobenzenesulfenyl chloride can give rise to the corresponding arenesulfenamide; details are given in Pasto and Johnson (p. 421):

$$2,4\text{-}(O_2N)_2C_6H_3\text{—}SCl \xrightarrow{2R_2NH} \underset{\text{arenesulfenamide}}{Ar\text{—}SNR_2} + R_2NH_2^+Cl^-$$

Occasionally the thiourea derivative reversibly undergoes a complicating disproportionation reaction with the original amine.

$$C_6H_5NH\overset{S}{\overset{\|}{C}}NHR + RNH_2 \rightleftharpoons RNH\overset{S}{\overset{\|}{C}}NHR + C_6H_5NH_2$$

$$C_6H_5NH\overset{S}{\overset{\|}{C}}NHR + C_6H_5NH_2 \rightleftharpoons C_6H_5NH\overset{S}{\overset{\|}{C}}NHC_6H_5 + RNH_2$$

This complication is averted if unduly long heating times are avoided.

Amines—Tertiary. Tertiary amines vary so greatly in nature that no type of derivative has been found to be generally applicable. Perhaps the most useful derivatives are the quaternary ammonium salts formed by the combination of the amine with methyl iodide (Procedure 20a), benzyl chloride (Procedure 20b), or methyl *p*-toluenesulfonate (Procedure 20b).

$$R_3N + CH_3I \rightarrow (R_3NCH_3)^+I^-$$
$$R_3N + C_6H_5CH_2Cl \rightarrow (R_3NCH_2C_6H_5)^+Cl^-$$
$$R_3N + CH_3C_6H_4SO_3CH_3 \rightarrow (R_3NCH_3)^+(CH_3C_6H_4SO_3)^-$$

The salts of halogen acids, picric acid (Procedure 22), *p*-toluenesulfonic acid, and chloroplatinic acid are also employed frequently. Picrolonates (salts of 3-methyl-4-nitro-1-*p*-nitrophenyl-5-pyrazolone) may be prepared by the same procedure given for picrates (Procedure 22). The picrolonates are among the most generally satisfactory derivatives of tertiary amines and are often formed readily when other derivatives offer difficulty. Reineckates [salts of the acid $H^+Cr(NH_3)_2(SCN)_4$] have also been employed.

Certain *N,N*-dialkylanilines react with nitrous acid to give *p*-nitroso derivatives, and these may sometimes serve as derivatives.

$$R_2N\text{-}C_6H_5 + HONO \rightarrow R_2N\text{-}C_6H_4\text{-}NO + H_2O$$

Procedure 20. Quaternary Ammonium Salts of Tertiary Amines

(a) $R_3N: + CH_3I \longrightarrow R_3\overset{+}{N}\text{—}CH_3\ I^-$

(b) $R_3N: \xrightarrow{C_6H_5CH_2Cl} R_3\overset{+}{N}CH_2C_6H_5\ Cl^-$

$R_3N: \xrightarrow{CH_3OSO_2\text{-}C_6H_4\text{-}CH_3} R_3\overset{+}{N}\text{—}CH_3\ ^-O_3S\text{-}C_6H_4\text{-}CH_3$

(a) A mixture of 0.5 g of the amine and 0.5 ml of methyl iodide is warmed in a test tube over a low flame for a few minutes and is then cooled in an ice bath. The tube is scratched with a glass rod to hasten crystallization. The product is purified by recrystallization from absolute ethyl or methyl alcohol or from ethyl acetate.

(b) One gram of the amine is added to a solution of 2 to 3 g of benzyl chloride or methyl *p*-toluenesulfonate in 10 ml of dry benzene.[43] The solution is boiled for 10 to 20 min and cooled. The products are recrystallized by dissolving them in the least possible amount of boiling ethanol; ethyl acetate is added

[43] Toluene, ethanol, acetone, or tetrachloroethane may be substituted for benzene solvent. The toxicity of benzene as well as amines and halides are discussed in Appendix IV.

until precipitation starts, and the mixture is cooled. The product is removed by filtration and quickly dried by being rubbed on a clay plate; the melting point is determined immediately.

Procedure 21. Nitroso Compounds from N,N-Dialkylanilines

$$C_6H_5-NR_2 \xrightarrow{HONO} ON-C_6H_4-NR_2 + H_2O$$

Two grams of the amine is dissolved in 20 ml of 10% hydrochloric acid, and the solution is cooled in a freezing mixture. A solution of 1.5 g of sodium nitrite dissolved in 2 ml of water is added slowly with vigorous stirring. The mixture is allowed to stand 15 min in the ice bath, and then the hydrochloride of the nitroso compound is collected on a filter. The yellow crystals are mixed with 3 ml of water and placed in an ice bath. Dilute sodium hydroxide solution is added until the solution is alkaline. The green nitroso compound is extracted with ether, and the ether is allowed to evaporate. The nitroso compound is deposited as brilliant green crystals. These are collected on a filter and dried on a porous plate.

Procedure 22. Picrates from Amines and from Aromatic Hydrocarbons

$$(O_2N)_3C_6H_2OH \xrightarrow{R_3N:} (O_2N)_3C_6H_2O^- \quad H\overset{+}{N}R_3$$

$$(O_2N)_3C_6H_2OH \xrightarrow[\text{(see Procedure 49)}]{ArH} (O_2N)_3C_6H_2OH \leftarrow ArH$$

(a) A sample of the compound (0.3 to 0.5 g) is added to 10 ml of 95% ethanol. If solution is not complete, the mixture is shaken until a saturated solution results and is then filtered. The filtrate is added to 10 ml of a saturated solution of picric acid in 95% ethanol, and the solution is heated to boiling. The solution is allowed to cool slowly, and the yellow crystals of the picrate are removed by filtration and recrystallized from ethanol. Certain picrates, especially those of hydrocarbons (Table, p. 556), dissociate when heated and consequently cannot be recrystallized. In such cases the original precipitate should be washed with a very small amount of ether and dried on a clay plate in preparation for the melting-point determination. (*Caution:* Some picrates explode when heated.)

(b) Equal amounts of the compound and picric acid are mixed in a test tube, which is then heated on the steam cone for 10 min or over a very low flame until the mixture melts. The solid is allowed to cool and, if sufficiently stable, is recrystallized from ethanol.

Picric acid is an oxidizing agent, as well as a *strong* acid. Addition complexes (π–π complexes) arise from the complexation of picric acid with some aromatic (usually π-electron rich) compounds. These "addition" compounds can be used to determine the molecular weight of the unknown compound as described below. Procedure 49 (Section 6.14) describes how picrates of aromatic ethers can be prepared.

A potentially invaluable use of picrate complexes of amines is their use for microdetermination of molecular weight.[44] These picrates have a uv maximum near 350 nm. The molecular weight (mol wt) of the complex can be determined (±2%) from the equation after having determined the absorbance ($\log I_0/I$) of the complex at 350 nm:

$$\text{mol wt} = \frac{13{,}440 \times C \times n}{\log (I_0/I)}$$

where C = concentration of complex in grams per liter, and n = mole ratio of picric acid/amine (or aromatic hydrocarbon) in the complex. Utilizing the molecular weight of picric acid (229), one can calculate the molecular weight of the amine (or aromatic hydrocarbon).

Procedure 23. Chloroplatinate Salts from Amines

$$R_3N\!: + H_2PtCl_6 \cdot 6H_2O \xrightarrow[H_2O]{HCl} R_3\overset{+}{N}H\ HPtCl_6^-$$

To a solution of 0.5 g of the amine in 10 ml of 10% hydrochloric acid is added slowly with shaking 10 ml of a 25% aqueous solution of chloroplatinic acid ($H_2PtCl_6 \cdot 6H_2O$). The crystalline chloroplatinate that separates should be collected on a filter and washed with 10% hydrochloric acid. It may be recrystallized from ethanol containing a drop of concentrated hydrochloric acid to prevent hydrolysis. Contact of chloroplatinates with metal spatulas must be avoided, since it leads to the formation of metallic platinum.

Another procedure to derivatize amines involves formation of tetraphenylborate salts:

$$RNH_2 + HCl \rightarrow R\overset{+}{N}H_3Cl^- \xrightarrow{NaB(C_6H_5)_4} R\overset{+}{N}H_3B(C_6H_5)_4^-$$

$$R_4N^+X^- + NaB(C_6H_5)_4 \rightarrow R_4N^+B(C_6H_5)_4^-$$

---

[44] L. F. Fieser and M. Fieser, *Reagents for Organic Synthesis*, Vol. 1 (Wiley, New York, 1967), p. 884; K. G. Cunningham, W. Dawson, and F. S. Spring, *J. Chem. Soc.*, 2305 (1951).

This procedure, which applies as well to quaternary ammonium salts, is described by Pasto and Johnson (p. 424).

A number of additional tests for amines from Cheronis, Entrikin, and Hodnett are listed:

Reagent	Page in C, E, and H
Fluorescein chloride	379
Basicity test[a]	380
Quinhydrone	380
N-Halosuccinimide	380
3,3′,5,5′-Tetrabromophenolphthalein	381
Lignin	384
Chloranil	385
2,4-Dinitrofluorobenzene	385

[a] Basicity classifications allow possible identification of, e.g., amines by their pK_b values.

As described in Chapter 5, the basicity of amines is quite sensitive to structure (compare the basicity of anilines to aliphatic amines). The basicity of amines also frequently causes these compounds to be amenable to lanthanide shift reagents (Chapter 8) as an aid to nmr studies.

6.8 AMINO ACIDS

One can gain a preliminary notion as to the chemistry of amino acids by studying the amine (Section 6.7) and carboxylic acid (Section 6.12) portions of this chapter; one must keep in mind, however, that the zwitterion (polar) form of amino acids heavily influences the chemical and physical properties of these compounds:

$$\text{R—CH(CO}_2\text{H)(NH}_2) \rightleftharpoons \text{R—CH(CO}_2^-)(\text{NH}_3^+) \quad \text{zwitterion}$$

Examples of how the zwitterion influence can cause amino acids to differ from the usual organic compound include the facts that their melting points are usually decomposition points and that they are often water-soluble.

A variety of chemical tests are available that are specific for some amino acids (Table 6.2). These tests are only infrequently used, since the *amino acid analyzer* (Chapter 8) allows facile qualitative, as well as quantitative, instrumental detection of such compounds. Tryptophan (Try) might be difficult to perceive instrumentally due to its sensitivity to acid; the chemical test (Table 6.2) might be more appropriate here. The ninhydrin test, described in Experiment 20, is the chemical basis for the amino acid analyzer.

Representative ir (Fig. 6.19) and nmr (Fig. 6.20) spectra of amino acids have been provided here. It is especially important to note the nature of these sampling

Table 6.2. Chemical Tests[a] for Amino Acids

Amino acid detected	Name of Reaction	Reagents	Color
Arginine	Sakaguchi reaction	α-Naphthol and sodium hypochlorite	Red
Cysteine	Nitroprusside reaction	Sodium nitroprusside in dil. NH_3	Red
Cysteine	Sullivan reaction	Sodium 1,2-naphthoquinone-4-sulfonate and sodium hydrosulfite	Red
Histidine, tyrosine	Pauly reaction	Diazotized sulfanilic acid in alkaline solution	Red
Tryptophan	Ehrlich reaction	*p*-Dimethylaminobenzaldehyde in conc. HCl	Blue
Tryptophan	Glyoxylic acid reaction (Hopkins-Cole reaction)	Glyoxylic acid in conc. H_2SO_4	Purple
Tyrosine	Folin-Ciocalteu reaction	Phosphomolybdotungstic acid	Blue
Tyrosine	Millon reaction	$HgNO_3$ in nitric acid with a trace of nitrous acid	Red
Tyrosine, Tryptophan, Phenylalanine	Xanthoproteic reaction	Boiling conc. nitric acid	Yellow

[a] A number of the experimental procedures for these tests are described in J. P. Greenstein and M. Winitz, *Chemistry of the Amino Acids* (Wiley, New York, 1961).

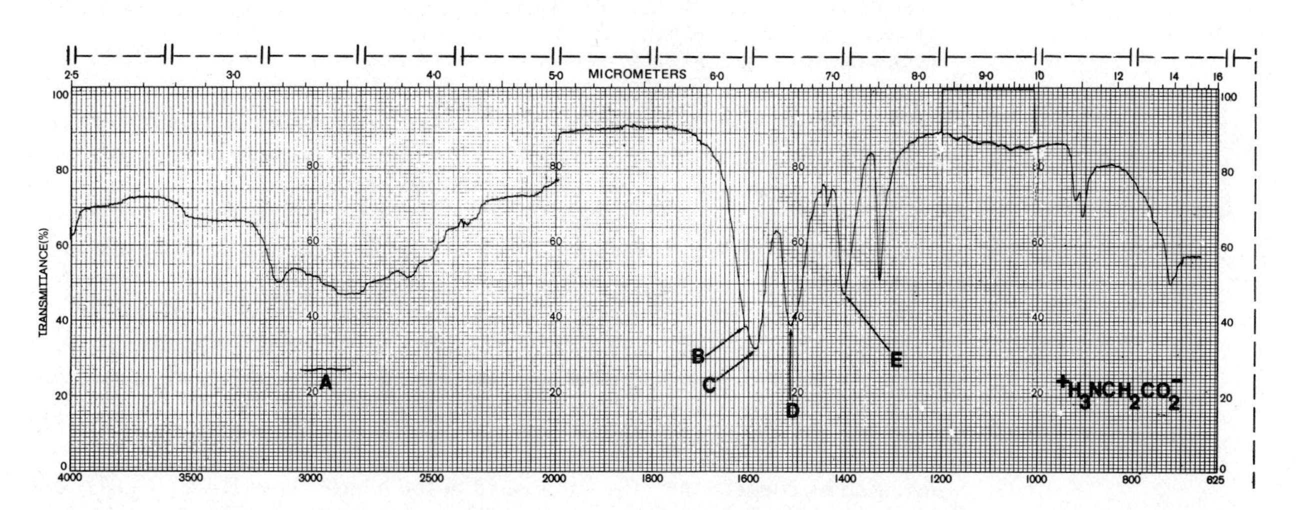

Figure 6.19. Infrared spectrum of glycine: KBr pellet. N—H and carboxylate vibrations: **A**, 3250–2300 cm^{-1} (3.08–4.35 μm); ν_{N-H} of $—NH_3^+$ group (C—H stretch is obliterated by **A**); **B**, 1608 cm^{-1} (6.22 μm) asymmetric N—H bend of $—NH_3^+$ group; **C**, 1570 cm^{-1} (6.37 μm), ν_{asym} $C(\overset{...}{—}O)_2^-$; **D**, 1530 cm^{-1} (6.54 μm), ν_{sym}, N—H bend of NH_3^+ group; **E**, 1403 cm^{-1} (7.13 μm), ν_{sym} $(C\overset{...}{—}O)_2^-$.

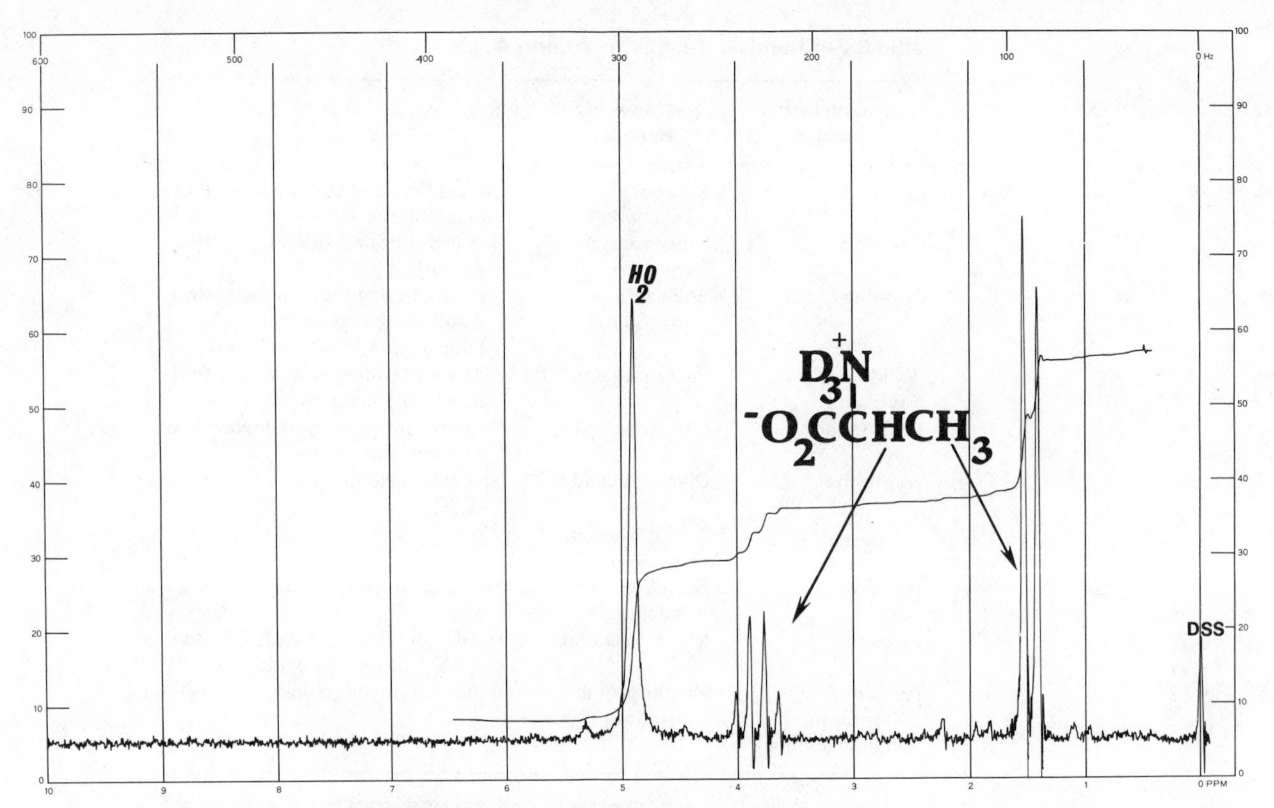

Figure 6.20. Nuclear magnetic resonance (proton) spectrum of L-alanine: 60 MHz, 600-Hz sweep width, D_2O solvent. The water peak is due largely to exchange of D_2O with NH_3^+ protons in alanine.

techniques used; amino acids usually have solubility properties very different from typical organic compounds.

Unless an aromatic substituent is present in the amino acid structure, the mass spectrum of an amino acid will show little or no molecular ion; this is especially so when standard ionization potentials are used because the fragmentation of amino acids is very facile under such conditions. Mass spectral analysis of amino acids is further limited by the low volatility of such zwitterionic compounds. Careful vacuum sublimation at temperatures of 150 to 240°C, which may result in decomposition of the pot residue, often will allow gaseous introduction of the sample into the spectrometer; corresponding sublimates are usually pure. Often, however, one is restricted to direct introduction of the solid sample into the spectrometer. Volatility, and hence sample manageability, is often greatly increased by conversion of the amino acid to the hydrohalide ($R\text{—}\overset{NH_3^+X^-}{\overset{|}{C}}HCO_2H$) salt, acetamides, trifluoroacetamides or various carboxylic acid ester derivatives. Trimethylsilyl derivatives can also be used.[45] Sodium salts do not lend themselves to vaporization.

[45] B. J. Mitruka, *Gas Chromatographic Applications in Microbiology and Medicine* (Wiley, New York, 1975), p. 152.

Amino acids can be divided into four general categories; this division is based on their acid-base and charge properties:

1. *Hydrophobic:* Amino acids that are substantially less water-soluble than class 2, for example, phenylalanine,

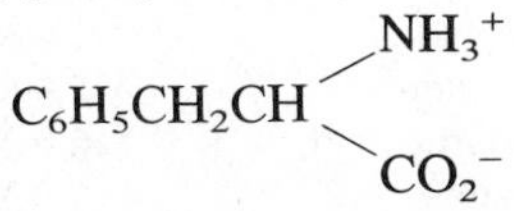

$$C_6H_5CH_2CH\begin{matrix} NH_3^+ \\ CO_2^- \end{matrix}$$

which is in solubility class $A_2(B)$, Chapter 5, is virtually insoluble in water.

2. *Hydrophilic* (*polar uncharged*): These amino acids have polar functional groups (OH) that, despite not having a positive or negative charge, are reasonably soluble in water, for example, threonine,

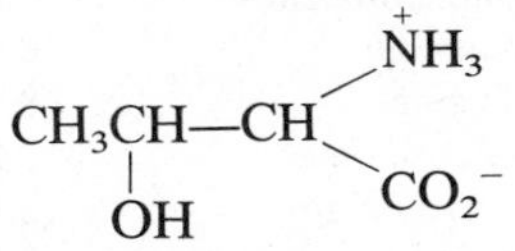

$$\underset{\displaystyle OH}{CH_3\underset{|}{C}H}—CH\begin{matrix} \overset{+}{N}H_3 \\ CO_2^- \end{matrix}$$

These compounds are differentiated from those in succeeding classes in that the polar groups are not appreciably acidic or basic (in the proton-transfer sense).

3. *Positively charged* (*basic*): Such amino acids have basic, usually nitrogenous, functions that are protonated at intracellular[46] pH. Lysine,

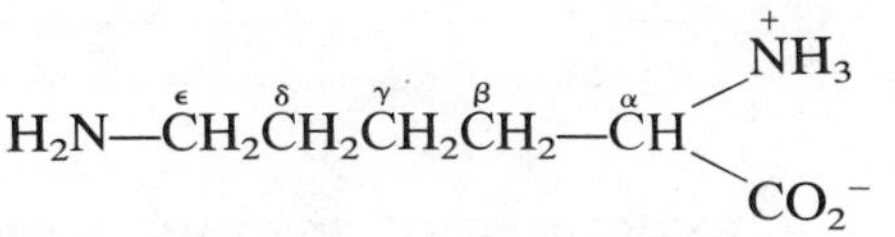

$$H_2N—\overset{\epsilon}{C}H_2\overset{\delta}{C}H_2\overset{\gamma}{C}H_2\overset{\beta}{C}H_2—\overset{\alpha}{C}H\begin{matrix} \overset{+}{N}H_3 \\ CO_2^- \end{matrix}$$

would also have its ε-amino group protonated under these conditions.

4. *Negatively charged* (*acidic*): These amino acids have an acidic, thus ionized, carboxyl group at intracellular pH. For example, the β-carboxyl group of aspartic acid,

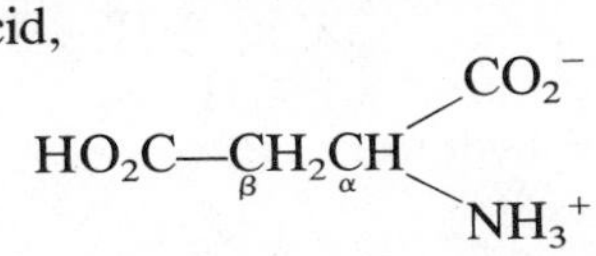

$$HO_2C—\underset{\beta}{C}H_2\underset{\alpha}{C}H\begin{matrix} CO_2^- \\ NH_3^+ \end{matrix}$$

would also be ionized to a carboxylate group under these conditions.

The melting points or decomposition points of amino acids are not exact. The values depend on the rate of heating. Hence, in using these constants to prepare a list of possibilities, allowance must be made for their inaccuracy.

The α-amino acids occurring naturally in plants and animals or obtained from the acid or enzymatic hydrolysis of proteins and peptides are optically active (except glycine) and belong to the *configurational* L-series. The specific rotations are valuable constants for identification. A list is given in Table 6.3. Specific rotations in other solvents and references to the literature may be found in recent editions of *Tables for Identification of Organic Compounds* (Chemical Rubber Co., Cleveland, Ohio).

---

[46] Intracellular pH: 6.0–7.0.

Table 6.3. Specific Rotations of α-Amino Acids[a]

Amino acid	Solvent	C (g/100 ml)	Temp. (°C)	$[\alpha]_D$ (deg.)	*R/S* configuration
L-Alanine	1.0 *N* HCl	5.8	15	+14.7	2(*S*)
L-Arginine	6.0 *N* HCl	1.6	23	+26.9	2(*S*)
L-Aspartic acid	6.0 *N* HCl	2.0	24	+24.6	2(*S*)
L-Cystine	1.0 *N* HCl	1.0	24	−214.4	2(*S*),2′(*S*)
L-Glutamic acid	6.0 *N* HCl	1.0	22	+31.2	2(*S*)
L-Histidine	6.0 *N* HCl	1.5	22	+13.0	2(*S*)
Hydroxy-L-proiine	1.0 *N* HCl	1.3	20	−47.3	2(*S*), 4(*R*)
Allohydroxy-L-proline	Water	2.6	18	−58.1	2(*S*), 4(*S*)
L-Isoleucine	6.0 *N* HCl	5.1	20	+40.6	2(*S*), 3(*S*)
L-Alloisoleucine	6.0 *N* HCl	3.9	20	+38.1	2(*S*), 3(*R*)
L-Leucine	6.0 *N* HCl	2.0	26	+15.1	2(*S*)
L-Lysine	6.0 *N* HCl	2.0	23	+25.9	2(*S*)
L-Methionine	0.2 *N* HCl	0.8	25	+21.2	2(*S*)
L-Phenylalanine	5.4 *N* HCl	3.8	20	−7.1	2(*S*)
L-Proline	0.5 *N* HCl	0.6	20	−52.6	2(*S*)
L-Serine	1.0 *N* HCl	9.3	25	+14.4	2(*S*)
L-Threonine	Water	1.3	26	+28.4	2(*S*), 3(*R*)
L-Allothreonine	Water	1.6	26	+9.6	2(*S*), 3(*S*)
L-Tryptophane	Water	1.0	22	−31.5	2(*S*)
L-Tyrosine	6.3 *N* HCl	4.4	20	−8.6	2(*S*)
L-Valine	6.0 *N* HCl	3.4	20	+28.8	2(*S*)

[a] A leading reference for amino acids is *The Merck Index*, 9th ed. (Merck & Co., Rahway, N.J., 1976).

EXPERIMENT 20. THE NINHYDRIN TEST

$$\text{ninhydrin} + R-\underset{|}{\overset{\overset{+}{N}H_3}{CH}}-CO_2^- \longrightarrow (\text{indanedione})=N-\overset{R}{CH}-CO_2H$$

$$(\text{indanedione})=N-\overset{R}{CH}-CO_2H \xrightarrow{H_2O} (\text{indanedione})-NH_2 + RCHO + CO_2$$

$$(\text{indanedione})-NH_2 + \text{ninhydrin (OH, OH)} \longrightarrow (\text{indanedione})-NH=(\text{indanedione})$$

Add about 2 mg of an α-amino acid to 1 ml of a solution of 0.2 g of ninhydrin (1,2,3-indanetrione monohydrate) in 50 ml of water. The test mixture is heated to boiling for 15 to 30 sec; a blue to blue-violet color is given by α-amino acids and β-amino acids and constitutes a positive test. Other colors (yellow, orange, red) are negative.

Discussion. This reaction is important not only because it is a qualitative test, but also because it is the source of the absorbing material that can be measured quantitatively by an automatic amino acid analyzer (Chapter 8). This color reaction is also used to detect the presence and position of amino acids after paper chromatographic separation (Chapter 7).

Proline, hydroxyproline, and *o*-, *m*-, and *p*-aminobenzoic acids fail to give a blue color; a yellow color is produced instead. Ammonium salts give a positive test and some amines (aniline) give orange to red colors, which are classed as negative.

An amino acid is usually converted, by base treatment, to a derivative via the amino-carboxylate salt; this enhances the rate of nucleophilic reaction of the free amino group.

$$\text{R—CH}\begin{matrix}\diagup NH_3^+\\ \diagdown CO_2^-\end{matrix} \xrightarrow{B^-} \text{R—CH}\begin{matrix}\diagup \ddot{N}H_2\\ \diagdown CO_2^-\end{matrix} + BH$$

Solid derivatives of the amino acids are usually obtained by reference to derivatives of the type used for amines rather than those described for acids. The Schotten-Baumann method (Procedure 17c, Experiment 18) yields the benzoyl

$$\text{R—CH}\begin{matrix}\diagup CO_2^-\\ \diagdown NH_3^+\end{matrix} \xrightarrow[\text{base}]{R'—\overset{O}{\overset{\|}{C}}—Cl} \text{R—CH}\begin{matrix}\diagup CO_2^-\\ \diagdown NH\underset{\underset{O}{\|}}{C}R'\end{matrix}$$

derivatives, and the same general procedure using acetic anhydride in place of benzoyl chloride leads to the acetyl derivatives (Procedure 17). Phenyl isocyanate reacts with amino acids to produce the corresponding substituted phenylureas (Procedure 1). The Hinsberg reaction, using *p*-toluenesulfonyl chloride (Procedure 18), furnishes good derivatives for a considerable number of the amino acids. The *N*-substituted 2,4-dinitroanilines obtained by the action of 2,4-dinitrofluorobenzene amino acids, peptides, and proteins. These dinitroaryl derivatives respond to the color tests with alkali [Experiment 29, p. 320].

Question

1. Propose a mechanism for the ninhydrin test.

Procedure 24. p-Toluenesulfonyl Derivatives of Amino Acids

$$R\text{—}CH(NH_3^+)(CO_2^-) \xrightarrow[\text{(2) } p\text{-TsCl/ether; (3) } H_3O^+]{\text{(1) NaOH}} R\text{—}CH(NHSO_2C_6H_4CH_3)(CO_2H)$$

$$p\text{-Ts} = CH_3\text{—}C_6H_4\text{—}SO_2\text{—}$$

About 1 g of the amino acid is dissolved in 20 ml of 1 *N* sodium hydroxide solution, a solution of 2 g of *p*-toluenesulfonyl chloride in 25 ml of ether is added, and the mixture is shaken mechanically or vigorously stirred for 3 to 4 hr. The ether layer is separated, and the aqueous layer is acidified to Congo red or to a pH of *ca.* 2 with pH paper using dilute hydrochloric acid. The derivative usually separates as a solid, which is removed by filtration and recrystallized from 4 to 5 ml of 60% ethanol. If an oil is obtained upon acidification, the mixture is placed in a refrigerator overnight to induce crystallization.

The sodium salts of the derivatives of phenylalanine and tyrosine are sparingly soluble in water and separate during the initial reaction. The resulting suspension is acidified, and the salts go into solution. The *p*-toluenesulfonyl derivatives then crystallize from the ether layer and are removed by filtration.

The derivatives of glutamic and aspartic acids, arginine, lysine, tryptophane, and proline crystallize with difficulty; other derivatives should be tried in the event that oils are produced.

Procedure 25. 2,4-Dinitrophenyl Derivatives of Amino Acids

$$R\text{—}CH(NH_3^+)(CO_2^-) \xrightarrow[\text{(2) } O_2N\text{—}C_6H_3(NO_2)\text{—}F]{\text{(1) } NaHCO_3/H_2O} \xrightarrow{\text{(3) } H_3O^+} R\text{—}CH(NH\text{—}C_6H_3(NO_2)\text{—}NO_2)(CO_2H) + NaF$$

To a solution or suspension of 0.5 g of the amino acid in 10 ml of water and 1.0 g of sodium bicarbonate is added a solution of 0.8 g of 2,4-dinitrofluorobenzene in 5 ml of ethanol. The mixture is shaken very vigorously and allowed to stand at room temperature for an hour with occasional vigorous shaking. Then 5 ml of saturated salt solution is added, and the mixture extracted twice with 10 ml of ether to remove unchanged reactants. The aqueous layer is then poured *into* 25 ml of cold 5% hydrochloric acid with vigorous stirring. This mixture should be distinctly acid to Congo red indicator. The product sometimes separates as an oil; a solid is induced to form by stirring or by scratching. The derivative is collected on a filter and recrystallized from 1:1 ethanol-water. The melting points are recorded in late editions of *Tables for Identification of Organic Compounds*, (Chemical Rubber Co., Cleveland, Ohio).

This procedure may also be used for amines and for proteins or peptides with uncombined primary amino groups on the chain or at the end.

Amino acids with functional groups on the side chain, such as OH, SH, and so on, can often be analyzed by derivatization at the side-chain functional group; consult the appropriate procedure in this chapter.

Paper and thin-layer chromatography (see Chapter 10 for references and Chapter 7 for descriptions) are excellent for characterizing amino acids. Lanthanide *salts* (e.g., europium nitrate) are easily utilized for shifting the nmr signals of amino acids (see Chapters 8 and 10).

6.9 AROMATICS

The results of preliminary chemical tests that lead to postulating aromatic character can suggest a range of approaches to chemically characterizing this class of organic compound. Specifically, we are often faced with introducing new substituents onto the aromatic ring, or modifying existing substituents, such that we may more readily characterize the compounds. If the molecule already contains reactive chemical substituents, the chemist is referred to other sections of this chapter for that group (e.g., carboxylic acids, amines, anilines, esters, ethers, carbonyl compounds, etc.). In addition, certain aromatic compounds are given special coverage in this book (aromatic nitro compounds, Section 6.17.1; aryl halides, Section 6.10; phenols, Section 6.19).

We shall consider the chemical tests beginning with the most vigorous conditions and, roughly speaking, continue on to tests in decreasing order of vigor of conditions. Some of the most inert aromatics may remain unchanged after even the most vigorous tests; characterization of such inert compounds may rely on spectral (ir, nmr, uv, and mass) and physical (mol wt, b.p., m.p., sp gr, [α], etc.) tests to a greater degree than usual.

EXPERIMENT 21. FUMING SULFURIC ACID

$$C_6H_6 \xrightarrow{H_2SO_4 \cdot SO_3} C_6H_5SO_3H$$

CAUTION: *Use this reagent with relatively inert compounds only. Compounds for which preliminary tests indicate highly activating groups (OH, NH_2, etc.) may be decomposed by fuming sulfuric acid.*

Procedure. Place 2 ml of 20% fuming sulfuric acid in a clean, dry test tube, and add 1 ml of aromatic. Shake the mixture vigorously, and allow it to stand for a few minutes. Note whether solution has been effected. Repeat the experiment using (1) benzene, (2) bromobenzene, (3) ethylene bromide, and (4) cyclohexane as model compounds.

Discussion. Concentrated sulfuric acid is a remarkable solvent in two respects. Its dielectric constant appears to be very much greater than that of many other compounds for which this property has been measured.[47] Thus, forces of attraction between dissolved ions are so small in dilute solution that activity coefficients may be taken as unity. The second unusual property is that, in addition to the

$$2H_2SO_4 \rightleftharpoons H_3SO_4^+ + HSO_4^-$$

autoprotolysis like that found in hydroxylic solvents (such as water), there is a self-dissociation resulting initially in the formation of sulfur trioxide and water. However, at the concentrations concerned, each of these reacts essentially completely with sulfuric acid so that the overall equilibrium is as follows:

$$(1)\ H_2SO_4 \rightleftharpoons H_2O + SO_3$$

$$(2)\ SO_3 + H_2SO_4 \rightleftharpoons HSO_3^+ + HSO_4^-$$

$$(3)\ H_2O + H_2SO_4 \rightleftharpoons H_3O^+ + HSO_4^-$$

$$1+2+3 = 3H_2SO_4 \rightleftharpoons HSO_3^+ + H_3O^+ + 2HSO_4^-$$

Also:

$$H_2SO_4 + HSO_3^+ + HSO_4^- \rightleftharpoons H_3SO_4^+ + HS_2O_7^-$$

Concentrated (100%) sulfuric acid reacts with ethylene to form ethyl hydrogen sulfate, but sulfuric acid containing added sulfur trioxide (fuming sulfuric acid) yields instead ethionic acid. The reason for the difference between the two sets of conditions is understood if it is realized that 100% sulfuric acid contains sulfonating species such as SO_3 or HSO_3^+ in small concentration and therefore the sulfonating reagent fails to compete with proton addition, the first step in alkyl sulfate formation:

$$CH_2{=}CH_2 + H_2SO_4 \rightarrow CH_3CH_2^+ + HSO_4^-$$

However, in fuming sulfuric acid, the addition of HSO_3^+, SO_3, or some other sulfonating agent becomes important, so that the principal reaction is the following:

$$CH_2{=}CH_2 + HSO_3^+ \rightarrow {}^+CH_2{-}CH_2SO_3H$$

In each case the second step is the same: the reaction of the first-formed carbonium ion with bisulfate ion.

$$CH_3CH_2^+ + HSO_4^- \rightarrow CH_3CH_2{-}O{-}SO_3H \qquad \text{ethyl hydrogen sulfate}$$

$${}^+CH_2CH_2{-}SO_3H + HSO_4^- \rightarrow HO_3SOCH_2CH_2{-}SO_3H \qquad \text{ethionic acid}$$

The first step in the formation of ethionic acid is, at least formally, like the first step in aromatic sulfonation. The exact nature of the electrophile is probably complex.

The action of fuming sulfuric acid on 1,2-dihalogen compounds is complex. The mixture turns dark, and some free halogen is liberated. It seems probable that loss of hydrogen halide occurs followed by polymerization of the vinyl halide. For example, ethylene bromide probably undergoes the following changes.

---

[47] R. J. Gillespie, E. D. Hughes, and C. K. Ingold, *J. Chem. Soc.*, 2473 (1950).

$$CH_2(Br)—CH_2(Br) \rightarrow CH_2{=}CHBr + HBr$$

$$n\,CH_2{=}CHBr \rightarrow —(CH_2—CH(Br)—)_n$$

$$2HBr + H_2SO_4 \rightarrow Br_2 + SO_2 + 2H_2O$$

Questions

1. Note that this test is useful only for compounds insoluble in sulfuric acid. Why?
2. Write equations for the reactions involved. Compare the products of this reaction with those formed in your solubility test with concentrated sulfuric acid and an olefin such as 1-hexene (see Section 5.1).

EXPERIMENT 22. AZOXYBENZENE AND ALUMINUM CHLORIDE

$$ArH \xrightarrow[AlCl_3]{C_6H_5N(\rightarrow O){=}N—C_6H_5} [Ar—C_6H_4—N{=}N—C_6H_5]\,AlCl_3$$

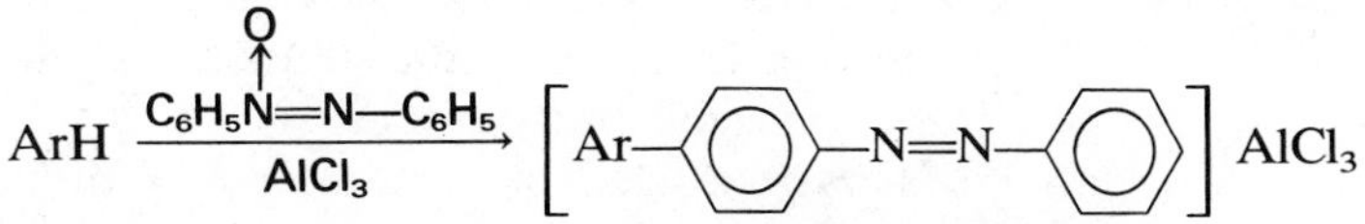

colored complex

Place 2 ml of dry aromatic in a clean, dry test tube; add one or two crystals of azoxybenzene and about 0.1 g of anhydrous aluminum chloride. Note the color. If no color is produced immediately, warm the mixture for a few minutes. Try the test on petroleum ether, chlorobenzene, ethyl bromide, naphthalene, and benzene. If the hydrocarbon is a solid, a solution of 0.5 g of it in 2 ml of dry carbon disulfide may be used.

Discussion. This test should be applied only to compounds that are insoluble in sulfuric acid. The color produced is due to an addition compound formed from the *p*-arylazobenzene and aluminum chloride.

The route to a *p*-arylazobenzene may be formulated as an aromatic substitution of the arene (ArH) by the aluminum chloride adduct of azoxybenzene followed by the net elimination of the elements of H_2O and $AlCl_3$:

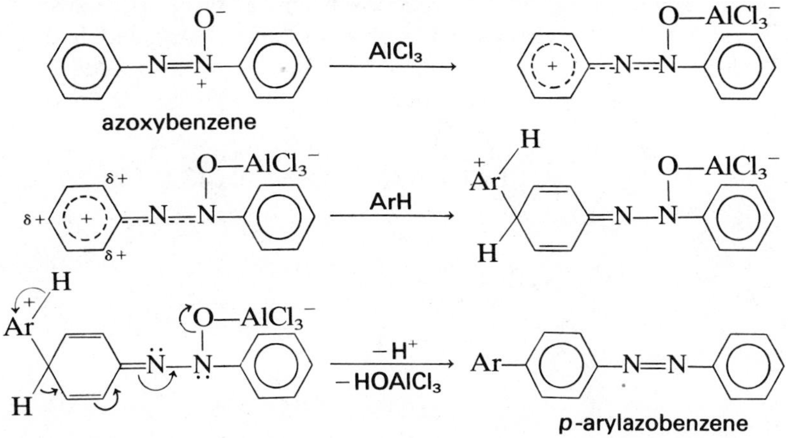

$$C_6H_5—N{=}N^+(O^-)—C_6H_5 \xrightarrow{AlCl_3} [C_6H_5^{(+)}{\cdots}N{\cdots}N(O—AlCl_3^-)—C_6H_5]$$

azoxybenzene

$$[\delta+\ C_6H_5{\cdots}N{\cdots}N(O—AlCl_3^-)—C_6H_5] \xrightarrow{ArH} (H)(Ar^+)C_6H_4{=}N—N(O—AlCl_3^-)—C_6H_5$$

$$(H)(Ar^+)C_6H_4{=}\ddot{N}—\ddot{N}(O—AlCl_3^-)—C_6H_5 \xrightarrow[-HOAlCl_3]{-H^+} Ar—C_6H_4—N{=}N—C_6H_5$$

p-arylazobenzene

Aromatic hydrocarbons derived from benzene and their halogen derivatives produce a deep orange to dark red color in solution or give a precipitate. Condensed polynuclear hydrocarbons such as naphthalene, anthracene, and phenanthrene produce brown colors. Aliphatic hydrocarbons give no color or, at most, a pale yellow.

EXPERIMENT 23. CHLOROFORM AND ALUMINUM CHLORIDE

$$3ArH + CHCl_3 \xrightarrow{AlCl_3} Ar_3CH$$

$$Ar_3CH + R^+ \longrightarrow \underset{\text{colored cation}}{Ar_3\overset{+}{C}} + 3HCl + RH$$

To 2 ml of dry chloroform in a test tube add 0.1 ml (or 0.1 g) of aromatic compound. Mix thoroughly, and incline the test tube so as to moisten the wall. Then add 0.5 to 1.0 g of anhydrous aluminum chloride so that some of the powder strikes the side of the test tube. Note the color of the powder on the side, as well as the color of the solution. Try the test on petroleum ether, chlorobenzene, biphenyl, and benzene.

Discussion. The colors produced by the reaction of aromatic compounds with chloroform and aluminum chloride are quite characteristic. Aliphatic compounds insoluble in sulfuric acid give no color or only a very light yellow. Typical colors produced are the following:

Compound	*Color*
Benzene and its homologs	Orange to red
Aryl halides	Orange to red
Naphthalene	Blue
Biphenyl	Purple
Phenanthrene	Purple
Anthracene	Green

With time the colors change to various shades of brown. Similar colors are obtained when chloroform is replaced by carbon tetrachloride.

The reaction begins as three successive (1–3 below) Friedel-Crafts reactions between the aromatic and chlorinated hydrocarbons; these alkylation reactions are promoted by the Lewis acid, $AlCl_3$, and facilitated by positive charge delocalization by the chlorine and the aryl groups:

$$CHCl_3 + AlCl_3 \longrightarrow \left[\overset{\delta+}{C}HCl_2\text{---}Cl\text{---}\overset{\delta-}{Al}Cl_3\right] \longrightarrow \left[H\text{—}\overset{\delta+}{C}\begin{matrix}Cl^{\delta+}\\ \\ Cl^{\delta+}\end{matrix}\right]^+ + AlCl_4^-$$

$$HCCl_2^+ + ArH \xrightarrow{(1)} \overset{+}{Ar}\begin{matrix}H\\ \\ CHCl_2\end{matrix} \xrightarrow{-H^+} ArCHCl_2 \xrightarrow{AlCl_3}$$

$$Ar\overset{+}{C}HCl \underset{(2)}{\xrightarrow{ArH}} \underset{\displaystyle Cl}{ArCH}\text{—}\overset{\displaystyle H}{\overset{+}{Ar}} \xrightarrow{-H^+}$$

$$\underset{\substack{|\\ \mathrm{Cl}}}{\mathrm{ArCH}}\mathrm{—Ar} \xrightarrow{AlCl_3} \underset{+}{\mathrm{ArCH}}\mathrm{—Ar} \xrightarrow[(3)]{ArH}$$

$$\underset{\substack{|\\ \mathrm{Ar}}}{\mathrm{ArCH}}\mathrm{—}\overset{\substack{\mathrm{H}\\ +|}}{\mathrm{Ar}} \xrightarrow{-H^+} Ar_3CH$$

Partially substituted chlorides (e.g., Ar_2CHCl or $ArCHCl_2$) may react with aluminum chloride to give mono or diaryl cations:

$$Ar_2CHCl + AlCl_3 \rightarrow Ar_2CH+ + AlCl_4^-$$

These cations abstract hydride from the triarylmethanes to give rise to stable triaryl cations:

$$Ar_3CH + Ar_2CH+ \rightarrow \underset{\text{colored}}{Ar_3C+} + Ar_2CH_2$$

Question

1. Propose a reason why tetraarylmethanes are not formed.

Typical ir (Fig. 6.21) and nmr (Fig. 6.22) spectra of aromatic compounds are shown. The chemistry of functional groups included in the aromatic structure may be learned and used for characterization after consulting the section of this chapter discussing that functional group. The number and relative positions of nonpolar substituents on a benzene ring may often be deduced by use of Table 6.4 and Fig. 6.23. Substitution of a polar group on an aliphatic side chain normally does not substantially alter the chemical shift of low-field aromatic protons. If the polar group is, however, substituted directly onto the aromatic nucleus, very substantial shifts in the position of the aromatic protons are observed (see Table 6.4).

The mass spectra of aromatic compounds are usually characterized by very intense molecular ion peaks. Thus, assuming no peaks due to fragments or impurities in the vicinity of the molecular ion, the M, M+1, and M+2 peaks are usually of sufficient intensity to allow determination of the molecular formula (see Chapter 4).

Procedure 26. Aromatic Nitration

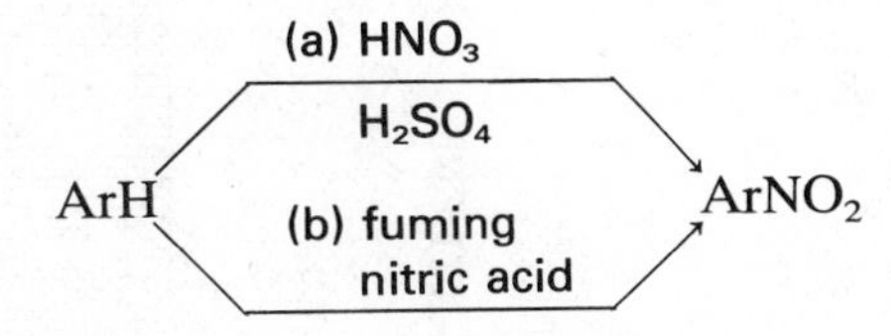

The nitration of an aromatic compound, especially if it is an unknown substance,

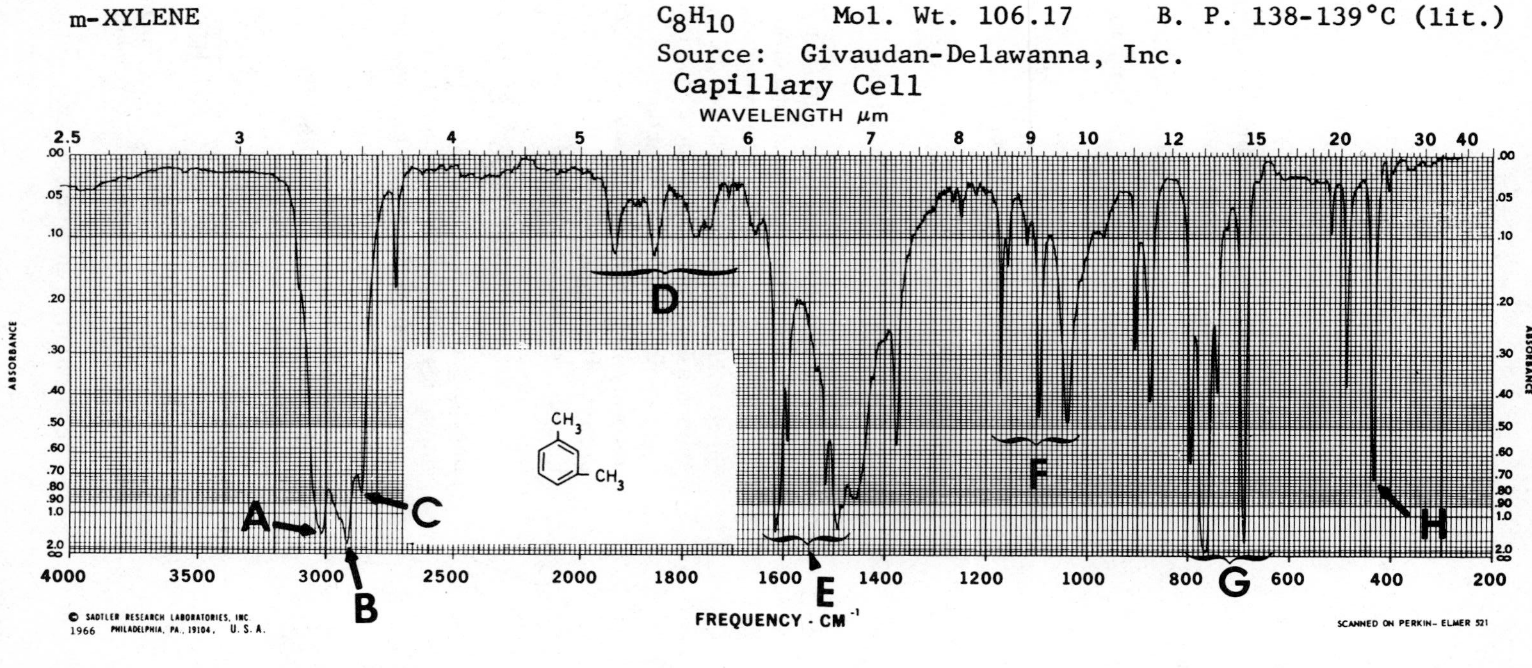

Figure 6.21. Infrared spectrum of *m*-xylene: neat. C—H stretch: **A**, 3010 cm^{-1} (3.32 μm), aromatic $\nu_{C—H}$; **B**, 2915 cm^{-1} (3.43 μm), ν_{asym} CH_3 (aliphatic); **C**, 2860 cm^{-1} (3.50 μm), ν_{sym} CH_3 (aliphatic). Overtone bands: **D**, 2000–1667 cm^{-1} (5.00–6.00 μm), indicative of meta substitution (see Fig. 6.23). C⚌C stretch: **E**, 1618, 1520, 1492 cm^{-1} (6.18, 6.70 μm), $\nu_{C—C}$. C⚌H bend: **F**, 1170, 1095, 1040 cm^{-1} (8.55, 9.13, 9.62 μm), δ_{CH_3}; **G**, 767, 690 cm^{-1} (13.04, 14.49 μm), out-of-plane ⚌C—H bend. C⚌C bend: **H**, 434 cm^{-1} (23.0 μm), out-of-plane C⚌C bend.

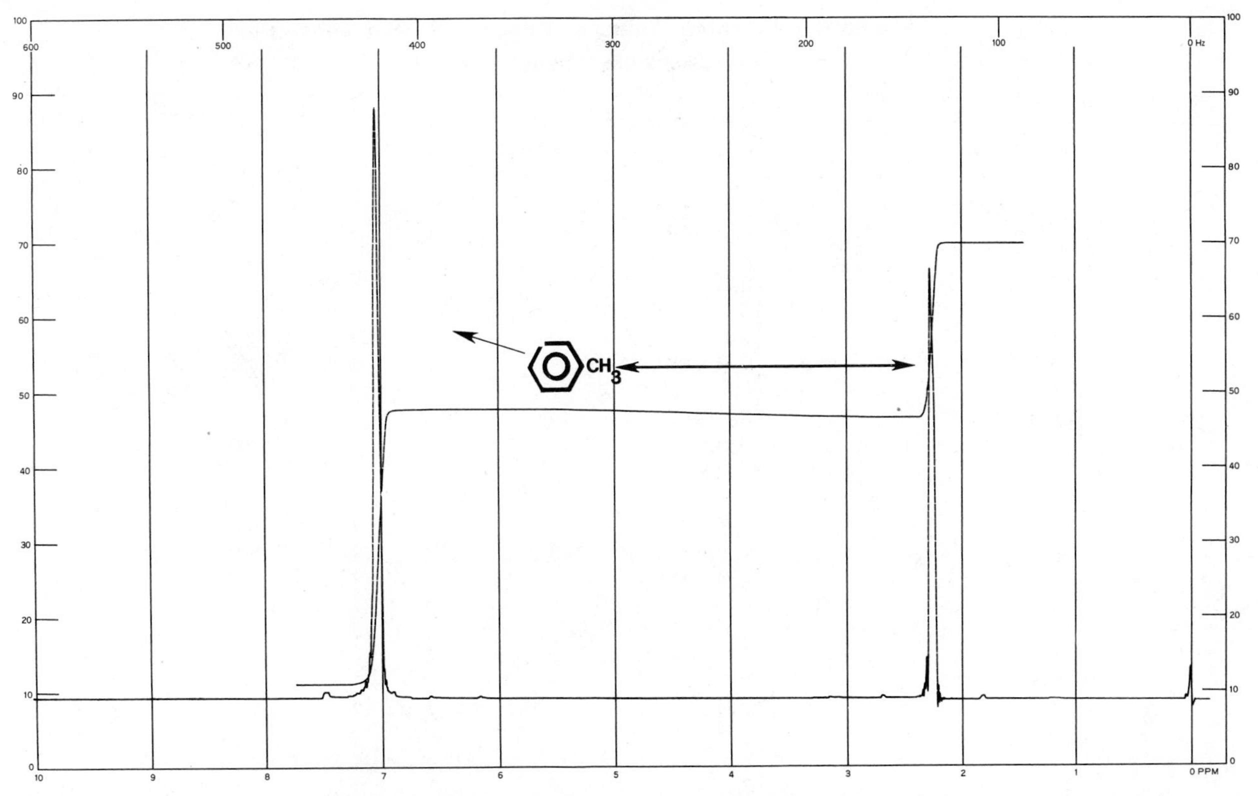

Figure 6.22. Nuclear magnetic resonance (proton) spectrum of toluene: 60 MHz, 600-Hz sweep width, $CDCl_3$ solvent.

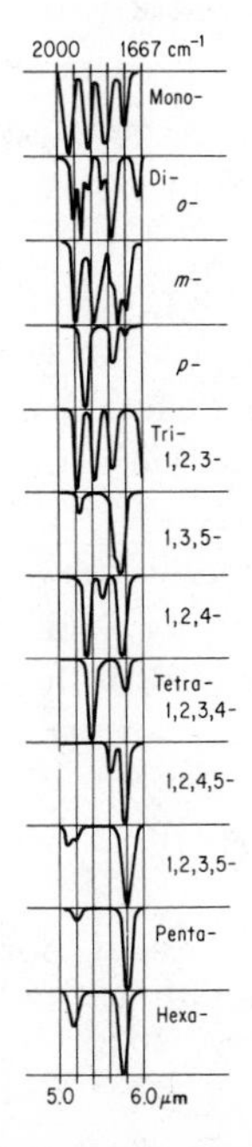

Figure 6.23. Schematic representation of the 5–6 μm IR region to be anticipated for benzenoid compounds of all substitution types. (Source: J. R. Dyer, *Applications of Absorption Spectroscopy of Organic Compounds*, © 1965, p. 52. Reprinted by permission of Prentice-Hall, Inc., Englewood Cliffs, N.J.)

Table 6.4. Chemical Shifts of Protons of Monosubstituted Benzenes (Benzene, Singlet, $\delta = 7.27$)

	Compound	Ortho	Meta	Para
	$C_6H_5CH_3$[a]	7.10	7.18	7.10
	$C_6H_5—C_6H_5$	7.45	7.27	7.35
	$C_6H_5—SR$	7.25	7.3	7.3
Electron-withdrawing substituents	$C_6H_5—NO_2$	8.22	7.44	7.60
	$C_6H_5—CHO$	7.85	7.48	7.50
	$C_6H_5—(C{=}O)R$	7.9	7.45	7.55
	$C_6H_5—CO_2H$	8.17	7.41	7.47
	$C_6H_5—CO_2R$	8.2	7.35	7.5
	$C_6H_5—(C{=}O)Cl$	8.10	7.43	7.6
	$C_6H_5—CN$	7.54	7.38	7.57
	$C_6H_5—NH_3^+$	7.62	—	—
Electron-donating substituents	$C_6H_5—NH_2$	6.52	7.03	6.64
	$C_6H_5—OH$	6.77	7.13	6.87
	$C_6H_5—OR$	6.75	7.2	6.75
	$C_6H_5—O(C{=}O)CH_3$	7.15	7.26	7.26
	$C_6H_5—OSO_2C_6H_4CH_3$-p	7.01	7.22	7.22
	$C_6H_5—NH(C{=}O)CH_3$	7.52	—	—
Halides	$C_6H_5—R$	6.97	7.25	7.04
	$C_6H_5—Cl$	7.29	7.21	7.21
	$C_6H_5—Br$	7.49	7.14	7.24
	$C_6H_5—I$	7.67	7.00	7.24

[a] Toluene will show only a slightly broadened singlet for the aromatic protons at 60 MHz, 600 Hz sweep width. A separation of greater than 0.1–0.2 Hz is necessary for splitting of the aromatic signals under these conditions.

should be carried out with special precautions, because many of these compounds react violently. A small-scale spot test behind a shield should be tried first.

(*a*) About 1 g of the compound is added to 4 ml of concentrated sulfuric acid. Four milliliters of concentrated nitric acid is added drop by drop, with shaking after each addition. The flask is connected to a reflux condenser and kept in a beaker of water at 45°C for 5 min. The reaction mixture is poured on 25 g of cracked ice and the precipitate collected on a filter. It may be recrystallized from dilute ethanol.

(*b*) The procedure outlined above is followed except that 4 ml of fuming nitric acid[48] is used instead of concentrated nitric acid, and the mixture is warmed on the steam cone for 10 min. Occasionally, with compounds that are difficult to nitrate, fuming sulfuric acid may be substituted for the concentrated sulfuric acid.

---

[48] This corresponds to highly concentrated (ca. 90%) nitric acid, sometimes called "white" or "yellow" fuming nitric acid; this is a powerful nitrating agent. It is not to be confused with "red" nitric acid, which contains dissolved NO_2 and is a vigorous oxidizing agent.

Discussion. Nitration introduces a functional group that provides synthetic access to a number of other substituted aromatics. For example, highly alkylated benzenes have been identified by nitration followed by reduction of the nitro compounds to the amines, which are then acetylated or benzoylated to give mono- or diacetamido or benzamido derivatives.

$$ArH \rightarrow ArNO_2 \xrightarrow{(1)} ArNH_2 \xrightarrow{(2)} ArNHCOR$$

(1) Procedure 50
(2) Experiment 18 and Procedure 17

Procedure (a) yields *m*-dinitrobenzene from benzene or nitrobenzene and the *p*-nitro derivative of chlorobenzene, bromobenzene, benzyl chloride, or toluene. Phenol, acetanilide, naphthalene, and biphenyl yield dinitro derivatives. It is best to employ Procedure (b) for halogenated benzenes because it produces dinitro derivatives that are easier to purify than the mononitro derivatives formed in Procedure (a). Mesitylene, the xylenes, and pseudocumene yield trinitro derivatives.

Procedure 27. Oxidation of a Side Chain of an Aromatic Compound

$$ArR \xrightarrow[H_2SO_4,\ \Delta]{(a)\ Na_2Cr_2O_7} ArCO_2H$$

$$ArR \xrightarrow[\Delta,\ OH^-]{(b)\ KMnO_4} ArCO_2^- \xrightarrow{H^+} ArCO_2H$$

(a) ***Dichromate Oxidation.*** In a small flask are placed 15 ml of water, 7 g of sodium dichromate, and 2 to 3 g of the compound to be oxidized. Ten milliliters of concentrated sulfuric acid is added, the flask is attached to a reflux condenser, and the mixture thoroughly shaken. The flask is heated carefully until the reaction starts; then the flame should be removed and the flask cooled if necessary. After the mixture has ceased to boil from the heat of the reaction it is heated under reflux for 2 hr. The contents of the flask are poured into 25 ml of water, and the precipitate is collected on a filter. The precipitate is mixed with 20 ml of 5% sulfuric acid, and the mixture is warmed on a steam cone with vigorous stirring. It is cooled, and the precipitate is separated and washed with 20 ml of water. The residue is dissolved in 20 ml of 5% sodium hydroxide solution, and the solution is filtered. The filtrate is poured, with vigorous stirring, into 25 ml of 10% sulfuric acid. The precipitate is collected on a filter, washed with water, and purified by recrystallization from ethanol or benzene.

(b) ***Permanganate Oxidation.*** One gram of the compound is added to 80 ml of water containing 4 g of potassium permanganate. One milliliter of 10% sodium hydroxide solution is added, and the mixture is heated under reflux until the purple color of the permanganate has disappeared (0.5 to 3 hr). At the end of this time the mixture is allowed to cool and is acidified carefully with sulfuric acid. The mixture is heated for 0.5 hr and cooled. Any excess manganese dioxide is removed by the addition of a little sodium bisulfite solution. The precipitated acid is collected on a filter and recrystallized from benzene or ethanol. If the acid is appreciably soluble in water, it may not separate from this dilute acid solution. In this event the acid may be extracted by means of chloroform, ether, or benzene. A slight precipitate of silicic acid sometimes appears on acidification; hence it is important to recrystallize the acid before taking the melting point.

Discussion. Aromatic hydrocarbons that have side chains may be oxidized to the corresponding acids. If there is only one such chain, this is usually an excellent method. Permanganate oxidation (Procedure 27b) is best for this purpose. Aromatic acids having several carboxyl groups are sometimes difficult to handle, and for this reason the utility of the oxidation method is limited. If there are two side chains situated in adjacent positions on the ring, oxidation is recommended because the resulting acid (phthalic acid) is easy to identify. *o*-Dialkylbenzenes, as well as compounds substituted on the *o*-dialkyl chains, undergo complete oxidation with Cr(VI) (Procedure 27a). Since this oxidizing agent may give misleading results, the permanganate oxidation should always be used (Procedure 27b). The melting points of the acids obtained by oxidation of the alkylbenzenes may be obtained by reference to the tables in the Appendix to this book.

Aromatic rings that are substituted with electron-withdrawing groups (nitro, halo, etc.) easily survive even the more vigorous oxidation procedures, whereas rings substituted with electron-donating groups may be oxidized on the ring more readily than on the side chain. For example, the oxidation of a substitited phenol can give the aliphatic acid in sufficient yield to be characterized:

$$\text{HO–C}_6\text{H}_4\text{–R} \xrightarrow[\text{H}_2\text{O, 25°C}]{\text{KMnO}_4} \text{RCO}_2\text{H} \quad \text{(characterize as per Section 6.12)}$$

It is clear that hydrogens α to the ring (benzylic hydrogens) facilitate this oxidation:

$$\text{C}_6\text{H}_5\text{–C(CH}_3)_3 \xrightarrow[\text{OH}^-,\ \Delta]{\text{KMnO}_4} \text{no reaction}$$

Substitution of a second aromatic ring onto the alkyl group can decrease the degree of oxidation:

$$\text{ArCH}_2\text{Ar}' \xrightarrow{\text{oxidation}} \text{Ar–C(=O)–Ar}'$$

Recently very good procedures for oxidation of side chains using crown ethers have appeared.[49] Similarly, an oxidation on a preparative scale using quaternary ammonium salts as the phase-transfer reagent (see Experiment 15b) could be carried out.

Procedure 28. Aroylbenzoic Acids

$$\text{ArH} + \text{phthalic anhydride} \xrightarrow[\text{CS}_2,\ \Delta]{\text{AlCl}_3} o\text{-}(\text{ArC(=O)})\text{C}_6\text{H}_4\text{CO}_2\text{H}$$

To a solution of 1 g of the dry aromatic hydrocarbon and 1.2 g of phthalic anhydride in 10 ml of dry carbon disulfide is added 2.4 g of anhydrous aluminum chloride. The mixture is heated under a reflux condenser in a boiling water bath for 30 min and cooled. The carbon disulfide layer is decanted, and 10 ml of concentrated hydrochloric acid and 10 ml of water are added to the residue. The acid should be added slowly at first, with cooling by ice if necessary, and the final mixture should be thoroughly shaken. If the aroylbenzoic acid separates as a solid, it is immediately collected on a filter and washed with cold water. If an oil separates, the mixture is cooled in an ice bath for some time to induce crystallization. If the product remains oily, the supernatant liquid is decanted and the oil is washed with cold water. The crude product, whether a solid or an oil, is boiled for 1 min with 30 ml of 10% ammonium hydroxide solution to which has been added about 0.1 g of Norite. The hot solution is filtered and cooled; 25 g of crushed ice is then added, and the solution is acidified with concentrated hydrochloric acid. The aroylbenzoic acid is removed by filtration and is recrystallized from dilute ethanol (30 to 80%). Sometimes it is necessary to allow the product to stand overnight in order to obtain crystals.

Discussion. Aromatic hydrocarbons and their halogen derivatives undergo the Friedel-Crafts reaction with phthalic anhydride, producing aroylbenzoic acids in good yield. This procedure produces a derivative with a functional group (CO_2H) that can be easily characterized (see Section 6.12). For example, the neutralization equivalent (p. 268) of the acid may be determined; this will yield the molecular weight of the keto acid derivative. The molecular weight (mol wt) of the original aromatic compound may be calculated:

$$\text{mol wt of ArH} = (\text{mol wt of Acid}) - 148$$

(Explain the origin of the value 148 in the preceding formula.)

---

[49] For the preparation of 18-crown-6 see L. F. Fieser and K. Williamson, *Organic Experiments*, 3rd. ed. (Heath, Lexington, Mass, 1975), pp. 386–388. For oxidations and other applications of crown ethers, see G. Gokel and H. Durst, *Synthesis*, 168 (1976). The latter is an extensive review with 190 references and detailed experimental procedures.

Procedure 29. Aryl Methyl Ketones

$$\text{ArH} \xrightarrow[\text{AlCl}_3]{\text{CH}_3\overset{\text{O}}{\overset{\|}{\text{C}}}\text{Cl}} \text{Ar}\overset{\text{O}}{\overset{\|}{\text{C}}}\text{CH}_3 \left(\longrightarrow \text{Ar}\overset{\text{NNHR}}{\overset{\|}{\text{C}}}\text{CH}_3 \right)$$

In a 50-ml conical flask, fitted with a calcium chloride drying tube, are placed 1.4 g of powdered anhydrous aluminum chloride, 5 ml of carbon disulfide, and 0.8 ml of acetyl chloride. After 5 min a solution of 2 ml of the aromatic in 5 ml of carbon disulfide is introduced with swirling. The mixture is allowed to stand at room temperature until all the aluminum chloride dissolves (10 to 15 min) and is then poured on a mixture of ice and 5 ml of concentrated hydrochloric acid. An additional 5 ml of carbon disulfide is added. The organic layer is washed successively with 3 ml of 10% hydrochloric acid, water, 5% sodium bicarbonate, and water. The final washing is repeated until the wash water is neutral to litmus. The organic layer is dried over calcium chloride, and the carbon disulfide is evaporated. The residue is dissolved in ethanol, and the 2,4-dinitrophenylhydrazone or semicarbazone is prepared by the usual Procedures (8 or 7, respectively).

Discussion. This Friedel-Crafts procedure works well with aromatic compounds of moderate reactivity. Compounds that are inactive toward electrophilic aromatic substitution often cannot be characterized by this procedure; for example, nitrobenzene will not undergo acetylation. Highly reactive aromatics (e.g., amines or phenols) may be substituted on the ring substituent (e.g., on the NH_2 or OH group); these active compounds often lead to intractable or otherwise uncharacterizable reaction products.

Other procedures suitable for characterizing aromatic compounds are described elsewhere in this book. Picrate derivatives (Procedure 22) provide the possibility of determining the molecular weight of the aromatic compound. Aniline and ring-substituted aniline derivatives can be characterized by nitrous acid (Experiment 19) and by the Hinsberg test (Procedure 16). Phenolic compounds can be characterized by treatment with bromine water (Procedure 60).

A procedure has been described by Pasto and Johnson (pp. 343–344) that can be used for preparing charge-transfer complexes of electron-rich aromatic compounds; the complex is formed from interaction of the aromatic compound with 2,4,7-trinitrofluorenone (TNF):

[Structure: 2,4,7-trinitrofluorenone with NO_2, O_2N, NO_2, and C=O labels] TNF

A procedure has been described by Cheronis, Entrikin, and Hodnett (p. 351) that differentiates aromatic compounds from more inert organic compounds:

$$\text{ArH} \xrightarrow[\text{H}_2\text{SO}_4]{\text{HCHO, H}_2\text{O}} \text{colored, polymeric carbocations}$$

The mechanism likely involves successive Friedel-Crafts alkylation of aromatic systems; reactive compounds (polyenes as well as aromatics) give various colored solutions and/or precipitates. Saturated compounds usually do not react.

The ultraviolet (uv) spectra of aromatic compounds offer the possibility for additional criteria for structural deduction or confirmation. Aromatic systems provide intense uv bands, the maxima of which often can be associated with specific substituents on the aromatic nucleus. This subject is discussed in Chapter 8.

In the case of highly unreactive aromatic compounds (e.g., fluorobenzene), one may have to rely on very accurate measurement of simple physical properties (b.p., m.p., index of refraction); these measurements would be followed by scrutiny of a number of tables until a compound is found that has the correct value for all of these constants.

6.10 ARYL HALIDES

Aryl halides are defined as compounds in which the halogen is bonded directly to the aromatic nucleus. The aryl halide class of compound is actually one particular subcategory of the preceding section (aromatics); it is, however, instructive to contrast this class of compound to the *alkyl* halide class. Aliphatic halides are usually much more reactive toward halogen displacement than are the corresponding aryl halides.

The sodium fusion test for qualitative halogen detection and mass spectral analysis discussed in Chapter 4 are both quite useful for structural diagnosis of aryl halides. Mass spectral analysis is very frequently diagnostic for the number of bromine and chlorine atoms in such compounds. Additional chemical analysis for the aryl halogen should be contemplated only after one is aware that aryl halides will usually resist typical displacement reaction tests. These include ethanolic silver nitrate (Experiment 16) and sodium iodide in acetone (Experiment 17), both of which have been discussed above.

Typical ir (Fig. 6.24) and nmr (Fig. 6.25) spectra of aryl halides are provided here. Halogen (X) can often be detected in an ir spectrum by observing the C—X stretch (see Section 6.6). In addition, ring halogens often cause substantial nonequivalence of the nmr signals of ring protons; in contrast, a simple alkyl group usually does not cause observable nonequivalence of the ring protons of a benzenoid ring. The effect of halogen, as well as other ring substituents, on the chemical shift of aromatic ring protons is described in Section 6.9.

Mass spectra of aryl halides are usually quite useful. The aromatic character frequently is sufficient to impart substantial stability to the molecular ion; the resulting molecular ions yield an intense peak that affords qualitative and quantitative bromine and chlorine determination. Less frequently, the cluster of peaks in the molecular ion region of bromides and chlorides can give rise to quantitative C, H, N, and O determination.

A number of methods outlined below are suitable for derivatizing aryl halides and have been described in other parts of this book; although many of these

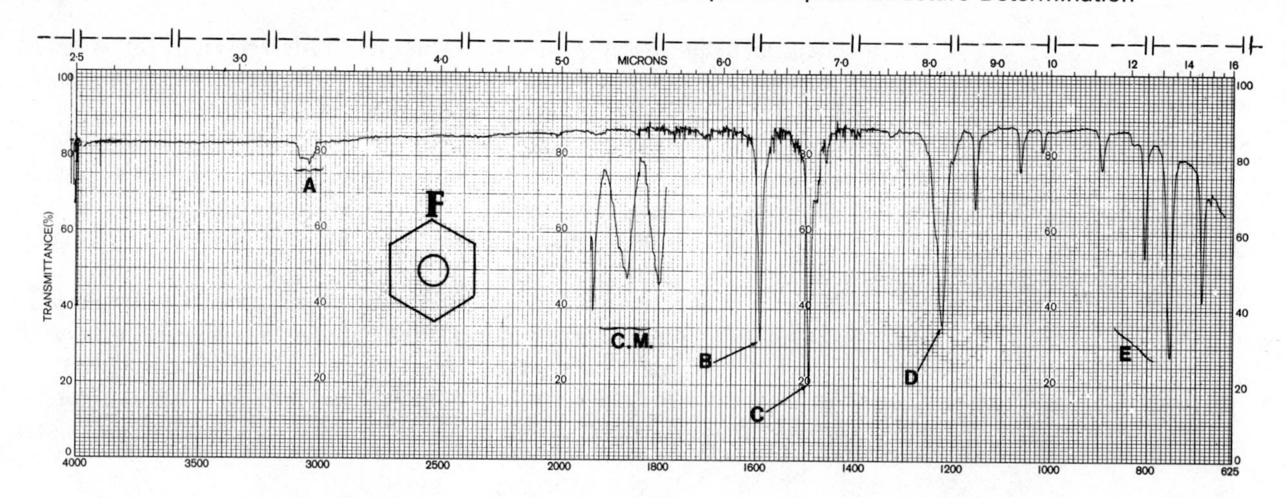

Figure 6.24. Infrared spectrum of fluorobenzene: neat. C—H stretch: **A**, 3090–3020 cm^{-1} (3.24–3.31 μm), $\nu_{C—H}$ (aromatic). C≕C stretch: **B**, 1583 cm^{-1} (6.32 μm), $\nu_{C≕C}$ (aromatic); **C**, 1486 cm^{-1} (6.73 μm), $\nu_{C≕C}$ (aromatic). C—F stretch: **D**, 1226 cm^{-1} (8.16 μm), $\nu_{C—F}$ (aromatic). Aromatic: **E**, 850–625 cm^{-1} (11.76–16.00 μm), ≕C—H out-of-plane bend and/or C≕C ring breathing. C.M. = calibration marker, 1944, 1871, 1802 cm^{-1} (5.14, 5.34, 5.55 μm). Average correction = +4 cm^{-1} (employed in 2000–625 cm^{-1} region).

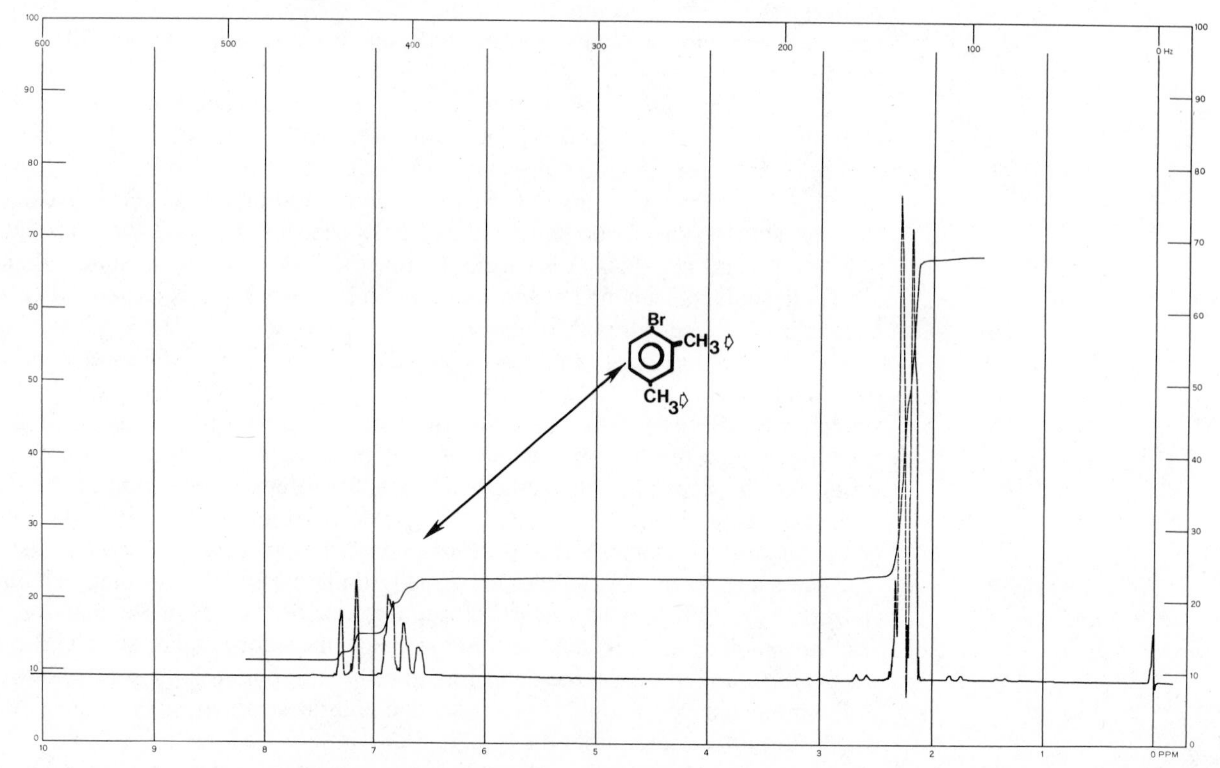

Figure 6.25. Nuclear magnetic resonance (proton) spectrum of 4-bromo-*m*-xylene: 60 MHz, 600-Hz sweep width, $CDCl_3$ solvent.

procedures may give some ortho-, and even a little meta-substituted compound, we have for convenience listed them as yielding the para isomer. A very general method is nitration (Procedure 26). However, bromine and chlorine derivatives having one side chain are frequently oxidized to the corresponding acids (Procedures 27a and 27b). For aryl iodides and bromides conversion to the corresponding carboxylic acids or anilides (Procedure 13) is recommended.

X = halogen, R = alkyl

HNO_3, Procedure 26 → X–C$_6$H$_3$(R)–NO_2

$K_2Cr_2O_7$ or $KMnO_4$, Procedure 27 → X–C$_6$H$_4$–CO_2H

(X = Br, I) Mg/dry ether, Procedure 12, X = Br → R–C$_6$H$_4$–MgX

(1) ArNCO (2) H_2O → R–C$_6$H$_4$–NH–C(=O)–Ar

CO_2, Pasto and Johnson, p. 352 → R–C$_6$H$_4$–CO_2H

Aryl halides react readily with chlorosulfonic acid to produce sulfonyl chlorides (Procedure 30), which yield sulfonamides when treated with ammonia (Procedure 31).

Procedure 30. Sulfonyl Chlorides

$$X\text{–}C_6H_5 \xrightarrow{2ClSO_3H} X\text{–}C_6H_4\text{–}SO_2Cl + HCl + H_2SO_4$$

A solution of 1 g of the aryl halide in 5 ml of dry chloroform in a clean, dry test tube is cooled in a beaker of ice to 0°C. About 5 ml of chlorosulfonic acid is added dropwise, and after the initial evolution of hydrogen chloride has subsided the tube is removed from the ice bath and allowed to warm to room temperature (about 20 min). The contents of the tube are poured into a 50-ml beaker full of cracked ice. The chloroform layer is removed and washed with water, and the chloroform is evaporated. The residual crude sulfonyl chloride may be recrystallized from low-boiling petroleum ether, benzene, or chloroform.

Note. The above conditions are satisfactory for most of the simple aryl halides and for alkylbenzenes. For the halotoluenes it is desirable to warm the chloroform solution to 50°C for 10 min and then pour the mixture on cracked ice.

Polyhalogen derivatives require more drastic treatment. For such compounds the sulfonation is carried out without any solvent, and the sulfonation mixture is warmed to 100°C for 1 hr under a reflux condenser.

Discussion. Two "abnormal" reactions may take place during the sulfonation. Fluorobenzene, iodobenzene, and *o*-dibromobenzene yield the corresponding sulfones when treated with chlorosulfonic acid at 50°C in the absence of a solvent.

$$2\,C_6H_5X \xrightarrow[50^\circ]{ClSO_3H} X{-}C_6H_4{-}SO_2{-}C_6H_4{-}X$$

These sulfones are solids and will serve as derivatives. In other cases (e.g., Procedure 31), small amounts of sulfones may be produced but are separable from the sulfonamide because they are insoluble in alkali. A second abnormal reaction is nuclear chlorination. This reaction takes place with *p*-diiodobenzene and 1,2,4,5-tetrachlorobenzene. Unsatisfactory results are frequently obtained with aryl iodides.

Aryl halides in which the halogen atom is activated by nitro groups react readily with piperidine to form useful piperidyl derivatives.

Aryl iodides take up chlorine to form iodoaryl dichlorides, which may serve as derivatives or chlorination reagents:

$$ArI + Cl_2 \rightarrow ArICl_2$$

References. M. K. Deikel, Piperidyl Derivatives of Aromatic Halogeno-Nitro Compounds, *J. Amer. Chem. Soc.*, **62,** 750 (1940); H. J. Lucas and E. R. Kennedy, Iodoaryl Dichlorides, *Org. Syn. Coll. Vol. III*, p. 482.

Procedure 31. Sulfonamides

$$X{-}C_6H_4{-}SO_2Cl \xrightarrow{NH_4OH} X{-}C_6H_4{-}SO_2NH_2$$

About 0.5 g of the sulfonyl chloride is boiled with 5 ml of concentrated ammonium hydroxide for 10 min. The mixture is diluted with 10 ml of water, cooled, and filtered.

The crude sulfonamide is dissolved in 10 ml of 5% sodium hydroxide solution, with gentle heating if necessary, and the solution is filtered to remove any sulfone or chlorinated products. The filtrate is acidified with dilute hydrochloric acid, and the sulfonamide is removed by filtration. It is purified by recrystallization from dilute ethanol and dried at 100°C.

Discussion. This procedure, or the Hinsberg procedure (Procedure 16), can be used to prepare *N*-substituted sulfonamides by using primary and secondary amines instead of ammonia.

6.11 CARBOHYDRATES AND SUGARS (SACCHARIDES)

Carbohydrates are polyhydroxy aldehydes and ketones; within this broad category are simpler compounds called sugars (or saccharides). The name carbohydrates is derived historically from the idea "hydrates of carbon," that is, the general formula $C_n(H_2O)_n$; for example, glucose has the molecular formula $C_6H_{12}O_6$. Sucrose (common table sugar) is consistent with another general formula, $C_n(H_2O)_m$, because it has the formula $C_{12}H_{22}O_{11}$. Carbohydrates do have uniformly recognizable properties. These compounds are usually water-soluble solids that melt with decomposition; these characteristics correctly point toward the presence of a number of highly polar functional groups in these molecules. Thus here we shall deal partly with functional group chemistry that is outlined elsewhere in this book (see Table 6.5). Often, as, for example, when derivatizing saccharides, the interaction of substituents in polyfunctional compounds causes results to be markedly different from those for monofunctional compounds; formation of osazones from saccharides, contrasted to phenylhydrazones from simple carbonyl compounds (Experiment 8), is a good example.

EXPERIMENT 24. OSAZONES

$$\text{RCHOH–CH=O} \;\; \textit{or} \;\; \text{RC(=O)–CH}_2\text{OH} \; + 3\,H_2NNHC_6H_5 \rightarrow \text{RC(=NNHC}_6\text{H}_5\text{)–CH=NNHC}_6\text{H}_5 + C_6H_5NH_2 + NH_3 + 2H_2O$$

Table 6.5. Tests Frequently Used to Characterize Saccharides[a]

Functional group (name)	Test reference	Tests
OH (alcoholic hydroxyl)	Procedure 5	Acetylation
C=O (keto carbonyl)	Experiment 8	Phenylhydrazones (osazones, Experiment 24)
CHO (aldehydic carbonyl)[b]	Experiment 11	Tollens test (ammoniacal silver nitrate)
$-\overset{\text{OH}}{\underset{\vert}{\overset{\vert}{C}}}-\overset{\text{OH}}{\underset{\vert}{\overset{\vert}{C}}}-$ (vicinal diol)	Experiment 5	Periodate oxidation
CHO (aldehydic carbonyl)[b]	Experiment 13	Benedict's reagent (and Fehling's solution)
$>C(OR)(OR')$ (acetal, ketal, hemiacetal)	Section 6.2.2	

[a] Other tests described in Sections 6.1 and 6.2 may be used; for derivatives of saccharides (e.g., glyconic acids), simply use tests for the new functional group (e.g., Section 6.12 for carboxylic acids).

[b] Sugars responding positively to these tests are described as "reducing" sugars.

Prepare 0.2-g samples of glucose, maltose, sucrose, and galactose. Test the four sugars simultaneously in the following manner. Place the samples in separate test tubes, and to each test tube add 0.4 g of phenylhydrazine hydrochloride, 0.6 g of crystalline sodium acetate, and 4 ml of distilled water. Stopper the test tubes with one-holed corks, and place them together in a beaker of boiling water. Make a note of the time of immersion and of the time of precipitation of each osazone.[50] It is necessary to shake the tubes occasionally to avoid supersaturation.

After 20 min, remove the tubes from the hot water bath and set them aside to cool. After they are cool, pour a small amount of the liquid and crystals on a watch glass. Tip the watch glass from side to side to spread out the crystals, and absorb some of the mother liquor with a piece of filter paper, taking care not to crush or break up the clumps of crystals. Examine the crystals under a low-power microscope (about 80 to 100×), and compare with photomicrographs.

The formation of tarry products due to oxidation of the phenylhydrazine may be prevented by the addition of 0.5 ml of saturated sodium bisulfite solution. This should be done if it is desired to isolate the osazone and determine its melting point.

Discussion. The time required for the formation of the osazone can be a valuable aid in distinguishing among various sugars. Mulliken gives the following figures for the time required for the osazone to precipitate from the hot solution: fructose, 2 min; glucose, 4 to 5 min; xylose, 7 min; arabinose, 10 min; galactose, 15 to 19 min; raffinose, 60 min; lactose, osazone soluble in hot water; maltose, osazone soluble in hot water; mannose, 0.5 min (hydrazone); sucrose, 30 min (owing to hydrolysis and formation of glucosazone).

A number of other tests, mostly involving color change as a positive result, are listed here:

Test	Reagents	References
Molisch	α-Naphthol, sulfuric acid	Pasto and Johnson, p. 396; Cheronis, Entrikin, and Hodnett, p. 390
Chromatographic spray	*p*-Anisidine hydrochloride, 1-butanol	Pasto and Johnson, p. 397
Anthrone	Anthrone, sulfuric acid	Cheronis, Entrikin, and Hodnett, p. 388
Resorcinol	Resorcinol, sulfuric acid	Cheronis, Entrikin, and Hodnett, p. 390

Solubility properties frequently suggest that the ir (see Fig. 6.26) and nmr (see Fig. 6.27) spectra of simple saccharides be run, respectively, as KBr pellets and heavy water (D_2O) solutions. Derivatization, such as formation of acetonides and glycosides, of saccharides often allows different approaches to their spectral analyses.

In the characterization of saccharides and their derivatives (e.g., glycosides), a common question is whether the compound is in the cyclic (or acyclic) form and

[50] W. Z. Hassid and R. M. McCready, *Ind. Eng. Chem., Anal. Ed.*, **14**, 683 (1942).

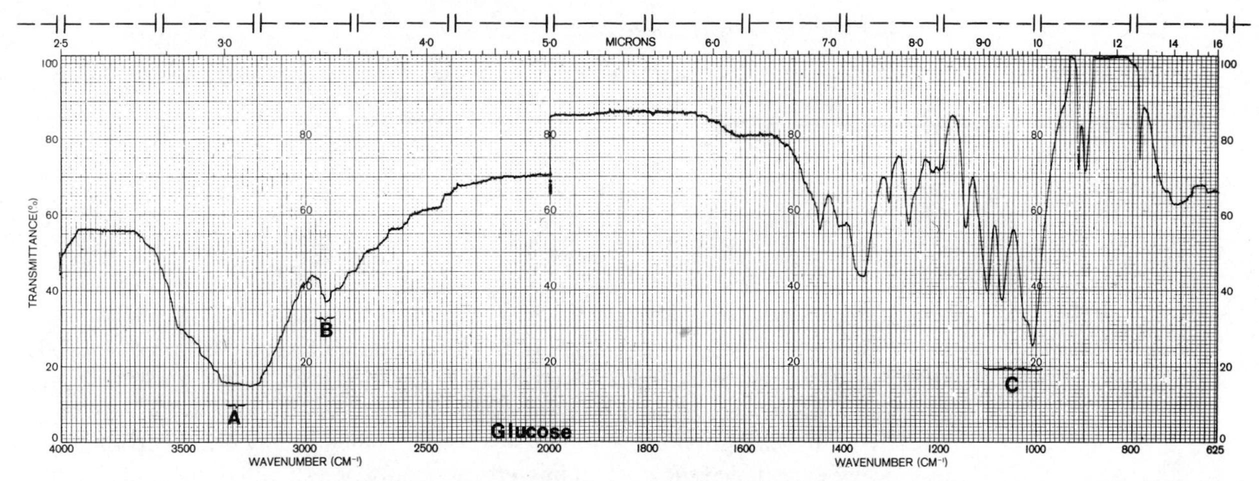

Figure 6.26. Infrared spectrum of β-D(+) glucose: KBr pellet. O—H stretch: **A**, 3600–3000 cm^{-1} (2.78–3.33 μm), $\nu_{O—H}$ (largely hydrogen bonded). C—H stretch: **B**, 2915 cm^{-1} (3.43 μm), $\nu_{C—H}$. C—O stretch: **C**, 1103, 1073, 1008 cm^{-1} (9.07, 9.32, 9.92 μm), $\nu_{C—O}$ (primary, secondary alcohols); **i**, 2000 cm^{-1} (5.00 μm), grating change.

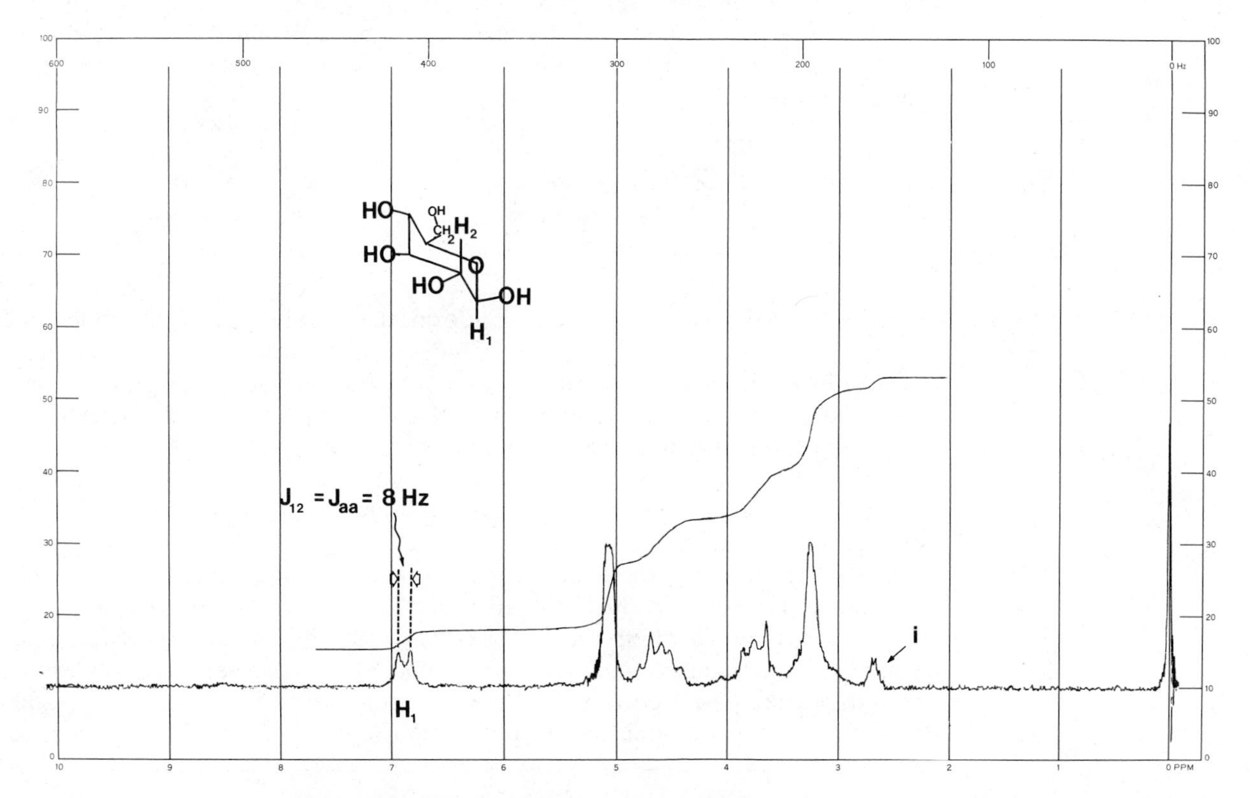

Figure 6.27. Nuclear magnetic resonance (proton) spectrum of β-D(+) glucose; 60 MHz, 600-Hz sweep width, DMSO-d_6 solvent. i = residual solvent signal.

whether the cyclic form is α or β:

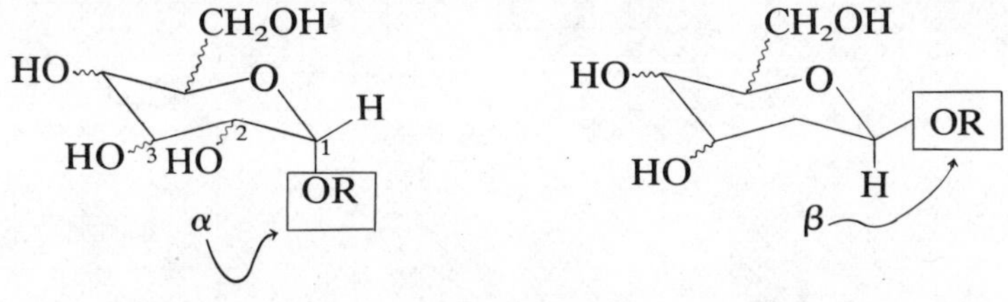

CH₂OH HO O H HO 3 HO 2 1 OR α — CH₂OH HO O OR HO H β

hexopyranoses

The α and β isomers are structurally differentiated by the position of the OR substituent attached to the anomeric (C-1) position. Since the proton at this position experiences the deshielding effect of the *two* oxygen atoms attached directly to this carbon atom, while the protons attached to any other carbon atom in the ring experience the deshielding effect of only one directly attached oxygen, the anomeric proton is at a well-separated and lower field position. The α and β isomers can be spectrally differentiated, because the anomeric proton of the α-isomer is equatorial and thus usually at a slightly lower field position than the anomeric proton of the β-isomer.[51] It is generally observed in six-membered rings that equatorial protons are at 0.3 to 0.9 ppm lower field than axial protons.

Additional configurational information can be ascertained from the splitting of the anomeric proton; this again is consistent with generalizations appropriate for six-membered rings:

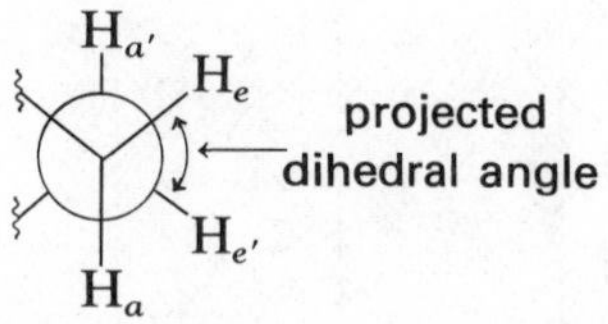

$H_{a'}$, H_e, $H_{e'}$, H_a — projected dihedral angle

$$J_{ee'} \cong J_{ea'} \cong J_{ae'} = 2\text{–}6\ \text{Hz}$$
$$J_{aa'} = 8\text{–}14\ \text{Hz}$$

As shown above in Newman projection form, there are both 180° (a–a') and 60° (a–e', e–a', e–e') dihedral angle magnitudes relating vicinal protons; the multiplicity of the anomeric proton signal may identify the configurational position (e or a) of the C-2 proton. For example, an axial anomeric proton split to the extent of ca. 8 Hz implies that the hydroxyl group at C-2 is equatorial.

Since the saccharides possess ordinary organic functional groups, as listed in Table 6.5 (p. 261), one should consult appropriate sections to obtain details regarding ir absorptions for these groups.

Mass spectra of carbohydrates are obtained only after utilization of special procedures largely associated with the very low volatility of these compounds. Direct ionization is often used to analyze such carbohydrates because they typically resist sublimation. In addition, commercially available reagents can be used to prepare polyesters or polyethers (e.g., polysiloxanes) that are more readily volatilized for normal ionization. Another difficulty associated with mass spectral

---

[51] R. M. Silverstein, G. C. Bassler, and T. C. Morrill, *Spectrometric Identification of Organic Compounds*, 3rd ed. (Wiley, New York, 1974), chap. 4, pp. 190–191.

analysis is the rather complex fragmentation pattern that most carbohydrates and their derivatives provide.

Since many saccharides and other carbohydrates and their derivatives are chiral, the specific rotation, [α], is often extremely useful; experimental measurement of [α] is described in Chapter 8. Lanthanide shift reagents (Chapters 8 and 10) can be used to aid nmr analyses of carbohydrates.

The osazone procedure, described in Experiment 24, can be used to prepare a derivative; polyacetates, as well as other polyesters, can also be used, and the procedure should be used as discussed in Section 6.1.1.

6.12 CARBOXYLIC ACIDS

Usually carboxylic acids (RCO_2H) will be found in solubility group A_1. Alternatively, water-soluble acids can be detected by an indicator. Two common types of indicators are used: Congo red and Universal. They are both available in solution and paper form. Acid character does not, however, prove that the compound is a carboxylic acid; sulfonic acids, certain nitro-substituted phenols, and other organic compounds will also give a positive tests for acid character.

Indicator	Color observation
Universal (Hydrion)	Variable, depends on pH range
Congo red	Blue color indicates acid

In a very few cases acid character will be difficult to detect; in such cases the following tests, as described in Cheronis, Entrikin, and Hodnett, can be used to probe for acidity:

Test	Page
Iodate-iodide reagent	359
Rhodamine-B/uranyl acetate	359
Liberation of nitrous acid	360

The detection of the carboxyl (CO_2H) group in an organic compound is greatly aided by spectral analysis. Perhaps the most useful is ir analysis; the ir spectrum of a typical carboxylic acid is shown in Fig. 6.28. The broad band observed in the 3.00–4.00 μm (3333–2500 cm^{-1}) region, often showing multiple maxima, is diagnostic of the carboxyl group. This broadness is at least partially due to substantial contributions of the dimeric, hydrogen-bonded form of these acids:

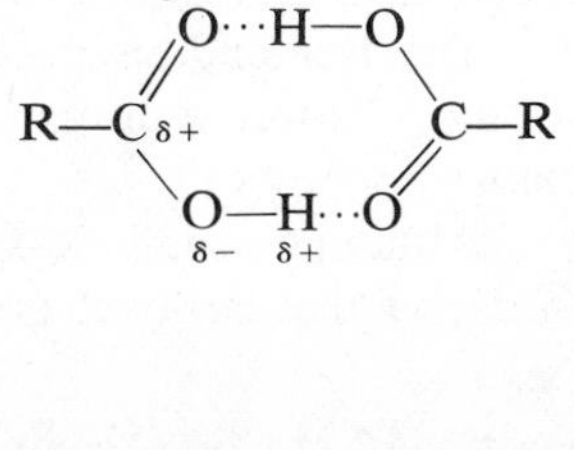

```
         δ−
         O···H—O
        //       \
R—C δ+           C—R
        \       //
         O—H···O
         δ−  δ+
```

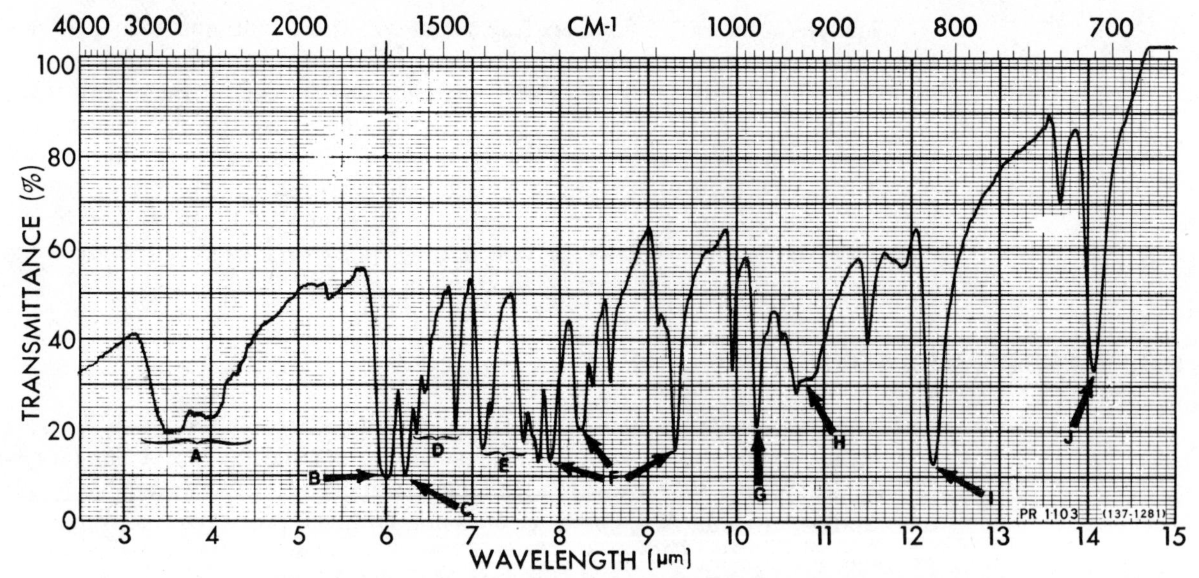

Figure 6.28. Infrared spectrum *p*-chlorocinnamic acid; KBr pellet.*

Cl—C_6H_4—CH═CH—CO_2H

O—H stretch: **A**, 3226–2326 cm^{-1} (3.10–4.30 μm),† $\nu_{O—H}$ (associated). C═O stretch: **B**, 1695 cm^{-1} (5.90 μm), $\nu_{C═O}$ (conjugated). C═C stretch: **C**, 1634 cm^{-1} (6.12 μm), $\nu_{C═C}$ (conjugated). C⩦C stretch; **D**, 1600, 1577, 1493 cm^{-1}, (6.25, 6.35, 6.70 μm), $\nu_{C⩦C}$. C—O stretch, O—H bend (in plane): **E**, 1425, 1307 cm^{-1} (7.02, 7.65 μm), $\nu_{C—O}$, $\delta_{O—H}$, coupled. ⩦C—H bend (in-plane); **F**, 1282, 1227, 1087 cm^{-1} (7.80, 8.15, 9.20 μm), $\delta_{⩦C—H}$ (in-plane); ═C—H bend (out-of-plane, olefinic): **G**, 988 cm^{-1} (10.12 μm), $\delta_{═C—H}$ (out-of-plane) O—H bend (out-of-plane); **H**, *ca.* 930 cm^{-1} (*ca.* 10.75 μm), $\delta_{O—H}$ (out-of-plane); ⩦C—H bend (out-of-plane): **I**, 823 cm^{-1} (12.15 μm), δ⩦C—H (out-of-plane); **J**, 715 cm^{-1} (13.98 μm), δ⩦C—H (out-of-plane).

* Calibration: since the tracing is skewed to slightly long wavelengths (μm), all band positions on spectrum decreased by 0.1 μm on tabulation. † Typical broad band corresponding to carboxylic acid dimer.

In addition, the C═O (5.81–5.86 μm, 1720–1706 cm^{-1}) and C—O (7.58–8.26 μm, 1320–1210 cm^{-1}) stretching bands are strong and useful in detecting the carboxyl group; the former band is especially useful because its position responds to π-conjugation with the carbonyl group very similarly to the responses described above for aldehydes and ketones (Section 6.2); specifically, the C═O stretch of α,β-unsaturated carboxylic acids,

$$\left(>\overset{\beta}{C}=\overset{\alpha}{C}(-)-\overset{O}{\overset{\|}{C}}OH \right)$$

occurs in the 5.85–5.95 μm (1710–1680 cm^{-1}) region.

The tremendous polarity of the carboxyl group is emphasized by the very strong resistance of carboxylic acids to movement through chromatographic media.

Although usually less useful than the ir spectrum, the nmr spectrum can help to detect the carboxyl group. The nmr spectrum of a typical carboxylic acid is

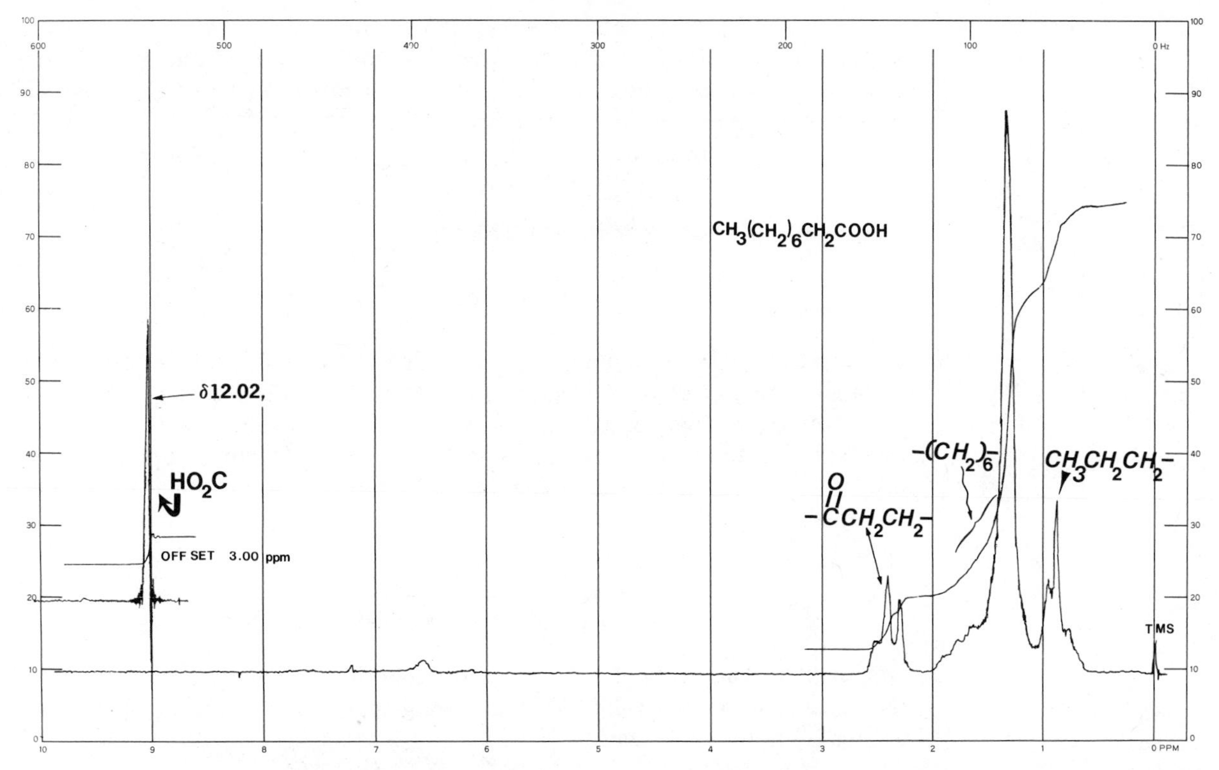

Figure 6.29. Nuclear magnetic resonance (proton) spectrum of nonanoic acid. 60 MHz, 600-Hz sweep width, $CDCl_3$ solvent. The insert at the upper left (carboxyl proton, singlet, δ12.02) has been offset upfield 3.00 ppm.

shown in Fig. 6.29. The carboxyl proton is usually found in the region δ10.0–13.3; in addition, since the signal's position is dependent on concentration and temperature, the student may be initially dismayed to find that it will be at different positions for various solutions of the same acid. Solutions of acids in dimethyl sulfoxide-d_6 may show the carboxyl proton at a substantially higher field than δ10.0. An absorption suspected to be due to a carboxyl proton can be checked by addition of deuterium oxide to the sample; the nmr spectrum of the resulting sample should no longer display the carboxyl OH peak and should now show a substantial water resonance at ca. δ4.8.

Mass spectrometry offers only an outside hope of providing the molecular formula as well as the molecular weight of a carboxylic acid, because the molecular ion peak is often very weak. Fragmentation of the carboxylic acid may provide useful information; common fragments are listed below.

Fragment	Mass	Comments
R—C≡O+	MW−17	Prominent for smaller R groups
R+	MW−45	Prominent for smaller R groups
$R_1R_2C{=}C(OH)_2^+$	$58+R_1+R_2$	McLafferty rearrangement

$M^{+\cdot} = RCO_2H^{+\cdot}$

One of the most useful pieces of information about a carboxylic acid is its *neutralization equivalent* (neut eq), which is obtained by quantitative titration with a standardized base (Procedure 32). The molecular weight of the acid, within experimental error, is an integral (usually simply 1, 2, 3, etc.) multiple of the neutralization equivalent.

Procedure 32. Neutralization Equivalents

$$RCO_2H + NaOH \rightarrow RCO_2Na + H_2O$$

A sample of the acid (about 0.2 g) is weighed accurately[52] and dissolved in 50 to 100 ml of water or ethanol. The mixture may be heated if necessary to dissolve all the compound. This solution is titrated with a previously standardized sodium hydroxide solution having a normality (N) of about 0.1, phenolphthalein being used as the indicator. The neutralization equivalent of the acid is calculated according to the formula

$$\text{neut eq} = \frac{\text{weight of sample} \times 1000}{\text{volume of alkali (ml)} \times N}$$

Discussion. The molecular weight (mol wt) of an acid may be determined from the neutralization equivalent by multiplying that value by the number of acidic groups (x) in the molecule:

$$\text{mol wt} = x(\text{neut eq})$$

The change in medium, even from pure water to pure ethanol, affects the pK of both the organic acid and the indicator. For this reason best results are obtained in water or aqueous ethanol with only enough ethanol to dissolve the organic acid. In absolute or 95% ethanol it is often impossible to obtain a sharp end point with phenolphthalein. In such cases bromothymol blue should be employed as the indicator. Acids may also be titrated in a solvent composed of ethanol and benzene or toluene.

A blank must always be run on the solvent; the same amount of phenolphthalein that was used in the titration must be employed in the blank. In ordinary work the neutralization equivalents should check the calculated values within ±1%. By using carefully purified and dried samples and good technique, the error may be reduced to ±0.3%.

In order to give an accurate neutralization equivalent, the substance titrated must be pure and anhydrous. If the value obtained for the neutralization equivalent does not check the theoretical value, the compound should be recrystallized from a suitable solvent and carefully dried.

Amine salts of strong acids may be titrated by the same procedure.

---

[52] Measured accurately to at least three figures on an analytical balance.

Questions

1. Calculate the neutralization equivalent of benzoic acid and of phthalic acid.
2. If an acid is imperfectly dried, will the neutralization equivalent be high or low?
3. Would the presence of an aromatic amino group interfere in the determination of the neutralization equivalent? What would be the effect of an aliphatic amino group?
4. What types of phenols may be titrated quantitatively?
5. From a theoretical point of view, what should the ionization constant of an acid be in order that the acid may be titrated with phenolphthalein?

Derivatives of carboxylic acids can be made by conversion of the acid to esters or amides; the text sections on these compounds, as well as sections on intermediates that can be converted to these compounds, should be consulted.

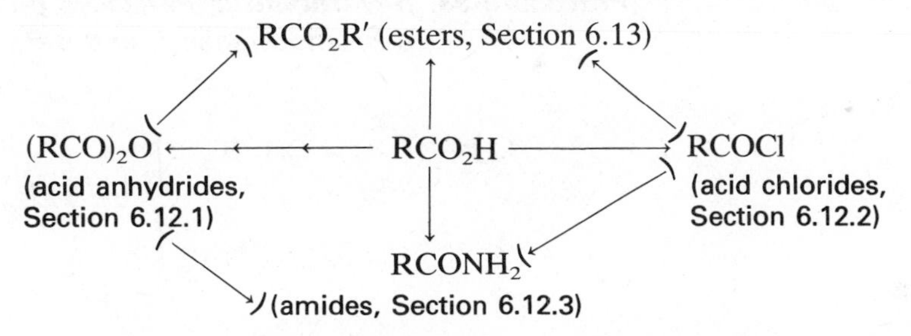

Solid esters furnish a useful means for characterizing the acids. Some methyl esters are solid, but in most cases one must use *p*-nitrobenzyl, phenacyl, *p*-chlorophenacyl, *p*-bromophenacyl, or *p*-phenylphenacyl esters. These are prepared by treating the salts of the acids with the corresponding halide (Procedure 33).

$$RCOONa + ClCH_2C_6H_4NO_2(p) \rightarrow RCOOCH_2C_6H_4NO_2(p) + NaCl$$

$$RCOONa + BrCH_2COAr \rightarrow RCOOCH_2COAr + NaBr$$

This method is particularly advantageous because it does not require an anhydrous sample of the acid.

The reaction products of phenylhydrazine and acids are also good derivatives. At the boiling point (243°C) of phenylhydrazine, simple unsubstituted aliphatic mono- and dibasic acids form phenylhydrazides (Procedure 34).

$$RCO_2H + H_2NNHC_6H_5 \rightarrow RCONHNHC_6H_5 + H_2O$$

Salts of phenylhydrazine are obtained from sulfonic acids, α-chloro aliphatic acids, and aliphatic dibasic acids when warmed with a benzene solution of phenylhydrazine.

Acids may be converted to amides by treatment with thionyl chloride and then with ammonia.

$$RCOOH + SOCl_2 \rightarrow RCOCl + SO_2 + HCl$$

$$RCOCl + 2NH_3 \rightarrow RCONH_2 + NH_4Cl$$

This method is particularly suitable if the amide is insoluble in water; it may be made by adding the acid chloride to concentrated aqueous ammonia (Procedure 35).

Anilides, toluidides, and *p*-bromoanilides are excellent derivatives because of the ease with which they may be made and purified. They may be made either from the free acid or from its salt (Procedure 36).

Salts derived from organic acids and strongly basic amines such as benzylamine, α-phenylethylamine, and piperazine are also good derivatives.

$$C_6H_5CH_2NH_2 + RCO_2H \longrightarrow C_6H_5CH_2NH_3^+ \; RCO_2^-$$

Carboxylic acids, or their derivatives, can be converted to thiuronium derivatives (Procedure 39).

Procedure 33. p-Nitrobenzyl, Phenacyl, p-Chlorophenacyl, p-Bromophenacyl, and p-Phenylphenacyl Esters

$$\left.\begin{matrix} O_2N\text{-}C_6H_4\text{-}CH_2Cl \\ \text{or} \\ ArCOCH_2Br \end{matrix}\right\} \xrightarrow{RCO_2^-Na^+} \left\{\begin{matrix} RC(=O)OCH_2Ar' \\ \text{or} \\ RC(=O)O\text{—}CH_2C(=O)Ar \end{matrix}\right. + NaX$$

Ar = phenyl or substituted phenyl groups

One gram of the acid is added to 5 ml of water in a small flask and is neutralized carefully with 10% sodium hydroxide solution. A little of the acid is added, the addition being continued until the solution is just acid to litmus. If the original acid is obtained as the sodium salt, 1 g of the salt is dissolved in 5 to 10 ml of water. If this solution is alkaline, a drop or two of dilute hydrochloric acid is added. Ten milliliters of alcohol and 1 g of the halide[53] are added; the mixture is then heated under reflux for 1 hr if the acid is monobasic, 2 hr if dibasic, and 3 hr if tribasic. Occasionally the addition of a few more milliliters of alcohol may be necessary if a solid separates during the refluxing. The solution is allowed to cool, and the precipitated ester is purified by recrystallization from alcohol.

In preparing derivatives with these reagents, care must be taken that the original reaction mixture is not alkaline. Alkalies cause hydrolysis of the phenacyl halides to phenacyl alcohols. In addition, *p*-bromophenacyl bromide should not be used if considerable amounts of sodium chloride are present in the sodium salt of the acid.

Crown ethers can be used to improve the ease with which esters are formed. For example potassium acetate will convert *n*-heptyl bromide to *n*-heptyl acetate in near quantitative yield.[54]

---

[53] *Caution!* Phenacyl halides are lachrymators.
[54] G. Gokel and H. Durst, *Synthesis*, 178 (1976).

Procedure 34. Phenylhydrazides and Phenylhydrazonium Salts

$$\text{(a)}\quad RCO_2H + C_6H_5NHNH_2 \xrightarrow[\Delta]{\text{heat}} R\overset{\overset{\displaystyle O}{\|}}{C}NHNHC_6H_5$$

$$\text{(b)}\quad \left.\begin{array}{c} RCO_2H + C_6H_5NHNH_2 \\ \text{or} \\ RSO_3H + C_6H_5NHNH_2 \end{array}\right\} \xrightarrow[(\Delta)]{\text{benzene}} \left\{\begin{array}{c} R\overset{\overset{\displaystyle O}{\|}}{C}NHNHC_6H_5 \\ \text{or} \\ RSO_3^-\ \overset{+}{N}H_3NHC_6H_5 \end{array}\right.$$

(*a*) One gram of the acid is dissolved in 2 ml of phenylhydrazine, and the solution is boiled gently for 30 min. The crystalline product that separates when the solution cools is isolated by suction filtration and washed with small quantities of benzene* or ether until the crystals are white. When a large excess of phenylhydrazine is used, it is sometimes necessary to dilute the mixture with benzene* in order to bring about precipitation of the product. The derivatives of the lower monobasic acids are recrystallized from hot benzene,* whereas those of the higher acids and dibasic acids are best recrystallized from ethanol or ethanol-water mixtures. The derivatives obtained from dibasic acids by this method are bis-β-phenylhydrazides.

(*b*) One gram of the acid is mixed with 2 ml of phenylhydrazine dissolved in 5 ml of benzene. Sometimes a white solid precipitates immediately; it is recrystallized from ethanol. If no solid separates, the mixture is heated under reflux for 30 min, and the product that precipitates upon cooling is collected on a filter, washed with ether, and recrystallized from benzene or ethanol. Sulfonic acids, halogen-substituted aliphatic acids, and aliphatic dibasic acids yield salts by this procedure, whereas simple unsubstituted aliphatic acids give phenylhydrazides.

Procedure 35. Amides from Acids

$$RCO_2H + SOCl_2 \xrightarrow[\Delta]{} R\overset{\overset{\displaystyle O}{\|}}{C}Cl \xrightarrow{NH_3} R\overset{\overset{\displaystyle O}{\|}}{C}NH_2$$

$$(RCOCl + R'NH_2 \longrightarrow RCONHR')$$

One gram of the acid is heated under reflux with 5 ml of thionyl chloride for 15 to 30 min. The mixture is poured cautiously into 15 ml of ice-cold, concentrated aqueous ammonia. The precipitated amide is collected on a filter and purified by recrystallization from water or dilute ethanol.

---

* See footnote 43, Chapter 6.

Procedure 36. Anilides, p-Toluidides, and p-Bromoanilides

(a) $$RCO_2H \xrightarrow{SOCl_2} RCOCl \xrightarrow{ArNH_2} RCONHAr$$

(b) $$RCO_2^-Na^+ \xrightarrow{HCl/ArNH_2} RCONHAr$$

(*a*) One gram of the acid or its sodium salt is mixed with 2 ml of thionyl chloride in a test tube, and the mixture is heated under a small reflux condenser for 30 min. The mixture is cooled, a solution of 1 to 2 g of the amine (aniline, *p*-toluidine, or *p*-bromoaniline) in 30 ml of benzene is added, and the mixture is warmed on the steam bath for 2 min. The benzene solution is decanted into a separatory funnel and washed successively with 2 ml of water, 5 ml of 5% hydrochloric acid, 5 ml of 5% sodium hydroxide solution, and 2 ml of water. The benzene is evaporated and the amide recrystallized from water or ethanol.

(*b*) A mixture of 0.4 g of the dry powdered sodium salt of the acid, 1 ml of the arylamine, and 0.3 ml of concentrated hydrochloric acid is placed in a test tube. The test tube is placed in an oil bath which is then heated, the temperature being kept between 150 and 160°C for 45 to 60 min. At the end of this time, the test tube is removed and the product purified by one of the following methods.

1. If the acid under consideration has fewer than six carbon atoms, the crude product is boiled with 5 ml of 95% ethanol and the solution decanted into 50 ml of hot water. The resulting solution is evaporated to a volume of 10 to 12 ml and cooled in an ice bath. The crystals are removed by filtration and recrystallized from a small amount of water or dilute ethanol.
2. If the acid contains six or more carbon atoms, the crude reaction product is powdered and washed with 15 ml of 5% hydrochloric acid and then with 15 ml of cold water. The residue is boiled with 30 to 40 ml of 50% ethanol, and the solution is filtered. The filtrate is chilled in an ice bath, and the crystals of the substituted amide are removed by filtration. The product may be recrystallized from dilute ethanol.

Carboxylic acids can be analyzed by nmr using lanthanide shift reagents; one must use shift reagents with more highly fluorinated ligands: $Eu(fod)_3$ and $Eu(tfn)_3$ (Chapter 8).

6.12.1 Acid Anhydrides

Solubility classification may be useless and even misleading in characterizing anhydrides; these compounds are rapidly converted by water to the corresponding acids, especially when mineral acid catalysis is employed.

$$(RCO)_2O + H_2O \xrightarrow{H^+} 2RCO_2H$$

In many cases the solubility results may thus actually reflect the properties of the acids.

Spectral analyses should be carried out with sampling techniques that carefully keep the anhydride as dry as possible. The ir spectrum of a typical anhydride is shown in Fig. 6.30. The most striking aspect of the ir spectrum is the appearance of *two* carbonyl stretching bands. This is due to the coupled interaction of the two carbonyl groups giving rise to both the symmetrical and asymmetrical interactions:[55]

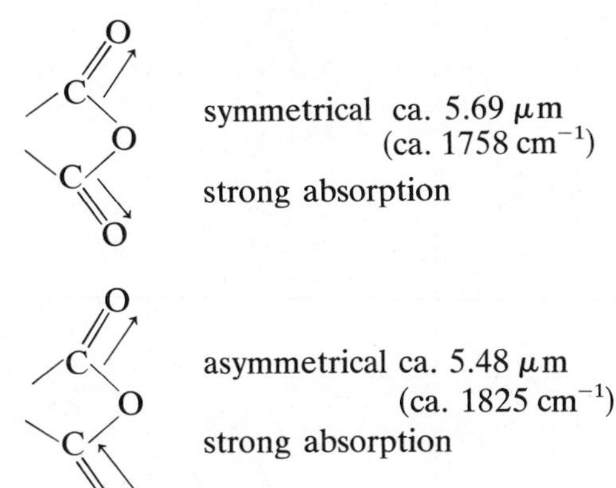

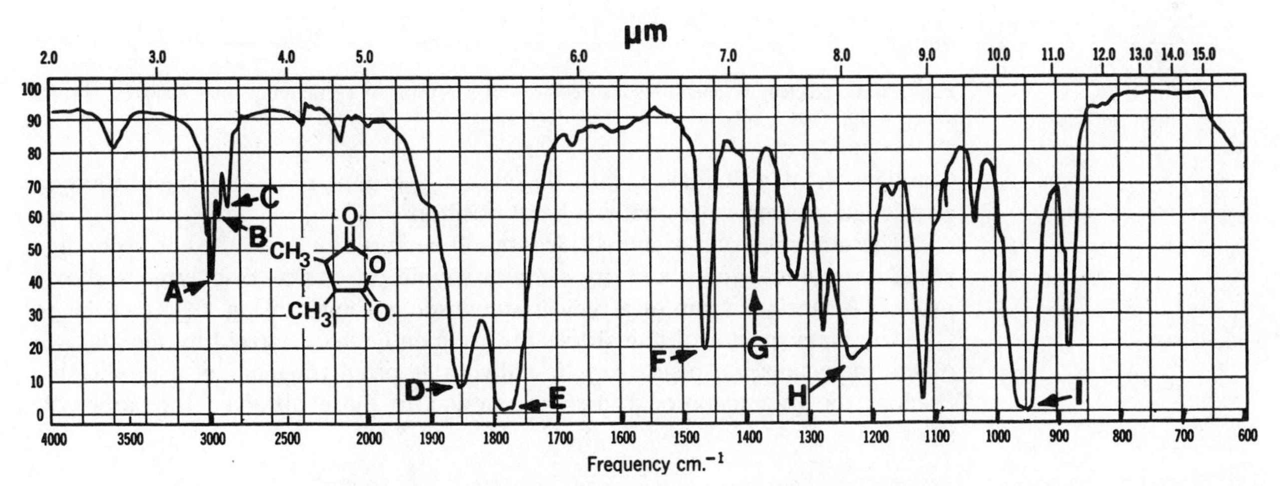

Figure 6.30. Infrared spectrum of α,α′-dimethylsuccinic anhydride. C—H stretch: **A**, 2980 cm^{-1} (3.36 μm), ν_{asym} CH_3; **B**, 2910 cm^{-1} (3.44 cm^{-1} (3.44 μm), $\nu_{C—H}$ (methine characteristically weak); **C**, 2860 cm^{-1} (3.50 μm), ν_{sym} CH_3. C═O stretch: **D**, 1850 cm^{-1} (5.40 μm), ν_{asym} C═O; **E**, 1775 cm^{-1} (5.63 μm), ν_{sym} C═O, coupled. C—H bend: **F**, 1470 cm^{-1} (6.80 μm), δ_{asym} CH_3; **G**, 1380 cm^{-1} (7.25 μm), δ_{sym} CH_3; C—CO—O—CO—C stretch (coupled): **H**, 1230 cm^{-1} (8.13 μm), $\nu_{C—O}$; **I**, 950 cm^{-1} (10.53 μm), $\nu_{C—O}$.

[55] The asymmetrical band is slightly stronger in aliphatic acyclic systems; the symmetrical band is stronger in cyclic anhydrides with 5-membered rings.

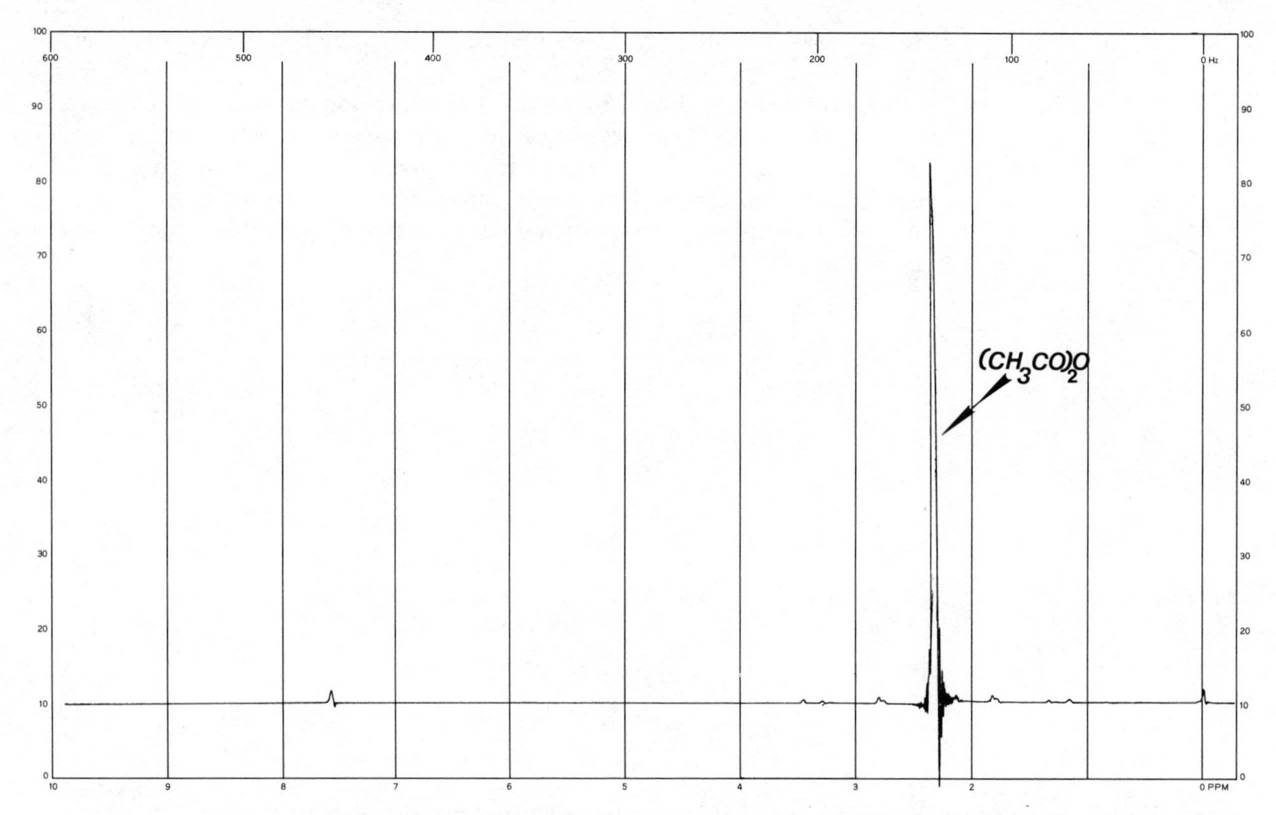

Figure 6.31. Nuclear magnetic resonance (proton) spectrum of acetic anhydride: 60 MHz, 600-Hz sweep width, $CDCl_3$ solvent.

Conjugation of a π system with the carbonyl groups results in slightly longer wavelength (lower wave number) band positions.

The nmr spectrum of anhydrides should be very similar to the corresponding acids; scrupulously pure and dry samples should, of course, display no carboxyl proton. A nmr spectrum of a typical anhydride is shown in Fig. 6.31.

The mass spectra of anhydrides are frequently characterized by the absence of the molecular ion peak. Acyclic anhydrides often fragment in a predictable fashion; the fragmentation routes are similar to those observed for carboxylic acids:

$$\left[\left(\overset{O}{\overset{\|}{RC}}\!\!\vdots\right)_2 O\right]^+ \longrightarrow RCO^+ \xrightarrow{-CO} R^+ \xrightarrow{-H_2} [R-H_2]^+$$

(RCO^+: often the base peak fragment)

Cylic anhydrides fragment via much more complex routes.[56]

[56] See H. Budzikiewicz, C. Djerassi, and D. H. Williams, *Mass Spectrometry of Organic Compounds* (Holden-Day, San Francisco, 1967).

Acid anhydrides of carboxylic acids, $(RCO)_2O$, can be derivatized by the same procedures used for carboxylic acids; simply substituting the anhydride for the carboxylic acid sample in the following procedures should lead to the desired derivative:

Derivative	Procedure
RCONHAr (anilide)	36(b)[a]
$RCONH_2$ (amide)	35[b]

[a] The *N*-phenylimide can arise from treatment with aniline, as in the case of phthalic anhydride:

$$\text{Phthalic anhydride} \xrightarrow{C_6H_5NH_2} C_6H_4(CONHC_6H_5)(CO_2H) \xrightarrow{\Delta} \text{N-}C_6H_5\text{ phthalimide}$$

[b] Thionyl chloride is *not* omitted from the procedure.

The Schotten-Baumann procedure (Experiment 18e) can be used for conversion of anhydrides to esters; just as is described for acid chlorides, one simply treats the anhydride with an alcohol:

$$RCOCl \text{ or } (RCO)_2O \xrightarrow[\text{(or ArOH)}]{R'OH} RCO_2R' \quad (RCO_2Ar)$$

Alcohols react with acid chlorides of dibasic acids to produce normal esters, but with an anhydride of a dibasic acid the acid ester is the chief product. Phthalic anhydride combines with alcohols to give alkyl hydrogen phthalates. The diester is produced only if a catalyst is used and an excess of the alcohol is present.

$$\text{Phthalic anhydride} + RCH_2OH \rightarrow C_6H_4(CO_2CH_2R)(CO_2H)$$

6.12.2 Acid Halides

Acid halides may give misleading solubility results similar to those described in the preceding section for acid anhydrides, because they may very readily hydrolyze to the carboxylic acid and mineral acid:

$$R\overset{O}{\overset{\|}{C}}Cl + H_2O \longrightarrow RCO_2H + HCl$$

Thus solubility "results" may actually correspond to characteristics of the hydrolysis products. Acid chlorides, the most common member of this halide class, can be detected by the "labile" halide test (ethanolic silver nitrate, Experiment

16). The more volatile acid halides can be detected merely by their obnoxious, lachrymatory odor.

As with the anhydrides, extreme care must be taken to exclude moisture when spectrally analyzing acid halides. The ir spectrum of a typical acid chloride is shown in Fig. 6.32. The carbonyl stretch of aliphatic acid chlorides occurs at ca. 5.51–5.60 μm (1815–1785 cm^{-1}); conjugation of a π system induces the usual 0.05-μm wavelength increase (15-cm^{-1} wave number decrease). These carbonyl bands often appear to be doublets due to the occurrence of a second maximum at slightly longer wavelength.

Nuclear magnetic resonance spectra of acid halides should be very similar to those of carboxylic acids; the nmr spectrum of a typical acid chloride is shown in Fig. 6.33. One should avoid hydroxylic solvents for nmr analysis, as these will usually react with and thus consume the acid halide. Scrupulously pure and dry samples of acid halides should show no carboxyl proton in the nmr spectrum and no O—H stretch in the ir spectrum.

Acid halides are often identified by reference to anilides, which they form when treated with aniline (Procedure 36). If the corresponding amide is water-insoluble, the acid halide may be treated with concentrated aqueous ammonia in order to produce the amide.

Acid chlorides of dibasic acids react with an excess of aniline at ordinary temperatures to produce the dianilides. If an anhydride of a dibasic acid is heated with aniline, the *N*-phenylimide may be produced. This frequently serves as a derivative.

$C_6H_4(CO_2H)_2 \xrightarrow{SOCl_2} C_6H_4(COCl)_2 \xrightarrow{C_6H_5NH_2} C_6H_4(CO)_2N\text{—}C_6H_5$

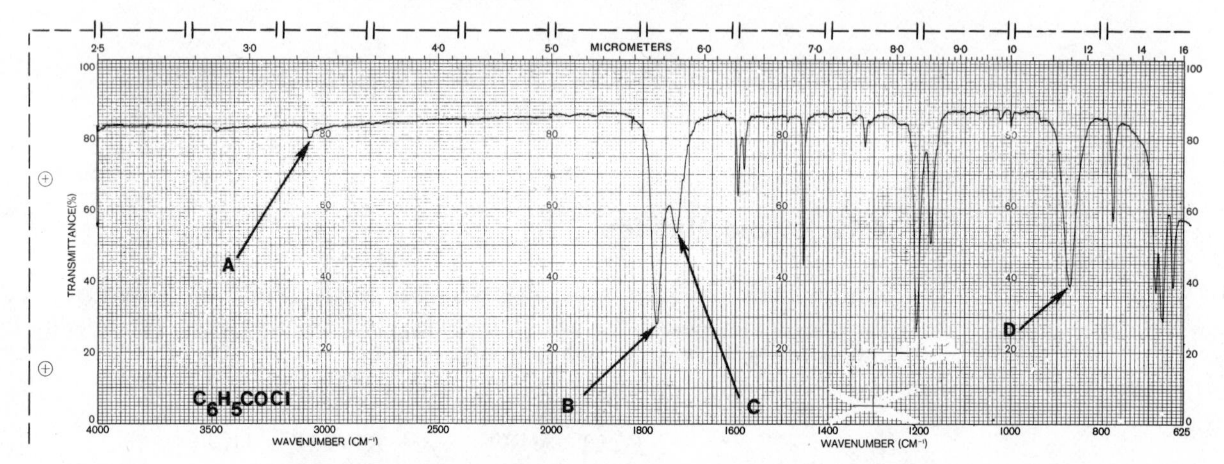

Figure 6.32. Infrared spectrum of benzoyl chloride; neat. C—H stretch: **A**, 3070 cm^{-1} (3.26 μm); ν_{C-H} (aromatic). C=O stretch: **B**, 1770 cm^{-1} (5.65 μm); $\nu_{C=O}$; **C**, 1727 cm^{-1} (5.79 μm); Fermi resonance band for resonance between B and overtone of 872 cm^{-1} (11.47 μm) band]. C—Cl stretch: **D**, 872 cm^{-1} (11.47 μm); ν_{C-Cl}.

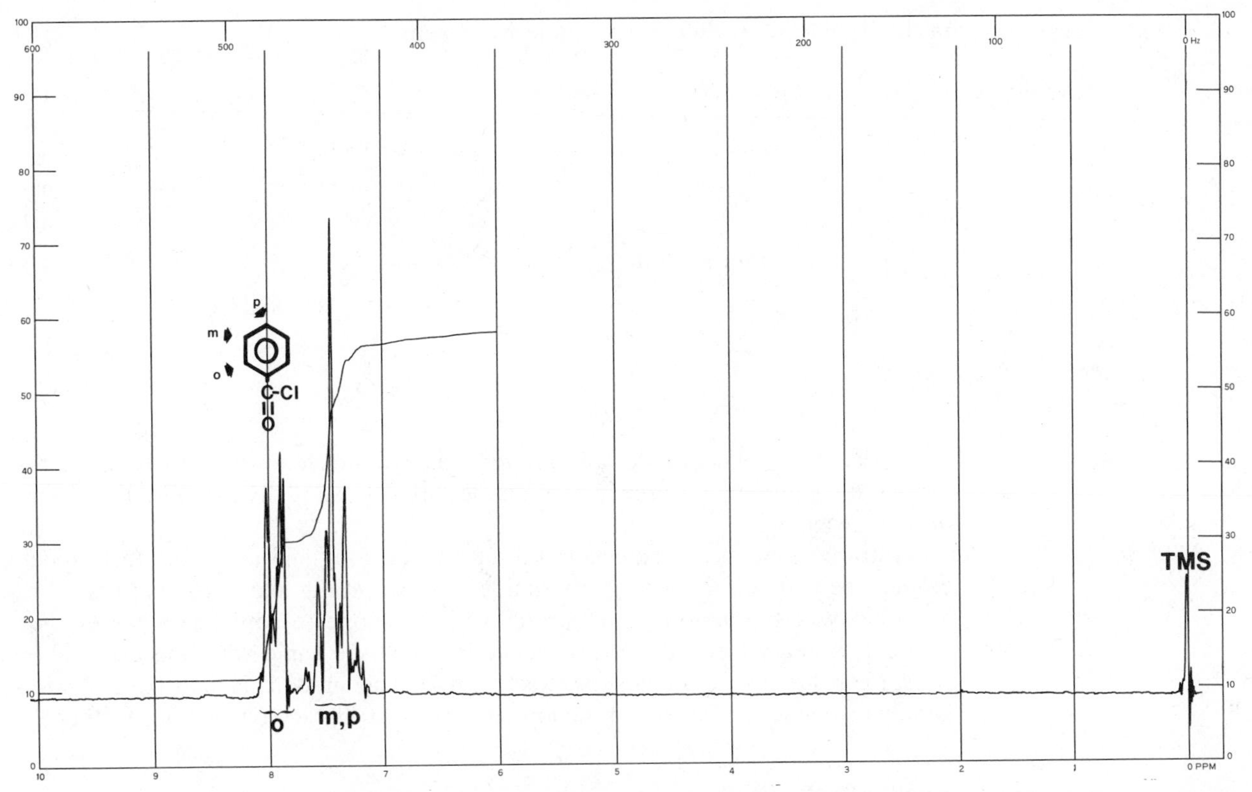

Figure 6.33. Nuclear magnetic resonance (proton) spectrum of benzoyl chloride: 60 MHz, 600-Hz sweep width, $CDCl_3$ solvent.

Hydrolysis converts acid halides (or anhydrides) into the corresponding acids. If the acid is a solid, it frequently will serve as a derivative. In other cases the acid halide (or anhydride) may be hydrolyzed with dilute alkali, and the solution neutralized with hydrochloric acid (phenolphthalein) and evaporated. The mixture of the sodium salt of the acid and sodium chloride may be used for preparing solid esters such as those described under acids in Procedure 33.

Acid halides (and anhydrides) react with alcohols and phenols to produce esters (Experiment 18d), which can be used for further characterization (Section 6.13).

6.12.3 Acid Amides

Because of the presence of nitrogen in amides and the neutral response of amides to acid and base, amides are usually found in solubility class MN. Exceptions occur when sufficiently strong electron-withdrawing groups on the amide nitrogen heighten the acidity of the otherwise neutral amide proton. An example is found in the significant acidity of the imide hydrogen:

$$\text{amides:}\quad R{-}\overset{\displaystyle O}{\overset{\|}{C}}NH_2 \xrightarrow{OH^-}\!\!\!\!\times\ RCONH^-$$

$$\text{imides:}\quad R\overset{\displaystyle O}{\overset{\|}{C}}NH\overset{\displaystyle O}{\overset{\|}{C}}R \xrightarrow{OH^-} \begin{matrix} R{-}\overset{O^{\delta-}}{\overset{\vdots}{C}} \\ \quad\quad N^{\delta-} \\ R{-}\underset{O_{\delta-}}{\underset{\vdots}{C}} \end{matrix} + H_2O$$

Usually, however, amides are not appreciably reactive toward 5% HCl or 5% NaOH at room temperature. In fact, amides are often initially diagnosed by their *lack* of reactivity.

Infrared spectra are especially useful for detecting and analyzing the amide group. The ir spectrum of a typical amide is shown in Fig. 6.34. The existence or multiplicity of the amide N—H stretch can allow one to conclude whether the amide is primary, secondary, or tertiary in much the same way one can make similar conclusions for amines (see Section 6.7). The amide I, II, and III ir bands, largely involving the carbonyl function, allow one to characterize amides further.

Amide structures:

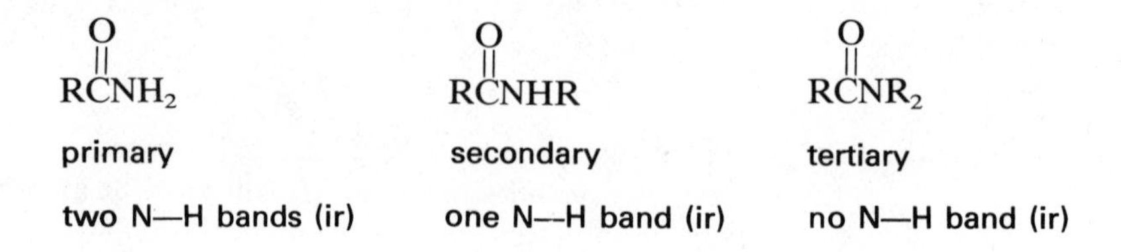

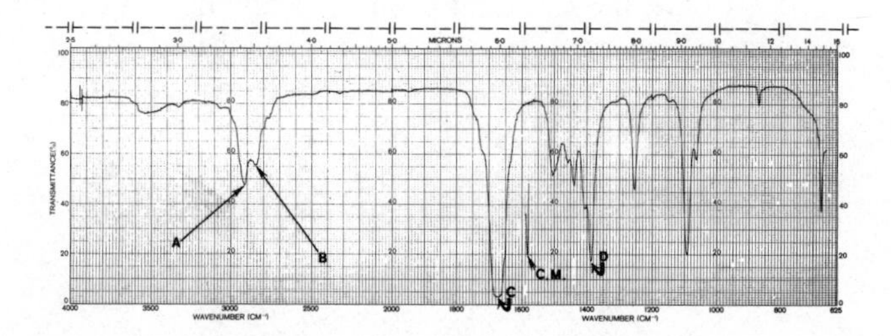

Figure 6.34. Infrared spectrum of *N,N*-dimethylformamide (DMF): neat. C—H stretch: **A**, 2915 cm^{-1} (3.43 μm); ν_{C-H} (aliphatic); **B**, 2850 cm^{-1} (3.51 μm); ν_{C-H} of CHO group. Amide I and II bands: **C**, 1675 cm^{-1} (5.97 μm) $\nu_{C=O}$ overlap of $\nu_{C=O}$ (Amide I) with N—H bend (Amide II). C—N band: **D**, 1387 cm^{-1} (7.22 μm); ν_{C-N}. C.M. = calibration marker: 1583 cm^{-1} (6.33 μm) band of the polystyrene window (no adjustments needed in the reading of band positions).

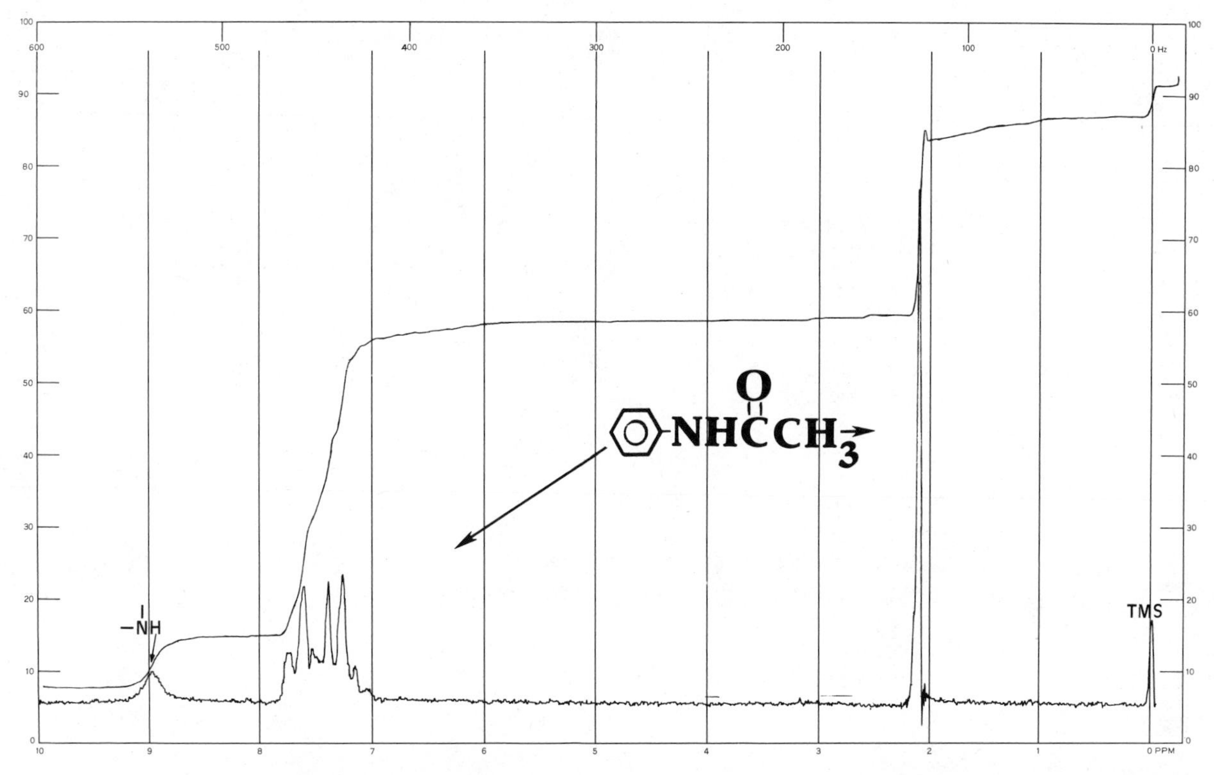

Figure 6.35. Nuclear magnetic resonance (proton) spectrum of acetanilide: 60 MHz, 600-Hz sweep width, $CDCl_3$ solvent.

The nmr spectrum of a typical amide is shown in Fig. 6.35. A table of nmr signals useful for characterizing amides is as follows:

Range (δ, ppm)	Structural feature
δ2.0–2.4[a]	—C<u>H</u>—CONR$_2$
δ1.95–2.85[a]	RCONHC<u>H</u>—
δ5.0–8.5	RCON<u>H</u>R′

[a] One usually finds methyl protons in the higher-field (lower δ) portion, methylene protons near the center portion, and methine protons in the lower-field portion of such ranges.

Although amide N—H protons do not exchange nearly as rapidly as amine N—H protons, the amide N—H signal still occurs as a broadened singlet in the nmr; this is due to the quadrupole moment of nitrogen. The broadened NH singlet thus does not have a discernible multiplicity that allows identification of the hydrogens on the carbon adjacent to nitrogen. The signals of the hydrogens on the carbon can, however, show splitting due to the N—H proton:

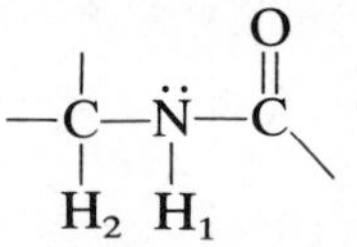

H_1 = broadened singlet
H_2 = definable multiplet

Typically J_{12} = ca. 5 Hz.

In view of the large contribution of resonance form B, there is substantial restricted rotation in acid amides:

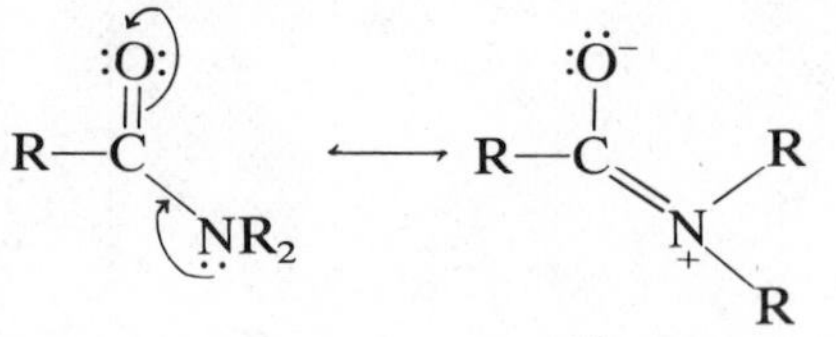

B

Thus, for example, the nmr spectrum of *N,N*-dimethylformamide shows separate signals for the nonequivalent methyl groups.

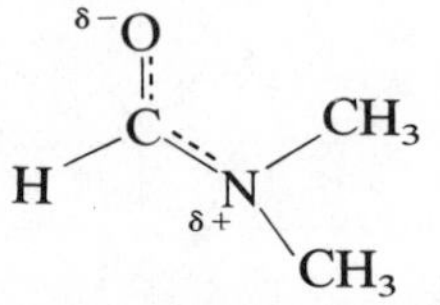

Mass spectra of amides usually provide discernible molecular ions. The molecular weights of amides are extremely useful, because compounds with even molecular weights must have an even number of or no nitrogen atoms in the molecular formula and those compounds with odd molecular weights must have an odd number of nitrogen atoms.

A number of the more important mass spectral fragments for amides are tabulated here:

Fragment	Mass	Comments
$[O{=}C{-}NR_2]^+$	42+2R	Strong for primary amides (R=H); useful for determination of mass of R groups for secondary and tertiary amides
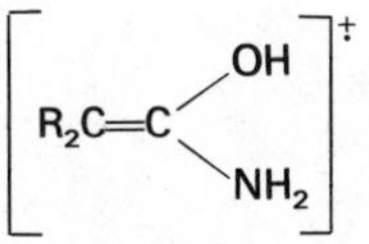 $[R_2C{=}C(OH)(NH_2)]^{+\cdot}$	57+2R	Results from McLafferty cleavage. For R=H, mass=59; base peak for primary amides with γ-hydrogen

The McLafferty cleavage fragment described above illustrates an important numerical relationship between the masses of molecular ions and of fragments from a given compound. If the compound has an even molecular weight, simple cleavage usually results in charged fragments of an odd mass. Similarly, simple cleavage of a compound with an odd molecular weight should give rise to fragments that have even masses. Simple cleavage is defined as the breaking of one single bond in the molecule. Since butyramide (mass 87) gives rise to its base peak at *m/e* 59, it is clear that this fragment is not the result merely of simple cleavage. The cleavage, in fact, involves rearrangement with a hydrogen transfer (McLafferty rearrangement):

$$\left[CH_3CH_2CH_2\overset{:O:^+}{\overset{\|}{C}}NH_2\right] \longrightarrow \text{(cyclic H transfer: } CH_2\text{–}CH_2\text{–}CH_2\text{–}C(=O^+)NH_2\text{)} \xrightarrow{-(CH_2{=}CH_2)} CH_2{=}C(\overset{+}{O}H)NH_2$$

m/e 87 → *m/e* 59

More complex fragmentation pathways are encountered when the amides are secondary or tertiary, due to alkyl substituents on nitrogen that are two carbon chains or longer; references described in Chapter 10 should be consulted.

The most general method for chemically characterizing amides consists in hydrolyzing them with acids or alkalies (Procedure 37):

$$RCONH_2 + H_2O + HCl \rightarrow RCOOH + NH_4Cl$$

$$RCONH_2 + NaOH \rightarrow RCO_2^-Na^+ + NH_3$$

Hydrolysis of substituted amides yields primary or secondary amines instead of ammonia.

$$RCONHR' + NaOH \rightarrow RCO_2^{-+}Na + R'NH_2$$

$$RCONR_2' + NaOH \rightarrow RCO_2^{-+}Na + R_2'NH$$

The identification is made by characterizing the hydrolysis products, namely, the acid, ammonia, or primary or secondary amine. One should consult the sections on carboxylic acids (6.12), acid salts (6.12.4), and amines (6.7) to aid such characterization.

Xanthydrol reacts readily with unsubstituted amides and imides to form 9-acylamidoxanthenes, which are good derivatives (Procedure 38).

A number of procedures useful for derivatizing amides have been described in other references. Cheronis, Entrikin, and Hodnett have outlined a number of spot tests:

$$3RCONH_2 \xrightarrow[(2)\ FeCl_3]{(1)\ 3NH_2OH\cdot HCl} \left(R\text{—}\overset{NOH}{\overset{\|}{C}}\text{—}O\text{—}\right)_3 Fe$$

(red-violet color)

(ferric hydroxamate test, Cheronis, Entrikin, and Hodnett, p. 373; *N*-substituted amides do not react);

$$ArCONH_2 \xrightarrow{6\%\ H_2O_2} [ArCONHOH] \xrightarrow{FeCl_3} (Ar\overset{NOH}{\overset{\|}{C}}O)_3Fe$$

(blue-red color)

(Cheronis, Entrikin, and Hodnett, p. 374; aliphatic amides are not oxidized to hydroxamic acids with hydrogen peroxide).

Phthalyl chloride reacts with unsubstituted amides to produce *N*-acylphthalimides, which are useful solid derivatives:

$$C_6H_4(COCl)_2 + RCONH_2 \rightarrow C_6H_4(CO)_2NCOR + 2HCl$$

[*J. Amer. Chem. Soc.*, **51,** 3651 (1929)].

Mercuric oxide reacts with certain amides to form *N,N'*-mercuribisacylamides, which are high-melting solid derivatives:

$$HgO + 2RCONH_2 \rightarrow Hg(NHCOR)_2 + H_2O$$

[*J. Amer. Chem. Soc.*, **64,** 1738 (1942)].

A few amides form stable salts with oxalic acid, which serve for identification:

$$RCONH_2 + (CO_2H)_2 \rightarrow \text{salt}$$

[*Ind. Eng. Chem., Anal. Ed.*, **12,** 737 (1940)].

Barbiturates may be converted to *N,N'*-*p*-nitrobenzyl derivatives by reaction with *p*-nitrobenzyl bromide by a procedure similar to that for the preparation of *p*-nitrobenzyl esters of acids (Procedure 33).

$$R_2C(CONH)_2C{=}O \xrightarrow[Na_2CO_3]{p\text{-}O_2NC_6H_4CH_2Br} R_2C[CON(CH_2C_6H_4NO_2(p))]_2C{=}O$$

[*J. Amer. Chem. Soc.*, **66,** 1440 (1944)].

Procedure 37. Hydrolysis of Amides, Nitriles, and Nitroanilines

(a) *Basic Conditions.*

$$RCONH_2 + NaOH \rightarrow RCOONa + NH_3$$
$$RCONHR' + NaOH \rightarrow RCOONa + R'NH_2$$
$$RCONR'_2 + NaOH \rightarrow RCOONa + R'_2NH$$
$$RCN + NaOH + H_2O \rightarrow RCOONa + NH_3$$

$$O_2N\text{-}C_6H_3(NO_2)\text{-}NH_2 + NaOH \rightarrow O_2N\text{-}C_6H_3(NO_2)\text{-}ONa + NH_3$$

Treat 5 ml of 10% sodium hydroxide solution in a test tube with 0.2 g of urea. Shake the mixture, and note if ammonia is evolved. Repeat the test with (1) benzamide; (2) acetanilide; (3) benzonitrile; (4) 2,4-dinitroaniline. Then heat each of the mixtures to boiling and note the odor. Test the action of the vapors on moist pink litmus paper.

Cool the above solutions and acidify each with hydrochloric acid. Note the result. What determines if a precipitate forms?

(*b*) *Acidic Conditions.* Many substituted amides are hydrolyzed more easily by heating them under reflux with 20% sulfuric acid.

$$2RCONHR' + H_2SO_4 + 2H_2O \rightarrow 2RCOOH + (R'NH_3)_2SO_4$$

$$2RCONR_2' + H_2SO_4 + 2H_2O \rightarrow 2RCOOH + (R_2'NH_2)_2SO_4$$

The acid may be separated by distillation if volatile or removed by filtration if insoluble in water. The amine may be liberated by the addition of alkali and characterized by conversion to its arenesulfonyl derivative.

Nitriles, particularly cyanohydrins, are frequently hydrolyzed by acids. Treatment with 90 to 95% sulfuric acid or concentrated hydrochloric acid at temperatures ranging from 10 to 50°C converts nitriles to amides. Amides may be hydrolyzed further by diluting the mixture with water and heating under reflux for 0.5 to 2 hr.

$$RCN \xrightarrow[H_2SO_4]{H_2O} RCONH_2 \xrightarrow[H_2SO_4]{H_2O} RCOOH + NH_4HSO_4$$

Discussion. The ammonia or amine that is the product of alkaline hydrolysis may be characterized by the Hinsberg test (Procedure 16). For this purpose it is best to use a larger sample (1 g) and distill the ammonia or volatile amine from the alkaline solution into a receiver containing dilute hydrochloric acid. This solution may then be neutralized and tested as directed in Procedure 16.

A water-insoluble amine may be removed by extraction with ether. The ether can then be distilled and the amine characterized by the usual tests. A water-soluble, nonvolatile amine may also be converted to an arenesulfonyl derivative and separated from the organic acid (which is the other product of hydrolysis) by taking advantage of solubility behavior or differences in volatility.

Aryl amines with nitro groups ortho or para to the amino group are hydrolyzed to the corresponding nitrophenols and ammonia or amines by the action of hot alkalies. The nitroso group resembles the nitro group in its labilizing effect on

$$p\text{-}ON{-}C_6H_4{-}NR_2 \xrightarrow{NaOH} R_2NH + \left[{}^{-}O{-}C_6H_4{-}N{=}O \longleftrightarrow O{=}C_6H_4{=}N{-}O^{-} \right] Na^+$$

groups ortho or para to it. For example, *p*-nitrosodialkylanilines are hydrolyzed by alkalies to the secondary amines and the sodium salt of *p*-nitrosophenol (benzoquinone monoxime). Part of the driving force for such reactions is the fact that nitro and nitroso groups ortho and para to nitrogen functions activate these functions via Meisenheimer complex formation; these complexes provide a low-energy, facile route to nucleophilic substitution:

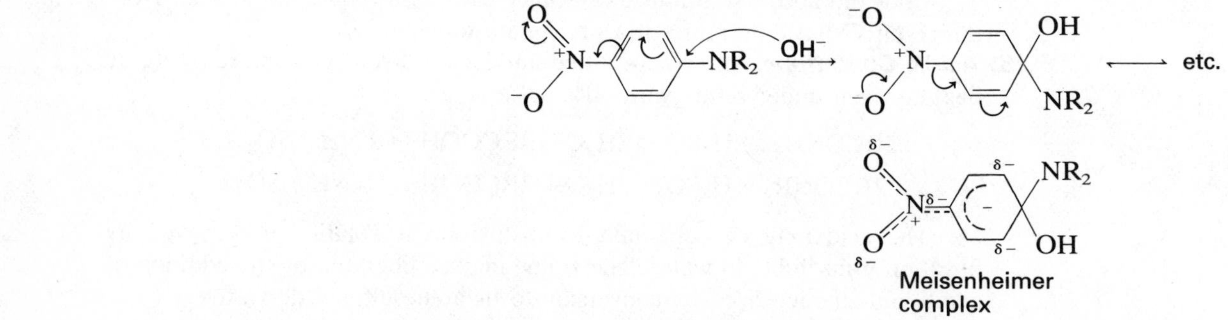

$$\text{O}_2\text{N}\text{-C}_6\text{H}_4\text{-NR}_2 \xrightarrow{OH^-} \text{Meisenheimer complex} \longrightarrow \text{etc.}$$

Meisenheimer
complex

Procedure 38. 9-Acylamidoxanthenes

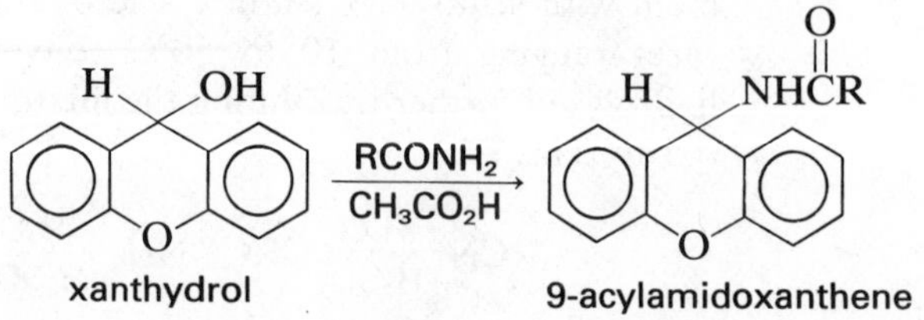

$$\text{xanthydrol} \xrightarrow[CH_3CO_2H]{RCONH_2} \text{9-acylamidoxanthene}$$

xanthydrol 9-acylamidoxanthene

About 0.5 g of xanthydrol is dissolved in 7 ml of glacial acetic acid. If the solution is not clear (because of disproportionation products of xanthydrol), it is allowed to stand a few minutes or is centrifuged and the clear solution decanted into a clean test tube. To this solution is added 0.5 g of the amide, and the mixture is warmed at 85°C in a beaker of water for 20 to 30 min. Upon cooling, the acylamidoxanthene is collected on a filter and recrystallized from a mixture of 2 parts of dioxane and 1 part of water.

Some amides fail to dissolve in the acetic acid and may be converted to the derivative by using a mixture of 5 ml of ethanol, 2 ml of glacial acetic acid, and 3 ml of water as the solvent for the reaction.

Amides can be analyzed by nmr facilitated by lanthanide shift reagents (Chapter 8); the binding site is very likely the carbonyl oxygen. Resonance forms such as those below emphasize the Lewis basicity of the oxygen atom; this oxygen can then interact with the Lewis acidic lanthanide atom of the shift reagent.

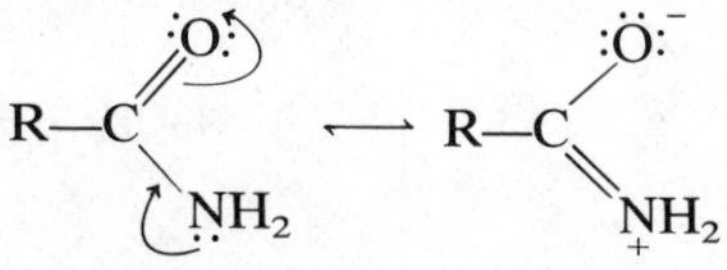

$$R\text{—}C(=O)NH_2 \longleftrightarrow R\text{—}C(O^-)=\overset{+}{N}H_2$$

6.12.4 Salts of Carboxylic Acids

Salts of carboxylic acids ($RCO_2^-Na^+$), because of their ionic character, have properties that are drastically different from the usual organic compound composed solely of covalent bonds. Such carboxylate salts are usually water-soluble but ether-insoluble; this not only places such compounds in solubility class S_2, it also relates to the basis for solvent extraction schemes (Chapter 7) for carboxylic acids. The ionic character of these salts is also reflected in their immobility, relative to completely covalent organic compounds, in chromatographic analyses. Additionally, these salts usually show very high melting points and low volatility, and leave a residue after ignition. The nature of the metal atom can be determined by utilizing procedures described in inorganic and analytical textbooks (Chapter 10).

The ir and nmr spectra of typical carboxylate salts are shown in Figs. 6.36 and 6.37, respectively; note the spectral sampling techniques, which reflect the physical properties of these ionic salts.

Characteristic of the ir spectra of carboxylate salts are the $C(\overset{...}{—}O)_2^-$ stretching bands. Since the carbon-oxygen bond of such salts is intermediate in bonding order between the C═O bond in ketones and the C—O bond in alcohols, the ir absorption due to these salts occurs at a position that is between such C═O and C—O absorptions. Also, the $C(—O)_2^-$ group results in two absorptions due to the coupled interaction of the carbon-oxygen bonds:

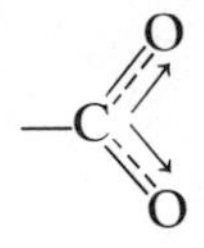

symmetrical stretching ca. 1400 cm^{-1} (7.15 μm) (the slightly weaker band of the two substantial bands)

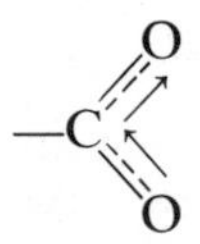

asymmetrical stretching ca. 1600 cm^{-1} (6.25 μm) (strong absorption)

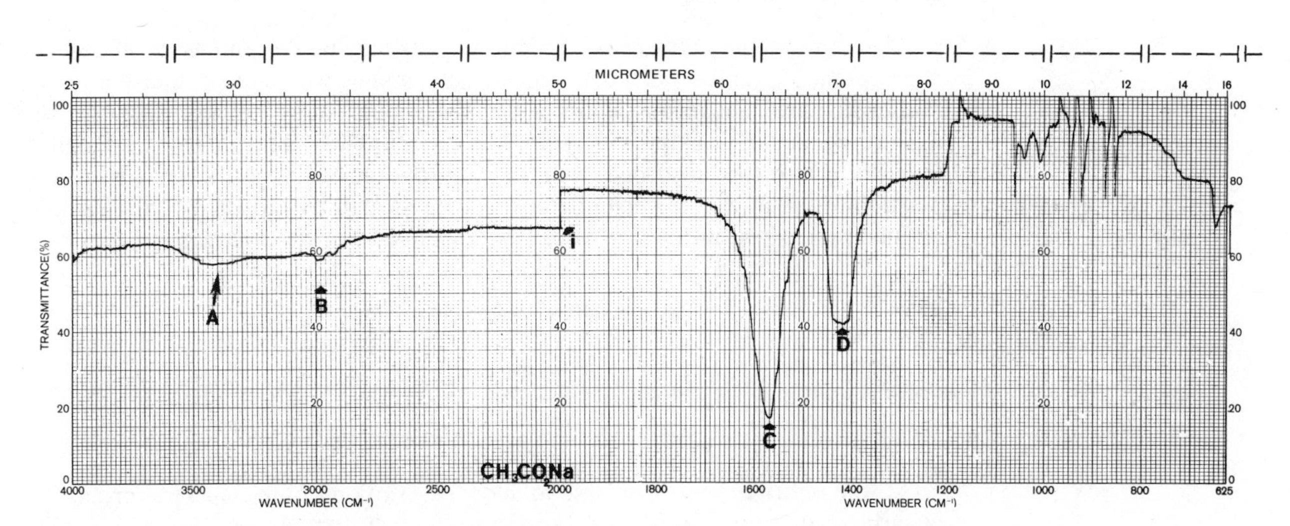

Figure 6.36. Infrared spectrum of sodium acetate: KBr pellet. O—H stretch: **A**, ca. 3450 cm^{-1} (2.90 μm), ν_{O-H} (O—H of acetic acid impurity). C—H stretch: **B**, 2990, 2930 cm^{-1} (3.34, 3.41 μm), ν_{C-H} (CH_3 group). $C(\overset{...}{—}O)_2$ stretch: **C**, 1570 cm^{-1} (6.37 μm), $\nu_{(C\overset{...}{—}O)_2}$, asym; **D**, 1420 cm^{-1} (7.04 μm), $\nu_{C(\overset{...}{—}O)_2}$, sym. i = grating change.

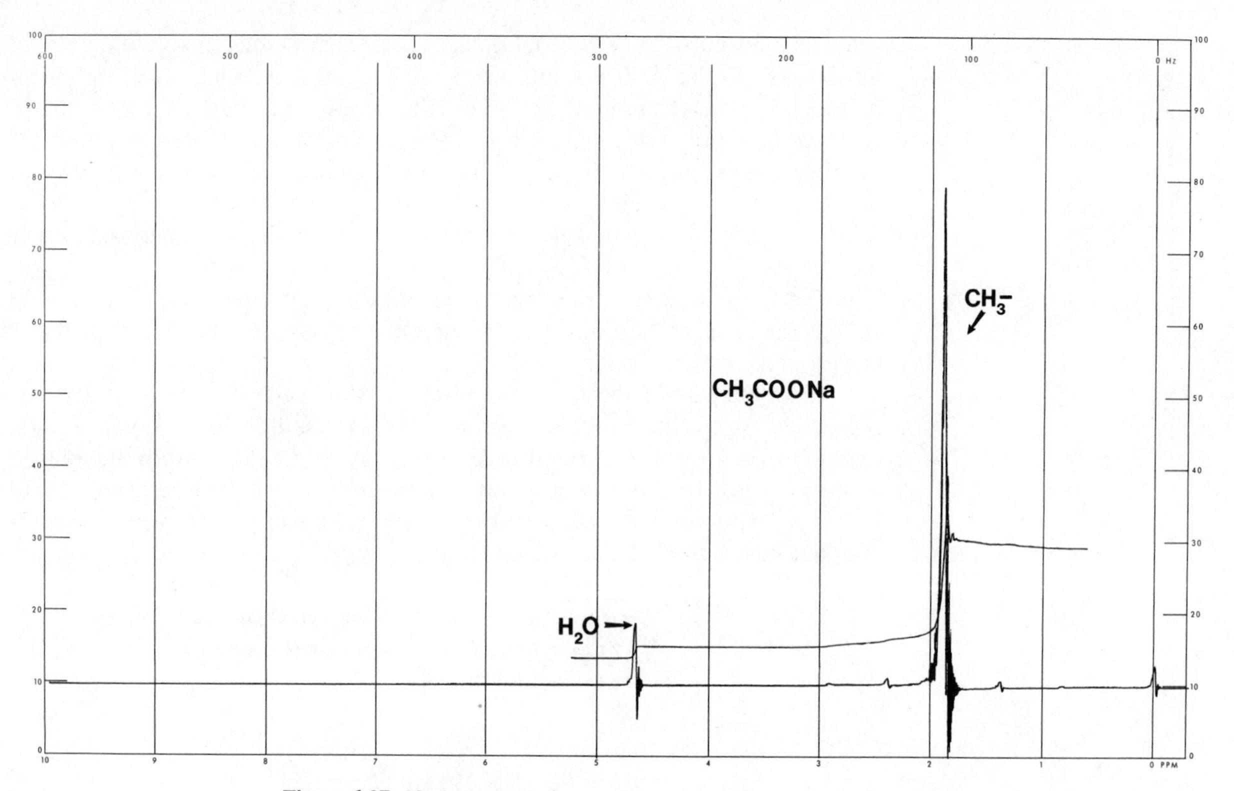

Figure 6.37. Nuclear magnetic resonance (proton) spectrum of sodium acetate: 60 MHz, 600-HZ sweep width, D_2O solvent. The relative areas of the two singlets indicates a ratio of 90 : 10 = sodium acetate : water (or acetic acid).

Nuclear magnetic resonance spectra of salts may be determined in DMSO-d_6 and acetone-d_6 (and other polar organic solvents) as well as in deuterium oxide. Although pure, anhydrous samples of salts should show no nmr signal in the δ10–15 region (CO_2H proton region), it is frequently difficult to avoid signals induced by moisture absorbed from the atmosphere. Although the exact structural position of the proton in such mixtures is complicated by proton exchange among various basic sites, water signals can be anticipated in the following ranges:

Solvent	Chemical shift of hydroxylic proton
Deuterium oxide	ca. 4.5–5.0
DMSO	3.2–3.4
Acetone	2.6–2.9

Little information is available to allow description of mass spectral properties of carboxylate salts; this is probably due to the fact that the very limited volatility of these salts inhibits mass spectral analysis. The salts can be converted to the corresponding acid (see below); mass spectral analysis of carboxylic acids is straightforward (Section 6.12).

Addition of a strong mineral acid to the metal salt of a carboxylic acid liberates the free organic acid:

$$2RCOONa + H_2SO_4 \rightarrow 2RCOOH + Na_2SO_4$$

The free organic acid may be isolated by distillation, extraction, or filtration. Since the metal salts melt only at very high temperatures and usually decompose, it is necessary to use the physical constants of the acid to prepare the list of possibilities. The acid may then be converted to a suitable derivative, or the salt may be used directly for the preparation of the anilide, toluidide (Procedure 36), or *p*-nitrobenzyl or *p*-bromophenacyl ester (Procedure 33).

S-(*p*-Bromobenzyl)-thiuronium bromide, prepared by heating *p*-bromobenzyl bromide with thiourea, reacts with sodium or potassium salts of carboxylic acids to yield the corresponding *S*-(*p*-bromobenzyl)-thiuronium salts (Procedure 39).

Procedure 39. Benzyl-, p-Chlorobenzyl-, and p-Bromobenzylthiuronium Salts

(a) Reagent Preparation

$$Br\text{—}C_6H_4\text{—}CH_2Br + H_2N\text{—}C(=S)\text{—}NH_2 \longrightarrow \left[Br\text{—}C_6H_4\text{—}CH_2SC(NH_2)_2\right]^+ Br^-$$

(see Procedure 15) *S*-(*p*-bromobenzyl)thiuronium bromide

(b) Derivative Preparation

$$\left[Br\text{—}C_6H_4\text{—}CH_2SC(NH_2)_2\right]^+ Br^- + RCO_2H \text{ (or } RCO_2^-Na^+) \longrightarrow$$

$$\left[Br\text{—}C_6H_4\text{—}CH_2SC(NH_2)_2\right]^+ \; ^-O_2CR$$

S-(*p*-bromobenzyl)thiuronium carboxylate

About 0.3 g of the acid (or 0.5 g of the sodium or potassium salt) is added to 3 or 4 ml of water, a drop of phenolphthalein indicator solution is added, and the solution is neutralized by the dropwise addition of 5% sodium hydroxide solution. An excess of alkali must be avoided. If too much is used, dilute hydrochloric acid is added until the solution is just a pale pink. To this aqueous solution of the sodium salt is added a hot solution of 1 g of the aralkylthiuronium chloride or bromide in 10 ml of 95% ethanol. The mixture is cooled, and the salt is collected on a filter. A few salts (e.g., from formic acid) fail to precipitate, and part of the ethanol must be evaporated to obtain the salt.

These thiuronium salts of organic acids separate in a state of high purity and usually do not require recrystallization. If necessary they may be recrystallized from a small amount of dioxane.

The melting points of many of these salts are close together, a large number of them melting within a narrow temperature range. Hence, it is always best to confirm the identification by some additional criterion such as neutralization equivalent.

S-Benzyl- and *S*-(*p*-chlorobenzyl)thiuronium salts are made in a similar way. The salts form in nearly pure condition but possess melting points that are very close together.

Ammonium carboxylate salts show essentially the same chemical and spectral features due to the carboxylate group as described above for sodium salts. One can detect ammonium (and amine) salts by treating the salt with sodium hydroxide (Experiment 25), which frees ammonia (or the amine). Additionally, the ir spectra of ammonium and amine salts show very distinct and strong bands due to N—H stretching and bending:

Salt	Vibration	cm^{-1}	μm
NH_4^+	N—H stretch	3300–3030 (broad)	3.0–3.3
	N—H bend	1430–1390	7.0–7.2
RNH_3^+	N—H stretch	ca. 3000 (very broad)	ca. 3.3
	asym. N—H bend	1600–1575	6.25–6.35
	sym. N—H bend	1490	ca. 6.7
$R_2NH_2^+$	coupled N—H stretch	2700–2250 (broad)[a]	3.7–4.4
	N—H bend	1600–1575[b]	6.25–6.35
R_3NH^+	N—H stretch	2700–2250[c]	3.7–4.45

[a] This band may show multiple maxima.
[b] This band is usually of medium intensity; all other bands listed in this table are normally strong.
[c] The N—H *bend* of R_3NH^+ groups is weak and is not of diagnostic value.

EXPERIMENT 25. SODIUM HYDROXIDE TREATMENT OF AMMONIUM SALTS AND AMINE SALTS

$$RCOONH_4 + NaOH \rightarrow RCOONa + NH_3 + H_2O$$

$$RNH_3Cl + NaOH \rightarrow RNH_2 + NaCl + H_2O$$

Place 5 ml of 10% sodium hydroxide solution in a test tube, add 0.2 g of ammonium benzoate, and shake the mixture vigorously. Note the odor of ammonia.

To 0.4 g of aniline hydrochloride in a test tube add 5 ml of 10% sodium hydroxide solution. Shake the mixture, and allow it to stand a few minutes. Note the separation of the oily layer of the amine.

6.13 ESTERS

Organic esters are normally soluble in organic solvents, (Classes N_1 and S_1) but they are water-soluble (class S_1) only if they contain five or fewer carbon atoms.

Solubility results must be viewed cautiously, because esters hydrolyze (often easily) to carboxylic acids and alcohols. The solubility of such hydrolysis products

$$RCO_2R' \xrightarrow[\text{hydrolysis}]{H_3O^+} RCO_2H + R'OH$$

$$RCO_2R' \xrightarrow[H_2O]{OH^-} RCO_2^- + R'OH \quad \text{(saponification)}; \quad RCO_2^- \xrightarrow{H^+} RCO_2H$$

must then be considered. The (sometimes sickly) sweet odor of esters is often very useful in their detection.

The ir (see Fig. 6.38) and nmr spectra (see Fig. 6.39) of esters are usually critical for structure determination. Infrared spectra usually show intense bands for C═O stretch and for "C—O" stretch; the latter is diagnostic for differentiation of esters from aldehydes and ketones.

Carbonyl band positions in ir spectra of esters are especially sensitive to other structural features, such as unsaturation α,β to the carbonyl group. There are two "C—O" bands for esters because there are two C—O single bonds; the C—O bond in the acid portion of the ester is strongly coupled[57] to the rest of that portion of the molecule, and the "alcohol C—O" bond is strongly coupled to the rest of the alcohol portion of the molecule. thus these bands have been called the C—(C═O)—O and O—C—C bands, respectively. Important ir bands for esters are listed in Table 6.6.

Nuclear magnetic resonance analysis of esters can yield three very interpretable types of information:

1. The chemical shift of α-protons in the "alcohol" portion of the molecule:

$$R{-}\overset{\overset{\large O}{\|}}{C}{-}O{-}\underset{\underset{\large H}{|}}{\overset{|}{C}}{-} \quad H \leftarrow \delta 3.2\text{–}4.5^{58}$$

2. The chemical shift of α-protons in the acid portion of the molecule:

$$-\underset{\underset{\large H}{|}}{\overset{|}{C}}{-}\overset{\overset{\large O}{\|}}{C}{-}OR' \quad H \longleftarrow \delta 2.0\text{–}2.5^{58}$$

---

57 For a more extensive discussion of coupling, see R. M. Silverstein, G. C. Bassler, and T. C. Morrill, *Spectrometric Identification of Organic Compounds*, 3rd ed. (Wiley, New York, 1974), chap. 3.

58 In these ranges, α-protons that are part of methyl groups are near the high-field end of the range, methylene protons are near the center-field portion, and methine protons are near the low-field end of the range.

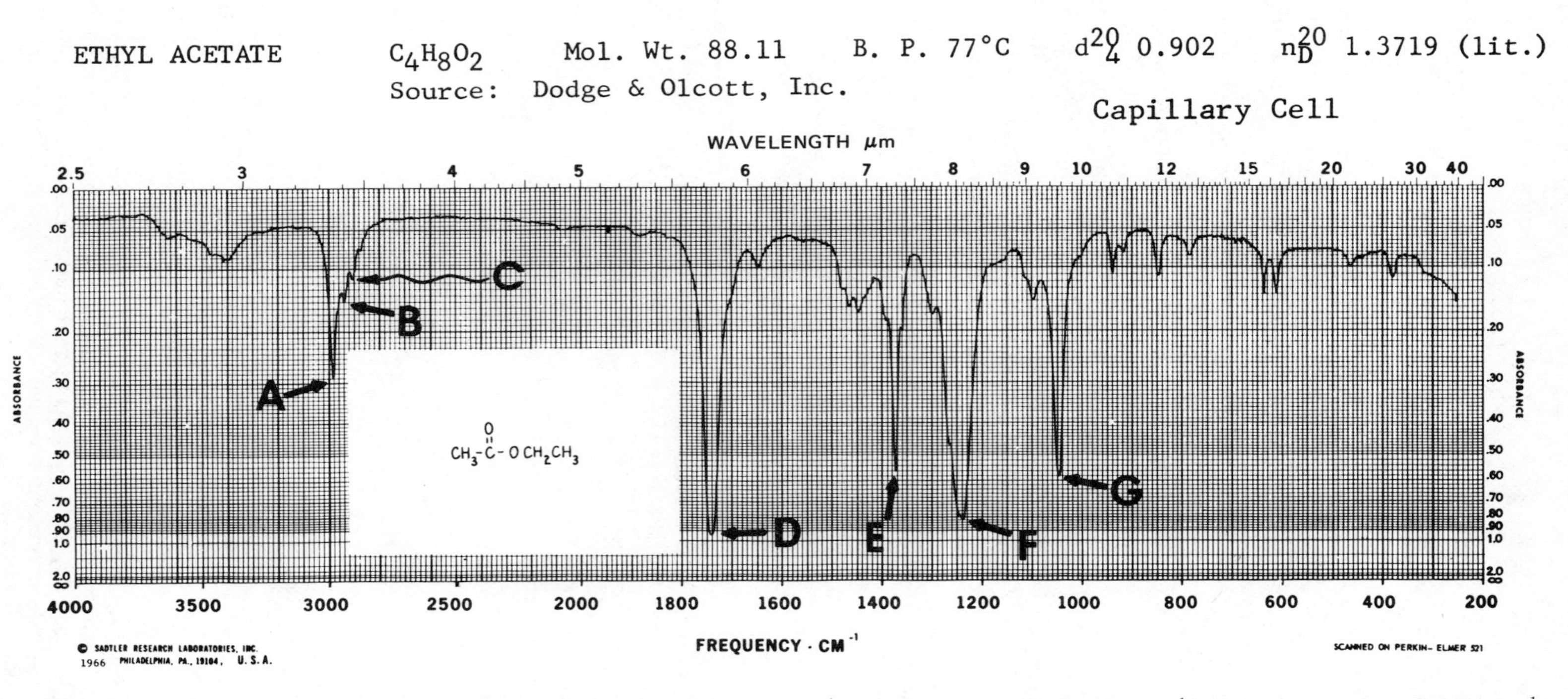

Figure 6.38. Infrared spectrum of ethyl acetate: thin film. C—H stretch: **A**, 2980 cm^{-1} (3.36 μm), ν_{asym} CH_3; **B**, 2930 cm^{-1} (3.41 μm), ν_{asym} CH_2; **C**, 2910 cm^{-1} (3.45 μm), ν_{sym} CH_3. C=O stretch: **D**, 1740 cm^{-1} (5.75 μm), $\nu_{C=O}$. C—H bend: **E**, 1375 cm^{-1} (7.27 μm), δ_{sym} CH_3. C—O stretch: **F**, 1240 cm^{-1} (8.07 μm), C(C=O)—O stretch; **G**, 1048 cm^{-1} (9.54 μm), O—C—C stretch.

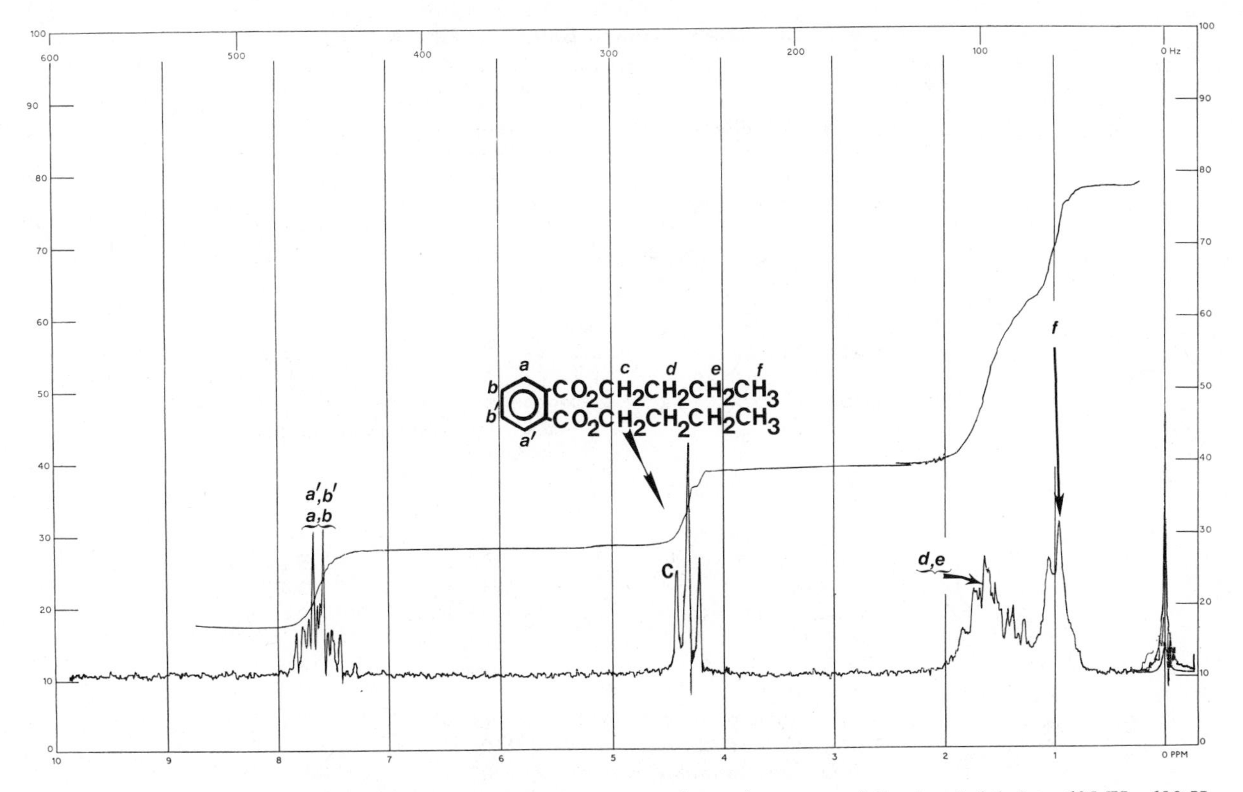

Figure 6.39. Nuclear magnetic resonance (proton) spectrum of di-*n*-butyl phthalate: 60 MHz, 600-Hz sweep width, $CDCl_3$ solvent.

3. The pattern of aromatic protons (especially ortho protons) as influenced by ester substituents (compare benzene, singlet, $\delta 7.27$)

ca. $\delta 7.6$ → (para H of C_6H_5–C(=O)OR′); ca. $\delta 7.5$ (meta H); ca. $\delta 8.2$ (ortho H)

ca. $\delta 7.3$ (meta and para H of C_6H_5–O–C(=O)–R); ca. $\delta 7.05$ (ortho H)

(see Section 6.9).

Ester analysis by nmr can be facilitated by lanthanide shift reagents (see Chapter 8); $Eu(fod)_3$ is usually the best reagent for these substrates.

The molecular ion of esters is usually distinct. Two general types of cleavage are very common for esters. One involves McLafferty rearrangement and the other involves cleavage of one of the groups attached to the carbonyl group.

McLafferty rearrangement, which involves hydrogen transfer, requires a

Table 6.6. Important Infrared Bands for Esters

Vibration	Band position cm^{-1}	μm
C=O stretch		
RCO_2R' (simple, aliphatic)	ca. 1750–1735	ca. 5.71–5.76
$ArCO_2R'$, $>C=C(-)-CO_2R'$	1730–1715	5.78–5.83
$R-C(=O)-O-C(-)=C<$	ca. 1776	ca. 5.63
Lactones	Depend on conjugation, ring size, etc. (see ir references listed in Chapter 10 for details)	
"C—O" stretch		
C—(C=O)—O stretch		
RCO_2R' (simple, aliphatic)	1210–1163	8.26–8.60
Acetates	1240	8.07
RCO_2Ar, vinyl acetates	1190–1140	8.40–8.77
$ArCO_2R'$	1310–1250	7.63–8.00
Lactones	1250–1111	8.00–9.00
O—C—C stretch		
RCO_2R' (R′ is primary, simple)	1064–1031	9.40–9.70
RCO_2R'(R′ is secondary, simple)	1100	9.09
$ArCO_2R'$ (R′ is primary, simple)	1111	9.00

γ-hydrogen on the acyl portion of the ester:

$$\text{RO—C(=}\ddot{\text{O}}^{+}\text{)—C—C—C(—)—H} \rightarrow \text{RO—C(—}\ddot{\text{O}}^{+}\text{—H)=C<} + >\text{C=C}<$$

Peaks due to these fragments can be very distinctive because they are often intense; in addition, whereas simple cleavage gives fragments of an odd-numbered mass from molecular ions of an even-numbered mass, the same molecular ions give rise to McLafferty fragments at even-numbered masses. For example, methyl caprylate ($M = m/e$ 158) displays its base peak at m/e 74 due to McLafferty cleavage.

There are four possible ions that can arise from cleavage of a bond α to the carbonyl group and these are listed in the Table on p. 293.

The most fundamental reaction of esters is the saponification reaction. Saponification converts an ester to an alcohol and the salt of a carboxylic acid.

Molecular ion	Observed fragment (charged)	Side product (radical)	Comments
$R—C(=\overset{+\cdot}{O}:)—OR'$ (cleavage of R—C bond)	$:O{\equiv}\overset{+}{C}—OR'$	R·	Infrequent; discernible for $R' = CH_3$ (*m/e* 59)
$R—C(=\overset{+\cdot}{O}:)—OR'$ (cleavage of R—C bond)	R+	$\ddot{O}{=}\dot{C}—OR'$	Usually of low intensity
$R—C(=\overset{+\cdot}{O}:)—OR'$ (cleavage of C—OR' bond)	R—C≡O+	R'O·	Common if R is short
$R—C(=\overset{+\cdot}{O}:)—OR'$ (cleavage of C—O bond)	$+\ddot{O}—R'$	$R—\dot{C}{=}\ddot{O}$	Infrequent

Note that although carboxylic acids and alcohols (from esters) can each be characterized, experience has shown that direct derivatization of the ester is a superior approach (see pp. 298–301).

Procedure 40. Saponification and Hydrolysis of Esters

Method A: Refluxing Water Solvent

Method B: Refluxing Diethylene Glycol

Method C: Concentrated Sulfuric Acid

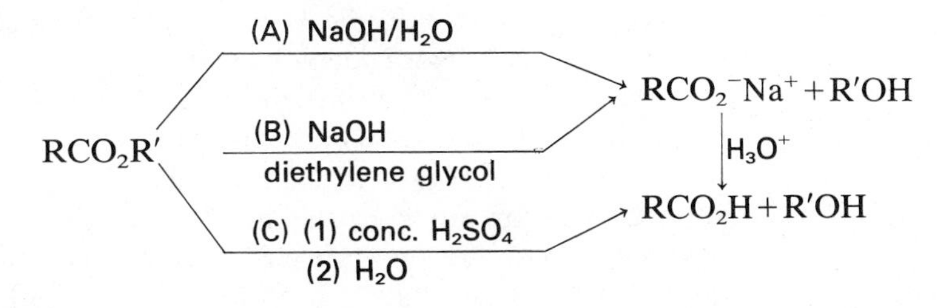

Method A. In a round-bottomed flask fitted with an efficient reflux condenser place 40 ml of a 25% sodium hydroxide solution. Add 5 ml of ethyl benzoate (or another ester), and heat to boiling. A piece of porous plate or a boiling tube should be placed in the flask to prevent bumping. Continue the boiling until the ester layer or the characteristic odor disappears (about 0.5 hr). Reverse the condenser, distil about 5 ml, and saturate the distillate with potassium carbonate.

Note the formation of two layers. (Explain.) The amount of sample required will obviously depend on the molecular weight of the alcohol to be isolated as well as on the molecular weight of the original ester.

Cool the residue in the flask, and acidify with dilute phosphoric acid. What is the solid that separates? How can you be sure that it is not sodium phosphate? If the acid were a volatile liquid it could be separated by distillation.

Liquid acids can be distilled as follows. The solution is placed in a distilling flask, and a few boiling chips are added. A thermometer should not be used, but the top of the distilling flask should be closed. The flask is connected to a short water-cooled condenser in such a way that the side arm of the distilling flask extends well into the narrow portion of the condenser tube. The apparatus should be set up so that the condenser makes about a 45° angle with the desk top. The distilling flask is heated so that drops come from the end of the condenser at a constant rate. The distillation should not be so rapid that a steady stream flows from the condenser or so slow that there is an appreciable time interval between drops.

Discussion. Esters vary considerably in the ease with which they may be saponified. Most simple esters boiling below 110°C will be saponified completely by refluxing with 25% sodium hydroxide solution for 0.5 hr as described in Method A. Esters boiling between 110 and 200°C require a longer time (1 to 2 hr) for complete saponification.

The hydrolysis of water-insoluble esters may be markedly accelerated by the addition of 0.1 g of sodium lauryl sulfate (Gardinol) to the alkali and the ester. The mixture is shaken vigorously to emulsify the ester and is then heated to refluxing. A large flask must be used because the emulsifying agent causes considerable foaming.

Very high-boiling esters (above 200°C) that are insoluble in water hydrolyze slowly, and prolonged refluxing may result in loss of a volatile alcohol. Procedure B utilizes a solution of potassium hydroxide in diethylene glycol (b.p. 244°C). Diethylene glycol is not only an excellent solvent for esters but also permits the use of a higher reaction temperature, and all but high-boiling alcohols can be distilled from the reaction mixture in a pure state.

Method B. In a 10- or 25-ml distilling flask place 3 ml of diethylene glycol, 0.5 g of potassium hydroxide pellets, and 0.5 ml of water. Heat the mixture over a low flame until the alkali has dissolved, and cool; add 1 to 2 g of the ester, and mix thoroughly. Use a thermometer at the still head and arrange a test tube cooled by a beaker of ice as a receiver. The flask is heated over a small flame at first, and the contents are mixed by shaking. When only one liquid phase or one liquid and one solid phase are present, the mixture is heated more strongly so that the alcohol distils. The distillate is used for the preparation of a solid derivative such as the 3,5-dinitrobenzoate (Procedure 2).

The residue in the flask is either a solution or a suspension of the potassium salt of the acid derived from the ester. Add 10 ml of water and 10 ml of ethanol to the residue, and shake thoroughly. Add dilute sulfuric acid (6 *N*) until the solution is slightly acid to phenolphthalein. Allow the mixture to stand about 5 min and then filter. The filtrate is used directly for the preparation of a derivative. It may be treated with p-nitrobenzyl bromide or p-phenylphenacyl bromide in order to obtain the corresponding solid ester. If the original ester was so high-boiling that a good boiling point could not be obtained, it may be desirable to divide the filtrate in half and make two derivatives.

Discussion. Saponification represents the most useful procedure for characterizing esters. However, it must be remembered that hot concentrated alkali also affects other functional groups. Aldehydes that have α-hydrogen atoms undergo the aldol condensation and resinification. Aldehydes that have no α-hydrogen atoms undergo Cannizzaro's reaction and form an alcohol and the sodium salt of an acid.

$$2R_3CCHO + NaOH \rightarrow R_3CCH_2OH + R_3CCOONa$$

Polyfunctional compounds such as β-diketones and β-keto esters undergo cleavage under the influence of hot alkalies. The possibility of such interfering reactions is detected by means of the other classification reagents mentioned in this chapter and emphasizes the fact that *a single classification reagent cannot be taken as proof of the presence of a certain functional group.* It is important to correlate all the tests in attempting to draw conclusions concerning the structure of an unknown compound.

Method C. Esters of sterically hindered acids, such as alkyl 2,4,6-trialkylbenzoates, are very difficult to hydrolyze with alkali. However, these esters may be hydrolyzed rapidly by dissolving them in 100% sulfuric acid and diluting the solution with ice water. This result is due to the fact that esters of sterically hindered acids undergo the following reaction when dissolved in 100% sulfuric acid:

$$2,4,6\text{-}(CH_3)_3C_6H_2COOR + 2H_2SO_4 \rightarrow 2,4,6\text{-}(CH_3)_3C_6H_2\overset{+}{C}O + ROH_2^+ + 2HSO_4^-$$

When water is added, the intermediate ion forms the acid:

$$2,4,6\text{-}(CH_3)_3C_6H_2\overset{+}{C}O + 2H_2O \rightarrow 2,4,6\text{-}(CH_3)_3C_6H_2COOH + H_3O^+$$

Conversely, the sterically hindered acid may be converted readily to an ester by dissolving it in 100% sulfuric acid and treating the solution with the alcohol.

Unhindered esters do not undergo these reactions. They dissolve in 100% sulfuric acid and are recovered unchanged when the solution is poured into ice water.

Esters of sterically hindered acids and alcohols such as *tert*-butyl 2,4,6-trimethylbenzoate, though extremely resistant to hydrolysis by alkalies, may be hydrolyzed readily by boiling for 1 hr with 18% hydrochloric acid.[59]

[59] S. G. Cohen and A. Schneider, *J. Amer. Chem. Soc.*, **63**, 3382 (1941).

Another saponification procedure, utilizing ethanol as solvent, has been described by Pasto and Jonhson (p. 412).

Question

1. Show by equations the products formed by the alkaline hydrolysis of (a) *p*-phenylphenacyl acetate; (b) ethylene glycol dibenzoate; (c) *n*-butyl oxalate; (d) the polyester of glycerol and phthalic acid. Devise suitable procedures for detecting the products formed in these reactions.

Saponification equivalents of esters (Procedure 41) are extremely useful, especially for samples where previous molecular weight determinations were unsuccessful. The saponification equivalent is simply the equivalent weight of the ester determined by a titrimetric procedure; the procedure is conceptually similar to that used for carboxylic acids above. Either the saponification equivalent or a small integral (x) multiple of the saponification equivalent, within experimental error, will be the molecular weight of the ester.

Procedure 41. Saponification Equivalents of Esters

(a) ***Diethylene Glycol Method. The Preparation of a Solution of Potassium Hydroxide in Diethylene Glycol.*** The reagent is made by dissolving 3 g of C.P. potassium hydroxide pellets in 15 ml of diethylene glycol. It is necessary to warm the mixture gently to effect solution. A thermometer should be used for stirring, and the mixture should not be heated above 130°C; higher temperatures may cause the reagent to be colored. After all the solid has dissolved, the warm solution is poured into 35 ml of diethylene glycol contained in a glass-stoppered bottle. The solution is mixed thoroughly and allowed to cool. It is approximately 1.0 N and is standardized[60] by pipetting 10 ml into a flask, adding 10 ml of water, and titrating with a previously standardized 0.25 N hydrochloric acid solution.

The Saponification of the Ester. Exactly 10 ml of the reagent is pipetted into a small glass-stoppered Erlenmeyer flask. The ester is placed in a small weighing bottle fitted with a small pipet. The weight of the bottle, ester, and pipet is determined, and a sample of 0.4 to 0.6 g of the ester is transferred to the reagent in the Erlenmeyer flask by means of the pipet, which is then returned to the weighing bottle. The loss in weight of the bottle represents the weight of the sample.

The ester is mixed with the reagent by a rotary motion of the flask. The stopper is held firmly in place and the mixture is heated in an oil bath so that a temperature of 70 to 80°C is reached in 2 to 3 min. The liquid is agitated by a whirling motion during heating[61]. At this point the flask is removed from the heating

---

[60] Three-figure accuracy on this Normality is useful; "0.25 N" means normality in the 0.230–0.270 range.

[61] Care should be exercised, because thermometer bulbs are fragile.

bath and shaken vigorously. The liquid is allowed to drain, and the stopper is carefully loosened to allow air to escape (*caution!*). The stopper is replaced, and the temperature is raised to 120 to 130°C. With very high-boiling esters the stopper may be removed and a thermometer inserted.

After 3 min at this temperature the flask and its contents are cooled to 80 to 90°C and the stopper is removed and washed with distilled water so that the rinsings drain into the flask. About 15 ml of distilled water is added; the contents are mixed and then titrated with 0.25 N hydrochloric acid, using phenolphthalein as the indicator. The saponification equivalent is calculated according to the following equation:

$$\text{saponification equivalent} = \frac{\text{weight of sample (mg)}}{[\text{volume of alkali (ml)} \times N_{KOH}] - [\text{volume of acid (ml)} \times N_{HCl}]}$$

where the volume of alkali is 10.0 ml.

Discussion. These procedures give complete saponification of esters that are insoluble in water. Esters such as benzyl acetate, butyl phthalate, ethyl sebacate, butyl oleate, and glycol and glycerol esters are completely saponified.

(b) ***Alcoholic Sodium Hydroxide Method. The Preparation of an Alcoholic Sodium Hydroxide Solution.*** Eight grams of sodium is dissolved in 250 ml of absolute ethanol, and, after solution is complete, 25 ml of water is added. This solution is standardized by titration against a weighed sample of pure potassium acid phthalate.

The Saponification of the Ester. The ester is placed in a weighing bottle containing a small pipet. The weight of the bottle and pipet is determined, and a sample of 0.2 to 0.4 g of the ester is transferred to a 150-ml Erlenmeyer flask by means of the pipet, which is then returned to the weighing bottle and the two are again weighed. The loss in weight[62] is the weight of the sample. Fifteen milliliters of the above alcoholic sodium hydroxide solution, measured from a buret, is then added to the flask containing the ester. The flask is attached to an efficient reflux condenser and the mixture is heated gently under reflux for $1\frac{1}{4}$ to $1\frac{1}{2}$ hr. At the end of this time it is allowed to cool slightly. The flask is loosened from the condenser, and the connector and condenser tube are washed with a stream of water from a wash bottle, the washings being allowed to run into the saponification mixture. Two drops of phenolphthalein solution are added, and the excess alkali is titrated by means of 0.25 N hydrochloric acid. The end point should be a faint pink. It is best to titrate the solution until the phenolphthalein is colorless and then back-titrate with the original alkali.

Discussion. The ester must be *pure* and *anhydrous* in order to give an accurate saponification equivalent. The following precautions must be observed in order to obtain accurate results.

1. The alcoholic sodium hydroxide solution should be standardized immediately before use and its normality (N) recorded.

---

[62] Determined on an analytical balance.

2. The standard solutions should be measured accurately from a buret, because a slight error in the amount of alkali will cause a large error in the saponification equivalent. This is especially noticeable with high-molecular-weight esters.
3. Heating for $1\frac{1}{2}$ hr will saponify most esters. For some, a longer time (2 to 24 hr) may be necessary.
4. Corks or rubber stoppers must not be used to attach the flask to the condenser, because the alcohol vapors extract substances that lower the strength of the alkali. Glass stoppers, cleaned by thorough washing with distilled water, should be used.
5. The end point should be a *faint* pink. This is the color assumed by phenolphthalein at pH 9.0, which represents the hydrogen ion concentration of solutions of the sodium salts of most organic acids.
6. The molecular weight of the ester is equal to x times the saponification equivalent, where x is the number of ester groups in the molecule.

Questions

1. What saponification value would be obtained for ethyl acetoacetate? *n*-Butyl β-bromopropionate? Ethyl hydrogen phthalate? Ethyl cyanoacetate? *n*-Butyl phthalate?
2. What would happen if Procedure 41 were applied to benzaldehyde?
3. If an ester had already partially hydrolyzed, what effect would this have on the saponification equivalent?

Because of difficulties involved in the separation and purification of the hydrolysis products, it is best to prepare ester derivatives by reactions on the original ester. Esters containing other functional groups may often be identified by reference to solid derivatives obtained by reactions such as halogenation, nitration, and acylation.

Derivatives of the acyl part of the ester may be obtained by the following methods.

1. Reaction with benzylamine in the presence of a little ammonium chloride yields *N*-benzylamides, which serve as derivatives (Procedure 42).

$$RCOOR' + C_6H_5CH_2NH_2 \rightarrow RCONHCH_2C_6H_5 + R'OH$$

The reaction proceeds well when R′ is methyl or ethyl. Esters of higher alcohols should be subjected to a preliminary methanolysis.[63]

$$RCOOR' + CH_3OH \xrightarrow{CH_3ONa} RCOOCH_3 + R'OH$$

The methyl ester thus obtained may be used for aminolysis.

---

[63] This transesterification procedure is described at the end of Procedure 42.

2. Hydrazine reacts readily with methyl and ethyl esters to produce acid hydrazides, which serve as satisfactory derivatives (Procedure 43).

$$RCOOCH_3 + H_2NNH_2 \rightarrow RCONHNH_2 + CH_3OH$$

Esters of the higher alcohols should be converted to the methyl esters.

3. Some simple esters react with aqueous or alcoholic ammonia to produce amides, which serve as derivatives.

$$RCOOR' + NH_3 \rightarrow RCONH_2 + R'OH$$

However, most esters must be heated under pressure in order to effect this reaction.

4. The *p*-toluidide of the acidic portion of the ester may be obtained by means of the following reactions:

$$C_2H_5MgBr + ArNH_2 \rightarrow ArNHMgBr + C_2H_6$$

$$RCOOR' + ArNHMgBr \rightarrow \underset{\displaystyle |\atop OMgBr}{RC(NHAr)_2} + R'OMgBr$$

$$\downarrow 2HCl$$

$$Ar = CH_3\text{-}C_6H_4\text{-} \qquad RCONHAr + Ar\overset{+}{N}H_3{}^{-}Cl + MgBrCl$$

A solid derivative of the alcohol portion of a simple ester may be obtained by effecting an interchange reaction between 3,5-dinitrobenzoic acid and the ester in the presence of concentrated sulfuric acid (Procedure 44).

$$ArCO_2H + RCO_2R' \xrightarrow{H_2SO_4} ArCO_2R' + RCO_2H$$

The method is applicable to a large number of simple esters but may not be used if either the R or R′ group of the ester reacts with concentrated sulfuric acid. High-molecular-weight esters (>250) also fail to react.

If it is not possible to obtain derivatives of the acid and alcohol portions directly from the ester, recourse must be had to hydrolysis, which is best accomplished by saponification with alkali according to Procedure 40, p. 293. The exact procedure to be followed in working up the saponification mixture depends on the nature of the acid and hydroxy compound. The acid may be monobasic or polybasic, and soluble or insoluble in water. The hydroxy compound may be a monohydroxy, polyhydroxy, or a phenol. Hence the separation and characterization of the products of saponification must be planned carefully in the light of the suggestions obtained from the list of possibilities and the results from classification reactions.

Procedure 42. N-Benzylamides from Esters

$$C_6H_5CH_2NH_2 \xrightarrow[NH_4Cl]{RCO_2R'} C_6H_5CH_2NHC(=O)R \qquad R' = CH_3, C_2H_5$$

$$\text{(includes } RCO_2R'' \xrightarrow[CH_3ONa]{CH_3OH} RCO_2CH_3\text{)}$$

A mixture of 1 g of the ester, 3 ml of benzylamine, and 0.1 g of powdered ammonium chloride is heated for 1 hr in a Pyrex test tube fitted with a small finger condenser. After being cooled, the reaction mixture is washed with water to remove excess benzylamine and to induce crystallization. Often the addition of a little dilute hydrochloric acid will promote crystallization. An excess of acid must be avoided, because it dissolves *N*-benzylamides. Occasionally the presence of unchanged ester may prevent crystallization. In that case it is best to boil the solution for a few minutes with water in an evaporating dish to volatilize the ester. The solid amide is collected on a filter, washed with a little ligroin, and recrystallized from a mixture of ethanol and water or of acetone and water.

Esters of alcohols higher than ethanol should be heated for 30 min with 5 ml of absolute methanol in which a small piece of sodium (0.1 g) has been dissolved. At the end of the reflux period, the methanol is evaporated and the residue treated by the above procedure.

Procedure 43. Acid Hydrazides from Esters

$$RCO_2R' + 85\%\ NH_2NH_2 \rightarrow RCONHNH_2 + R'OH$$
$$R' = CH_3 \text{ or } C_2H_5$$

One gram of the methyl or ethyl ester and 1 ml of 85% hydrazine hydrate are mixed, and the mixture is heated under reflux for 15 min. Just enough absolute ethanol is then added, through the top of the condenser, to obtain a clear solution. After the mixture has been heated under reflux for 2 hr, the alcohol is evaporated and the residue cooled. The crystals of the hydrazide are collected on a filter and recrystallized from water or a mixture of water and ethanol.

Higher esters must be subjected to methanolysis, as described above, before treatment with hydrazine.

Procedure 44. Alkyl 3,5-Dinitrobenzoates from Esters

$$3,5\text{-}(O_2N)_2C_6H_3CO_2H + RCO_2R' \xrightarrow[\Delta]{\text{conc. } H_2SO_4} 3,5\text{-}(O_2N)_2C_6H_3CO_2R' + RCO_2H$$

About 2 ml of the ester is mixed with 1.5 g of 3,5-dinitrobenzoic acid, and 2 drops of concentrated sulfuric acid is added. If the original ester boiled below 150°C, the mixture is heated gently under reflux. If the ester boiled above 150°C, the mixture is heated in an oil bath at 150°C, with frequent stirring. The time required varies from 30 min to 1 hr, the longer time being used in those cases in which the 3,5-dinitrobenzoic acid fails to dissolve in about 15 min. After the mixture has been cooled, 25 ml of absolute ether is added, and the solution is extracted twice with two portions of 15 ml of 5% sodium carbonate solution to remove the sulfuric and 3,5-dinitrobenzoic acids. The ether layer is washed with 10 ml of water, and the solvent is evaporated. The residue (usually an oil) is dissolved in 5 ml of boiling ethanol. After the solution has been filtered, water is added to incipient cloudiness. The mixture is cooled and stirred to induce crystallization of the derivative.

6.14 ETHERS

Ethers are only a little more polar and slightly more reactive than either saturated hydrocarbons or alkyl halides. The ether oxygen can be protonated by concentrated sulfuric acid:

$$R_2\ddot{\underset{\cdot\cdot}{O}} + H_2SO_4 \rightleftharpoons R_2\ddot{O}H^+HSO_4^-$$

This reaction shows why ethers are in solubility class N_1 (as contrasted to class I for inert saturated hydrocarbons and alkyl halides).

NOTE: Ethers form extremely explosive peroxides upon standing, especially when exposed to air and/or light. Liquid ether that shows solid precipitates should not be handled at all.

Peroxides can be detected by treating the ether with starch-iodide paper that has been moistened with dilute hydrochloric acid; peroxides will cause the paper to turn blue. Methods of removing peroxides and hydroperoxides have been described by Pasto and Johnson (p. 371).

Pure ethers are compounds that are more likely to be initially diagnosed by their failure to undergo reactions rather than by their ability to undergo chemical reactions. Thus, spectra of ethers become extremely important for structure determination. The ir spectrum of a typical ether is shown in Fig. 6.40. Bands due to C—O—C stretching are the most important for characterizing ethers and are tabulated below; these band positions are sensitive to conjugation of the nonbonded pair of electrons on oxygen with a π system:

$$\left(-\ddot{\underset{\cdot\cdot}{O}}-\overset{|}{C}=C\langle\right)$$

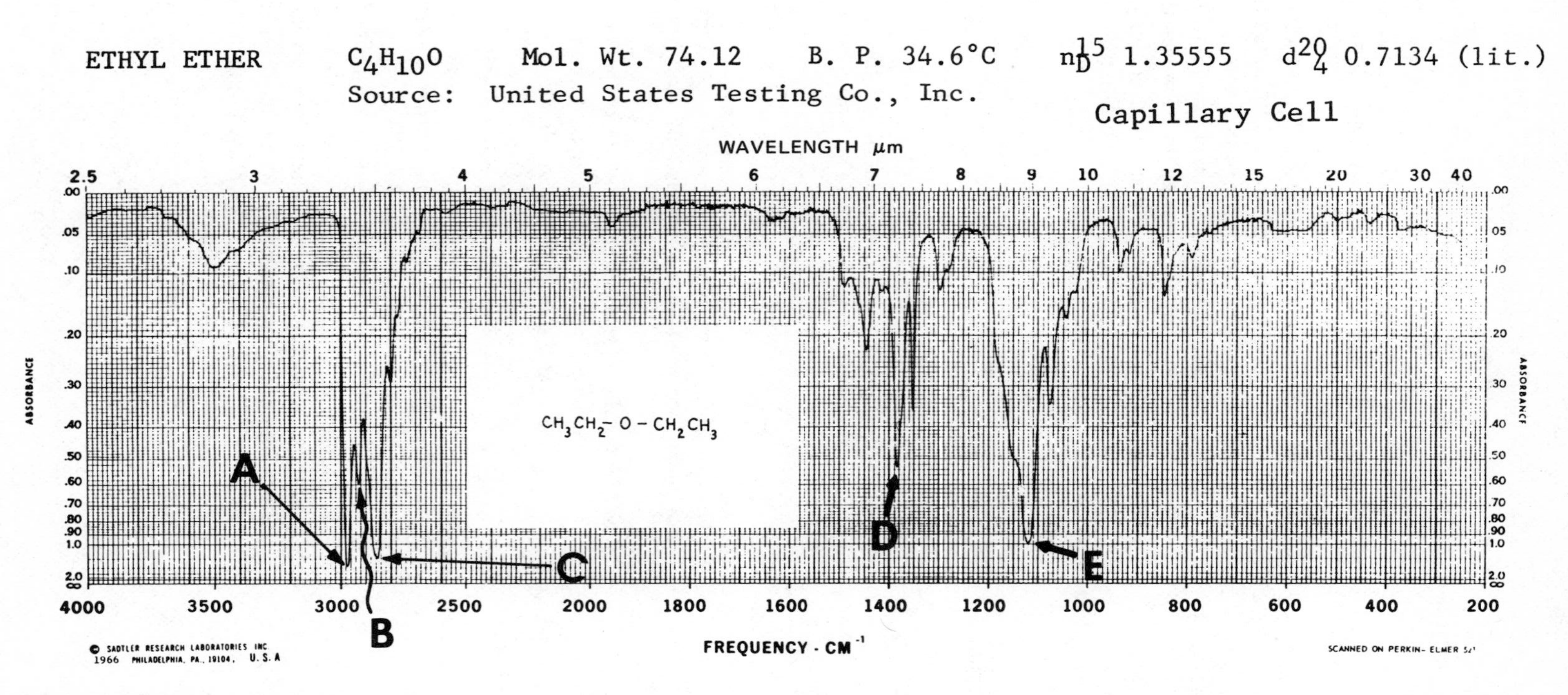

Figure 6.40. Infrared spectrum of ethyl ether: neat. C—H stretch; **A**, 2975 cm^{-1} (3.36 μm), ν_{asym} CH_3; **B**, 2930 cm^{-1} (3.41 μm), ν_{asym} CH_2; **C**, 2860 cm^{-1} (3.50 μm), ν_{sym} CH_3. C—H bend: **D**, 1383 cm^{-1} (7.23 μm), δ_{sym} CH_3 (bend). "C—O" stretch: **E**, 1120 cm^{-1} (8.93 μm), ν_{asym} C—O—C (stretch).

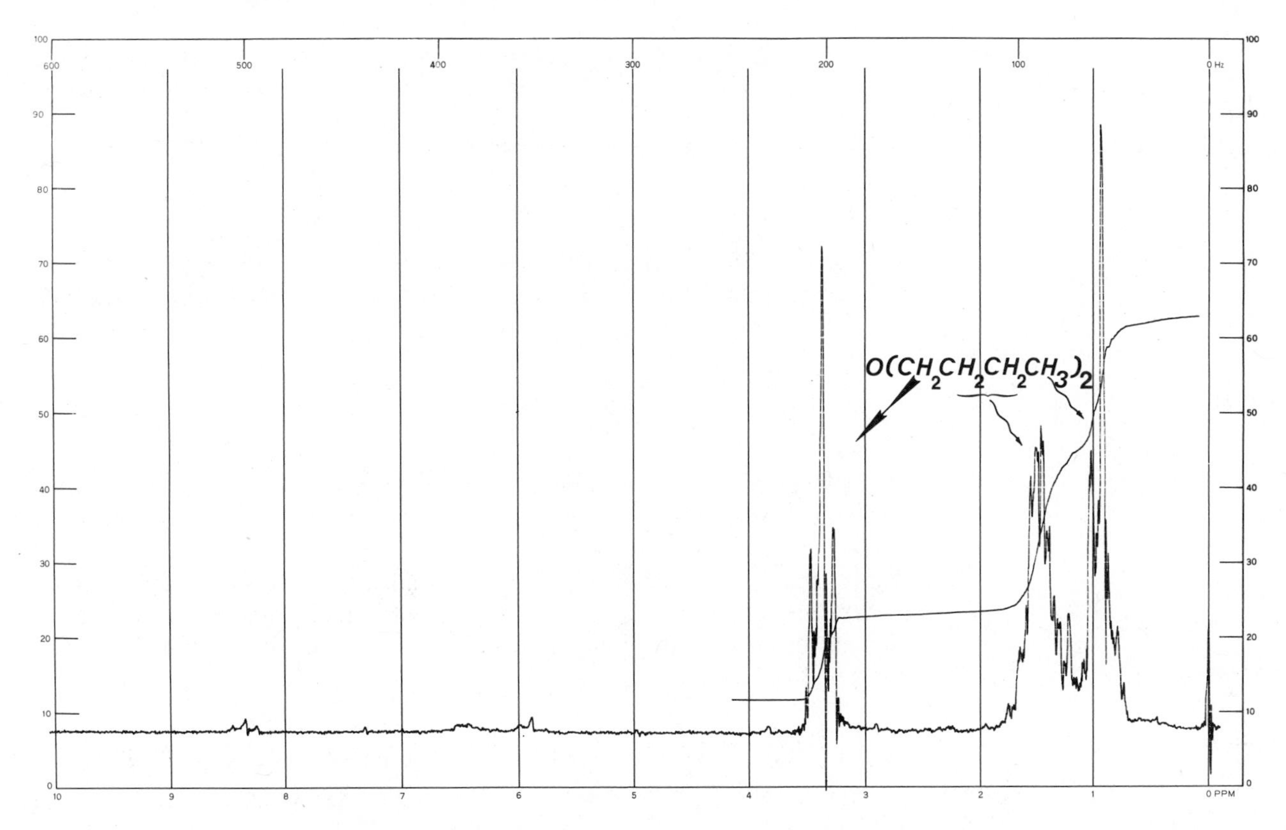

Figure 6.41. Nuclear magnetic resonance (proton) spectrum of *n*-butyl ether: 60 MHz, 600-Hz sweep width, $CDCl_3$ solvent.

C—O—C Infrared Bands of Ethers

Structural feature	cm^{-1}	μm
ROR		
(Simple aliphatic)		
Asymmetrical C—O—C stretch	1150–1085 (s)[a]	8.70–9.23
Symmetrical C—O—C stretch	ca. 1125 (w)	ca. 8.89
ROAr		
Asymmetrical C—O—C stretch	1275–1200 (s)	7.84–8.33
Symmetrical C—O—C stretch	1075–1020 (s)	9.30–9.80
Vinyl ethers		
Asymmetrical C—O—C stretch	1225–1220 (s)	8.16–8.33
Symmetrical C—O—C stretch	1075–1020 (s)	9.30–9.80

[a] s = strong, w = weak.

An nmr spectrum of a typical ether is shown in Fig. 6.41. The most easily interpreted feature is usually any signals for protons on carbon bearing the oxygen (α-protons):[64]

$$\underset{\text{ca. } \delta 3.2}{CH_3OR} \qquad \underset{\text{ca. } \delta 3.4}{R'CH_2OR} \qquad \underset{\text{ca. } \delta 3.6}{R_2'CHOR}$$

These chemical shift positions are very similar to the shifts of analogous protons in alcohols and esters. When R is replaced by Ar, each signal is moved downfield by a few tenths of a ppm:

$$\underset{\text{ca. } \delta 3.85}{CH_3OAr} \qquad \underset{\text{ca. } \delta 4.0}{R'CH_2OAr} \qquad \underset{\text{ca. } 4.55}{R'_2CHOAr}$$

An alkoxy group on an aromatic ring usually shifts the aromatic protons to a higher field than the protons of benzene ($\delta 7.27$); Section 6.9 should be consulted for more details.

Mass spectral analysis of ethers can be difficult, because the molecular ion peak (M) is usually weak. If, however, the sample concentration is increased, the $M+1$ peak often becomes more intense; this is a result of H· transfer during ion–molecule collisions:

$$RCH_2\ddot{O}R' \xrightarrow{e^-} \underset{M}{RCH_2\ddot{O}R'^{\cdot +}} \xrightarrow[\sim H\cdot]{RCH_2\ddot{O}R} \underset{M+1}{RCH_2-\overset{+}{O}(H)-R'} + RCH\ddot{O}R$$

A very common fragmentation pathway for ethers involves cleavage of the bond between the α and β carbon atoms:

$$-\underset{\beta}{C}-\underset{\alpha}{CH_2}-\ddot{O}^{\cdot +}-R \longrightarrow -\dot{C}\cdot + CH_2{=}\overset{+}{\ddot{O}}-R' \longleftrightarrow \overset{+}{C}H_2-\ddot{O}-R$$

This is an example where cation stabilization (by resonance) is a factor that greatly influences the position of cleavage when more than one cleavage pattern is possible. Another factor that influences the direction of such cleavage is the size of R· group lost; usually the larger R· group is removed more readily.

Treatment of an ether with hydriodic acid results in cleavage of the ether. Hydriodic acid is sufficiently strong as an acid that the ether is protonated; the resulting iodide ion is sufficiently nucleophilic to cause the required displacement reaction:

$$R_2\ddot{O} \xrightarrow{HI} R_2\overset{+}{\ddot{O}}H \xrightarrow{I^-} RI + ROH \xrightarrow{HI}$$

$$ROH_2^+ \xrightarrow{I^-} RI + H_2O$$

$$ROH_2^+ \longrightarrow R^+ \xrightarrow{I^-} RI$$

[64] Note that the protons in each set maintain the same, regular order: methyl protons are at higher field than methylene, which are at higher field than methine.

EXPERIMENT 26. HYDRIODIC ACID (ZEISEL'S ALKOXYL METHOD)

$$ROR' + 2HI \rightarrow R'I + RI + H_2O$$
$$HOR' + HI \rightarrow R'I + H_2O$$
$$ArOR' + HI \rightarrow R'I + ArOH$$
$$RCOOR' + HI \rightarrow R'I + RCOOH$$
$$RCH(OR')_2 + 2HI \rightarrow 2R'I + RCHO + H_2O$$

Place about 0.1 g (or 0.1 ml) of the compound in a 16-by-150 mm test tube. Carefully add, by means of a pipet, 1 ml of glacial acetic acid and 1 ml of 57% hydriodic acid (sp gr 1.7). Add a small piece of unglazed porcelain, and insert into the mouth of the test tube a gauze plug prepared as described below. The gauze plug is twisted so as to make a good fit and pushed down so that it is 4 cm from the mouth of the test tube. A small piece of nonabsorbent cotton is gently tamped on top of the plug by means of a glass rod so as to make a disk of cotton 2 to 3 mm thick. A piece of filter paper about 2 by 10 cm is folded longitudinally, moistened with a solution of mercuric nitrate, and placed on the cotton disk. The test tube is immersed to a depth of 4 to 5 cm in an oil bath kept at 120 to 130°C.

When the reaction mixture boils, vapors rise through the porous plug, which usually turns gray. The volatile alkyl iodide, rising through the plugs, reacts with the mercuric nitrate to produce a light orange or vermilion color due to the formation of mercuric iodide. A positive test consists in the formation of an orange or vermilion color on the test paper within a 10-min heating period. A yellow color constitutes a negative or doubtful test.

Try this test on (1) anisole; (2) methyl benzoate; (3) α-methyl glucoside.

Gauze Plugs. A solution of 1 g of lead acetate in 10 ml of water is added to 60 ml of 1 *N* sodium hydroxide solution and stirred until the precipitate dissolves. To this sodium plumbite solution is added a solution of 5 g of hydrated sodium thiosulfate in 10 ml of water. About 1 ml of glycerol is added, and the solution is diluted to 100 ml. About 5 ml of this solution is pipetted on strips of double cheesecloth 2 by 45 cm. The strips of cloth are dried and rolled to fit the test tube.

Mercuric Nitrate Solution. A saturated solution of mercuric nitrate is prepared in 49 ml of distilled water to which has been added 1 ml of concentrated nitric acid.

Discussion. This test is based on the classic Zeisel method for estimating quantitatively the percentage of methoxyl or ethoxyl groups. Functional groups containing methyl, ethyl, *n*-propyl, or isopropyl radicals attached to oxygen are cleaved by the hydriodic acid with the formation of a volatile alkyl halide. Alkoxy derivatives in which the group is *n*-butyl or larger are difficult to cleave, and the iodide is too high-boiling to be volatilized. Some *n*-butoxy compounds give a positive test, but the procedure is not reliable (the boiling point of *n*-butyl iodide is 131°C).

This class reaction is most useful for ethers, esters, and acetals in which the groups are methyl or ethyl. Methanol, ethanol, the two propyl alcohols, and even higher alcohols such as *n*-butyl and isoamyl alcohol will also give a positive test.

The test has been applied to numerous alkaloids and methylated sugars. The chief interference is caused by the presence of a sulfur-containing functional group that liberates hydrogen sulfide when heated with hydriodic acid.

Some ethers may require a more vigorous reagent for cleavage. It is sometimes advantageous to use 2 ml of the hydroiodic acid, 0.1 g of phenol, and 1 ml of propionic anhydride for 0.1 g of the sample.

A method suitable for the preparation of solid derivatives for the identification of small amounts of aliphatic ethers involves the cleavage of the ethers and the formation of the corresponding 3,5-dinitrobenzoyl chloride in the presence of zinc chloride (Procedure 45).

$$ArCOCl + ROR \rightarrow ArCOOR + RCl$$

This method is not suitable for mixed ethers in which both the groups of the ether are aliphatic. Aliphatic ethers and aryl alkyl ethers may be cleaved by hydriodic acid (Experiment 26, p. 305).

$$R_2O + 2HI \rightarrow 2RI + H_2O$$

$$ArOR + HI \rightarrow ArOH + RI$$

By use of a large sample (5 to 10 g), the alkyl iodide may be isolated and converted to a derivative (Procedures 12–15). The phenol from an aryl alkyl ether may be isolated also and transformed into a derivative (Procedures 59–61).

Procedure 45. Alkyl 3,5-Dinitrobenzoates from Ethers

$$3,5\text{-}(O_2N)_2C_6H_3COCl \xrightarrow[ZnCl_2]{ROR} 3,5\text{-}(O_2N)_2C_6H_3COOR$$

A mixture of 1 ml of the ether, 0.15 g of anhydrous zinc chloride, and 0.5 g of 3,5-dinitrobenzoyl chloride is placed in a test tube which is then attached to a reflux condenser. The mixture is boiled for 1 hr and cooled. To it is then added 10 ml of sodium carbonate solution. The mixture is warmed to 90°C in a water bath, cooled, and filtered. The precipitate is washed with 5 ml of 5% sodium carbonate solution and 10 ml of distilled water. The residue is dissolved in 10 ml of hot carbon tetrachloride, and the solution is filtered while hot; the filtrate is then cooled in an ice bath. If the ester does not separate, the carbon tetrachloride is evaporated. The residue is dried on a porous plate, and the melting point is determined.

Ethers can be cleaved by reagents that provide an active acyl electrophile that attacks the ether oxygen. In one procedure, an ether can be cleaved to an acetate by treatment with anhydrous ferric chloride in acetic anhydride:[65]

$$ROR' + (CH_3CO)_2O \xrightarrow{FeCl_3} RO\overset{O}{\overset{\|}{C}}CH_3 + R'O\overset{O}{\overset{\|}{C}}CH_3$$

In another procedure, the active acylium ion is obtained from a mixed anhydride;[66] this approach requires that one first prepare the mixed anhydride from the corresponding sulfonic acid or acid derivatives. The acetate is the product that is usually more readily characterized.

$$R_2O + CH_3\overset{O}{\overset{\|}{C}}OTs \longrightarrow RO\overset{O}{\overset{\|}{C}}CH_3 + ROTs$$

$$Ts = CH_3-C_6H_4-\underset{O}{\underset{\|}{\overset{O}{\overset{\|}{S}}}}-$$

$$CH_3-C_6H_4-SO_3H + CH_3\overset{O}{\overset{\|}{C}}Cl \quad \text{or} \quad CH_3-C_6H_4-SO_3H + (CH_3CO)_2O \longrightarrow TsO\overset{O}{\overset{\|}{C}}CH_3$$

The derivatives of aromatic ethers that are employed most frequently are those obtained by bromination (Procedure 46) and nitration (Procedures 26a and 26b). Picrates (Procedure 49) are also used. All of these procedures take advantage of the ability of the aromatic ring to undergo attack by electron-deficient species.

Aromatic ethers containing an alkyl side chain may be oxidized to substituted benzoic acids (Procedures 27*a* and 27*b*).

$$CH_3O-C_6H_4-CH_2CH_3 \xrightarrow{KMnO_4} CH_3O-C_6H_4-CO_2H$$

Aromatic ethers react smoothly with chlorosulfonic acid at 0°C to produce sulfonyl chlorides (Procedure 47):

$$ROC_6H_5 + ClSO_3H \rightarrow p\text{-}RO-C_6H_4-SO_2Cl$$

The latter are oils or low-melting solids and are treated with ammonia or ammonium carbonate (Procedure 48). The sulfonamides obtained in this way are useful derivatives.

$$p\text{-}RO-C_6H_4-SO_2Cl + (NH_4)_2CO_3 \rightarrow p\text{-}RO-C_6H_4-SO_2NH_2$$

---

[65] B. Ganem and V. R. Small, jr. *J. Org. Chem.*, **39**, 3728 (1974).
[66] M. H. Karger and Y. Mazur, *J. Amer. Chem. Soc.*, **90**, 3878 (1968).

Procedure 46. Bromination

$$C_6H_5OR \xrightarrow[\text{(or } CCl_4)]{Br_2 \atop CH_3CO_2H} p\text{-}BrC_6H_4OR$$

One gram of the compound is dissolved in 15 ml of glacial acetic acid, and 3 to 5 g of liquid bromine is added. The mixture is allowed to stand for 15 to 30 min and is then poured into 50 to 100 ml of water. The bromo compound that separates is removed by filtration and purified by recrystallization from dilute ethanol. In some cases carbon tetrachloride may be substituted for the acetic acid as the solvent: The carbon tetrachloride is distilled, and the residue is recrystallized.

Procedure 47. Sulfonyl Chlorides

$$C_6H_5OR \xrightarrow[CHCl_3]{2ClSO_3H} p\text{-}RO{-}C_6H_4{-}SO_2Cl + HCl + H_2SO_4$$

A solution of 1 g of the compound in 5 ml of dry chloroform in a clean, dry test tube is cooled in a beaker of ice to 0°C. About 5 g of chlorosulfonic acid is added dropwise, and after the initial evolution of hydrogen chloride has subsided the tube is removed from the ice bath and allowed to warm up to room temperature (about 20 min). The contents of the tube are poured into a 50-ml beaker full of cracked ice. The chloroform layer is removed and washed with water. The chloroform is evaporated, and the residual sulfonyl chloride is recrystallized from low-boiling petroleum ether, benzene, or chloroform.

Procedure 48. Sulfonamides

$$p\text{-}RO{-}C_6H_4{-}SO_2Cl \xrightarrow[\Delta]{(NH_4)_2CO_3} p\text{-}RO{-}C_6H_4{-}SO_2NH_2 + NH_4Cl + H_2O + CO_2$$

The sulfonyl chloride (0.5 g) is mixed with 2.0 g of dry powdered ammonium carbonate and heated at 100°C for 30 min. The mixture is cooled and washed with three 10-ml portions of cold water. The crude sulfonamide is dissolved in 10 ml of 5% sodium hydroxide solution, gentle heating being used if necessary, and the solution is filtered to remove any sulfone or chlorinated products. The filtrate is acidified with dilute hydrochloric acid, and the sulfonamide is removed by filtration. It is purified by recrystallization from dilute ethanol and dried at 100°C.

Procedure 49. Picrates of Phenolic Ethers

$$C_6H_5OR + 2,4,6\text{-}(NO_2)_3C_6H_2OH \xrightarrow{CHCl_3} C_6H_5OR \cdot 2,4,6\text{-}(NO_2)_3C_6H_2OH$$

A solution of 1 to 2 g of the ether in the smallest possible amount of chloroform (2 to 10 ml) is added to a separately prepared solution of 2 g of picric acid in 10 ml of boiling chloroform. The mixture is stirred thoroughly, set aside, and allowed to cool. The picrate crystallizes when the mixture is allowed to stand. It may be purified by recrystallization from the smallest possible amount of boiling chloroform. The melting point should be determined as soon as possible, because some picrates decompose.

These picrate complexes can often be used to determine the molecular weight of the ether (see Procedure 22, Section 6.9).

6.15 EPOXIDES

Epoxides can usually be expected to fall into the same solubility class (N_1) as ethers; the question as to whether or not the epoxides will chemically survive the solubility tests must be carefully considered. Epoxides will undergo carbon–oxygen bond cleavage much more readily than will ethers; these reactions are both acid- and base-catalyzed:

$$R\text{—}CH(\text{—}O\text{—})CH_2 \xrightarrow{OH^-} RCH(O^-)CH_2OH \xrightarrow[\text{reagent}]{\text{protic}} RCH(OH)\text{—}CH_2OH$$

$$R\text{—}CH(\text{—}O\text{—})CH_2 \xrightarrow{H^+} RCH(\text{—}\overset{+}{O}H\text{—})CH_2 \longrightarrow R\overset{+}{C}HCH_2OH \xrightarrow{H_2O} RCH(OH)\text{—}CH_2OH$$

$$RCH(\text{—}\overset{+}{O}H\text{—})CH_2 \xrightarrow{H_2O} RCH(OH)\text{—}CH_2OH_2^+ \xrightarrow{-H^+} RCH(OH)\text{—}CH_2OH$$

In such cases, solubility as well as chemical character of the diol products must be considered.

Epoxides can be analyzed by ir and nmr spectral methods; interpretation of the spectral bands is normally a complex task.

The strained three-membered ring of the epoxide contributes directly to the features of their ir spectra:

Group	Vibrational mode	cm^{-1}	μm
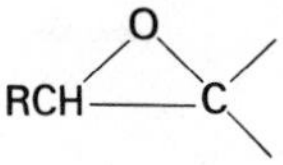 RCH—C (epoxide, O bridging)	Methine C—H stretch	3040–3000	3.99–3.33
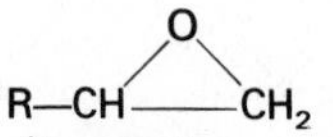 R—CH—CH_2 (epoxide, O bridging) (terminal epoxide)	CH_2 stretch	3050	3.28
Epoxy	Sym. ring breathing[a]	1250	8.00
Epoxy	Asym. ring bending[b]	950–810	10.53–12.35
Epoxy	—[c]	840–750	11.90–13.33

[a] Commonly called the "8μ band."
[b] Commonly called the "11μ band."
[c] Commonly called the "12μ band."

The nmr data for ethylene oxide and propylene oxide are shown below:

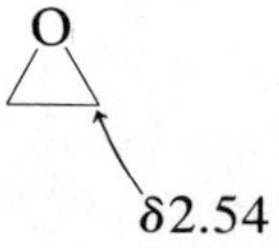

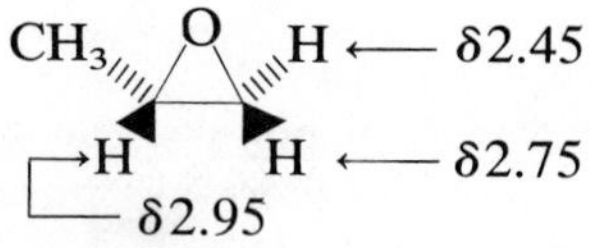

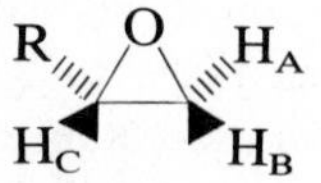

δ2.54 (ethylene oxide)

CH_3 (propylene oxide): H ← δ2.45; H ← δ2.75; H δ2.95

R, O, H_A, H_C, H_B

One should note that epoxides can, in principle, display complex nmr patterns; this is because such ABC systems are often composed of protons (ABC) of only slightly different chemical shifts; thus:

1. The geminal protons (A and B) are shift-nonequivalent and frequently display **geminal coupling.**
2. The coupling of each of B and of A with C are nonidentical.
3. The multiplicities of A, B, and C very possibly cannot be analyzed by first-order rules.

Mass spectra of epoxides should give rise to molecular ion peaks; fragmentation patterns are very complex.

A number of procedures and reactions for the chemical characterization of epoxides have been described by Pasto and Johnson.

(Procedure, p. 376, Pasto and Johnson):

$$\text{R-epoxide (terminal)} \xrightarrow{H_3O^+} R\text{—}\overset{OH}{\underset{|}{C}}H\text{—}\overset{OH}{\underset{|}{C}}H_2 \xrightarrow{HIO_4} RCHO + HCHO + IO_3^-$$

$$IO_3^- + Ag^+ \longrightarrow \underline{AgIO_3} \text{ (white ppt.)}$$

(Procedure, p. 376, Pasto and Johnson):

$$\text{R,R-epoxide} \xrightarrow{HCl} R\text{—}\overset{OH}{\underset{|}{C}}H\text{—}\overset{Cl}{\underset{|}{C}}H\text{—}R$$

chlorohydrin

(Procedure, p. 378, Pasto and Johnson):

$$R_2C\overset{O}{\text{—}}CH_2 \xrightarrow[(2)\ H_2O]{(1)\ LiAlH_4} R_2\overset{OH}{\underset{|}{C}}\text{—}CH_3$$

6.16 MISCELLANEOUS COMPOUNDS

It would be too ambitious to describe characterization procedures for all types of organic compounds. In this section are tabulations of references to chemical and spectral analyses (Table 6.7) too specialized to be described in this book. We have also included a table of ir bands for these compounds (Table 6.8).

Table 6.7

Compound class	General structure	Characterization references[a]
Amine oxides	$R_3N\rightarrow O$ (or $R_3\overset{+}{N}—\overset{-}{O}$)	P & J, p. 447
Azides	RN_3	P & J, p. 445–446; C, pp. 123–124
Azo	RN=NR	C & M, p. 262; P & J, p. 443; J & S, p. 224
Azoxy	R—N=N(O)R	C & M, p. 262; P & J, p. 443; J & S, p. 178
Carbodiimides	R_2C=N—N=CR_2	P & J, pp. 442–444
Diazo	RN_2	C & M, p. 266; C, pp. 123–124
Hydrazines	$RNHNH_2$	P & J, p. 443; C & M, p. 289; C, E, & H, p. 339; J & S, p. 217
Hydrazo	RNHNHR	P & J, p. 443
Hydrazones	R_2C=NNHAr	C, E, & H, p. 406; P & J, p. 445; J & S, pp. 226, 344
Imides, imines	R_2C=NR	C & M, p. 229; P & J, pp. 432–435; C, p. 139
Isocyanides	R—NC	C & M, p. 254; P & J, p. 446; C, pp. 124–126; J & S, pp. 164, 354

Table 6.7 (Continued)

Compound class	General structure	Characterization references[a]
Isocyanates	R—NCO	C & M, p. 229; P & J, pp. 442–445; C, pp. 125–126
Isothiocyanates	R—NCS	C & M, p. 297; P & J, pp. 463; C, pp. 124–126; J & S, p. 178
Lactams	$(CH_2)_x$ ring with C=O and NH	P & J, pp. 432–435; C, pp. 152, 167, 171, 180; J & S, p. 198, 276, 287, 362
Lactones	$(CH_2)_x$ ring with C=O and O	C & M, p. 180; C, pp. 152, 161, 177; J & S, pp. 188, 198, 248, 317
Nitramines	$R_2N—NO_2$	P & J, pp. 446–447
Nitrates	$RONO_2$	P & J, p. 447; C, E, & H, p. 401
Nitrites	RONO	C, E, & H, p. 401
Nitrones	$R_2C{=}N(O)R$	P & J, p. 448
Nitroso	R—NO	P & J, p. 447; C, E, & H, p. 405; C & M, p. 301; C, 184, 186; J & S, p. 226
Oximes	$R_2C{=}NOH$	C, E, & H, p. 406; P & J, p. 445; J & S, pp. 191, 216, 226
Perfluoroalkanes	C_xF_y	P & J, p. 353
Peroxides	R—O—O—R	C & M, p. 204; P & J, pp. 408–409
Quinones	etc.	C & M, p. 212; P & J, pp. 394–395; J & S, p. 248
Semicarbazones	$R_2C{=}NNH(C{=}O)NH_2$	P & J, p. 445; J & S, p. 336
Sulfenyl compounds	$RSX, RSNH_2, RSOH$	P & J, p. 457
Thiocyanates	RSCN	C & M, p. 348; J & S, p. 177
Thiocarbonyl	$R(C{=}O)SH, R_2C{=}S, RCH{=}S, R(C{=}O)SR', R(C{=}S)OR', R(C{=}S)NH_2$	P & J, pp. 462–463; C & M, pp. 297, 355; J & S, p. 177

[a] C = R. T. Conley, *Infrared Spectroscopy*, 2nd ed. (Allyn & Bacon, Boston, 1972); C, E, & H = N. Cheronis, J. B. Entrikin, and E. M. Hodnett, *The Systematic Identification of Organic Compounds*, 3rd ed. (Wiley-Interscience, New York, 1965); C & M = N. Cheronia, and T. S. Ma, *Organic Functional Group Analysis* (Wiley-Interscience, New York, 1964); J & S = L. M. Jackman, and S. Sternhell, *Nuclear Magnetic Resonance in Organic Chemistry*, 2nd ed. (Pergamon, Elmsford, N.Y., 1969) P & J = D. Pasto, and C. Johnson, Organic Structure Determination (Prentice-Hall, Englewood Cliffs, N.J., 1969).

Table 6.8. Infrared Data for Miscellaneous Compounds[a]

Class of compound	Compound or general structure[b]	Vibration[c]	Band or range [wave number, cm^{-1} (wavelength, μm)]	Comments[d]
Amine oxides	R_2NO	ν_{NO}	970–950 (10.31–10.53)	Hydrogen bonding reduces magnitude in wave numbers; usually s
	(N→O (aromatic)	ν_{NO}	1300–1200 (7.69–8.33)	See R_2NO comments
	Picoline *N*-oxide	ν_{NO}	1250 (8.00)	See R_2NO comments
Azides	RN_3	$\nu_{N_3}^{asym}$	ca. 2100 (4.76)	s
		$\nu_{N_3}^{sym}$	1350–1170 (7.41–8.55)	m–w
	Cyclohexyl azide	$\nu_{N_3}^{asym}$	2110 (4.76)	s, neat
Azo	RN=NR	$\nu_{N=N}$	1380–1570 (7.25–6.37)	w, not diagnostic
	trans-$CH_3CH_2\ddot{N}=NCH_2CH_2CH_3$	$\nu_{N=N}$	1563 (6.40)	w
	trans-$C_6H_5N=NC_6H_5$	$\nu_{N=N}$	1455, 1395 (6.87, 7.17)	w, KBr, position varies little with substitution on ring
	cis-$C_6H_5\underset{..}{N}=\underset{..}{N}C_6H_5$	$\nu_{N=N}$	1511 (6.62)	
	trans-$C_6H_5\underset{..}{N}=\ddot{N}C_6H_5$	$\nu_{N=N}$	1442 (6.93)	Raman, s, diagnostic
	ArN=NAr	$\nu_{N=N}$	1463–1380 (6.83–7.25)	Raman, s, diagnostic
Carbodiimides	R—N=C=N—R	$\nu_{N=C=N}$	2260–2100 (4.42–4.76)	s
	$CH_3CH_2N=C=NCH_2CH_3$	$\nu_{N=C=N}$	2128 (4.70)	s, neat
Diazo	$R_2C=N_2$	$\nu_{C=N_2}$	2200–2000 (4.54–5.00)	s, conjugation places band in upper end of wave number range
	CH_2N_2	$\nu_{C=N_2}$	2075 (4.82)	s, CCl_4
	$N_2CHCO_2CH_2CH_3$	$\nu_{C=N_2}$	2101 (4.76)	s, CH_2Cl_2
	$N_2C(CN)_2$	$\nu_{C=N_2}$	2140, 2225 (4.67, 4.49)	s, KBr
Fluorides	RF	ν_{C-F}	1350–1120 (7.41–8.93)	s
	ArF	ν_{C-F}	1100–1270 (9.09–7.87)	s
	Vinyl fluorides	$\nu_{C=CF}$	1340–1300 (7.46–7.69)	
Imines	$R_2C=NH$	ν_{N-H}	3400–3000 (2.94–3.33)	s, not diagnostic
	$R_2C=NR'$, R = alkyl, H	$\nu_{C=N}$	1675–1620 (5.97–6.17)	s, m, w

Table 6.8 (Continued)

Class of compound	Compound or general structure[b]	Vibration[c]	Band or range [wave number, cm^{-1} (wavelength, μm)]	Comments[d]
	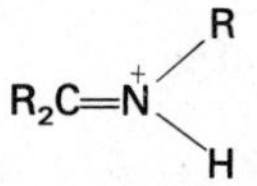 $R_2C{=}\overset{+}{N}(R)H$	$\nu_{C{=}N}$	1650–1705 (6.06–5.87)	s
	$CH_3CH_2CH{=}NCH_2CH_2CH_3$	$\nu_{C{=}N}$	1673 (5.98)	s, CCl_4
Isocyanides RNC (isonitriles)		$\nu_{N{=}C}$	2180–2100 (4.59–4.76)	s, conjugation places band in lower end of wave number region
	$CH_3CH_2CH_2CH_2NC$	$\nu_{N{=}C}$	2146 (4.66)	s, neat
	C_6H_5NC	$\nu_{N{=}C}$	2117 (4.72)	s, neat
Isocyanates	R—NCO	ν_{NCO}^{asym}	2280–2250 (4.39–4.44)	s
		ν_{NCO}^{sym}	1460–1340 (6.84–7.46)	w, not diagnostic
	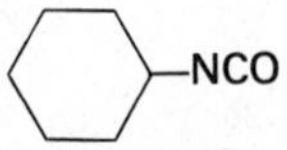 cyclohexyl—NCO	ν_{NCO}^{asym}	2257 (4.43)	s, CCl_4
	C_6H_5—NCO	ν_{NCO}^{asym}	2267 (4.41)	s, CCl_4
	$C_6H_5(C{=}O)NCO$	ν_{NCO}^{asym}	2225 (4.49)	s, neat
Isothiocyanates	RNCS, ArNCS	ν_{NCS}	2220–1975 (4.50–5.06)	w–s, multiple bands
	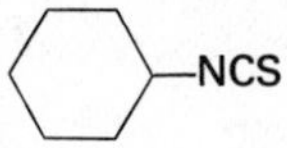 cyclohexyl—NCS	ν_{NCS}	2180–2053 (4.59–4 87)	Four bands, CCl_4
Lactams	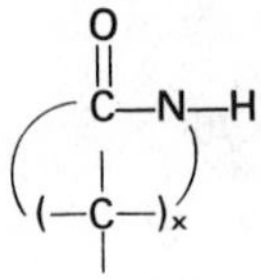 cyclic —C(=O)—N—H with $(—C—)_x$	ν^{assoc}	3375–3000 2.96–3.33)	m–s, multiple bands hydrogen bonded
		ν_{NH}^{free}	3450–3400 (2.90–2.94)	s, no hydrogen bonding
		ν_{CONH}^{comb}	ca. 3200 (ca. 3.12)	m, combination band
	x = 2	ν_{CO}	1760–1730 (5.68–5.78)	s, carbonyl stretch occurs at higher wave numbers for smaller rings
	x = 3	ν_{CO}	1750–1700 (5.71–5.88)	s
	x = 4, 5	ν_{CO}	ca. 1650 (ca. 6.06)	s
Lactones	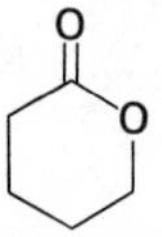 six-membered lactone (δ-valerolactone)	ν_{CO}	1748 (5.72)	s, CCl_4

Table 6.8 (Continued)

Class of compound	Compound or general structure[b]	Vibration[c]	Band or range [wave number, cm^{-1} (wavelength, μm)]	Comments[d]
	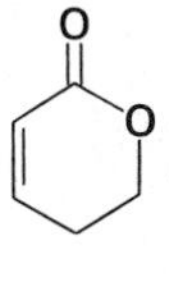 (six-membered lactone structure, O / O)	ν_{CO}	1743 (5.73)	s, CCl_4, position of conjugated unsaturation affects the magnitude of carbonyl stretch
	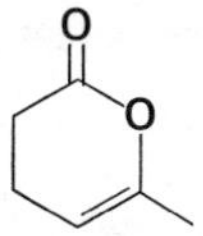 (six-membered lactone structure, O / O)	ν_{CO}	1770 (5.65)	s, CCl_4
	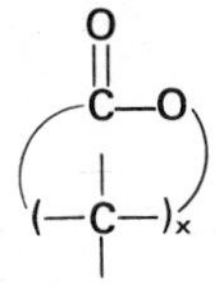 (C(=O)—O ring, (—C—)$_x$)			
	$x = 2$, β-lactones	ν_{CO}	ca. 1818 (ca. 5.50)	s, ring size decrease means an increase in magnitude of wave number for carbonyl stretch
	$x = 3$, γ-lactones	ν_{CO}	1795–1760 (5.57–5.68)	s
	x=4, δ-lactones	ν_{CO}	1750–1735 (5.71–5.76)	s
Nitrates	$RONO_2$	$\nu_{NO_2}^{asym}$	1660–1625 (6.02–6.15)	s
		$\nu_{NO_2}^{sym}$	1285–1270 (7.78–7.87)	s
	$CH_3CH_2ONO_2$	$\nu_{NO_3}^{asym}$, $\nu_{NO_3}^{sym}$	1634, 1282 (6.12–7.80)	s, neat
	NO_3^-	$\nu_{NO_3}^{asym}$	1049 (9.53)	Inorganic salts in Nujol
Nitrites	RONO	$\nu_{N=O}^{trans}$	1681–1648 (5.95–6.07)	s
		$\nu_{N=O}^{cis}$	1625–1605 (6.15–6.23)	s
		$\nu_{N—O}$	814–751 (12.29–13.32)	s
Nitroso compounds	R—NO	$\nu_{N=O}$	1621–1539 (6.17–6.50)	s
	Ar—NO	ν_{NO}	1513–1488 (6.61–6.72)	s
	$(R—NO)_2$	ν_{NO}^{trans}	1290–1176 (775–8.50)	s

Table 6.8 (Continued)

Class of compound	Compound or general structure[b]	Vibration[c]	Band or range [wave number, cm^{-1} (wavelength, μm)]	Comments[d]
		ν_{NO}^{cis}	1420–1330 (7.04–7.52) 1344–1323 (7.44–7.56)	s
	$(Ar—NO)_2$	ν_{NO}^{trans}	1299–1253 (7.70–7.98)	s
		ν_{NO}	ca. 1409, 1397–1389 (ca. 7.10, 7.16–7.20)	s
Oximes	$R_2C{=}N—OH$	ν_{OH}	3300–3130 (3.03–3.19)	s
		$\nu_{C{=}N}$	1690–1620 (5.92–6.17)	w
		$\nu_{N—O}$	ca. 930 (ca. 10.75)	s
Peroxides	$(RCO)_2O_2$	$\nu_{C{=}O}$	1816–1787 (5.51–5.60) 1790–1765 (5.59–5.67)	s
	RO_2H	$\nu_{C—O}$ primary	ca. 1465 (ca. 6.83)	s; these C—O stretches are about 30 cm^{-1} higher than the C—O stretch of the corresponding alcohols
		$\nu_{C—O}$ secondary	1352–1334 (7.40–7.50)	m–s
	RO_2R	$\nu_{C—O}$	779–769 (12.84–13.00)	Raman active
	RCO_3H	ν_{OH}	ca. 3280 (3.05)	Gas
		$\nu_{C{=}O}$	ca. 1760 (5.68)	Gas
		δ_{OH}	ca. 1450 (6.90)	Gas
		$\nu_{C—O}$	ca. 1175 (8.51)	Gas
		$\nu_{O—O}$	ca. 865 (11.56)	Gas
	$C_6H_5CO_3H$	$\nu_{C{=}O}$	1775 (5.63)	s, CCl_4
	m-chloroperbenzoic acid	$\nu_{C{=}O}$	1735 (5.76)	s, CCl_4
Quinones		$\nu_{C{=}O}$	1680–1640 (5.96–6.10)	s
	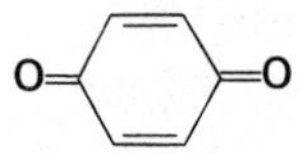 O= =O	$\nu_{C{=}O}$	1669 (5.99)	s, CS_2

Table 6.8 (Continued)

Class of compound	Compound or general structure[b]	Vibration[c]	Band or range [wave number, cm^{-1} (wavelength, μm)]	Comments[d]
		$\nu_{C=C}$	1600 (6.25)	m, CS_2
		ν_{CH}	3077 (3.25)	w, CS_2
Thiocyanates	R—SCN	ν_{CN}	2170–2135 (4.61–4.68)	m–s
Thiocarbonyl	CS_2	ν_{CS}	1510 (6.62)	s, neat
	RCS_2R'	$\nu_{C=S}$	1200–1170 (8.3–8.55)	s, neat
	$CH_3CH_2C(=S)—SCH_2CH_3$	$\nu_{C=S}$	1187 (8.42)	s, neat
	$R(C=S)NH_2$	$\nu_{C=S}$	1290–1020 (9.75–9.80)	s, neat

[a] Collected from ir the references listed in Chapter 10 and from D. Dolphin and A. Wick, *Tabulation of Infrared Spectral Data* (Wiley-Interscience, New York, 1977).
[b] R = alkyl, Ar = aromatic (aryl).
[c] ν = stretch, δ = bend.
[d] Intensity notation: s = strong, m = medium, w = weak; solvent or mulling medium listed for specific compounds.

6.17 NITROGEN COMPOUNDS (NITRO COMPOUNDS, NITRILES)

Amines have been described in Section 6.7; anilines are special cases of amines and are also described in Section 6.7. Here we discuss nitro compounds (RNO_2, Section 6.17.1) and nitriles (RCN, less frequently called organic cyanides, Section 6.17.2). Many other nitrogen compounds are briefly mentioned in the preceding section (6.16).

6.17.1 Nitro Compounds

Nitro compounds are usually in solubility class MN. Unusual solubility characteristics of certain nitro compounds have been described in Chapter 5.

The ir spectrum of a typical nitro compound is shown in Fig. 6.42; nitro groups show two intense bands due to the coupled interaction of the highly polar nitrogen–oxygen bonds:

		cm^{-1}	μm
$—N(O)_2$	asymmetric $N(\overset{...}{=}O)_2$ stretch	1661–1499	6.02–6.67
$—N(O)_2$	symmetric $N(\overset{...}{=}O)_2$ stretch	1389–1259	7.20–7.94

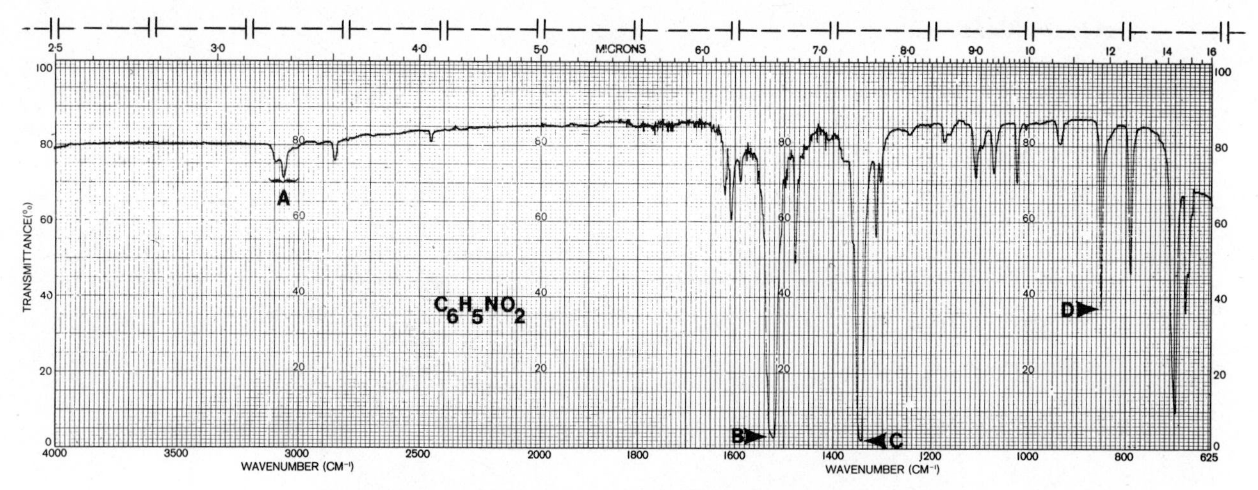

Figure 6.42. Infrared spectrum of nitrobenzene: neat. C—H stretch: **A**, 3100, 3080 cm^{-1} (3.23, 3.25 μm), $\nu_{C—H}$ (aromatic). N(—O)$_2$ stretch: **B**, 1520 cm^{-1} (6.58 μm), $\nu_{N(\cdots O)_2}$, asymmetric; **C**, 1345 cm^{-1} (7.44 μm), $\nu_{N(\cdots O)_2}$, symmetric. C—N stretch: **D**, 850 cm^{-1} (11.76 μm), $\nu_{C—N}$ (aromatic). The low-frequency (<800 cm, >12.5 μm) bands are of little use here for determination of the number and relative positions of ring substituents. This is because many of these bands are due to the interaction of nitro group vibrations with the out-of-plane bending vibrations of the C—H groups.

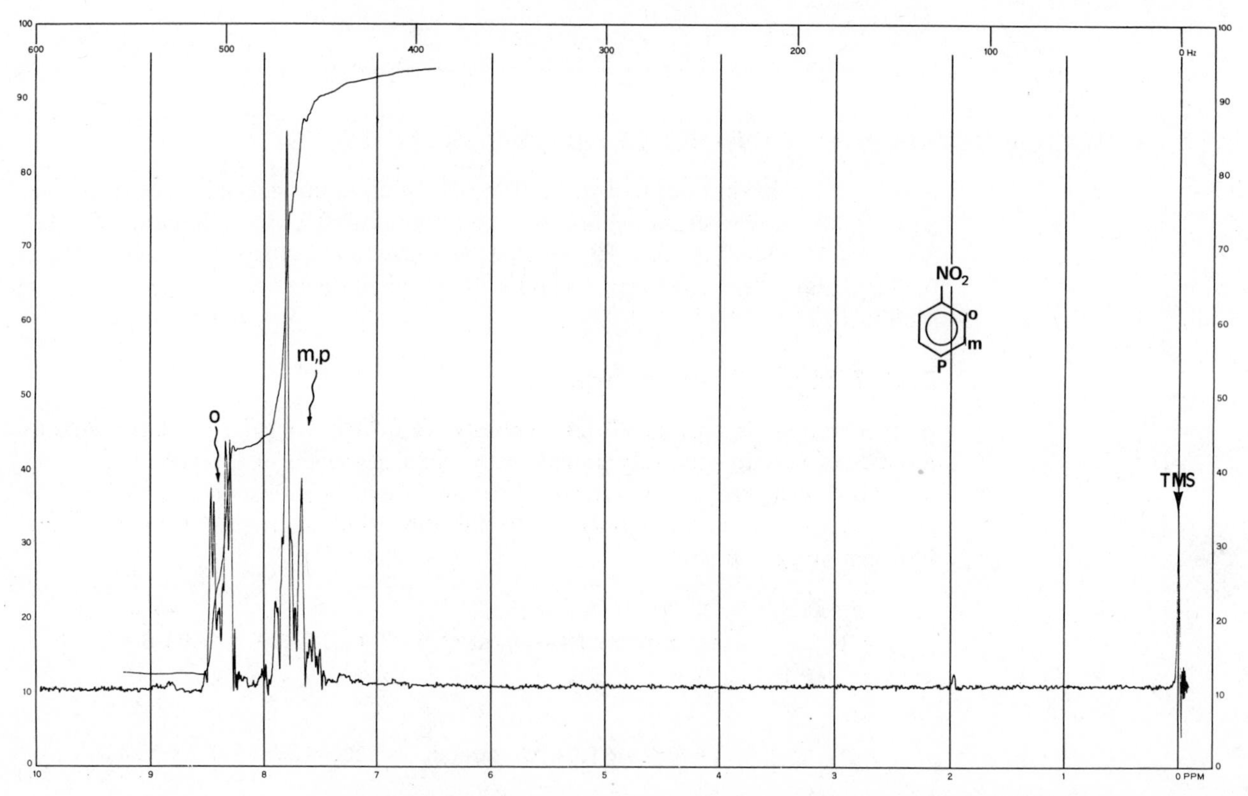

Figure 6.43. Nuclear magnetic resonance (proton) spectrum of nitrobenzene: 60 MHz, 600-Hz sweep width, $CDCl_3$ solvent.

The nmr spectrum of a typical aromatic nitro compound is shown in Fig. 6.43. The nitro substituent causes the aromatic protons to be several tenths of a ppm downfield of the protons of benzene; more examples of this effect are described in Section 6.9. The chemical shifts of protons α to nitro groups follow the usual order, albeit in unusually small increments, as a function of carbon branching at that α-proton (see p. 304):

CH_3NO_2 — $\delta 4.3$ $\qquad$ RCH_2NO_2 — ca. $\delta 4.35$ $\qquad$ R_2CHNO_2 — ca. $\delta 4.6$

Molecular ion peaks of aliphatic nitro compounds are usually weak or nonexistent. Aromatic rings induce sufficient stability to cause the molecular ion of aromatic nitro compounds to be quite stable. The use of the nitrogen rule (Chapter 4) is often indispensable in the mass spectral analysis of nitro compounds. The mass spectral fragments of nitro compounds are often very useful; the $(M - NO_2)$ peak can be used in place of the molecular ion peak (M) of aliphatic nitro compounds. Important fragments are tabulated here:

Class of nitro compound	Fragment	Mass
Aliphatic	$[M\text{-}NO_2]^+$	—
Aliphatic	NO^+	30
Aliphatic	NO_2^+	46
Aromatic	$[M\text{-}NO_2]^+$	(77 for nitrobenzene)
Aromatic	$[M\text{-}NO]^+$	(93 for nitrobenzene)
Aromatic	$\{M\text{-}[NO_2+C_2H_2]\}^+$	—
Aromatic	$\{M\text{-}[NO+CO]\}^+$	—
Aromatic	NO^+	30

Both chemical tests for and derivatization of nitro compounds focus on the reduction of the nitro group to amino or hydroxylamino groups. The number of nitro groups present can be deduced by the response to sodium hydroxide treatment (Experiment 29).

EXPERIMENT 27. FERROUS HYDROXIDE REDUCTION

$$RNO_2 + 6Fe(OH)_2 + 4H_2O \rightarrow RNH_2 + 6Fe(OH)_3$$

Add a small amount (about 10 mg) of the compound to be tested to 1 ml of the ferrous sulfate reagent (A) in a test tube, and then add 0.7 ml of the alcoholic potassium hydroxide solution (B). Insert a glass tube so that it reaches the bottom of the test tube, and pass a stream of inert gas through the tube for about 30 sec in order to remove air. Stopper the tube quickly, and shake. Note the color of the precipitate after 1 min. Try the test on (1) nitrobenzene; (2) *m*-nitroaniline; (3) ethyl alcohol; (4) isopropyl alcohol.

Reagents. (A) to 500 ml of recently boiled, distilled water are added 25 g of ferrous ammonium sulfate crystals and 2 ml of concentrated sulfuric acid. An iron nail is introduced to retard oxidation by the air.

(B) Thirty grams of stick potassium hydroxide is dissolved in 30 ml of distilled water, and this solution is added to 200 ml of 95% ethanol.

Discussion. A positive test is indicated by the formation of a red-brown to brown precipitate.[67] This is ferric hydroxide formed by oxidation of ferrous hydroxide by the nitro compound, which in turn is reduced to the amine. A negative test is indicated by a greenish precipitate. In some cases partial oxidation may cause a darkening of the ferrous hydroxide.

Practically all nitro compounds give a positive test in about 30 sec. The speed with which the nitro compound is reduced depends on its solubility. *p*-Nitrobenzoic acid, which is soluble in the alkaline reagent, gives a test almost immediately, whereas α-nitronaphthalene must be shaken for about 30 sec.

A positive test is also given by other compounds that oxidize ferrous hydroxide. Nitroso compounds, quinones, hydroxylamines, alkyl nitrates, and alkyl nitrites are in this group. Highly colored compounds cannot be tested.

EXPERIMENT 28. ZINC AND AMMONIUM CHLORIDE REDUCTION

$$C_6H_5NO_2 + 4[H] \xrightarrow[NH_4Cl]{Zn} C_6H_5NHOH + H_2O$$

Dissolve 0.5 ml of nitrobenzene in 10 ml of 50% ethanol, and add 0.5 g of ammonium chloride and 0.5 g of zinc dust. Shake, and heat to boiling. Allow to stand for 5 min, filter, and test the action of the filtrate on Tollens reagent (Experiment 11).

This test depends on the reduction of the unknown to a hydrazine, a hydroxylamine, or an aminophenol; all these compounds are oxidized by Tollens reagent, which is reduced to metallic silver.

$$C_6H_5NHOH + 2Ag(NH_3)_2OH \rightarrow C_6H_5NO + 2H_2O + 2Ag + 4NH_3$$

This test cannot be applied if the original compound reduces Tollens' reagent.

The reduction of nitro compounds to azo compounds with lithium aluminum hydride has also been suggested as a qualitative test for the nitro function.[68]

EXPERIMENT 29. TREATMENT OF AROMATIC NITRO COMPOUNDS WITH SODIUM HYDROXIDE

To 5 ml of 20% sodium hydroxide solution add 3 ml of ethanol and a drop of nitrobenzene, and shake vigorously. Compare the color of the solution with that produced by a drop of nitrobenzene and 5 ml of water plus 3 ml of ethanol. Repeat the test with *p*-nitrophenol and *p*-nitroaniline.

---

[67] W. M. Hearon and R. G. Gustavson, *Ind. Eng. Chem., Anal. Ed.*, **9,** 352 (1937).
[68] H. Gilman and T. N. Goreau, *J. Amer. Chem. Soc.*, **73,** 2939 (1951).

Dissolve 0.1 g of *m*-dinitrobenzene in 10 ml of acetone, and add 2 to 3 ml of 10% sodium hydroxide solution, with shaking. Note the color produced. Try the test with nitrobenzene.

Discussion. Mononitro benzenoid compounds give no color (or a very light yellow) with these reagents. If two nitro groups on the same ring are present, a bluish purple color develops; the presence of three nitro groups produces a blood red color. The presence of an amino, substituted amino, or hydroxyl group in the molecule inhibits the formation of the characteristic red and purple colors.

Polynitro compounds can form Meisenheimer complexes, which may lead to colored solutions:

$$O_2N\text{-}C_6H_3(NO_2)\text{-}G \xrightarrow{OH^-} O_2N\text{-}[C_6H_3(NO_2)(OH)(G)]^-$$

Nitrophenols can form highly conjugated and stable phenoxide anions that may be a source of color:

$$HO\text{-}C_6H_4\text{-}NO_2 \xrightarrow{NaOH} \left[{}^-O\text{-}C_6H_4\text{-}N^+(=O)O^- \longleftrightarrow O{=}C_6H_4{=}N^+(O^-)_2 \right] Na^+$$

Nitro compounds can be classified by their reaction with nitrous acid; for any reaction to occur there must be a proton α to the nitro group. Primary nitro compounds undergo removal of the remaining α-proton with base:

$$\text{primary:}\quad R\overset{\alpha}{C}H_2NO_2 \xrightarrow{HONO} R\text{—}CH(NO)NO_2 \xrightarrow{OH^-} [RC(NO)NO_2]^- \text{ (red)}$$

A procedure is described in Pasto and Johnson (p. 441).

$$\text{secondary:}\quad R_2\overset{\alpha}{C}HNO_2 \xrightarrow{HONO} R_2C(NO)NO_2 \text{ (blue)}$$

$$\text{tertiary:}\quad R_3\overset{\alpha}{C}NO_2 \xrightarrow{HONO} \text{no reaction (no color)}$$

The reduction of nitro compounds in acidic media leads to the formation of primary amines (Procedure 50).

$$RNO_2 + 6[H] \rightarrow RNH_2 + 2H_2O$$

These may be converted into suitable derivatives such as the *N*-substituted benzamides, acetamides, and arenesulfonamides (Section 6.7).

Aromatic mononitro compounds may often be characterized by conversion to the corresponding dinitro or trinitro derivatives (Procedures 26*a* and 26*b*).

In some cases bromination of the ring (Procedure 46) or oxidation of an alkyl side chain [Procedures 27*a* and 27*b*) offers the most suitable means of obtaining a derivative.

The nitroalkanes may be characterized by reduction with zinc and hydrochloric acid to the corresponding primary amines (Experiment 28), which may then be converted to solid derivatives. Recently a procedure has been developed that allows reduction of one nitro group in the presence of another (Procedure 50a).

Procedure 50. Reduction with Tin and Hydrochloric Acid

$$RNO_2 \xrightarrow[\text{HCl}]{\text{Sn}} RNH_3^+Cl^- \xrightarrow{\text{NaOH}} RNH_2$$

One gram of the nitrogen-containing compound (nitro, nitroso, azo, azoxy, or hydrazo) is added to 2 g of granulated tin in a small flask. The flask is connected to a reflux condenser, and 20 ml of 10% hydrochloric acid is added, in small portions, with vigorous shaking after each addition. Finally the mixture is warmed on the steam bath for 10 min. The solution is decanted while it is still hot into 10 ml of water, and sufficient 40% sodium hydroxide solution is added to dissolve the tin hydroxide. The solution is extracted several times with 10-ml portions of ether. The ether extract is dried and the ether distilled.

With very insoluble nitro compounds the addition of 5 ml of ethanol will hasten the reduction.

The primary amine remaining after distillation of the ether is converted to one of the derivatives described on pp. 230–234.

Procedure 50a. Selective Reduction of Polynitro Aromatic Compounds to Nitroanilines

$$C_6H_4(NO_2)_2 \text{ (m-dinitrobenzene)} \xrightarrow[\text{CH}_3\text{OH}]{\text{NaSH}} C_6H_4(NH_2)(NO_2) \text{ (m-nitroaniline)} + H_2S\uparrow$$

Two grams of the polynitro compound is dissolved in 30 to 50 ml of methanol in a 200-ml round-bottomed flask fitted with a reflux condenser. Since hydrogen sulfide is evolved during the reduction reaction, the apparatus should be placed under a good hood, or a tube from the top of the condenser should lead to a 500-ml filter flask containing 200 ml of 20% sodium hydroxide solution. A boiling stone is added and the mixture brought to gentle reflux (water bath). If the nitro compound does not dissolve in 50 ml of methanol, 5 to 10 ml of toluene should be added to bring all the compound into solution. The water bath is removed and

40 ml of the methanolic sodium hydrogen sulfide reagent (below) is added with shaking over a 5-min period. The mixture is refluxed for 20 min and then filtered hot into a distilling flask and the methanol removed by distillation from a water bath. The residue is treated with 50 ml of warm water. Products that do not contain an acidic group (OH, CO_2H) may be collected on a filter, recrystallized from hot aqueous isopropyl alcohol, and dried before the melting point is taken. Products that have an acidic group (nitroaminophenols, or nitroamino acids) will dissolve in the alkaline aqueous solution and are precipitated by making the solution just acid to litmus with acetic acid. The precipitate is collected on a filter, and recrystallized from aqueous isopropyl alcohol. The product must be dried before taking the melting point.

Reagent. The methanolic sodium hydrogen sulfide reagent is prepared by dissolving 40 g of $Na_2S \cdot 9H_2O$ in 100 ml of water. The solution is cooled below 20°C and 14 g of powdered sodium bicarbonate is added slowly with stirring and cooling. Keeping the temperature below 20°C, 100 ml of methanol is added, the mixture is stirred for 30 min and the precipitate of hydrated sodium carbonate is removed by filtration. The filtrate is slightly alkaline to phenolphthalein, and keeps for about a week in a brown bottle just below 20°C (not in a refrigerator). Use of ca. 40 ml of this reagent is needed for selective reduction of one nitro group of a 2-g sample of substrate.

Discussion. Di- and trinitro aromatic compounds are selectively reduced to aromatic nitroamines in 70 to 93% yields by treatment with methanolic sodium hydrogen sulfide. Some examples are

$$\text{1,3-dinitrobenzene} \xrightarrow{\text{NaSH}} \text{3-nitroaniline}$$

$$\text{2,4-dinitrotoluene} \xrightarrow{\text{NaSH}} \text{4-amino-2-nitrotoluene}$$

$$\text{2,4-dinitrobenzoic acid} \xrightarrow{\text{NaSH}} \text{4-amino-2-nitrobenzoic acid}$$

$$\text{2,4-dinitrophenol} \xrightarrow{\text{NaSH}} \text{2-amino-4-nitrophenol}$$

$$\text{2,4,6-trinitrotoluene} \xrightarrow{\text{NaSH}} \text{4-amino-2,6-dinitrotoluene}$$

$$\text{1,3,8-trinitronaphthalene} \xrightarrow{\text{NaSH}} \text{2-amino-4,5-dinitronaphthalene}$$

$$\text{3,4'-dinitrobiphenyl} \xrightarrow{\text{NaSH}} \text{4'-amino-3-nitrobiphenyl}$$

Such aromatic nitroamines are solids and serve as useful derivatives. The amines can be further characterized by application of procedures outlined in the amine (Section 6.7) and aniline (Section 6.7) portions of this chapter.

Since the mechanism of the reaction has not been completely elucidated, it is not clear as to how one predicts which nitro group (in sets of nonequivalent nitro groups) will be reduced.

The above procedure may also be used for the reduction of aromatic mononitro compounds, for example, *p*-nitrophenylacetic acid to *p*-aminophenylacetic acid. However, reduction with tin (Procedure 50), zinc, or iron (Experiment 27) with hydrochloric acid is usually preferable.

Some ortho-dinitro compounds undergo displacement of one of the nitro groups:

$$o\text{-}C_6H_4(NO_2)_2 \xrightarrow{NaSH} o\text{-}C_6H_4(S^-Na^+)(NO_2)$$

References. J. P. Idoux, *J. Chem. Soc.* (*C*), 435 (1970); H. H. Hodgson and E. R. Ward, *J. Chem. Soc.*, 794 (1945).

Aliphatic nitro compounds can be derivatized by condensation with aldehydes if the nitro compound has two acidic hydrogens on the α-position. The nitro group stabilizes the anion resulting from removal of an α-proton:

$$H_\alpha{-}\overset{|}{\underset{|}{C}}{-}\overset{+}{N}(=O)O^- \xrightarrow{base} \ddot{C}^-{-}\overset{+}{N}(=O)O^- \longleftrightarrow C{=}\overset{+}{N}(O^-)O^-$$

This anion can then attack a carbonyl compound. Loss of water from the intermediate β-nitro alcohol results in the formation of a nitro olefin; a typical reaction sequence is shown in Procedure 50b.

Procedure 50b. Base-Induced Condensation of Nitromethane and Benzaldehyde

$$CH_3NO_2 + C_6H_5CHO \xrightarrow[CH_3OH]{NaOH} \left[C_6H_5{-}\overset{OH}{\overset{|}{C}}HCH_2NO_2 \right] \xrightarrow{HCl} C_6H_5CH{=}CHNO_2$$

A mixture of 3 ml of nitromethane, 5.3 ml of benzaldehyde, and 10 ml of methanol, in a small flask, is immersed in an ice-salt bath. When the mixture is well cooled, a chilled solution of 2.1 g of sodium hydroxide in 7 ml of water is added *very slowly* (e.g., via a capillary pipette), keeping the temperature below 10°C during the addition. After 15 min, 30 ml of crushed ice and water is added and the slightly cloudy solution is poured into a previously chilled solution of 10 ml of concentrated hydrochloric acid and 15 ml of water in a small beaker. The cloudy supernatant liquid is decanted and the solid collected on a filter, washing with 40 to 50 ml of water. Most of the residual water is removed by melting under warm (60°C) water, cooling to resolidify, and decanting. The crude product, in a 30-ml beaker, is crystallized from 5 ml of ethanol on a steam bath. The yield is 2 to 3 g of fluffy, pale yellow needles, m.p., 56–58°C. (*Caution:* β-Nitrostyrene is a skin irritant!)

6.17.2 Nitriles

Nitriles are usually in solubility class MN. These compounds have a somewhat sweet odor, often slightly resembling the corresponding aldehyde; for example, both benzaldehyde (C_6H_5CHO) and benzonitrile (C_6H_5CN) have odors that are almondlike. Impure nitriles may contain traces of isocyanides (RNC), which are foul-smelling, and thus odor for characterization should be interpreted cautiously.

Infrared analysis of nitriles is usually definitive; the cyano group shows a sharp band of moderate to weak intensity at 2260–2210 cm^{-1} (4.42–4.52 μm) due to C≡N stretch. Very few bands due to other groups appear in this region; the C≡C and —NCO groups are two other possibilities in this region. The C≡N band is very weak and possibly undetectable in some cases; α-hydroxy and α-amino nitriles are examples of compounds that often are ir-transparent in this region. The ir spectrum of a typical nitrile is shown in Fig. 6.44.

The nmr spectrum of a typical nitrile is shown in Fig. 6.45. Protons on the carbon bearing the cyano group* are somewhat deshielded.

CH_3CN — $\delta 2.15$ RCH_2CN — ca. $\delta 2.45$ R_2CHCN — ca. $\delta 2.90$

Special mass spectral techniques (see Chapter 3) normally must be used to locate the molecular-ion peak of nitriles, because this peak can be very weak. The nitrogen rule, in relation to the molecular weight, as described in Chapter 4 is

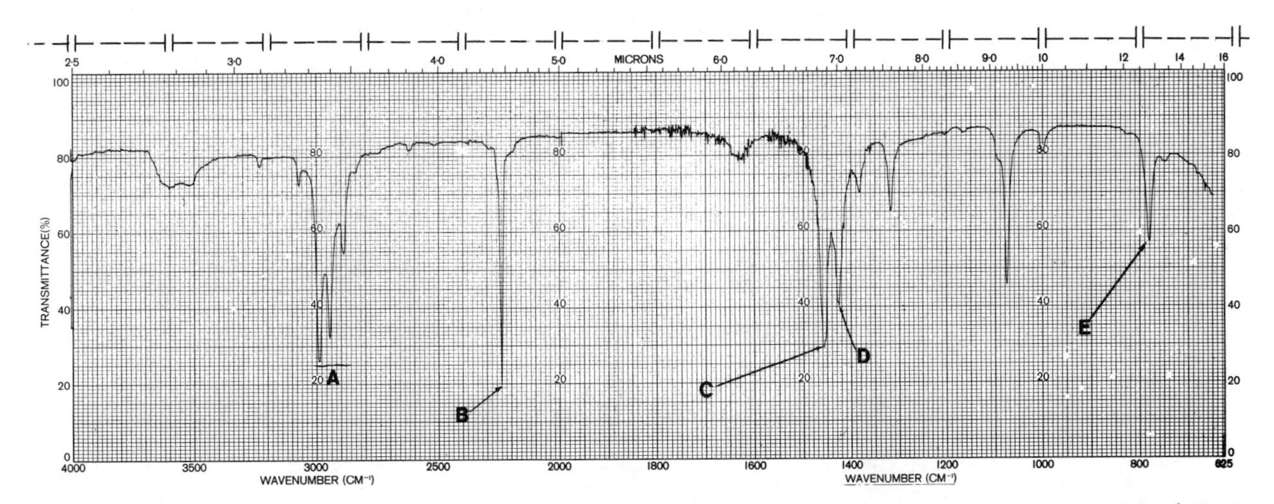

Figure 6.44. Infrared spectrum of propionitrile: neat. C—H stretch: **A**, 3010–2890 cm^{-1} (3.32–3.46 μm), ν_{C-H} (aliphatic). C≡N stretch: **B**, 2240 cm^{-1} (4.46 μm), $\nu_{C\equiv N}$ (aliphatic). C—H bend: **C**, 1458 cm^{-1} (6.86 μm), δ_{asym} CH_3C (aliphatic); **D**, 1427 cm^{-1} (7.01 μm), δ_{sym} CH_2CN: **E**, 781 cm^{-1} (12.80 μm), methylene rock.

* See footnote 58, p. 289 6.

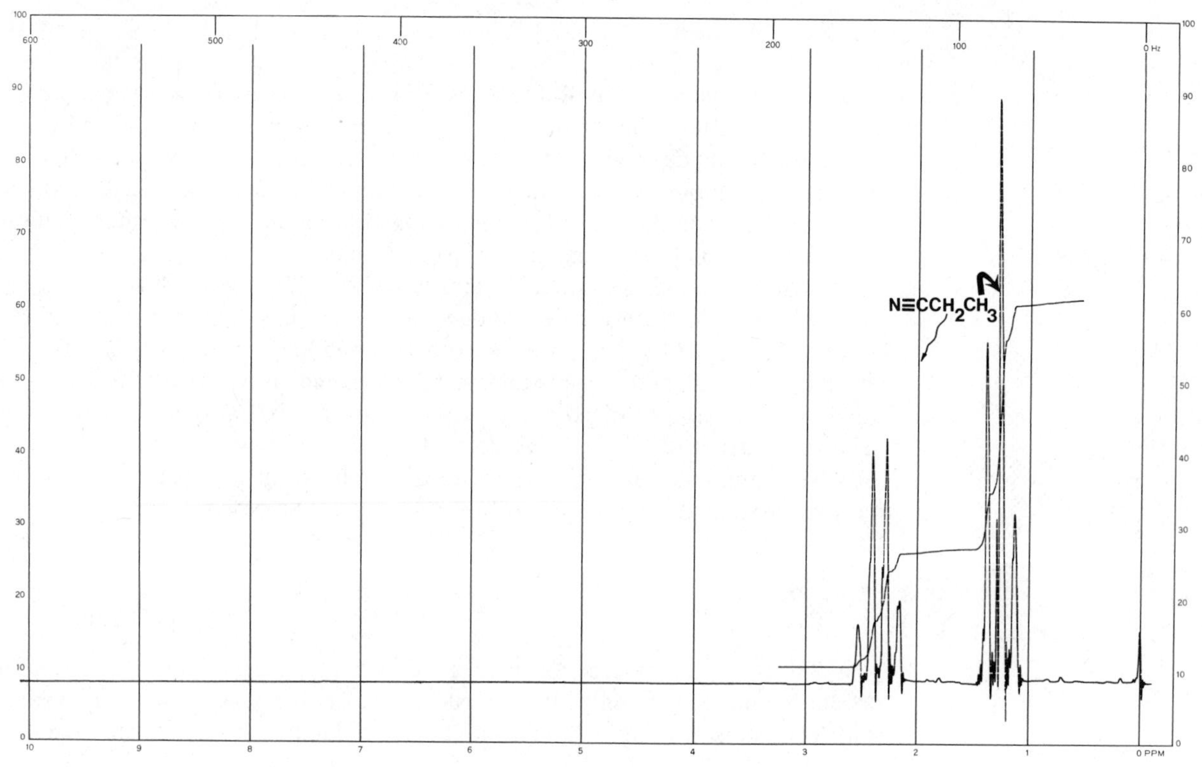

Figure 6.45. Nuclear magnetic resonance (proton) spectrum of propionitrile: 60 MHz, 600-Hz sweep width, $CDCl_3$ solvent.

frequently useful for nitriles. Nitriles with at least one α-hydrogen have the possibility of displaying an $M-1^+$ peak due to the loss of α-hydrogen; this peak can also be difficult to detect because of its low intensity:

$$R_2CH{-}CN{:} \xrightarrow{+e^-} \underset{M^{+\cdot}}{R_2CH{-}CN^{+\cdot}} \longrightarrow \left[\underset{H}{R_2\overset{|}{C}}{-}C{\equiv}N^{+\cdot} \right] \longrightarrow R_2C{=}C{=}N^{+\cdot}$$

McLafferty rearrangement can occur for nitriles with at least one γ-hydrogen:

$$\underset{M^{+\cdot}}{\text{(cyclic } \dot{N}^{+}{\equiv}C{-}C{-}C{-}C{-}H \text{)}} \xrightarrow{\sim H\cdot} H{-}\dot{N}^{+}{=}C{=}C\Big\langle \; + \; \Big\rangle C{=}C\Big\langle$$

Chemical characterization of nitriles focuses on converting the only moderately reactive cyano group to the more reactive carboxyl or primary amino groups. These products can be used as derivatives or can be derivatized themselves.

The nitriles may be hydrolyzed to the corresponding carboxylic acids by means of mineral acids (Procedures 51 and 37b) or alkalies (Procedure 37a, p. 282).

$$RCN + 2H_2O + HCl \rightarrow RCOOH + NH_4Cl$$

$$RCN + H_2O + NaOH \rightarrow RCOONa + NH_3$$

If the resulting acid is a solid, it serves as an excellent derivative. If the acid is a liquid or is water-soluble it is difficult to separate in pure form and is best characterized by means of its *p*-bromophenacyl ester (Procedure 33).

Controlled hydrolysis (Procedure 51a, p. 329) results in the corresponding amide.

$$RCN + H_2O \longrightarrow \left[R{-}\overset{\overset{\displaystyle OH}{|}}{C}{=}NH \right] \longrightarrow R{-}\overset{\overset{\displaystyle O}{\|}}{C}NH_2$$

Reduction of nitriles with sodium and an alcohol forms primary amines (Procedure 52), which may be identified by direct conversion to substituted phenylthioureas:

$$RCN + 4[H] \xrightarrow[\text{ROH}]{\text{Na}} RCH_2NH_2$$

Treatment of the nitriles with Grignard reagent and subsequent hydrolysis produce ketones[69] that can be characterized, for example, by their semicarbazones:

$$RCN + R'MgX \rightarrow R\overset{\overset{\displaystyle R'}{|}}{C}{=}NMgX$$

$$R\overset{\overset{\displaystyle R'}{|}}{C}{=}NMgX + H_2O + 2HX \rightarrow R\overset{\overset{\displaystyle R'}{|}}{C}{=}O + NH_4X + MgX_2$$

Certain nitriles, however, undergo an alternative reaction in which the cyano group is replaced by alkyl or aryl.[70] Compounds of this type include α-keto, α-hydroxy or alkoxy, and α-amino nitriles.

Mercaptoacetic acid (thioglycolic acid) condenses with nitriles in the presence of hydrogen chloride to produce α-(imidoylthio)acetic acid hydrochlorides (Procedure 53):

$$RCN + HSCH_2COOH + HCl \rightarrow RC\begin{matrix} \nearrow\!\!\!\!= NH \cdot HCl \\ \searrow SCH_2COOH \end{matrix}$$

These salts may be characterized by their decomposition points and neutralization equivalents. They act as dibasic acids when titrated with alkali, thymol blue being used as the indicator.

---

[69] R. L. Shriner and T. A. Turner, *J. Amer. Chem. Soc.*, **52,** 1267 (1930). The preparation of Grignard reagent is described in Procedure 12 of this text.

[70] See M. S. Kharasch and O. Reinmuth, *Grignard Reactions of Nonmetallic Substances* (Prentice-Hall, Englewood Cliffs, N.J., 1954), p. 776.

All these methods involve addition to the cyano group and fail with ortho-substituted benzonitriles such as *o*-tolunitrile or *o*-chlorobenzonitrile. Sterically hindered nitriles may be hydrolyzed to acids by heating with 75% sulfuric acid (Procedure 51).

Procedure 51. Acid Hydrolysis of Nitriles

$$C_6H_5\text{—CN} \xrightarrow[\Delta]{H_2SO_4/H_2O/NaCl} C_6H_5\text{—}CO_2H$$

benzonitrile → benzoic acid

About 25 ml of 75% sulfuric acid and 1 g of sodium chloride are placed in a 100-ml round-bottomed flask fitted with a reflux condenser. The flask is heated to 150–160°C by means of an oil bath, and 5 g of the nitrile is added in 0.5-ml portions through the top of the condenser, with vigorous shaking after the addition of each portion. The mixture is heated, with stirring, at 160°C for 30 min and at 190°C for another 30 min. It is then cooled and poured on 100 g of cracked ice in a beaker, and the precipitate is collected on a filter. The precipitate is treated with a slight excess of 10% sodium hydroxide solution, and any insoluble amide is removed by filtration. Acidification of the filtrate yields the carboxylic acid, which may be purified by recrystallization from benzene* or an acetone-water mixture.

Hydrochloric acid is more effective than sulfuric for some nitriles. However, for nitriles that are difficult to hydrolyze, it is advisable to choose sulfuric acid because of its higher boiling point. The addition of a small amount of hydrochloric acid (as sodium chloride) increases the rate of the reaction.

Alternatively, nitriles can be hydrolyzed to carboxylic acid precursors by treatment under strongly basic conditions; a procedure has been described herein (Procedure 37a, p. 282) and elsewhere (Pasto and Johnson, p. 436):

Nitriles in which the cyano group is hindered, for example by α-branches, are more difficult to hydrolyze. A higher boiling solvent may be used:

$$R\text{—CN} \xrightarrow[\text{a glycol solvent}]{KOH/\Delta} RCO_2^-K^+ \xrightarrow{H_3O^+} RCO_2H$$

Acid-catalyzed hydrolysis of nitriles can be stopped at the primary amide stage for those primary amides that are solid and water insoluble (procedure on p. 437 of Pasto and Johnson):

$$RCN \xrightarrow{\text{conc. } H_2SO_4} \xrightarrow{\text{cold } H_2O} RCONH_2 \quad (\xrightarrow{HONO} RCO_2H)$$

* See footnote 43, Chapter 6.

Procedure 51a. Controlled Hydration of a Nitrile to an Amide

$$R{-}C{\equiv}N \xrightarrow[BF_3 \cdot CH_3CO_2H]{H_2O} \left[R{-}\overset{\overset{\displaystyle OH}{|}}{C}{=}NH \right] \longrightarrow R{-}\overset{\overset{\displaystyle O}{\|}}{C}{-}NH_2$$

A mixture of 3 g of the nitrile, 4 g of water, and 20 g of boron trifluoride/acetic acid complex[71] is placed in a 200-ml flask fitted with a reflux condenser. The mixture is heated to 115–120°C in an oil bath, held at this temperature range for 10 min, and then cooled in an ice bath to 15–20°C. A solution of 6 *M* sodium hydroxide is added slowly, keeping the temperature below 20°C until the mixture is just alkaline to litmus paper. About 90 to 100 ml of the sodium hydroxide will be needed. The cold solution is extracted with three 100-ml portions of 1 : 1 ethyl ether/ethyl acetate solvent. The combined extracts are dried with 5 g of anhydrous sodium sulfate and filtered into a dry distilling flask, and the ether/ethyl acetate solvent is removed by distillation from a water bath. The residual amide is purified by recrystallization from water or a mixture of water and methanol. The amide is dried on a clay plate in a vacuum desiccator before the melting point is taken.

This controlled hydration of a nitrile to an amide may be accomplished in a short time in yields of 90 to 95% by use of boron trifluoride/acetic acid catalyst using a limited amount of water.[72]

Diamides (from dinitriles) may be extracted from the cold alkaline mixture more efficiently by use of three 100-ml portions of methylene chloride or chloroform.

Procedure 52. Reduction of Nitriles to Amines and Conversion to Substituted Phenylthioureas

$$\underset{\text{butyronitrile}}{CH_3CH_2CH_2C{\equiv}N} \xrightarrow{Na/C_2H_5OH} \xrightarrow[R = n\text{-butyl}]{HCl} RNH_3^+Cl^-$$

$$\xrightarrow{NaOH} \underset{n\text{-butyl amine}}{CH_3CH_2CH_2CH_2NH_2}$$

$$\left(RNH_2 \xrightarrow{C_6H_5NCS} RNH\underset{\underset{\displaystyle S}{\|}}{C}NHC_6H_5 \right).$$

In a clean, dry, 200-ml, round-bottomed flask, fitted with a reflux condenser, are placed 20 ml of absolute ethanol and 1 g of an aliphatic nitrile (or 2 g of an aromatic nitrile). Through the top of the condenser 1.5 g of finely cut sodium slices is added as rapidly as possible without causing the reaction to become too

---

[71] The liquid complex, $BF_3 \cdot 2CH_3CO_2H$ is used; this complex is commercially available from Harshaw Chemical Co. (1945 E. 97th St., Cleveland, Ohio). The preparation is described in L. F. Fieser and M. Fieser's *Reagents for Organic Synthesis* (Wiley, New York, 1967, p. 69) and *Org. Syn., Coll.* Vol. II, 586 (1943).

[72] C. R. Hauser and D. S. Hoffenberg, *J. Org. Chem.*, **20,** 1448 (1955).

vigorous. When the reduction is complete (10 to 15 min), the mixture is cooled to 20°C, and 10 ml of concentrated hydrochloric acid is added through the condenser, dropwise and with vigorous swirling of the contents of the flask. The reaction mixture is tested to make certain that it is acid to litmus and transferred to a 200-ml distilling flask connected to a condenser. About 20 ml of ethanol and water is removed by distillation. The flask and contents are cooled, and a small dropping funnel containing 15 ml of 40% sodium hydroxide solution is fitted to the top of the distilling flask. An adapter is attached to the end of the condenser and arranged so that it dips into 3 ml of water in a 50-ml Erlenmeyer flask. The alkali is added drop by drop, with shaking. The reaction is vigorous, and care must be exercised to avoid adding the alkali too fast. After all the alkali has been added, the mixture is heated until the distillation of the amine is complete. Distillation is stopped when the contents of the flask are very viscous.

To the distillate[73] is added 0.5 ml of phenyl isothiocyanate, and the mixture is shaken very vigorously for 3 to 5 min. If the derivative does not separate immediately, crystallization is induced by cooling the flask in an ice bath and scratching the walls. The precipitate is collected on a filter, washed with a little cold 50% ethanol, and recrystallized twice from a hot ethanol-water mixture.

Alternatively, a nitrile can be reduced to the corresponding primary amine by treatment with lithium aluminum hydride (procedure on p. 438 of Pasto and Johnson):

$$R{-}C{\equiv}N \xrightarrow[(2)H_2O]{(1)LiAlH_4} RCH_2NH_2$$

CAUTION:* *Note that lithium aluminum hydride violently reacts with water and other protic reagents; work-up of reactions with residual lithium aluminum hydride should be carried out with extreme care (use small-scale reactions, and add protic reagents very slowly with cooling).

Reduction of nitriles is much less sensitive to steric hindrance than is hydrolysis. Thus, with a compound such as *iso*-butyronitrile, reduction would be the more desirable method for characterization.

Procedure 52a. Reduction-Hydrolysis of Nitriles to the Corresponding Aldehyde[74]

$$R{-}C{\equiv}N \xrightarrow[\text{75\% formic acid}]{\text{Raney Ni}} \left[R{-}\overset{H}{\overset{|}{C}}{=}NH\right] \xrightarrow{H_2O} R{-}\overset{H}{\overset{|}{C}}{=}O \xrightarrow{ArNHNH_2} RCH{=}NNHAr)$$

[73] A portion of this distillate should be saved for spectral and other chemical tests.
[74] T. van Es and B. Staskun, *J. Chem. Soc.*, 5775 (1965).

(a) *For Nitriles Boiling Above 125°C.* In a 100-ml round-bottomed flask is placed 12 ml of 90% formic acid, 3 ml of water, 1 g (or 1 ml) of the nitrile, and 1 g of Raney nickel alloy (Al:Ni).[75] A reflux condenser is attached, the contents of the flask is thoroughly mixed by shaking, and the mixture is allowed to reflux gently. The reaction mixture foams as a result of the evolution of hydrogen and carbon dioxide. Hydrogen is formed by the reaction of the aluminum with formic acid and the finely divided nickel catalyzes the decomposition of formic acid to hydrogen and carbon dioxide.

After 1 hr the hot mixture is filtered (*Caution:* Avoid contact with formic acid or its vapors) and the residue of nickel washed with 3 to 4 ml of isopropyl alcohol. The original filtrate and alcohol washings are combined. One half can be used for conversion of the aldehyde to the 2,4-dinitrophenylhydrazone by Procedure 9. If necessary, the other half may be used for the preparation of the methone derivative by Procedure 10.

(b) For Nitriles Boiling Below 125°C. In a 100-ml round-bottomed flask is placed 16 ml of 90% formic acid, 9 ml of water, 1.5 g of the nitrile, and 1.5 g of Raney nickel alloy (Al:Ni). The contents of the flask are thoroughly mixed and allowed to stand 5 min with occasional shaking. The flask is then attached to a still head with the condenser set for normal distillation. The adapter on the condenser is arranged so that the end of the tube just dips into 2 ml of water in a short test tube. The mixture is heated gently to boiling and the flame adjusted so that refluxing occurs in the still head. After about 15 min the heating is increased so that distillation occurs slowly. The aldehyde plus some water and a little formic acid distills over into the 2 ml of water in the receiver. About 3 ml of distillate is collected and one half is used for the preparation of the 2,4-dinitrophenylhydrazone by Procedure 9. The other half may be used for preparing the methone derivative by Procedure 10.

Procedure 53. α-(Imidioylthio)acetic Acid Hydrochlorides

$$R{-}C{\equiv}N \xrightarrow{HSCH_2CO_2H} \left[R{-}\overset{\overset{\large NH}{\|}}{C}{-}SCH_2CO_2H\right] \xrightarrow{\text{dry HCl}} R{-}\overset{\overset{\large NH\cdot HCl}{\|}}{C}{-}SCH_2CO_2H)$$

One gram of the nitrile and 2.0 g of mercaptoacetic acid (thioglycolic acid) are dissolved in 15 ml of absolute ether in a clean, dry test tube. The solution is cooled in an ice bath and thoroughly saturated with dry hydrogen chloride. The tube is tightly stoppered and kept in the ice bath or refrigerator until crystals of the derivative separate. Aliphatic nitriles form the addition compound in 15 to 30 min, whereas aromatic nitriles usually have to stand overnight in a refrigerator.

The crystals are removed by filtration, washed thoroughly with absolute ether, and placed in a vacuum desiccator containing concentrated sulfuric acid in the bottom and small beakers of potassium hydroxide pellets and paraffin wax in the top. The decomposition point is determined in the usual melting-point

---

[75] The 1:1 alloy is used here. Raney nickel catalysts are available from W. R. Grace—Raney Catalyst Division, Chattanooga, Tenn.

apparatus and, if necessary, the neutralization equivalent by titration with standard alkali, thymol blue being used as the indicator.

Determination of the neutralization equivalent (p. 268) can lead to the mass of the R group; one should keep in mind that there are two acid groups in such a derivative.

Nitriles are only weakly basic toward lanthanide shift reagents (Chapter 8); thus only the more highly fluorinated shift reagents offer even the possibility of aiding nmr analysis of nitriles.

6.18 ORGANOSULFUR COMPOUNDS

Both chemical (Na fusion) and mass spectral ($M+2$ peak intensity) determinations ordinarily can easily detect sulfur in organic compounds (Chapter 4); this sulfur atom causes most of these compounds to fall in solubility class MN.

Since sulfur can form bivalent compounds as well as those of higher oxidation states, a large number of classes of compound are possible. Classes that are more important to organic chemists are listed in Table 6.9. Since ir analysis is usually important in pinpointing the class of organosulfur compound, significant ir bands have also been listed on Table 6.9.

6.18.1 Mercaptans

Mercaptans (or thiols) can be initially detected by their fantastically foul odor, provided that the mercaptan is of sufficient volatility. More definitive chemical proof involves chemical and spectral substantiation of the SH functional group. Treatment of a few drops of mercaptan with saturated lead(II) acetate in ethanol usually results in a yellow solid:

$$RSH + Pb(II) + 2H_2O \rightarrow 2H_3O^+ + \underset{\text{(a yellow ppt.)}}{Pb(SR)_2}$$

An important reagent for differentiating mercaptans from other sulfur compounds, n-butyl nitrite, is described below (Experiment 29a). This reagent can also differentiate tertiary from primary and secondary mercaptans. In addition, Benedict's solution can be used as a test for mercaptans (Experiment 29b).

EXPERIMENT 29A. NITROSATION OF MERCAPTANS

$$RSH + n\text{-}C_4H_9ONO \xrightarrow[\text{ether}]{\text{HCl}} \underset{\text{(red)}}{RSNO} + n\text{-}C_4H_7OH$$

Add about 10 mg of the sulfur-containing compound to 1 ml of ether. Then add 1 drop of 6 N hydrochloric acid and 2 drops of n-butyl nitrite[76] (or ethyl nitrite).

[76] *Caution: n*-Butyl nitrite, like other alkyl nitrites, is physiologically potent; they are vasodilators.

Table 6.9. Classes of Organosulfur Compounds and Related Infrared Data

Section	Compound	General structure	Important ir vibrations		cm^{-1}	μm
6.18.1	Mercaptans	RSH	S—H stretch		2600–2550 (w)[a]	3.85–3.92
	(thiophenols)	ArSH	S—H stretch		2560–2550 (w)	3.91–3.92
6.18.2	Sulfides	RSR	C—S stretch		Extremely weak,	usually not useful
6.18.3	Disulfides	RSSR	C—S, S—S stretch		Usually not useful	
6.18.4	Sulfoxides	R—S(=O)—R	S=O stretch		1070–1030 (s)	9.35–9.71
6.18.5	Sulfones	R—S(=O)₂—R	$S(=O)_2$ stretch,	asym	1350–1300 (s)	7.41–7.70
				sym	1160–1120 (s)	8.62–8.93
6.18.6	Sulfinic acids	R—S(=O)—OH	O—H stretch,	free	ca. 3650 (s)	ca. 2.74
			S=O stretch		ca. 1050 (s)	ca. 9.52
6.18.7	Sulfonic acids	RS(=O)₂OH	O—H stretch		3500–3100 (m)	2.86–3.23
			$S(=O)_2$ stretch,	asym	1350–1342 (s)	7.41–7.45
				sym	1165–1150 (s)	8.58–8.69
		$R—SO_3^-M^+$ (salts)	$S(=O)_2$ stretch,	asym	ca. 1175 (s)	ca. 8.5
				sym	ca. 1055 (s)	ca. 9.5
6.18.8	Sulfonyl chlorides	R—S(=O)₂—Cl	S—O stretch		700–610 (s)	14.28–16.39
			$S(=O)_2$ stretch	asym	1410–1380 (s)	7.09–7.25
				sym	1204–1177 (s)	8.31–8.49
6.18.9	Sulfonamides	R—S(=O)₂—NH_2	$S(=O)_2$ stretch	asym	1370–1320 (s)	7.30–7.57
				sym	1170–1140 (s)	8.55–8.77
6.18.10	Sulfonate esters	R—S(=O)₂OR′ (covalent)	$S(=O)_2$ stretch	asym	1372–1375 (s)	7.29–7.49
				sym	1195–1168 (s)	8.37–8.56
6.18.11	Sulfates	R—O—S(=O)₂—O—R	$S(=O)_2$ stretch	asym	1415–1380	7.06–7.24
	(covalent)			sym	1200–1185	8.33–8.44

[a] Band intensities: s = strong, m = medium, w = weak.

Shake the mixture gently, and note the color produced. Primary and secondary alkyl mercaptans give a red color due to the reactions:

$$RCH_2SH + C_4H_9ONO \xrightarrow{HCl} RCH_2SNO + C_4H_9OH$$

$$R_2CHSH + C_4H_9ONO \xrightarrow{HCl} R_2CHSNO + C_4H_9OH$$

Tertiary alkyl mercaptans, $R_3C—SH$, thio ethers, and disulfides give a pale green to yellow color in the ether layer. Sulfoxides, sulfones, sulfinates, and sulfonates likewise give a negative test.

Discussion. Nitrous acid (Experiment 19) will also produce such derivatives. Any reagent that will produce an effective transfer of NO^+ can cause the reaction to occur. Other reagents include N_2O_4, NOBr, and NOCl.

$$\mathrm{RSH} \xrightarrow{\mathrm{HONO}} \mathrm{RSNO} + \mathrm{H_2O}$$

Mercaptans will also decolorize bromine water; one should consult the method and discussion as outlined in Experiment 30.

EXPERIMENT 29B. TREATMENT OF MERCAPTANS AND THIOPHENOLS WITH BENEDICT'S SOLUTION

$$\mathrm{RSH} + \text{Benedict's solution} \rightarrow \mathrm{(RS)_2Cu}$$

$$\mathrm{ArSH} + \text{Benedict's solution} \rightarrow \begin{cases} \mathrm{(ArS)_2Cu} \\ \mathrm{Ar{-}S{-}S{-}Ar} \end{cases}$$

To a solution or suspension of 0.2 g of the compound in 3 ml of 10% sodium bicarbonate solution add 2 ml of Benedict's solution.[77] Note the color and whether a precipitate is formed; then boil the mixture for 1 min and note the result obtained. Try this test on (1) lauryl mercaptan; (2) thiophenol; (3) thiophene; (4) di-*n*-butyl disulfide; (5) L-cysteine hydrochloride; (6) thiourea.

Discussion. Aliphatic and aromatic mercaptans give a yellow precipitate that collects as a greasy yellow ball on boiling. Tertiary alkyl mercaptans merely give a blue solution, as do compounds containing the sulfide or disulfide grouping. A few thiol-containing compounds (thioglycolic acid, mercaptoethanol, and thiosalicyclic acid) yield no precipitates but give nearly colorless solutions.

L-Cysteine and L-cystine undergo extensive degradation, forming black cupric sulfide. Dialkyl sulfones, diaryl sulfones, and sulfonates do not react with Benedict's reagent.

The SH stretch of a mercaptan can be detected by ir; one should be certain, however, that the sample is sufficiently concentrated to allow observation of the usually weak S—H stretch. An ir spectrum for a typical mercaptan is shown in Fig. 6.46. Infrared data for mercaptans are listed above in Table 6.9 (p. 333).

Nuclear magnetic resonance data for appropriate model mercaptans are shown below; the nmr spectrum of a typical mercaptan is shown in Fig. 6.47. The SH proton of mercaptans, unlike the proton on the heteroatom of amines or alcohols, does not undergo rapid exchange; thus, the chemical shift usually occurs as a sharp, well-defined multiplet in a relatively narrow range. The SH proton couples to the α-protons on carbon with $J =$ ca. 8 Hz.

RSH	ArSH	CH_3SH	RCH_2SH	R_2CHSH
δ1.2–1.6	δ2.8–3.6	δ2.1	ca. δ2.6	ca. δ3.05

Unfortunately, the SH proton signal for aliphatic mercaptans is frequently obliterated by the signal due to the aliphatic (C—H) protons.

[77] Preparation of the reagent is discussed in Experiment 13.

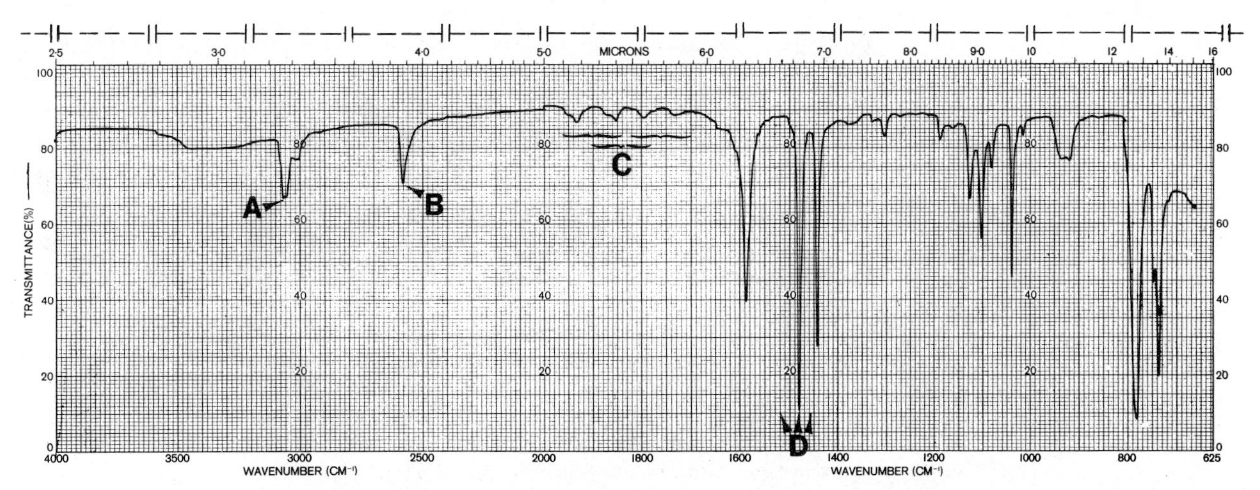

Figure 6.46. Infrared spectrum for benzenethiol (also called thiophenol or phenyl mercaptan): neat. C—H stretch: **A**, 3060 cm^{-1} (3.27 μm), $\nu_{C—H}$ (aromatic). S—H stretch: **B**, 2580 cm^{-1} (3.73 μm), $\nu_{S—H}$ (this band can often be difficult to detect in the spectra of other thiols, because the band is often much weaker than shown here). Aromatic overtones: **C**, 2000–1667 cm^{-1} (5.00–6.00 μm); the pattern here indicates a monosubstituted aromatic (see Fig. 6.23). C═C stretch: **D**, 1600–1400 cm^{-1} (6.25–7.15 μm), $\nu_{C=C}$.

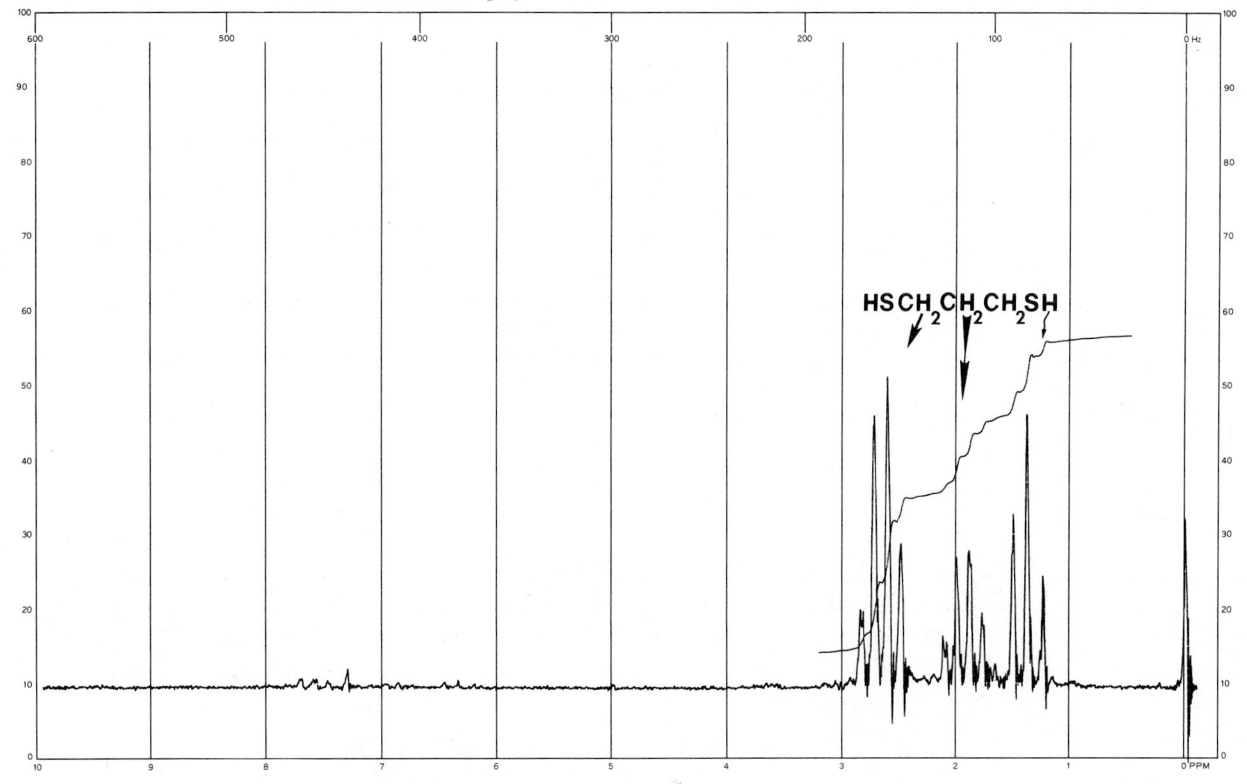

Figure 6.47. Nuclear magnetic resonance (proton) spectrum of 1,3-propanedithiol: 60 MHz, 600-Hz sweep width, $CDCl_3$ solvent.

Aliphatic mercaptans usually yield mass spectra with sufficiently strong molecular ion (M) peaks and $M+2$ peaks such that these compounds can be characterized. The $M+2$ peak intensity can be used for the detection of sulfur in view of the contribution of ^{34}S to the $M+2$ peak intensity (see Chapter 4, p. 89). The $M+2$ peak can also lead to the molecular formula (see Chapter 4). A few useful fragments arising from mercaptans are the following:

Fragment	Mass (m/e)	Comment
[CH_2SH]	47	Implies RCH_2SH structure
[$M-34$]	—	Intense; due to loss of H_2S

A number of derivatization procedures illustrate the substantial nucleophilicity of mercaptans and the mercaptide anion (RS^-).

Mercaptans and thiophenols may be converted to the 2,4-dinitrophenylthioethers by reaction with 2,4-dinitrochlorobenzene:

$$RSNa + Cl\text{-}C_6H_3(NO_2)_2 \rightarrow RS\text{-}C_6H_3(NO_2)_2 + NaCl$$

(for procedure see Pasto and Johnson, p. 451).

3,5-Dinitrobenzoyl chloride may be used to prepare the thio esters:

$$RSH + 3,5\text{-}(O_2N)_2C_6H_3COCl + C_5H_5N \rightarrow 3,5\text{-}(O_2N)_2C_6H_3C(=O)SR + C_5H_5N\cdot HCl$$

(for procedure see Pasto and Johnson, p. 451).

Mercaptans also react with 3-nitrophthalic anhydride:

$$RSH + \text{3-nitrophthalic anhydride} \rightarrow C_6H_3(NO_2)(CO_2H)(C(=O)SR)$$

This acidic thiol-ester derivative allows the possibility of establishing the mass of the R group via the neutralization equivalent (Procedure 32).

Mercaptans can be easily oxidized to disulfides (or to higher oxidation states); mild conditions, such as treatment with iodine, are sufficient:

$$2RSH + I_2 \rightarrow RSSR + 2HI$$

Their great ease of formation and high stability make such disulfides suspect as contaminants in mercaptans and in other reasonably reactive sulfur compounds.

6.18.2 Sulfides

Sulfides (thioethers) are to be suspected when organic molecules containing one sulfur atom and no oxygen atoms show no evidence of a mercapto group. Infrared analysis is not very useful in characterizing organic sulfides. The following nmr data can, however, be useful in characterizing them:

$(CH_3)_2S$ — δ2.05
$(RCH_2)_2S$ — ca. δ2.55
$(R_2CH)_2S$ — ca. δ3.0
$CH_3SC_6H_5$ — δ2.45

CH_3CH_2SR — ca. δ1.25
$R'CH_2CH_2SR$ — ca. δ1.6
R'_2CHCH_2SR — ca. δ1.9

Addition of a small amount of sulfide to a saturated solution of mercuric chloride in ethanol results in a precipitate:

$$R_2\ddot{\underset{..}{S}} + xHgCl_2 \rightarrow \underset{\text{ppt.}}{R_2\ddot{S}{:}(HgCl_2)_x}$$

Sulfides may be oxidized to sulfoxides (R_2SO, Section 6.18.4) or to sulfones (R_2SO_2, Section 6.18.5); reactions and references are as follows:

$$R_2S \xrightarrow{KMnO_4/CH_3CO_2H} R_2SO_2 \qquad \text{(Pasto and Johnson, p. 453)}$$

$$R_2S \xrightarrow{V_2O_5} R_2SO_2 \qquad \text{(Pasto and Johnson, p. 453)}$$

$$R_2S \xrightarrow{H_2O_2/CH_3CO_2H} R_2SO_2 \qquad \text{(Pasto and Johnson, p. 453)}$$

$$R_2S \xrightarrow[C_2H_5OH]{NaIO_4} R_2SO \qquad [\textit{J. Org. Chem.}, \textbf{27}, 282 (1962)]$$

The nmr analysis of sulfides can be supplemented by the use of lanthanide shift reagents (Chapter 8); it is clear, however, that only the more highly fluorinated reagents will provide substantial shifts.[78]

6.18.3 Disulfides

Disulfides can be directly implicated by the presence of two sulfur atoms (elemental analysis, especially mass spectrometry, Chapter 4) in the molecule and indirectly suggested by the absence of evidence for a mercapto (thiol) group (see Section 6.18.1). Infrared spectrometry is of little use in detecting disulfides. Analysis of disulfides by nmr is very similar to that discussed above for sulfides; some useful data are listed here:

CH_3SSR — ca. δ2.35
RCH_2SSR — ca. δ2.70

---

[78] T. C. Morrill et al., *Tetrahedron Lett.*, 3715 (1973), 397 (1975).

Although the disulfide sulfur–sulfur bond is readily formed, this bond can also be readily cleaved with common reagents. Such disulfide cleavages can involve oxidation to sulfonates or reduction to mercaptans:

$$RSSR \xrightarrow{KMnO_4} 2RSO_3H \qquad \text{(Procedure 27)}$$

$$RSSR \xrightarrow{Zn/H^+} 2RSH$$

The resulting products can then each be characterized by procedures described elsewhere in this book (sulfonic acids, Section 6.18.7; mercaptans, Section 6.18.1).

6.18.4 Sulfoxides

Organic sulfoxides (R_2SO), especially those of lower molecular weight, often are very hygroscopic compounds; this can easily lead to spurious O—H bands in their ir spectra. Sulfoxides often have a foul odor; this odor can be due to contamination by an organic sulfide. Both the sulfur–oxygen bond and the nonbonded electron pair of sulfoxides are configurationally stable; for example, there are geometric isomers possible for this *bis*-sulfone:

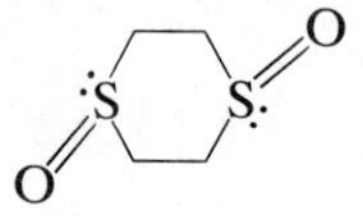

[1,4-dithiane 1,4-dioxide structure: ring with two S atoms, each bearing =O and a lone pair]

Common nmr solvents are

$$CH_3\overset{O}{\overset{\|}{S}}CH_3 \qquad\qquad CD_3\overset{O}{\overset{\|}{S}}CD_3$$

DMSO (dimethyl sulfoxide) — DMSO-d_6

NOTE: Sulfoxides react explosively with perchloric acid! DMSO or DMSO-d_6 should not be used as an nmr solvent for studies using $HClO_4$ catalysis.

Infrared spectra, in view of the intense S═O stretching band (see Table 6.9, p. 333), are invaluable for the detection of the sulfinyl group. The following chemical shifts should assist in interpreting nmr spectra:

$$\underset{\uparrow \text{ ca. } \delta 2.6}{CH_3}—\overset{O}{\overset{\|}{S}}—R \qquad\qquad R'\underset{\uparrow \text{ ca. } \delta 3.0}{CH_2}\overset{O}{\overset{\|}{S}}R$$

The methylene protons in the second structure above have nonequivalent chemical shifts; such a system of protons can result in a surprisingly complex nmr signal.

Sulfoxides can be oxidized to sulfones (R_2SO_2) via the procedures referred to in Section 6.18.3 (sulfides).

6.18.5 Sulfones

Sulfones are usually odorless, colorless, and chemically somewhat inert. The strong pair of ir bands due to the coupled sulfur–oxygen stretching vibrations (see

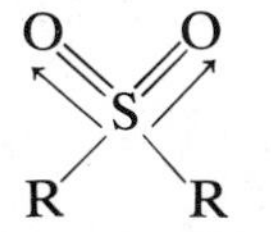

$$\underset{\text{symmetric}}{R_2S(=O)_2} \qquad \underset{\text{asymmetric}}{R_2S(=O)_2}$$

Table 6.9) are normally valuable for the detection of the sulfonyl group. Analysis by nmr is often useful; the following model compound data are provided:

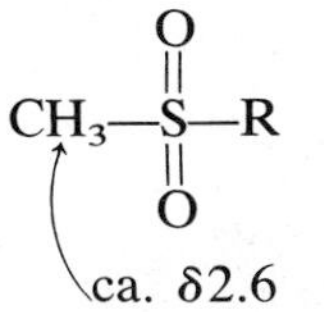

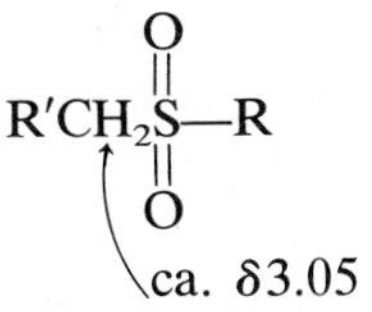

$$\underset{\text{ca. } \delta 2.6}{CH_3}\text{—}SO_2\text{—}R \qquad R'\underset{\text{ca. } \delta 3.05}{CH_2}SO_2\text{—}R$$

One of the chemical reactions of sulfones is their reduction to sulfides by lithium aluminum hydride.[79] Another reaction is the formation of anionic conjugate bases of the sulfones and the subsequent trapping of this anion with electrophiles:

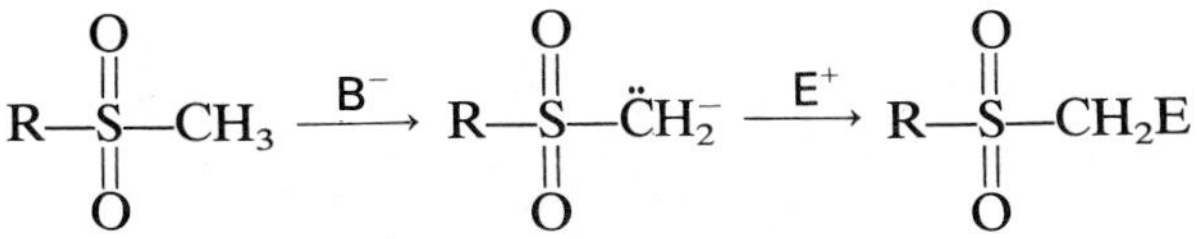

$$R\text{—}SO_2\text{—}CH_3 \xrightarrow{B^-} R\text{—}SO_2\text{—}\ddot{C}H_2^- \xrightarrow{E^+} R\text{—}SO_2\text{—}CH_2E$$

6.18.6 Sulfinic Acids

A number of aromatic sulfinic acids ($ArSO_2H$) have been prepared, but the free acids are unstable, changing to the corresponding sulfonic acids and thiosulfonates. The thiosulfonates undergo hydrolysis to yield thiophenols and sulfonic acids.

$$3ArSO_2H \rightarrow ArSO_3H + ArS\text{—}SO_2Ar + H_2O$$

$$ArS\text{—}SO_2Ar \rightarrow ArSH + ArSO_3H$$

Aliphatic sulfinic acids are even less stable than aromatic ones; hence both types usually are met as the sodium, potassium, or magnesium salts, which are more stable. The best derivatives are those produced by alkylation, which forms sulfones rather than esters. Methyl iodide, for example, reacts with sodium

---

[79] F. G. Bordwell and W. H. McKellin, *J. Amer. Chem. Soc.*, **73**, 2251 (1951).

arenesulfinates to yield methyl sulfones:

$$ArSO_2Na + CH_3I \rightarrow ArSO_2CH_3 + NaI$$

Ethylene bromide has been used to prepare solid 1,2-dialkylsulfonylethanes from salts of aliphatic sulfinic acids.[80]

$$BrCH_2CH_2Br + 2RSO_2Na \rightarrow \begin{array}{c} CH_2SO_2R \\ | \\ CH_2SO_2R \end{array} + 2NaBr$$

The sodium salts of arene- and alkanesulfinates react with mercuric chloride to form the corresponding aryl- and alkylmercuric chlorides, which are crystalline solids.[81]

The most useful technique for quantitative analysis of sulfinic acids is titration with standard sodium nitrite in an acid solution:[82]

$$2RSO_2H + HNO_2 \rightarrow (RSO_2)_2NOH + H_2O$$

Starch-iodide paper is used as an external spot indicator. No color will be observed on the test paper until excess nitrite is present; purple color indicates excess nitrite. Sulfinic acids can be easily identified by dissolving the sample in cold, concentrated sulfuric acid and adding one drop of anisole (or phenetole). A blue color is formed in positive tests.[83]

Other sulfinic acid analyses are as follows:

ir, uv:	S. Detoni and D. Hadzi, *J. Chem. Soc*, 3163 (1955); H. Bredereck, G. Brod, and G. Höschele, *Chem. Ber.*, **88**, 438 (1955)
purification via Fe(III) salts:	C. G. Overberger and J. J. Godfrey, *J. Polymer Sci.* **40**, 179 (1959)
analysis via Fe(III) salts:	S. Krishna and H. Singh, *J. Amer. Chem. Soc*, **50**, 792 (1928)
quantitative analysis via hypochlorite oxidation:	L. Ackerman, *Ind. and Eng. Chem, Anal. Ed*, **18**, 243 (1946)

6.18.7 Sulfonic Acids

Since sulfonic acids are structurally different from sulfuric acid only in that an organic group has been substituted for one hydroxyl group, the high acid strength

$$\begin{array}{ccc} HO{-}\overset{\overset{O}{\|}}{\underset{\underset{O}{\|}}{S}}{-}OH & R{-}\overset{\overset{O}{\|}}{\underset{\underset{O}{\|}}{S}}{-}OH & Ar{-}\overset{\overset{O}{\|}}{\underset{\underset{O}{\|}}{S}}{-}OH \end{array}$$

sulfuric acid — alkanesulfonic acid — arenesulfonic acid

---

[80] C. F. H. Allen, *J. Org. Chem.*, **7**, 23 (1942).
[81] C. S. Marvel, C. E. Adams, and R. S. Johnson, *J. Amer. Chem. Soc.*, **68**, 2735 (1946).
[82] J. L. Kice and K. W. Bowers, *J. Amer. Chem. Soc.*, **84**, 605 (1962); B. Lindberg, *Acta Chem. Scand.*, **17**, 383 (1963); C. S. Marvel and R. S. Johnson, *J. Org. Chem.*, **13**, 822 (1948).
[83] A. I. Vogel, *Practical Organic Chemistry*, **3rd ed.** (Longmans, London, 1956), p. 1078.

of sulfonic acids is not surprising. Sulfonic acids and their metal salts are usually soluble in water but not in organic solvents. Infrared analysis may be complicated by the fact that sulfur–oxygen vibration may result in a multiplicity of bands rather than three well-defined bands as described in Table 6.9. Analysis by ir may also be complicated by the fact the sulfonic acids are very hygroscopic.

Sulfonic acids and their salts, like carboxylic acids and their salts, readily form characteristic derivatives with *S*-benzylthiuronium chloride. This reaction represents the shortest and most direct method for obtaining derivatives of these compounds (Procedure 53a).

Sulfonic acids and their salts are converted into sulfonyl chlorides by heating with phosphorus pentachloride (Procedure 54):

$$RSO_3Na + PCl_5 \rightarrow RSO_2Cl + POCl_3 + NaCl$$

The chloride is then treated with ammonia or an amine to obtain the amide (Procedures 54, 31):

$$RSO_2Cl + 2NH_3 \rightarrow RSO_2NH_2 + NH_4Cl$$

Since the sulfonic acids are strong, it is not possible to obtain them by acidification of their salts unless the sulfonic acid is very insoluble in water or hydrochloric acid. Sulfonic acids combine with amines to produce salts that have definite melting or decomposition points and hence may be used as derivatives.

By treating a concentrated solution of a sodium salt of a sulfonic acid with hydrochloric acid and *p*-toluidine it is possible to cause the *p*-toluidine salts to separate (Procedure 55). These are useful derivatives.

$$RSO_3Na + HCl + ArNH_2 \rightarrow ArNH_3^+RSO_3^- + NaCl$$

The phenylhydrazine salts are useful for aliphatic sulfonic acids.[84]

When silver sulfonates are allowed to react with *p*-nitrobenzyl chloride and pyridine, the corresponding *p*-nitrobenzylpyridinium salts are formed.[85] These salts crystallize readily, have sharp and characteristic melting points, and are suitable as derivatives. Aminosulfonic acids in the benzene and naphthalene series may be characterized by replacement of the amino group by chlorine through the Sandmeyer reaction, followed by conversion of the sulfonic acid to a sulfonamide or sulfonanilide (Procedure 56).

$$RSO_3^-\overset{+}{A}g + ArCH_2Cl + C_5H_5N \rightarrow C_5H_5\overset{+}{N}\text{—}CH_2Ar\ RSO_3^- + AgCl$$

Procedure 53a. Benzylthiuronium Sulfonates

$$RSO_3H + NaOH \rightarrow RSO_3^-Na^+$$

$$RSO_3^-Na^+ + C_6H_5CH_2SC(NH_2)_2^+Cl^- \rightarrow C_6H_5CH_2SC(NH_2)_2^+RSO_3^- + NaCl$$

About 1 g of the sodium or potassium salt of the sulfonic acid is dissolved in the smallest amount of water, heat being used if necessary to effect solution. If the

---

[84] P. H. Latimer and R. W. Bost, *J. Amer. Chem. Soc.*, **59**, 2500 (1937).
[85] E. H. Huntress and G. L. Foote, *ibid.*, **64**, 1017 (1942).

free sulfonic acid is the starting material, it is dissolved in 2 *N* sodium hydroxide solution, and any excess alkali is neutralized with hydrochloric acid, phenolphthalein being employed as the indicator. A preparation of the thiuronium reagent is included in Procedure 39 (p. 287).

A solution of 1 g of benzylthiuronium chloride is dissolved in the smallest possible amount of water. This solution and that of the sulfonate are chilled in an ice-water bath, mixed, and shaken thoroughly. Occasionally it is necessary to scratch the tube and cool in an ice bath to induce crystallization. The benzylthiuronium sulfonate crystals are collected on a filter, washed with a little cold water, and recrystallized from hot 50% ethanol.

Procedure 54. Sulfonyl Chlorides and Sulfonamides from Salts of Sulfonic Acids

$$C_6H_5SO_3^-Na^+ \xrightarrow{PCl_5} C_6H_5SO_2Cl \xrightarrow{\text{conc. } NH_4OH} C_6H_5SO_2NH_2$$

Two grams of the salt are mixed with 5 g of phosphorus pentachloride in a clean, dry flask. A reflux condenser is attached, and the flask is heated in an oil bath at 150°C for 30 min. The mixture is cooled, and 20 ml of dry benzene* is added. The mixture is then warmed on a steam cone, the solid mass being stirred thoroughly. The solution is filtered through a dry filter paper and the filtrate is washed with two 15-ml portions of water. The benzene is removed by distillation from a steam bath. The residual sulfonyl chloride may be recrystallized from petroleum ether or chloroform. Since the sulfonyl chlorides usually are low-melting compounds, it is best to prepare the amide by adding the benzene solution to 20 ml of concentrated ammonia, with vigorous stirring. Occasionally the sulfonamide precipitates and may be removed by filtration; otherwise it is obtained by evaporation of the benzene layer. The sulfonamides may be recrystallized from ethanol.

Procedure 55. p-Toluidine Salts of Sulfonic Acids

(a) $C_6H_5SO_3H + p\text{-}CH_3C_6H_4NH_2$

(b) $C_6H_5SO_3^-Na^+ \xrightarrow[\text{conc. HCl}]{CH_3-C_6H_4-NH_2}$

$$\longrightarrow C_6H_5SO_3^- \; H_3\overset{+}{N}-C_6H_4-CH_3$$

(a) *From Free Sulfonic Acids.* One gram of the sulfonic acid is dissolved in the minimum amount of boiling water and 1 g of *p*-toluidine is added. More

* See footnote 43, p. 235.

water or an additional portion of the sulfonic acid is added to obtain a clear solution. The solution is cooled and the flask scratched to induce crystallization of the salt. The salt is removed by filtration and recrystallized from the minimum amount of boiling water.

(*b*) ***From Soluble Salts of Sulfonic Acids.*** About 2 g of the sodium, potassium, or ammonium salt of the sulfonic acid is dissolved in the minimum amount of boiling water, and 1 g of *p*-toluidine and 2 to 4 ml of concentrated hydrochloric acid are added. If a precipitate separates or if the *p*-toluidine is not completely dissolved, more hot water and a few drops of concentrated hydrochloric acid are added until a clear solution is obtained at the boiling point. The solution is cooled, and the walls of the flask are scratched to induce crystallization of the salt. The product is removed by filtration and recrystallized from a small amount of water or dilute ethanol.

Procedure 56. Chloroarenesulfonamides and Chloroarenesulfonanilides from Aminosulfonic Acids

(a) SO_3H / NH_2 (G) $\rightleftharpoons$ SO_3^- / NH_3^+ (G) $\xrightarrow{Na_2CO_3}$ $SO_3^-Na^+$ / NH_2 (G)

$\xrightarrow{HONO}$ $SO_3^-Na^+$ / $N_2^+Cl^-$ (G)

$\xrightarrow[HCl]{CuCl}$ SO_3H / Cl (G)

(b) Cl—Ar(G)—SO_3H $\xrightarrow{PCl_5}$ Cl—Ar(G)—SO_2Cl

$\xrightarrow{NH_4OH}$ Cl—Ar(G)—SO_2—NH_2

(c) Cl—Ar(G)—SO_2—Cl $\xrightarrow{C_6H_5NH_2}$ Cl—Ar(G)—SO_2—NH—C_6H_5

(a) About 1.5 g of the aminoarenesulfonic acid is dissolved in 10 ml of water containing 0.5 g of sodium carbonate. Diazotization is effected by adding 2 ml of concentrated hydrochloric acid and then, quickly, about 5 ml of 10% sodium nitrite solution, the temperature being maintained at 10 to 15°C by the addition of ice.

Meanwhile, cuprous chloride is prepared by mixing a solution of 2.16 g of copper sulfate and 0.56 g of sodium chloride in 10 ml of water with a solution of 0.46 g of sodium bisulfite and 0.33 g of sodium hydroxide in 10 ml of water. The precipitated cuprous chloride is then dissolved in 10 ml of concentrated hydrochloric acid. The solution is cooled in ice to 5°C, and the diazonium solution is added rapidly, with stirring. The temperature is allowed to rise slowly to room temperature, stirring being continued for 1 hr, and the solution is then heated at 60 to 70°C for 30 min on a steam bath. The copper is precipitated by hydrogen sulfide, and the resulting copper sulfide is removed by filtration. The crude chloroarenesulfonic acid is obtained by evaporating the filtrate to dryness on the steam bath.

(b) The crude acid is then mixed with double its weight of phosphorus pentachloride in a small beaker. When the vigorous reaction has ended, the beaker is heated for a short time in an oil bath at 130 to 140°C (under a hood) to expel the phosphorus oxychloride. After being cooled, the chloride is washed by decantation with cold water. The resulting oil is added to 45 ml of concentrated ammonium hydroxide, and the solution is evaporated to dryness on the steam bath. The crude sulfonamide is recrystallized, with the addition of Norit, from ethanol or water.

(c) In order to prepare the chloroarenesulfonanilide, a solution of the crude sulfonyl chloride in 10 ml of benzene* is mixed with 2.5 g of aniline, and the resulting solution is heated under reflux for 1 hr. It is then concentrated to half its volume and chilled. The solid that separates is collected on a filter, washed thoroughly with warm water, and recrystallized from chlorobenzene.

6.18.8 Sulfonyl Chlorides

In addition to having distinctive ir bands (Table 6.9, p. 333), volatile sulfonyl halides have a penetrating, unpleasant odor. Because of their reactivity, sulfonyl halides are frequently characterized by conversion to other compounds such as sulfonamides that are less reactive and thus more easily completely characterized.

Sulfonyl chlorides are readily converted to amides by treatment with aqueous ammonia (Procedure 54) or with ammonium carbonate (Procedure 48). The Hinsberg reaction (Procedure 16, p. 230) serves to produce sulfonanilides or sulfontoluidides, which are good derivatives.

6.18.9 Sulfonamides

Most sulfonamides show distinct $S(=O)_2$ and N—H stretching bands. These amides are usually stable solids; it is normally best to analyze theese compounds chemically by cleaving the sulfur–nitrogen bond and then characterize the cleavage products.

---

* See footnote 43, p. 235.

A fundamental reaction of sulfonamides is cleavage by hydrolysis (Procedure 57); the amines (Section 6.7) and sulfonic acids (Section 6.18.7) can be characterized by procedures described in the sections cited.

Procedure 57. Hydrolysis of Sulfonamides

$$RSO_2NHR' + H_2O + HCl \rightarrow RSO_3H + R'NH_3Cl$$

$$RSO_2NR'_2 + H_2O + HCl \rightarrow RSO_3H + R'_2NH_2Cl$$

The sulfonamide is hydrolyzed by heating 10 g of it with 100 ml of 25% hydrochloric acid under reflux. Sulfonamides of primary amines require 24 to 36 hr refluxing, whereas sulfonamides of secondary amines may be hydrolyzed in 10 to 12 hr. After solution is complete, the mixture is cooled, made alkaline with 20% sodium hydroxide solution, and extracted with three 50-ml portions of ether. The ether solution is dried, and, after the ether has been driven off, the amine is distilled. With certain very low- or very high-boiling amines it is often more convenient to recover them as hydrochlorides by passing dry hydrogen chloride gas into the dry ether solution.

The amine may be separated by the addition of alkali and characterized by a suitable derivative (p. 227ff). If necessary, the sulfonic acid may be recovered in the form of its sodium salt (after removal of the amine) and converted to a derivative (Procedure 54 or 55).

A method that may be more satisfactory consists in treating the amide with 48% hydrobromic acid and phenol (Procedure 16, p. 231):

$$2ArSO_2NHR + 5HBr + 5C_6H_5OH \rightarrow ArSSAr + 2RNH_2 + 5BrC_6H_4OH + 4H_2O$$

Not only can the amine be characterized, but also the diaryl disulfide can be isolated and can serve as a derivative (Section 6.18.3).

The amide proton is labile, and can be removed from the nitrogen atom to form sulfonamide anions which will nucleophilically attack alkyl halides and carbonyl carbon atoms; the remaining reactions are examples of such nucleophilic attack.

Sulfonamides that have at least one hydrogen atom on the nitrogen atom can be alkylated by treatment with base and reactive halides or alkyl sulfates:

$$RSO_2NH_2 + NaOH + R'X \rightarrow RSO_2NHR' + NaX + H_2O$$

$$RSO_2NHR' + NaOH + R''X \rightarrow RSO_2NR'R'' + NaX + H_2O$$

These reactions are useful if the alkyl group is so chosen that a known sulfonamide is produced.

Primary sulfonamides react with phthaloyl chloride to produce *N*-sulfonylphthalimides:

$$C_6H_4(COCl)_2 + RSO_2NH_2 \longrightarrow C_6H_4(CO)_2NSO_2R + 2HCl$$

Primary amides of sulfonic acids, like those of carboxylic acids (Section 6.12, Procedure 38), react with xanthydrol to form *N*-xanthylsulfonamides, which are satisfactory derivatives (Procedure 58).

Procedure 58. N-Xanthylsulfonamides

$$RSO_2NH_2 + \text{xanthydrol (HO–CH)} \xrightarrow{CH_3CO_2H} \text{N-xanthylsulfonamide (RSO}_2\text{NH–CH)} + H_2O$$

xanthydrol

About 0.2 g of xanthydrol is dissolved in 10 ml of glacial acetic acid. If the mixture is not clear, it is filtered or centrifuged, and to the clear solution is added 0.2 g of the sulfonamide. The mixture is shaken and allowed to stand at room temperature until the derivative separates; this may require as long as 1.5 hr. The *N*-xanthylsulfonamide is removed by filtration and recrystallized from a dioxane-water mixture (3:1).

Sulfonamides can be readily cleaved by reduction with sodium naphthalene anion radical. This provides, via a simple procedure,[86] the sulfinate anion and the salt of the amine:

$$RSO_2NHR \xrightarrow[-70°C]{[C_{10}H_8]^{\dot{-}}\,Na^+} \xrightarrow{H_2O} RSO_2^-Na^+ + R'NH_2$$

6.18.10 Sulfonic Acid Esters (sulfonates)

This section discusses sulfonate esters, RSO_3R', which contain a covalent carbon–oxygen bond to the R′ group. These are to be contrasted with sulfonate salts, $RSO_3^-M^+$, which contain an ionic metal–oxygen bond (Section 6.18.7).

Sulfonate esters are often quite stable compounds; these esters, in fact, serve as alternatives to sulfonamides as derivatives of sulfonic acids. Esters of phenols, glycols, and primary alcohols are best for these purposes.

$$\left.\begin{array}{l} RSO_3H \\ RSO_3^-Na^+ \end{array}\right\} \xrightarrow{PCl_5} RSO_2Cl \xrightarrow[C_5H_5N]{R'OH} R{-}\overset{O}{\underset{O}{\overset{\|}{\underset{\|}{S}}}}{-}OR' + C_5H_5NH^+Cl^-$$

(a sulfonate ester)

[86] W. D. Closson, S. Ji and S. Schulenberg, *J. Amer. Chem. Soc.*, **92**, 650, (1970); sulfonamides and sulfonates can also be cleaved by processes described in references 1 and 2 of this paper, The gas chromatographic analysis of amines, often a difficult task, is discussed in the Closson reference.

Such reactions are carried out by simply treating the sulfonyl chloride in pyridine (C_5H_5N) solvent with an equivalent of alcohol; an ice bath should be applied if the reaction at room temperature is fast enough to cause observable decomposition. The pyridine solvent serves to act as a scavenger base for the mineral acid side product; the pyridine hydrochloride may crystallize from solution to serve as a visual monitor of reaction progress.

Alkyl arenesulfonates ($ArSO_3R$) are very common examples of this molecular class. Methyl "tosylate" (methyl *p*-toluenesulfonate) is an excellent alkylating reagent:

$$R_2NH + \underbrace{CH_3\text{—}C_6H_4\text{—}\overset{O}{\underset{O}{\overset{\|}{\underset{\|}{S}}}}\text{—}}_{TsO\text{—}}OCH_3 \longrightarrow R_2\overset{H}{\underset{+}{N}}CH_3 \quad {}^{-}OTs$$

p-Bromobenzenesulfonates ("brosylates"), as well as other substituted benzenesulfonates, have been used to study carbocations and carbocation character of the R group:

$$Br\text{—}C_6H_4\text{—}\overset{O}{\underset{O}{\overset{\|}{\underset{\|}{S}}}}\text{—}O\text{—}R \longrightarrow Br\text{—}C_6H_4\text{—}SO_3^{-}\,R^{+}$$

The R group can be trapped with metal hydrides:

$$ArSO_3R \xrightarrow[NaBH_4]{LiAlH_4} RH + ArSO_3^{-}$$

The resulting sulfonate salt can be characterized as described in Section 6.18.7.

6.18.11 Sulfates

Organic sulfates (esters) are usually much more reactive than, for example, sulfonate esters; dimethyl sulfate for example, is a common methylating reagent:

$$RCO_2H + \underset{[(CH_3)_2SO_4]}{CH_3O\overset{O}{\underset{O}{\overset{\|}{\underset{\|}{S}}}}OCH_3} \longrightarrow RCO_2CH_3 + CH_3OSO_3H$$

CAUTION: ***Alkyl sulfates are highly toxic and should be handled with adequate ventilation and with extreme caution. Skin contact is to be prevented, as such contact is a method of toxification***

6.19 PHENOLS

Solubility of phenols in sodium hydroxide and the insolubility in sodium bicarbonate of water-insoluble phenols are the criteria for their classification as weak acids (class A_2).

$$ArOH + NaOH \rightarrow ArO^-Na^+ + H_2O$$

Exceptions of this have been discussed in Chapter 5; picric acid, for example, is soluble in the weaker bicarbonate base:

$$\text{2,4,6-}(NO_2)_3C_6H_2\text{—}OH + NaHCO_3 \longrightarrow \text{2,4,6-}(NO_2)_3C_6H_2\text{—}O^-Na^+ + H_2O + CO_2$$

Phenols can be detected by treatment with ferric chloride (Experiment 29c). The procedure using pyridine solvent has resulted in accurate results in 90% of the phenolic substrates tested; previous procedures using water or alcohol–water solvents have had only a 50% success rate.

EXPERIMENT 29C. FERRIC CHLORIDE-PYRIDINE REAGENT

$$3ArOH + 3\,C_5H_5N + FeCl_3 \longrightarrow Fe(OAr)_3 + 3\,C_5H_5\overset{+}{N}H\ \bar{Cl}$$

Add 30 to 50 mg of the solid unknown (or 4 to 5 drops of a liquid) to 2 ml of pure chloroform in a clean, dry test tube. Stir to effect solution. If the unknown does not seem to dissolve (even partially), add an additional 2 to 3 ml of chloroform and warm gently. Cool to 25°C and add 2 drops of a 1% solution of anhydrous ferric chloride in chloroform followed by 3 drops of pyridine. Shake the tube to effect mixing and note the color produced *immediately*. A positive test is shown by production of a blue, violet, purple, green, or red-brown solution. Frequently the colors change in a few minutes. Try the test on (1) *p*-cresol; (2) salicylaldehyde; (3) *o*-nitrophenol; (4) *m*-bromophenol.

Reagents. Pure chloroform free from ethanol should be used as the solvent and for preparing the ferric chloride solution. The latter is made by adding 1 g of the black crystals of *anhydrous* ferric chloride to 100 ml of pure chloroform in a 150-ml bottle. The mixture is shaken occasionally for about an hour, and allowed to stand to permit the insoluble material to settle. Decant the pale yellow solution into a screw-cap bottle fitted with a medicine dropper. Pure analytical reagent-grade pyridine is also placed in a screw-cap bottle fitted with a medicine dropper.

Discussion. This reagent is useful for detecting compounds containing a hydroxyl group directly attached to an aromatic nucleus. Treatment of chloroform solutions of phenols, naphthols, and their ring-substituted derivatives with a chloroform solution of anhydrous ferric chloride and pyridine produces characteristic blue, violet, purple, green, or red-brown colored complexes.

Alcohols, ethers, aldehydes, acids, ketones, hydrocarbons, and their halogen derivatives give negative results, that is, colorless, pale yellow, or tan solutions.

This method is especially valuable for substituted phenols and naphthols that are very insoluble in water. Even 2,4,6-trichlorophenol, 2,4,6-tribromophenol, nonylphenol, phenolphthalein, and thymolphthalein give positive tests provided that sufficient chloroform is used (about 5 ml) to get them into solution.

Phenolic compounds that have failed to give positive tests are picric acid, 2,6-di-*tert*-butylphenols, phenol- and naphtholsulfonic acids, hydroquinone, dl-tyrosine, *p*-hydroxyphenylglycine, and *p*-hydroxybenzoic acid. The latter gives a distinct yellow color (which would be classed as negative), whereas salicylic acid gives a violet color. The esters of *p*-hydroxybenzoic acid give purple colors and *p*-hydroxybenzaldehyde a violet-purple color.

It is of interest that 5,5-dimethyl-1,3-cyclohexandione (dimedon, methone) gives a beautiful purple color. Resorcinol gives a blue-violet color. Note that

dimedon: CH_3 / CH_3 (O, O) $\rightleftharpoons$ CH_3 / CH_3 (OH, O) $\rightleftharpoons$ CH_3 / CH_3 (OH, OH)

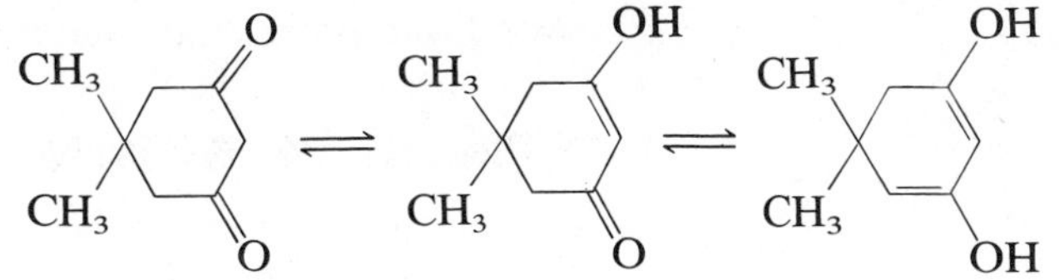

resorcinol: OH / OH $\rightleftharpoons$ OH / O $\rightleftharpoons$ O / O

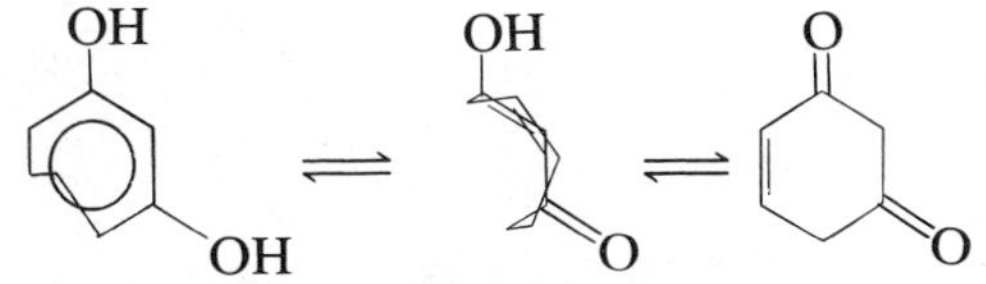

several of the tautomeric forms of these compounds are similar in structure to tautomeric forms of phenols. Salicylaldehyde forms a highly colored complex with ferric chloride:

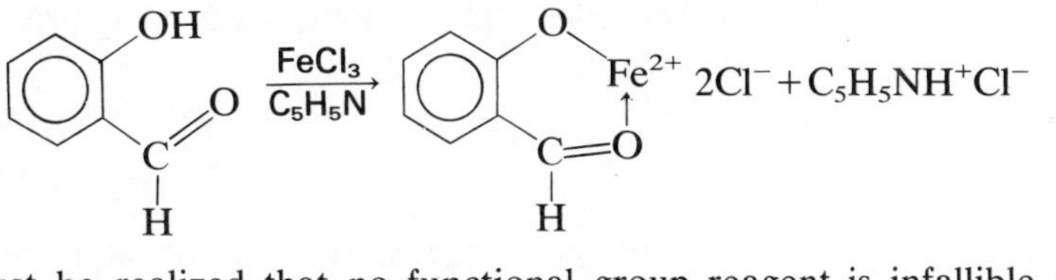

$$\text{OH, C(H)=O} \xrightarrow[C_5H_5N]{FeCl_3} \text{O–}Fe^{2+}\text{←O=C(H)}\ 2Cl^- + C_5H_5NH^+Cl^-$$

It must be realized that no functional group reagent is infallible and it is frequently necessary to use a second or third reagent. For example, *p*-hydroxybenzoic acid does not give a positive ferric chloride test for the phenolic hydroxyl group. However, bromine water (Experiment 30) readily yields 2,4,6-tribromophenol:

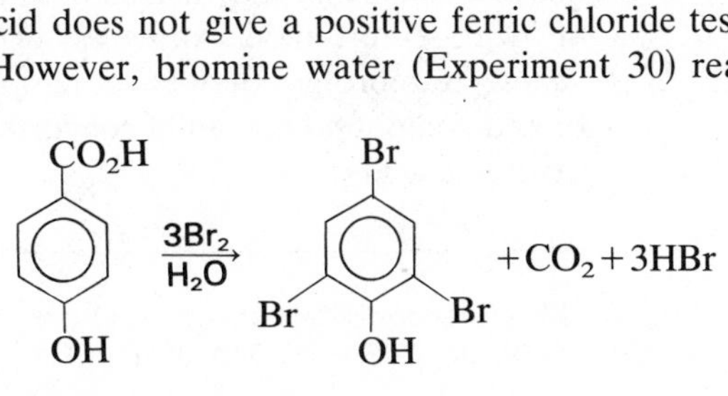

$$\text{(}CO_2H\text{, OH)} \xrightarrow[H_2O]{3Br_2} \text{(Br, Br, Br, OH)} + CO_2 + 3HBr$$

The 2,4,6-tribromophenol gives a blue color in the ferric chloride test. Also, an alkaline solution of *p*-hydroxybenzoic acid readily couples with a diazonium salt to give an orange-red dye (Experiment 19).

$$p\text{-HOC}_6\text{H}_4\text{CO}_2\text{H} \xrightarrow[H_2O]{NaOH} p\text{-Na}^+\,{}^-\text{OC}_6\text{H}_4\text{CO}_2^-\text{Na}^+ \xrightarrow{ArN_2^+Cl^-} \text{(OH)(N{=}N{-}Ar)C}_6\text{H}_3\text{CO}_2^-\text{Na}^+ + \text{NaCl}$$

In aqueous or aqueous-alcoholic solutions, some enols, oximes, and hydroxamic acids produce red, brown, or magenta colored complexes with *aqueous* ferric chloride. However, in this anhydrous chloroform test, these compounds give yellow or pale tan solutions quite different from the phenols.

Since the aromatic nucleus of a phenol is substantially more reactive toward electrophilic aromatic substitution than benzene, bromination of phenols should be carried out under mild conditions (Experiment 30).

EXPERIMENT 30. BROMINE WATER

$$\text{C}_6\text{H}_5\text{OH} + 3\text{Br}_2 \longrightarrow 2,4,6\text{-Br}_3\text{C}_6\text{H}_2\text{OH} + 3\text{HBr}$$

Prepare 1% aqueous solutions of (1) phenol, (2) aniline, (3) salicylic acid, and (4) *p*-nitrophenol; to each solution add bromine water drop by drop until the bromine color is no longer discharged.

Discussion. It has been shown that, in the bromination of benzene and *o*-nitroanisole with bromine water, the brominating agent operates by complex mechanisms.[87]

Mercaptans are converted readily by bromine water to disulfides:

$$2\text{RSH} + \text{Br}_2 \rightarrow \text{RSSR} + 2\text{HBr}$$

The advantage of bromine in water over bromine in carbon tetrachloride is that the more polar solvent greatly increases the rate of bromination by the ionic mechanism. Of course, it is impossible with this solvent to observe the evolution of hydrogen bromide. An excess of bromine water converts tribromophenol to a yellow tetrabromo derivative, 2,4,4,6-tetrabromocyclohexadienone. The tetrabromo compound is readily converted to the tribromophenol by washing with 2% hydriodic acid.

---

[87] C. K. Ingold, *Structure and Mechanism in Organic Chemistry*, 2nd ed. (Cornell University Press, Ithaca, N.Y., 1969), pp. 345–346, 349–351.

OH, Br, Br, Br (2,4,6-tribromophenol) $\underset{\text{2\% HI}}{\overset{Br_2}{\rightleftharpoons}}$ O, Br, Br, Br, Br

Questions

1. Why is tribromoaniline insoluble in dilute hydrobromic acid? Could the decolorization of the bromine water result from the presence of an inorganic compound? Give examples.
2. Is bromine hydrolyzed in water? What effect would bromine water have on a water-soluble salt of a water-insoluble acid?

For more discussion of the reaction of bromine with organic compounds, consult Experiment 14.

The ir absorptions of phenols are generally very similar to those of alcohols; both classes contain a C—O—H unit. Important absorption ranges are as follows:

Vibration	cm^{-1}	μm
O—H stretch	ca. 3610 (s, b)[a]	ca. 2.77
O—H bend	1410–1310 (m, b)	7.09–7.63
C—O stretch	ca. 1230 (s, b)	ca. 8.13

[a] s = strong (intensity), m = moderate (intensity); b = broad.

An infrared spectrum of a typical phenol is shown in Fig. 6.48.

Interpretation of nmr spectra of phenols (e.g., Fig. 6.49) involves consideration of the position of the hydroxyl proton. In addition, the aromatic proton signals are influenced by the hydroxyl substituent; this has been described earlier (Section 6.9).

Ultraviolet spectrometry (Chapter 8) of phenols is especially useful; of particular use is the shift of major maxima to longer wavelengths (bathochromic shift) upon conversion of phenols to the corresponding oxy anions; for example:

	$\lambda_{max}(\epsilon_{max})$		
	E_2 band	B Band	Solvent
OH (phenol)	210.5 (6200)	270 (1450)	Water (pH = 3)
O^- (phenolate anion)	235 (9400)	287 (2600)	Water (pH = 11)

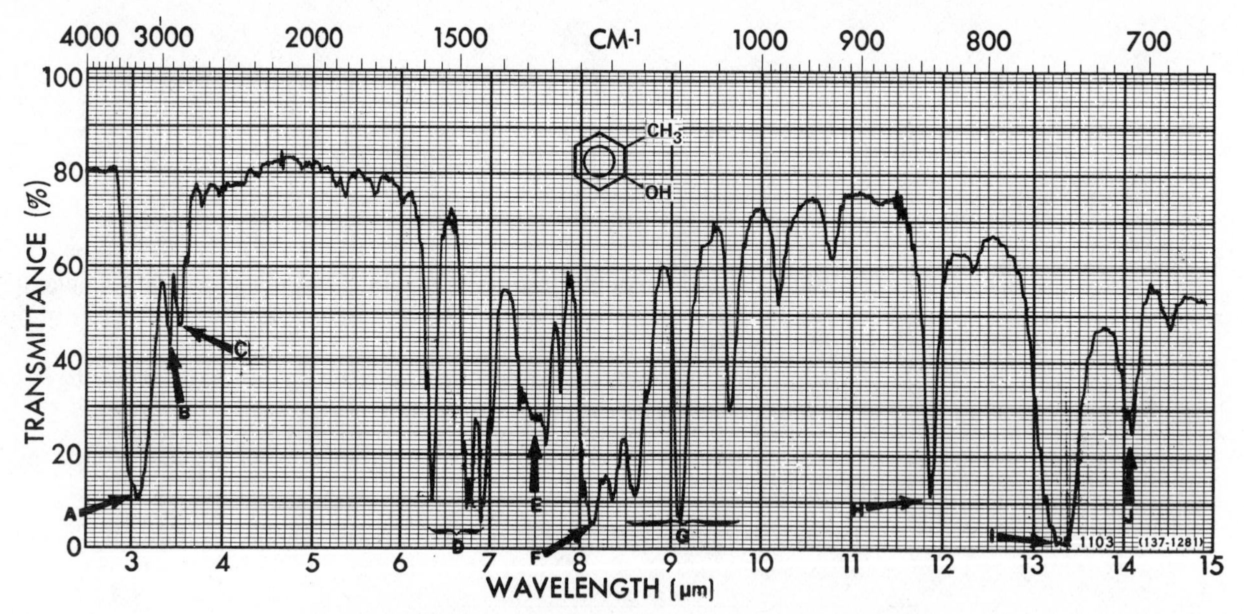

Figure 6.48. Infrared spectrum of *o*-cresol: neat. O—H stretch: **A**, 3390 cm^{-1} (2.95 μm), ν_{O-H} (associated). C—H stretch: **B**, 3021 cm^{-1} (3.31 μm), ν_{asym} CH_3; **C**, 2924 cm^{-1} (3.42 μm), ν_{sym} CH_3. C⩦C stretch: **D**, 1587, 1493, 1460 cm^{-1} (6.30, 6.70, 6.85 μm), νC⩦C. O—H bend: **E**, ca. 1333 cm^{-1} (ca. 7.5 μm), δ_{O-H} (out of plane). C—O stretch: **F**, 1235 cm^{-1} (8.10 μm), ν_{C-O} (aryl-oxygen). ⩦C—H bend (in plane): **G**, 1205, 1170, 1105, 1042 cm^{-1} (8.30, 8.55, 9.05, 9.60 μm). ⩦C—H bend (out of plane): **H**, 848 cm^{-1} (11.80 μm); **I**, 752 cm^{-1} (13.3 μm). C⩦C bend (out of plane): **J**, 714 cm^{-1} (14.0 μm). Calibration: bands on the spectrum in 3.00–4.00 μm region are all decreased by 0.10 μm upon tabulation. Bands on the spectrum in the 6.00–16.00 μm region are all decreased by 0.05 μm upon tabulation.

Because of stabilization of the molecular ion by the aromatic ring, the molecular-ion peak of phenols is normally the base peak in the mass spectrum. The following are major fragment peaks for typical phenols; these are tabulated by reference to the atoms lost in the fragmentation process:

Fragment	Mass	Fragment arising from phenol formula, mass
M − CO	M − 28	$C_5H_6^+$, *m/e* 66
M − CHO	M − 29	$C_5H_5^+$, *m/e* 65

Most phenols do not show intense M − 1 peaks; an exception is the intense M − H peak of cresols resulting from facile cleavage of the benzylic carbon–hydrogen bond.

In the presence of alkali, phenols react readily with chloroacetic acid to give aryloxyacetic acids. These derivatives crystallize well from water and have proved to be exceedingly useful in characterization work (Procedure 59):

$$ArONa + ClCH_2COOH \rightarrow ArOCH_2COOH + NaCl$$

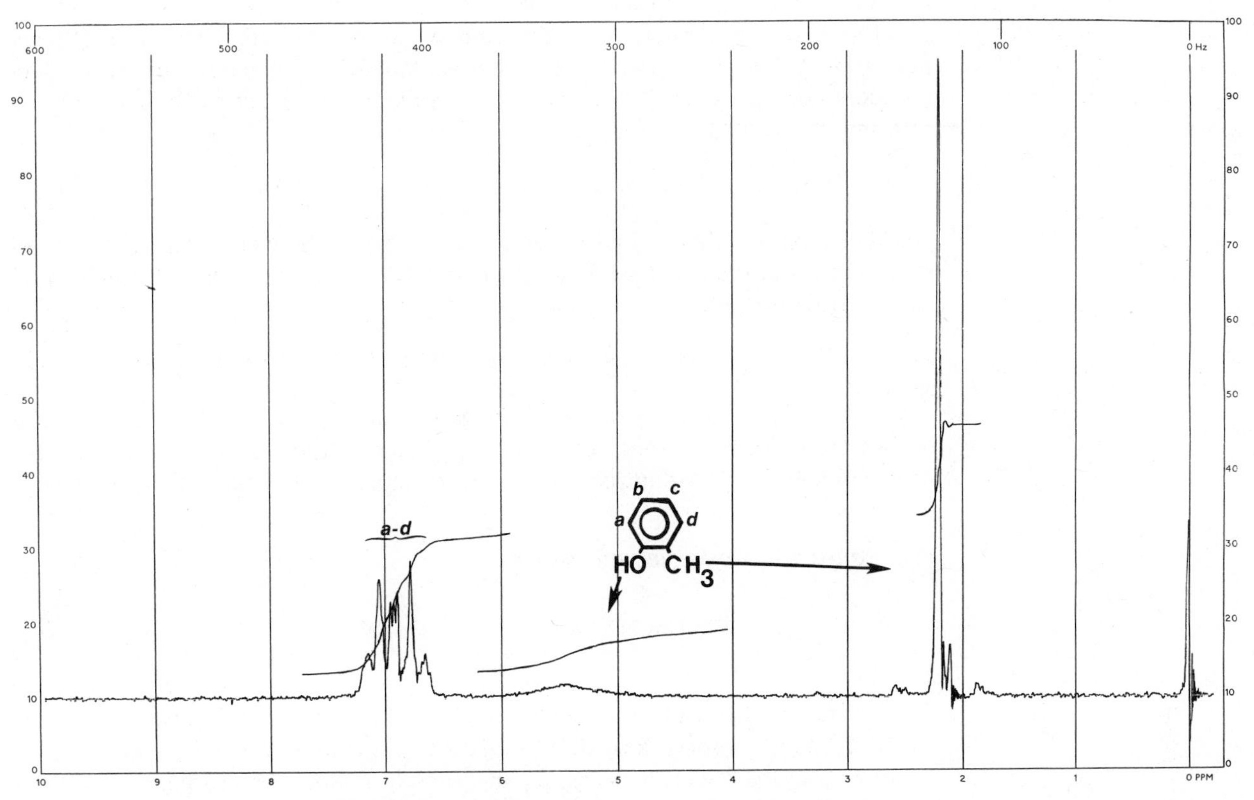

Figure 6.49. Nuclear magnetic resonance (proton) spectrum of *o*-cresol: 60 MHz, 600-Hz sweep width, $CDCl_3$ solvent.

The aryloxyacetic acids can be compared not only by melting-point determinations but also by reference to their neutralization equivalents. Accurate determination of the neutralization equivalent can lead to an estimate of the total mass of substituents on the aryl ring.

Similar to the use of bromine as a qualitative test for phenols (Experiment 30), bromine can be used to form derivatives of phenols (Procedure 60). Since the phenolic ring is reactive toward such electrophilic reagents, every proton atom in an ortho or para position is displaced by bromine; in fact, bromine often substitutes for groups other than protons. For example, bromine reacts with both phenol and *p*-hydroxybenzoic acid to form 2,4,6-tribromophenol:

Phenol (OH) $\xrightarrow{3Br_2}$ 2,4,6-tribromophenol (OH; Br, Br, Br)

p-Hydroxybenzoic acid (OH; CO_2H) $\xrightarrow{3Br_2}$ 2,4,6-tribromophenol

Phenols, like alcohols, yield urethans when treated with isocyanates (Procedure 61). Among the latter, α-naphthyl isocyanate is a generally useful reagent for identifying phenols (Procedure 1a). This reaction is catalyzed by the addition of a few drops of dry pyridine.

$$C_{10}H_7N{=}C{=}O + ArOH \rightarrow C_{10}H_7NHCOOAr$$

Another type of urethan that is often used is the diphenylurethan; it is derived from diphenylcarbamyl chloride, $(C_6H_5)_2NCOCl$, according to the following equation (Procedure 1b):

$$(C_6H_5)_2NCOCl + ArOH \rightarrow (C_6H_5)_2NCO_2Ar + HCl$$

Other types of derivatives of phenols that can be made are the nitro derivatives (Procedure 26a), picrates (Procedures 22a and 22b), acetates (Procedures 5a and 5b), 3,5-dinitrobenzoates (Procedure 2), and sulfonic acid esters.

Procedure 59. Aryloxyacetic Acids

$$ArOH \xrightarrow{\text{33\% NaOH}} ArO^-Na^+ \xrightarrow{ClCH_2CO_2H} ArOCH_2CO_2H$$

To a mixture of 1 g of the phenol with 5 ml of a 33% sodium hydroxide solution is added 1.5 g of chloroacetic acid. The mixture is shaken thoroughly, and 1 to 5 ml of water may be added if necessary in order to dissolve the sodium salt of the phenol. The test tube containing the mixture is then kept in a beaker of boiling water for 1 hr. The solution is cooled, diluted with 10 to 15 ml of water, acidified to Congo red with dilute hydrochloric acid, and extracted with 50 ml of ether. The ether solution is washed with 10 ml of cold water and is then shaken with 25 ml of 5% sodium carbonate solution. The sodium carbonate solution is acidified with dilute hydrochloric acid; the aryloxyacetic acid is then collected on a filter and recrystallized from hot water.

Procedure 60. Bromination of Phenols

$$ArOH \xrightarrow{Br_2/KBr} Ar'(Br)_xOH$$

A brominating solution is prepared by dissolving 15 g of potassium bromide in 100 ml of water and adding 10 g of bromine. This solution is added slowly, with shaking, to a solution of 1 g of the phenol dissolved in water, ethanol, acetone, or dioxane. Just enough of the brominating solution is added to impart a yellow color to the mixture. About 50 ml of water is then added, and the mixture is shaken vigorously to break up the lumps. The bromo derivative is removed by filtration and washed with a dilute solution of sodium bisulfite. It is recrystallized from ethanol or a water-ethanol mixture.

Procedure 61. Phenylurethans

$$\xrightarrow{C_5H_5N} ArO\overset{\overset{\displaystyle O}{\|}}{C}NHC_6H_5$$

To a mixture of 0.5 g of the dry phenol and 0.5 ml of phenyl isocyanate in a dry 25-ml flask is added 1 drop of dry pyridine. The flask is loosely stoppered with a plug of cotton and heated on a steam cone for 15 min. If separation of the derivative does not occur during this time, crystallization is induced by cooling the flask and scratching the walls. When crystals have formed, 10 ml of dry ethyl acetate (or dry benzene*) is added. The mixture is heated on a steam cone and filtered through a fluted filter. Hexane is now added until a turbidity or crystals are obtained. Crystallization is allowed to proceed overnight; the product is then removed by filtration, washed with hexane containing a little benzene, and air dried.

If any water is present in the original sample or reagents, the product will be contaminated with *s*-diphenylurea (m.p. 238°C). Purification may be effected by warming the product with 10 ml of carbon tetrachloride and removing the insoluble diphenylurea by filtration. The phenylurethan may be obtained by cooling the filtrate. Occasionally the filtrate may have to be evaporated to 2 to 3 ml to cause the crystallization to occur.

The nmr spectra of phenols can be simplified by the use of lanthanide shift reagents (Chapter 8). The more highly fluorinated shift reagents must be used in order to avoid neutralization of the shift reagent ligands by the moderately acidic phenols.

---

* See footnote 43, Chapter 6.

CHAPTER SEVEN

separations

7.1 INTRODUCTION

The identification of the components of a mixture involves first a separation into individual compounds and second the characterization of each of the latter according to the procedures in Chapter 6. It is very rarely possible to identify the constituents of a mixture without previous separation. The separation of the compounds in a mixture should be as nearly quantitative as possible in order to give some idea of the actual percentage of each component. It is far more important, however, to carry out the separation in such a manner that each compound is obtained in the pure state, because this renders the individual identification much easier.

The method of separation chosen should be such that the compounds are obtained as they existed in the original mixture. Derivatives of the original compounds are not very useful unless they may be reconverted readily into the original compounds. This criterion of separation is necessary because the identification of a compound rests ultimately on agreement between physical constants of the original and of a derivative with similar data obtained from the literature.

The history of a mixture will frequently furnish sufficient information to indicate the group to which the mixture belongs and hence the general mode of separation to be used.

In recent years the field of analytical separation has been extensively developed and widely applied by organic chemists. It is thus necessary to be aware of the nature of the many techniques available in order to be able to choose the technique that is most appropriate for the mixture in hand. There are a number of separation problems that frequently occur for the organic chemist. One common situation is that in which the mixture is comprised of a number of components,[1] all of reasonable purity and all of substantial proportion. Another type of separation is the isolation of a single component from large amounts of unreacted starting material or from undesired side products; these side products frequently are simply intractable tars or polymeric materials. One should try to place each new separation problem into one of these two categories in order to be able to select the most efficient separation approach.

Before selecting a separation procedure, the preliminary tests outlined below should be carried out. As these tests are performed, one should constantly be concerned with the following:

1. Will the sample survive the separation procedure? That is, are the components of the mixture stable under the conditions of the procedure?
2. Is this the easiest and most efficient way to carry out the separation?

Stability of the sample under the conditions of the separation procedure may not be known until the separation is attempted. Thermal stability is always of concern. Samples that are thermally unstable to the heat required for distillation at atmospheric pressure can be distilled at reduced pressure. Chromatographic separations (other than gas chromatography) normally do not involve heat application and thus may be very appropriate for samples that cannot be distilled; it is possible, however, that samples subjected to chromatographic separation may decompose as a result of chemical reactions with the chromatographic packing or support. A quick TLC test (see pp. 33–37) is a good check for sample durability under chromatographic conditions.

7.1.1 Preliminary Examination of Mixtures

1. The physical state is noted. If a solid is suspended in a liquid, the solid is removed by filtration and is examined separately. If two immiscible liquids are present, they also are separated and examined individually.

2. The solubility or insolubility of the mixture in water is determined.

3. With liquid mixtures, 2 ml of the solution is evaporated to dryness on a watch crystal or porcelain crucible cover and the presence or absence of a residue

---

[1] In mixtures assigned for laboratory exercises, typically two to five components are present in the mixture.

noted. The ignition test is applied to the residue. For a solid mixture the ignition test is applied directly.

4. In liquid mixtures the presence or absence of water is detected by (a) determining the miscibility of the solution with ether; (b) anhydrous copper sulfate; (c) distillation test for water.* The distillation test is the most reliable and is carried out in the following manner. Five milliliters of the liquid mixture and 5 ml of anhydrous toluene are placed in a small distilling flask. The mixture is heated gently until distillation occurs, and 2 ml of distillate is collected. About 5 to 10 ml of anhydrous toluene is added to the distillate. The presence of two layers or distinct drops suspended in the toluene indicates water. If the solution is only cloudy, traces of water are indicated.

5. If water has been found to be absent, the presence or absence of a volatile solvent in a liquid mixture is determined by placing 10 ml of the mixture in a 25-ml distilling flask. The flask is placed in a beaker of cold water that is heated to boiling. Any liquid that distils under these conditions is classified as a volatile solvent. The distillate, which may be a mixture of readily volatile compounds, and the residue in the flask are examined separately.

It frequently happens that distillation of a mixture originally water-soluble yields a volatile solvent and a water-insoluble residue. The separation of such a mixture is therefore carried out by removing all the volatile solvent. The residue is then treated as a water-insoluble mixture.

If the residue after distillation is a water-soluble liquid, it is best not to remove the solvent at this stage because the separation is usually not quantitative.

If, however, the residue after distillation is a water-soluble solid and the removal of the solvent seems to be quantitative, then it is desirable to remove all the volatile solvent and to examine the distillate and residue separately.

It is to be noted that if water is present, no such separation should be attempted.

6. The reaction of an aqueous solution or suspension of the mixture to litmus and phenolphthalein is determined. If the mixture is distinctly acid, 5 ml should be titrated with 0.1 *N* sodium hydroxide solution in order to determine whether considerable amounts of free acid are present or whether the acidity is due to traces of acids formed by hydrolysis of esters. The titration must be performed in an ice-cold solution, and the first pink color of phenolphthalein taken as the end point.

7. Two milliliters of the mixture is acidified with hydrochloric acid, and the solution is cooled. The evolution of a gas or the formation of a precipitate is noted. Dilute sodium hydroxide solution is now added, and the result is noted.

8. Two milliliters of the mixture is made distinctly alkaline with sodium hydroxide solution. The separation of an oil or solid, the liberation of ammonia, and any color changes are noted. The solution should be heated just to boiling and then cooled. The odor is now compared with that of the original mixture. The

---

* The Karl Fischer test, referenced in Appendix I, can be used to quantitatively determine the amount of water present.

presence of esters is often indicated by a change in odor. Dilute hydrochloric acid is now added, and the result is noted.

9. In the case of water-insoluble mixtures, an elemental analysis should be made. If water or a large amount of a volatile solvent is present in a water-soluble mixture, the elemental analysis of the mixture is omitted. If the water-soluble mixture is composed of solids, an elemental analysis is made.

10. If water is absent, the effect of the following classification reagents is cautiously determined: (a) metallic sodium; (b) acetyl chloride.

11. The action of the following classification reagents should be determined on an aqueous solution or suspension of the original mixture: (a) bromine water; (b) potassium permanganate solution; (c) ferric chloride solution; (d) alcoholic silver nitrate solution; (e) fuchsin-aldehyde reagent; (f) phenylhydrazine.

At this stage of the examination the results of the foregoing tests are summarized and as much information as possible is deduced from the behavior of the mixture. The preliminary study will show the group in which the mixture should be classified and will, therefore, indicate which of the following procedures should be used in its separation.

REPORT FORM

Preliminary Examination of Mixtures

Student Name: Date:
Mixture No.:
Components:

1. (a) Physical state(s):
   (b) Mechanical separation procedure(s):

2. Water solubility:
3. (a) Dryness test:
   (b) Ignition test:
4. Test(s) for presence of water:

5. Test for volatile solvent(s):

6. Tests for acid/base character:

7. Tests for gaseous components liberated by acid or by base:

8. Test for involatile components liberated by acid or by base:

9. Elemental analysis on mixture: Water-insoluble?
   F Cl Br I
   N S Metals
10. Classification tests: Water present?
    (a) Metallic sodium: (b) Acetyl chloride:

11. Classification tests:
    (a) Bromine water: (b) Potassium permanganate:

    (c) Ferric chloride: (d) Alcoholic silver nitrate:

    (e) Fuchsin-aldehyde reagent: (f) Phenylhydrazine:

12. Recommendations for separation technique:

Distillation: Crystallization: TLC:
G.C: Column chromatography: HPLC:

Comments and references:

7.2 DISTILLATION AND SUBLIMATION

7.2.1 Distillation

An introduction to simple distillation has been given on pp. 46–47 (in conjunction with boiling-point determination) and on pp. 48–52 (Chapter 3); in these earlier treatments we were concerned with simple sample purification and with b.p. determination. Since distillation is a technique of ancient origin, it is a technique for which there has been extensive development. We shall now consider more sophisticated distillation techniques.

Figure 7.1 shows an apparatus useful for a (small-scale) short-path distillation. Although the evaporation-condensation process is inefficient, such a short-path process allows distillation of materials (e.g., low-melting solids) for which long exposure to elevated temperature could be hazardous.

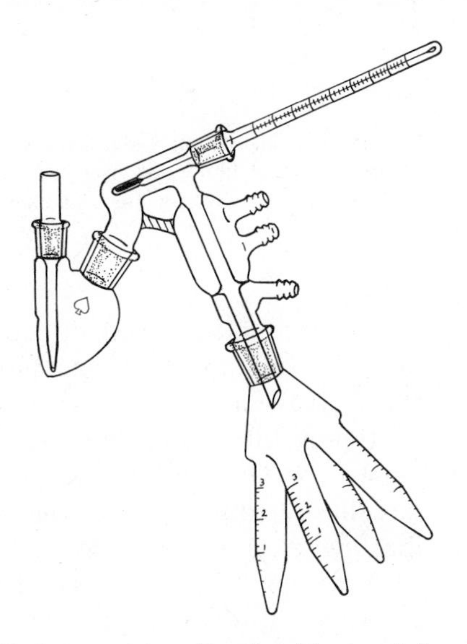

Figure 7.1. Short-path distillation apparatus. (Reprinted by permission of Ace Glass, Inc., Vineland, N.J.)

In order to improve the efficiency of a distillation, a column can be placed between the vessel to be heated and the condenser tube. This is illustrated in Fig. 7.2, which shows columns for which the internal surface area has been increased (e.g., Vigreux columns) and columns that are actually modified condensers and provide increased surface area and/or increased cooling surfaces. A complete apparatus utilizing an intermediate column is shown below (Fig. 7.6). The spinning band apparatus (see Fig. 7.3) allows a very efficient distillation because of the large number of theoretical plates[2] provided for the distillate.

[2] Other books should be consulted for the theory of distillation (see Chapter 10).

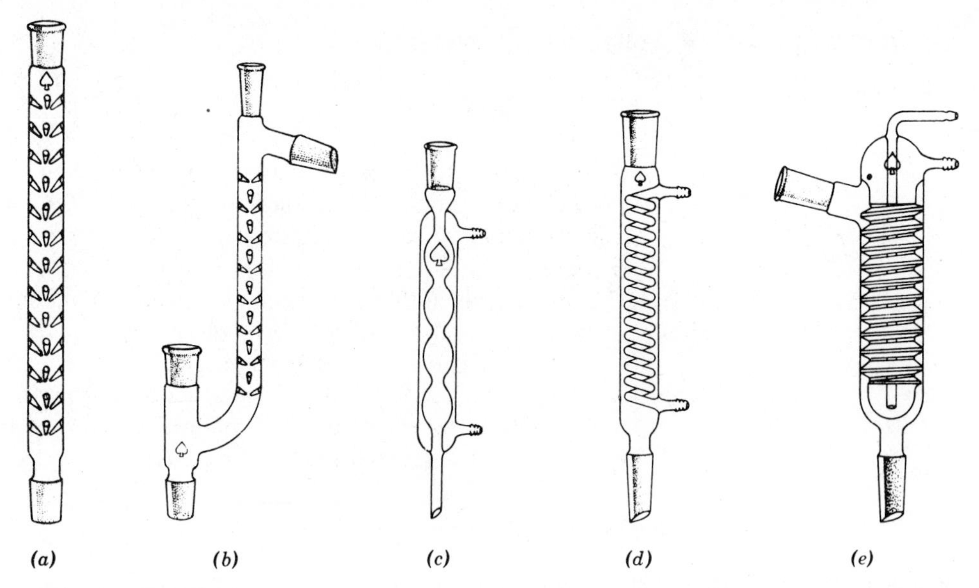

Figure 7.2. Distillation Columns and condensers; (*a*) Vigreux column, (*b*) Vigreux column with adapters, (*c*) Allihn condenser, (*d*) coiled condenser, (*e*) Friederichs condenser. (Reprinted by Permission of Ace Glass, Inc., Vineland, N.J.)

The frequent result of the use of more efficient distillation apparatus is a longer "hang-up" time for materials on the columns of the given apparatus. In order to avoid heat loss, the column should be externally insulated with asbestos[2a] or (Pyrex) glass wool. The user also must be prepared to vary the amount of heat applied in different places during the distillation. Two commonly used techniques are illustrated in Fig. 7.4. The use of a heating tape (Fig. 7.4*a*) provides uniform heating; electrical power to the tape must be regulated by, e.g., a Powerstat (Fig. 7.5). A heat gun (Fig. 7.4*b*) can be used to drive remarkably involatile materials through a distillation apparatus. This heat gun is analogous in appearance and use to a hand-held hair dryer: One merely points the heat at places in the apparatus that need a small temperature increase to drive over the distillate.

In order to distill liquids and solids of low volatility that might be somewhat heat-sensitive, a vacuum distillation apparatus (Fig. 7.6) is normally used. Use of a laboratory aspirator (not shown) can provide a working vacuum of ca. 15 Torr (mm Hg). A good vacuum pump, modified by a bleed apparatus (e.g., the gas inlet regulator of a bunsen burner), can provide vacuums in the range 0.01 to 15 Torr. The apparatus in Fig. 7.6 shows a useful inlet device (b) which, when fitted with, for example, a clamped rubber tube, allows regulation of admitted air; this air provides a bubbling action that should prevent bumping during the distillation process. Note also the other features described in the legend of Fig. 7.6.

[2a] Asbestos presents a toxicity hazard and should be avoided.

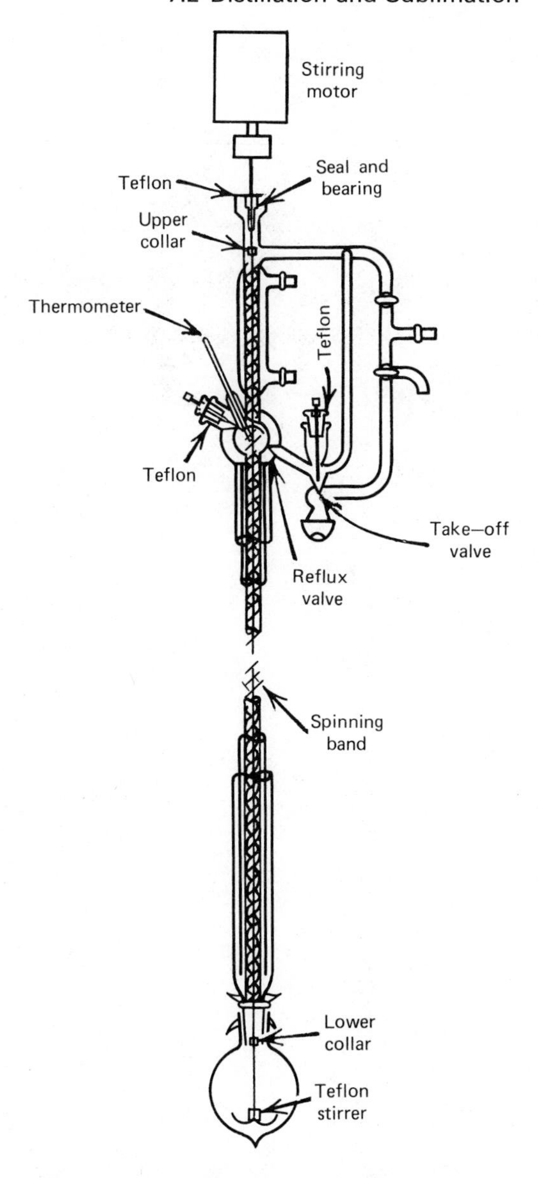

Figure 7.3. Spinning band distillation apparatus. (Source: R. B. Bates and J. P. Schaefer, *Research Techniques in Organic Chemistry*, Copyright © 1971, p. 50. Reprinted by permission of Prentice-Hall, Inc., Englewood Cliffs, N.J.)

An important aspect in distillations is the method of heating the pot (for example, application of heat at point a in Fig. 7.6). A steam bath nicely provides moderate heat (up to the vicinity of the b.p. of water), but this heat source is often an uneven supplier. Baths containing oils or other involatile, inert substances can be used; the application of heat by such hot liquids is a very even method of heat application and can be used to higher temperatures (ca. 250 to 300°C). Where

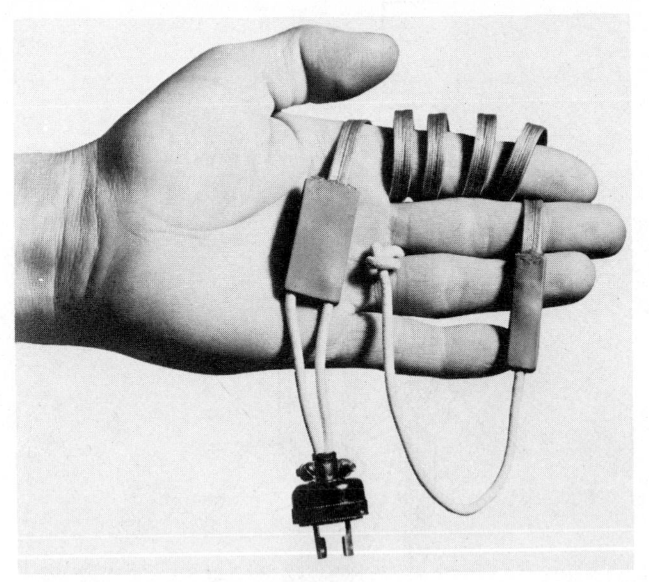

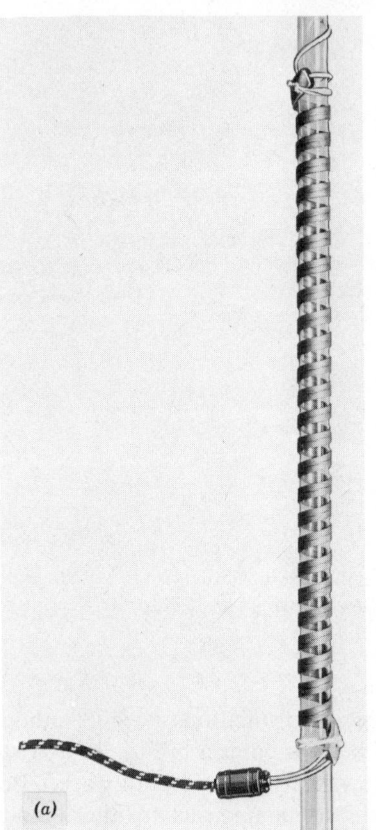

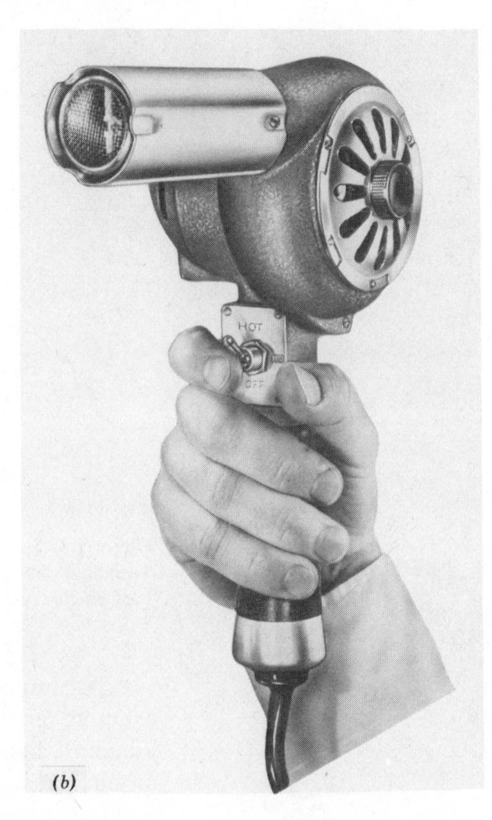

Figure 7.4. (*a*) Column heating tape (Courtesy of Glas-Col Apparatus Co., Terre Haute, Ind.). (*b*) Heat gun. (Courtesy of Master Appliance Corp, Racine, Wisc.).

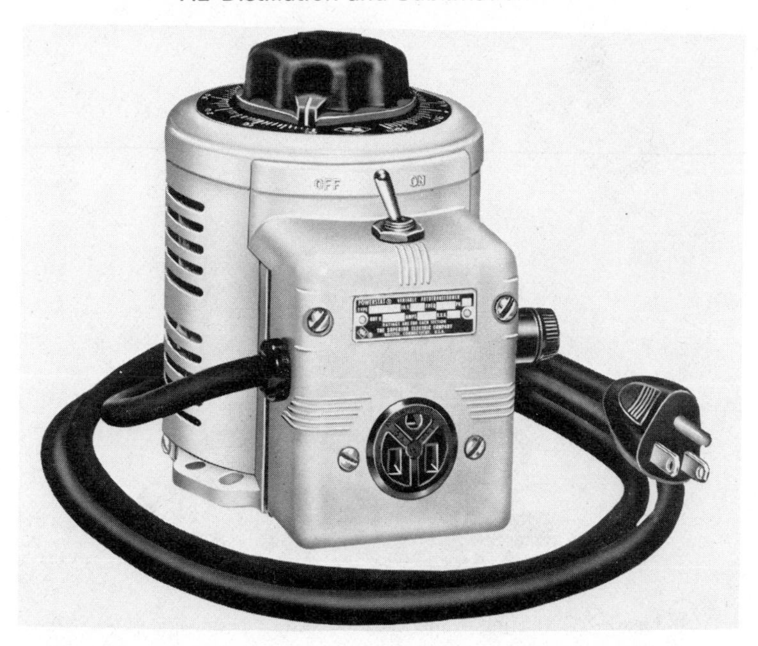

Figure 7.5. Powerstat (variable voltage transformer). (Superior Electric Co., Bristol, Conn.)

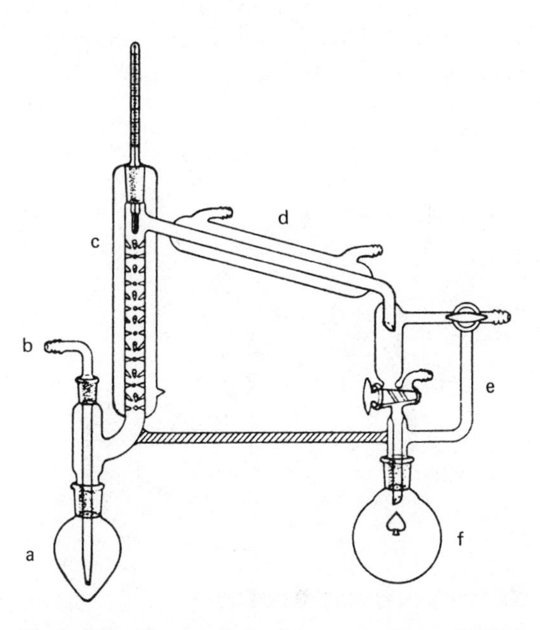

Figure 7.6. Vacuum distillation apparatus: *a*, distillation pot (temperature should be measured and controlled here), see the text discussion of Fig. 7.1 as to how the pot can be modified to include magnetic stirring (this removes the necessity for an air or gas inlet system); *b*, air inlet system; *c*, Vigreux column; *d*, water-jacketed condenser; *e*, system for separation and regulation of pressure in the receiver; *f*, receiver. (Reprinted by permission of Ace Glass, Inc., Vineland, N.J.)

Figure 7.7. Heating mantle (Reproduced by permission of Ace Glass, Inc., Vineland, N.J.)

available, heating mantles (Fig. 7.7) can be used for heat application; use of such mantles allows one to avoid the use of messy oils for external heating. Power to the heating mantle must be controlled by, e.g., a Powerstat (Fig. 7.5).

As a sidelight to distillation, it should be pointed out that the organic chemist is often concerned simply with the removal of volatile solvent from, for example,

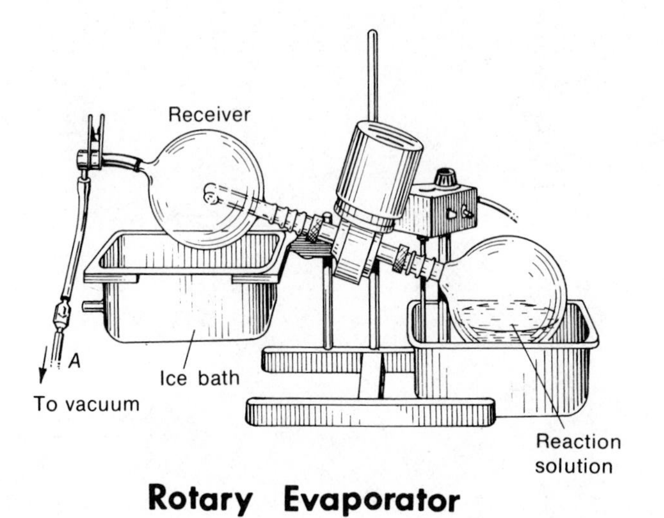

Figure 7.8. (*a*) Rotary evaporator. (Source: J. Landgrebe, *Theory and Practice* in the Organic Laboratory, 2nd. ed. D. C. Heath, Lexington, Mass., © 1977. Reprinted with permission.)

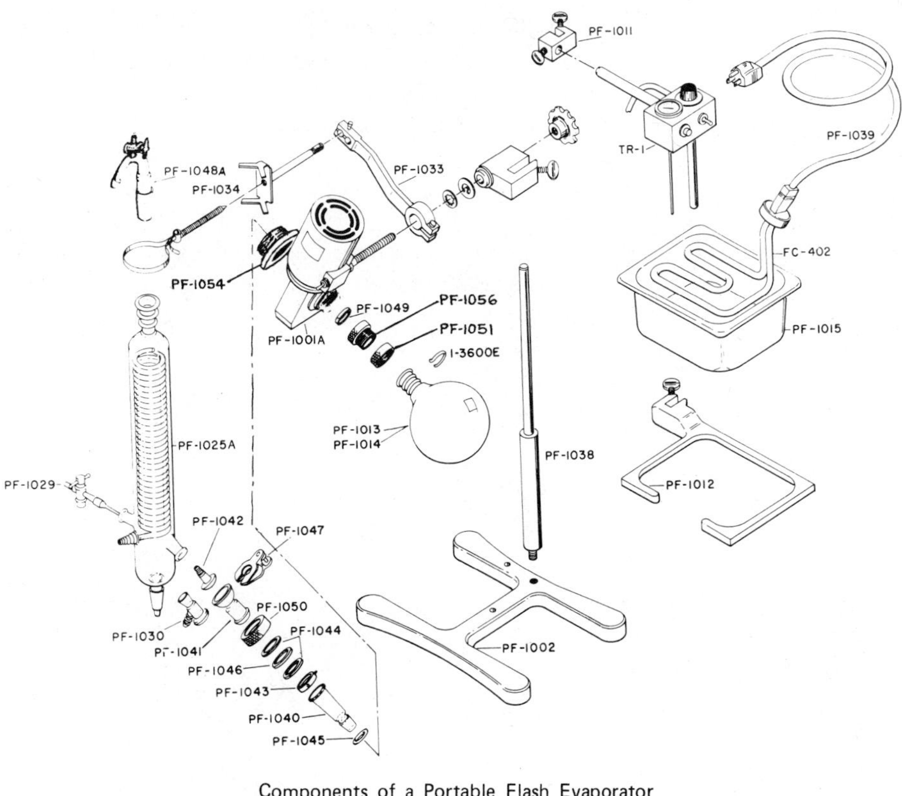

Components of a Portable Flash Evaporator

(*b*)

Figure 7.8. (*Continued*) (*b*) Components of a portable flash evaporator.

a solution resulting from the work-up of a reaction. Here a rotary solvent evaporator (solvent "stripper") is useful (see Fig. 7.8*a*). In order to keep such equipment running efficiently, the O-rings (part PF-1049, Fig. 7.8*b*) should be periodically checked and replaced.

A class of distillation apparatus that is quite popular is illustrated by the Kugelrohr apparatus of Fig. 7.9. Less sophisticated modifications (e.g., without

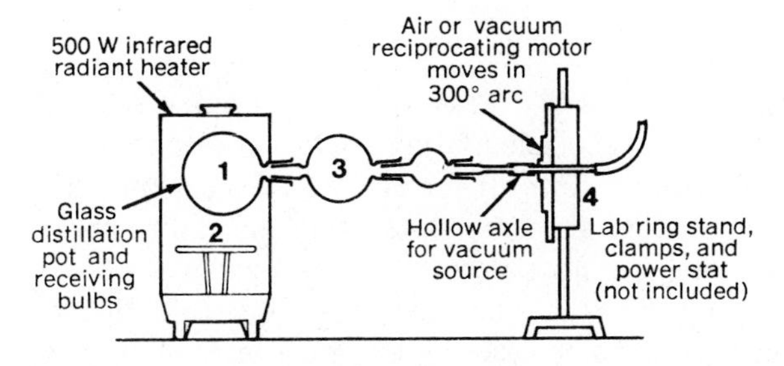

Figure 7.9. Kugelrohr distillation apparatus.

the rotational accessory) can be purchased. A great advantage of this type of apparatus is its ability to apply good vacuums (0.1 Torr), especially when the glassware is composed of only one piece, rather than of a number of fitted pieces with many possibilities for leaks.

7.2.2 Sublimation

Sublimation can be utilized for purification employing the apparatus shown in Fig. 7.10. The inner chamber is a cold finger that may be lifted out of the apparatus from the top. This cold finger is charged (on the inside) with a coolant, usually

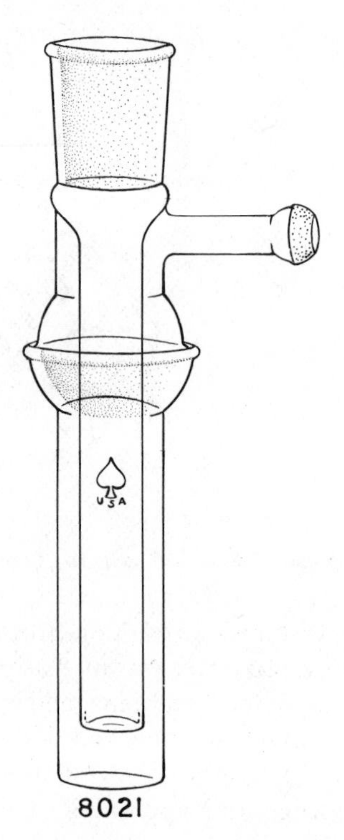

Figure 7.10. Sublimation apparatus. Reprinted by permission of Ace Glass, Inc., Vineland, N.J.

ice, Dry Ice, or Dry Ice-acetone. The material to be sublimed is placed within the outer unit of the sublimation apparatus on the bottom. Heat is applied externally, usually in the form of an oil bath. Successful sublimation of material from the crude mixture at the bottom will result in formation of solid on the cold finger. It may well be necessary to interrupt the sublimation periodically and scrape collected solid off the surface of the cold finger. The entire process can be enhanced in some cases by pumping the system down to reduced pressure via the side arms shown in Fig. 7.10.

7.2.3 Steam Distillation

Steam distillation is a technique whereby a compound of relatively low volatility can be purified by co-distilling it with water. This distillation occurs because both of the liquid components contribute to the vapor pressure and thus the distillation can be carried out at temperatures slightly less than 100°C (1 atm). The distillation is actually carried out by simply forcing steam through a vessel containing the mixture and collecting the distillate with a water-cooled condenser. Procedures and diagrams are found in introductory organic laboratory manuals.[3]

A difference in polarity sufficient to permit separation by steam distillation is generally provided by a second functional group in the molecule. Thus, monohydroxy alcohols can be separated from dihydroxy and polyhydroxy alcohols by this scheme. Similarly, simple acids, amines, and many other volatile compounds can be separated from the corresponding di- and polyfunctional compounds. Moreover, the additional group or groups need not be the same as the original. Amino acids, hydroxy acids, nitro acids, keto acids, keto alcohols, and cyano ketones are rarely volatile with steam. In fact, it is a general rule that the presence in a molecule of two or more functional (polar) groups will render a compound nonvolatile. Table 7.1 shows which types of compounds are volatile with steam and which are not. Acetic acid and oxalic acid, ethyl alcohol and ethylene glycol, benzoic acid and phthalic acid are mixtures that illustrate the point. In each pair the first named can be removed by steam distillation whereas the other remains behind.

A very interesting group of exceptions to the multiple-function rule is found in the aromatic series. *o*-Nitrophenol, salicylaldehyde, and many other ortho-disubstituted benzene derivatives are volatile with steam. The explanation for this apparently anomalous loss of polar character is found in the observation that *all these exceptional compounds are capable of intramolecularly hydrogen-bonded forms.* These forms tend not to associate with the water and are thus relatively volatile. The hydrogen-bonded structures of *o*-nitrophenol and salicylaldehyde are shown below.

Another valuable use of steam distillation is the separation of reaction products from solvents such as *N,N*-dimethylformamide (DMF) and dimethyl sulfoxide (DMSO). DMF and DMSO are good solvents for carrying out many

---

[3] For example, see J. Landgrebe, *Theory and Practice in the Organic Laboratory,* 2nd ed. (D. C. Heath, Lexington, Mass., 1977), pp. 85–87. Steam can also be generated within the vessel.

Table 7.1. Solubility and Steam Distillation

Solubility	Types of compounds	Volatility	Volatility with steam
Soluble in water and ether	Low-molecular-weight alcohols, aldehydes, ketones, acids, esters, amines, nitriles, acid chlorides	Readily distil. Many compounds boil below 100°C	Volatile with steam
Soluble in water but insoluble in ether	Polyhydroxy alcohols, diamines, carbohydrates, amine salts, metal salts, polybasic acids; hydroxyaldehydes, -ketones, and -acids; amino acids	Low volatility. With certain exceptions these compounds cannot be distilled at atmospheric pressure	Not volatile with steam
Insoluble in water but soluble in NaOH and $NaHCO_3$	High-molecular-weight acids; negatively substituted phenols	Low volatility	Usually not volatile, but there are some exceptions
Insoluble in water and $NaHCO_3$ but soluble in NaOH	Phenols, sulfonamides of primary amines, primary and secondary nitro compounds; imides, thiophenols	High boiling points; many cannot be distilled	Usually not volatile
Insoluble in water but soluble in dilute HCl	Amines containing not more than one aryl group attached to nitrogen; hydrazines	High boiling points	Many are volatile with steam
Insoluble in water, dilute NaOH, and HCl, but contain elements other than carbon, hydrogen, oxygen, and the halogens	Nitro compounds (tert), amides, negatively substituted amines; sulfonamides of secondary amines; azo and azoxy compounds; alkyl or aryl cyanides, nitrites, nitrates, sulfates, phosphates	High boiling points; many cannot be distilled	Some are volatile with steam
Insoluble in water, dilute NaOH, and HCl, but soluble in H_2SO_4	Alcohols, aldehydes, ketones, esters, unsaturated compounds	High-boiling compounds	Usually volatile with steam
Insoluble in water, dilute NaOH, dilute HCl, and H_2SO_4	Aromatic and aliphatic hydrocarbons and their halogen derivatives	Volatile	Volatile with steam

reactions, but their very high boiling points (as well as other properties) make their removal from the reaction mixture a very difficult process when conducted by other procedures. Neither DMSO nor DMF is volatile in a steam distillation. Thus in many cases one may merely dilute a reaction mixture with water and remove the products or the unreacted starting materials by steam distillation.

7.3 RECRYSTALLIZATION

Recrystallization depends on the decreased solubility of a solid in a solvent, or mixture of solvents, at lower temperature. Thus, one should be familiar with the theory of solubility (pp. 95–99) in order to understand better the theory of recrystallization.

In the simplest cases, recrystallization is accomplished by dissolving a solid, oil, or semisolid material in a solvent; the procedure is often most effective when heating is necessary to dissolve the material completely. The warm solution is then allowed to cool slowly down to room temperature (or lower). In the "ideal" case, uniform crystals slowly appear; frequently, an overnight wait or longer is required for complete recrystallization. After the crystals have been isolated (by filtering and air drying on, e.g., a Büchner funnel), they are checked for purity (m.p., gc, nmr or tlc, etc.).

Actual recrystallization procedures can be quite complex; standard laboratory manuals for organic chemistry normally have exercises to introduce these procedures. Impure acetanilide is often recrystallized as a standard laboratory exercise. In this section we shall deal with some principles that may be useful in solving recrystallization problems in the more unpredictable situations involved in organic qualitative analysis.

The first consideration in planning a recrystallization is the choice of solvent. A rule of thumb suggests that the solid sample should be five times as soluble in the hot solvent as in the cold solvent. It may be very convenient to test the solubility by the use of gas chromatography described in Section 7.4.5. Table 7.2 lists a variety of recrystallization solvents. Table 7.3 lists solvent mixtures that can be employed. Even if the precise solute-solvent pair cannot be found on Table 7.2 or 7.3, the tables should at least give an idea as to what general class of solute-solvent pairs are appropriate. For example, if a phenylurethane derivative does not recrystallize from petroleum ether, then petroleum ether/benzene might work. Or, if you have a *m*-nitrobenzyl ester, the characterization of which is not published, you may very likely find that it recrystallizes from methanol-water (why?).

A common technique for crystallization from a solvent pair involves dissolving the sample with warming in the solvent in which the sample is more soluble, then slowly adding the second solvent until the point at which it is expected that cooling the solution should cause crystallization of the sample. When addition of the second solvent results in a clouding of the solution that disappears *only slowly* with stirring, the proper solvent combination may be in hand. Any approach will normally involve a number of trial-and-error sequences.

All too frequently recrystallization attempts will result in oil formation rather than the desired solids. The formation of an oil may very possibly be due to the fact that the sample is impure. If there is reason to believe this, the sample may be dissolved in a solvent in which it is readily soluble—for example, ether—and this solution can be treated with decolorizing carbon, followed by drying agent (e.g., magnesium sulfate). The solvent is removed (rotary evaporator, Fig. 7.8*a*) and recrystallization attempts are repeated, if necessary.

Table 7.2. Common Solvents for Recrystallization of Standard Functional Classes

	Sample to be recrystallized	Solvent[a]	Solvent b.p. (°C)	Co-solvent possibilities
1.	Acid anhydrides	Carbon tetrachloride	76.5	Ether, benzene, hydrocarbons
2.	Acid chlorides	Carbon tetrachloride	76.5	see line 1
3.	Acid chlorides	Chloroform	61.7	Hydrocarbons
4.	Amides	Acetic acid	118	Water
5.	Amides	Dioxane	102	Water, benzene,[b] hydrocarbons
6.	Amides	Water	100	Acetone, alcohols, dioxane, acetonitrile
7.	Aromatics	Benzene[b]	80	Ether, ethyl acetate, hydrocarbons
8.	Bromo compounds	Acetone	56	Water, ether, hydrocarbons
9.	Bromo compounds	Ethyl alcohol	78	Water, hydrocarbons, ethyl acetate
10.	Carboxylic acids	Acetic acid	118	See line 4
11.	Carboxylic acids	Water	100	See line 6
12.	Complexes	Benzene[b]	80	See line 7
13.	Esters	Ethyl acetate	77	Ether, hydrocarbons, benzene[b]
14.	Esters	Ethyl alcohol	78	See line 9
15.	General	Acetone	56	See line 8
16.	General	Chloroform	61.7	See line 3; also ethyl alcohol
17.	General	Ethyl acetate	77	See line 13
18.	General	(Ethyl) ether	34.5	Acetone, hydrocarbons, ethyl acetate, benzene, carbon tetrachloride
19.	General	Methylene chloride	40	Ethyl alcohol, hydrocarbons
20.	General	Ethyl alcohol	78	See line 9; bromo compounds
21.	Hydrocarbons	Benzene[b]	80	See line 7
22.	Hydrocarbons	*n*-Hexane	69	Any but acetronitrile, acetic acid, water
23.	Low-melting compounds	Ether	34.5	See line 18
23.	Low-melting compounds	Methylene chloride	40	See line 19
25.	Nitro compounds	Acetone	56	See line 8
26.	Nitro compounds	Ethyl alcohol	78	See line 9
27.	Nonpolar compounds	Carbon tetrachloride	76.5	See line 1
28.	Osazones	Acetone	56	See line 8
29.	Polar compounds	Acetonitrile	81.6	Water, ether, benzene[b]
30.	Salts	Acetic acid	118	See line 4
31.	Salts	Water	100	See line 6
32.	Sugars	Methyl cellosolve		Water, benzene,[b] ether

[a] More details on these and other solvents, especially with regard to solvent toxicity, flammability, and practical handling comments, may be found in A. J. Gordon and R. A. Ford, *The Chemists' Companion* (Wiley, New York, 1972), pp. 442–443. **Caution: Remember that many of these solvents, of which benzene is an important example, have very significant toxicity characteristics. Consult Appendix IV on this subject.**

[b] The toxicity of benzene should be kept in mind (see footnote 43, p. 235 and Appendix IV).

Oils may persist, even after repeated purifications. This may be due to the fact that the sample is inherently difficult to crystallize or to the fact that last traces of impurity must be removed by recrystallization. The following techniques may be tried:

1. During a recrystallization, add a small *seed crystal* of the pure sample desired. This should be added after the solution has been supersaturated;

Table 7.3. Solvents and Solvent Pairs for Recrystallization of Common Derivatives

Derivative	Solvent or solvent system[a]
Acetates	Methanol; ethanol
Amides	Methanol; ethanol
Anilides	Methanol/water; ethanol
Benzoates	Methanol; ethanol
Benzyl esters	Methanol/water; ethanol
Bromo compounds	Acetone/Alcohol; methanol; ethanol
3,5-Dinitrobenzoates	Methanol; ethanol
3,5-Dinitrophenylurethans	Petroleum ether/benzene
Esters	Ethyl acetate; methanol; ethanol
Hydrazones	Methanol/water; ethanol
α-Naphthylurethans	Petroleum ether
p-Nitrobenzyl esters	Methanol/water; ethanol
Nitro compounds	Methanol; ethanol; acetone/alcohol
p-Nitrophenylurethans	Petroleum ether/benzene[b]
Osazones	Acetone/alcohol
Phenylurethans	Petroleum ether
Picrates	Benzene;[b] ethanol; methanol/Water
Quaternary ammonium salts	Ethyl acetate; isopropyl ether
Semicarbazones	Ethanol; methanol/water
Sulfonamides	Methanol/water; ethanol
Sulfonyl chlorides	Chloroform; carbon tetrachloride
p-Toluidides	Methanol; ethanol
Xanthylamides	Dioxane/water

[a] Specific information about these solvents can be found in Table 7.1, and in the reference cited on that table. Refer to the toxicty footnote on Table 7.2.

[b] The toxicity of benzene should be kept in mind (see footnote 43, p. 235 and Appendix IV).

for example, add a seed crystal to a solution that has been allowed to cool to room temperature but has as yet not produced crystals.

2. A site for crystal nucleation can initiate the crystallization process. A glass rod, a scratched surface on the inside of the flask, a wooden stick, or a boiling stone may provide the surface necessary to initiate solid formation.

3. Lower temperatures may be necessary. The purpose is to decrease the solubility of the sample, *not* to freeze the solvent or sample. An increase in the solvent volume by 20 to 30% can be used effectively to require a lower temperature for crystal formation. The following conditions produce increasingly lower temperatures (see Appendix I for more details):

Room temperature
Refrigeration
Ice water bath
Refrigerator freezer or salted ice bath
Dry Ice-acetone
Liquid nitrogen

Be careful not to confuse frozen solvent or frozen amorphous oils with crystals; frozen oils will melt and form oils at room temperature.

Sometimes it is necessary to mash the neat oil sample, for example, with a stirring rod, to induce crystallization. This mashing can also be done in contact with the mother liquor. The other techniques mentioned above, such as seeding, can be used with this. In any case, it may take a very long period of mashing and grinding to induce crystallization. A detailed flow procedure for recrystallizing oils has been described.[4]

7.4 CHROMATOGRAPHY

7.4.1 Introduction and General Comments

Chromatography may be defined[5] as the science of separation techniques involving a mobile phase (e.g., the solvent in column chromatography) passing by a stationary phase (e.g., the alumina in column chromatography). The ability to separate various components of a mixture of organic compounds is based on selective and preferential absorption of these components in the mobile phase by the stationary phase.

Organic chemists are interested in two major classes of chromatography: gas chromatography (gc) and liquid chromatography (lc). Gas chromatography is useful for relatively volatile and thermally stable organic compounds; this method involves a gaseous mobile phase (helium, or, less frequently, nitrogen) and a liquid stationary phase. The stationary phase is usually spread evenly over a solid support, such as crushed brick. Gas chromatography has already been introduced in this text (Chapter 3, pp. 61–67).

Liquid chromatography (lc) involves a liquid mobile phase (usually common organic solvents, see p. 34) and either solid stationary phases (e.g., the alumina or silica gel of column and thin-layer chromatography) or stationary phases of liquids spread over solids (e.g., as used in high-pressure liquid chromatography). Thin-layer chromatography (an example of liquid chromatography) was introduced earlier in this text (pp. 33–37).

Most organic chemists are concerned with subdividing a chromatographic method into analytical and preparative classes. Techniques developed on an analytical scale usually involve handling quite small amounts of material; care must be exercised in extrapolating the methods to the different equipment and larger samples used in preparative-scale work. Preparative separations require working samples large enough that a number of chemical and spectral analyses and chemical reactions can be carried out.

In choosing between gc and lc, the following facts should be recalled

[4] J. S. Swinehart, *Organic Chemistry, An Experimental Approach* (Appleton-Century-Crofts, New York, 1969), pp. 81–84.

[5] The linguistic origin of the word chromatography is based on color (Greek; chromatismos, meaning color); early chromatography was carried out on paper using colored derivatives of naturally occurring compounds.

before making a choice:

Gas chromatography:

1. The sample should be at least moderately volatile and reasonably stable to heat. Specifically, the compound must be stable enough to survive the conditions necessary to convert it to the gas phase.
2. Simple gc instruments are inexpensive, easy to operate, and usually give results rapidly; instruments are now available for very modest prices.

Liquid chromatography:

1. Procedures are usually time-consuming, especially for classical gravity-flow conditions and preparative-scale work; rapid analyses can, however, be carried out with dry column chromatography (section 7.4.3, pp. 380–389) and high-performance (high-speed) liquid chromatography (hplc, Section 7.4.4, pp. 389–395).
2. Slow procedures mean that careful consideration must be made of the possible chemical reactivity of the sample of the column (for example, a sample of alcohol + $Al_2O_3 \rightarrow$?). Proper choice of conditions, however, allows virtually any organic compound (other than, for example, salts) to be analyzed by lc.
3. High-performanance liquid chromatography is somewhat more expensive, in terms of initial cost, because it is a newer technique than gc and because of the high-quality pumps and column packings that are necessary.

The chromatography possibilities discussed above are routinely used by most organic chemists. Purity checks on gc and tlc are routine (see Chapter 3). Even tlc can be done on a scale large enough to provide preparative-scale samples.

7.4.2 Liquid Chromatography

Column Chromatography. Column chromatography is directly applicable to preparative-scale separations and purifications because, in principle at least, one can simply choose the size of the column and its contents to fit the dimensions of the sample to be fractionated. It should be recalled, however, that this approach may very well involve a time commitment of several hours or even days.[6] *It is imperative that a knowledge of the tlc* (pp. 33–37) *characteristics of a sample be known before column chromatography is employed.*

A number of commercial setups are available (Fig. 7.11). One can, however, assemble a simple setup (for ca. 5-g samples) from miscellaneous items available in any chemistry laboratory.

---

[6] These difficulties can often be overcome by using dry column chromatography (p. 380) or hplc (Section 7.4.3, p. 389).

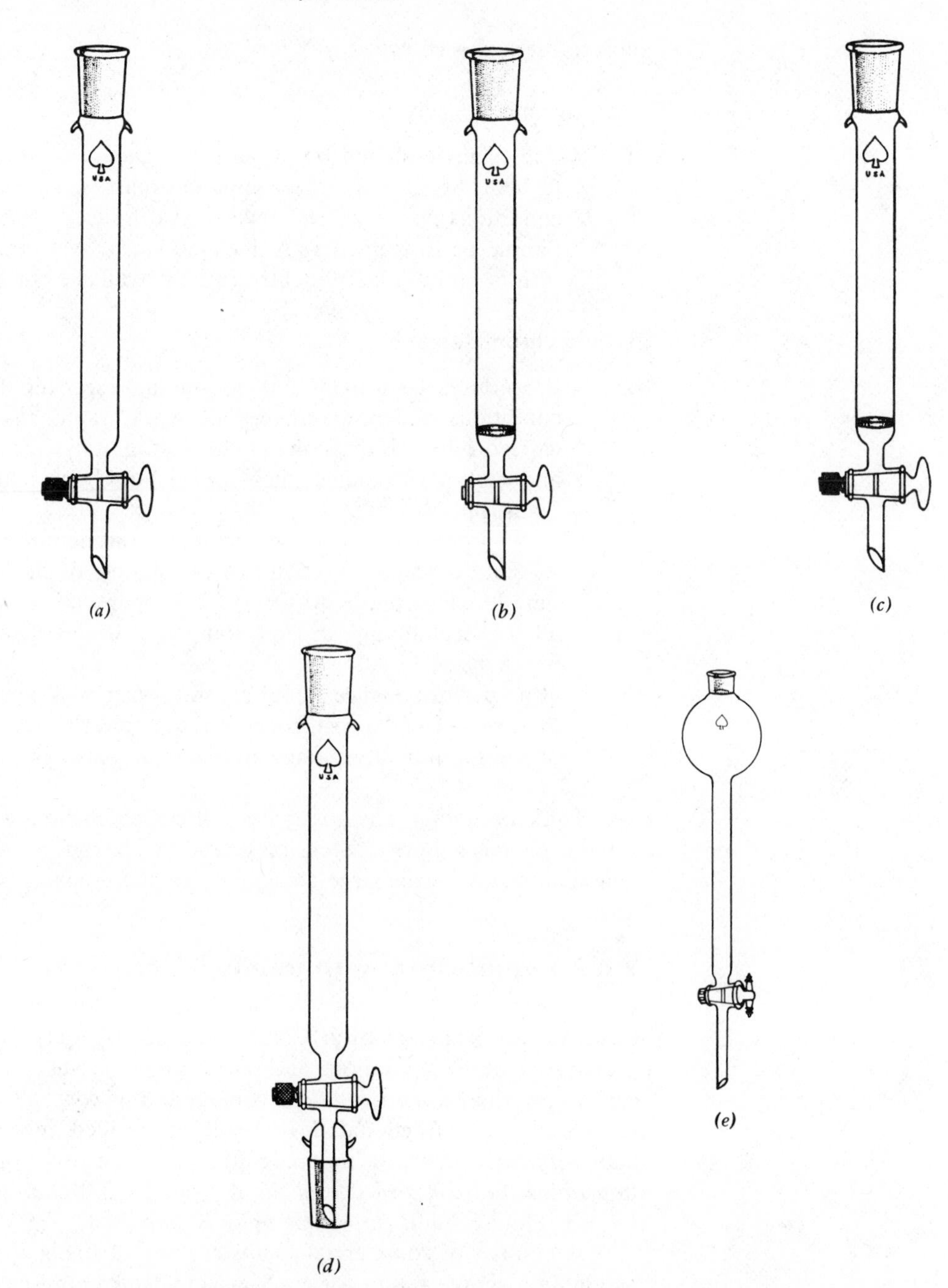

Figure 7.11. Chromatography Columns: (*a*) Ground glass opening at top. (*b*) Ground glass top. A safety clamp should also be used to keep the stopcock from slipping out. This column is fitted with a fritted disk near the bottom that allows passage of solvent, but prevents passage of column packing. (*c*) This has the characteristics of (*b*) plus a built-on safety holder for the stopcock. (*d*) This has the characteristics of (*c*) plus a ground-glass joint at the bottom to adapt to a receiving flask. (*e*) A built-in reservoir has been included as part of the column.

Column Construction. Fit a 2-cm-diameter piece of glass tubing (ca. 45 cm long) with a glass wool plug at one end. Attach a snug-fitting short rubber tube to the same end. Clamp the glass tube in an upright position (precisely parallel to gravity; use a plumb line) with the rubber tube at the bottom. Put a few centimeters of sand on the top of the glass wool plug, level the sand by tapping the column, and pinch a clamp onto the rubber tube. Charge the column with a good grade[7] of petroleum ether (b.p. 40–60°C). Slowly funnel 140 g of adsorbent (see Table 7.4, below) into the column with care; *packing technique is crucial.* All adsorbent used should be kept scrupulously dry or maintained at a proper activity grade. The adsorbent should be carefully and gently packed by lightly and continuously tapping the column (e.g., with a rubber stopper on the end of a glass rod handle). *It is crucial that the absorbent never be added to a level higher than the solvent and that solvent never be allowed to drop below the level of the adsorbent during packing or during the chromatographic analysis.* After all of the adsorbent has been added to the column (and it has been leveled by tapping), gently place 2 to 3 cm of sand on the top of the adsorbent. Level the sand. Note that this sand should normally be covered with solvent.

If it becomes necessary to leave a column for an extended period of time, charge the void space above the adsorbent with enough solvent to leave just enough room for a (*cork*) stopper. (Why are stoppers made of plastic or rubber not acceptable?) Leave the column tightly stoppered to avoid evaporation losses.

A 50-ml buret can be used instead of the glass tube; the stopcock removes the necessity for a rubber tube/wire clamp outlet. Use 115 g of adsorbent for such a buret.

Table 7.4 lists common adsorbents used to pack chromatographic columns. *A ratio of ca.* 30 g *of adsorbent per gram of sample is a good rule of thumb.*

Table 7.4. Chromatographic Adsorbents[a]

Adsorbent	Comments
Carbon black	Rated weaker for hydrogen bonding compounds
Magnesium silicate	
Alumina (Al_2O_3)	Commonly used for stable organic compounds
Silica gel	Commonly used, especially for more sensitive organic compounds
Calcium sulfate	
Sugar	
Powdered cellulose	Very weak adsorbent

[a] Listed in order of decreasing adsorbing power; alumina and silica gel are by far the most commonly used. Note that so-called reverse-phase adsorbents are available; for such adsorbents the more polar sample components are usually more quickly eluted.

[7] A poor grade of solvent should be washed with aqueous potassium permanganate and/or sulfuric acid, followed by water wash, drying, and distilling.

Table 7.5. Standard Elution Order of Organic Compounds[a]

Alkanes
Alkenes
Aromatics
Ethers
Esters
Ketones
Aldehydes
Alcohols
Amides (having at least one N—H bond)
Diols
Carboxylic acids
Polyfunctional compounds

[a] Listed in order of decreasing elution speed; exceptions to this order occur as a result of large differences in other properties, such as molecular weight. The elution order would essentially be reversed if reverse-phase adsorbent (see footnote, Table 7.4) is used.

A variety of chromatography columns that are commercially available are shown in Fig. 7.11. A common alternative to the reservoir (shown in Fig. 7.11*e*) is a separatory funnel (see Fig. 7.17). The separatory funnel, filled with solvent and closed both at the stopcock and stoppered at the top, is hung (via a ring attached to the ring stand holding the column) such that the lower tip is under the surface of the solvent in the column (see Fig. 7.17). If the stopcock of the *stoppered* separatory funnel is opened, solvent will bubble out of the funnel into the column whenever the solvent in the column drops below the tip of the funnel. This will supply a constant solvent head on the column until the contents of the separatory funnel are depleted.

Table 7.5 describes the normal elution order of standard organic compounds. Alkanes are eluted most rapidly, because they are usually least strongly attracted to the adsorbent. Highly polar compounds, such as carboxylic acids, are eluted much more slowly because they strongly adhere to the adsorbent (e.g., by hydrogen bonding or by polar interactions).

Reasons for the elution power of solvents based on polarity have been discussed in Chapter 3 (see pp. 33–34; Table 3.1, p. 34).

Use of the Packed Chromatography Column. Weigh the (dried) mixture that is to be analyzed. Dissolve this dried mixture in a minimum of low-boiling petroleum ether; use heat and/or methylene chloride to assist solution *only* if necessary. Drain the column until the solvent is about halfway down the sand layer at the top. Transfer the solution mixture onto the surface of the sand as gently and evenly as possible (e.g., with a disposable pipet). Carefully fill the rest of the column with petroleum ether; install a ring-supported solvent reservoir in place (as described above and in Fig. 7.17) containing petroleum ether; maintain a constant supply of solvent on top of the column via this reservoir. Liquid should drip from the bottom of the column at a rate of ca. one drop per second. Allow solvent to pass through the column until all pertinent sample bands are eluted.

In those cases where it is very difficult to dissolve the sample in petroleum ether, the sample may be dissolved in a polar but volatile solvent and dispersed on to a small amount of adsorbent (this procedure is discussed on p. 385).

Collect the eluted liquid in regular volume increments (e.g., collection of 50-ml portions in 100-ml flasks may be convenient). Carefully evaporate off the solvent (e.g., using a rotary evaporator, p. 366) by reduced pressure using as little heat as possible.[8] Care should be exercised to crush the sample and remove the trapped solvent by continued evaporation. The total mass of all compounds eluted should be monitored[9] and compared to the mass of the mixture originally placed on the column. The total mass of collected samples should be equal to the mass of the original mixture placed on the column. An exception to this mass balance occurs in those cases where substantial amounts of intractable tars are held behind near the top of the column.

During the course of the chromatographic elution,[10] the solvent composition will usually be varied through a range of increasing polarities, for example:

Petroleum ether (methylene chloride)	
100% (0%)	
96% (4%)	
90% (10%)	
80% (20%)	
50% (50%)	
20% (80%)	
0% (100%)	% by volume

Preliminary tlc analysis should be useful in determining the choice of elution solvents. Care should be exercised to avoid streaking of the bands: streaking will occur if solvent polarity is increased too abruptly.[11]

Elution of colored samples may be monitored visually. Colorless samples must be monitored by an analytical technique such as gc, nmr, tlc, ir, or uv. These techniques can also be used to establish the identity and purity of colored or colorless components that have been eluted. Compounds that fluoresce when

---

[8] Flasks used on the rotary evaporator could possibly implode; round-bottomed or conical flasks should always be used for flasks of greater than 50 ml volume.

[9] The tare weights for the collection flasks should be determined after they have been numbered.

[10] As a rule of thumb, a column should be treated with 10 to 20 ml of solvent per gram of alumina before changing solvent polarity; this normally amounts to two to four fractions.

[11] Band resolution problems (streaking, unevenness) can be caused by many factors: impure solvents; channeled columns resulting from uneven adsorbent packing; contaminated (e.g., wet) adsorbent; decomposition of the sample on the column; etc. If the column has been packed too tightly, the elution flow may not reach the desired flow rate of one drop per second.

irradiated with uv light can be followed down the column; a hand-held "black light" lamp and a darkened room should be employed.

In summary, it should be pointed out that the various lc concepts (tlc and column chromatography) interrelate. Results of tlc should be used to plan column construction and the choice of solvents.

7.4.3 Dry Column chromatography

A technique called dry column chromatography has now been described in detail;[12] this powerful technique provides several improvements over traditional column chromatography. These include degree and speed of separations and a high degree of component resolution. Another important characteristic of the dry column procedure is the near-quantitative applicability of tlc results (pp. 33–37) to dry column analysis.

The tremendous potential of this technique is indicated by the report[13] that isomeric compounds A and B could be separated on silica gel using a dry column

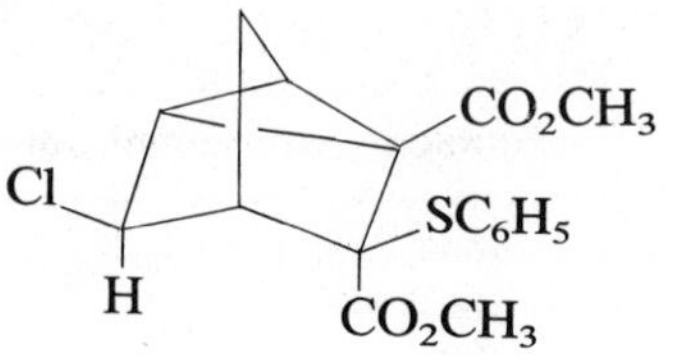

Cl H CO_2CH_3 SC_6H_5 CO_2CH_3

A

Cl H CO_2CH_3 CO_2CH_3 SC_6H_5

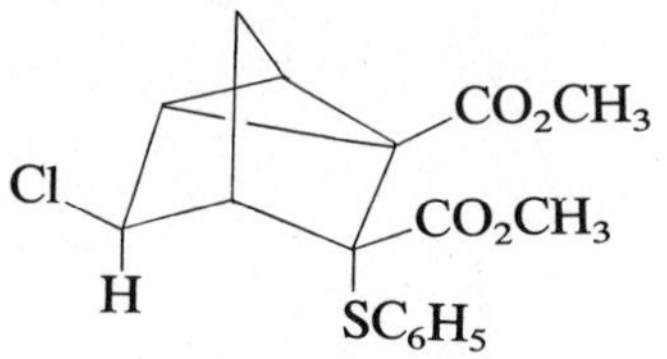

B

approach. This is quite impressive when one realizes that compounds A and B have only very slight structural differences; replacement of the chlorine atom by hydrogen in each of these structures would produce an enantiomeric pair. Thus only the slightly different dipolar attractions of A and B to the silica gel of the dry column results in their separation.[14]

In brief, in this approach the column is prepared dry, that is, without the use of solvent. Rather than eluting with several volumes of solvent, the dry column is developed just once with a single volume of solvent. The packing of dry columns is in contrast with traditional (or "wet") column chromatographs; these wet columns are often prepared by a slurry technique (see p. 385). Wet columns usually require a long and laborious series of solvent elutions in order to separate the components of a mixture.

---

[12] B. Loev and M. M. Goodman, *Chem. Ind.*, 2026 (1967); also consult the references listed at the end of this section.

[13] T. C. Morrill, S. Malasanta, K. Warren, and B. Greenwald, *J. Org. Chem.*, **40,** 3032 (1975).

[14] That is, the diastereomeric differences between A and B are imposed by the chlorines; these chlorines are remote from C-3, the site of diastereomeric differentiation. Note that the nortricyclene hydrocarbon skeleton has a $\mathbf{C}_{3v}$ axis of symmetry that is positioned coincidental to the (front) methine C—H bond and that punctures the center of the face of the cyclopropane ring.

A dry column can be prepared by charging a glass column[15] (such as those shown above in Fig. 7.11) with dry alumina (or dry silica gel). The mixture to be analyzed should be dissolved in a minimum of the eluting solvent. This mixture should be placed carefully on the top of the freshly packed dry column. Columns made of pliable nylon, rather than glass, are often used. Eluting solvent is added only until the solvent front reaches the bottom of the column. The developed column is then fractionated, for example, by slicing with a knife when a nylon sleeve is used.[15] An analyst should be careful to use any (visual, etc.) and all guides to determine band positions and thus to direct the position for the cuts. All organic compounds are then extracted (usually with methyl alcohol or ether) from the slices.

Procedure. Each of the following topics should be considered carefully, in order, before attempting dry column chromatography:

TLC behavior
Adsorbent
Column preparation (glass, nylon)
Column loading
Column development

TLC Behavior. The mixture to be treated should be analyzed by tlc, as has been described earlier (pp. 33–37); if at all convenient, good tlc development should be carried out employing only *one solvent.* If solvent mixtures are required, the adsorbent (see below) must be pretreated with the mixed solvent. *It is common for the separation to be more efficient on the dry column than on tlc.*

Adsorbent. Careful control of the moisture content of the adsorbent is crucial to dry column as well as other types of chromatography. Commercial and well-defined grades of adsorbent must be deactivated to match tlc conditions according to the following table:

tlc	Dry column
Silica gel (anhydrous)	15% water
Alumina (anhydrous)	3–6% water

Various packing qualities and useability factors of the adsorbent have been detailed (see the reference cited in footnote 12 on p. 380).

Deactivated adsorbent is prepared by adding the appropriate quantity of water to the adsorbent and rotating this mixture in a rotary evaporator (or ball

[15] Glass columns are not as easily fractionated as nylon columns. Glass columns must have their contents removed by inversion, followed by carefully sliding the column contents (by tapping or air pressure through the spout) onto a tray. The adsorbent should then be carefully cut into segments.

Table 7.6. Activity of Alumina (Standard Dye: *p*-Aminoazobenzene)

R_f of dye	Brockmann activity grade	Percent water
0.00	I	0
0.12	II	3
0.24	III	6
0.46	IV	8
0.54	V	10

mill) for ca. 3 hr. Note that it is very useful to use a fluorescent indicator (see p. 383) at this point.

Activity of adsorbents has been measured by extensively detailed procedures.[16] We shall describe here only the simpler Loev-Goodman procedure using a single dye standard and capillary minicolumns.

A 1×75 mm sealed capillary (e.g., m.p. tube) is filled with adsorbent and one drop of benzene is placed on the open end. The tube is inverted (open end down) and the closed tip is broken off the top. The damp end is then touched to the surface of a standard (e.g., 0.5% in benzene) dye solution which has been prepared in a small vial. This charged capillary is transferred to another vial containing a little pure benzene. This tiny column is allowed to develop and the position of the dye when the benzene solvent reaches the column top is used to calculate the R_f value (see formula, p. 37); this allows deduction of the activity of the alumina using Tables 7.6 and 7.7.

A fluorescent column adsorbent can be easily prepared as described here; this adsorbent is extremely useful for monitoring the development of bands of

Table 7.7. Activity of Silica Gel (Standard Dye: *p*-Dimethylaminoazobenzene[a])

R_f of dye	Brockmann activity grade	Percent water
0.15	I	0
0.22	—	3
0.33	—	6
0.44	—	9
0.55	II	12
0.65	III	15

[a] This compound has been listed as "hazardous" in OSHA Act 1910.93c. It can be replaced by 1,4-di-p-toluidinoanthraquinone (Allied Chemical, Special Chemicals Div., New York, called "D & C Green 6").

[16] H. Brockmann and H. Schodder, *Ber.*, **74b**, 73 (1941); also consult the references listed at the end of this section.

colorless compounds. One simply disperses enough inert fluorescent material[17] in the adsorbent to cause the fluorescent material to be present in ca. 0.5% (w/w) concentration; this may be done conveniently at the point described above at which water deactivation is being carried out. Band progress can now be monitored by observing the column under a hand-held uv lamp and noting those areas of the column that have the fluorescence *blocked out.* Ideally, these blocked-out regions are due to bands of a single compound. Since water content is crucial to this method of analysis, one should be very careful to check the activity of commercial adsorbents before adding the water necessary to prepare adsorbent of a certain reduced activity

Column Preparation

Glass Column. Standard glass columns with a fritted disk (for example, see Fig. 7.11*b*) can be used (Fig. 7.12).

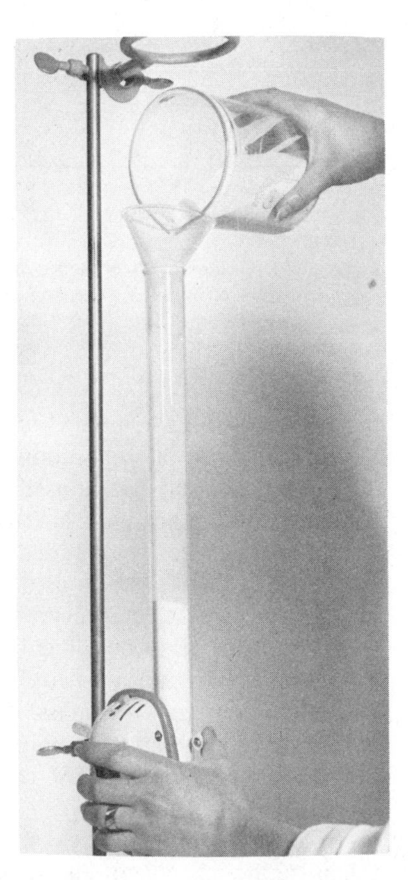

Figure 7.12. Packing a glass column with the aid of a vibrator (From B. Loev and M. Goodman, *Chem. Ind.*, 2026, 1967; with permission.)

[17] Fluorescent materials are available from du Pont (Photo Products Dept.), Towanda, Pa. (No. 609 Luminescent Chemical), or from Alupharm Chemicals (Woelm fluorescent green indicator), New Orleans, La.

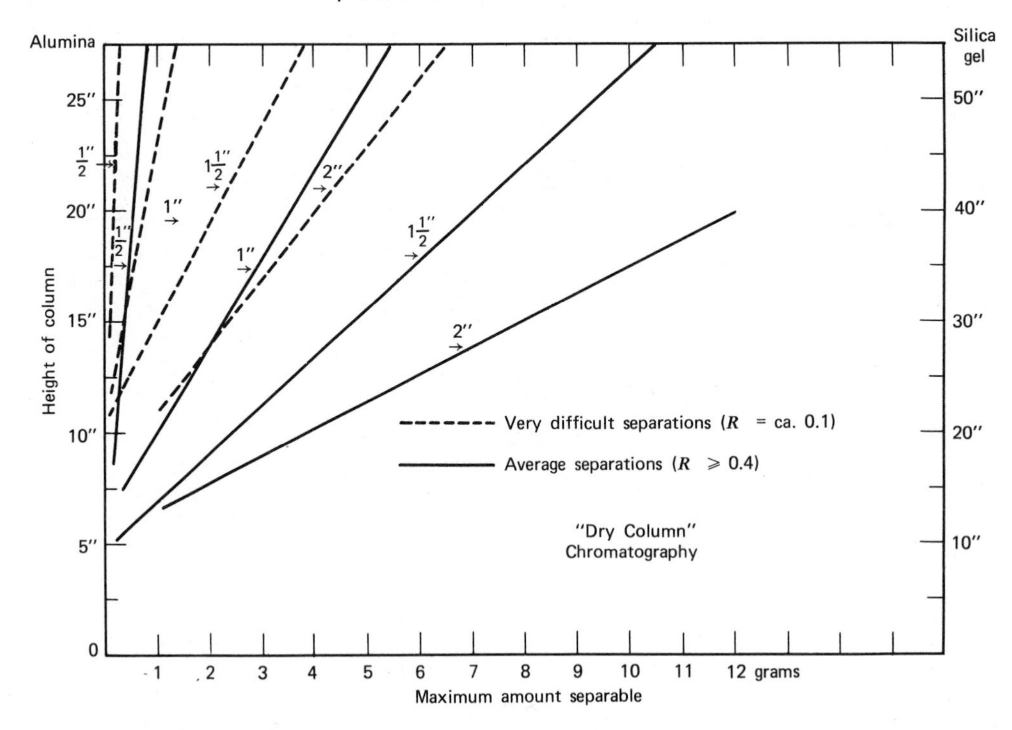

Figure 7.13. Graph for determining diameter and height of a dry column necessary to separate specified masses of mixtures. (The experimental procedure on which this graph is based is described in Loev and Goodman, *Chem. Ind.*, 2026 (1967); reproduced with permission.

The size of the column chosen depends directly on the preliminary tlc results; the more difficult the separation, the larger the column that must be used. Figure 7.13 is a graph that can be used to choose the required depth of adsorbent in the column. Compounds that are reasonably mobile (R_f = ca. 0.4) are usually involved in "average" separations. If an alumina column is, for example, 2 in. thick, the graph of Fig. 7.13 tells us that we must use a column that is about 12 in. high in order to separate the "average" components of a 6.0 g sample. If the 6.0 g of compounds prove to be difficult to separate (R_f only ca. 0.1), the graph of Fig. 7.13 predicts that we must use a column that is about 25 in. high. Note that the graph sets the *maximum* mass of mixture that can be analyzed. More efficiency is gained by using samples of only 50 to 75% of the column capacity; for example, 3.0 to 4.5 g of the "average" mixture described above would be more efficiently separated than would 6.0 g.

The column's stopcock should be left open during packing in order to minimize the trapping of air in the absorbent. Dry adsorbent is *slowly* poured into the column while gently tapping the column (or while holding an electric vibrator against the column; see Fig. 7.12).

The data in Fig. 7.13 correspond to a requirement of ca. 70 g of adsorbent per gram of mixture for "average" separations and of ca. 300 g of adsorbent per gram of mixture for "difficult" separations.[18] This can be contrasted to the ca. 30 g of adsorbent per gram of mixture mentioned earlier as a rule of thumb for classical ("wet") column chromatography.

Nylon Columns. Nylon tubing is available[19] that is transparent to uv light; this tubing is thus very useful for dry column chromatography. Columns developed in this tubing can be sliced with a knife to allow division of the portions of the adsorbent containing chromatographically separated compounds. The nylon must be strong and inert to the solvents used.

Flat diameter (in.)	Column diameter (in.)
1.0	$\frac{3}{4}$
1.5	1
2.0	1.5
3.0	2.0

Roll length; 100 or 1000 ft.; nylon thickness 1.6 mils.

To pack the nylon column, a nylon tube is chosen with a width based on application of Fig. 7.13; this tube is cut to a length based on consideration of Fig. 7.13. One end is heat-sealed,[20] and this end is used as the bottom. A pad of Pyrex glass wool is placed at the bottom of the tube. Two or three small holes are made (Fig. 7.14) at the bottom (to prevent air pocket formation during packing). About one-third of the adsorbent is added rapidly; this partially filled tube should be compacted by dropping it two or three times on to the bench top from a height of about 6 in. (Fig. 7.15). This packing process is carried out a second and third time; a properly prepared nylon column is quite sturdy and can be supported by a single clamp (Fig. 7.16).

Column Loading. Mixtures could be placed on the dry column in a minimum volume of the eluting solvent as has been described above for traditional column chromatography. A much superior method involves *direct deposition* of the mixture onto some adsorbent before placing the mixture on the column. The mixture is dissolved in a minimum of a low-boiling solvent, such as ether or methylene chloride. This low-boiling solvent need not be the eluting solvent. A quantity of adsorbent, approximately five times as heavy as the mixture to be

[18] The decreased capacity (grams of mixture/column height) for silica gel compared to alumina (Fig. 7.13) is due largely to the fact that silica gel is about one-half as dense as alumina.

[19] "C"-guage nylon tubing is available from Walter Coles and Co. Ltd., Backhouse Works, Surrey Square, Walworth, London, S.E., U.K.

[20] Heat sealing may be done with a hot flat iron or with a hand sealer (item S-2020, Bel-Art Products, Pequannock, N.J.; use the maximum setting). Alternatively, the column can be folded over a few times and stapled shut.

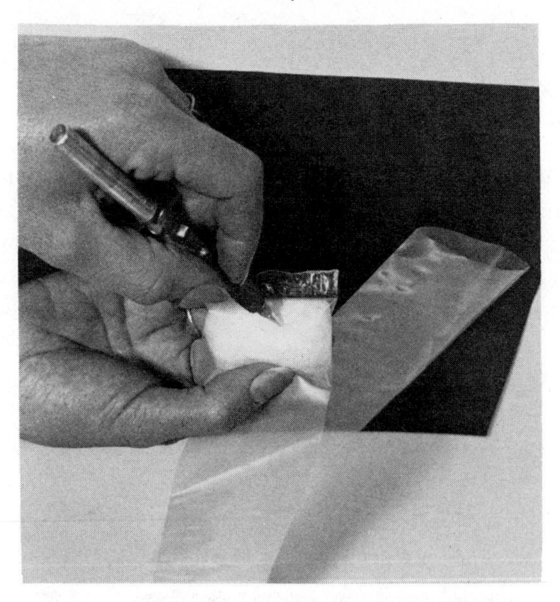

Figure 7.14. Puncturing the bottom of the nylon sleeve for a dry column chromatography. [From B. Loev and M. Goodman, *Chem Ind.*, 2026 (1967); with permission.]

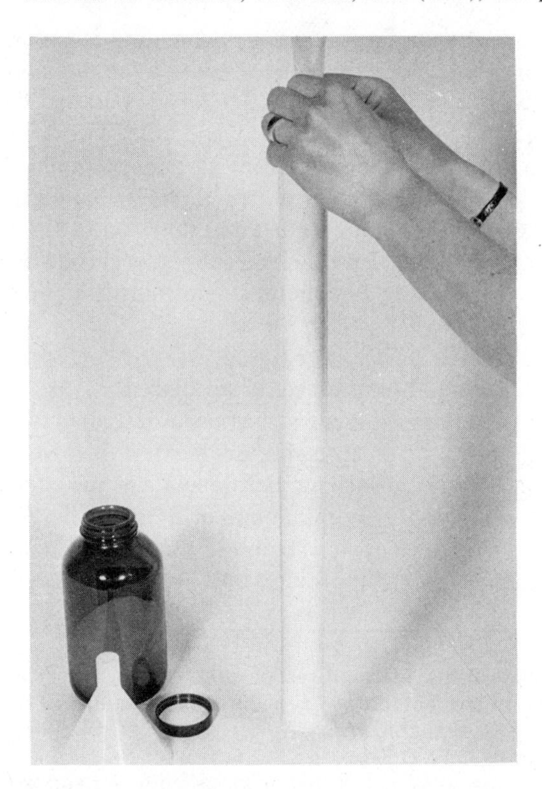

Figure 7.15. Compacting the filled nylon column. [From B. Loev and M. Goodman, *Chem. Ind.*, 2026 (1967); with permission.]

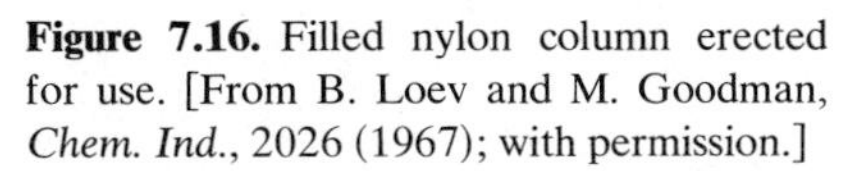

Figure 7.16. Filled nylon column erected for use. [From B. Loev and M. Goodman, *Chem. Ind.*, 2026 (1967); with permission.]

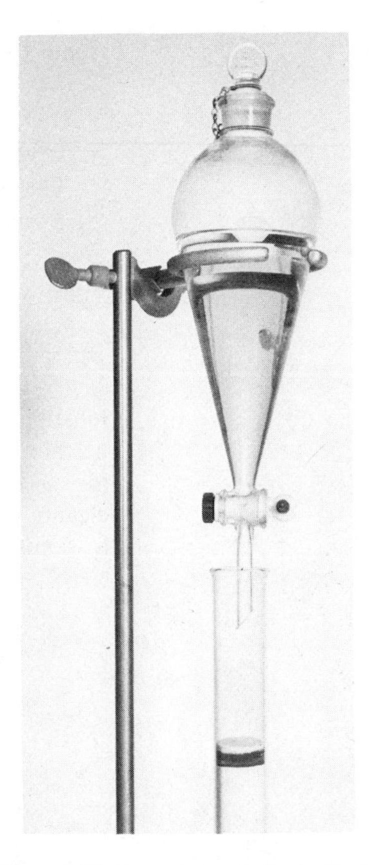

Figure 7.17. Dry column charged with dye mixture and ready for development. [From B. Loev and M. Goodman, *Chem. Ind.*, 2026 (1967); with permission.]

analyzed, is dispersed in this solution. After thoroughly mixing this dispersion, the solvent is carefully removed (rotary evaporation at 30 to 40°C). This dried dispersion is distributed evenly onto the top of the partially packed column; this packed column is covered with a small layer (ca. $\frac{1}{8}$ in.) of sand or small glass beads (Fig. 7.17).

Column Development. Note carefully that only a limited quantity of solvent is to be added to the column. This quantity should be just sufficient to reach the bottom. Remember that ideally the solvent should *not* be a mixed solvent (see p. 381) if at all possible. Approximate solvent volumes may be found in Table 7.8, but a little additional solvent should be on hand in case of a shortage. Solvent should be added such that a constant head of 3 to 5 cm is maintained on top of the adsorbent; this is nicely automated by using a stoppered separatory funnel with an open stopcock (see Fig. 7.18 and also the discussion of p. 378). *As soon as the solvent has reached the bottom, solvent addition should be halted*; *development is complete.* Positions at which the nylon column will be cut are marked with a

Table 7.8. Volume of Solvent for Development of Various Sizes of Dry Chromatographic Columns

Column size (in.)	Solvent volume (ml)
20×0.5	20
20×1.0	90
20×1.5	300
20×2.0	500

felt-tipped marking pen. If the compounds are colorless and a fluorescent mixture has been used for an adsorbent, these marked divisions should be positioned with the aid of a hand-held uv light. The nylon column and its contents are then sliced cleanly and the separated units of adsorbent are extracted with methanol or ether to remove the desired components.

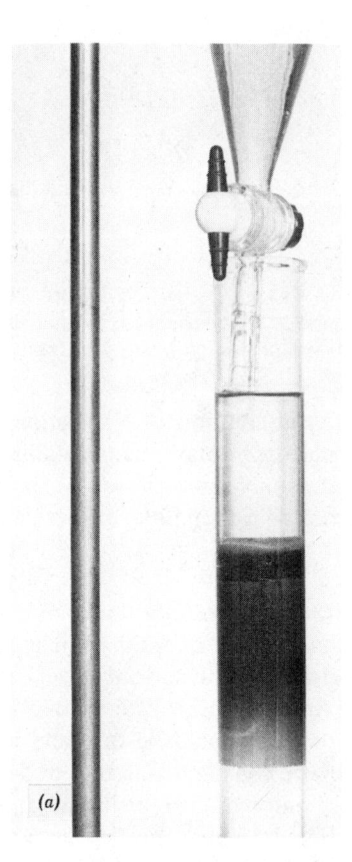

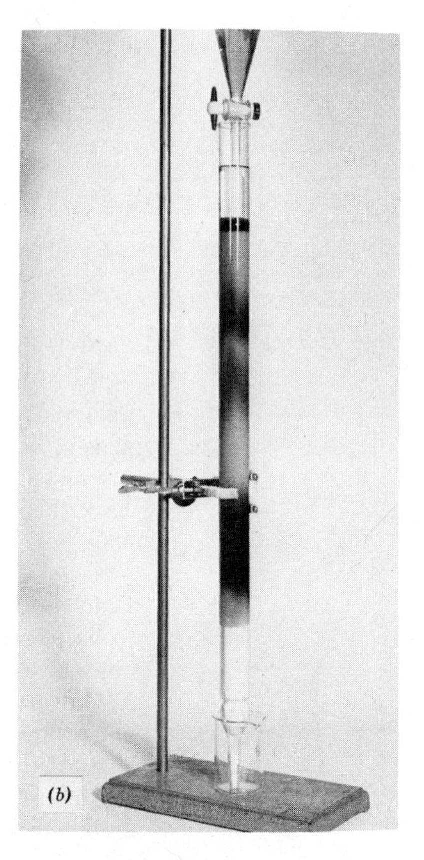

Figure 7.18. (*a*) Early stages of development of a dry column (glass column). (*b*) Dry column nearly completely developed. [From B. Loev and M. Goodman, *Chem. Ind.*, 2026 (1967); with permission.]

If a glass column is used, the column is inverted and the contents are carefully removed by opening the stopcock and tapping the sides of the column and by adding air pressure through this stopcock where necessary.[21] The contents should be carefully laid out (e.g., on a pan) in order to avoid mixing the separated components. The adsorbent is sliced into separate pieces and these pieces are extracted as described above in the nylon tube procedure.

Using the dry column technique, columns as long as 6 ft have been packed and mixtures of mass of up to 50 g have been separated.

Additional Dry Column Chromatography References

B. Loev and M. M. Goodman, *Progress in Separation and Purification*, Vol. 3, edited by E. S. Perry and C. J. Van Oss (Wiley, New York, 1970), "Dry-Column Chromatography," pp. 73–95.

J. M. Bohen, M. M. Joullié, F. A. Kaplan, and B. Loev, *J. Chem. Educ.*, **50,** 367 (1973).

B. Loev, P. E. Bender, and R. Smith, *Synthesis*, 362 (1973); this article describes the separation of mixtures of alkenes using a dry column treated with silver nitrate.

7.4.4 High-Performance Liquid Chromatography

High-performance liquid chromatography (hplc), sometimes called high-pressure liquid chromatography (and less frequently, high-speed liquid chromatography), attacks problems associated with traditional column chromatography with great success.

Flow rates in traditional lc columns are usually low compared to, for example, those used in gc, because the diffusion of the sample into the lc stationary phase is relatively slow. This occurs because traditionally lc chromatographic procedures involved stationary phases of a particle size that is relatively large (much like the sizes used in gc). In order to speed the diffusion of sample into the stationary phase, *very fine particles* of stationary phase are used in hplc. The small size of such fine particles produces a new problem; pressures well above atmospheric pressure are necessary to push the mobile phase through tightly packed columns of these fine particles. Since 1968, rapid technical developments have taken place. Pumps delivering thousands of pounds per square inch push mobile phases through columns of very fine particles. These particles are of the order of a few micrometers in size. Special stationary phases have been developed which possess a hard core that allows organic compounds to diffuse only into the skin. This limits the extent of the diffusion process and expedites the elution procedure. Detectors that differentiate samples by refractive index, by uv absorptions, and by fluorescence are commonly used. An entire subject field has rapidly

---

[21] The glass column procedure should be avoided, as tightly packed columns often require tremendous force in order to be removed from the glass.

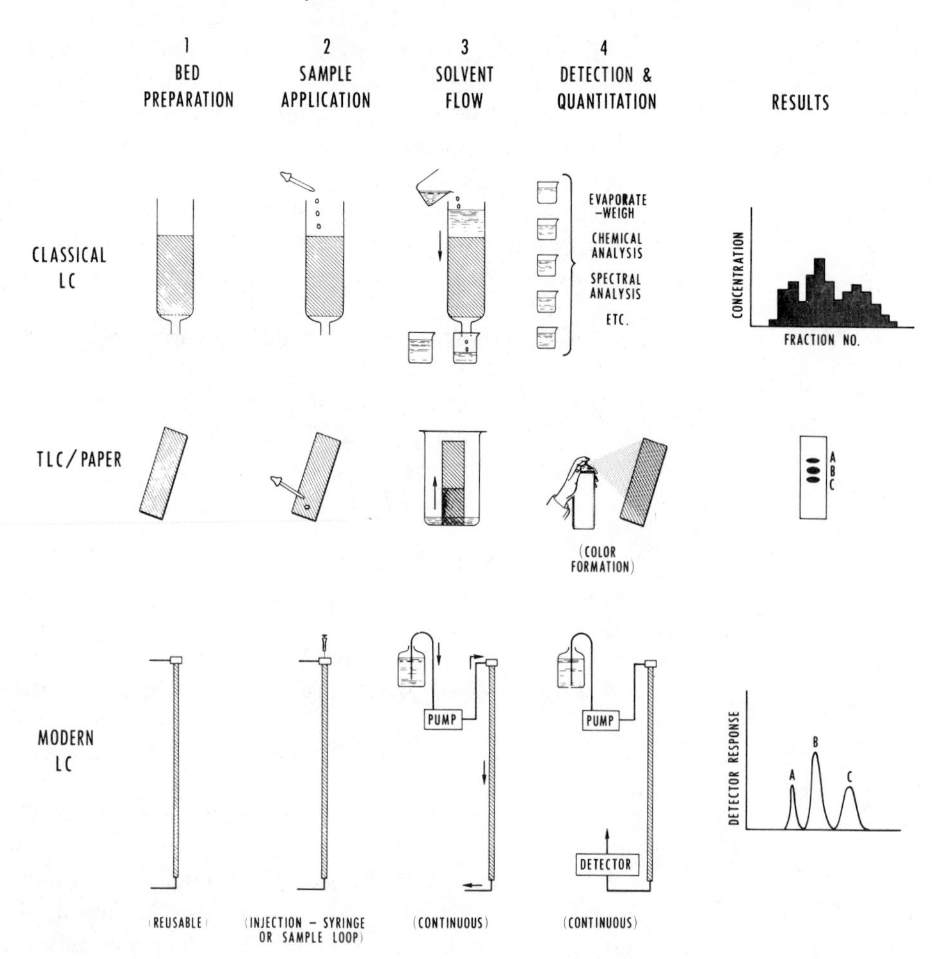

Figure 7.19. Schematic representation of lc, tlc, and modern lc chromatographic methods. (From J. J. Kirkland and L. Snyder, *Modern Liquid Chromatography*, Wiley-Interscience, New York, Copyright © 1974, with permission.

evolved, and this highly developed technique is now available to the organic chemist.

An overview of how hplc fits into the general chromatography picture can be gained by examining Fig. 7.19. High-pressure liquid chromatography offers the advantages of high speed, reusable columns, automatic and continuous solvent addition, reproducible programed gradients of solvents, and automatic and continuous monitoring of the samples that are eluted. The disadvantages of hplc include the fact that the system demands high-quality equipment (pumps, columns, etc.). Recently the technology of hplc has been extended to the preparative scale; thus large quantities of compounds can be isolated, and this technique has become an alternative to gc and traditional column chromatography. Despite its limitations, hplc is an outstanding technique and is becoming routinely available to most laboratories.

The hplc Instrument. In order to appreciate how to apply hplc, the chemist should have some understanding of the components of the standard types of hplc equipment.[22] Figure 7.20 outlines schematically the components of a standard hplc setup. We shall discuss these basic components, keeping in mind that various commercial instruments will emphasize certain specific choices in these areas (e.g., the type of pump employed). Figure 7.21 is an illustration of a typical commercial unit for hplc.

First let us consider the equipment associated with the solvent reservoir (Fig. 7.20, parts 1–3). The reservoir itself is made either of glass or of stainless steel. Attachments that provide heating or stirring of the solvent are available; these are very useful, because the eluting solvent should be degassed. Inlets can also be incorporated that allow nitrogen purging for solvent degassing. Solvents used as a mobile phase should be degassed, because oxygen in the solvent can react with the mobile or stationary phase of the system; in addition, bubbles of oxygen can easily disrupt the operation of the detector. It should be kept in mind that hplc involves demanding conditions and thus careful techniques and highly stable components are necessary.

It is clear that the quality of the pump is one of the most important factors in determining the quality of the entire hplc apparatus. One must have a pump that can drive the mobile phase through long, narrow columns packed with very fine

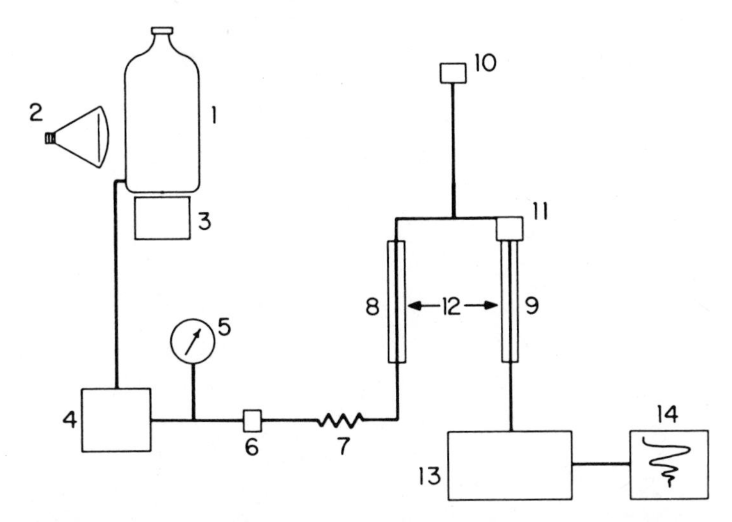

Figure 7.20. Schematic of hplc instrument components: 1, solvent reservoir; 2, heating unit; 3, magnetic stirrer; 4, high-pressure pump; 5, pressure guage; 6, line filter; 7, restrictor (for pulse dampening): 8, precolumn (optional); 9, analytical column; 10, pressure sensor–pump detection device (optional); 11, sample introduction system; 12, thermostatted oven area (optional); 13, detector (e.g., refractive index or uv); 14, recorder of data handling system. (From J. J. Kirkland and L. Snyder, *Modern Liquid Chromatography*, Wiley-Interscience, New York, copyright © 1974; with permission.

[22] Commercial instruments are available from Waters Associates, Varian Associates, and du Pont Instruments Division.

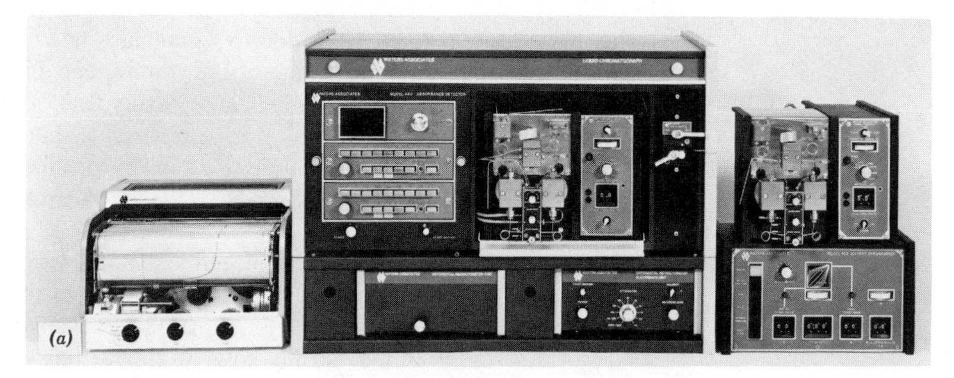

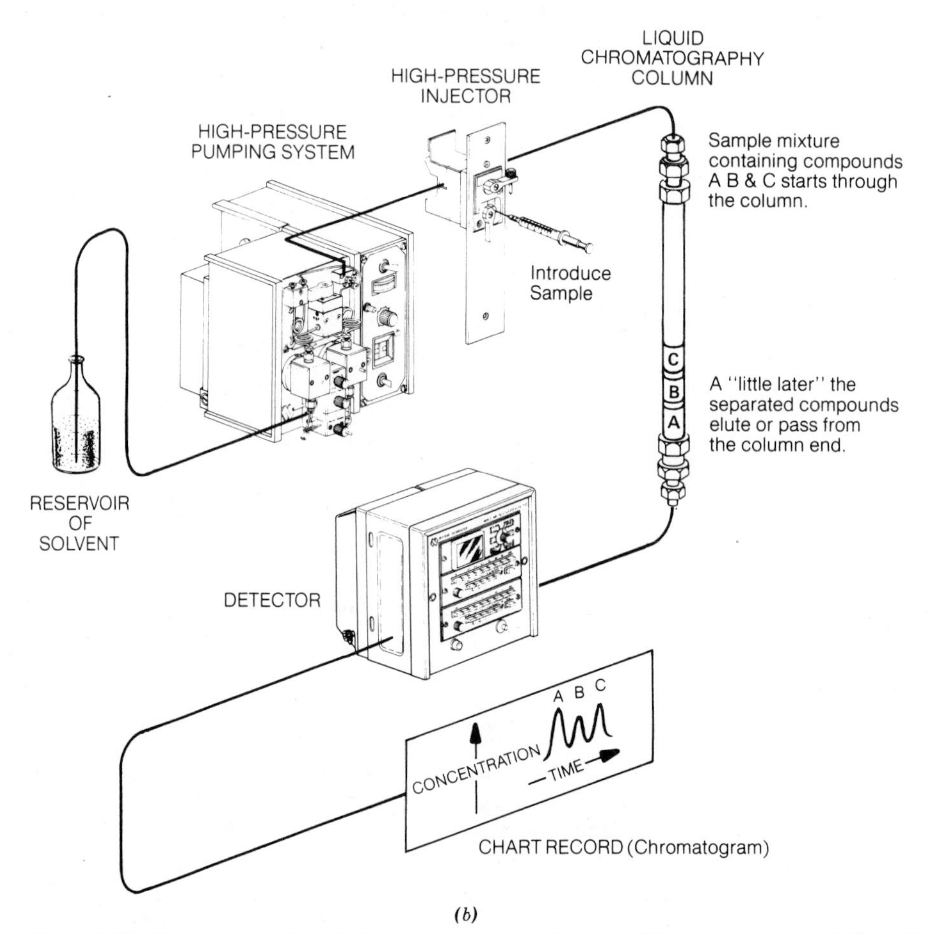

(b)

Figure 7.21. Commercial hplc unit for analytical determinations: Waters Associates, Milford Mass. Both a photograph (a) and a schematic diagram (b) of the same instrument are shown.

particles that have been highly compacted. In addition, it is crucial that pulses in the flow be kept to a minimum, because pulses will induce gradients in the sample, which will disrupt automatic detection procedures, especially when refractive index detectors are used. A flow rate of ca. 1 to 6 ml/min is needed for analytical work. The exact flow rate depends on the column diameter and the packing material (adsorbent).

There are two general classes of "pulse-free" pumps: mechanical pumps and pneumatic pumps. The former maintain constant pressure and the latter maintain constant flow rate.

Commercial instruments with mechanical pumps contain either a syringe type or a reciprocating type of pump. The syringe-type, or screw-driven pump, maintains constant displacement by screw feed. This pump was used in many of the earlier instruments. Two pumps in tandem can be used for gradient elution (mixed solvents), but any of these constant-displacement systems must be stopped for refilling.

Most newer instruments use the reciprocating type of pump. In this pump, each stroke of a plunger pushes some mobile phase into the system *and* on the reverse stroke the piston cavity is refilled from the reservoir via a ball-check valve system. In such a system, a constant flow rate can be maintained, independent of variations in the back pressure of the column. An example of a reciprocating pump is shown in Fig. 7.22.

Some general comments about pump systems are in order. Pulse-damping devices external to the pumping system should be avoided if possible, because they increase the dead volume of the system; this increase in dead space will be antagonistic toward gradient elution and recycling procedures. Also, Teflon or

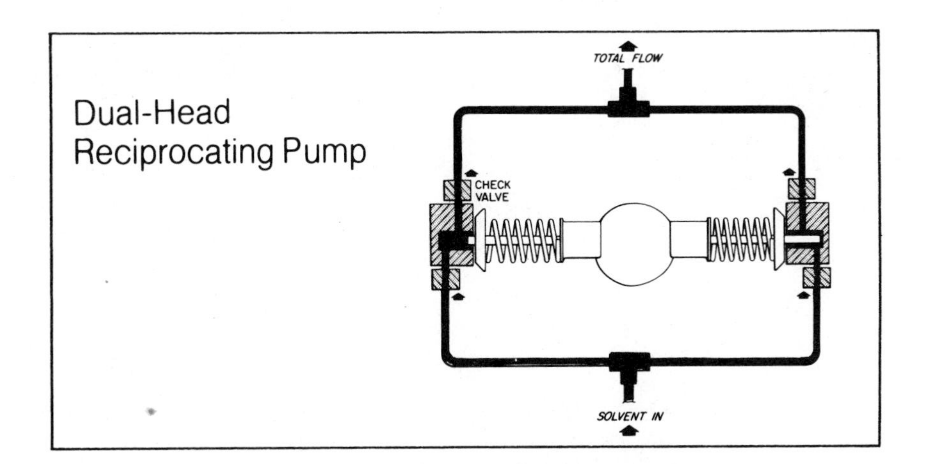

Figure 7.22. Schematic diagram of the Perkin-Elmer dual-head reciprocating pump. Each 100-μl side alternately fills while the opposing piston is pumping. The low-mass, double-ball check valves assure that eluting solvent can flow only toward the column. Phasing of the opposing pistons causes combined flow output to be virtually pulse-free. Flow is digitally controlled by varying the motor speed electronically. (Reproduced with permission of Perkin-Elmer Corp, Norwalk, Conn.)

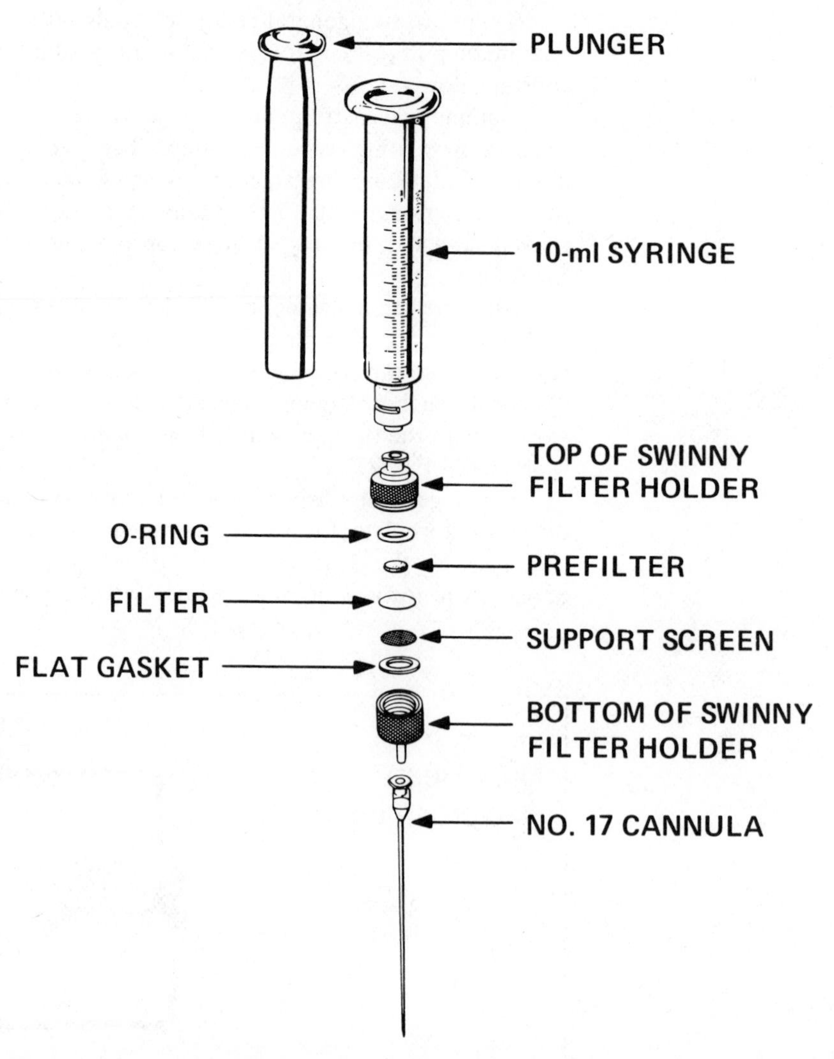

Figure 7.23. A "sample clarification" kit, one of the ways to clean samples of particulates. If samples are not filtered to remove particulate matter, the column, injector or syringe may become plugged. The "sample clarification kit" consists of a Swinney adaptor, which will hold a very low porosity filter. For organic solvents a 0.45-μm Teflon filter is recommended. For aqueous solvents a 0.45-μm cellulose acetate filter should be used. It is of the utmost importance that samples be filtered before injecting into an instrument. (Courtesy of Waters Associates, Milford, Mass.)

comparably inert fittings for many pump parts (gaskets, O-rings, etc.) are necessary, because the high pressures involved increase the possibility of chemical and physical destruction of such parts. A filter between the pump and the column is essential so that the column, which is packed with fine particles, will not be plugged by an undesirable particle. Filters made of sintered stainless steel which are impermeable to particles larger than 2 μm are frequently the choice.

Sample introduction in hplc is even more sensitive than the careful procedures required for traditional chromatography. A very narrow band of sample should be introduced onto the chromatographic bed. A microsyringe that will tolerate pump pressures up to 1500 psi can be used; above that pressure an off-line injection procedure should be used. Stop-flow injection should be avoided, as it disturbs both the column and the detector. Samples should be introduced using the apparatus shown in Fig. 7.23.

Columns for hplc are constructed from stainless steel or glass; it must be kept in mind that these columns must withstand pressures of up into the thousands of pounds per square inch. Fittings and plugs must be inert and should not detract from the homogeneity of flow. Columns are usually 25 to 150 cm long. Analytical columns are usually 1 to 8 mm in inside diameter; columns of less than 8 mm inside diameter are difficult for a novice to pack.

A list of column packings commonly used for hplc is shown in Fig. 7.24. Typical results using a μ-Porasil column are shown in Fig. 7.25; note that an elegant separation of positional isomers has been done in less than 1 min. A very useful approach in hplc is shown in Fig. 7.26; this so-called reverse-phase type of adsorbent allows compounds to be separated on the basis of their lack of polarity.

Recently preparative-scale lc instruments have become fairly common; an example is shown in Fig. 7.27. Their great power is due largely to the fact that large samples may be placed on the column and, although the peaks may overlap (see Fig. 7.28) under these conditions, the samples may be repeatedly recycled through the very large columns and thus excellent separations may be obtained in relatively short times.

7.4.5 Gas Chromatography

Gas chromatography (gc) has been introduced as an analytical and a purity probe in Chapter 3, pp. 61–67. We shall now expand upon the earlier introduction and develop the idea of preparative-scale chromatography. If the sample to be analyzed can survive the conditions of the gc instrument, gas chromatography is one of the simplest, quickest, and most useful analytical methods that can be used.

Columns for analytical gc studies are either *capillary* columns (small capillary tubing packed only with a stationary phase) or larger-diameter columns packed[23] with a stationary phase dispersed over an inert support (e.g., pulverized firebrick). Such analytical columns can vary in length from a few inches to as long as 1 mile;

---

[23] Commercially prepared columns are available, but substantial savings may be gained by using "home-made" columns (see book by Bates and Schaefer described on p. 363, Fig. 7.3).

Packing Structures:	Analytical & Semi-Preparative Separations	Analytical Separations	Preparative Separations
	(Schematic of Packing Structures)		
Definition:	• Small Diameter Particles • Fully Porous	• Pellicular Particles (Thin layer silica fused to solid glass bead)	• Large Diameter Particles • Fully Porous
Characteristics:	• Very High Efficiency • Moderate Capacity • High Speed	• High Efficiency • Low Capacity • Moderate Speed	• Moderate Efficiency • High Capacity • Moderate Speed

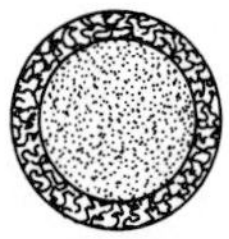

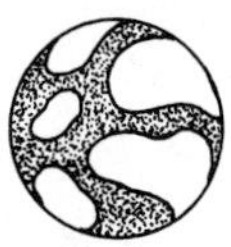

Functional Groups	Polarity	Separations Mode	Products Available		
Silica-OH	High	• Adsorption, Normal Phase	μPORASIL (Micro particles of fully porous silica)	CORASIL (Silica fused to Solid Glass Core)	PORASIL (Fully porous silica)
Amino — NH_2	High	• Adsorption, Normal Phase • Ion Exchange • Partition, Reverse Phase	μBONDAPAK NH_2	—	—
Cyano — CN	Intermediate	• Adsorption, Normal Phase, • Partition, Reverse Phase	μBONDAPAK CN	—	—
Phenyl	Low	• Partition, Reverse Phase	μBONDAPAK Phenyl	BONDAPAK Phenyl/ CORASIL	BONDAPAK Phenyl/ PORASIL
C_{18}	Low	• Partition, Reverse Phase	μBONDAPAK C_{18}	BONDAPAK C_{18}/CORASIL	BONDAPAK C_{18}/PORASIL
—	—	• Fatty Acid Analysis	Fatty Acid Analysis Column	—	—
—	—	• Carbohydrate Analysis	Carbohydrate Analysis Column	—	—
—	—	• Triglyceride Analysis	Triglyceride Analysis Column	—	—
Styrene, Di-Vinyl Benzene Copolymer	—	• Size Exclusion Gel Permeation	μSTYRAGEL	—	STYRAGEL
Ether $(RO)_x CH_3$	Intermediate	• Size Exclusion	μBONDAGEL	—	—

Figure 7.24. Column packing commonly used for hplc. (Courtesy of Waters Associates, Milford, Mass.)

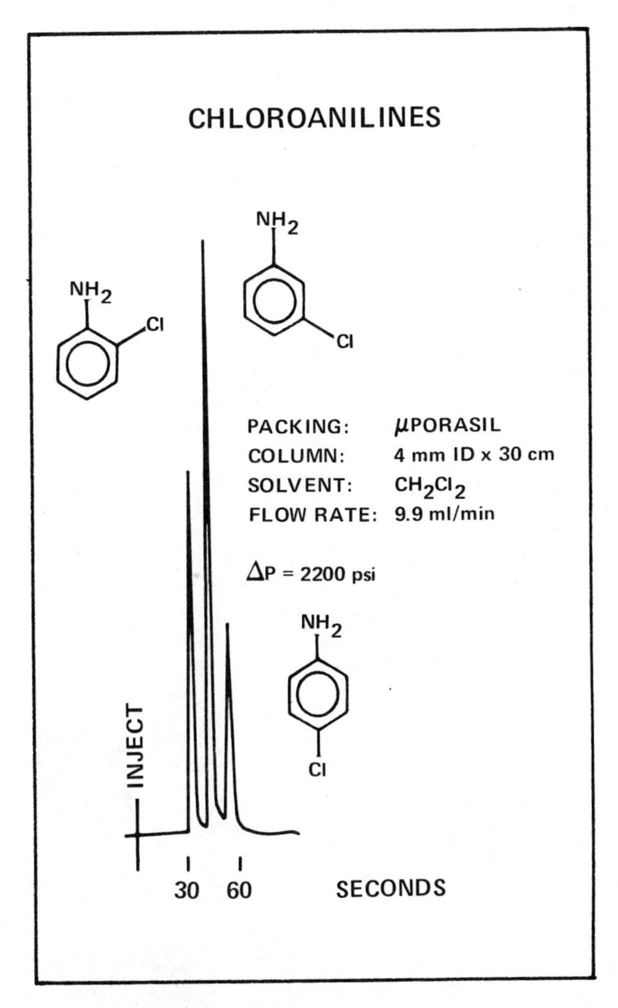

Figure 7.25. Chromatogram of chloro-substituted anilines separated by hplc. (Courtesy of Waters Associates, Milford, Mass.)

5 to 10 ft is typical. In order to accommodate preparative-scale samples, columns of moderate length (5 to 10 ft) are usually preferable in order to avoid unrealistically long retention times. Although it is possible to use apparatus containing much wider columns, it is usually most convenient to use instruments, typically with thermal conductivity detectors, that are equipped with columns of a diameter of $\frac{1}{4}$ to $\frac{3}{8}$ in. for preparative work.[24] If the gc instruments are equipped with detectors (e.g., flame ionization detectors) that destroy the sample, *splitters* must be used within the "plumbing" to divert part of the sample into an outlet port at which preparative collections can be carried out.

---

[24] Columns of $\frac{1}{4}$ to $\frac{3}{8}$ in. diameter are handy for both analytical and preparative-scale work.

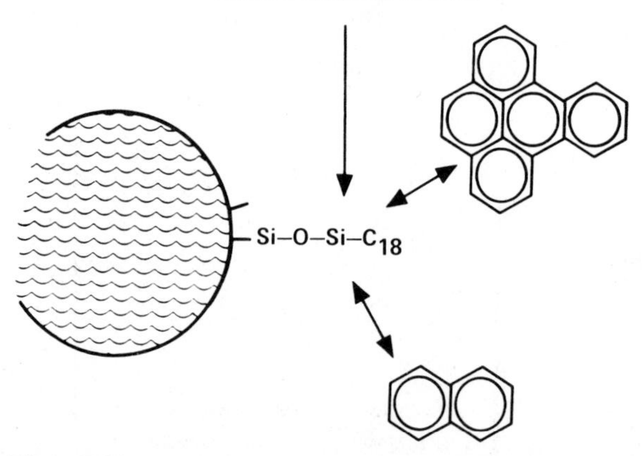

Figure 7.26. Bonded packing for Reverse-Phase lc. Bonded packings, such as C_{18} or phenyl, where the bonded material is very nonpolar, are used in reverse-phase separations. These are actually liquid/solid separations. Normally a reverse-phase packing is run with a carrier such as a water/acetonitrile mixture or water/methanol mixtures. For the separation of naphthalene from benzopyrene, since both the packing and the sample are nonpolar, the packing does not reject the samples, whereas the carrier (the water/acetonitrile in this case) is very polar and does reject them. This forces the samples into the packing. The actual mechanisms are very involved. (Courtesy of Waters Associates, Milford, Mass.)

Figure 7.27. Preparative-scale hplc unit. (Courtesy of Waters Associates, Milford, Mass.)

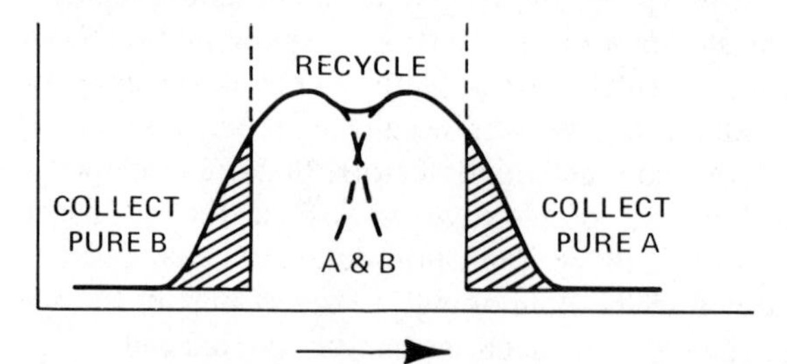

Figure 7.28. Preparative scale-up for two major components. Chromatogram of two poorly resolved components. The shaded areas correspond to material that is "shaved" or saved. The material from the center is recycled and, thus, successive passes are lighter loads and lead to improved resolution. This procedure is especially useful in prep-scale hplc. (Courtesy of Waters Associates, Milford, Mass.)

The desire to expedite preparative work should not be allowed to result in the use of overly large samples. As one reaches the point where sample sizes are so large that the column is overloaded, peaks observed on the gc recorder will begin to have other than a sharp spike-like appearance or a regular, bell-shaped appearance.

A number of definitions and formulas are presented at this time such that a better understanding of gc can be gained. It should be noted that many of these definitions and formulas are also applicable to other chromatographic procedures (tlc, hplc, etc.). The general theory here thus is of use for various types of chromatography. A good understanding of this theory is very useful and will serve as a foundation for the planning and implementation of successful chromatographic analyses.

Let us define the velocity (cm/sec) of the mobile phase as v and the velocity of a given band in the mobile phase as v_x. We can now relate these velocities by the factor R, the fraction (e.g., mole fraction) of molecules of x in the mobile phase, where

$$v_x = vR$$

Note that if $R = 0$, then $v_x = 0$ (that is, the x molecules are immobile); and if $R = 1.00$, then $v_x = v$ (that is, the x molecules are as mobile as the solvent). Thus the velocity at which a band proceeds within a mobile phase is decreased by stationary phases that decrease the number of molecules in the mobile phase.

It is also useful to define k', the capacity or retention factor, as the ratio of n_s to n_m, which are the number (n) of molecules of x in, respectively, the stationary (s) and mobile (m) phases. Thus the equation

$$k' = \frac{n_s}{n_m}$$

implies that k' will increase for chromatography in which there is an increase in the attraction of the sample for the stationary phase.

From the development of R above, we can deduce that

$$R = \frac{v_x}{v} = \frac{n_m}{n_m + n_s}$$

Thus, using equations from above,

$$v_x = \frac{v}{k' + 1}$$

Experimentally, we could measure v_x if we could determine L, the length (cm) of the column, and t_R, the retention time (sec) of x on this column:

$$v_x = \frac{L}{t_R}$$

In like manner, we can evolve a relationship for the velocity (v_0) and retention time (t_0) of a reference standard:[25]

$$v_0 = \frac{L}{t_0}$$

It is now important to realize that *ratios* of retention times under comparable condition do not depend on the length (L) of the column:

$$\frac{t}{t_0} = \frac{v}{v_x}$$

Rearranging this ratio of times and disposing of the velocity terms by using the equations relating v and k', we find that

$$k' = \frac{t_R - t_0}{t_0}$$

This means that the capacity factor (k') can be determined by dividing the increase in retention time of x over the retention time of the standard by the retention time of the standard (t_0). If $t_R = t_0$, then $k' = 0$; that is, the column has no capacity. If $t_R = 2t_0$, then $k' = 1.00$, and if $t_R = 3t_0$, then $k' = 2.00$, and so on. Thus k', at reasonable conditions, should be constant for a given component. The quantities t_0 and t_r are illustrated in Fig. 7.29.

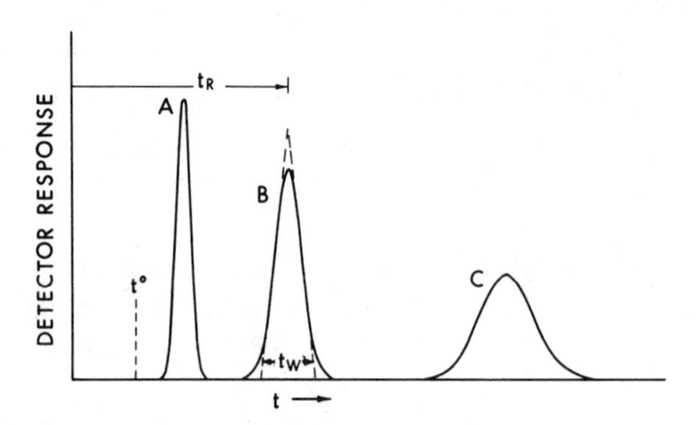

Figure 7.29. Typical chromatogram (gc, tlc, or hplc): t = time; t_R = retention time of component B; t_0 = retention time of reference standard (frequently a peak will show here) t_w = time for width of peak B. (From J. J. Kirkland and L. Snyder, *Modern Liquid Chromatography*, Wiley-Interscience, New York, copyright © 1974; with permission.)

Peak	k' (see text)
A	0.7
B	2.2
C	5.7

[25] This is usually a relatively weakly retained component, e.g., air or the solvent.

The selectivity, α, is measured by taking the ratio of retention factors (k'):

$$\alpha = \frac{k_2'}{k_1'} = \frac{V_2 - V_0}{V_1 - V_0}$$

where

$$V = \text{retention volume}$$

The magnitude of retention times described just above depends on the flow rate employed for the measurement of t_R, t_0, and so on. Retention volumes (V_R) can be measured and are more inherent characteristics of the sample than retention times, because retention volumes, in principle, are independent of the flow rate used:

$$V_R = t_R F$$

where F = the flow rate of the mobile phase (ml/sec), and V_R is the retention volume (ml) of component x of retention time t_R (sec). If t_0 is the retention time of the solvent, the retention volume of the solvent (i.e., the volume of mobile phase that the empty column will hold) is V_m and is found from

$$V_m = t_0 F$$

and the ratio of retention volumes can be related to the capacity factor, k', by

$$\frac{V_R}{V_m} = 1 + k'$$

Magnitudes of V_R are often preferred to values of t_R. Since V_R values ostensibly are independent of the flow rate, they are extremely useful in those cases where the flow rate can vary (e.g., in hplc) or where the flow rate is difficult to measure.

From the definition of k' (the capacity factor), we can deduce that k' is proportional to V_s, the volume of the stationary phase, because

$$k' = \frac{n_s}{n_m} = \frac{[x]_s V_s}{[x]_m V_m}$$

The ratio of concentrations is defined as K, the distribution constant:

$$K = \frac{[x]_s}{[x]_m}$$

The magnitude of K is a measure of the equilibrium distribution of molecules of x between the stationary and the mobile phases. Note that the magnitude of K is independent of V_s.

Since the capacity factor, k', is directly proportional to the volume of the stationary phase, V_s, the magnitude of k' will increase with an increased loading of support by a stationary phase. Specifically, pellicular or surface-coated glass beads commonly used in hplc will usually have lower values of V_s and k' for a given column size and a given set of experimental conditions.

An important measure of the efficiency of a chromatographic column is the number of theoretical plates, N. This can be determined from the equation

$$N = 16\left(\frac{t_R}{t_w}\right)^2$$

where t_w = is the bandwidth in seconds. The quantity t_w has been defined graphically for component B in Fig. 7.29. In principle, the quantity N should be the same for all bands on a given column under given conditions. In such cases, the bands with longer retention times will be correspondingly broader (see Fig. 7.29). We can see that the ratio of t_R/t_w, and thus N, is essentially a measure of the resolving power of the column; that is, larger values of N mean that we can deal with a larger range of retention times (t_r) for bands that are relatively narrow (small t_w).

The number of theoretical plates, N, is directly proportional to the length of the column, L. We can avoid this proportionality by developing H, the height equivalent of a theoretical plate (*hetp*), which measures the efficiency of the column per unit length.

$$H = \frac{L}{N} = \frac{\text{length in cm}}{\text{no. of theoretical plates}}$$

The primary objective of chromatography is to minimize H, that is, to have as short as possible lengths to achieve as many theoretical plates as possible.

There are five major types of factors that control H and N and thus that control the efficiency of the chromatographic procedure; these must be reviewed in order to understand and apply chromatography:

1. Eddy diffusion.
2. Mobile-phase mass transfer.
3. Longitudinal diffusion.
4. Stagnant mobile-phase mass transfer.
5. Stagnant stationary-phase mass transfer.

Eddy diffusion arises from the fact that the mobile phase is broken into a number of streams which each flow through different sections of a column at different flow rates, much as part of a river will flow rapidly in streams in open channels whereas other areas of a river will have restricted, slower streams. The greater the range of streams with various flow rates, the more the band will be diffused on a column and the greater the possibility that a set of bands will overlap and be difficult to separate. This problem can be minimized by using homogeneously packed columns of small, uniform (e.g., spherical) particles so that the eddies are as similar as possible in flow rate.

Mobile-phase mass transfer refers to the fact that in a single stream, the solvent molecules closer to the "river bed" (the stationary phase) flow more slowly than those molecules in the center of the stream. This leads to essentially the same dispersion problem as was described above for eddy diffusion. This problem can be minimized by using small particles so that the channels are small.

Longitudinal diffusion refers to the tendency of molecules to diffuse randomly away from the band center. This is greater for those mobile phases that move more slowly and that have larger diffusion coefficients for the samples in the mobile phase. Fast flow rates minimize this problem.

Stagnant mobile-phase mass transfer refers to the fact that some of the molecules of a band can be temporarily trapped in pores of the particles of column packing. As a result, these molecules are delayed in their progress through the column and this, in turn, results in band spreading. Use of either hard-core particles (or small particles) of stationary phase minimize this problem by minimizing the number of pools that lead to stagnation.

Stagnant stationary-phase mass transfer is essentially an extrapolation of the effect described just above. After the molecules of x have penetrated into a pore of the stationary phase, they actually penetrate the stationary phase. Use of thin films of stationary phase on the support can minimize this problem.

Most of the preceding discussion has centered on a model for gc that results in ideal Gaussian-shaped curves (see Fig. 7.29. When gc analyses are carried out on a preparative scale, extensive departure of peak shapes from Gaussian is common. In order to understand the problems involved, we should first consider the concept of "tailing" (Fig. 7.30).

There are a number of factors that cause a gc peak to "tail," that is, to be an unsymmetrical peak and skewed toward the side of the peak corresponding to a greater retention time; these are illustrated in Fig. 7.30. Chemical tailing (Fig.

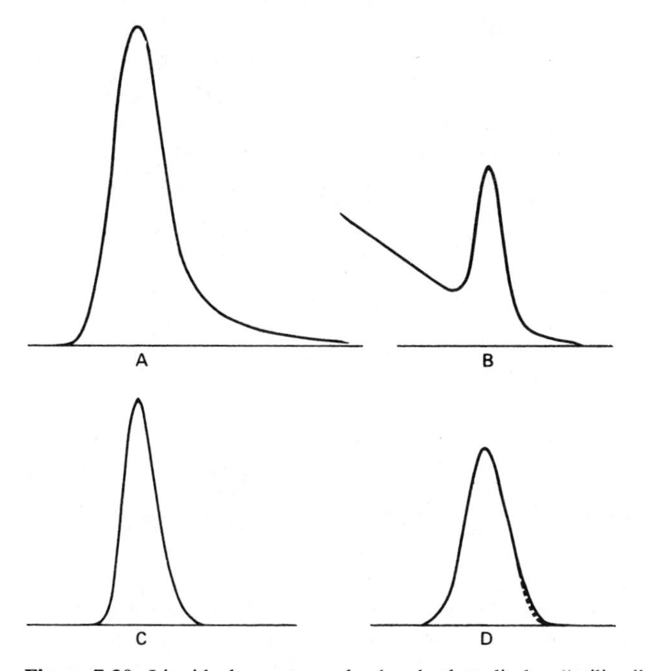

Figure 7.30. Liquid chromatography bands that display "tailing." A, chemical tailing; B, solvent tailing; C, Poisson tailing; D, exponential tailing. (From J. J. Kirkland and L. Snyder, *Modern Liquid Chromatography*, Wiley-Interscience, New York, copyright © 1974; with permission.)

7.30, A) usually occurs because of a mismatch between the sample and, for example, the stationary phase. When this tailing, typified by a slow return to baseline, is so severe as to cause poor peak separations, one should resort to another chromatographic technique. Solvent tailing (Fig. 7.30, B) occurs when the band of interest is overlapped by a band due to another, very large component; very frequently the large component is a solvent. Poisson and exponential tailings (Fig. 7.30, C and D, respectively) are usually not a problem; these occur because of inefficient columns and normal departure from symmetry, respectively. A more detailed discussion of tailing has been published.[26]

Resolution (R_s) is the goal of the separations chemist. This can be determined numerically from

$$R_s = \frac{t_2 - t_1}{(\frac{1}{2})(t_{w_1} - t_{w_2})}$$

where t_2 and t_1 are the retention times of components 2 and 1, respectively, and t_{w_2} and t_{w_1} are the respective times for the widths of these two components (see Figs. 7.29 and 7.31). The fraction equal to R_s is thus the ratio of the retention time difference to one-half the peak width difference; thus R_s gives us some idea of the potential for overlap of the two peaks (Fig. 7.31). Figure 7.32 shows us the appearance of peaks due to various degrees of resolution (R_s) as a function of the relative proportions of a two-component system.

Another parameter that is of interest is time (t). Often we are simply trying to expedite analysis by minimizing t. We must realize, however, that separation time and resolution are strongly interrelated. The concept can be illustrated by considering a phenomenon observed in liquid chromatography. Separation time is a function of the viscosity of the mobile phase, total column porosity, column length, and the pressure drop across the column. Intensive study of hplc (Section 7.4.4) has meant that there are many reports of how these factors influence separation time. Figure 7.33 shows the relationship of the retention time of the

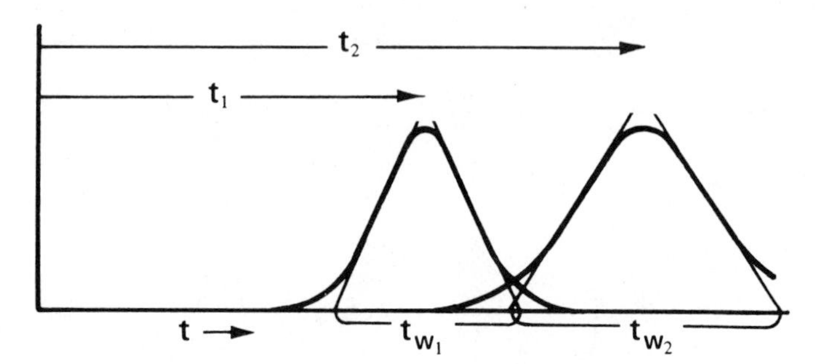

Figure 7.31. Two peaks that overlap slightly. An increase in any one of t_{w_1}, t_{w_2}, or t_1 (or a decrease in t_2) will decrease the resolution R_s. (From J. J. Kirkland and L. Snyder, *Modern Liquid Chromatography*, Wiley-Interscience, New York, copyright © 1974; with permission.)

[26] L. R. Snyder, *J. Chromatog. Sci.*, **10**, 200 (1972).

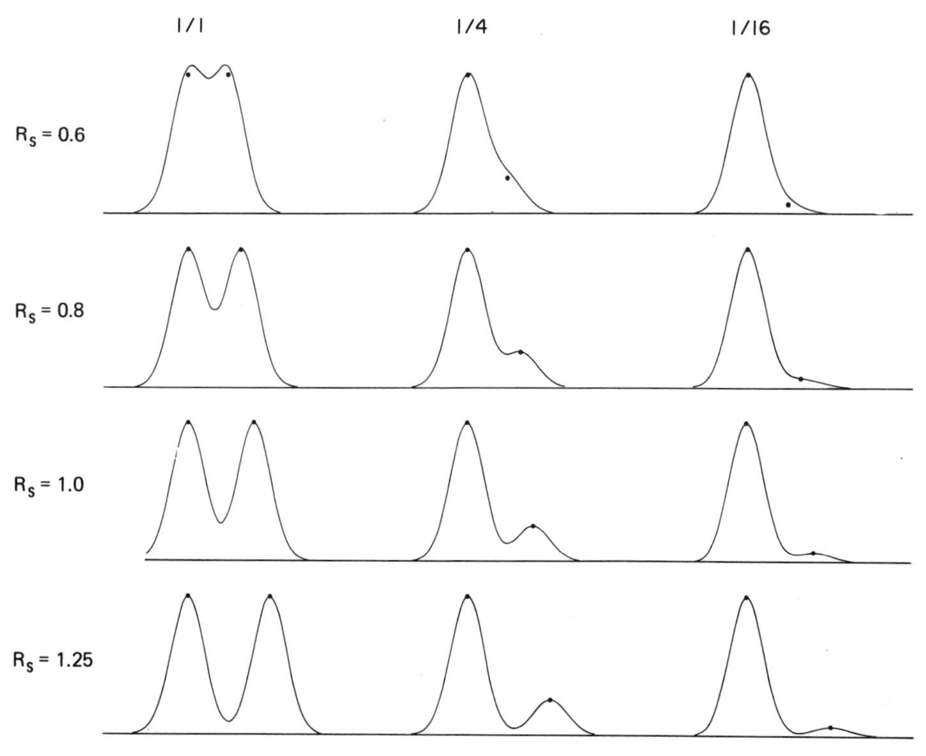

Figure 7.32. Chromatographic separations. The column of figures down the left is a variation of the magnitude of resolution (R_s), the axis along the bottom is time (three segments), and the ratios along the top are relative amounts of the two components. (From J. J. Kirkland and L. Snyder, *Modern Liquid Chromatography*, Wiley-Interscience, New York, copyright © 1974; with permission.)

reference peak (t_0) to the column length for hplc columns of various particles sizes. It is clear that small particles, a subject of extensive hplc study, expedite lc analysis.[27] Figure 7.33 is the result of data obtained from analyses using well-packed columns of porous, irregular particles. Porous, spherical particles are more permeable and thus give t_0 times that are about one-half those in Fig. 7.33; pellicular particles give times that are about one-fourth of those in Fig. 7.33. Gels are more deformable, less permeable, and thus result in higher t_0 times.

A great concern in preparative-scale chromatography is analytical usefulness as a function of sample size. Figure 7.34 shows the variation in chromatogram appearance as a function of sample size. Note in Fig. 7.34, case b, that a decrease in retention time has occurred for both components, the component of longer retention time showing the greater decrease. Thereafter, increased sample size (cases c and d) results in a poorer degree of separation; that is, the peaks are clearly less resolved. This occurs because at certain sample sizes the linear

---

[27] The advantage of small particle size is at a maximum at 10 μm. Smaller particles, say, 5 μm, give better resolution but are very susceptible to plugging problems.

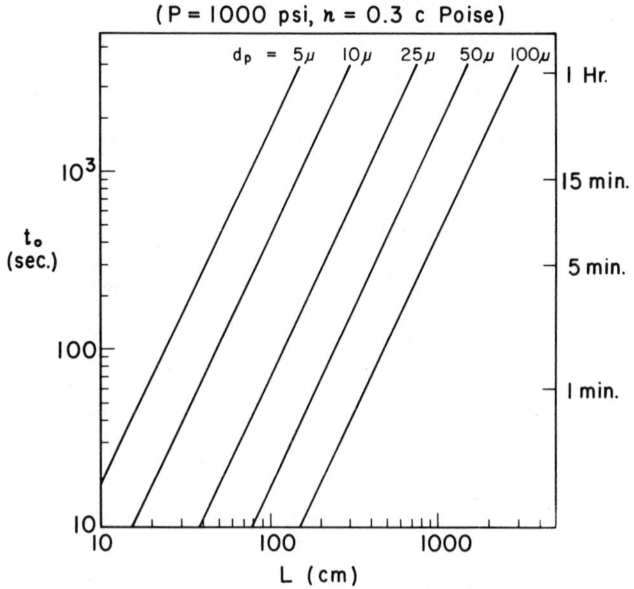

Figure 7.33. The log/log plot of the retention time of a reference standard versus column length (L) for a variety of particle sizes (d_p). Conditions: pressure = 1000 psi; viscosity = 0.3 cP. (From J. J. Kirkland and L. Snyder, *Modern Liquid Chromatography*, Wiley-Interscience New York, copyright © 1974; with permission.)

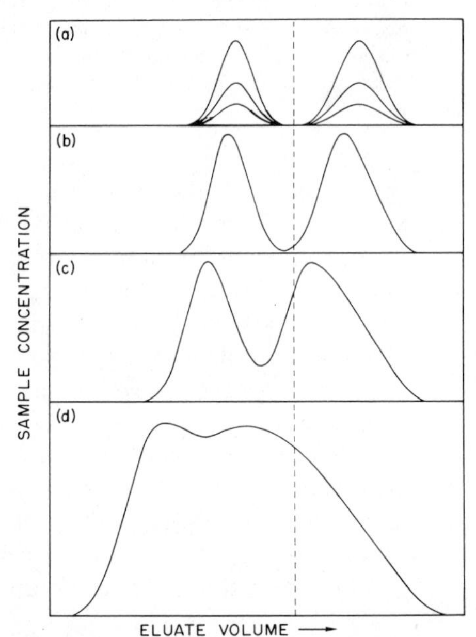

Figure 7.34. Effect on appearance of two peaks as both sample sizes are increased (case a, smallest sample size, to case d, largest sample size). (From J. J. Kirkland and L. Snyder, *Modern Liquid Chromatography*, Wiley-Interscience, New York, copyright © 1974; with permission.)

capacity of the column is exceeded; thereafter the previously gradual decrease in the height equivalent of a theoretical plate (hetp) takes place very rapidly. This decrease in plate height means that a given column contains fewer plates and thus separations will be less efficient.

In order to carry out a preparative-scale collection from a gas chromatograph, a proper collection device must be at hand. A number of commercial, automatic devices are available; some of these are built in to the chromatograph or are available as separate units. The turntable device that biochemists use to collect fractions of substances such as amino acids can be used.

Most practicing organic chemists find it convenient and sufficient to use a collection device such as shown in Fig. 7.35. It should be pointed out that care must be exercised in the choice of coolant; some compounds, when collected at condensation temperature conditions that are far too low, will form an aerosol that cannot be easily condensed in the collection trap. Centrifuge tubes are handy for collection (Fig. 7.35*c*) when small samples require spinning down after collection.

Samples collected directly from gc are usually very pure and can be analyzed directly; for example, the sample can be washed with $CDCl_3$ solvent directly into a nmr tube. For many compounds it is *necessary* that they be collected from the gas chromatograph and analyzed quickly so that physical constants and other characteristics determined from these samples will be accurate. The samples collected can be checked for purity by reinjecting a small sample onto the gc instrument. Sometimes the column adsorbent will bleed off with the sample; in these cases the collected sample should be redistilled in a microapparatus.

High-pressure liquid chromatography has become very useful for preparative-scale collections. This procedure has advantages over traditional lc and gc in the speed, ease, and nondestructive nature of these collections. A typical instrument is shown in Fig. 7.27.

Determination of the solubility class (Chapter 5) may be difficult to carry out. These difficulties may be due to factors such as the inability to note visibly any change as a result of the solubility test or the development of emulsions or color changes that obscure the degree to which solution takes place. Gas chromatography can be used to aid such solubility studies; for example, gc will detect small amounts of solute and thus aid determinations in those cases where solubility is limited.

Gas chromatography can be used to carry out solubility analyses as follows: The solute and solvent are prepared in the amounts as have been described earlier (p. 93) or as in the following table:

	Mass of solute/volume of solvent	
Scale	**solids**	**liquids**
Macro	100 mg/3 ml	0.2 ml/3 ml
Semimicro	30 mg/1 ml	1 drop/15 drops

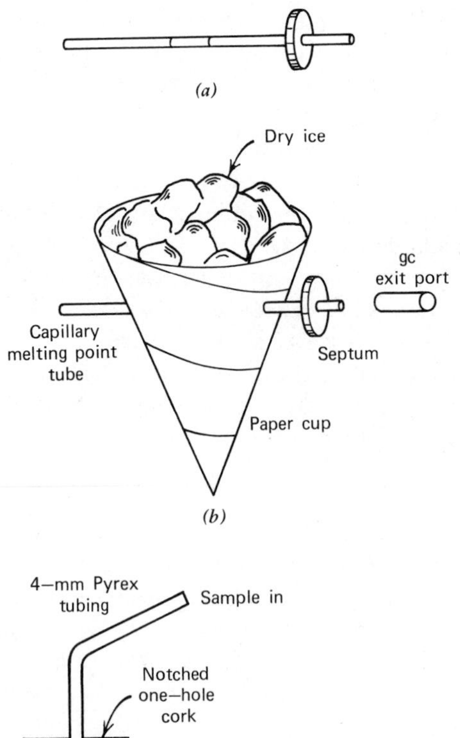

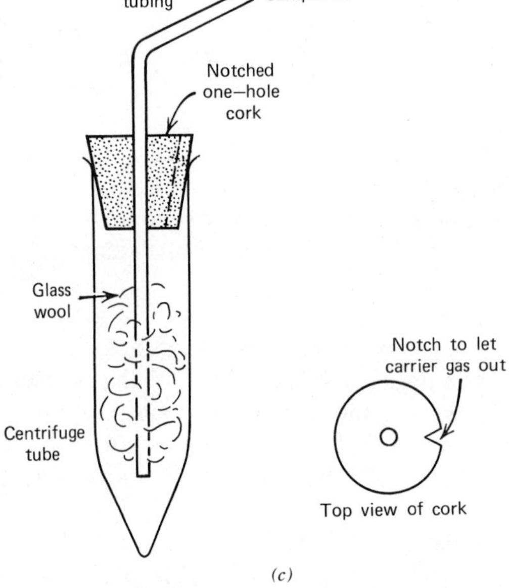

Figure 7.35. Collection devices for gc. (*a*) Capillary tube; no cooling. (*b*) Capillary tube; cooling added. (*c*) Centrifuge tube (glass-wool can be omitted if no aerosol problems occur). [Sources: Parts (*a*) and (*b*) are from R. C. Crippen, *The Identification of Organic Compounds with the Aid of Gas Chromatography*, McGraw-Hill, New York, copyright © 1973. Part (*c*) is from R. B. Bates and J. P. Schaefer, *Research Techniques in Organic Chemistry*, copyright © 1971; by permission of Prentice-Hall, Inc., Englewood Cliffs, N.J.]

Then one should try very hard to dissolve the solute in the solvent (normally by extensive shaking). If the solute is highly soluble in the solvent (and thus the solubility classification can be easily made by visual observation), one need not use gc to aid the solubility analyses. If not, determine the gc retention times and molar responses (see below), under the same instrument conditions, for each of the solvent and the solute. Then extract 2 to 10 μl of the liquid layer arising from the solubility test and inject this onto the chromatograph.

Instrument conditions and sample sizes should be adjusted so that the areas of both the solvent and solute peaks can be measured under the same conditions. Molar responses (in cm^2/mole) can be measured by chromatographing known volumes of pure solvent (of known density) and standard solutions of the solute. The area of the peak (in cm^2) can then be divided by the number of moles of sample causing that peak to obtain the molar response. If the molar response of the solute is equal to that of the solvent, the ratio of the areas of the solute and solvent peaks allows direct determination of the amount of solute extracted. If the molar responses are not equal, the peak areas must be converted to sample amounts by use of the molar response of each component: If P.A. is the peak area for a component of interest and M.R. is the molar response for a given component, the following equation allows conversion of ratios of peak areas to mole ratios of these components:

$$\frac{\text{moles of component 1}}{\text{moles of component 2}} = \frac{(\text{P.A.})_1(\text{M.R.})_1}{(\text{P.A.})_2(\text{M.R.})_2}$$

For example, if component 1 is the compound whose solubility is being determined and component 2 is the solvent, the preceding formula provides the solubility of component 1.

Water in an organic compound can be detected by using gas chromatography.[28] This is done by injecting solutions of solids onto a nonpolar gc column (for example, DC-710 silicone oil, or a hydrocarbon polymer such as Porapak Q) and looking for the water peak. One must use thermal conductivity, rather than flame ionization, as the detection system for gc water tests.

In choosing a gc column for analysis of a specific class of compound, the following factors should be considered. Retention times (and thus retention volumes) on relatively inert columns (such as SE-30 columns, see Fig. 7.36) usually are longer for compounds with higher boiling points. This is shown to be so in Fig. 7.36 for alkanes, alkylbenzenes, and alcohols. Polar gc columns, such as Carbowax 20M, will cause peak shapes to vary due to association (hydrogen bonding) and molecular weight changes: Note in Fig. 7.37 that benzyl alcohol, and to even greater extent hydroxycinnamic acid, show broad peaks due to these two factors and due to longer retention times.

---

[28] Water can be detected in liquids by adding white, anhydrous copper sulfate to the liquid and using the appearance of the blue color of the hydrated copper sulfate as a positive test for water. Water can be removed from liquids by using molecular sieves, or anhydrous calcium sulfate or anhydrous magnesium sulfate. Desiccators or drying pistols (p. 84) can be used for solids.

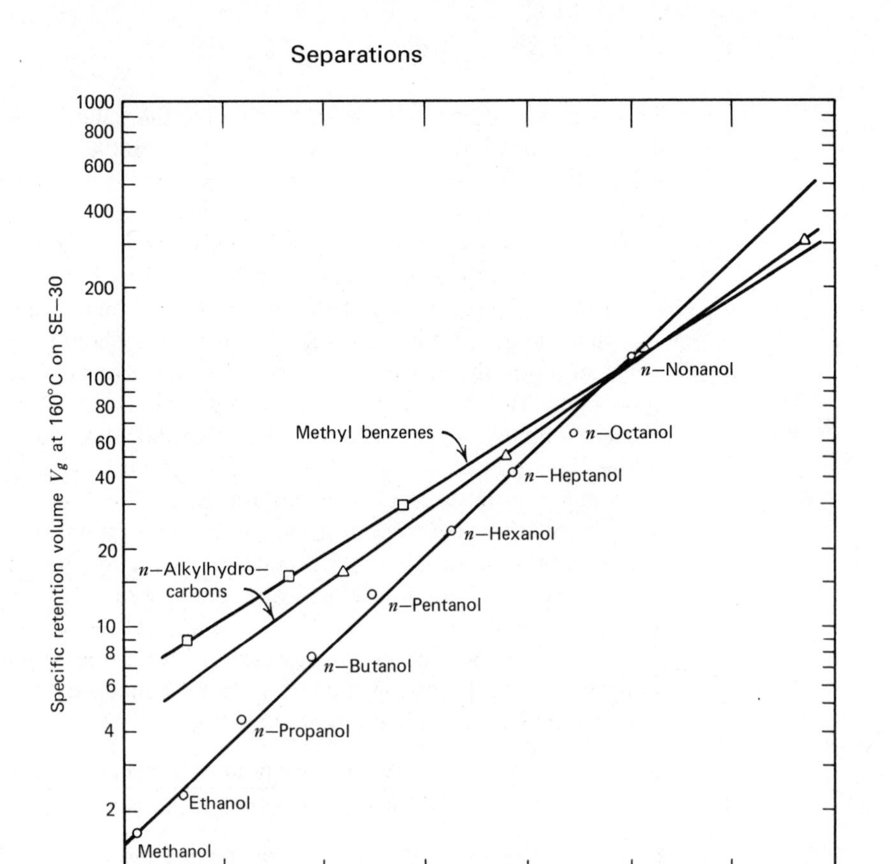

Figure 7.36. Hydrocarbons and alcohols: relationships of boiling points to specific gc retention volumes at 160°C on an SE-30 column (20%). From R. C. Crippen, *The Identification of Organic Compounds with the Aid of Gas Chromatography*, McGraw-Hill, New York, copyright © 1973.

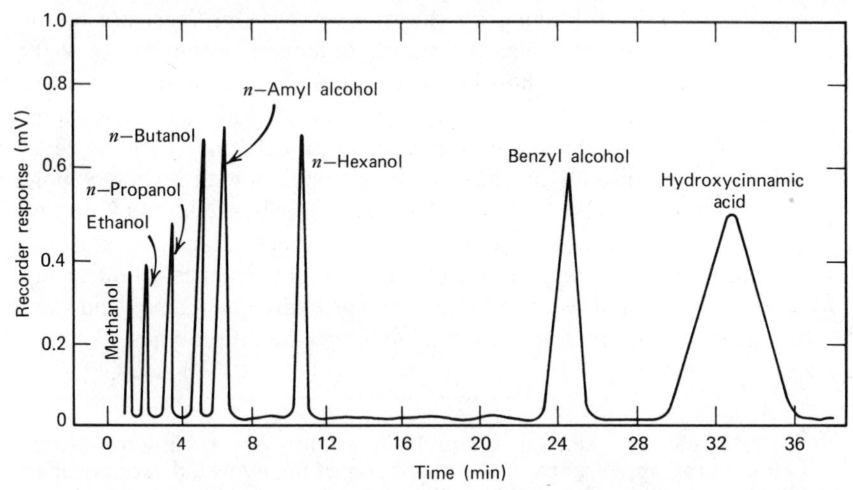

Figure 7.37. Chromatogram of n-alkyl alcohols, benzyl alcohol, and hydroxycinnamic acid on $12\,\text{ft}\times\frac{1}{4}$ in. carbowax 20 M column (20%) at 225° C. From R. C. Crippen, *The Identification of Organic Compounds with the Aid of Gas Chromatography*, McGraw-Hill, New York, copyright © 1973.

7.5 OPTICAL RESOLUTION

When an unknown is deduced to have a structure that possesses chirality, the optical activity of the sample should be measured (see pp. 412–416). The specific rotation, when measured in the same solvent and at reasonably similar concentrations, should be compared to literature values as a measure of the identity and purity of a compound. If the compound shows no optical activity, this should not be taken as an indication of a lack of chirality, because the compound may exist as a racemic mixture, or the compound may possess chirality that produces negligible optical activity. Separation of the enantiomers of a racemic mixture (resolution) will allow for confirmation of the chiral nature of an unknown.

A procedure for the resolution of α-phenylethylamine has been published.[29]

CH_3
$C_6H_5CH—NH_2$ $\xrightarrow{\text{resolution}}$
racemic

C_6H_5 H C CH_3 NH_2
R(+)
$[\alpha]_D^{25°C} = +40°$ (neat)

H C_6H_5 C H_2N CH_3
S(−)
$[\alpha]_D^{25°C} = -40°$ (neat)

This procedure involves treatment with (+)-tartaric acid, a reagent that is readily available in most organic laboratories. In Table 7.9 are outlined resolution procedures for a variety of functional groups.

Table 7.9. Reagents and References to Procedures for Optical Resolution

Functional group	Reagent used	Reference[a]
Alcohols	Amines, via phthalate esters of alcohols	Boyle[b]
Aldehydes, ketones	Amine tartrate and amine sulfate complexes	Ault[c]
	Menthyl *N*-aminocarbonate	Boyle
Alkenes	Pt complexes	Boyle
Amines	(+)-Tartaric acid	Ault
Amino acids	α-Phenylethylamine	Ault
	Paper chromatography	Boyle
Aromatic (biphenyls)	Column chromatography (on lactose or potato starch)	Boyle
Carboxylic acids	α-Phenylethylamine	Ault
Esters of amino acids	gc (peptide adsorbents)	Boyle
Salts of acids	Ion exchange chromatography	Boyle

[a] Boyle is not a primary reference but rather a lead to primary references. Ault describes a detailed procedure for the resolution of α-phenylethylamine.

[b] P. H. Boyle, *Quart. Rev.* (London), 323 (1971).

[c] A. Ault, *Org. Syn., Vol. V,* edited by H. E. Baumgarten (Wiley, New York, 1973), p. 932.

See also S. H. Wilen, *Resolving Agents and Resolutions in Organic Chemistry* (University of Notre Dame Press, Notre Dame, Ind., 1971).

[29] A. Ault, *Org. Syn, Coll. Vol. V,* edited by H. E. Baumgarten (Wiley, New York, 1973), p. 932. Since this procedure appears in *Org. Syn., it has been verified independently.* Most procedures that appear in standard journals, however, have not been verified.

CHAPTER EIGHT

special characterization techniques

In this chapter we outline characterization techniques that are not usually part of the standard analysis of an organic compound. One usually will consult the procedures in this chapter only after having analyzed the compound by the procedures described in Chapter 6. Results of the tests in Chapter 6 and a general knowledge of the value of the procedures in this chapter will provide a guide to choosing specific procedures described below.

8.1 OPTICAL ROTATION

The optical rotation is determined only if the list of possible compounds contains optically active substances.

8.1.1 Preparation of the Solution

Procedure. An accurately weighed sample (about 0.1 to 0.5 g) of the compound is dissolved in 25 ml of solvent in a volumetric flask. The solvents commonly used are water, ethanol, and chloroform. The solution should be clear; it must contain no suspended particles of dust or filter paper. It should also be colorless if possible. If the solution is not clear, either the original compound should be recrystallized or else 50 ml of the solution made up and filtered through a small dry filter paper. The first 25 ml of filtrate is discarded; the last 25 ml is used for the determination.

8.1.2 Filling the Polarimeter Tube

Procedure. The cap is screwed on the small end of the polarimeter tube (T, Fig. 8.1), the tube is held vertically, and the solution is poured in until the tube is full and the rounded meniscus extends above the top end of the tube. The glass plate is caused to slide over the end of the tube so that no air bubbles are imprisoned. The brass cap is then screwed on.

Precautions

1. A rubber washer should be placed between the glass plate and the brass cap. There is no washer between the glass end plate and the glass tube. This is a glass-to-glass contact.

2. The ends should not be screwed on too tightly. They should be turned up enough to make a firm, leak-proof joint. If the ends are screwed on too tightly, the glass end plates will be strained and a rotation will be observed with nothing in the tube at all. For substances with very low rotations it is advisable to loosen the caps and tighten them again between readings.

3. If the brass ends come off the glass tube, they may be cemented in place again by means of litharge-glycerol cement. In putting an end back on, care must be taken that the glass part extends 1 mm beyond the brass end.

8.1.3 The Use of the Polarimeter

One form of the polarimeter is an instrument of the Lippich double-field type. A schematic diagram of the working parts is shown in Figs. 8.1 and 8.2.

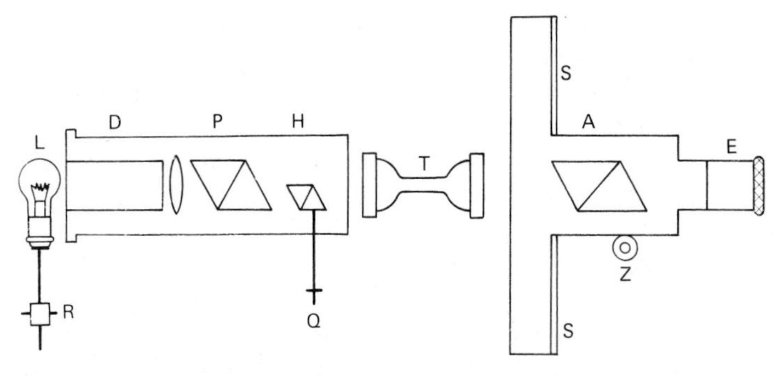

Figure 8.1. Schematic diagram of polarimeter (longitudinal section). L, light (may be replaced by sodium flame or electric sodium lamp); D, dichromate filter; P, polarizing Nicol prism; Q, half-shadow adjustment, H, half-shadow Nicol prism; T, tube containing solution; A, analyzing Nicol prism; E, eyepiece; S, scale; R, switch.

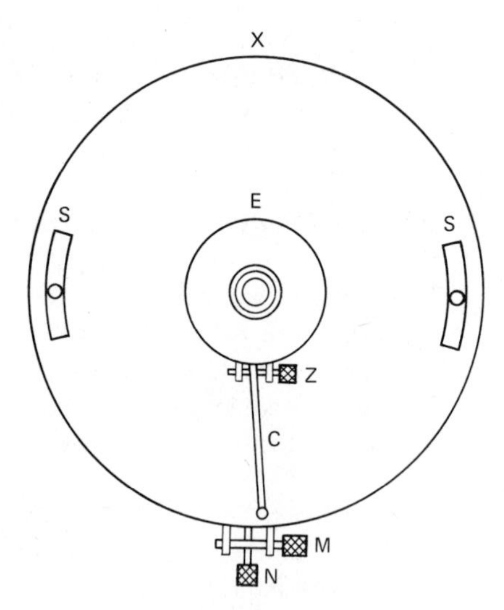

Figure 8.2. Polarimeter (end view). E, eyepiece; Z, zero adjustment, S, scale; C, coarse adjustment; N, locking nut; M, micrometer adjustment; X, lamp.

Procedure. The light (L) is turned on by means of snap switch R. The eyepiece (E) is focused by turning to right or left until the line dividing the two fields is sharp.

To determine the *zero reading*, the locknut (N) is loosened and the two fields are matched approximately by means of the coarse adjustment lever (C). To use C it is necessary first to loosen N and then to tighten C by turning the white knob on the end of C. After approximate equality of the fields is obtained, N is tightened and the fields are matched exactly by turning the micrometer adjustment (M) to the right or to the left. The lamp (X) is turned on, and the scale (S) is read. The main circle is divided into degrees and 0.25 degree. The vernier or outside scale is divided into 25 divisions, enabling the reading to be made to 0.01 degree. At least five readings should be made and the values averaged. For very exact work, both scales (S) should be read and the readings averaged in order to correct for any lack of centering of the scale with reference to the eyepiece and Nicols.

To get the *rotation* of the solution, the polarimeter tube is placed in the trough. The cover is closed, and the procedure that was followed for the zero reading is repeated. The average of at least five readings should be taken. The observed rotation is the difference between this value and the zero reading.

Notes and Precautions

1. The instrument is usually set up for use with yellow light corresponding to the sodium D line. Light of this wavelength is most readily obtained by means of electrically heated sodium lamps, which produce a brilliant yellow glow. A sodium flame may also be used, but it is not as brilliant. A white light with the following solution in a 3-cm filter cell gives good results for compounds possessing low

observed rotations. The filter solution is made of 8.9 g of hydrated copper sulfate, 9.4 g of potassium dichromate, and 300 g of water. The solution is filtered and allowed to stand in order to permit dust particles to settle.

A mercury arc makes it possible for the green mercury line to be used. To use the green mercury line, the light (L) is removed and the dichromate filter (D) is unscrewed. The arc is substituted for L, and the above procedure is repeated.

2. The instrument is set up and adjusted for all ordinary work. The half-shadow lever (Q) or the screw (Z) should not be changed.

3. In making readings, it is best to start with the scale (S) adjusted nearly to zero. If the arm (C) is moved very much, reversal of the fields will occur.

4. The lever (C) and locknut (N) should not be screwed up too tightly. It is only necessary to turn them firmly.

8.1.4 Expression of Results

The specific rotation of a substance is calculated by one of the following formulas:

For pure liquids: *For solutions:*

$$[\alpha]_{D}^{25^\circ} = \frac{\alpha}{ld} \qquad [\alpha]_{D}^{25^\circ} = \frac{100\alpha}{lc}$$

where

$[\alpha]_{D}^{25^\circ}$ = specific rotation at 25°C (using the D line of sodium)
α = observed rotation
l = length of tube (decimeters)
d = density in g/ml
c = grams in 100 ml of solution

It should be noted that the specific rotation may be quite sensitive to the nature of the solvent and, in certain cases, even to the concentration of the substance being examined. The wavelength of the light used for measurement can also affect not only the magnitude but also the sign of rotation. Attention should be paid, therefore, to the exact conditions under which a rotation reported in the literature was measured.

The following is the correct way to report specific rotation:[1]

$$[\alpha]_{546}^{25^\circ} = -40 \pm 0.3^\circ \ (c = 5.44 \text{ g/100 ml, water})$$

The preceding relationship refers to a specific rotation determined at 25°C, in water, with light of a wavelength of 546 nm, at a concentration of 5.44 g/100 ml of solution. It is necessary to determine the observed rotation, α, at two different

---

[1] All of these items should be reported. It will be common, however, to find data reported that omit the concentration and/or the solvent. The method by which the error range is determined (here ±0.3°) should be reported. For example, is the error plus or minus one standard deviation, or is it an estimated error based on the measurement methods? The error shown in this example is large compared to that which applies to modern instrumentation.

concentrations. In the simplest cases, observed rotations will be decreased by the same factor as the concentration decrease; for example,

α	**Concentration**
−50°	x
−5.0°	0.1x
0.50°	0.01x

Thus in such a case the $[\alpha]$ determined from all three experiments will be the same, and one has in hand a value of $[\alpha]$ that can be safely compared to literature values of $[\alpha]$ determined at other concentrations.[2] In this way the value of $[\alpha]$ in such simple cases can be used to confirm the identity of the compound of interest. If the value of $[\alpha]$ has been determined upon a liquid sample using no solvent, the specific rotation should be reported as follows:

$$[\alpha]_D^{25°} = +40° \text{ (neat)}$$

8.2 ULTRAVIOLET SPECTROSCOPY

The use of ultraviolet spectroscopy by organic chemists is made clearer by relating ir, nmr, and uv. To characterize an organic compound, the chemist would normally use ir to detect and identify functional groups and nmr to determine the structural arrangement of protons and carbons. After having established at least a preliminary idea of the structure of the compound, the chemist can decide if uv would be useful for characterizing the compound. The existence of multiple bonds, especially when conjugated, is an indication that uv could be useful. A clearer idea of when and how to use uv can be obtained upon study of the remainder of this section.

Although a comprehensive discussion of ultraviolet spectroscopy is beyond the scope of this book we shall indicate briefly the types of structural problems that it might be expected to solve. The following general references are recommended and should be consulted by anyone interested in using the method for structure determination.

H. H. Jaffé and M. Orchin, *Theory and Applications of Ultraviolet Spectroscopy* (Wiley, New York, 1962).
A. E. Gillam and E. S. Stern, *An Introduction to Electronic Absorption Spectroscopy in Organic Chemistry*, 2nd ed. (Edward Arnold, London, 1957).

---

[2] In a few special cases the magnitude of the specific rotation will depend on concentration. For example, concentration-dependent hydrogen bonding of an alcoholic solvent to a polar chiral substrate could cause this dependence. Thus, it is important to include the concentration of the sample.

E. S. Stern and T. C. J. Timmons, *Electronic Absorption Spectroscopy in Organic Chemistry* (St. Martin's Press, New York, 1971).

These books provide the most comprehensive and generally useful sources of information for the organic chemist.

E. A. Braude, *Determination of Organic Structures by Physical Methods,* edited by E. A. Braude and F. C. Nachod (Academic Press, New York, 1955), chap. 4.
S. F. Mason, *Quart. Rev.*, **15,** 287 (1961).

The two reviews provide a good introduction to the theory of ultraviolet and visible light absorption.

H. M. Hershenson. *Ultraviolet and Visible Absorption* (Academic Press, New York, 1955). This work provides a relatively simple route to spectra published between 1930 and 1957.
A. Streitwieser, jr., *Molecular Orbital Theory for Organic Chemists* (Wiley, New York, 1961). Ultraviolet spectrometry is related to orbital theory.

The ultraviolet region is commonly divided for practical reasons into the "near" or "quartz" ultraviolet (the region in which air and quartz are transparent), which includes the region 200–750 nm (2000–7500 Å), and the "far" or "vacuum" ultraviolet. The Cary and Beckmann spectrophotometers conveniently measure spectra in the near region, although measurements at wavelengths of 200–220 nm are reliable only under special circumstances. Instruments are available that may be employed down to 175 nm. The far ultraviolet, because air in the path of the light absorbs strongly in this region, requires special techniques.

The ultraviolet spectrum, quite unlike the infrared or Raman, arises from *electronic excitation* of the molecule by the irradiating light. The important consequence of this fact, from the point of view of the structural chemist, is that spectra in the ultraviolet region are diagnostic of unsaturation or nonbonded electrons in the absorbing molecule. This comes about because, with few exceptions, only molecules containing multiple bonds have sufficiently stable excited states to give rise to absorption in the near ultraviolet. Thus saturated hydrocarbons, alcohols, and ethers are transparent in this region. Furthermore, monofunctional olefins, acetylenes, carboxylic acids, esters, amides, and oximes have absorption maxima just outside the near region (at about 200 nm) and, in general, show only end absorption.

A third class of compounds comprising aldehydes, ketones, aliphatic nitro compounds, and nitrate esters is characterized by absorption maxima in the near ultraviolet, but they have intensities so low that they are useful only under special circumstances. For example, acetone has a molar absorptivity (ε, the measure of the absorption intensity) of 16 at its maximum at 270 nm. By way of comparison,

Table 8.1.

Functional group	λ_{max}(nm)	ε
RCH=CHR	185	8,000
$R_2C{=}O$	270	16
RCOOH	204	41
RCOCl	235	53
RNO_2	275	12
R_2S	210	1,200
$R_2C{=}NNHCONH_2$	229	10,900
RN=NR	350	140
RN_3	210	500
	270	60
RN=O	300	100
	670	20

a strong band would have an ε of 1000–10,000; and values as high as 100,000 are not uncommon. Nevertheless, ultraviolet spectroscopy has sometimes been quite useful for studying such weakly absorbing functional groups. In Table 8.1 are listed some representative data giving the positions of the absorption maxima and molecular extinction coefficients of a few isolated (nonconjugated) functional groups.

It can be seen that, if the use of ultraviolet spectroscopy were confined to monofunctional molecules, it would be severely limited. Fortunately, however, even functional groups such as the olefinic, acetylenic, and carboxyl (and, in general, any function containing a double or triple bond) give rise to strong absorption in the near ultraviolet when they are conjugated with one another. Thus, ultraviolet spectroscopy has proved to be primarily a tool for the study of conjugated systems.[3] It is obvious from this discussion that the use of ultraviolet spectra in structural determination is rather different from that of infrared techniques; in fact, the two methods often supplement each other.

The modern double-beam ultraviolet spectrophotometer generally plots absorbance (optical density) against wavelength in nanometers. Since the absorbance (A) of a given solution has quantitative significance in ultraviolet spectroscopy, and since the solutions employed are sufficiently dilute that in general Beer's law holds, it is almost always desirable to convert absorbance to molar absorptivity (ε) in any curve that is to be published.[4] These quantities are related to each other by the equation

$$\varepsilon = \frac{A}{bC}$$

[3] L. N. Ferguson and J. C. Nnadi, *J. Chem. Educ.*, **42,** 529 (1965).

[4] Note that the molar absorptivity, ε, has units corresponding to area per mole. Thus this quantity has been referred to as the cross section of the molecule with respect to photons of the energy corresponding to this wavelength.

Table 8.2. Ultraviolet Solvents

Solvent	Lower wavelength limit (nm)
Water	205
Ethanol (95% or absolute)	210
Hexane	210
Cyclohexane	210
Methanol	210
Diethyl ether	210
Acetonitrile	210
Tetrahydrofuran	220
Dichloromethane	235
Chloroform	245
Carbon tetrachloride	265
Benzene	280

where b = length of cell in centimeters (generally 1 cm), and C = concentration in moles per liter. It is seen that ε is a measure of the absorbance of a solution at a concentration of 1 mole/liter in a 1-cm cell, and ε values of different spectra can be compared even though somewhat different concentrations may have been used in making the two curves.

The most common solvents for ultraviolet spectroscopy are cyclohexane, 95% ethanol, benzene-free absolute-ethanol, and dioxane. A list of uv solvents, in decreasing order of useable range, is given in Table 8.2. The approximate concentration to use may be estimated if any information is available as to the expected value of the maximum value of ε. Thus the Cary absorbance scale runs from 0.0 to 2.4 absorbance units, and, for a plot such that a maximum with $\varepsilon = 10^4$ has an absorbance of 2.0, it is seen that the correct concentration is $2/10^4$ or 2×10^{-4} molar when a 1-cm cell is employed.

8.2.1 Identification of Functional Groups from Knowledge of Positions of Absorption Maxima

The use of spectroscopy to identify functional groups from knowledge of positions of absorption maxima, which is a common infrared technique, is rare in the ultraviolet for two reasons. First, most simple functional groups absorb weakly or not at all, and second, the spectra of most molecules are relatively simple; that is, they have only one or two maxima instead of the 10 or 20 common in an infrared spectrum. Necessarily, therefore, many types of functional groups have absorption in the same region. Nevertheless, inspection of the ultraviolet spectrum by an experienced observer has occasionally suggested the presence of a previously unsuspected functional group. Probably such inspections are most commonly made to reveal the class of aromatic or heterocyclic ring present in a natural product of unknown structure. As another example, the presence of a nitrophenyl group in chloromycetin was suggested on the basis of the ultraviolet spectrum. In general, however, some information about possible functional groups is required before the ultraviolet spectrum is useful. Of course, the absence of functional groups known to absorb in the ultraviolet region may often be unambiguously inferred.

8.2.2 Use of Model Compounds

Even though it might at first appear to be a disadvantage that only rather specific functional groups give rise to ultraviolet absorption and that the spectrophotometer is blind to changes not affecting the absorbing unit, this narrowness of vision is largely responsible for the usefulness of ultraviolet spectroscopy in structural work. For, although it is almost never possible to predict on the theoretical grounds the spectrum of a molecule, the spectrum of a "model," which is a simple substance that differs from the unknown in a way that should not affect the absorbing unit, can often be obtained.

An example of the use of model substances to approximate ultraviolet spectra is provided by the structure proof of the quinoline dicyanides.[5] Quinoline reacts with cyanogen bromide and hydrogen cyanide to form two dicyanides, which were believed to have the structures I and II. Confirmation was necessary,

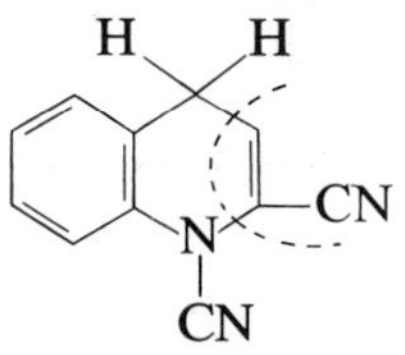

H H
CN
N
CN
I

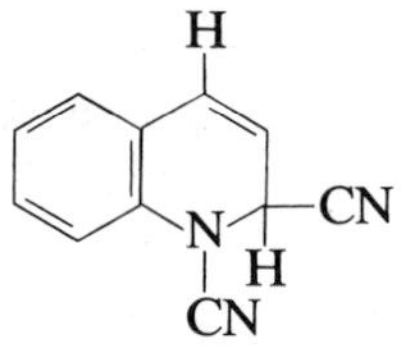

H
CN
N
H
CN
II

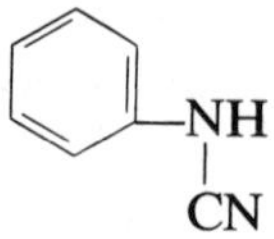

NH
CN
III

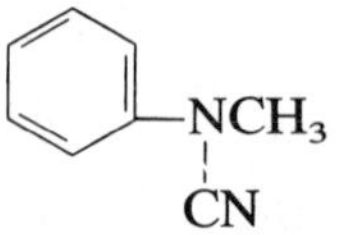

NCH_3
CN
IV

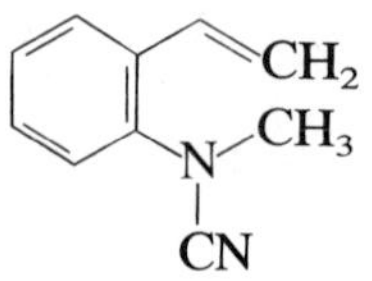

CH_2
CH_3
N
CN
V

and furthermore, it was unsettled as to which quinoline dicyanide was I and which was II. Since no unambiguous synthetic or degradative method was available, it was necessary to resort to a study of the ultraviolet spectra. As model substances for "quinoline dicyanide" I were chosen phenylcyanamide (III) and *N*-methylphenylcyanamide (IV). In Fig. 8.3 the general shapes of the curves, the positions and heights of the maxima, are seen to be in good agreement. It may be noted that quinoline dicyanide I has an α,β-unsaturated nitrile unit that does not appear in the model substances. This grouping could be ignored because its absorption is very much smaller than that of the other absorbing unit.

[5] M. G. Seeley, R. E. Yates, and C. R. Noller, *J. Amer. Chem. Soc.*, **73**, 772 (1951).

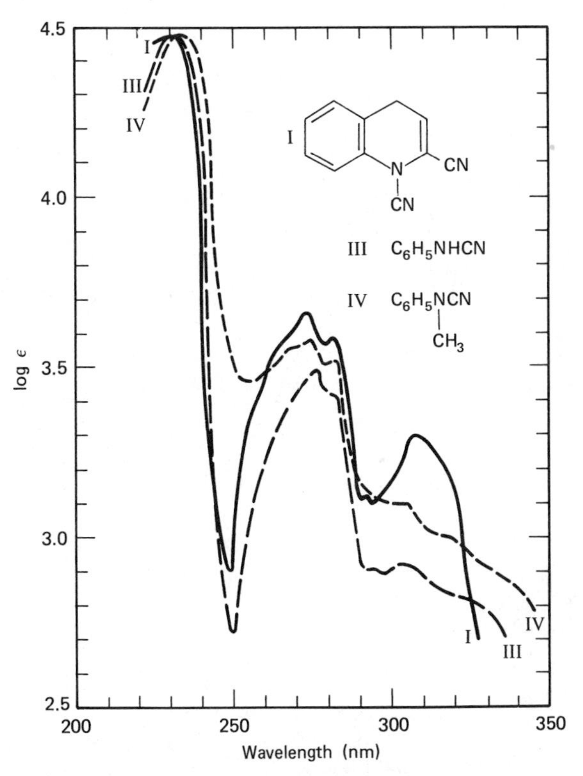

Figure 8.3. Absorption spectra of low-melting quinoline dicyanide (I), phenylcyanamide (III), and *N*-methylphenylcyanamide (IV). (Reproduced from the *Journal of the American Chemical Society* through the courtesy of the Society and Professor Carl R. Noller.)

As a model for the other quinoline dicyanide (II) was chosen *N*-methyl-*o*-styrylcyanamide (V). The spectra of II and V are shown together in Fig. 8.4 and are seen to agree well with each other and to be different from the spectra in Fig. 8.3. The ultraviolet spectra of the quinoline dicyanides thus provide strong evidence that they have the structures indicated.

8.2.3 The Additivity Principle

The aspect of ultraviolet spectroscopy that gives it greatest flexibility in structural problems is the additivity principle, which can be stated as follows. The spectrum of a molecule containing two absorbing units separated by one or more insulating atoms (a —CH_2— group, for instance) is approximated closely by addition of the spectra of the absorbing units.[6,7] Some examples of the use of the additivity principle follow.

[6] See E. A. Braude, *J. Chem. Soc.*, 1902 (1949).

[7] Although one methylene group is sufficient insulation between two chromophores as a first approximation, for really satisfactory additivity two or more methylene groups are required (see References on pp. 416–417).

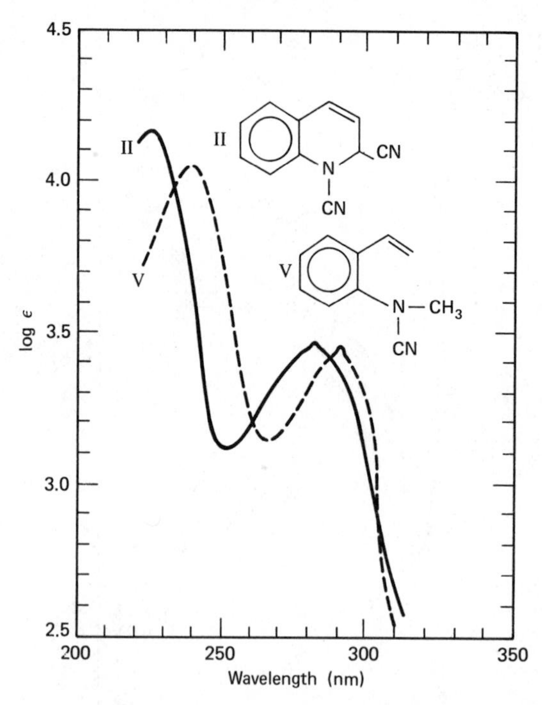

Figure 8.4. Absorption spectra of high-melting "quinoline dicyanide" (II) and N-methyl-σ-styrylcyanamide (V). (Reproduced from the *Journal of the American Chemical Society* through the courtesy of the Society and Professor Carl R. Noller.)

In the course of the structure determination of terramycin,[8] an important degradation product was VI, of which that part of the structure indicated was

CH_3
$CHOH[C_6H_2(OCH_3)_3]$
CH_2OH
CH_3O OCH_3

VI

established but that in square brackets was undetermined. The investigators suspected that the residue $[C_6H_2(OCH_3)_3]$ was one of the trimethoxybenzenes. It will be seen that the two chromophores, the naphthalene system and the benzene system, are insulated (nonconjugated) by the —CHOH—, and therefore the spectrum of the molecule may be approximated by adding the spectra of the two halves. The authors, therefore, added the spectrum of substance VII, which was available, in turn to those of 1,2,3-, 1,2,4-, and 1,3,5-trimethoxybenzene. Only

[8] C. R. Stephens, L. H. E. Conover, R. Pasternack, F. A. Hochstein, W. T. Moreland, P. P. Regna, F. J. Pilgrim, K. J. Brunings, and R. B. Woodward, *J. Amer. Chem. Soc.*, **76**, 3568 (1954).

for the 1,2,4-derivative did the summation curve coincide reasonably well with the spectrum of VI.

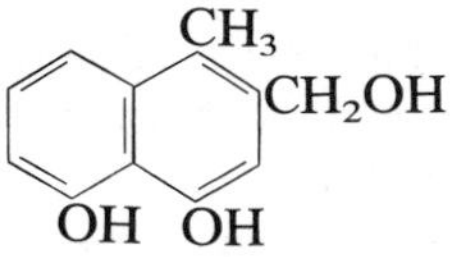

CH_3 CH_2OH
OH OH
VII

MacKenzie,[9] in a study of the structure of usnic acid, examined the spectra of the degradation products, VIII and IX. He noted that if the proposed structures are correct, structure VIII has the chromophore IX attached to the insulated chromophore of acetylacetone (enol form) (X). He therefore subtracted the spectrum of IX from that of VIII and obtained a curve that was essentially superimposable with that of acetylacetone. The proposed structures were thus confirmed.

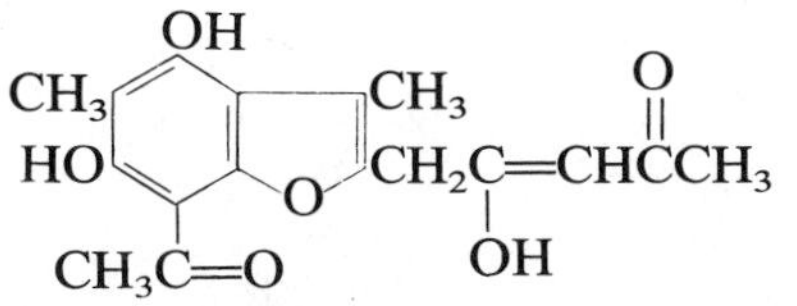

OH CH_3 HO CH_3 $CH_2C{=}CHCCH_3$ (O) OH $CH_3C{=}O$
VIII

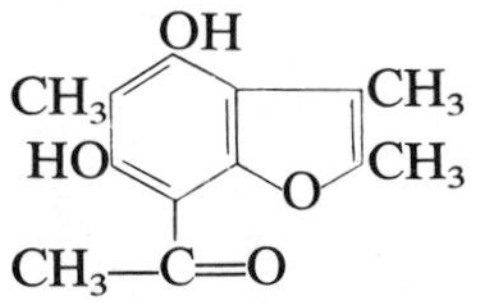

OH CH_3 HO CH_3 CH_3 $CH_3{-}C{=}O$
IX

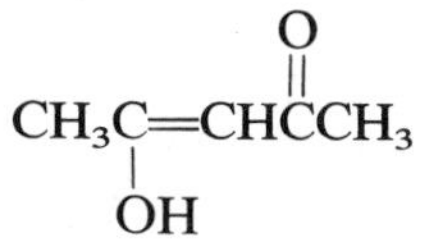

$CH_3C{=}CHCCH_3$ (O) OH
X

8.2.4 Effects of Molecular Geometry on Ultraviolet Absorption Spectra

An ever-present difficulty in selecting model compounds for comparison of their absorption spectra with spectra of unknowns is the possibility of unexpected changes in spectrum because of special steric effects in one or the other of the compounds to be compared. The most general source of such difficulties is steric inhibition of resonance. For example, the spectrum of biphenyl is not at all the sum of the spectra of two benzene rings. This is true because resonance interaction of the two conjugated phenyl rings in biphenyl stabilizes the excited states resulting from light absorption. It has been shown, on the other hand, that such resonance interaction requires that the two rings of biphenyl be approximately coplanar. Dimesityl (XI), which has a bulky methyl group in each of the four

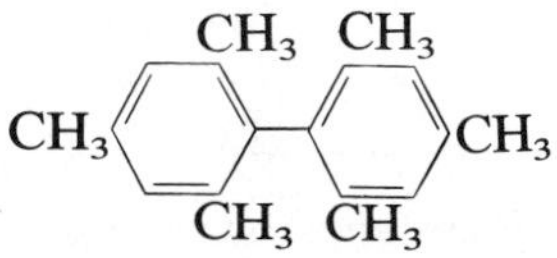

CH_3 CH_3
CH_3 CH_3
CH_3 CH_3
XI

ortho positions, is unable to assume the conformation with the two phenyl rings coplanar. Resonance interaction between the two rings is therefore inhibited, and,

[9] S. MacKenzie, *J. Amer. Chem. Soc.*, **74,** 4067 (1952).

as a consequence, the spectrum of dimesityl is very similar to the sum of the spectra[10] of the two isolated mesitylene rings and different from that of biphenyl.[11]

An examination of the spectra of ortho-substituted nitrobenzenes has shown that, as the size of the ortho substituent increases and the nitro group is forced more and more out of the plane of the benzene ring, the spectrum changes gradually; finally the maximum at 265 nm, which had an ε of more than 10,000 in *p*-*t*-alkylnitrobenzenes, has almost completely disappeared in *o*-nitro-*t*-butylbenzene.[12]

Similar steric inhibition of resonance has been observed in studies of the effect of ortho substituents on the spectra of aniline derivatives.[13] An extreme example is provided by benzoquinuclidine (XII), in which the nitrogen atom is held by the rigid carbon skeleton at an angle that effectively inhibits resonance interaction between the amino group and the aromatic ring. As might be expected, the ultraviolet spectrum of benzoquinuclidine resembles that of benzene very much more than that of dimethylaniline.[14]

N

XII

In addition to steric inhibition of resonance, ultraviolet spectra may be markedly changed by steric factors that alter the positions of dipoles in the molecule with respect to each other. An example is given later in the discussion (p. 429) of the effect of position of the chlorine atom in α-chloro ketones upon the ultraviolet spectrum of the carbonyl function. A study of α-diketones[15] has shown that their spectra are drastically influenced by the orientation of the carbonyl groups with respect to each other. In general, then, caution must be exercised in the choice of model substances to be used in ultraviolet spectroscopy.

Another effect that depends on molecular geometry to a large extent is the interaction of nonconjugated chromophores that are suitably located in space. For example, the diketo olefin (XIII) shows an absorption maximum at 223 nm

O

O

XIII

---

[10] Remember that normally the intensity of absorption of a uv-active compound is directly proportional to the number of isolated chromophores causing that absorption. For example, a compound with four isolated benzene rings would have twice the intensity of benzenoid absorptions as a compound with two isolated benzene rings.

[11] L. W. Pickett, G. F. Walter, and H. France, *J. Amer. Chem. Soc.*, **58**, 2296 (1936); M. T. O'Shaughnessy and W. H. Rodebush, *ibid.*, **62**, 2906 (1940).

[12] W. G. Brown and H. Reagan, *J. Amer. Chem. Soc.*, **69**, 1032 (1947).

[13] W. R. Remington, *J. Amer. Chem. Soc.*, **67**, 1838 (1945).

[14] B. M. Wepster, *Rec. Trav. Chim.*, **71**, 1159 (1952).

[15] N. J. Leonard and P. M. Mader, *J. Amer. Chem. Soc.*, **72**, 5388 (1950).

(log ε 3.36), as well as absorption maxima at 296 and 307 nm (log ε 2.41). The infrared spectrum shows that the compound is not enolized, and the unusually intense spectrum has been attributed to transannular interaction of the olefinic double bond with the carbonyl groups.[16]

Other interactions of nonconjugated chromophores have been reviewed and interpreted.[17]

8.2.5 More Refined Correlations of Structure with Ultraviolet Spectra

A number of studies have been made of the effect on the ultraviolet spectrum of increasing the number of double or triple bonds in systems of the type X—(CH═CH)$_n$—Y and X—(C)$_n$—Y. Since these have been summarized by Gillam and Stern,[18] they will not be discussed further here.

The most important quantitative correlation of spectra with structure for the natural products chemist is that of Woodward,[19] who formulated an empirical treatment of the position of the absorption maximum of conjugated dienes and α,β-unsaturated ketones (Tables 8.3 and 8.4). In the preceding discussion we have assumed that replacement of hydrogen atoms, even those attached directly to a chromophore, by alkyl groups leaves the spectrum unchanged; this is true

Table 8.3. Woodward's Rules for Conjugated Dienes (in Ethanol)

Acyclic and heteroannular dienes	215 nm
Homoannular dienes	253 nm
Addition for each substituent	
—R alkyl (including part of a carbocyclic ring)	+5 nm
—OR alkoxy	+6 nm
—SR thioether	+30 nm
—Cl, —Br	+5 nm
—OC(═O)R acyloxy	+0 nm
—CH═CH— additional conjugation	+30 nm
If one double bond is exocyclic to one ring	+5 nm
If exocyclic to two rings simultaneously	+10 nm
Solvent shifts are negligible	
ε_{max} 6,000–35,000	

[16] C. A. Grob and A. Weiss, *Helv. Chim. Acta,* **43,** 1390 (1960).
[17] C. F. Wilcox, S. Winstein, and W. G. McMillan, *J. Amer. Chem. Soc.,* **82,** 5450 (1960).
[18] See General References on p. 416.
[19] See L. F. Fieser and M. Fieser, *Steroids* (Reinhold, New York, 1959), pp. 15 ff, or any of the general references mentioned on pp. 416–417.

Table 8.4. Woodward's Rules for α,β-Unsaturated Carbonyl Compounds (in Ethanol)

Ketones, $-\overset{\beta}{\underset{\vert}{C}}=\overset{\alpha}{\underset{\vert}{C}}-CO-$ acyclic or 6-membered ring	215 nm 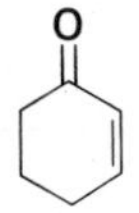
5-membered ring	202 nm 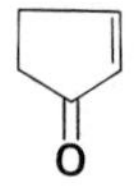
Aldehydes, $-\overset{\vert}{C}=\overset{\vert}{C}-CHO$	207 nm
Acids and esters, $-\overset{\vert}{C}=\overset{\vert}{C}-CO_2H(R)$	197 nm

Additional conjugation

($-\overset{\delta}{C}=\overset{\gamma}{C}-\overset{\beta}{C}=\overset{\alpha}{C}-CO-$ etc.	+30 nm
(If the second double bond is homoannular with the first,	+39 nm

Addition for each substituent

	α	β	γ	δ
—R alkyl (including part of a carbocyclic ring)	+10 nm	12 nm	17 nm	
—OR	35 nm	30 nm	17 nm	31 nm
—OH	35 nm	30 nm	30 nm	50 nm
—SR	—	80 nm	—	—
—Cl	15 nm	12 nm	12 nm	12 nm
—Br	25 nm	30 nm	25 nm	25 nm
—OCOR (acyloxy)	6 nm	6 nm	6 nm	6 nm
$-NH_2$, —NHR, $-NR_2$	—	95 nm	—	—
If one double bond is exocyclic to one ring		5 nm		
If exocyclic to two rings simultaneously		10 nm		
ε_{max} 4,500–20,000				

Solvent shifts

Solvent	Shift
Water	+8 nm
Methanol	0 nm
Chloroform	−1 nm
Dioxane	−5 nm
Diethyl ether	−7 nm
Hexane	−11 nm
Cyclohexane	−11 nm

only as a first approximation, however. In system XIV, the absorption maximum of the parent unsubstituted compound in the solvent ethanol may be taken as 215 nm. The maxima of a wide variety of α- and β-alkyl substituted relatives can be predicted by using the generalizations that for each α-substituent 10 nm is

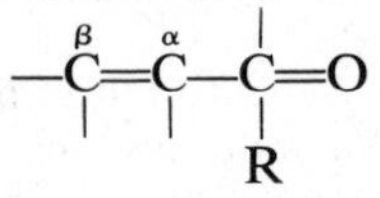

$$-\overset{\beta}{C}=\overset{\alpha}{C}-\underset{R}{C}=O$$

XIV

added to the value of 215 nm, for each β-substituent 12 nm is added, and for each ring to which the carbon–carbon double bond is exocyclic 5 nm is added (see Table 8.4). Application of the method is illustrated by the following example.

Problem. Predict the position of the ultraviolet maximum of the following substance (in ethanol solution):

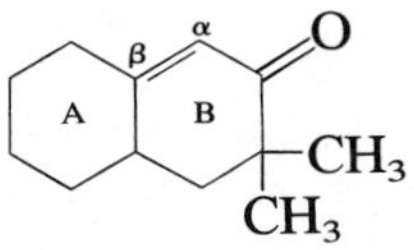

An exocyclic double bond is one attached to a carbon atom in a ring as shown in the formula on the left below. The double bond in the second example is exocyclic to ring A but endocyclic to ring B. There are no substituents on the α-carbon atom but two on the β, for which 24 nm (2×12) must be added to the base value of 215 nm. Also, the double bond is exocyclic to ring A so that a total of $5 + 24 = 29$ nm is added to 215 nm to give the predicted value of 244 nm for the position of the maximum for the compound above.

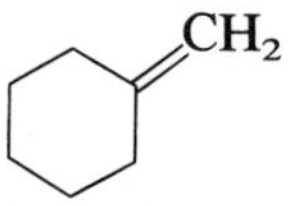

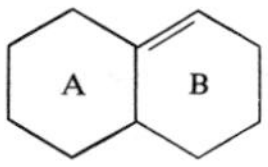

The treatment of conjugated dienes is similar. The references cited should be consulted for the details. In view of the simplicity of the method, its extensive agreement with experiment is remarkable; it has been responsible for the revision of incorrect structures of a number of natural products.

A second correlation, important for aromatic chemistry, is that of Scott[20] who found that the effect of substituents on the spectrum of benzene could be correlated with structure (Table 8.5). Use of this table is illustrated by the following example:

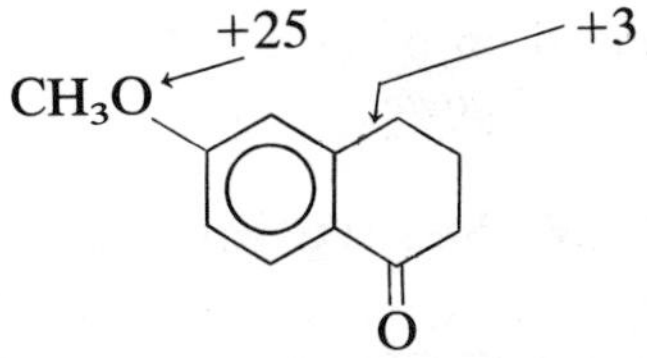

6-methoxytetralone

Calc: $\lambda_{max}^{EtOH} = 246 + 3 + 25$ (values from Table 8.5) $= 274$ nm

Obs: $\lambda_{max}^{EtOH} = 276$ nm

[20] A. I. Scott, *Interpretation of the Ultraviolet Spectra of Natural Products* (Pergamon Press, Elmsford, N.Y., 1964).

Table 8.5. Data for Calculation of the Principal Band of Substituted Benzene Derivatives, Ar—COG (in EtOH)

ArCOR/ArCHO/$ArCO_2H$/$ArCO_2R$		λ_{max}^{EtOH} (nm)
Parent chromophore: Ar = C_6H_5—		
G = Alkyl or ring residue, (e.g., ArCOR)		246
G = H, (ArCHO)		250
G = OH, OAlk, ($ArCO_2H$, $ArCO_2R$)		230
Increment for each substituent on Ar:		
—Alkyl or ring residue	*o*-, *m*-	+3
	p-	+10
—OH, —OCH_3, —OR	*o*-, *m*-	+7
	p-	+25
—O^- (oxyanion)	*o*-	+11
	m-	+20
	p-	+78[a]
—Cl	*o*-, *m*-	+0
	p-	+10
—Br	*o*-, *m*-	+2
	p-	+15
—NH_2	*o*-, *m*-	+13
	p-	+58
—NHAc	*o*-, *m*-	+20
	p-	+45
—$NHCH_3$	*p*-	+73
—$N(CH_3)_2$	*o*-, *m*-	+20
	p-	+85

[a] This value may be decreased markedly by steric hindrance to coplanarity.

Other systematic studies on the variation of spectrum with structural change of such types of compounds as conjugated polyenes and acetylenes, semicarbazones, 2,4-dinitrophenylhydrazones, and oximes—conjugated and unconjugated—are discussed in the books by Gillam, Stern, and Timmons referred to on p. 416. A further extensive study of 2,4-dinitrophenylhydrazones has been reported.[21]

8.2.6 Absorption by Monofunctional Compounds

A few groups such as the azo and nitroso have intense absorption in the near ultraviolet. The absorption maxima of other compounds, such as aldehydes, ketones, and amines, in the near ultraviolet region are, however, of very low intensity. The danger inherent in using such bands is clear when it is considered that a concentration of 0.1% of an impurity with an ε of 15,000 would give rise to an absorption maximum with the same intensity as a pure ketone, for example,

[21] G. D. Johnson, *J. Amer. Chem. Soc.*, **75**, 2720 (1953).

with an ε of 15. Nevertheless, valuable information has been obtained from such spectra.

Since the region of the ultraviolet accessible to the organic chemist is being extended to include absorption due to many isolated functional groups, the subject is reviewed briefly here.

Unconjugated Olefins. Although the principal absorption maxima of unconjugated olefins lie in the vacuum ultraviolet (160–200 nm), the number of alkyl substituents on the carbon atoms of the double bond can be correlated with the spectrum,[22] such correlations are valuable because this region of the ultraviolet is more generally accessible to the organic chemist.[23] Furthermore, it has been possible to use the end absorption in the near ultraviolet (205–225 nm) to determine the degree of substitution of carbon–carbon double bonds.[24]

Carbonyl Compounds. The weak carbonyl absorption at 260–300 nm might well be a useful supplement to the infrared spectrum in certain cases, because it can be used to distinguish between ketones or aldehydes and esters. For example, both five-membered cyclic ketones and aliphatic esters have infrared absorption at about 1740 cm^{-1}, whereas only the former have appreciable ultraviolet absorption above 210 nm. A further use of the carbonyl maximum has been found.[25] Its position is shifted by the presence of an α-chlorine or bromine atom, and the amount of the shift depends on whether the halogen atom is equatorial or axial when a substituted cyclohexanone is examined. Similar shifts have been observed in the spectra of α-hydroxy and α-acetoxy ketones.[26] The technique therefore supplements the infrared as a means of assigning configurations of such substituted ketones.

Amines. Examination of the spectra of simple amines in the vacuum ultraviolet[27] indicates that extension of the practical region may make this a valuable tool for distinguishing between primary, secondary, and tertiary aliphatic amines. Since saturated tertiary amines have an absorption maximum just inside the near ultraviolet region (at about 214 nm), which is shifted to longer wavelengths in vinyl amines with the structure $R_2\ddot{N}—CH═CHR$ (228–238 nm), ultraviolet spectroscopy has proved valuable in distinguishing this functional group from either saturated amines or unconjugated amines in which the amine and olefinic double bond are isolated.[28]

---

[22] See D. Semenow, A. J. Harrison, and E. P. Carr, *J. Chem. Phys.*, **22**, 638 (1954); or D. W. Turner, *J. Chem. Soc.*, 30 (1959), for a discussion and references.
[23] R. A. Micheli and T. H. Applewhite, *J. Org. Chem.*, **27**, 345 (1962).
[24] P. Bladon, H. B. Henbest, and G. W. Wood, *J. Chem. Soc.*, 2737 (1952).
[25] R. C. Cookson, *J. Chem. Soc.*, 282 (1954).
[26] R. C. Cookson and S. H. Dandegaonker, *J. Chem. Soc.*, 352 (1955).
[27] E. Tannenbaum, E. M. Coffin, and A. Harrison, *J. Chem. Phys.*, **21**, 311 (1953).
[28] N. J. Leonard and D. M. Locke, *J. Amer. Chem. Soc.*, **77**, 437 (1955).

Sulfur-Containing Functional Groups. Mercaptans, thioethers, and di- and polysulfides have characteristic spectra in the near ultraviolet region.[29] Sulfones and sulfoxides also have characteristic absorption.[30]

Questions

How would you distinguish the following compounds by using uv methods in as many ways as possible? Give a brief discussion of the differences you anticipate.

1. $CH_3CH_2CH{=}CH{-}CH{=}CHCH_2CH_3$ and $CH_3CH_2CH{=}CHCH_2CH{=}CHCH_3$.

2. 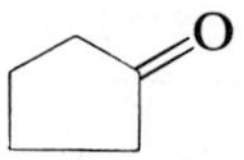 [cyclopentanone] and [γ-butyrolactone]

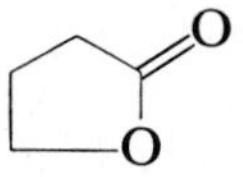

3. 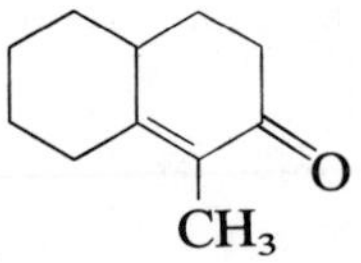 [bicyclic enone with CH_3 on the α-carbon] and [bicyclic enone with CH_3 on the ring carbon next to the carbonyl]

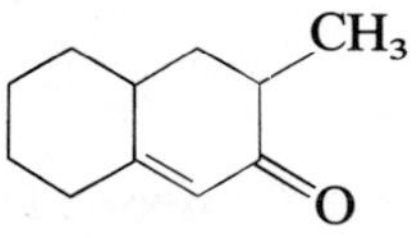

4. Calculate the λ_{max} in ethanol for each of the isomers below. If the predicted values can be assumed to be accurate within 5 nm, will this technique differentiate between the two compounds?

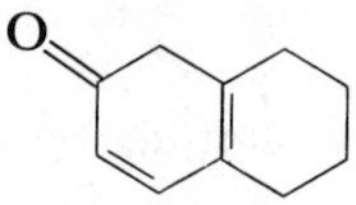

A 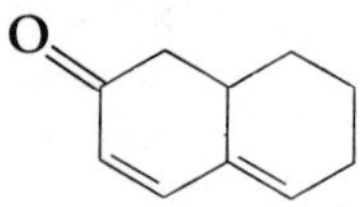 B

5. Match the table of uv data observed with the appropriate structure.

λ_{max}	ε
231	21,000
236	12,000
245	18,000
265	6,400
282	11,900

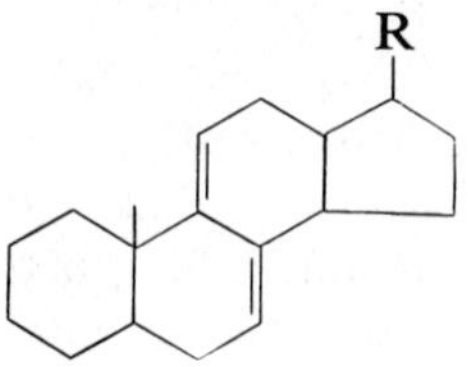

(a) 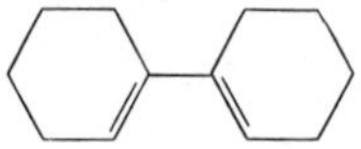 (b)  (c) 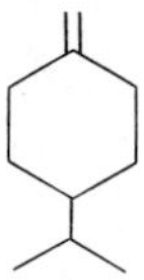 (d) 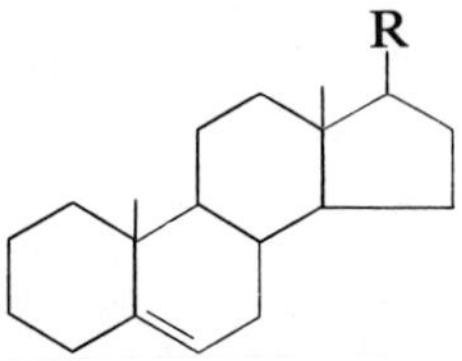 (e)

R = saturated alkyl group

---

[29] E. A. Fehnel and M. Carmack, *J. Amer. Chem. Soc.*, **71**, 84 (1949); J. E. Baer and M. Carmack, *ibid.*, **71**, 1215 (1949).

[30] E. A. Fehnel and M. Carmack, *J. Amer. Chem. Soc.*, **71**, 232 (1949); **72**, 1292 (1950).

6. Match the observed maxima (ethanol solvent) with the correct structure using Table 8.5.

λ_{max}	
245	*p*-aminobenzoic acid
253	*p*-bromobenzoic acid
288	*m*-methoxybenzoic acid

8.3 OPTICAL ROTATORY DISPERSION

The index of refraction of a substance depends on the wavelength of transmitted light. Similarly, the specific rotation, $[\alpha]$, of a substance depends on wavelength. The change of specific rotation with wavelength is called *optical rotatory dispersion* (ord).

Extensive studies have shown that a plot of the optical rotation against wavelength over the visible and ultraviolet region of the spectrum can be more valuable than the rotation at a single wavelength. Such optical rotatory dispersion curves have proved to be very valuable in the determination of configurations and conformations of optically active compounds.

In regions of absorption where electronic transitions occur, optical rotatory dispersions often behave quite anomalously. They often pass through one or more maxima ("peaks") and minima ("troughs"). Such anomalies are called the *Cotton effect* (See Fig. 8.5).

If the curve shows a single maximum and single minimum (Fig. 8.5), the curve is that of a "single Cotton effect." Curves with two or more peaks and corresponding troughs for an absorption band correspond to "multiple Cotton effects." The Cotton effect is positive when the maximum is at a longer wavelength than the minimum (see Fig. 8.5) and negative when the maximum is

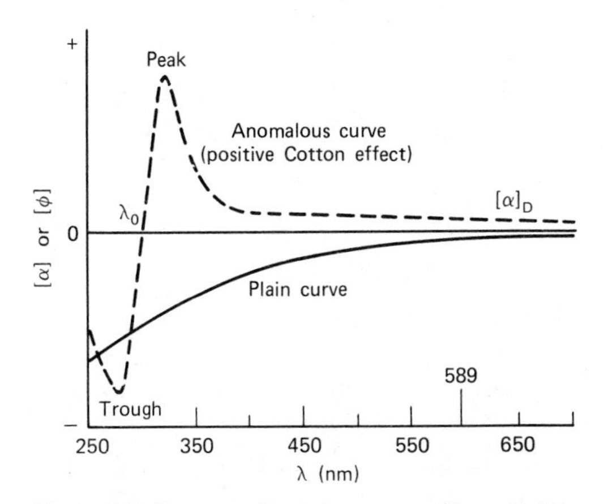

Figure 8.5. Rotatory dispersion curves. (From C. Djerassi, *Optical Rotatory Dispersion*, McGraw-Hill, New York, copyright © 1960.)

toward the lower wavelength. One enantiomorph will give a positive Cotton effect, and the other, having equal rotations always of opposite sign, a negative Cotton effect. These effects are related to the orientation of groups in the molecule near the chromophore.

Instrumentation for optical rotatory dispersion consists of a combination of the polarimeter to measure optical rotation, a source of monochromatic light (visible or ultraviolet) of known wavelength, and a photoelectric device to determine the angle of minimum transmittance: it is thus a *photoelectric spectropolarimeter.*

The most successful applications of the optical rotatory dispersion procedure have been to the stereochemistry of cyclohexanones and complex ring compounds containing the cyclohexanone ring. For this, the octant rule applies.

8.3.1 The Octant Rule

The octant rule, which has an empirical origin,[31] allows one to deduce the fourth of the following series when any one of the other three is known: structure, conformation, configuration, and sign of the Cotton effect.

When the molecule containing the cyclohexanone ring is oriented as in Fig. 8.6 with the carbonyl group on the origin of the axes, and the molecule is divided by planes and coordinates into octants and then projected on the frontal plane, C, perpendicular to the C=O bond, the octant diagram (Fig. 8.6) results.

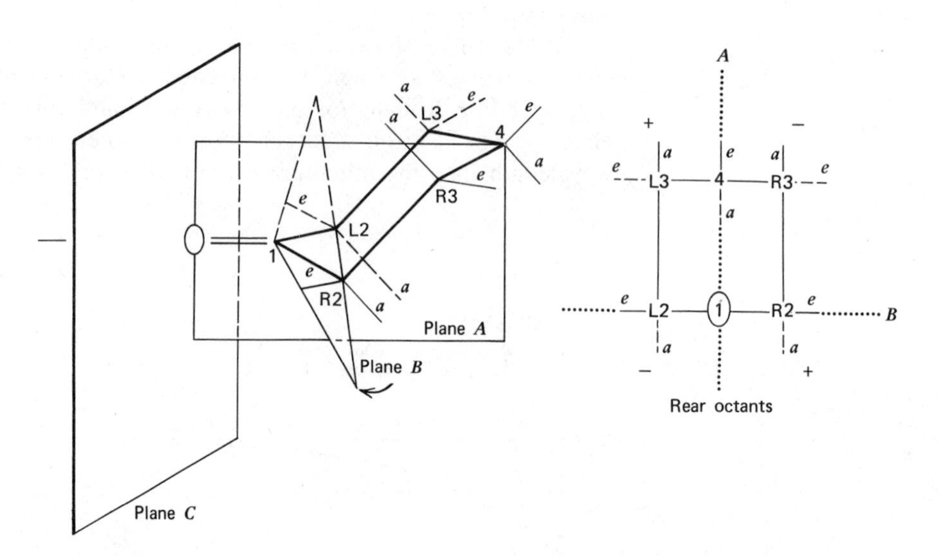

Figure 8.6. Octant rule. (From C. Djerassi, *Optical Rotatory Dispersion*, McGraw-Hill, New York, copyright © 1960.

[31] W. Moffitt, R. B. Woodward, A. Moscowitz, W. Klyne, and C. Djerassi, *J. Amer. Chem. Soc.*, **83,** 4013 (1961); K. Mislow, M. A. W. Glass, A. Moscowitz, and C. Djerassi, *ibid.*, **83,** 2771 (1961).

The octant rule for cases in which all ring substituents lie within the rear octants (remote to the eye relative to plane *C*) is comprised of the following:

Atoms lying in any of the dividing planes (*A*, *B*, *C*) make no contribution to the specific rotation. These include any directly connected atoms in the R2*e*, L2*e*, 4*a*, and 4*e* positions.

Atoms lying in the upper right octant (above plane *B* and to the right of *A*), and those lying in the lower left octant (below plane *B* and to the left of *A*) make negative contributions to the rotation. These include any groups on R3*a*, R3*e*, and L2*a*, as well as any attached to L2*e*, R2*e*, 4*a*, and 4*e* that lie in these octants.

Any atoms in the upper left octant (above plane *B* and to the left of *A*), and those lying in the lower right octant (below plane *B* and to the right of *A*) make positive contributions to the specific rotation. These include any on L3*a*, L3*e*, and R2*a*, as well as any attached to L2*e*, R2*c*, 4*a*, and 4*e* that lie in these octants.

When any cyclohexane ring substituents are in the front octants (to the left of plane *C*), the contributions of these groups to the specific rotation are opposite in sign to those in the octants behind them.

The signs apply to the long-wavelength sides of Cotton effect curves; hence, if the sum of contributions is positive, the Cotton effect is positive, and the Cotton effect is negative if the sum of contributions is negative. The predicted Cotton effect is that for the uv band due to the keto group.

Since spatially opposed like atoms and groups on opposite sides of a reference plane each cancel the contribution of the other, only groups that have no counterbalancing group need to be considered in evaluating the total specific rotation. These are called significant groups (or atoms).

Two applications of the rule illustrate its usefulness.

Configuration. Determination of configuration is illustrated by (+)-*trans*-10-methyl-2-decalone (Fig. 8.7). If this molecule has the configuration shown, carbons 8, 7, and 6, being equatorially attached to C_9, are in the upper left octant, whereas carbon 5 and the angular methyl group are in the *A* plane and make no contribution. No substituents are attached at C_1 or C_3. Since the only substituents outside the coordinate plane are thus in the upper left octant, the Cotton effect for the molecule should be positive. Since, in fact, the Cotton effect for (+)-*trans*-10-methyl-2-decalone is positive, the absolute configuration shown in Fig. 8.7 is the correct one.

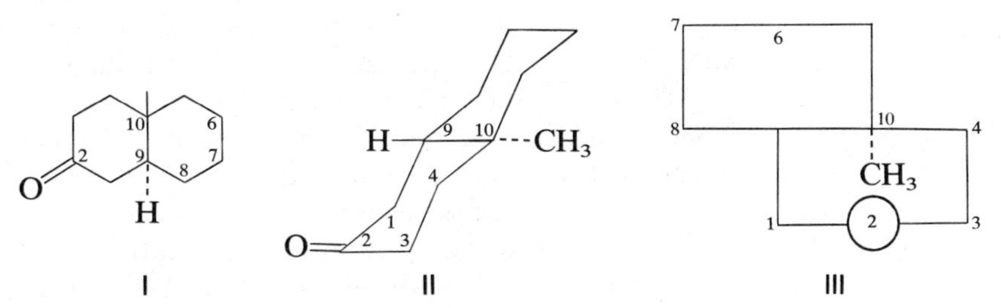

Figure 8.7. (+)-*trans*-10-Methyl-2-decalone. (From C. Djerassi, *Optical Rotatory Dispersion*, McGraw-Hill, New York, copyright © 1960.)

Conformation. (+)-3-Methylcyclohexanone is known to have the **R** configuration (Fig. 8.8). If the ketone is oriented appropriately for application of the octant rule, either the equatorial representation with methyl on the upper left (substituent at L3) or the axial representation with methyl on the upper right (at R3) is obtained (see Fig. 8.8). There are no other substituents and therefore the equatorial conformer should have a positive Cotton effect and the axial conformer, negative. The fact that the Cotton effect is indeed positive supports the conclusion (expected on conformational grounds) that the equatorial isomer predominates.

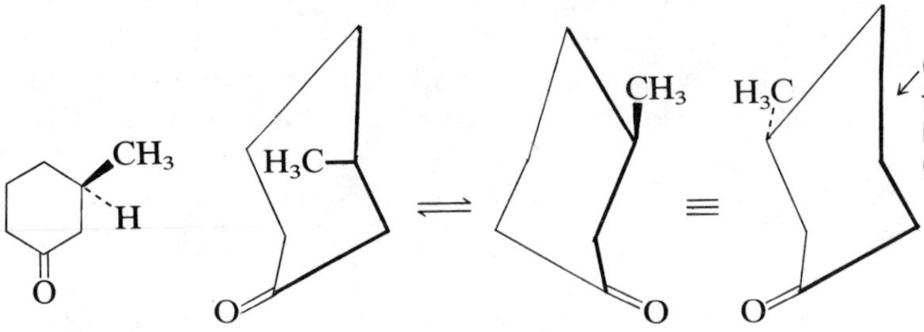

Figure 8.8. (R)-(+)-3-Methylcyclohexanone.

In summary, a major application of optical rotatory dispersion in organic structure determination is obtaining the result in one of the following three categories when answers are known for the other two: (1) functional group position(s), (2) configuration(s), and (3) molecular conformation.

8.4 CARBON-13 NUCLEAR MAGNETIC RESONANCE

The general principles that pertain to proton magnetic resonance also pertain to carbon magnetic resonance. Until recent developments, it has been much more difficult to analyze carbon by magnetic resonance because ^{13}C, the magnetically responsive isotope of carbon, is not the predominant isotope of carbon, and because the ^{13}C nucleus is inherently less sensitive than the proton to the magnetic resonance experiment due to the lower magnetogyric ratio for carbon-13. Modern ^{13}C instruments overcome these problems with computers which help to control procedures including pulsed excitation of the ^{13}C nuclei and Fourier transform analysis of the data obtained from the excited ^{13}C nuclei.

The value of the information available from carbon magnetic resonance (cmr) cannot be overemphasized. The large number of nonequivalent carbons in organic compounds and the fact that cmr allows identification of nearly all of these nonequivalent carbons makes cmr spectrometry a very powerful tool available to organic chemists for structure determination. Until recently, cmr had not been available to very many organic chemists. Now cmr instruments can be purchased at a cost of \$50,000–\$100,000. Since proton magnetic resonance (pmr) instruments are, however, much less expensive than cmr instruments, proton data are still more accessible to most organic chemists.[32] It is clear, though, that cmr analysis for organic structure determination will be every bit as useful as proton magnetic resonance and ir spectrometry.

Our approach to cmr spectrometry will focus on the use of this technique for organic structure determination. For a discussion of the theory related to this subject, we refer the reader to the cmr references listed in Chapter 10.

Solvents used for cmr analysis are usually those used for pmr (e.g., $CDCl_3$, etc., see Chapter 5). Certain instruments (e.g., the Varian CFT-20) require that at least part of the solvent be deuterated solvents ($CDCl_3$, D_2O, CD_3CN, etc.) for internal lock purposes.

It is now conventional to determine all chemical shifts in ppm relative to the ^{13}C signal of TMS (tetramethylsilane), with the more common positive values corresponding to downfield of TMS and negative values corresponding to upfield. Many earlier determinations involved different internal standards, the chemical shifts of which are listed in Table 8.6.

For example, CS_2 (chemical shift of 193.7 ppm downfield of TMS) was used quite often. If data are reported in ppm relative to CS_2, they are converted to ppm relative to TMS by adding 193.7 ppm. Thus:

$$\delta_c^{TMS} = \delta_c^{\,i} + F_i$$

Table 8.6. TMS-Based ^{13}C Chemical Shifts for Common Standards and Solvents

	Chemical shift[a]	
Compound, solvent	Protio compound	Deuterated compound[b]
Acetic acid (carbonyl carbon)	178.3	
Acetone (methyl carbon)	30.4	29.2
Benzene	128.5	128.0
Carbon tetrachloride	96.0	
Chloroform	77.2	76.9
CS_2[c]	192.8	
CS_2 capillary[c]	193.7	
Cyclohexane	27.5	26.1
Dimethyl sulfoxide	40.5	39.6
Dioxane	67.4	
Methylene chloride	54.0	53.6

[a] In δ ppm downfield from internal TMS (±0.05 ppm at 38°C); see G. C. Levy and J. D. Cargioli; J. *Mag. Res.* **6**, 143 (1972). Useful as F_i factors (see text).

[b] All protons have been replaced by deuteriums.

[c] Equivalent to the shifts utilized by J. B. Stothers in *Carbon-13 NMR Spectroscopy* (Academic Press, New York, 1972).

[32] The abbreviations pmr and cmr are used for ^1H and ^{13}C nmr, respectively. Our use of pmr is not to be mistaken for phosphorus magnetic resonance, and the use of pmr notation has not been approved by IUPAC.

When δ_c^{TMS} is the desired chemical shift relative to TMS, δ_c^{i} is the shift relative to some other internal standard and F_i is the factor taken from Table 8.6 corresponding to internal standard *i*.

For structure determination, we should be concerned with extracting the following three classes of information from cmr spectra:

1. Chemical shifts.
2. Signal multiplicity.
3. Signal intensity.

8.4.1 Chemical Shift

It is a relatively rare situation in which the cmr spectrum of even a complex organic molecule does not provide a separate, well-resolved signal for each type of carbon in the molecule (see Fig. 8.9). A typical spectrum, e.g., 200 ppm wide (when completely proton spin-decoupled) will normally provide a single line that can be assigned to each type of carbon (Fig. 8.9). It is very common to find that the pmr spectrum of the same compound displays proton resonances that are not as well resolved as the carbon resonances; thus the pmr spectrum frequently does not provide as much structural information as is provided by the cmr spectrum. This is only one of a number of ways in which we can learn of the great power of cmr for organic structural analysis.

Up to a point, the general principles that one uses to correlate structure with chemical shift in spectrometry can be used in cmr spectrometry. For example, it can be seen (Fig. 8.10) that alkanes give rise to carbon signals (at ca. $\delta -10$ to $\delta 35$ ppm) that are upfield of alkene carbons (at ca. $\delta 100$ to $\delta 165$ ppm); recall that it is well known that protons of alkanes are upfield of olefinic protons. In addition, electronegative atoms (e.g., N, O) cause the resonances of the carbons bearing these atoms to be downfield of alkane carbons; again, the same general trend holds for proton resonances.

If one has a rough idea as to the structure of an organic compound, it is often very easy to refine this idea using carbon resonances of the various types of carbon atoms in the organic compound. This is especially true of the carbonyl carbon; Fig. 8.10 indicates that the chemical shift of the carbonyl carbon depends on both the *type* of functional group containing the carbonyl group, such as aldehyde, ketone, ester, and so on, and on whether or not the carbonyl group is conjugated with another π system in the molecule. The relatively low-field position for carbonyl carbon resonances (ca. $\delta 145$ to 210 ppm, Fig. 8.10) is due, at least in part, to resonance contributions to the carbonyl group that place positive charge on carbon:

$$\rangle C{=}\ddot{O} \longrightarrow {}^{+}\rangle C{-}\ddot{O}{:}^{-}$$

The decreased charge density at the carbonyl carbon decreases the paramagnetic shielding, which, in turn, leads to lower field resonance for such carbon atoms.

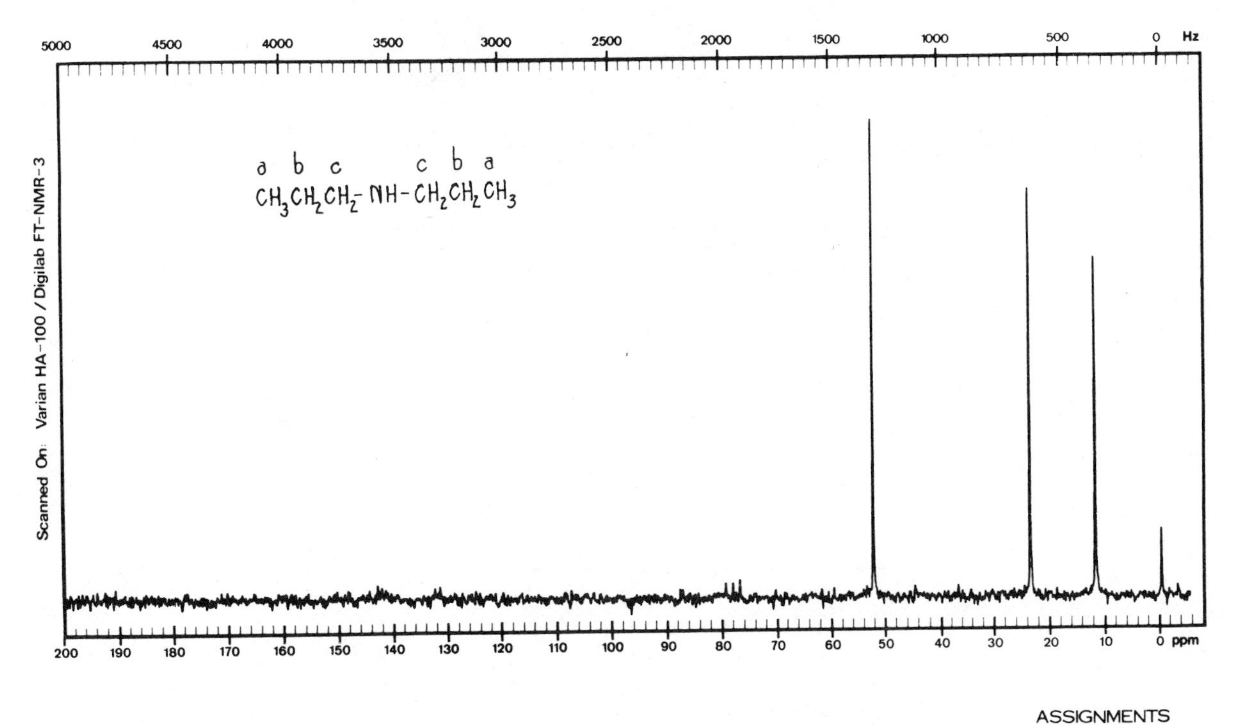

COMPOUND: DIPROPYLAMINE

SOURCE OF SAMPLE: Eastman Organic Chemicals, Rochester, New York

CHEMICAL FORMULA	$C_6H_{15}N$	SOLUTION CONC.	50% v/v
		SOLVENT	$CDCl_3$
MOLECULAR WT.	101.19	REFERENCE	TMS
PROTON NMR NO.	17051 M	TRANSIENTS	400
BOILING POINT	105-110°C	TIME	8 minutes
		SWEEP OFFSET	-

ASSIGNMENTS

A	11.9	N	
B	23.6	O	
C	52.3	P	
D		Q	
E		R	
F		S	
G		T	
H		U	
I		V	
J		W	
K		X	
L		Y	
M		Z	

Figure 8.9. ^{13}C magnetic resonance spectrum of di-*n*-propyl amine. (Published by permission of Sadtler Laboratories, Inc., Philadelphia, Pa.)

Rarely does the pmr spectrum directly provide information about the carbonyl groups. Clearly cmr spectra will often detect and identify carbonyl groups. Thus cmr and ir spectra can be used jointly to characterize carbonyl compounds.

The cmr spectrum of a simple hydrocarbon is illustrated in Fig. 8.11. Typically, the terminal carbon atom of an alkane shows a carbon resonance at higher field (such as C-1 of octane, Table 8.7) than an internal carbon (such as C-4 of octane, Table 8.7). This effect is general; for example, in many cases the

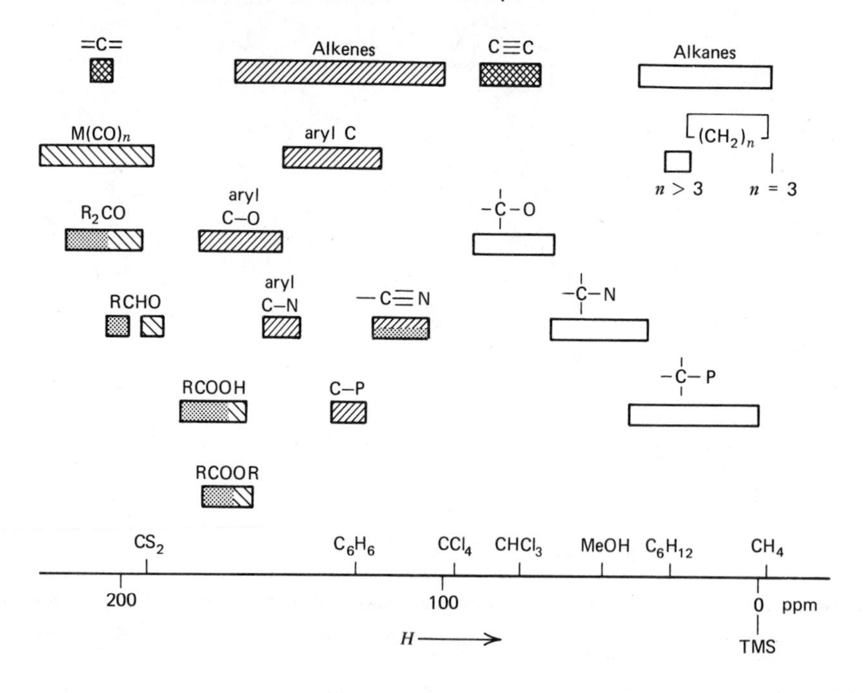

Figure 8.10. Chemical shift ranges for neutral compounds Carbon hybridization denoted as ▭ for sp^3, ▨ for sp^2, and ▩ for sp. Carbonyl carbons are distinguished as isolated ▒ and conjugated ▧; for example, the former notation would be appropriate for the carbonyl carbon of 2-pentanone and the latter for the carbonyl carbon of methyl vinyl ketone. (From "Carbon-13 NMR Spectroscopy," by J. B. Stothers, Academic Press, New York, copyright © 1972; with permission.)

introduction of a carbon α, β, or γ (etc.) to a given carbon atom causes the same degree of shift in that carbon's resonance (Table 8.8). Table 8.8 indicates that a carbon α or β to a given carbon usually shifts that carbon's resonance downfield by about 9 ppm, whereas introduction of a γ carbon into the skeleton causes a small (2.5-ppm) upfield shift. This upfield shift shows again that changes in proton resonances as a function of structure changes could be misleading if the same trends were used as guidelines for carbon resonances since such upfield shifts are not typical for pmr spectra.

It is very tempting to use data such as those of Table 8.8 to predict chemical shifts for alkanes. It is necessary, however, to use a much more involved set of parameters to predict such chemical shifts, because the number and type of branch sites affect these shifts. The parameters listed in Table 8.9, in conjunction with the following equation, can be used to calculate the carbon shifts expected for alkanes:

$$\delta_i = B_n + \sum_{M=2}^{M=4} (N_M A_{nM}) + \gamma_n N_{i3} + \Delta_n N_{i4}$$

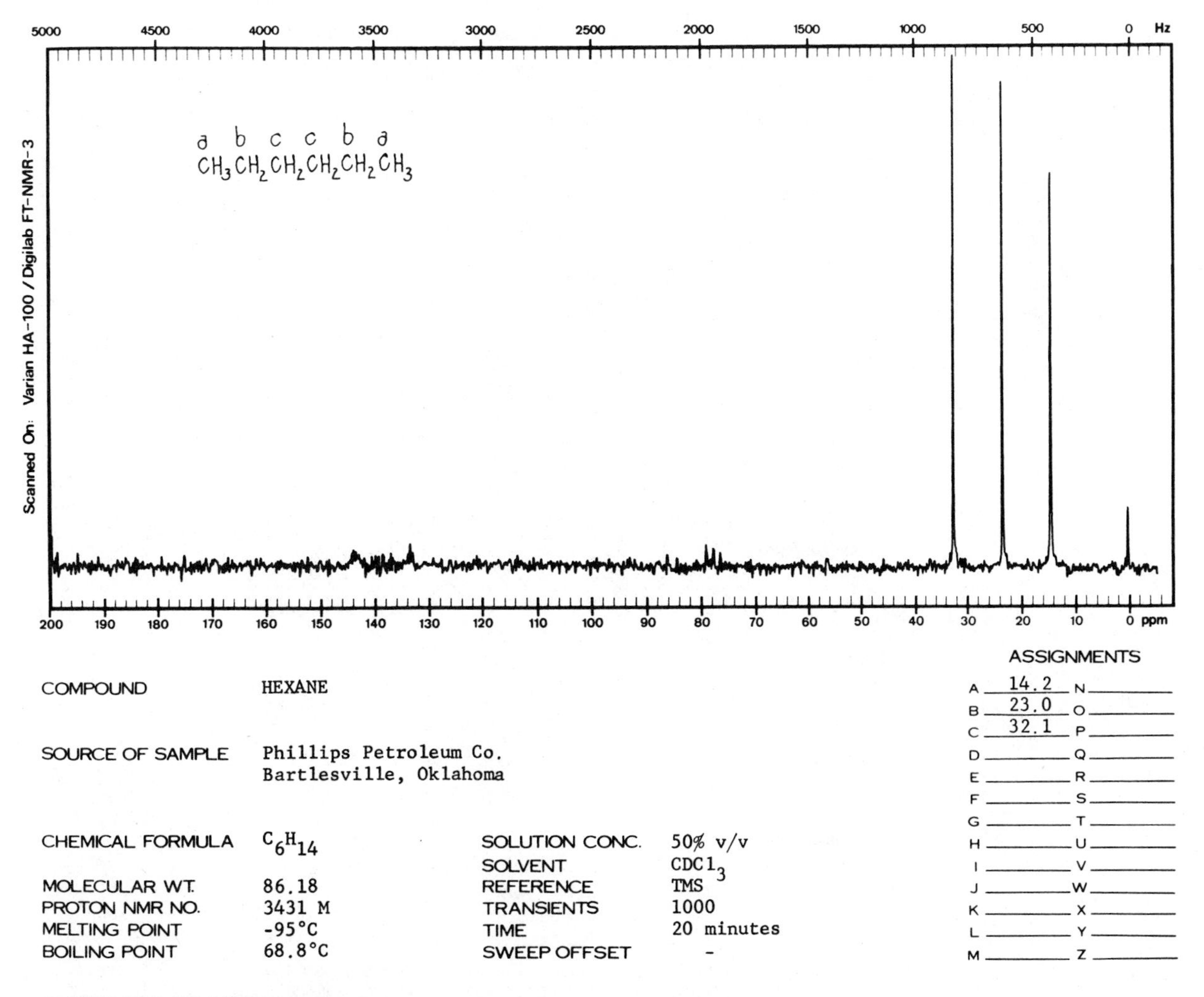

Figure 8.11. ^{13}C magnetic resonance spectrum of hexane. (Published with permission of Sadtler Laboratories, Inc., Philadelphia, Pa.)

where

δ_i = the calculated shift for carbon atom i

B_n = a parameter that depends on the number of carbons (n) that are attached to atom i

N_M = the number of carbon atoms connected to atom i having two or more attached carbons, where M is that number of attached carbons

A_{nM} = a parameter that depends on the number of carbon atoms (n) directly bonded to i and the number of carbon atoms (M) attached to the carbon that is directly bonded to atom i

Table 8.7. Chemical Shifts of Carbons in Straight- and Branched-Chain Alkanes[a]

Straight chains

Compound	C-1	C-2	C-3	C-4	C-5
Methane	−2.1				
Ethane	5.9	5.9			
Propane	15.6	16.1	15.6		
Butane	13.2	25.0	25.0	13.2	
Pentane	13.7	22.6	34.5	22.6	13.7
Hexane	13.9	22.9	32.0	32.0	22.9
Heptane	13.9	23.0	32.4	29.5	32.4
Octane	14.0	23.0	32.4	29.7	29.7
Nonane	14.0	23.1	32.4	29.8	30.1
Decane	14.1	23.0	32.4	29.9	30.3

Branched chains

Compound	C-1	C-2	C-3	C-4	C-5
Isobutane	24.3	25.2			
Isopentane	22.0	29.9	31.8	11.5	
Isohexane	22.5	27.8	41.8	20.7	14.1
Neopentane	31.5	27.9			
Neohexane	28.9	30.4	36.7	8.7	
3-Methylpentane	11.3	29.3	36.7 18.6[b]		
2,3-Dimethylbutane	19.3	34.1			
2,2,3-Trimethylbutane	27.2	32.9	38.1	15.9	
3,3-Dimethylpentane	6.8 4.4[b]	25.1	36.1		

[a] From D. G. Grant and E. G. Paul, *J. Amer. Chem. Soc.*, **86**, 2984 (1964); J. D. Roberts et al., *ibid.*, **92**, 1338 (1970); H. Spiescke and W. G. Scheider, *J. Chem. Phys.*, **35**, 722 (1961).

[b] Branch methyl carbon. Chemical shifts in ppm downfield from TMS. Since these are obtained from spectra using various internal standards (TMS, benzene, CS_2), the error in shift position is at least 0.2 ppm.

γ_n = a parameter that depends on the number of atoms (n) directly bonded to i

N_{iP} = the number of carbon atoms P bonds away from atom i (only those atoms that are three or more bonds away will contribute)

Δ_n = a parameter that depends on the number of atoms (n) directly bonded to atom i

Use of this equation and Table 8.9 is illustrated by the calculations for 2,2,3-trimethylbutane:

```
      CH3 CH3
       |   |
CH3—C——CH—CH3
(1)  |(2) (3)  (4)
      CH3
```

$$\delta_1 = B_1 + N_4A_{14} + \gamma_1N_3 + \Delta_1N_4$$

Table 8.8. Shift Induced by Introduction of Carbon at Various Positions in Alkanes[a]

Position	Induced shift
α	$+9.1 \pm 0.1$
β	$+9.4 \pm 0.1$
γ	-2.5 ± 0.1
δ	$+0.3 \pm 0.1$
ε	$+0.1 \pm 0.1$

[a] From D. M. Grant and E. G. Paul, *J. Amer. Chem. Soc.*, **86**, 2984 (1964).

where $B_n = B_1 = 6.80$ because $n = 1$ (there is one carbon attached to carbon 1 and $B_1 = 6.80$ on Table 8.9), and there is only one term involving $N_M A_{nM}$ because there is only one type of carbon (carbon 2) attached to carbon 1 that has two or more attached carbons and because there are four carbons attached to carbon 2, $N_M = N_4$; also because there is only one of this type, $N_4 = 1$. Also, carbon 1 has one attached carbon, and thus $n = 1$ in the subscript to A. Thus

$$\sum_{M=2}^{M=4} (N_M A_{nM}) = N_4 A_{14} = (1)(25.48)$$

$\gamma_n = \gamma_1$ (because there is one carbon attached to atom 1), and $N_3 = 1$, because there is one carbon atom (carbon 4) three bonds away from carbon 1. Finally, there are no carbons that are four bonds removed from carbon 1, so $N_4 = 0$.

Table 8.9. Carbon-13 Chemical Shift Parameters for Paraffins[a]

Parameter	Value (ppm)	Parameter	Value (ppm)
B_1	6.80	Δ_2	0.25
A_{12}	9.56	B_3	23.46
A_{13}	17.83	A_{32}	6.60
A_{14}	25.48	A_{33}	11.14
γ_1	−2.99	A_{34}	14.70
Δ_1	0.49	γ_3	−2.07
B_2	15.34	B_4	27.77
A_{22}	9.75	A_{42}	2.26
A_{23}	16.70	A_{43}	3.96
A_{24}	21.43	A_{44}	7.35
γ_2	−2.69	γ_4	0.68

[a] From the work of L. P. Lindeman and J. Q. Adams, *Anal. Chem.*, **43**, 1245 (1971).

Therefore

$$\delta_1^{\text{calc.}} = 6.80 + (1)(25.48) + (-2.99)(1) = 29.29$$

This is to be compared to $\delta_1^{\text{obs.}} = 27.2$.

Likewise, for the rest of the chain, one can calculate:

$$\begin{aligned}\delta_2^{\text{calc.}} &= B_4 + N_3 A_{43} + \gamma_4 N_3 + \Delta_4 N_4 \\ &= 27.77 + (1)(3.96) + \gamma_4(0) + \Delta_4(0) = 31.73\end{aligned}$$

Compare

$$\delta_2^{\text{obs.}} = 32.9$$

and

$$\delta_3^{\text{calc.}} = 23.46 + (1)(14.70) = 38.16$$

compare

$$\delta_3^{\text{obs.}} = 38.1$$

and

$$\delta_4^{\text{calc.}} = 15.66 \text{ ppm}$$

compare

$$\delta_4^{\text{obs.}} = 15.9 \text{ ppm}$$

The standard error for such calculations is 0.8 ppm.

Problem. *Outline the calculation for* δ_4.

The preceding calculations involving Table 8.9 do not work well with cyclic or polycyclic compounds.

A similar type of approach can be used when functional groups are part of aliphatic compounds. The data in Table 8.10 can be used to predict chemical shifts for straight- and branched-chain compounds. For example, compare the calculated and observed carbon shifts for 2-butanol:

	CH_3—	CH_2—	CH(OH)—	CH_3
$\delta_c^{\text{calc.}}$	8.2	33.0	66.0	21.2 ppm
$\delta_c^{\text{obs.}}$	10.0	32.0	69.2	22.7 ppm

The calculated shifts are determined by adding the appropriate parameter from Table 8.10 to the shift for the corresponding carbon in the unsubstituted alkane (Table 8.7). For example,

$$\delta_3^{\text{calc.}} = 25.0 + 8.0 = 33.0 \text{ ppm}$$

where 25.0 ppm is the shift for C-2 of butane (Table 8.7) and +8.0 is the shift

Table 8.10. Shift Effect Due to Replacement of H by Functional Groups (R) in Alkanes[a]

n = ...γ β α R and ...γ β α(R) β γ... = br

	α		β		γ
R	n	br	n	br	
CH_3	+9	+6	+10	+8	−2
COOH	+21	+16	+3	+2	−2
COO^-	+25	+20	+5	+3	−2
COOR	+20	+17	+3	+2	−2
COCl	+33	+28		+2	
COR	+30	+24	+1	+1	−2
CHO	+31		0		−2
Phenyl	+23	+17	+9	+7	−2
OH	+48	+41	+10	+8	−5
OR	+58	+51	+8	+5	−4
OCOR	+51	+45	+6	+5	−3
NH_2	+29	+24	+11	+10	−5
NH_3^+	+26	+24	+8	+6	−5
NHR	+37	+31	+8	+6	−4
NR_2	+42		+6		−3
NO_2	+63	+57	+4	+4	
CN	+4	+1	+3	+3	−3
SH	+11	+11	+12	+11	−4
SR	+20		+ 7		−3
F	+68	+63	+9	+6	−4
Cl	+31	+32	+11	+10	−4
Br	+20	+25	+11	+10	−3
I	−6	+4	+11	+12	−1

[a] n = straight-chain compounds, br = branched-chain compounds. The effect of the γ-position is virtually the same for straight- and branched-chain compounds.

Reprinted from F. W. Wehrli and T. Wirthlin, *Interpretation of Carbon-13 NMR Spectra*, Heyden, N.Y., 1976; with permission.

induced at a carbon β to a OH that branches from an alkane. Such calculations can be successfully applied to branched alkanes. For example, consider 1,3-butanediol:

	CH_2(OH)	CH_2	CH(OH)	CH_3
$\delta_c^{calc.}$	56.2	43.0	61.0	21.2 ppm
$\delta_c^{obs.}$	60.0	40.6	60.0	23.4 ppm

Sample calculation: $\delta_1 = 13.2 + 48.0 - 5.0 = 56.2$ ppm, where 13.2 is the shift for C-1 of butane (Table 8.7), +48.0 is the shift of an α-carbon caused by a OH attached to the end of an alkane, and −5.0 is the shift of a γ-carbon caused by a OH branched to the side of an alkane.

Examination of Tables 8.8, 8.9, and 8.10 reveals addition of either a carbon or a functional group to an alkane causes *deshielding* of the γ-carbon in that chain. This is due at least in part to a steric effect; for example, in hydrocarbons, the steric interaction of protons on the α-carbon with protons on the γ-carbon cause an upfield shift:

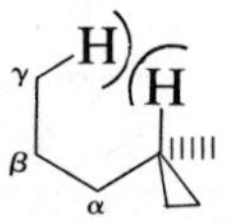

γ H H β α

This is referred to as the γ-*gauche effect.*

Empirical parameters for calculating chemical shifts in other types of systems (cycloalkanes, alkenes, alkynes, etc.) may be obtained from the references cited at the end of this section.

The spectrum of a typical benzenoid compound is illustrated in Fig. 8.12. Chemical shifts can be assigned in Fig. 8.12 using chemical shift/structure correlations and the intensity of the signals. This second approach involving intensity will be discussed later.

The use of empirical parameters to predict chemical shifts has been successfully applied to benzenoid compounds. Specifically, the data of Table 8.11 can be used to predict the chemical shifts of the benzene carbons. For example, Table 8.11 can be used to predict the chemical shifts of p-chloroanisole:

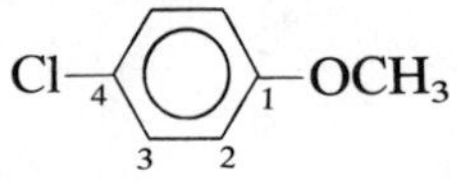

Cl 4 1 OCH₃ 3 2

$$\delta_1^{\text{calc.}} = 128.7 + 31.4 + (-1.9) = 158.2$$

Compare

$$\delta_1^{\text{obs.}} = 158.2$$

where, for $\delta_1^{\text{calc.}}$, 128.7 is the carbon shift for benzene and +31.4 is from Table 8.11 and is the change in shift caused by the directly attached methoxy group. Likewise, the value −1.9 is the shift induced at C-1 by the *para*-chloro group.

In like manner:

$$\delta_2^{\text{calc.}} = 128.7 + (-12.7) + 1.3 = 117.3$$

Compare

$$\delta_2^{\text{obs.}} = 115.2$$

To complete the comparison for p-chloroanisole:

$$\delta_3^{\text{calc.}} = 130.7 \qquad \delta_3^{\text{obs.}} = 129.2$$

$$\delta_4^{\text{calc.}} = 127.5 \qquad \delta_4^{\text{obs.}} = 125.4$$

The predicted and calculated shifts for a variety of benzenoid compounds usually agree within 2 ppm. This agreement, along with the fact that signals due to nonequivalent carbons of benzene rings show separate well-resolved signals, means that structures of many mono- and many disubstituted benzenoid compounds can be deduced by cmr analysis. Poor agreement often occurs for those

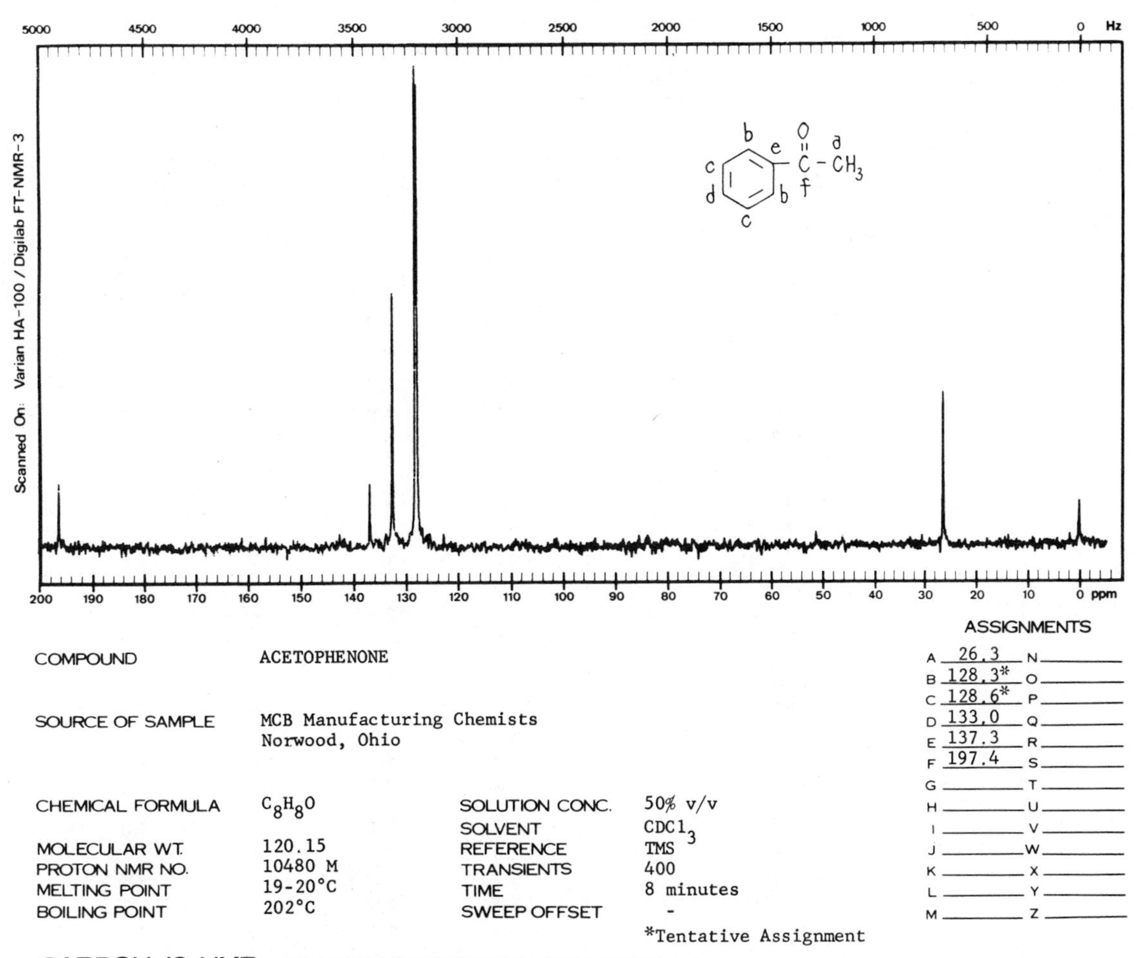

Figure 8.12. ^{13}C magnetic resonance spectrum of acetophenone. (Published with permission of Sadtler Laboratories, Inc., Philadelphia, Pa.)

compounds that bear two or more substituents in an ortho relationship. Specifically, the observed and calculated results for 2,6-dichlorobenzaldehyde agree poorly, especially at the carbon that has the most ortho neighbors (C_1):

(2,6-dichlorobenzaldehyde: CHO at C1, Cl at C2 and C6)	Carbon	$\delta^{calc.}$	$\delta^{obs.}$
	1	137.9	130.3
	2	137.3	136.6
	3	127.6	129.6
	4	136.6	133.5

It is generally accepted that these cases show poor agreement because of steric effects; the shifts listed on Table 8.11 are due largely to electronic effects. In

Table 8.11. Effect of Substituents on the C-13 Shift of Benzene Ring Carbons[a]

	Position			
Substituent[b]	C-1	Ortho	Meta	Para
—Br	−5.5	+3.4	+1.7	−1.6
$—CF_3$	−9.0	−2.2	+0.3	+3.2
$—CH_3$	+8.9	+0.7	−0.1	−2.9
—CN	−15.4	+3.6	+0.6	+3.9
—C≡C—H	−6.1	+3.8	+0.4	−0.2
1,4-di—C≡C—H	−5.6	+3.8	—	—
$—COCF_3$	−5.6	+1.8	+0.7	+6.7
$—COCH_3$	+9.1	+0.1	0.0	+4.2
—COCl	+4.6	+2.4	0.0	+6.2
—CHO	+8.6	+1.3	+0.6	+5.5
—COOH	+2.1	+1.5	0.0	+5.1
$—COC_6H_5$	+9.4	+1.7	−0.2	+3.6
—Cl	+6.2	+0.4	+1.3	−1.9
—F	+34.8	−12.9	+1.4	−4.5
—H	0.0	—	—	—
—NCO	+5.7	−3.6	+1.2	−2.8
$—NH_2$	+18.0	−13.3	+0.9	−9.8
$—NO_2$	+20.0	−4.8	+0.9	+5.8
$—OCH_3$	+31.4	−14.4	+1.0	−7.7
—OH	+26.9	−12.7	+1.4	−7.3
$—C_6H_5$	+13.1	−1.1	+0.4	−1.2
—SH	+2.3	+1.1	+1.1	−3.1
$—SCH_3$	+10.2	−1.8	+0.4	−3.6
$—SO_2NH_2$	+15.3	−2.9	+0.4	+3.3
$—O^-$ (oxyanion)	+39.6	−8.2	+1.9	−13.6
$—OC_6H_5$	+29.2	−9.4	+1.6	−5.1
$—OCOCH_3$ (acetoxy)	+23.0	−6.4	+1.3	−2.3
$—N(CH_3)_2$	+22.6	−15.6	+1.0	−11.5
$—N(CH_2CH_3)_2$	+19.9	−15.3	+1.4	−12.2
$—NHCOCH_3$	+11.1	−9.9	+0.2	−5.6
CH_2OH	+12.3	−1.4	−1.4	−1.4
I	−32.0	+10.2	+2.9	+1.0
$Si(CH_3)_3$	+13.4	+4.4	−1.1	−1.1
$CH{=}CH_2$	+9.5	−2.0	+0.2	−0.5
CO_2CH_3	+1.3	−0.5	−0.5	+3.5
COCl	+5.8	+2.6	+1.2	+7.4
CHO	+9.0	+1.2	+1.2	+6.0
$COCH_2CH_3$	+7.6	−1.5	−1.5	+2.4
$COCH(CH_3)_2$	+7.4	−0.5	−0.5	+4.0
$—COC(CH_3)_3$	+9.4	−1.1	−1.1	+1.7

[a] Data obtained in various solvents (CCl_4, neat, DMF = *N,N*-Dimethylformamide) with various internal standards (TMS, CS_2). A positive value means a downfield shift, negative means an upfield shift. Estimated error is ±0.5 ppm. Chemical shift of unsubstituted benzene is 128.7 ppm.

[b] —CO— = —C(=O)—

some cases, however, ortho-substituted compounds show good agreement; for example, for *o*-xylene,

	Carbon	$\delta^{calc.}$	$\delta^{obs.}$
o-xylene (CH_3, CH_3)	1	138.4	136.4
	2	129.1	129.9
	3	125.4	126.1

Problem. Two compounds, both of molecular formula C_7H_7Cl, show different cmr spectra:

Compound A	Compound B
21.0 ppm	20.7 ppm
125.5	128.3
127.1	130.4
129.1	131.2
129.3	136.2
134.0	
139.7	

Deduce the structure of each compound and assign all of the signals for each compound.

8.4.2 Signal Multiplicity

There are three types of cmr spectra that are of interest to organic chemists. These types vary as to the extent to which protons are magnetically decoupled:

1. Completely decoupled (no ^{13}C–^{1}H coupling)
2. Off-resonance decoupled
3. Completely coupled

An illustration of each of these three types is shown in Fig. 8.13.

When the organic chemist goes to the literature to obtain illustrations of spectra to compare to those of an unknown organic compound, spectra that result from complete proton decoupling will be encountered most frequently. Such spectra are obtained from samples that are irradiated with a broad band of noise spread across the proton region. This irradiation completely removes all proton to carbon coupling and thus the resulting cmr spectra (e.g., see Fig. 8.13*c*) consist of a singlet for each type of carbon atom. These spectra are often referred to as "completely decoupled" spectra. (Note: The ^{13}C nucleus, since it has a spin of $\frac{1}{2}$, can couple to other ^{13}C nuclei. In fact spectra determined from ^{13}C-enriched samples show that vicinal ^{13}C atoms coupling constants are 30 to 170 Hz. Spectra determined from samples that contain a natural abundance of ^{13}C, however, show no carbon–carbon coupling. *Explain.*)

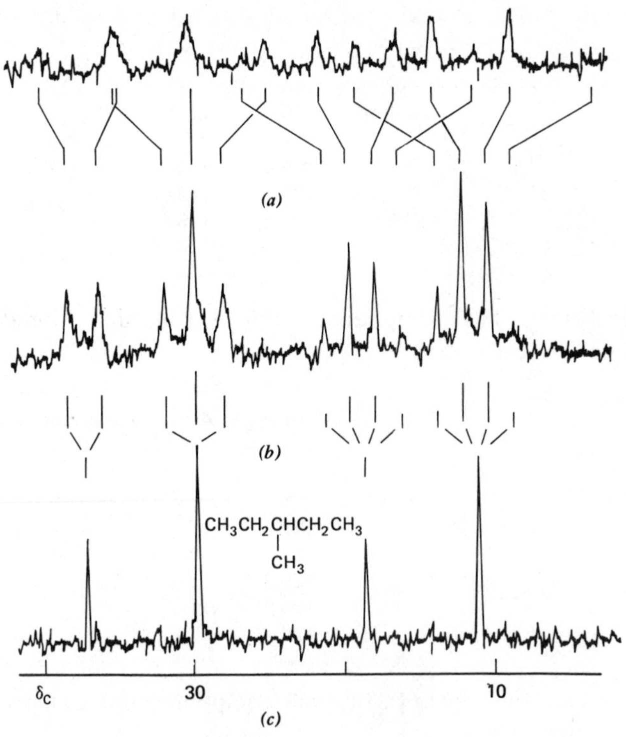

Figure 8.13. ^{13}C magnetic resonance spectra of 3-methylpentane. (*a*) No proton decoupling ("coupled" spectrum). (*b*) Protons decoupled at an off-resonance position ("off-resonance decoupled" spectrum). (*c*) Protons completely decoupled ("completely decoupled" spectrum). Spectra determined at 25.15 MHz. (From "Carbon-13 NMR Spectroscopy," by J. B. Stothers, Academic Press, New York, copyright © 1972; with permission.)

Another type of cmr spectrum that is of value is the "off-resonance decoupled" spectrum. In such a spectrum, the irradiation used to decouple the protons is offset a few hundred hertz; this offset introduces a small degree of splitting in the carbon signals. The splitting in a given signal is usually due only to those protons directly attached to the carbon giving rise to that signal. Thus this technique normally gives rise to a splitting (see Fig. 8.13*b*), which reveals the number of protons attached to a carbon; specifically:

Structural unit containing the carbon	Apparent multiplicity of a successful off-resonance decoupling experiment
Methyl (CH_3—)	Quartet
Methylene (—CH_2—)	Triplet
Methine (CH—)	Doublet
Quarternary (—C—) (with bonds above and below)	Singlet

If the off-resonance decoupling experiment gives rise to easily identifiable multiplets, it is convenient simply to label the singlets of the completely decoupled spectrum with a letter; specifically, q for "quartet," t for "triplet," and so on, and m for a "multiplet" of such complexity that no one of the q, t, d, or s designations can be assigned.

Problem. A compound of molecular formula $C_4H_6O_2$ gives rise to the following cmr data:

δ_c	Multiplicity
20.2	q
96.8	t
141.8	d
167.6	s
(singlet position, completely decoupled spectrum)	(apparent multiplicity, off-resonance decoupled spectrum)

Assume that no other carbon resonances exist for this compound and deduce the structure of the compound. Assign all signals.

An illustration of a cmr spectrum that was obtained in the absence of any proton decoupling is provided in Fig. 8.13*a*. This type of spectrum is rarely used for routine structure determinations, because the magnitude of carbon–hydrogen coupling constants (100 to 300 Hz) is of the order of magnitude of differences in carbon shift positions. Thus, in such "completely coupled" spectra the components of multiplets very frequently coincide or overlap with one another. This causes it to be very difficult to determine shift positions and signal multiplicities. In addition, "completely coupled" spectra display complexity due to the long-range coupling of carbons to protons that are more than one bond away. These long-range couplings cause only negligible splitting in "off-resonance decoupled" spectra. It is also clear that proton decoupling enhances the intensity of carbon signals for a number of reasons; thus, in comparison, the signals are very weak in "proton coupled" spectra determined under similar instrument conditions. Thus, for routine structure determinations spectra showing complete proton–carbon coupling are normally not used.

8.4.3 Signal Intensity

The intensity of a carbon resonance is not simply proportional to the number of equivalent carbons in a molecule that would be expected to give rise to that signal. There are a number of additional, more important factors that control the intensity of a carbon resonance. Many of these depend on instrument settings, such as pulse length, pulse delay, and so on, and others depend on the structural environment of the ^{13}C nuclei (e.g., as seen in terms of the spin-lattice relaxation time, T_1). It is useful to remember, however, that the spin-lattice relaxation time of a given carbon atom normally has a substantial dependence on the number of protons directly attached to that carbon. Thus the following intensity trends usually hold:

$$CH_3 \text{—} \sim \text{—} CH_2 \text{—} \sim \text{—} \overset{|}{\underset{|}{C}}H \text{ (methine)} > \text{—}\overset{|}{\underset{|}{C}}\text{—(quaternary)}$$

and

$$CH_3 \text{—} \sim \text{—} CH_2 \text{—} \sim \text{—} \overset{|}{\underset{|}{C}}H > \; \rangle C{=}O$$

Specifically, the preceding trends indicate orders of decreasing intensity of cmr signals corresponding to these groups. Sometimes it is even possible to differentiate these structural units even more selectively; for example, methyl can be differentiated from methine.

In some cases cmr intensities may seem to be misleading; for example, it is common to obtain the cmr spectrum of 1,4-cyclohexadiene under conditions that

[Structure: 1,4-cyclohexadiene, ring positions numbered 1–6]

cause the intensity of the olefinic carbon signal to be nearly equal to the intensity of the methylene carbon signal. (Explain this equality.) Normally, however, the intensity trends can be useful. Specifically, benzene ring carbons bearing substituents, and thus no protons, usually give rise to weak signals (see Fig. 8.12). Also, carbonyl carbon resonances (of other than aldehydes) are normally relatively weak, because they have no directly attached protons.

Problem. Deduce the structure of the compound corresponding to both the molecular formula $C_9H_{11}NO$ and the following cmr spectrum:

δ_c	Intensity[a]
39.7	s
110.8	s
124.9	w
131.6	s
154.1	w
189.7	m

[a] s = strong, m = medium, and w = weak.

Make use of intensity trends and the data of Table 8.11.

Under certain conditions that involve the addition of relaxation agents containing Cr(III) or Gd(III) ions to a solution of an organic compound, the cmr spectrum can give rise to signals of intensities proportional only to the number of equivalent carbons that correspond to that signal. The student is referred to the C-13 references listed in Chapter 10 (Section 10.2.10).

8.5 LANTHANIDE SHIFT REAGENTS

When carrying out proton magnetic resonance (pmr) analysis of organic compounds, it is very common to find that pmr signals are not well resolved. This frequently occurs when the environment of various types of protons in a molecule are only slighly dissimilar. This slight dissimilarity, in turn, results in only slightly different chemical shifts for the protons. In such cases, the overlap of the proton signals makes it very difficult either to correlate specific signals with appropriate protons in the organic compound or to extract coupling constant information from these signals.

It has now become common practice to simplify the pmr spectra of many organic compounds by treatment with a compound known as a *lanthanide shift reagent* (LSR). These LSRs are almost always β-diketonate complexes of certain paramagnetic lanthanide (Ln) ions:

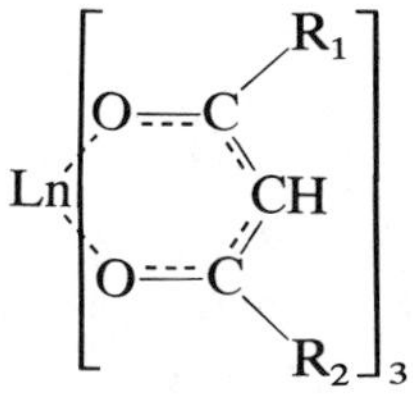

```
      ┌        R1 ┐
      │O---C      │
   Ln │       CH  │
      │O---C      │
      └        R2 ┘3
```

The central lanthanide ion bears a +3 charge and is usually either europium(III), praeseodymium(III), or ytterbium(III). Complexes containing any one of these three metal ions are commercially available (Aldrich Chemical Co., Eastman Organic Chemicals, Merck, Alpha-Ventron), most commonly with $R_1 = R_2 = t$-butyl or $R_1 = t$-butyl and $R_2 =$ perfluoro-n-propyl. Specific structures are shown here for the common europium reagents:

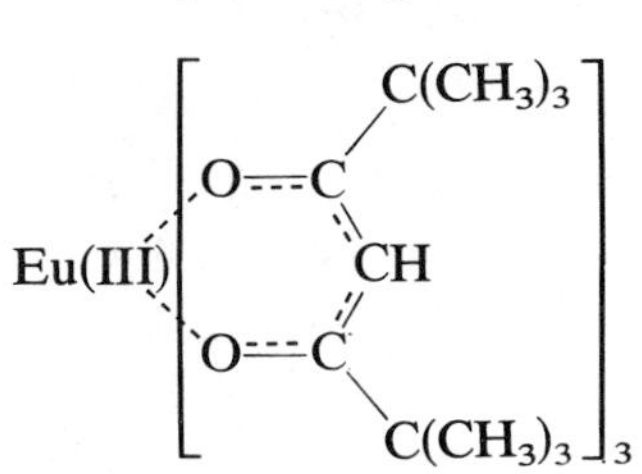

```
           ┌        C(CH3)3 ┐
           │O---C           │
   Eu(III) │       CH       │
           │O---C           │
           └        C(CH3)3 ┘3
```

tris-(**dip**ivalo**m**ethanato)europium

$Eu(dpm)_3$

These compounds are often listed by shortened names composed of initials taken from the ligand's more systematic name; thus $Eu(dpm)_3$, $Pr(dpm)_3$, $Yb(fod)_3$, and so on.

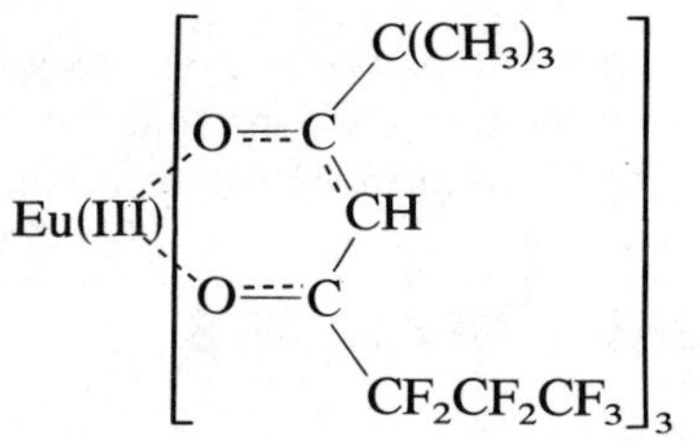

tris(1,1,1,2,2,3,3-heptafluoro-7,7-dimethyl-3,5-octanedionato)europium

$Eu(fod)_3$

Such shift reagents are added to samples of common organic compounds dissolved in the usual nmr solvents ($CDCl_3$, CCl_4, C_6D_6). This addition frequently results in an increased nonequivalence of the signals in the pmr spectrum of the organic compounds. Addition of shift reagent can result in the complete separation of the signals in the spectrum, and this separation can lead to first-order analysis of the separate signals. For example, in the absence of shift reagent, the pmr spectrum of benzyl alcohol at 60 MHz shows only a slightly broadened singlet for all of the aromatic protons. Upon addition of a ratio of 0.39 moles of

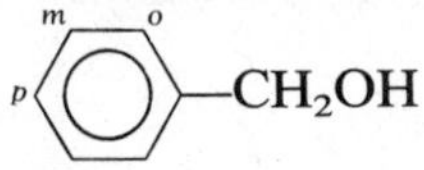

benzyl alcohol

$Eu(dpm)_3$ for each mole of alcohol, the aromatic protons separate into three separate signals.[33] In addition, the signal due to the proton para to the CH_2OH group appears as a triplet. This triplet appearance is consistent with first-order rules and with significant coupling only to the meta protons. It is also important to note that the extent to which these aromatic protons shift per increment of shift reagent follows the pattern

ortho > meta > para

This is consistent with the McConnell–Robertson equation, which is used to explain the shifts in proton signals for most organic compounds:

$$\frac{\Delta H}{H} = K\left(\frac{3\cos^2\theta - 1}{R^3}\right)$$

where $\Delta H/H$ is the magnitude of the induced shift and K is a constant that depends on the shift reagent and the organic substrate (e.g., benzyl alcohol) under study. The magnitudes of R and θ can be visualized by considering the dipole–dipole interaction of the Lewis acidic LSR with a Lewis basic site in the substrate:

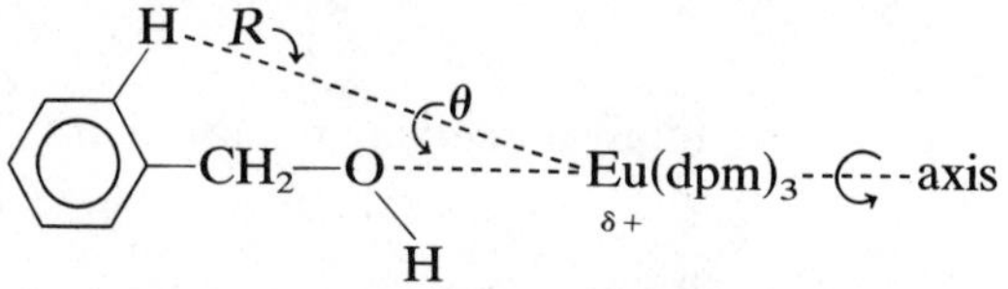

[33] J. K. M. Sanders and D. H. Williams, *J. Chem. Soc. D, Chem. Commun.*, 422 (1970).

Thus R is the distance from the center of the lanthanide atom to the proton of interest in this complex. The angle θ is rigorously described by the intersection of R with the principal magnetic axis of the complex. It is usually sufficient to assume that this magnetic axis is collinear with an axis from the center of the lanthanide atom to the Lewis basic atom in the organic substrate coordinated to this lanthanide (e.g., the oxygen of benzyl alcohol).

The McConnell–Robertson equation can be applied to the LSR analysis of benzyl alcohol. The term $(3\cos^2\theta - 1)$ changes very little upon passing from the ortho to the meta to the para protons. In view of the inverse cube dependence on distance, it is clear that the protons closest to the lanthanide atom in the complex should show the greatest shift per increment of shift reagent. This interpretation explains the order of LSR-induced shifts for the aromatic protons of benzyl alcohol listed above. Such an interpretation can also be used to aid in the assignment of proton signals of other organic compounds. In general, if one can conclude, for example, from molecular models, that the angle term $(3\cos^2\theta - 1)$ will undergo very little change from one proton to another near a functional group,[34] one may assume that those protons that shift more per increment of shift reagent are the protons closer to the lanthanide atom in the complex and thus closer to the functional group coordinated to that lanthanide atom.

Shift reagents are more useful when applied to organic substrates with relatively basic functional groups. Thus, the following sequence of decreasing affinity to the usual LSR is expected and normally observed:

$$\text{amines} > \text{alcohols} > \text{aldehydes} > \text{ketones} > \text{esters} > \text{nitriles}$$

Exceptions to this order are observed where other factors, such as steric effects, offset the basicity factors. For example, *t*-butyl alcohol associates more poorly with the LSR than do simple ketones. Shifts of a reasonable magnitude due to complexation with sulfoxides, sulfites, sulfines, *N*-oxides, *N*-nitroso compounds, and azoxy compounds have been observed. Carboxylic acids and phenols may be shifted, but only LSR reagents with less basic ligands (e.g., fod) should be used. Treatment of substrates as acidic as phenols with $Eu(dpm)_3$ results in the unwanted acid-base reaction of the dpm anion with the hydroxyl protons of the phenols.

The β-diketonate complexes of Pr(III) and Eu(III) provide complementary properties. Usually the Pr(III) complexes shift protons of organic substrates upfield and the Eu(III) complexes shift them downfield. Often the Eu(III) complexes are the reagent of choice because, in simple compounds, the protons that are downfield in the absence of shift reagent are often the protons closest to the functional group and thus are the protons that are shifted to the greatest extent upon addition of the shift reagent. For example, the pmr spectrum of n-butyl alcohol shows the protons on the carbon α to the OH group at lowest field in the absence of shift reagent. Addition of a Pr(III) diketonate complex

---

[34] The tacit assumption is also made that this functional group is the only one that is significantly coordinated to the LSR to a substantial degree.

would rapidly shift these α-protons to higher field and thus these protons would overtake the upfield protons; this clearly could cause signal overlap and an undesirable increase in the complexity of the signals. The Eu reagents would be desirable in such cases because they should *increase* the differences in chemical shifts of protons in substrates such as n-butyl alcohol, the protons at lower field being shifted to even lower field at faster rates.

In some cases it is desirable to avoid contact shifts so that the magnitude of LSR-induced shifts can be adequately interpreted by the McConnell–Robertson equation. In those cases where contact shifts might be a problem (e.g., in cmr in general and in the pmr of highly basic amines), $Yb(dpm)_3$ and other ytterbium shift reagents may be used, rather than Pr or Eu reagents.

Analysis of chiral organic compounds can be facilitated by use of chiral shift reagents. It has been found that derivatives of camphor are extremely useful for this purpose:

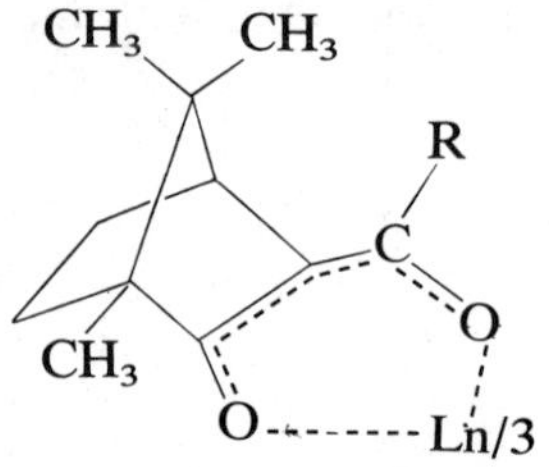

CH_3 CH_3 R C O CH_3 O Ln/3

Specific chiral shift reagents and the abbreviations under which commercial suppliers (e.g., Alpha-Ventron) list their reagents are given here:

Pra-Opt®:

$Ln = Pr, R = CF_3$

tris(3-trifluoromethylhydroxymethylene-*d*-camphorato)praseodymium(III)

Eu-Opt®:

$Ln = Pr, R = CF_3$

tris(3-trifluoromethylhydroxymethylene-*d*-camphorato)europium(III)

Application of chiral shift reagents can be visualized by considering a sample of racemic α-phenylethylamine containing both (+) and (−) enantiomers:

$$\underset{(\pm)}{C_6H_5\overset{CH_3}{CH}\text{—}NH_2} + \underset{(+)}{Eu\text{—}Opt} \longrightarrow \underset{(\substack{+\\+})}{C_6H_5\overset{CH_3}{C}HNH_2\cdot Eu\text{—}Opt} + \underset{(\mp)}{C_6H_5\overset{CH_3}{C}HNH_2\cdot Eu\text{—}Opt}$$

Racemic amine sample

The pmr signals of corresponding protons in each of the enantiomers of the $(\pm)$ amine reactant are identical, because enantiomers differ only in their ability to rotate plane polarized light. Treatment of the racemic amine with chiral, optically active Eu-Opt, here arbitrarily assumed to be $(+)$, can give rise to two diastereomeric complexes: $(^{+}_{+})$ and $(\mp)$. The pmr signals of corresponding protons in these diastereomers are nonidentical; integration of the areas due to these nonidentical signals yields the relative amounts of the $(^{+}_{+})$ and $(\mp)$ complexes and thus the relative amounts of $(+)$ and $(-)$ amine in the original sample.

Shift studies are usually most successful in solvents such as $CDCl_3$, CCl_4 and C_6D_6. These solvents are relatively weak as Lewis bases and thus do not fill the coordination sites of the shift reagent. The protons of the β-diketonate ligands in the shift reagents absorb in regions as follows:

$Eu(fod)_3$	0.4 to 2.0 ppm
$Eu(dpm)_3$	1.0 to −2.0 ppm
$Pr(dpm)_3$	3.0 to 5.0 ppm

These absorption regions depend to some extent on the solvent employed and on the organic substrate to which the shift reagent coordinates. Deuterated shift reagents are available (Aldrich, Alpha); these alleviate the interference of such ligand resonances.

Additional material on this subject may be found in the following references:

R. E. Sievers, *N.M.R. Shift Reagents* (Academic Press, New York, 1973).
O. Hofer, *Topics in Stereochemistry*, Vol. 9, edited by N. Allinger and E. L. Eliel (Wiley-Interscience, New York, 1976), p. 111ff.
M. R. Willcott, III and R. E. Davis, *Science*, **190,** 850 (1975).
A. Cockerill, *et al.*, *Chem. Rev.*, **73,** 553 (1973).
J. R. Campbell, *Aldrichimica Acta*, **4,** 55 (1971).

CHAPTER NINE

structural problems—solution methods and exercises

The laboratory examination of an organic compound results in the accumulation of data concerning the physical properties, elements present or absent, solubility, spectral data, and behavior toward certain suitable class reagents and in various special tests. All these observed facts must be correlated and interpreted in order to arrive at possible structural formulas for the compound in question. It is necessary to show what functional groups are present, to determine the nature of the nucleus to which they are attached, and to find the positions of attachment.

It is the purpose of this discussion to point out, by means of several specific examples, the mode of attack and reasoning involved in deducing information concerning the structure of a molecule from experimental data.

SAMPLE PROBLEMS

Part I

Compounds with Structures Previously Described in the Literature

The identification of these compounds does not require quantitative analysis for the elements present, molecular weight determination, or calculation of molecular formulas. The identification is based on the *matching* of the physical and chemical properties of the substance being studied and the data on its derivatives. The laboratory work in this course as described in the preceding chapters is concerned with these previously described compounds.

Two very helpful physical constants described in Chapter 6 are the neutralization equivalents of acids and bases and the saponification equivalents of esters. These numerical data, in conjunction with the solubility class and behavior toward reagents, frequently give valuable clues concerning the molecular structure of the compound. Their use may best be explained by reference to examples.

Example 1 *An organic acid has a neutralization equivalent of* 45 ± 1.

As pointed out on p. 268, the neutralization equivalent of an acid is dependent on the number of carboxyl groups in the molecule. If one carboxyl group is present, the neutralization equivalent is equal to the molecular weight. If the present compound is monobasic, its molecular weight[1] must be 44, 45, or 46. A carboxyl group weighs 45; hence if the molecular weight were 45, nothing could be attached to the carboxyl group. A molecular weight of 44 is obviously impossible, but a molecular weight of 46 leaves a residue of 1 after the weight of the carboxyl radical is subtracted. Only one element has the atomic weight of 1; hence, formic acid (HCOOH) is one possibility.

However, the compound might be dibasic, in which event the molecular weight would be 90 ± 2. Two carboxyl groups equals $2 \times 45 = 90$. Hence, a possible residue of 0, 1, or 2 units remains. There are no bivalent atoms of this atomic weight; hence the only possible dibasic acid is oxalic acid, in which the two carboxyl groups are united:

```
COOH
|
COOH
```

Thus, by assuming first a monobasic acid and then a dibasic acid, two possible structures have been deduced from the neutralization equivalent alone. In order to decide between the two, the physical state or the solubility class of the compound would serve. If this compound, with a neutralization equivalent of 45 ± 1, is a liquid, soluble in water and in pure ether (solubility group S_1), it must be formic acid. If it is a solid, soluble in water but insoluble in ether (group S_2), it is anhydrous oxalic acid.

Consideration of the molecular weights indicates in a similar fashion that the compound could not be tribasic (mol wt 135 ± 3) or tetrabasic (mol wt 180 ± 4).

---

[1] For purposes of illustration and calculation in this chapter, whole numbers have been used for the atomic weights of the elements carbon, hydrogen, oxygen, nitrogen, and bromine. The actual atomic weights (which must be used in all precise quantitative analyses) differ from these rounded-off values by an amount less than the experimental error involved in the determination of neutralization and saponification equivalents.

Example 2 *An acid* (A) *possessed a neutralization equivalent of* 136 ± 1. *It gave negative tests for halogen, nitrogen, and sulfur. It did not decolorize cold potassium permanganate solution; but when an alkaline solution of the compound was heated with this reagent for an hour and acidified, a new compound* (B) *was precipitated. This compound had a neutralization equivalent of* 83 ± 1.

First consider the compound B. Assume it to be monobasic.

$$\begin{aligned} \text{molecular weight} &= 83 \pm 1 \\ \text{less one —COOH} &= \underline{45} \\ \text{residue} &= 38 \pm 1 \end{aligned}$$

This residue to which the carboxyl group is attached must be made up of some combination of carbon, hydrogen, and perhaps oxygen that is stable to hot potassium permanganate solution. Examination shows that this is not possible.

$$\begin{aligned} \text{residue} &= 38 \pm 1 \\ \text{three carbon atoms} = 3 \times 12 &= \underline{36} \\ \text{remainder} &= 2 \pm 1 \end{aligned}$$

The residue might be C_3H, C_3H_2, or C_3H_3, none of which corresponds to a compound that would be stable to permanganate. The alkane would require C_3H_8 as the parent compound, and the alkyl group would have to be C_3H_7—; similarly, the cycloalkane would have to be C_3H_6 and the cyclopropyl group C_3H_5—.

The presence of oxygen in this residue is also excluded. If it is assumed to be present, the following figures are obtained:

$$\begin{aligned} \text{residue} &= 38 \pm 1 \\ \text{one oxygen atom} &= \underline{16} \\ & 22 \pm 1 \\ \text{one carbon atom} &= \underline{12} \\ \text{remainder} &= 10 \pm 1 \end{aligned} \qquad \textit{or} \qquad \begin{aligned} \text{residue} &= 38 \pm 1 \\ \text{two oxygen atoms} &= \underline{32} \\ \text{remainder} &= 6 \pm 1 \end{aligned}$$

Neither of these remainders corresponds to any atom or groups of atoms that correspond to a reasonable organic compound. For example, the former suggest $CH_{10}OCO_2H$ for the compound and the latter, $H_6O_2CO_2H$, both of which are unreasonable. Thus, it is now safe to conclude that compound B *cannot be monobasic.*

$$\begin{aligned} \text{molecular weight} = 2 \times 83 \pm 1 &= 166 \pm 2 \\ \text{two carboxyl groups} &= \underline{90} \\ \text{residue} &= 76 \pm 2 \end{aligned}$$

If this residue is saturated and aliphatic it must be made up of —(CH_2)— units.

$$\text{five —CH}_2\text{—} = 5 \times 14 = 70$$

$$\text{six —CH}_2\text{—} = 6 \times 14 = 84$$

Neither of these corresponds to the weight of the residual radical, 76 ± 2.

Another grouping that is stable to hot permanganate is the benzene nucleus. This is an arrangement of six CH groups or $6 \times 13 = 78 =$ molecular weight of benzene itself. If two carboxyl groups are present, two of the hydrogen atoms are displaced and the residue becomes $78 - 2 = 76$. This value checks that calculated above for the residue, and hence a possible structure for B is $C_6H_4(COOH)_2$; that is, it may be one of the phthalic acids.

The question now arises whether compound B could be tribasic. If so we have the following values:

$$\begin{aligned} \text{molecular weight} &= 3 \times 83 \pm 1 = 249 \pm 3 \\ \text{three carboxyl groups} &= 3 \times 45 \quad\; = 135 \\ \text{residue} &\quad\quad\quad\quad\;\; = 114 \pm 3 \end{aligned}$$

Inspection shows that this residue cannot be aromatic because it does not correspond to one or more benzene rings. A benzene ring plus a side chain is excluded because the side chain would be oxidized by the permanganate. The value 114 ± 3 does correspond, however, to eight CH_2 groups $(8 \times 14 = 112)$ within experimental error. Hence $C_8H_{15}(COOH)_3$, with a molecular weight of 246, falls within the limit of 249 ± 3. Although this tricarboxylic acid represents a possible structure for a compound with a molecular weight of 249 ± 3, it would be impossible to produce it from compound A, which has a neutralization equivalent of 136 ± 1.

Assume that A is monobasic.

$$\begin{aligned} \text{molecular weight A} &= 136 \pm 1 \\ \text{one —COOH} &= \;\;45 \\ \text{residue} &= \;\;91 \pm 1 \end{aligned}$$

Since B has a C_6H_4 grouping stable to permanganate, this same group must also be present in A.

$$\begin{aligned} \text{residue} &= 91 \pm 1 \\ C_6H_4 &= 76 \\ \text{remainder} &= 15 \pm 1 \end{aligned}$$

This remainder of 15 ± 1 corresponds to a methyl group that must be attached to the ring.

$$C_6H_4\begin{matrix} \diagup COOH \\ \\ \diagdown CH_3 \end{matrix} \longrightarrow C_6H_4\begin{matrix} \diagup COOH \\ \\ \diagdown COOH \end{matrix}$$

compound A neut equiv = 136 | compound B neut equiv = 83

The original must be *o*-, *m*-, or *p*-toluic acid, each of which would give the reactions cited. Additional data, such as a melting point, a derivative, or spectra are necessary to distinguish among them.

This example also illustrates the fact that oxidation almost invariably converts a compound with a given neutralization equivalent to a product that has a lower neutralization equivalent. This generalization follows naturally from the increase in the number of carboxyl groups or cleavage of the molecule into smaller fragments.

Example 3 *A colorless crystalline compound* (A) *gave a positive test for nitrogen but not for halogens or sulfur. It was insoluble in water, dilute acids, and alkalies. It produced a red-colored complex with ammonium hexanitratocerate reagent but did not react with phenylhydrazine. Compound* A *dissolved in hot sodium hydroxide solution with the liberation of ammonia and the formation of a clear solution. Acidification of this solution produced compound* B, *which contained no nitrogen and gave a neutralization equivalent of* 182±1. *Oxidation of* B *by hot permanganate solution produced* C, *which had a neutralization equivalent of* 98±1. *When either* A *or* B *was heated with hydrobromic acid for some time, a compound* D *separated. This compound contained bromine but no nitrogen. It gave a precipitate with bromine water and a violet color with ferric chloride, and it readily reduced dilute potassium permanganate. It was soluble in sodium bicarbonate solution.*

These reactions may be summarized in the following chart.

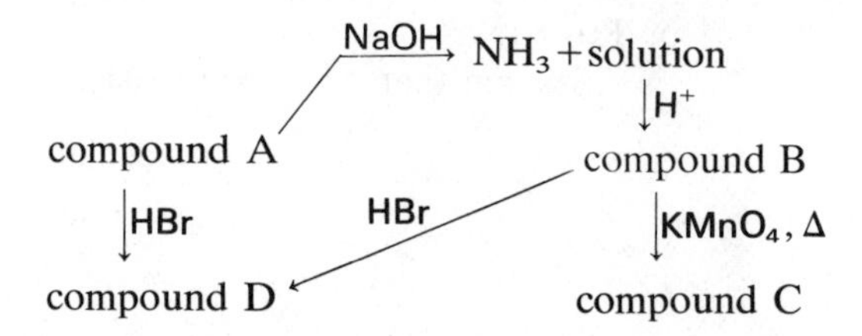

The elimination of nitrogen from compound A by alkaline hydrolysis suggests the presence of a nitrile or amide grouping, because these functional groups liberate ammonia when they undergo hydrolysis.

$$RCN + NaOH + H_2O \rightarrow RCOONa + NH_3$$

$$RCONH_2 + NaOH \rightarrow RCOONa + NH_3$$

The imide grouping, —CONHCO—, which also liberates ammonia, is excluded by the fact that compound A is not soluble in sodium hydroxide solution. Since compound B contained no nitrogen, negatively substituted amines, such as 2,4-dinitroaniline, are excluded. The absence of nitrogen in B and the fact that it was acidic and not basic likewise eliminate a substituted urea, which also liberates ammonia when hydrolyzed.

$$RNHCONH_2 + H_2O \rightarrow RNH_2 + CO_2 + NH_3$$

The positive red color with ammonium hexanitratocerate reagent suggests an alcohol; an amino or phenolic group is excluded by the fact that compound A is neutral. Further evidence for the presence of a hydroxyl group is furnished by the fact that compound D, produced from A by the action of hydrobromic acid, contained bromine.

$$ROH + HBr \rightarrow RBr + H_2O$$

The properties of compound D strongly suggest the presence of a phenolic hydroxyl group. Ease of bromination, sensitivity to permanganate, and color with ferric chloride are properties characteristic of substituted phenols. This phenolic hydroxyl group was produced by the action of hydrobromic acid on some functional group present in A and B, because neither of these originally contained the phenol grouping. One type of compound that produces a phenol when treated with hydrobromic acid is an aryl alkyl ether.

$$ArOR + HBr \rightarrow ArOH + RBr$$

If the alkyl group is small, it would be lost as alkyl bromide during the treatment with hydrobromic acid. Thus compounds A, B, and C probably contain such a mixed ether group and also an aromatic nucleus, because the substituted phenol D contains one. The solubility of D in sodium bicarbonate solution is probably due to the presence of a carboxyl group, because both the nitrile and amide groups are hydrolyzed to carboxyl groups by acids as well as alkalies. Hence compound D is a hydroxybenzoic acid with the bromine attached to a side chain. This side chain must also be present in compound A with an alcoholic group in place of the bromine atom.

The neutralization equivalents of compounds B and C may now be considered. It will be noted that the neutralization equivalent of compound C, produced by permanganate oxidation, is lower than that of compound B, which acquired its acidic properties by an hydrolysis reaction only. This oxidation obviously affects the side chain, and compound C must have more carboxyl groups than B.

Assume compound C to be dibasic.

$$\text{molecular weight} = 2 \times 98 \pm 1 = 196 \pm 2$$

$$\text{two carboxyl groups} = 2 \times 45 = \underline{90}$$

$$106 \pm 2$$

$$\text{one oxygen atom in ether linkage} = \underline{16}$$

$$90 \pm 2$$

$$\text{benzene minus three hydrogen atoms } (C_6H_3) = \underline{75}$$

$$\text{residue} = 15 \pm 2$$

This residue of 15 corresponds to a CH_3— group, and hence the ether grouping must have been CH_3O—.

It is now necessary to find the length of the side chain to which the alcohol group in A and B is attached.

If C is dibasic, B is monobasic.

$$
\begin{aligned}
\text{neut equiv} = \text{mol wt} &= 182 \pm 1 \\
\text{carboxyl group} &= \underline{\ \ 45\ \ } \\
&\ \ \ 137 \pm 1 \\
\text{methoxyl group} &= \underline{\ \ 31\ \ } \\
&\ \ \ 106 \pm 1 \\
\text{hydroxyl group} &= \underline{\ \ 17\ \ } \\
&\ \ \ 89 \pm 1 \\
\text{benzene nucleus } (C_6H_3) &= \underline{\ \ 75\ \ } \\
\text{residue} &= 14 \pm 1
\end{aligned}
$$

This residue represents the weight of the aliphatic side chain and obviously corresponds to —CH_2—. Hence, possible structures for A, B, C, and D are

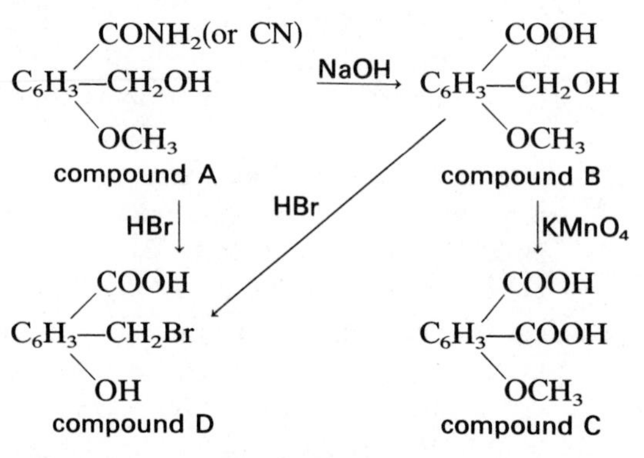

This example illustrates the fact that a given reagent may affect more than one functional group. Thus, boiling hydrobromic acid affected three functional groups in compound A and two in compound B.

Example 4 *A solid compound, melting at 60–61°C, gave no tests for sulfur, nitrogen, or halogen. It was insoluble in water, dilute acid, and alkali, but reacted with concentrated sulfuric acid and was placed in solubility group* MN. *A thin film (melt) was prepared and the following infrared spectrum was recorded:*

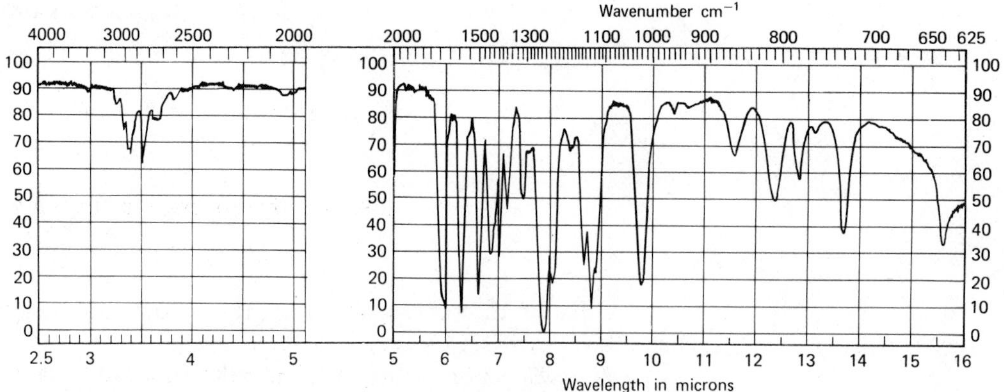

Example 4. Infrared spectrum. (Reprinted from "The Aldrich Collection of IR Spectra," Edited by C. J. Pouchert, copyright © 1971, 1975; with permission.)

The following infrared bands have immediate implications as to functional group features for this compound:

5.95 μm (1680 cm^{-1})	C=O stretch (conjugated with a π system)
6.3 μm (1590 cm^{-1})	aromatic C⩴C stretch
6.65 μm (1500 cm^{-1})	aromatic C⩴C stretch

These features are suggested by tables in Chapters 5 and 6 and by Chapter 3 of Silverstein, Bassler, and Morrill.[2] The strong bands at 7.9 μm (1270 cm^{-1}), 8.80 μm (1140 cm^{-1}), and 9.8 μm (1020 cm^{-1}) imply that the molecule possesses a C—O—C unit; either ether and/or ester functional groups would provide this unit.

A sample of the original compound gave an orange-red precipitate with the 2,4-dinitrophenylhydrazine reagent, which confirmed the presence of an aldehyde or ketone group, but did not exclude the possibility of an ester group *in addition* to the aldehyde or ketone group.

Ms. Mary Four, who was working on the identification of this compound, then consulted the tables of solid ketones, aldehydes, and esters with the melting-point range[3] 60 to 61 ±2°C. The following list of possible compounds was obtained:

Solid ketones

		m.p. (°C)			m.p. (°C)
(A)	$C_6H_5COCH_2C_6H_5$	60	(D)	$C_6H_5CH{=}CHCOC_6H_5$	62
(B)	CH_3— —COC_6H_5	60	(E)	CH_3O— —COC_6H_5	62
(C)	$C_6H_5COCH_2COCH_3$ ⇅ C_6H_5—C(OH)=$CHCOCH_3$	61			

Solid aldehydes

		m.p. (°C)			m.p. (°C)
(F)	CH_3O— —CHO, CH_3O	58	(I)	C_6H_5— —CHO	60
(G)	—CHO	59	(J)	CHO, OCH_3	60
(H)	CHO	60			

---

[2] R. M. Silverstein, G. C. Bassler, and T. C. Morrill, *Spectrometric Identification of Organic Compounds,* 3rd ed. (Wiley, New York, 1974).

[3] This overall range of 58 to 62°C was used to take into consideration problems of possible inpurities and the differences in techniques and thermometer use by prior chemists. By using this wide range, the chances of missing a compound are reduced.

Note that the above list contains some compounds that would not give the C—O—C stretching that the ir seems to display; it is, however, clearly better to list too many compounds initially than presumptuously to limit the list and inadvertently omit the proper structure.

A search of tables of solid esters (m.p. 58–63°C) that *also* had an aldehyde or keto group resulted in no additional possibilities. Ms. Four now reexamined the ir spectrum in conjunction with the above structural formulas. The moderate absorption bands at ca. 3.4 μm (2940 cm^{-1}) and 3.5 μm (2860 cm^{-1}) suggest an aromatic aldehyde (aldehydic C—H stretching bands); the ir bands for C—O—C stretch (see above) suggest aromatic methoxy groups. To confirm a reduced number of possibilities, Ms. Four carried out two additional tests. The hydriodic acid reagent (Zeisel test for CH_3O—) gave a positive test. This eliminated compounds (A), (B), (C), (D), (G), (H), and (I) from consideration and left only compounds (E), (F), and (J) containing methoxyl groups in addition to a carbonyl group.

Although the ir spectrum already favored an aromatic aldehyde, the ketone (E) was eliminated by showing that the original compound was oxidized by potassium permanganate solution and the acidic chromium trioxide reagent. To distinguish between the remaining possibilities [compounds (F) and (J)], the oxime derivative was prepared. It melted at 95°C, thus confirming the identity of the original as 3,4-dimethoxybenzaldehyde (F, veratraldehyde), since the literature reported the m.p. of the oxime of compound (J) as 102°C.

Note that although they were not used here, nmr and other chemical tests could have been helpful. A pair of nmr singlets at ca. δ3.8 would have supported the aromatic methoxyl groups, and a positive Tollens test would have supported the aldehyde group.

Example 5 *One of Mr. John Five's unknowns melted at* 60 to 61°C *and gave no tests for sulfur, nitrogen, or halogen. It was in solubility group* MN *and gave an orange-red precipitate with the* 2,4-*dinitrophenylhydrazine reagent. The infrared spectrum of his unknown gave the following:*

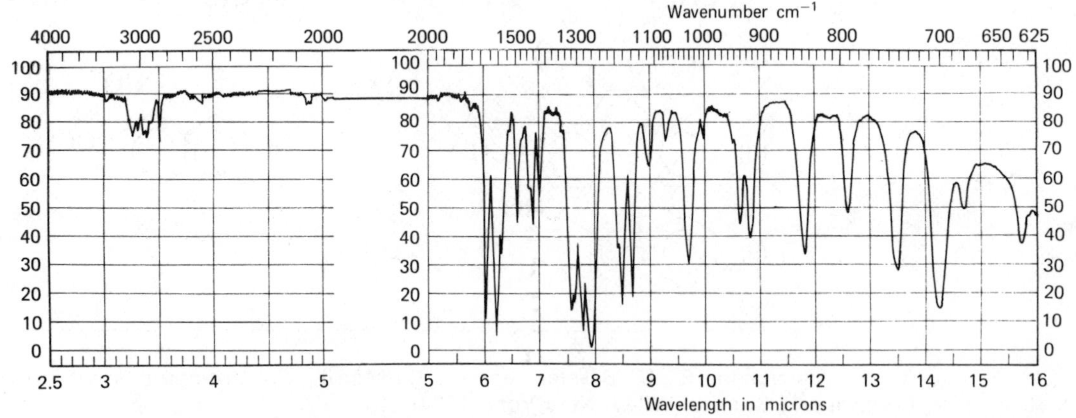

Example 5. Infrared spectrum. (Reprinted from "The Aldrich Collection of IR Spectra," Edited by C. J. Pouchert, copyright © 1971, 1975; with permission.)

The nmr spectrum of this unknown ($CDCl_3$, 60 MHz, TMS $\delta = 0.00$) showed:

δ3.83, 3H, sharp singlet
δ6.95, 2H, doublet ("leans" upfield)
δ7.3–8.0, 7H, multiplet

The first nmr signal strongly implies a methyl group; the chemical shift implies that the methyl is on oxygen. All remaining signals imply aromaticity.

Observing the carbonyl band at 6.0 μm (1667 cm^{-1}), Mr. Five prepared a list of possible solid aldehydes and ketones melting over the range 58 to 63°C, and his list was identical to that assembled by Ms. Mary Four shown on p. 463. His unknown did not reduce potassium permanganate or the chromium trioxide reagent, thus eliminating all the oxidizable compounds, (C), (D), (F), (G), (H), (I), and (J). The remaining compounds possible were (A), (B), and (E). Inspecting these structures, he noted that only compound (E) contained a methoxyl group and that the ir spectrum [asymmetric C—O—C stretch of aryl alkyl ether, 7.9 μm (1266 cm^{-1}), symmetric C—O—C stretch, 9.7 μm (1030 cm^{-1})] also indicated this group as a possible function. Hence, the compound was tested with hydriodic acid reagent and found to give a positive methoxyl test. Final identification was made by preparing the 2,4-dinitrophenylhydrazone which, after recrystallization, melted at 180 to 181°C, and this confirmed the original as 4-methoxybenzophenone.

Example 6 *Mr. Henry Six received a compound for identification that melted at* 60 *to* 61°; *it contained no nitrogen, sulfur, or halogen and was in solubility Group* MN. *It gave a precipitate with* 2,4-*dinitrophenylhydrazine reagent; the infrared spectrum of Mr. Six's compound looked like this:*

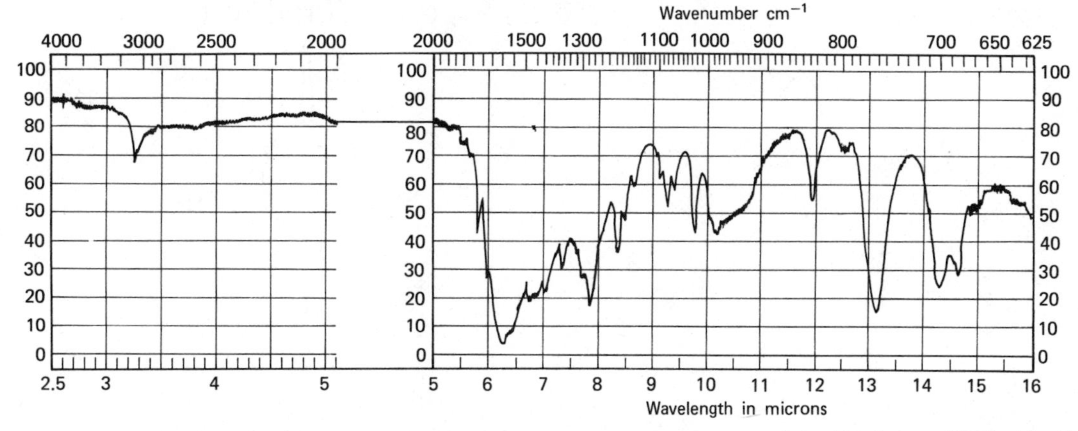

Example 6. Infrared spectrum. (Reprinted from "The Aldrich Collection of IR Spectra," Edited by C. J. Pouchert, copyright © 1971, 1975; with permission.)

The NMR spectrum of this compound (determined in deuteriochloroform, TMS = δ0.00, 60 MHz) was taken:

δ6.92, 2H, sharp singlet

δ7.3–7.6, 6H, multiplet

δ7.8–8.1, 4H, multiplet

The rather broad IR band at 6.0 to 6.5 μm (1667–1538 cm^{-1}) was puzzling, but Mr. Six prepared a list of solid aldehydes and ketones melting from 58 to 63°, and his list of possibilities was identical with those made (p. 463) by Ms. Mary Four and Mr. John Five. Mr. Six treated his unknown with potassium permanganate solution and chromium trioxide reagent and found that both these reagents were reduced. This eliminated nonoxidizable compounds (A), (B), and (E). Next, he tried the Zeisel test for methoxyl group and this was negative, thus eliminating compounds (F) and (J); also, (E) was eliminated (again). The lack of nmr evidence for a methoxyl group supports these eliminations. The remaining possibilities were compounds (C), (D), (G), (H), and (J). Mr. Six restudied his ir spectrum and compared the bands with the tables and discussion in Chapters 5 and 6. It appeared that the wide band at 6.2–6.5 μm suggested the C=C stretch of an enolized ketone. The O—H stretch band of enols is broad and here runs from 3.1 to 4.0 μm (3200 to 2500 cm^{-1}, low absorbance, because of broadening). Hence, he tested the reaction of the compound with the ferric chloride-pyridine reagent and noted that a bluish red color was produced. A sample of the 2,4-dinitrophenylhydrazone was prepared and found to melt at 150 to 151°C, a value that agreed with the literature value for that derivative of benzoylacetone, compound (C).

In retrospect, the nmr signals can be discussed further. The most upfield singlet is due to the methylene group and the most downfield aromatic signals are due to the protons ortho to the benzoyl groups.[4]

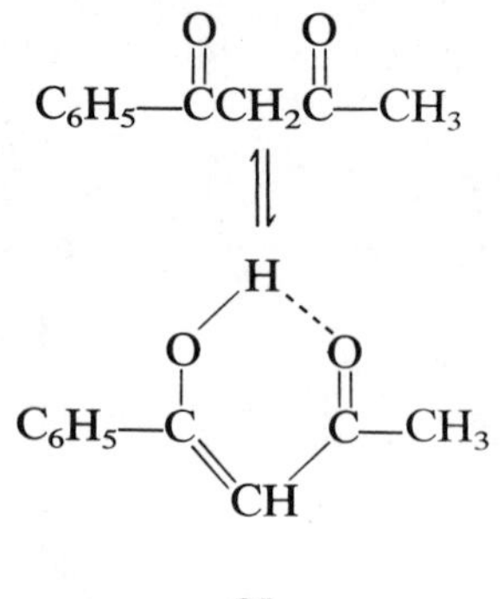

$$C_6H_5—\overset{O}{\overset{\|}{C}}CH_2\overset{O}{\overset{\|}{C}}—CH_3 \qquad (C)$$

$$\rightleftharpoons$$

$C_6H_5—C(—O—H\cdots O=)C—CH_3$, with C=CH (enol form, intramolecularly hydrogen-bonded)

or

$C_6H_5—C(=O\cdots H\cdots O=)C—CH_3$, with delocalized C⋯CH⋯C (resonance-hybrid form)

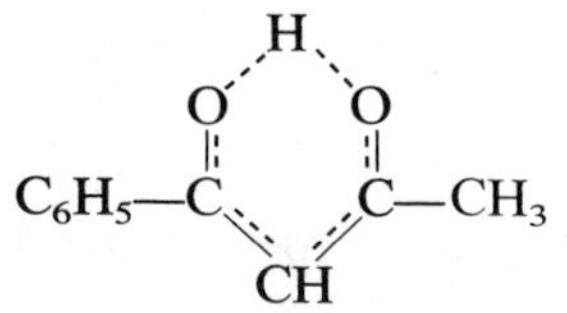

Example 7 *An ester, containing only carbon, hydrogen, and oxygen, possessed a saponification equivalent of* 74 ± 1.

---

[4] It is of interest that the nmr sample is not observably enolized and yet the ir sample is highly enolized.

The first step is to work out the possibilities on the assumption that the molecule contains only one ester group. In that event the molecular weight is equal to the saponification equivalent. The type of formula for an ester is R—COO—R′, and therefore the first step is to subtract the weight of —COO— from the molecular weight.

$$\begin{aligned} \text{molecular weight} &= 74 \pm 1 \\ \text{—COO—} &= \underline{44\quad\ } \\ \text{residue} &= 30 \pm 1 \end{aligned}$$

This residue represents the combined weight of R and R′. In saturated esters[5] containing only carbon, hydrogen, and oxygen, this residue must always be equal to C_nH_{2n+2} and, moreover, must always be an even number.[6] Thus residual weights of 31 and 29 are impossible, and the value 30 represents the molecular weight of C_nH_{2n+2}. Mere inspection in this case shows that the hydrocarbon residue is C_2H_6, but the logical approach is to solve for the value of n by multiplying its value by the atomic weights of the elements in the formula and setting this equal to the residual weight.

$$12n + 1(2n+2) = 30$$

$$14n = 28 \quad \text{and} \quad n = 2$$

This residue of C_2H_6 represents the sum of R and R′, and it is now necessary to write the possibilities.

$$R + R' = C_2H_6$$

$$H + C_2H_5$$

$$CH_3 + CH_3$$

Hence, on the assumption that the compound was a monoester, two possibilities are

$HCOOC_2H_5$ and CH_3COOCH_3

ethyl formate — methyl acetate

If two ester groups are present in the molecule, then the molecular weight is twice the saponification equivalent. Two ester groups will contain two —COO— combinations; hence

$$\begin{aligned} \text{molecular weight} &= 2 \times 74 \pm 1 = 148 \pm 2 \\ \text{two —COO—} &= 2 \times 44 \quad\ = \underline{\ 88\quad\ } \\ \text{residue} &= \quad\quad\quad\quad\ \ 60 \pm 2 \end{aligned}$$

The value of 60 ± 2 represents the summation of those portions of the molecule

---

[5] In an olefinic ester the residue is C_nH_{2n}; in an acetylenic ester, C_nH_{2n-2}.

[6] See footnote on p. 457.

other than the two —COO— groups. Again, $C_nH_{2n+2} = 60 \pm 2$ and

$$12n + 1(2n + 2) = 60 \pm 2$$
$$14n = 58 \pm 2$$

Since n must be an integer, the only value of the right-hand side of the equation that will fulfill this requirement is 56, and hence $n = 4$. The residue must be C_4H_{10} and has to be divided among the various hydrocarbon radicals present in the type formulas for a compound with two ester groups. Some of the possible type formulas are the following:

COOR	COOR	RCOO	RCOO
|	|	|	|
$(CH_2)_x$	$(RCH)_x$	$(CH_2)_x$	$(CHR)_x$
|	|	|	|
COOR′	COOR	RCOO	RCOO

Using these type formulas for esters of a dicarboxylic acid, or a dihydroxy alcohol, possible structures may now be written in which the four carbon atoms and ten hydrogen atoms are distributed properly in each of the above formulas.

This example illustrates the use of saponification equivalents in deducing possible structures. It also shows that a saponification equivalent is not quite so useful as the neutralization equivalent of an acid. It is always desirable to have some additional data concerning either the acid or alcohol or both produced by saponification of the ester in order to reduce the number of isomeric esters that possess the required saponification equivalent.

Part II

Determination of the Structure of New Compounds That Have Not Been Described in the Chemical Literature

Interpretation of Molecular Formulas

In research, quantitative analyses together with molecular weight determinations routinely yield the molecular formula of any unknown substance. Much information about possible functional groups can often be deduced from such information alone, and for this reason it is pertinent to consider the significance of the molecular formula.

A saturated hydrocarbon without any rings has the general formula C_nH_{2n+2}. Introduction of oxygen to give an alcohol, ether, acetal, or any other saturated acyclic compound does not change the carbon-to-hydrogen ratio, and the molecular formula is $C_nH_{2n+2}O_m$. The introduction of a double bond or a ring into a saturated molecule requires the removal of two hydrogen atoms, and the introduction of a triple bond involves the removal of four hydrogen atoms.

By an examination of the carbon-to-hydrogen ratio, then, it is possible to draw conclusions concerning the possible number of multiple bonds or rings in a molecule.

For example, the substance C_3H_6O must have either one double bond (either olefinic or carbonyl) or one ring but cannot have a triple bond because it is only two hydrogen atoms short of saturation. The compound $C_8H_{12}O$ must have three double bonds or three rings or some combination of double bonds, triple bonds, and rings that accounts for the three pairs of missing hydrogen atoms. It cannot contain a benzene ring, however, because such a ring requires a shortage from saturation of four pairs of hydrogen atoms. (Thus the possibility that the oxygen function is a phenolic hydroxyl group is immediately excluded.)

Furthermore, the introduction of a halogen atom into a saturated molecule necessitates the removal of one hydrogen atom, and consequently the general formula of a saturated acyclic monohalide is $C_nH_{2n+1}X$. On the other hand, introduction of a nitrogen atom to give an acyclic saturated amine requires also the addition of an extra hydrogen atom so that the formula is $C_nH_{2n+3}N$. A consequence of these generalizations, which is of value in deriving molecular formula from the analyst's carbon and hydrogen determination, is that a molecule with no elements other than carbon, hydrogen, and oxygen must contain an even number of hydrogen atoms; an odd number of halogen or nitrogen atoms requires an odd number of hydrogen atoms; and an even number of halogen or nitrogen atoms requires an even number of hydrogen atoms. Thus a molecular formula such as $C_5H_{11}O_3$, calculated from analytical data, is obviously incorrect; the correct formula must be either $C_5H_{10}O_3$ or $C_5H_{12}O_3$.

Sample Problems with Molecular Formulas

The preceding discussion can be summarized in equation form. For the generalized molecular formula $I_yII_nIII_zIV_x$ (e.g., $C_xH_yN_zO_n$), where

I can be H, F, Cl, Br, I, D, etc. (i.e., any monovalent atom),
II can be O, S, or any other bivalent atom,
III can be N, P, or any other trivalent atom,
IV can be C, Si, or any other tetravalent atom,

the index of hydrogen deficiency $= x - y/2 + z/2 + 1$.

This index represents "missing pairs" of monovalent atoms that correspond to double bonds, triple bonds, and/or cyclic features in the structure of interest. This formula should be used with care when dealing with other than simple oxidation states of elements in the second and lower rows of the periodic table. For example, the index for the formula C_3H_8OS is zero. Possible structures for this formula include the following:

$HOCH_2CH_2SCH_3$ $\quad$ $HOCH_2CH_2CH_2SH$ $\quad$ $CH_3S(=O)CH_2CH_3$

The last formula, a sulfoxide, seems to present a contradiction. We can, however, include this if we visualize it in terms of its polar-covalent resonance form:

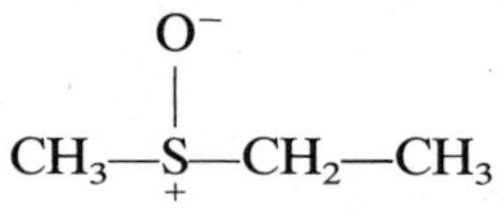

$$CH_3{-}\overset{\overset{\textstyle O^-}{|}}{\underset{+}{S}}{-}CH_2{-}CH_3$$

Thus, all three structures would have an index of zero.

Problem

Write all the structures possible for the formula $C_3H_9O_4P$.

Example 8 *A substance* (A) *has the formula* C_6H_{10}. *On hydrogenation over platinum under mild conditions it is converted to* B, C_6H_{14}. *When compound* A, C_6H_{10}, *was ozonized and the ozonide reductively cleaved, two products, acetaldehyde and glyoxal, were formed. These known compounds were identified by means of their* 2,4-*dinitrophenylhydrazones.*

The molecular formula shows that the maximum number of double bonds and rings is two, or else there is one triple bond. Uptake of two moles of hydrogen indicates that there are, in fact, either two double bonds or one triple bond.[7] Ozonolysis to give two-carbon fragments suggests that the six-carbon skeleton originally present must have been cleaved at two points to give three two-carbon fragments rather than at one. This means that two double bonds rather than one triple bond must have been present. Finally, the probable arrangement of the three two-carbon pieces is as shown.

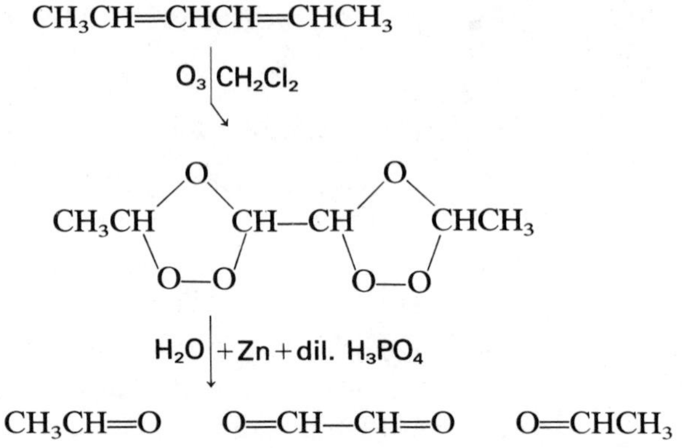

$$CH_3CH{=}CHCH{=}CHCH_3$$

$$\downarrow O_3 \mid CH_2Cl_2$$

$$CH_3CH\langle O, O{-}O\rangle CH{-}CH\langle O, O{-}O\rangle CHCH_3$$

$$\downarrow H_2O \mid +Zn+\text{dil. } H_3PO_4$$

$$CH_3CH{=}O \qquad O{=}CH{-}CH{=}O \qquad O{=}CHCH_3$$

By considering the molecular formulas of several compounds and the reactions that produced them, one is frequently able to deduce possible formulas. The following problem involves the deduction of a large amount of information from only a few clues.

---

[7] Less frequently, hydrogen will add across a *strained* ring of a compound containing no multiple bonds; we have assumed that this is not so here in order to allow initial hypotheses to be made.

Example 9 *A neutral compound* (A), $C_{15}H_{14}O$, *gives a negative Baeyer test and is not attacked by hydrogen bromide; it is oxidized to an acid* (B), $C_{14}H_{10}O_3$, *by hot chromic acid solution.*

First, note that the oxidation has caused a *loss* of one carbon atom and four hydrogen atoms but a *gain* of two oxygen atoms. It is necessary to find a functional group or several groups that will do this. In this connection it is useful to tabulate the behavior of the common functional groups on oxidation, making a note of the gain and loss in composition. From the table below it will be noted that the oxidation of an ethyl side chain corresponds exactly to the oxidation of A to B; in both cases there are gains of two oxygen atoms and a loss of four hydrogen atoms and one carbon atom.

Functional group	Oxidation product	Gain	Loss	
RCHO	→ RCOOH	1 O		
RCH_2OH	→ RCOOH	1 O	2 H	
R_2CHOH	→ R_2CO		2 H	
$R_2C(OH)CH_3$	→ R_2CO		4 H	1 C
$RCH{=}CH_2$	→ RCOOH	2 O	2 H	1 C
$RC{\equiv}CH$	→ RCOOH	2 O		1 C
$ArCH_2CH_2OH$	→ ArCOOH	1 O	4 H	1 C
$ArCOCH_3$	→ ArCOOH	1 O	2 H	1 C
$ArCH_3$	→ ArCOOH	2 O	2 H	
$ArCH_2CH_3$	→ ArCOOH	2 O	4 H	1 C
ArC_nH_{2n+1}	→ $ArCO_2H$	2 O	$2n$ H	$(n-1)$ C
Ar_2CHCH_3	→ Ar_2CO	1 O	4 H	1 C
Ar_2CH_2	→ Ar_2CO	1 O	2 H	
From Example 9:				
(A) $C_{15}H_{14}O$	→ (B) $C_{14}H_{10}O_3$	2 O	4 H	1 C

Subtracting an ethyl group from A ($C_{15}H_{14}O$) or a carboxyl group from B leaves a unit $C_{13}H_9O$. This unit is derived from some parent compound (C), $C_{13}H_{10}O$, which is stable to oxidation.

The next problem concerns the character of the functional group containing the oxygen atom. What functional groups containing only one oxygen atom are stable to oxidation? Consideration of various functional groups leads to the conclusion that the ether linkage is one possibility and that a properly substituted ketone is a second. Ketones with no hydrogen atoms on the α-carbon atoms are usually stable to oxidation. The commonest examples are diaryl ketones.

The next step consists of considering the ratio of carbon to hydrogen in the compounds A and B and the hypothetical parent compound (C), $C_{13}H_{10}O$. These carbon and hydrogen atoms must be combined so that the resulting compound will be stable to oxidation. A completely saturated compound, C_nH_{2n+2}, would require a formula $C_{13}H_{28}$ for a 13-carbon-atom compound (C). Even allowing two hydrogen atoms as equivalent to the oxygen atom, it is obvious that the compound has no such ratio of carbon and hydrogen atoms. An alicyclic compound would require C_nH_{2n} or $C_{13}H_{26}$. Ordinary olefinic compounds and acetylenic compounds with enough double or triple bonds to lower the ratio of hydrogen to carbon are excluded by the stability to oxidation.

The only large class of compounds with such a low ratio of hydrogen to carbon atoms is aromatic in nature. Since benzene has six carbon atoms whereas the parent compound (C) has 13, the possibility of two benzene rings is suggested. This leaves one carbon atom to be accounted for.

Subtracting two phenyl groups, $(C_6H_5—)_2$, from compound $C(C_{13}H_{10}O)$ leaves a residue of CO. It will be remembered that diaryl ketones are stable to oxidation. The parent compound (C) is evidently benzophenone, $C_6H_5COC_6H_5$, and the compounds A and B are probably

$$C_6H_5COC_6H_4CH_2CH_3 \longrightarrow C_6H_5COC_6H_4CO_2H$$

compound A	compound B
One of the isomers of ethylbenzophenone	One of the isomers of benzoylbenzoic acid

Part III

Problems

The following problems are designed to give the student added experience in the types of reasoning illustrated by the examples above. It is of the utmost importance to seek the answers by systematic procedures, and students are urged to avoid a random attack on the problems.

After determining structures, equations should be written for all reactions and all major spectral bands should be assigned to the appropriate structural features.

For spectra, the following abbreviations are used:

nmr

Integration, 1 H = one proton, 2 H = 2 protons, etc.
Multiplicity, s = singlet, d = doublet, t = triplet, q = quartet, b = broad, m = multiplet (undetermined).

All nmr spectra were determined at 60 MHz, unless otherwise noted.

ir

band intensities
s = strong, m = medium, w = weak, b = broad.

Set 1

In the investigation of unknown compounds, the following types of behavior are observed frequently. Indicate in each instance the deductions which may be made as to the nature of the compound.

1. A yellow, neutral compound containing only carbon, hydrogen, and oxygen was changed to an acid by the action of hydrogen peroxide. This yellow compound displayed a singlet in the nmr spectrum at ca. $\delta 9.0$ ppm, as well as other signals.
2. A neutral compound reacted with phenylhydrazine to yield a product that differed from the expected phenylhydrazone by the elements of ethanol; that is, the condensation involved the elimination not only of the elements of water but also of those of ethanol.
3. A compound containing only carbon, hydrogen, and oxygen reacted with acetyl chloride but not with phenylhydrazine. Treatment with periodic acid converted it into a compound that reacted with phenylhydrazine but not with acetyl chloride.
4. A yellow, neutral compound formed a derivative with *o*-phenylenediamine. The original compound showed an ir band(s) at 1710 cm^{-1} (5.85 μm).
5. An alcohol gave a positive iodoform test and a negative Lucas test.
6. A neutral compound containing only carbon, hydrogen, and oxygen reacted with acetyl chloride but not with phenylhydrazine. Heating with mineral acids converted it to a compound that failed to react with acetyl chloride but that gave positive tests with phenylhydrazine and bromine in carbon tetrachloride. The nmr spectrum of the original compound showed only two singlets.
7. A nitrogen-containing compound gave a positive nitrous acid test for secondary amines, but its derivative with benzenesulfonyl chloride was soluble in alkalies.
8. A compound, when treated with ethyl orthoformate and a trace of acid, was found to take up the elements of ethyl ether.

   One of the two possible classes of compounds that serves as an answer showed a proton nmr signal at ca. $\delta 9.0$–10.0 before orthoformate treatment; orthoformate treatment moved this signal to ca. $\delta 5.0$.
9. A neutral compound containing carbon, hydrogen, and oxygen underwent dimerization in an ethanolic solution of sodium cyanide. The ir spectrum of the reactant showed bands at ca. 2800 and 2700 cm^{-1} (3.57 and 3.71 μm).
10. A basic compound failed to react with benzenesulfonyl chloride but yielded a derivative with nitrous acid. The original basic compound showed no appreciable ir bands near 3333 cm^{-1} (3.00 μm). The product of nitrous acid treatment showed an ir band at ca. 1550 cm^{-1} (6.45 μm).
11. An ester had a saponification equivalent of 59 ± 1. The nmr spectrum showed two singlets with an integration ratio of 3:1.
12. A solid water-soluble acid had a neutralization equivalent of 54 ± 1. At temperatures above its melting point, the compound lost carbon dioxide and formed a new solid acid with a neutralization equivalent of 59 ± 1. The

nonexhangeable protons of the original compound resulted in a multiplet nmr signal; the product formed upon heating showed the nonexchangeable protons as a singlet.

13. A water-soluble acidic compound containing nitrogen and sulfur gave a neutralization equivalent of 142 ± 1. Addition of barium chloride to an aqueous solution produced a precipitate insoluble in acids. Alkali caused the separation of a basic compound whose hydrochloride had a neutralization equivalent of 130 ± 1. Both the first and last compounds (the acids) showed broad ir bands at ca. 3333–2500 cm^{-1} (3.00–4.00 μm).
14. A water-soluble compound containing nitrogen gave a neutralization equivalent of 73 ± 1 when titrated with standard hydrochloric acid and methyl red as the indicator. It formed a precipitate when treated with benzenesulfonyl chloride and sodium hydroxide solution. The original compound showed a quartet (4 H), a triplet (6 H), as well as a broadened singlet (1 H) in the nmr spectrum.
15. A 12-g sample of a compound, C_8H_8O, was treated with a carbon tetrachloride solution of bromine containing 60 g of bromine. Hydrobromic acid was evolved, and a quantitative determination showed that 16.2 g was liberated. After the reaction was complete the solution was still red, and it was found that 12 g of bromine remained unused. Calculate the number of atoms of bromine per molecule that were introduced by substitution and, if addition also took place, the number of atoms of bromine which were added. Use Br = 80, H = 1, C = 12, O = 16.

    Describe important nmr and ir features you would anticipate for both the reactant and the product; emphasize the important spectral changes expected as the result of the bromine treatment.

Set 2

In solving the problems of this set and those in Sets 3 and 4, follow the steps outlined above using the data cited in the problems. Make the customary allowances for experimental error in boiling points and melting points. Assign spectral data to structural features.

I. (1) White crystals; m.p. 117–118°C.
(2) Elemental analysis for X, N, S—negative.
(3) Solubility—water (+).
(4) Classification tests:

$C_6H_5NHNH_2$—negative	CH_3COCl—positive
$KMnO_4$—positive	HIO_4—positive
Br_2 in CCl_4—negative	
N.E. = 151 ± 1	

(5) Derivative:
p-Nitrobenzyl ester, m.p. 123°C.
(6) Spectra:
ir (mineral oil mull): 3333–2400 cm^{-1} (3.00–4.17 μm, s, b); 1706 cm^{-1}

(5.86 μm, s); 1449, 1379, 1300, 1200, 1190, 943, 865, 730 cm^{-1} (6.9, 7.25, 7.7, 8.2, 8.4, 10.6, 11.55, 13.7 μm, all m).
nmr (acetone): δ5.22, 1 H, s; δ6.93–7.88, 7 H, m.

II. (1) Colorless liquid; b.p. 259–261°C.
(2) Elemental analysis for Br—positive; for Cl, I, N, S—negative.
(3) Solubility—water (−), NaOH (−), HCl (−), H_2SO_4 (+).
(4) Classification tests:

$H_2NOH \cdot HCl$—negative $KMnO_4$—negative
CH_3COCl—negative Br_2 in CCl_4—negative
$AgNO_3$—negative   NaI—negative

Hot sodium hydroxide—clear solution which on acidification gave white crystals, m.p. 250°C, containing bromine. Saponification equivalent of the original compound = 229 ± 2.

(5) Derivatives:
(a) Treatment with hydrazine gave colorless crystals; m.p. 164°C.
(b) Treatment with 3,5-dinitrobenzoic acid and sulfuric acid gave pale yellow crystals; m.p. 92°C.

III. (1) Brown liquid; b.p. 198–200°C.
(2) Elemental analysis for X, N, S—negative.
(3) Solubility—water (−), NaOH (+), $NaHCO_3$ (−), HCl (−).
(4) Classification tests:

CH_3COCl—positive Br_2 in H_2O—precipitate
$C_6H_5NHNH_2$—negative $FeCl_3$—violet
Ce(IV)—positive

(5) Derivative: treatment with chloroacetic acid gave white crystals; m.p. 102–103°C.
(6) Spectra:
ir (neat): 3390 cm^{-1} (2.95 μm, s, b); 1163, 935, 780, 690 cm^{-1} (8.6, 10.75, 12.85, 14.5 μm, all s).
nmr ($CDCl_3$): δ2.25, 3 H, s; δ5.67, 1 H, s; δ6.5–7.3, 4 H, m.

IV. (1) Colorless liquid; b.p. 194–195°C.
(2) Elemental analysis for X, N, S—negative.
(3) Solubility—all (−).
(4) Classification tests:

$H_2SO_4 \cdot SO_3$—negative
$AlCl_3 + CHCl_3$—light yellow

(5) Derivatives: none.
Sp gr $\frac{20}{4}$ = 0.8963 n_D^{20} = 1.4811

V. (1) White crystals; m.p. 187–188°C.
(2) Elemental analysis for X, N, S—negative.
(3) Solubility—water (+). Aqueous solution is acid to litmus.
(4) Classification tests:

$KMnO_4$—negative $C_6H_5NHNH_2$—negative
Br_2 in CCl_4—negative CH_3COCl—negative
Neutralization equivalent = 59   Partition coefficient = 7.7

(5) Derivative: Heating with phenylhydrazine gave white crystals, m.p. 209–210°C.

(6) Spectra:

ir (Nujol mull): 3333–2222 cm^{-1} (3.0–4.5 μm, s, b); 1695, 1418, 1307, 1198, 913 cm^{-1} (5.90, 7.05, 7.65, 8.35, 10.95 μm, all s).

nmr (DMSO-d_6): δ2.43, s, δ11.80, bs; areas 2:1 (upfield singlet to downfield singlet).

VI. (1) Reddish-brown solid; m.p. 65–70°C.

(2) Elemental analysis for N—positive; for X, S—negative.

(3) Solubility—water (−), NaOH (−), HCl (+). A solution of the compound in dilute hydrochloric acid was decolorized with Norit and alkali added. The compound purified in this manner melted at 73°C.

(4) Classification tests:

$KMnO_4$—positive

Br_2 in CCl_4—precipitate

Bromine water—precipitate

$C_6H_5NHNH_2$—positive

$C_6H_5SO_2Cl + NaOH$—residue soluble in hydrochloric acid

$Fe(OH)_2$—negative

Tollens reagent—positive

Hot sodium hydroxide decomposed the compound. After removal of a dark brown solid by filtration, the filtrate was neutralized and a light tan precipitate obtained. After recrystallization from a water-alcohol mixture, it melted with decomposition at 236–240°C. It was soluble in HCl and NaOH, insoluble in sodium bicarbonate. The original compound reacted with acetone and sodium hydroxide to give a yellow precipitate; m.p. 134–135°C.

(5) Derivatives:

Oxime, m.p. 144°C.
Phenylhydrazone, m.p. 148°C.
Semicarbazone, m.p. 224°C.

(6) Spectra:

ir (Nujol mull): no absorption near 3333 cm^{-1} (3.00 μm); absorptions at 1653, 1587 cm^{-1} (6.05, 6.30 μm, m, s).

nmr ($CDCl_3$): δ3.05, 3 H, s; δ6.69, 2 H, d; δ7.71, 2 H, d; δ9.70, 1 H, s.

Set 3

1. A brown liquid (I) boiled at 193–195°C. It contained nitrogen but gave negative tests for sulfur and the halogens. It was insoluble in water but soluble in dilute acid. It did not react with acetyl chloride or benzenesulfonyl chloride. Treatment of a hydrochloric acid solution of the unknown compound with sodium nitrite, followed by neutralization, gave a compound (II) that melted at 83–84°C. Compound II was insoluble in alkalies but dissolved in boiling concentrated sodium hydroxide solution, with the liberation of a gas (III). This gas (III) was absorbed in water, and the aqueous solution was treated with phenyl isothiocyanate; there was formed a compound (IV) with a melting point of 134–135°C. Careful acidification of the above alkaline

solution, followed by extraction, gave a compound (V) that melted at 125–126°C.

The nmr spectrum of compound V (in acetone) showed δ6.63, 2 H, d; δ7.67, 2 H, d; ca. δ8.7, 1 H, bs. The δ8.7 signal was so broad as to be detectable only by electronic integration.

2. A colorless liquid was found to be soluble in water and in ether. It boiled at 94–96°C, and gave negative tests for the halogens, nitrogen, and sulfur. It reduced a dilute potassium permanganate solution, decolorized bromine in carbon tetrachloride, reacted with acetyl chloride, and liberated hydrogen upon treatment with sodium. It did not give iodoform when treated with sodium hypoiodite and did not react with phenylhydrazine. Treatment with 3,5-dinitrobenzoyl chloride transformed it into a compound melting at 47–48°C.

   The nmr spectrum of the original compound (in $CDCl_3$) showed δ3.58, s, 1 H; δ4.13, 2 H, m; δ5.13, 1 H, m; δ5.25, 1 H, m; ca. δ6.0, 10 lines, 1 H. The δ3.58 signal showed a chemical shift that was concentration-dependent.

3. A yellow solid (I), melting at 113–114°C, contained nitrogen but no halogens, sulfur, or metals. It was insoluble in water and alkalies but soluble in dilute acids. The acid solution of I was treated with sodium nitrite in the cold and then boiled. The product (II) of this reaction separated when the solution was cooled. It contained nitrogen and melted at 95–96°C; it was insoluble in acids and sodium bicarbonate solution but soluble in sodium hydroxide solution. The products obtained by treating compounds I and II with zinc and a boiling solution of ammonium chloride readily reduced Tollens reagent. The original compound (I) was treated with benzenesulfonyl chloride and alkali. Acidification of the resulting solution gave a compound (III) that melted at 135–136°C.

   The nmr spectrum of compound II (in DMSO-d_6) showed δ7.0–7.75, 4 H, m; δ9.8, 1 H, s.

4. A colorless crystalline compound (I) melted at 186–187°C; it contained nitrogen but no halogens or sulfur. It was insoluble in water and dilute acids but soluble in dilute sodium bicarbonate solution. It gave a neutralization equivalent of 180 ± 2 but did not react with bromine in carbon tetrachloride, dilute potassium permanganate solution, acetyl chloride, or phenylhydrazine. It was treated for some time with boiling hydrochloric acid. When this reaction mixture was cooled, a compound (II) separated that melted at 120–121°C and gave a neutralization equivalent of 121 ± 1. The filtrate remaining after the removal of II was evaporated to dryness, and the residue (III) was purified by recrystallization. It contained nitrogen and chlorine, was rather hygroscopic, and decomposed when an attempt was made to determine is melting point. It was insoluble in ether, and its aqueous solution gave a precipitate with silver nitrate. A solution of III was treated with nitrous acid in the cold. A vigorous evolution of a gas was observed. Compound III was treated with benzenesulfonyl chloride and sodium hydroxide solution. Acidification of the resulting solution gave a new product (IV) that melted at 164–165°C.

nmr spectra:

I (Sodium Salt, D_2O solvent): δ4.06, 2 H, s; δ4.77, HDO impurity, s; δ7.5–8.0, 5 H, *m*.

II (CCl_4 solvent): δ7.3–7.7, 3 H, m; δ8.0–8.25, 2 H, m; δ12.8, 1 H, s.

III (D_2O solvent): δ3.58, s; δ4.9, HDO.

5. A colorless crystalline compound (I) melted at 162–165°C, with decomposition. It gave negative tests for nitrogen, halogens, sulfur, and metals. It was soluble in water but insoluble in ether. It reacted with acetyl chloride, decolorized permanganate solution, reduced Fehling's solution and Tollens reagent, but gave no color with Schiff's reagent. It reacted with phenylhydrazine to give a product (II) that melted with decomposition, at 199–201°C. When I was warmed with concentrated nitric acid, a vigorous reaction took place, and a compound (III) separated when the reaction mixture was cooled. This compound (III) was insoluble in water but readily soluble in alkalies; it gave a neutralization equivalent of 104 ± 1. Compound III reacted with acetyl chloride but not with phenylhydrazine, and melted at about 212–213°C, with decomposition. If kept above its melting point for some time it was converted into a new compound (IV) that melted at 132–133°C, after recrystallization. Compound IV was insoluble in water but soluble in sodium bicarbonate solution, and it gave a neutralization equivalent of 111 ± 1. Treatment of the sodium salt of IV with *p*-bromophenacyl bromide gave a compound (V) melting at 137–138°C.

   The original compound (I) was optically active, having a specific rotation of +81°C, but the degradation products III, IV, and V were optically inactive.

   A solution of the potassium salt of IV (in D_2O) gave rise to these nmr signals: δ6.59, 1 H, m; δ7.05, 1 H, m; δ7.64, 1 H, m. A Nujol mull of IV gave these ir bands: 3100–2400 cm^{-1} (3.22–4.27 μm, s); 1675 cm^{-1} (5.97 μm, s); several bands in the 1667–1000 cm^{-1} (6.00–10.00 μm) region; 932 cm^{-1} (10.73 μm, m); 885 cm^{-1} (11.3 μm, m); 758 cm^{-1} (13.2 μm, m).

Set 4

1. A compound insoluble in water, sodium hydroxide, and hydrochloric acid but soluble in concentrated sulfuric acid and containing nitrogen melted at 68°C. When treated with tin and hydrochloric acid it yielded a substance that reacted with benzenesulfonyl chloride to give an alkali-soluble derivative. When the original compound was treated with zinc and a hot sodium hydroxide solution it was converted to a new substance melting at 130°C.

   The product of the reaction with tin and hydrochloric acid was neutralized with base and then distilled. This distillate (in CCl_4 solvent) gave rise to the following nmr spectrum: δ3.32, 2 H, s; δ6.44, 2 H, m; ca. δ6.6, 1 H, m; δ7.0, 2 H, m.

2. A compound boiled at 166–169°C and contained sulfur but no nitrogen or halogen. It was insoluble in water and dilute acids but dissolved in sodium hydroxide solutions. Its sodium derivative reacted with 2,4-dinitrochlorobenzene to give a compound melting at 118–119°C. When allowed to stand in air, the original compound was slowly oxidized to a derivative melting at 60–61°C.

   The nmr spectrum of the air oxidation product (in $CDCl_3$) showed only δ7.2–7.7, m. The nmr spectrum of the original compound showed δ3.39, 1 H, s; δ7.12, s, 5 H.

3. A compound melted at 141–142°C and contained nitrogen but no halogen or sulfur. It was insoluble in water, dilute acids, and dilute alkalies. It was unaffected by treatment with tin and hydrochloric acid. When treated for a long time with hot sodium hydroxide solution, it reacted, forming an insoluble oil (I). The oil was soluble in dilute hydrochloric acid and reacted with acetyl chloride to give a solid, melting at 111–112°C. Acidification of the alkaline solution from which I was removed gave a solid melting at 120–121°C, whose neutralization equivalent was 122 ± 1.

   An nmr spectrum of the 120–121°C m.p. solid (in $CDCl_3$) showed δ7.4–7.7, 3 H, m; δ8.0–8.3, 2 H, m; δ12.8, 1 H, s. The nmr spectrum of compound I (in $CDCl_3$) showed: δ2.15, 3 H, s; δ3.48, 2 H, bs; δ6.45–6.8, m, 2 H; δ6.8–7.15, m, 2 H.

4. A compound boiled at 159–161°C and contained chlorine but no nitrogen or sulfur. It was insoluble in water, in dilute acids and alkalies, and in cold concentrated sulfuric acid. It dissolved in fuming sulfuric acid. It gave no precipitate with hot alcoholic silver nitrate solution. Treatment with a hot solution of potassium permanganate caused the compound to dissolve slowly. The resulting solution, when acidified with sulfuric acid, gave a precipitate that melted at 138–139°C and had a neutralization equivalent of 157 ± 1.

   An nmr spectrum of the original compound (in $CDCl_3$) showed δ2.37, 3 H, s; δ7.0–7.35, 4 H, m. An ir spectrum of the material melting at 138–139°C (in a Nujol mull) showed 3333–2381 cm^{-1} (3.0–4.2 μm, s); 1678 cm^{-1} (5.96 μm, s); several bands in the region 1587–1389 cm^{-1} (6.3–7.2 μm, m); 1316 cm^{-1} (7.6 μm, s); 1053, 1042 cm^{-1} (9.5, 9.6 μm, m); 9.13 cm^{-1} (10.95 μm, m, broad); 742 cm^{-1} (13.47 μm, s).

5. A colorless liquid boiled at 188–192°C. It contained only carbon, hydrogen, and oxygen. It was insoluble in water, dilute acids, and alkalies, but dissolved readily in cold concentrated sulfuric acid. It did not react with phenylhydrazine or acetyl chloride and did not decolorize a carbon tetrachloride solution of bromine. Boiling alkalies dissolved it slowly. The resulting mixture was subjected to steam distillation. The distillate contained a compound that, when pure, boiled at 129–130°C and reacted with α-naphthyl isocyanate to give a derivative melting at 65–66°C.

   The alkaline residue, left after the steam distillation, was acidified with phosphoric acid and steam-distilled. The distillate contained an acid that yielded an anilide melting at 108–109°C.

   Treatment of the original compound with lithium aluminum hydride (followed by the usual careful work-up) resulted in only one compound; the resulting compound showed nmr and ir spectra identical to those spectra obtained from material of 129–130°C b.p. described above. The nmr spectrum showed: δ0.92, 3 H, d; δ1.0–2.0 3 H, m; δ2.13, 1 H, s; δ3.63, 2 H, d.

6. An unknown compound was a pink solid that melted at 109–112°C. Treatment with decolorizing carbon and recrystallization removed the color and brought the melting point to 112–114°C. The compound burned with a smoky flame and left no residue. Elemental analysis showed nitrogen to be present and sulfur and halogens to be absent.

   The compound was insoluble in water and dilute alkalies but dissolved in ether and dilute acids. It reacted with benzenesulfonyl chloride to give a derivative that was soluble in alkali and melted at 101–102°C. The acetyl derivative melted at 132°C.

   Infrared bands for the original compound (5000–1250 cm^{-1} in $CHCl_3$, 1250–650 cm^{-1} in CS_2) were found at 3400 cm^{-1} (2.94 μm), 3350 (2.99), 3200 (3.13), 3050 (3.28), 1640 (6.10), 1610 (6.21), 1520 (6.58), 1480 (6.75), 1400 (7.15), 1290 (7.75), 1280 (7.81), 1230 (8.13), 1190 (8.40), 1130 (8.85), 970 (10.31), 890 (11.24), 870 (11.50), 850 (11.76), 820 (12.19), 750 (13.33), and 720 (13.89).

Set 5

For each of the problems in this and the following sets, give the structural formula of an organic compound that will fulfill the conditions stated and show by equations the changes that it undergoes. Associate all major spectral bands with appropriate components of your answer structures.

1. An acid (A) containing only carbon, hydrogen, and oxygen had a neutralization equivalent of 103 ± 1. It gave a negative test with phenylhydrazine. Treatment with sulfuric acid converted it to a new acid (B) that decolorized permanganate and bromine solutions and had a neutralization equivalent of 87 ± 1. The original acid (A) was transformed by hypoiodite to iodoform and a new acid (C), the neutralization equivalent of which was 52 ± 1.

   The nmr spectrum of compound B (in $CDCl_3$) showed δ1.90, 3 H, d of d ($J = 8$, 2 Hz); δ5.83, 1 H, d of q ($J = 15$, 2 Hz); δ7.10, 1 H, m; δ12.18, 1 H, s.
2. An acid had a neutralization equivalent of 97. It could not be made to undergo substitution of bromine for hydrogen readily, even in the presence of phosphorus tribromide. Vigorous oxidation transformed it into a new acid whose neutralization equivalent was 83.

   The nmr spectrum of the second acid (in $CDCl_3$) showed δ8.08, s; δ11.0, s; relative areas = 2:1 for the δ8.08 to 11.0 signals.
3. An optically active hydrocarbon dissolved in cold, concentrated sulfuric acid, decolorized permanganate solutions, and readily absorbed bromine. Oxidation converted it to an acid that contained the same number of carbon atoms as the parent substance and had a neutralization equivalent of 66. Mass spectrometry indicated the molecular ion of the hydrocarbon to be m/e 68.

4. A compound had a neutralization equivalent of 66. The substance was not affected by bromine in carbon tetrachloride, but heat transformed it into an acid whose neutralization equivalent was 88. The nmr spectrum of the latter acid (in $CDCl_3$) showed $\delta 1.20$, 6 H, d; $\delta 2.57$, 1 H, septet; $\delta 12.4$, 1 H, bs.
5. A base had a neutralization equivalent of 121 ± 1. Vigorous oxidation converted it to an acid having a neutralization equivalent of 121 ± 1. The nmr spectrum of the acid (in CCl_4) showed $\delta 7.52$, 3 H, m; $\delta 8.14$, 2 H, m; $\delta 12.82$, 1 H, s. The nmr spectrum of the base (in $CDCl_3$) showed $\delta 0.91$, 2 H, s; $\delta 2.78$, 4 H, 8 lines; $\delta 7.23$, 5 H, s.
6. An acid whose neutralization equivalent was 166 was unaffected by bromine in carbon tetrachloride but gave a positive iodoform test.
7. A compound (A) gave negative tests for nitrogen, sulfur, and halogens. It was insoluble in water but dissolved in dilute sodium hydroxide solution. Compound A gave no color with ferric chloride and did not decolorize a solution of potassium permanganate.

   Treatment with concentrated hydrobromic acid converted A into two compounds. One was a compound (insoluble in all tests) containing bromine, which gave a precipitate with sodium iodide in acetone. The other was a new acid (B), which decolorized bromine solutions and gave a color with ferric chloride. Compound B contained no halogen. The neutralization equivalents of compounds A and B were, respectively, 180 ± 2 and 137 ± 1.

   The nmr spectrum of acid B (in $CDCl_3$ containing DMSO-d_6) showed $\delta 6.84$, 2 H, d ($J = 9$ Hz); $\delta 7.86$, 2 H, d ($J = 9$ Hz); $\delta 8.2$, 2 H, bs.
8. An optically active acid had a molecular weight of 98.
9. A compound (I) gave a precipitate with 2,4-dinitrophenylhydrazine. Compound I was heated with 25% aqueous sodium hydroxide, and the mixture was partially distilled. The distillate contained a compound that reacted with sodium and gave a positive iodoform test. It gave a negative Lucas test.

   The residue from the distillation was acidified with phosphoric acid and the mixture stream-distilled. A volatile acid was isolated. Its *p*-bromophenacyl ester had a saponification number of 257 ± 1.

   Mass spectral analysis of the original compound (I) gave rise to the following data (only peaks of intensity greater than or equal to 10% of the base peak are reported):

m/e	Percent of base	*m/e*	Percent of base
15	13.7	45	13.2
27	22.6	85	18.0
29	42.7	88	12.3
31	28.9	130	10.4
42	13.5	131	0.8
43	(100, base)		

   The ir spectrum of a neat sample of compound I showed 3000 cm^{-1} (3.33 μm, m); 2933 cm^{-1} (3.41 μm, w); 1715 cm^{-1} (5.83 μm, s); 1634 cm^{-1}

(6.12 μm, w); 1408 cm^{-1} (7.10 μm, w); 1364 cm^{-1} (7.33 μm, m); 1316 cm^{-1} (7.60 μm, s); 1250 cm^{-1} (8.0 μm, s); 1149 cm^{-1} (8.70 μm, s); 1042 cm^{-1} (9.60 μm, s).

Set 6

1. An acid (I) contained nitrogen and had a neutralization equivalent of 197 ± 2. Treatment with thionyl chloride followed by ammonium hydroxide converted I to a neutral compound (II) that reacted with an alkaline hypobromite solution to yield a basic compound (III). Hydrolysis of compound III produced a new compound (IV), soluble in both acids and bases and having a neutralization equivalent of 186 ± 2. Treatment of this compound with nitrous acid in the presence of sulfuric acid gave a clear solution. When cuprous cyanide was added to this solution, compound I was regenerated.

   Hydrolysis of I gave an acid (V) having a neutralization equivalent of 107 ± 1. Heat converted V into a neutral substance (VI), which could be reconverted to V by hydrolysis.

   Oxidation of compound IV gave a new acid (VII) having a neutralization equivalent of 71 ± 1.

   The nmr spectrum of compound V (in $CDCl_3$ containing DMSO-d_6) showed the following signals (of area 1:1:1:1): δ7.6, m; δ7.9, m; δ8.28, s; δ11.8, bs. The δ11.8 signal showed a chemical shift that was concentration-dependent. The two upfield multiplets, although complex, showed symmetry; the two signals were nearly mirror images of each other.
2. An unknown (I) was insoluble in water but soluble in both dilute acid and dilute alkali. It contained nitrogen and bromine. No satisfactory neutralization equivalent could be obtained. Treatment with acetic anhydride converted it into an acid (II), which gave a neutralization equivalent of 270 ± 3. When I was treated with cold nitrous acid, a gas was liberated and a compound (III) was produced that gave a neutralization equivalent of 230 ± 2. Compounds I, II, and III when vigorously oxidized gave the same product—a bromine-containing acid whose neutralization equivalent was found to be 199 ± 2.

   The nmr spectrum of the product of vigorous oxidation (in $CDCl_3$ containing DMSO-d_6) showed δ7.60, 2 H, d; δ7.95, 2 H, d; δ12.18, 1 H, bs. The δ12.8 signal showed a chemical shift that was concentration-dependent.[8]
3. A solid ester (I) was saponified (saponification equivalent 173 ± 2), and the alkaline aqueous solution was evaporated nearly to dryness. The distillate was pure water. The residue was acidified and distilled; a colorless oil (II) was isolated. Substance II reacted with acetyl chloride and gave a color with ferric chloride solution. The residue from the distillation yielded a solid (III) that was alkali-soluble. When III was heated above its melting point, it changed to

[8] The two doublets "leaned" toward one another so heavily that the signal could be mistaken for a quartet centered at ca. δ7.75.

IV; IV was dissolved by shaking with warm aqueous alkali, and the solution was acidified; the solid that separated was identical with III.

The nmr spectrum of compound II (in $CDCl_3$) showed δ2.25, 3 H, s; δ5.67, 1 H, s, δ6.5–7.3, 4 H, m. The δ5.67 signal had a concentration-dependent chemical shift.

The nmr spectrum of III (in $CDCl_3$) showed δ7.4–7.9, m, and δ12.08, s; the singlet had an area one-half that of the multiplet. In addition, the position of the chemical shift of the singlet was concentration-dependent.

4. A solid (I) giving a positive test for nitrogen was soluble in water and insoluble in ether. Addition of cold alkali liberated a water-soluble, ether-soluble compound (II) that possessed an ammoniacal odor and reacted with benzenesulfonyl chloride to give a derivative insoluble in alkali. Addition of hydrochloric acid to a dilute aqueous solution of I gave a solid acid (III) that possessed a neutralization equivalent of 167 ± 2. After it had been boiled with zinc dust and ammonium chloride solution, compound III reduced Tollens reagent.

   The nmr spectrum of II (in CCl_4) showed δ0.53, 1 H, s; δ0.90, 6 H, t, distorted; δ1.2–1.7, 8 H, m; δ2.52, 4 H, t.

   The nmr spectrum of III (in $CDCl_3/CF_3CO_2H$) showed δ7.77, 1 H, t ($J = 7$ Hz); ca. δ8.54, 2 H, d of m ($J = 7$ Hz, ca. 2 Hz and smaller splittings); δ8.96, 1 H, t ($J =$ ca. 2 Hz).

5. An optically active compound ($C_5H_{10}O$) was found to be slightly soluble in water. Its solubility was not increased appreciably by sodium hydroxide or hydrochloric acid. It gave negative tests with phenylhydrazine, Lucas reagent, and hypoiodite but decolorized solutions of bromine and permanganate. It reacted with acetyl chloride. Oxidation with permanganate converted it to an acid having a neutralization equivalent of 59 ± 1. When heated, the acid lost carbon dioxide and was converted to a new acid having a neutralization equivalent of 73 ± 1.

6. A compound (I) containing carbon, hydrogen, oxygen, and nitrogen was soluble in dilute sodium hydroxide solution and in dilute hydrochloric acid but insoluble in sodium bicarbonate solution. It reacted with an excess of acetic anhydride to give a product (II) that was insoluble in water, in dilute acids, and in dilute alkalies. Compound I decolorized bromine water; when I was dissolved in an excess of dilute hydrochloric acid and the cold solution was treated with sodium nitrite, a new product (III) separated without the evolution of nitrogen.

   The hydrogen sulfate salt of I, prepared using sulfuric acid, resulted in a nmr spectrum (CF_3CO_2H solvent) that showed δ3.31, 3 H, t ($J = 5$ Hz); δ7.12, 2 H, d ($J = 10$ Hz); δ7.49, 2 H, d ($J = 10$ Hz); δ8.74, 2 H, bs.

Set 7

1. When a solid, $C_9H_6O_2$ (I), was heated with a dilute solution of potassium carbonate, it was converted to the potassium salt of an acid, $C_9H_8O_3$ (II). The salt reverted to I when treated with acids. Compound I reacted with bromine

in carbon tetrachloride to yield $C_9H_6O_2Br_2$ (III). When heated with a solution of sodium hydroxide, compound III gradually went into solution. Acidification of the solution precipitated a compound, $C_9H_6O_3$ (IV).

The nmr spectrum of acid II (in $CDCl_3$ containing DMSO-d_6) showed δ6.4–8.2, m; δ10.3, very broad s. The δ10.3 signal showed a chemical shift that was dependent on sample concentration; the area of this signal was one-third that of the δ6.4–8.2 signal.

2. A compound, C_8H_8ONBr, when treated with boiling potassium hydroxide solution, gave the potassium salt of an acid, $C_8H_8O_3$, which was resolvable into *d* and *l* forms.

   An nmr spectrum of the acid of formula $C_8H_8O_3$ (in acetone) showed δ5.22, 1 H, s; δ7.2–7.7, 7 H, m. The chemical shift of the δ5.22 signal was not concentration-dependent.
3. A compound, $C_{11}H_{12}O_4$ (I), reacted with hot sodium hydroxide solution to yield a salt, $C_{11}H_{13}O_5Na$ (II). Treatment of the salt with hot dilute sulfuric acid converted it to an acid, $C_9H_8O_4$ (III). Heat converted compound III to a high-melting compound, $C_{18}H_{12}O_6$ (IV).
4. A compound, $C_5H_8O_2$ (I), was changed to $C_5H_9O_2Cl$ (II) by treatment with boiling concentrated hydrochloric acid. Compound I reacted slowly with a solution of potassium hydroxide to yield $C_5H_9O_3K$ (III). This compound was converted into $C_5H_8O_3$ (IV) by treatment with an alkaline permanganate solution. Compound IV, when treated with a solution of sodium hypochlorite, was transformed into $C_4H_6O_4$ (V), an acid whose neutralization equivalent was 58 ± 1. This acid, when heated, gave $C_4H_4O_3$ (VI). When the aqueous solution of II was made exactly neutral with sodium hydroxide solution, the original compound (I) was slowly regenerated.

   The nmr spectrum of V (in DMSO-d_6) showed only two singlets, one at δ2.43 and the other at δ11.8. The two singlets were of area 2:1, respectively, and the δ11.8 signal was very broad.
5. A liquid neutral compound had the formula $C_7H_7NO_3$. The nmr spectrum of this compound (in CCl_4) showed δ3.89, 3 H, s; δ6.91, 2 H, d ($J = 10$ Hz); δ8.12, 2 H, d ($J = 10$ Hz).
6. A compound, $C_{16}H_{13}N$, formed salts with strong mineral acids. The salts were hydrolyzed by water. The nmr spectrum of this compound (in CCl_4) showed δ5.61, 1 H, bs; δ6.60–7.55, 10 H, m; δ7.80, 2 H, m.
7. A compound, $C_{14}H_{10}O$, gave $C_{13}H_{10}O_3$ when oxidized by alkaline permanganate. The original compound reacted with sodium to give $C_{14}H_9ONa$.
8. An acid of neutralization equivalent 57 was unaffected by bromine in carbon tetrachloride.

Set 8

1. A liquid, $C_5H_4O_2$ (I), decolorized a permanganate solution and also reduced Benedict's solution. When heated with an alkali cyanide, compound I dimerized. Treatment of compound I with ethyl orthoformate in the presence of an acid catalyst converted it to a new substance, $C_9H_{14}O_3$.

The nmr spectrum of compound I (in $CDCl_3$) showed $\delta 6.63$, 1 H, d of d ($J = 5$, 2 Hz); $\delta 7.28$, 1 H, d ($J = 5$ Hz); $\delta 7.72$, 1 H, m; $\delta 9.67$, 1 H, s. The chemical shift position of the $\delta 9.67$ signal was not concentration-dependent.

2. A compound, $C_4H_4O_4$ (I), was soluble in dilute sodium hydroxide solution and when treated with bromine was converted into $C_4H_4O_4Br_2$ (II). Compound I was regenerated from II by treatment of the latter with zinc dust. Compound II was converted into $C_4H_6O_4$ (III) by heating with hydrogen iodide. Compound III possessed a neutralization equivalent of 59 and lost a molecule of water when heated. The anhydride thus produced reacted with benzene in the presence of aluminum chloride to give $C_{10}H_{10}O_3$, which was soluble in alkali, reacted with phenylhydrazine, and was converted into benzoic acid by vigorous oxidation.

   The nmr spectrum of the compound of formula $C_{10}H_{10}O_3$ (in $CDCl_3$) showed $\delta 2.8$, 2 H, t ($J = 7$ Hz); $\delta 3.3$, 2 H, t ($J = 7$ Hz); $\delta 7.2$–7.6, 3 H, m; $\delta 8.0$, 2 H, m; $\delta 11.7$, 1 H, s. The chemical shift of the $\delta 11.7$ signal was concentration-dependent.
3. A compound, $C_{11}H_{10}N_2$, was converted by vigorous oxidation to $C_{11}H_8N_2O$.
4. A compound, $C_5H_{10}O$, decolorizes an alkaline solution of potassium permanganate but is not affected by bromine in carbon tetrachloride. The mass spectrum of this compound (b.p., 75°C) results in the following peaks of intensity greater than or equal to 5% of the base peak:

m/e	Percent of base	*m/e*	Percent of base
86	34	41	83
57	(100, base)	39	17
55	7	29	40
43	18	27	11

5. A compound, $C_{14}H_{12}O$, is converted by chromic acid oxidation into an acid whose neutralization equivalent is 226. The nmr spectrum of the oxidation product (in $CDCl_3$ containing DMSO-d_6) shows $\delta 7.45$–7.95, 7 H, m; $\delta 8.19$, 2 H, d ($J = 7$ Hz); $\delta 10.8$, 1 H, bs.
6. A neutral compound, $C_{10}H_6O_4$ (I), when heated with a sodium hydroxide solution, was converted to the salt of an acid (II) having a neutralization equivalent of 209 ± 2. Compound I reacted with alkaline hydrogen peroxide to yield a new acid (III) having a neutralization equivalent of 111 ± 1.

   The nmr spectrum of compound I (in $CDCl_3$) showed $\delta 6.64$, 2 H, d of d ($J = 5$ Hz, 2 Hz); $\delta 7.63$, 2 H, d ($J = 5$ Hz); $\delta 7.78$, 2 H, d ($J = 2$ Hz).
7. A compound, $C_{10}H_7O_2N$ (I), when treated with iron and hydrochloric acid was converted into $C_{10}H_9N$ (II), a compound that was soluble in dilute hydrochloric acid. By vigorous oxidation of I and II, $C_8H_5O_6N$ and $C_8H_6O_4$, respectively, were produced. Both oxidation products were soluble in alkali.

   The nmr spectrum of I (in $CDCl_3$) showed a complex multiplet extending from $\delta 7.15$ to $\delta 8.5$. The product of oxidation of II, $C_8H_6O_4$, when heated, lost the elements of water.

Set 9

1. A naturally occurring compound, $C_{10}H_{10}O_2$ (I), was found to be unreactive toward acetyl chloride and phenylhydrazine. Heating with potassium hydroxide, however, brought about isomerization. The isomer (II) also failed to react with acetyl chloride and phenylhydrazine. Both isomers decolorized solutions of bromine and permanganate. Ozone converted isomer II to a compound (III) that formed a phenylhydrazone. Oxidation of compound II or III with alkaline potassium permanganate yielded an acid, $C_8H_6O_4$ (IV).

   The nmr spectrum of I (in $CDCl_3$) showed δ3.30, 2 H, d of m ($J=7$ Hz, additional small splittings); δ4.90, 1 H, m; δ5.15, 1 H, m; δ5.88, 2 H, s; ca. δ5.6–6.2, 1 H, m (a wide band, highly split); δ6.67, 3 H, s (with small splittings).
2. A compound, $C_8H_5O_2Cl$ (I), when heated with absolute alcohol, gave $C_{14}H_{20}O_4$ (II), which by oxidation with alkaline permanganate was converted into $C_8H_6O_4$. Treatment with aniline converted I into $C_{20}H_{16}ON_2$. The product of permanganate oxidation was converted (via treatment with excess thionyl chloride) to compound III, $C_8H_4Cl_2O_2$. Compound III (in $CDCl_3$) yielded a nmr spectrum that showed δ7.72, 1 H, t ($J=$ca. 7 Hz); δ8.33, 2 H, t of d ($J=7$, 2 Hz); δ8.82, 1 H, t ($J=$ca. 2 Hz).
3. A compound, $C_8H_{14}O_3$ (I), was soluble in dilute sodium hydroxide solution. Phenylhydrazine converted it into $C_{12}H_{14}ON_2$ (II), and heating with 20% hydrochloric acid transformed (I) into $C_5H_{10}O$ (III). Compound III gave a crystalline precipitate when treated with semicarbazide but, when treated with a solution of sodium hypoiodite, gave no iodoform.

   The compound arising from semicarbazide treatment of III, when dissolved in $CDCl_3$, yielded an nmr spectrum that showed δ1.09, 6 H, t; δ2.26, 4 H, q; δ5.89, 2 H, bs; δ8.59, 1 H, bs.
4. A compound, C_9H_7N (I), was converted by catalytic reduction into $C_9H_{11}N$ (II). When compound II was treated with an excess of methyl iodide followed by silver oxide and the reaction product was heated, a compound, $C_{11}H_{15}N$ (III), was produced. This compound was converted (a) by vigorous oxidation into $C_8H_6O_4$ (IV); (b) by treatment with ozone and hydrolysis of the reaction product into $C_{10}H_{13}ON$ (V). The ozonization product yielded $C_{20}H_{26}O_2N_2$ (VI) when heated with a dilute solution of potassium cyanide. Compound VI was converted into IV by vigorous oxidation.
   Compound I (in $CDCl_3$) yielded a nmr spectrum that showed δ7.25–8.1, 5 H, m; δ8.52, 1 H, d ($J=7$ Hz); δ9.26, 1 H, s.
5. A compound, $C_{10}H_{10}O$ (I), reacted with sodium to give a derivative, $C_{10}H_9ONa$, which was completely hydrolyzed by water. Treatment of compound I with cold 80% sulfuric acid in the presence of mercuric sulfate and then with water gave $C_{10}H_{12}O_2$ (II). This compound dissolved slowly in a solution of sodium hypochlorite, and by acidification of the solution $C_9H_{10}O_3$ (III) was obtained. Boiling 48% hydrobromic acid converted compound III into $C_8H_7O_2Br$ (IV), a compound that vigorous oxidation transformed into $C_8H_6O_4$.

The nmr spectrum of the compound of formula $C_8H_6O_4$ (in $CDCl_3$ containing DMSO-d_6) showed only two singlets of area 2:1 at, respectively, δ8.08 and δ11.0.

6. A compound having the molecular formula $C_{14}H_{12}$ is converted by permanganate oxidation into a derivative whose molecular formula is $C_{13}H_{10}O$ and that is not affected by further treatment with permanganate.

   The nmr spectrum of the original compound (in CCl_4, 100 MHz) shows δ5.35, 2 H, s; δ7.21, 10 H, s (slightly distorted near the base).
7. An optically active compound, $C_9H_{13}N$, was converted by vigorous oxidation into an acid whose neutralization equivalent is 83. A closely related isomer of the original compound (in $CDCl_3$) yielded an nmr spectrum that showed δ0.95, 2 H, s; δ2.30, 3 H, s; δ2.5–3.0, 4 H, m (8 lines, symmetrical); δ7.05, 4 H, s. Oxidation of the isomer resulted in the same acid as was obtained upon oxidation of the original compound.
8. A compound does not decolorize bromine water but reacts with sodium to give $C_8H_6O_3Na_2$. It has a neutralization equivalent of 152. This compound is not optically active, and vigorous oxidation converts it to an acid $C_8H_6O_4$ of neutralization equivalent 82 ± 2. A nmr spectrum of this acid (in $CDCl_3$ containing DMSO-d_6) shows δ8.08, 4 H, s; δ11.0, 2 H, bs.

Set 10

1. A colorless, crystalline solid (I) contained nitrogen but no halogen or sulfur. It was soluble in water but insoluble in ether. Its aqueous solution was acidic and gave a neutralization equivalent of 123 ± 1. It was treated with sodium nitrite and hydrochloric acid. A gas evolved, one-third of which was soluble in potassium hydroxide solution. The solution from the nitrous acid treatment was evaporated to dryness; only sodium chloride, sodium nitrate, and sodium nitrite were left. Compound I was heated with dilute sodium hydroxide solution; ammonia was evolved, and the solution on acidification gave off a gas. Evaporation of this solution left only inorganic salts.

   The original compound was carefully neutralized with base under mild conditions to give a solid of m.p. of 135°C. The uv spectrum of this solid, in the presence of sodium hydroxide, showed no features above 220 nm. The solid resulted in the following spectra.

   Mass spectrum (only peaks of intensity greater than or equal to 10% of base are reported):

m/e	Percent of base
60	(100, base)
44	60
43	18
28	17

Infrared spectrum: 3450 cm^{-1} (2.90 μm); 3350 cm^{-1} (2.99 μm); 1690 cm^{-1} (5.90 μm); 1640 cm^{-1} (6.10 μm); 1600 cm^{-1} (6.25 μm); 1470 cm^{-1} (6.80 μm); 1160 cm^{-1} (8.62 μm); 1050 cm^{-1} (9.52 μm); 1000 cm^{-1} (10.00 μm); 790 cm^{-1} (12.66 μm); 710 cm^{-1} (14.8 μm).

2. An unknown compound containing carbon, hydrogen, oxygen, and nitrogen was insoluble in water, in dilute acids, and in dilute alkalies. It did not react with acetyl chloride or phenylhydrazine and was not easily reduced. When treated with hot aqueous alkali, the substance slowly dissolved. Distillation of this alkaline solution gave a distillate containing a compound that was salted out by means of potassium carbonate. This compound gave the iodoform test but no reaction with a hydrochloric acid solution of zinc chloride. The original alkaline solution (residue from the distillation) was acidified with sulfuric acid, and the solution was again distilled; a volatile acid was obtained in the distillate. This distillate reduced permanganate. The residual liquor from this second distillation was exactly neutralized, and a solid was obtained. This solid contained nitrogen and possessed a neutralization equivalent of 137 ± 1.

   A 100-mg sample of the volatile acid (in 0.5 ml of CCl_4) yielded a nmr spectrum that showed δ5.8–6.75, 3 H, m; δ12.4, 1 H, s. The nmr spectrum of a 60-mg sample of the solid of neutralization equivalent 137 (dissolved in 0.5 ml of acetone) showed δ6.55, 3 H, s (shift position was concentration-dependent); δ6.75, 2 H, d ($J = 8$ Hz); δ7.83, 2 H, d ($J = 8$ Hz), The compound that salted out with potassium carbonate showed, among other bands, one nmr band that was concentration-dependent and that appeared as a singlet when the analysis was done on a $CDCl_3$ solution and as a triplet ($J =$ ca. 7 Hz) when done on a DMSO-d_6 solution.

3. An acid, $C_6H_8O_4$ (I), was converted into $C_6H_6O_2Cl_2$ (II) by phosphorus pentachloride. The chlorine derivative, when treated with benzene in the presence of anhydrous aluminum chloride, gave $C_{18}H_{16}O_2$ (III). This compound did not decolorize permanganate but readily formed a dibromide and a dioxime. The dioxime rearranged under the influence of phosphorus pentachloride to yield $C_{18}H_{18}O_2N_2$ (IV), a compound that gave the original acid (I) when hydrolyzed.

   The nmr spectrum of compound II (in CCl_4 solution) showed δ2.2–2.6, 4 H, m; δ3.90, 2 H, t (distorted, $J =$ ca. 7 Hz); δ11.9, 2 H, bs.

4. A compound (I) contained carbon, hydrogen, oxygen, nitrogen, and chlorine. It was soluble in water but insoluble in ether. The aqueous solution immediately gave a precipitate with silver nitrate. When the aqueous solution of I was exactly neutralized, a new compound (II) free from chlorine separated. It reacted with acetic anhydride to give an alkali-soluble, acid-insoluble compound (III) possessing a neutralization equivalent of 207 ± 2. Compound II reacted with benzenesulfonyl chloride to give an alkali-soluble product and with nitrous acid without evolution of any gas even when heated. The product from the latter treatment still contained nitrogen and was soluble in alkalies and insoluble in acids. Vigorous oxidation of either I, II, or III gave a nitrogen-free acid, insoluble in water, and possessing a neutralization equivalent of 82 ± 2.

The nmr spectrum of this nitrogen-free acid (in $CDCl_3$ containing DMSO-d_6) showed δ7.62, 1 H, t ($J = 7$ Hz); δ8.25, 2 H, distorted doublet (J = ca. 2 Hz); δ8.70, 1 H, distorted s; δ12.28, 2 H, s.

5. A compound, $C_{10}H_6O_3$ (I), decolorized alkaline permanganate and reacted with hydroxylamine. It decomposed when distilled at ordinary pressure to give $C_9H_6O_2$ (II), a compound that yielded a monosodium derivative and was readily oxidized to $C_8H_6O_4$ (III). Compound III was an acid that, when heated with soda lime, was converted into $C_7H_6O_2$ (IV). Compound IV was decomposed by heating with dilute hydrochloric acid under pressure and yielded a weakly acidic compound having the formula $C_6H_6O_2$.

   The nmr spectrum of compound IV (in $CDCl_3$) showed δ5.90, 2 H, s; δ6.83, 4 H, s (slight distortions at the bottom of this signal). The nmr spectrum of compound III (in $CDCl_3$) showed δ6.00, 2 H, s; δ6.8, 1 H, d ($J = 7$ Hz); δ7.38, 1 H, d ($J = 2$ Hz); δ7.55, 1 H, d of d ($J = 7$, 2 Hz); δ7.6, 1 H, bs. The chemical shift of the δ7.6 signal was concentration-dependent.

Set 11

1. A compound (A) contained only carbon, hydrogen, and oxygen. It was insoluble in water, dilute acids, and dilute alkalies but dissolved in cold concentrated sulfuric acid. It gave negative tests with phenylhydrazine and acetyl chloride. When heated with aqueous sodium hydroxide it yielded an oil (B) and a volatile acid having a neutralization equivalent of 59 ± 1. Compound B reacted with acetyl chloride but not with phenylhydrazine. Treatment with a concentrated solution of hydrogen bromide transformed it into a bromine-containing compound (C) that gave positive tests with silver nitrate, ferric chloride, and bromine water.

   Mild oxidation converted B to an acid (D) having a neutralization equivalent of 164 ± 2.

   Treatment of acid D with ethyl alcohol and a trace of sulfuric acid resulted in compound E. The nmr spectrum of compound E (in CCl_4) showed δ1.35, t ($J = 7$ Hz); δ1.40, t ($J = 7$ Hz); δ3.8–4.5, 4 H, 6 lines (relative intensities 1:3:4:4:3:1, spaced by ca. 7 Hz each); δ6.79, 2 H, d ($J = 9$ Hz); δ7.89, 2 H, d ($J = 9$ Hz). The signals at δ1.35 and δ1.40 had a combined area of 6 H.

2. A compound (I), giving positive tests for nitrogen and chlorine, was insoluble in water and hydrochloric acid but soluble in sodium bicarbonate solution. A neutralization equivalent of 210 ± 2 was obtained. Compound I reacted with acetyl chloride but not with hot alcoholic silver nitrate. The acetyl derivative (II) had a neutralization equivalent of 253 ± 2. Boiling alkali liberated ammonia from I, and acidification of the resulting solution precipitated a new acid (III), which had a neutralization equivalent of 115 ± 1. Compound III contained chlorine but no nitrogen. When compound III was boiled with potassium permanganate solution, a new compound having the same solubility characteristics as I was produced that still contained chlorine and gave a neutralization equivalent of 81 ± 1.

Compound IV (neutralization equivalent 71) could be chlorinated to give the compound of neutralization equivalent 81; compound IV yielded an nmr spectrum ($CDCl_3$/DMSO-d_6 solvent) displaying δ7.57, 1 H, t ($J = 7$ Hz); δ8.18, 2 H, d ($J = 7$ Hz); δ11.2, 3 H, bs.

The ir spectrum of I (in Nujol) showed, among other bands, strong absorptions at 1779 cm^{-1} (5.62 μm) and 1712 cm^{-1} (5.84 μm).

3. A neutral compound (A) was a colorless solid that gave negative tests for nitrogen, sulfur, and the halogens. It gave positive tests with phenylhydrazine and acetyl chloride. Mild oxidation converted it to a new compound that was a yellow solid (B). The new compound reacted with phenylhydrazine to give the same derivative that was obtained from A. Vigorous oxidation of B converted it into an acid (C) that had a neutralization equivalent of 121 ± 1.

   The nmr spectrum of compound A (in $CDCl_3$) showed δ4.5, 1 H, bs; δ5.9, 1 H, bs; δ7.2–7.5, 8 H, m (the major part of which was a sharp singlet); δ7.85, 2 H, d of d.

4. A liquid (I) containing chlorine was insoluble in water, dilute hydrochloric acid, and dilute sodium hydroxide. It dissolved in cold concentrated sulfuric acid but not in phosphoric acid. It gave no precipitate with warm alcoholic silver nitrate and did not react with acetyl chloride. It gave a precipitate with phenylhydrazine but no color with Schiff's reagent. When boiled with concentrated sodium hydroxide solution, the compound (I) dissolved. The distillate from this alkaline solution gave an iodoform test. Acidification of the alkaline solution precipitated a compound (II) that contained chlorine and had a neutralization equivalent of 156 ± 1. Compound II was not affected by permanganate solution. After removal of compound II, a portion of the acidic filtrate was distilled, and the distillate was found to be acid to litmus. The original compound (I) had a saponification equivalent of 113 ± 1. The acidic filtrate yielded a compound that had a neutralization equivalent of 61 ± 1.

   The nmr spectrum of compound II (in $CDCl_3$/DMSO-d_6) showed δ7.1–7.5, 3 H, m (primarily a large singlet); δ7.82, 1 H, m; δ9.0, 1 H, bs.

5. A liquid had a saponification equivalent of 163 ± 1. Saponification yielded an oil, which gave a positive ferric chloride test, and an acid of neutralization equivalent 60 ± 1.

   The nmr spectrum of 49 mg of the oil in 0.5 ml of CCl_4 showed δ2.18, 6 H, s; δ5.73, 1 H, s; δ6.33, 2 H, s; δ6.45, 1 H, s. Upon addition of more CCl_4, the δ5.73 signal moved upfield.

Set 12

1. An ether-insoluble compound (I), containing nitrogen, dissolved in water, giving an alkaline solution. Titration of this solution with standard acid gave a neutralization equivalent of 37 ± 1. Treatment of a cold solution of I with sodium nitrite and hydrochloric acid liberated a gas. The resulting solution was made distinctly alkaline, and benzoyl chloride was added. A neutral compound (II) separated. Compound II gave a saponification equivalent of 142 ± 1.

The nmr spectrum of compound I (in $CDCl_3$) showed $\delta 1.08$, s; $\delta 1.58$, 5 lines ($J = 7$ Hz, relative intensities 1:4:6:4:1); $\delta 1.75$, t. The relative signal areas were, from high to low field, 2:1:2.

2. A colorless liquid (I) gave no tests for halogen, nitrogen, sulfur, or metals. It was insoluble in water, dilute hydrochloric acid, dilute sodium hydroxide, and phosphoric acid, but soluble in cold concentrated sulfuric acid. It did not react with acetyl chloride or phenylhydrazine and was not affected by heating with sodium hydroxide. It was boiled with dilute phosphoric acid, and an oil (II) separated when the solution was cooled. Compound II gave a precipitate with phenylhydrazine and with sodium bisulfite solution but did not react with acetyl chloride When II was vigorously shaken with strong alkali, a compound (III) separated from the alkaline solution. This product (III) reacted with acetyl chloride but not with phenylhydrazine. Acidification of the alkaline solution gave IV, which had a neutralization equivalent of 136 ± 1. Strong oxidation of IV gave an acid (V) with a neutralization equivalent of 82 ± 1.

   The phosphoric acid solution, from which II was separated, was distilled. The distillate was saturated with potassium carbonate, and a compound (VI) was obtained. This compound reacted with sodium and acetyl chloride and gave a yellow precipitate with sodium hypoiodite. It did not react with Lucas reagent.

   The nmr spectrum of II (in $CDCl_3$) showed $\delta 2.42$, 3 H, s; $\delta 7.18$, 2 H, d ($J = 8$ Hz); $\delta 7.66$, 2 H, d ($J = 8$ Hz); $\delta 9.81$, 1 H, s.

3. A colorless liquid (I) gave no tests for nitrogen, sulfur, or halogen. It was soluble in water and ether. It did not react with sodium, acetyl chloride, phenylhydrazine, or dilute permanganate solution. It did not decolorize bromine in carbon tetrachloride and was unaffected by boiling alkalies. When compound I was heated with an excess of hydrobromic acid, an oil (II) separated. This oil (II) contained bromine and readily gave a precipitate with alcoholic silver nitrate. It was insoluble in water, acids, and alkalies. After II was dried and purified, it was treated with magnesium in pure ether. A reaction occurred with the liberation of a gas (III). No Grignard reagent could be detected. Treatment of II with alcoholic potassium hydroxide liberated a gas (IV) that gave a precipitate when passed into ammoniacal silver nitrate. Compound III did not give a precipitate with ammoniacal cuprous chloride. Both III and IV decolorized bromine water and reduced permanganate solutions. A careful examination of the action of hydrobromic acid on compound I showed that II was the only organic compound produced and that no gases were evolved during this reaction.

   The mass spectrum of compound I showed peaks at m/e values (relative abundance) as follows: m/e 88 (650); m/e 89 (32); m/e 90 (3); and no peaks at higher mass. The nmr spectrum of I (in $CDCl_3$) showed only a singlet at $\delta 3.69$.

4. A compound (I) containing carbon, hydrogen, nitrogen, and oxygen was insoluble in dilute alkalies and acids. When heated for some time with hydrochloric acid, compound I yielded a solid acid (II) having a neutralization

equivalent of 180 ± 1. If compound I was oxidized with potassium dichromate and sulfuric acid, it yielded a solid, nitrogen-containing acid (III) with a neutralization equivalent of 166 ± 1. Compound I reacted with benzaldehyde in the presence of alkalies to give a benzal derivative.

If compound I was treated with stannous chloride and hydrogen chloride in dry ether and the resulting mixture was treated with water, a new compound (IV), whose molecular formula was C_8H_7N, resulted. A Zerewitinoff determination showed IV to possess one active hydrogen atom. Compound IV was feebly basic and was resinified by acids.

The nmr spectrum of I (in $CDCl_3$) showed resonances at δ4.25, 2 H, s; δ7.5–8.0, 3 H, m; δ8.25, 1 H, d (distorted slightly). The nmr spectrum of II (in $CDCl_3$ containing DMSO-d_6) showed δ4.02, 2 H, s; δ7.3–7.65, 3 H, m; δ8.0, 1 H, d (distorted); δ11.2, 1 H, bs. The nmr spectrum of III (in $CDCl_3$ containing DMSO-d_6) showed δ7.5–8.0, 4 H, m; δ12.15, 1 H, s. The nmr spectrum of IV (160 mg in 0.5 ml of CCl_4) showed δ6.38, 1 H, m; δ6.76, 1 H, t (distorted); δ6.95–7.10, 4 H, m; δ7.4–7.65, 1 H, m.

Set 13

1. A colorless, oily liquid (I) having an agreeable odor and containing only carbon, hydrogen, and oxygen reacted with phenylhydrazine but not with acetyl chloride, and it readily decolorized potassium permanganate solution. When it was treated with a concentrated sodium hydroxide solution and the reaction mixture acidified, two products were obtained: an oxygen-containing compound (II) that reacted with acetyl chloride, and an acid (III) that reacted with fuming sulfuric acid and that at 200°C lost carbon dioxide to yield an oxygen-containing compound (C_4H_4O). The latter decolorized permanganate solutions but did not react with sodium or phenylhydrazine. When compound I was warmed with potassium cyanide, a compound (IV) was produced that yielded an osazone and reduced alkaline copper solutions and that, on oxidation with periodic acid, was converted to the original compound (I) and to the acid (III).

   The nmr spectra of some of the above compounds showed the following signals (see p. 493).
2. A weakly acidic compound (I) contained nitrogen but no sulfur or halogens. It reacted with acetyl chloride but not with phenylhydrazine. When compound I was heated with dilute acids, a new compound (II) was obtained; it was isolated by distillation from the acid solution and saturation of the distillate with potassium carbonate. Compound II did not react with acetyl chloride or Schiff's reagent but gave positive tests with phenylhydrazine and sodium bisulfite.

   When compound I was warmed with phosphorus pentachloride and poured into water, a new compound (III) was obtained. Compound III still contained nitrogen but no halogen, was neutral, and was decomposed by alkalies. By the addition of benzoyl chloride to this alkaline solution an acid (IV) was obtained. It gave a neutralization equivalent of 220 ± 2. When

Chemical shift	No. of protons	Multiplicity
Compound I (in $CDCl_3$) Problem 1 (set 13)		
δ6.63	1 H	d of d ($J=5$ Hz, 2 Hz)
δ7.28	1 H	d ($J=5$ Hz)
δ7.72	1 H	m ($J=$ ca. 2 Hz and smaller splittings)
δ9.67	1 H	s
Compound II (in $CDCl_3$) Problem 1 (set 13)		
δ2.83	1 H	s (chemical shift concentration-dependent)
δ4.57	2 H	s
δ6.33	2 H	m[a]
δ7.44	1 H	m[a]
Compound III (DMSO-d_6 solvent) Problem 1 (set 13)		
δ6.88	1 H	m[a]
δ7.78	1 H	m[a]
δ8.32	1 H	m[a]
δ11.88	1 H	bs
Compound of formula C_4H_4O (in $CDCl_3$) Problem 1 (set 13)		
δ6.37	2 H	t
δ7.42	2 H	t

[a] All of these so-marked signals showed only small splittings (ca 0–2 Hz).

compound IV was heated for some time with dilute acids and distilled, two products resulted. One (V) contained no nitrogen, was acidic, and gave a neutralization equivalent of 120 ± 2. The other proved to be identical with III. When III was treated with cold concentrated hydrochloric acid, a compound (VI) was obtained that contained nitrogen and chlorine and was soluble in water. It liberated a gas when treated with cold sodium nitrite solution.

The following nmr spectra were determined:

Chemical shift	No. of protons	Multiplicity
Compound I (in $CDCl_3$) Problem 2 (set 13)		
δ1.77	4 H	m
δ2.40	4 H	m
δ9.12	1 H	bs
Compound III (in $CDCl_3$) Problem 2 (set 13)		
δ1.80	4 H	m
δ2.38	2 H	m
δ3.32	2 H	m
δ7.60	1 H	bs

3. A compound (I) soluble in water but not in ether was decomposed by heat into a compound (II) insoluble in all tests, and a basic compound (III). When II and III were heated together, I was reformed. Both I and II gave a precipitate immediately when treated with silver nitrate. Compound III did not give a benzenesulfonamide but yielded a nitroso derivative.

   The following nmr spectra were determined:

Chemical shift	No. of protons	Multiplicity
Compound I (D_2O; 80 mg/0.5 ml)		
δ3.70	9 H	s
δ7.4	5 H	m
Compound II ($CDCl_3$ solvent)		
δ2.20	—	s
Compound III ($CDCl_3$ solvent)		
δ2.85	6 H	s
δ6.4–6.7	3 H	m
δ7.10	2 H	m

4. A neutral compound (I) gave positive tests for chlorine, bromine, and iodine. Alcoholic silver nitrate gave a white precipitate that was readily soluble in ammonia. Phenylhydrazine produced a precipitate, but acetyl chloride failed to react. Permanganate was slowly decolorized, as was bromine, in carbon tetrachloride. When compound I was shaken with cold dilute alkali for some time, it dissolved. Acidification of the alkaline solution produced a compound (II) that gave positive tests for bromine and iodine and possessed a neutralization equivalent of 369 ± 3. When compound I was boiled with dilute alkali and then acidified, a compound (III) precipitated that gave a test for iodine. Compound III possessed a neutralization equivalent of 306 ± 3 and reacted with both phenylhydrazine and acetyl chloride. Treatment of I or II with sodium hypochlorite and acidification yielded an acid containing iodine and having a neutralization equivalent of 146 ± 1.
5. A neutral compound (A) contained bromine but no nitrogen or other halogens. It did not react with hot alcoholic silver nitrate solution, acetyl chloride, phenylhydrazine, or bromine in carbon tetrachloride. It dissolved in boiling sodium hydroxide solution, but the distillate from this alkaline solution contained no organic compounds. Acidification of the alkaline solution with phosphoric acid caused the precipitation of a compound (B) that contained bromine and had a neutralization equivalent of 200 ± 2. Steam distillation of the acid solution gave a distillate that was repeatedly extracted with chloroform. Removal of the chloroform left a colorless liquid (C) that was purified by distillation. Compound C contained no bromine and was soluble in sodium bicarbonate solution. It had a neutralization equivalent of 102 ± 1. After removal of C, the solution remaining in the steam distillation

flask was made distinctly alkaline, benzoyl chloride added, and the mixture shaken vigorously. A new compound (D) separated from the alkaline solution. Compound D contained no bromine and was neutral. It had a saponification number of 135 ± 1.

The nmr spectra of compounds B and C showed the following:

Chemical shift	No. of protons	Multiplicity
Compound B (in DMSO-d_6)		
δ7.3	1 H	bs (very broad)
δ7.71	2 H	d (distorted)
δ7.90	2 H	d (distorted)
Compound C (in CCl_4)		
δ0.93	3 H	t (distorted)
δ1.2–1.8	4 H	m
δ2.31	2 H	t
δ11.7	1 H	s

Set 14

1. A compound (I) that contained only carbon, hydrogen, and oxygen had a neutralization equivalent of 179 ± 1. When it was heated with aqueous sodium hydroxide and the reaction mixture was acidified with sulfuric acid, a solid (II) separated that was soluble in alkalies and had a neutralization equivalent of 138 ± 1. Compound II decolorized bromine water and gave a color with ferric chloride. Distillation of the filtrate from II yielded an acid (III) with a neutralization equivalent of 60 ± 1.

   The nmr spectrum of compound II (in DMSO-d_6 plus $CDCl_3$) showed: δ6.7–7.0, 2 H, m; δ7.35, 1 H, t of d ($J = 9$, 3 Hz); δ7.75, 1 H, d of d ($J = 9$, 3 Hz), δ11.55; 2 H, s.

2. A compound (I) containing nitrogen was insoluble in water and dilute alkalies but soluble in dilute hydrochloric acid. Treatment with benzenesulfonyl chloride and alkali gave a clear solution. Acidification of this solution produced a precipitate that dissolved in an excess of the acid. The original compound did not react with phenylhydrazine but was decomposed by boiling with hot sodium hydroxide solution. An oil (II) was separated from the alkaline solution (III). The oil still contained nitrogen and was soluble in dilute hydrochloric acid. Compound II gave a precipitate with bromine water and reacted with acetyl chloride and sodium. When compound II was treated with benzenesulfonyl chloride and alkali, an oil remained that proved to be soluble in hydrochloric acid. Compound II was dissolved in ether and the solution saturated with hydrogen chloride; a solid compound (IV) separated. Compound IV was soluble in water and had a neutralization equivalent of 187 ± 1.

   Acidification of the above alkaline solution (III) produced a precipitate

(V) that dissolved in an excess of acid. Addition of sodium nitrite to the ice-cold acid solution of V gave a clear solution without the evolution of nitrogen. Addition of this solution to a solution of sodium β-naphthoxide gave a red solution. Compound V gave a neutralization equivalent of 137 ± 1.

Methylation (twice) of V followed by mild reduction gave II. The nmr spectrum of V (in acetone) showed: δ6.55, 3 H, bs; δ6.76, 2 H, d ($J = 8$ Hz); δ7.83, 2 H, d ($J = 8$ Hz).

3. A solid (I) gave tests for nitrogen, sulfur, and bromine. It was insoluble in ether but dissolved in water to give an acid solution. It gave a neutralization equivalent of 221 ± 1. Addition of cold alkali caused an oil (II) to separate, which contained bromine and nitrogen but no sulfur. Compound II reacted with benzenesulfonyl chloride and alkali to give a clear solution from which acid (III) precipitated, which contained bromine, nitrogen, and sulfur. Compound II did not give a precipitate with silver nitrate but decolorized both bromine water and dilute permanganate solution. Treatment of I with nitrous acid in the cold gave a solution without the evolution of any gas. This solution was poured into cuprous cyanide solution, and a compound (IV) separated. Compound IV still contained bromine and nitrogen but was insoluble in dilute acids and alkalies. When IV was boiled with dilute sulfuric acid for some time, it was converted to V. This compound no longer contained nitrogen but did contain bromine. It (V) was insoluble in water and dilute hydrochloric acid but soluble in sodium bicarbonate. A neutralization equivalent of 200 ± 2 was obtained. Compound V did not react with silver nitrate or permanganate.

   The nmr spectrum of II (in $CDCl_3$) showed δ3.53, 2 H, bs; δ6.57, 2 H, d ($J = 9$ Hz); δ7.21, 2 H, d ($J = 9$ Hz).

4. A pale yellow crystalline compound (I), giving tests for nitrogen and bromine, was insoluble in water, dilute acids, and alkalies. It did not react with phenylhydrazine, acetyl chloride, or cold dilute permanganate. Cold alcoholic silver nitrate did not react, but when the solution was boiled for some time a precipitate of silver bromide formed. When compound I was heated with zinc and ammonium chloride solution and the mixture fitered, it was found that the filtrate reduced Tollens reagent. Vigorous oxidation of I produced a new compound (II), which still contained bromine and nitrogen and was insoluble in hydrochloric acid but soluble in sodium bicarbonate solution. It had a neutralization equivalent of 145 ± 1.

   Treatment of compound I with tin and hydrochloric acid followed by alkali gave a new compound (III) that contained nitrogen and bromine and was soluble in dilute hydrochloric acid. Compound III gave a precipitate with bromine water, reacted with acetyl chloride, and gave a clear solution when treated with benzenesulfonyl chloride and alkali. Vigorous oxidation of III produced a white crystalline acid (IV). Compound IV gave no tests for bromine or nitrogen and had a neutralization equivalent of 82 ± 1.

   Compound IV, when heated, lost the elements of water. The nmr spectrum of III (in $CDCl_3$) showed δ3.96, 2 H, bs; δ6.45, 1 H, d ($J = 8$ Hz); δ7.25–7.75, 4 H, m; δ8.0–8.2, 1 H, m.

5. A light yellow neutral solid (I) contained chlorine but not nitrogen. It did not react with hot alcoholic silver nitrate, acetyl chloride, or bromine in carbon tetrachloride. It gave a precipitate with phenylhydrazine. Compound I was not attacked by cold alkalies, but when it was heated for some time with concentrated sodium hydroxide a clear solution resulted. The distillate from this alkaline solution contained no organic compounds. Acidification of the alkaline solution with phosphoric acid gave a precipitate (II), which was removed by filtration. No organic compounds could be obtained from the filtrate by distillation or evaporation to dryness.

   Compound II contained chlorine, had a neutralization equivalent of 297 ± 2, and reacted with acetic anhydride to produce a compound (III) that had a neutralization equivalent of 340 ± 3. Compound III did not react with bromine water, bromine in carbon tetrachloride, or phenylhydrazine.

   Vigorous oxidation of I with alkaline permanganate gave a very good yield of a product, IV, which contained chlorine and possessed a neutralization equivalent of 156 ± 1. No other oxidation product could be found.

   Vigorous oxidation of II or III with potassium dichromate and sulfuric acid also produced IV but in very poor yield.

   The nmr spectrum of IV (in $CDCl_3$ containing DMSO-d_6) showed a 3 H signal composed of d, $J = 9$ Hz at $\delta 7.37$ with a bs at $\delta 7.45$; in addition, a 2 H signal appeared at $\delta 7.81$ (d, $J = 9$ Hz). Treatment of the solution with deuterium oxide caused the $\delta 7.45$ signal to disappear, leaving the two doublets.

Set 15

1. An acid (I) was found to have a neutralization equivalent of 151 ± 2. Treatment in the cold with acetyl chloride converted it to a new acid (II), which had a neutralization equivalent of 193 ± 2. Gentle oxidation of I with cold potassium permanganate solution transformed it to an acid (III) having a neutralization equivalent of 149 ± 2. Compound III yielded a derivative with phenylhydrazine. Vigorous oxidation of I, II, or III yielded an acid (IV) having a neutralization equivalent of 122 ± 1.

   The nmr spectrum of compound I (in acetone) showed $\delta 5.22$, 1 H, s; $\delta 6.93$–7.88, 7 H, m. The ir spectrum of compound I (in a mineral oil mull) showed, among other bands, a strong band at 1706 cm^{-1} (5.86 μm) and a broad band at 3333–2400 cm^{-1} (3.00–4.17 μm).
2. A compound (I) had a neutralization equivalent of 223. Vigorous oxidation converted it to a new acid (II) having a neutralization equivalent of 167. Treatment of I with zinc and hydrochloric acid gave an acid (III) with a neutralization equivalent of 179. Compounds I, II, and III were found to contain nitrogen.

   The nmr spectrum of compound II (in $CDCl_3$/DMSO-d_6) showed: $\delta 7.72$, 1 H, t ($J = 8$ Hz); $\delta 8.2$–8.55, 2 H, m; $\delta 8.71$, 1 H, t ($J = 2$ Hz); $\delta 12.98$, 1 H, s.
3. A sulfur-containing compound was soluble in strong alkalies but not in sodium bicarbonate solutions. Vigorous oxidation gave a sulfur-containing

acid having a neutralization equivalent of 102 ± 1. When this compound was treated with superheated steam, a sulfur-free acid was formed.

The nmr spectrum of 100 mg of the sulfur-free acid (in 0.5 ml of acetone-d_6) showed δ7.00–7.75, 4 H, m; δ8.00, 2 H, s. The δ8.00 signal position had a chemical shift that was concentration-dependent. The nmr spectrum of the original compound (in $CDCl_3$) showed δ2.29, 3 H, s; δ3.37, 1 H, s; δ7.02, 4 H, s (somewhat broadened at the base).

4. A compound (I) containing chlorine gave positive tests with alcoholic silver nitrate, sodium hypoiodite solution, and acetyl chloride. Hydrolysis with sodium bicarbonate yielded a chlorine-free compound (II) that gave a positive iodoform test and was oxidized by periodic acid to a single compound (III). Compound III also gave a positive iodoform test.
5. A compound (A), containing only carbon, hydrogen, and oxygen, was found to react with acetyl chloride but not with phenylhydrazine. Oxidation with periodic acid converted it to a new compound (B), which reduced Tollens but not Fehling's reagent. Treatment of B with potassium cyanide in aqueous ethanol converted it to a new compound (C), which gave positive tests with acetyl chloride and phenylhydrazine. Oxidation of C with Fehling's solution or nitric acid converted it to a yellow compound (D) that yielded a derivative with *o*-phenylenediamine. When D was treated with hydrogen peroxide it yielded an acid (E) having a neutralization equivalent of 135 ± 1. Catalytic hydrogenation of C or D produced the original compound (A).

   The nmr spectrum of compound B (in $CDCl_3$) showed δ2.42, 3 H, s; δ7.18, 2 H, d ($J = 10$ Hz); δ7.56, 2 H, d ($J = 10$ Hz); δ9.81, 1 H, s.
6. A solid, neutral substance (A) contained nitrogen. It was recovered from attempted hydrolysis with dilute acids and bases. It gave negative tests with acetyl chloride, bromine in carbon tetrachlorde, sodium hypoiodite solution, and Tollens reagent. It reacted slowly with dinitrophenylhydrazine and, after treatment with zinc and ammonium chloride, reduced Tollens reagent.

   When A was treated with hydroxylamine hydrochloride in pyridine, it was slowly converted to a new compound B. Substance B was treated with phosphorus pentachloride, which changed it to C. Boiling with acid converted C to D, a nitrogen-containing acid of neutralization equivalent 168 ± 2, and the salt of a base E. The hydrochloride of E had a neutralization equivalent (by titration with alkali) of 195 ± 2.

   Acid D was treated with thionyl chloride, and the product was added to aqueous ammonia. The neutral substance so obtained was treated with bromine and sodium hydroxide solution. The product (F) was an acid-soluble substance that, after treatment with hydrochloric acid and sodium nitrite, gave a color with β-naphthol. Treatment of F with tin and hydrochloric acid produced G, a substance readily attacked by oxidizing agents. Compound G did not give a crystalline derivative when treated with benzil.

   Base E, obtained in the reaction with phosphorus pentachloride, reacted with sodium nitrite and dilute sulfuric acid. The resulting solution gave a color with β-naphthol. Boiling the aqueous solution produced a nitrogen-free substance H, which dissolved in aqueous alkali but not in aqueous sodium bicarbonate. Oxidation of either E or H produced an acid (neutralization equivalent, 83) which was readily converted to an anhydride. Substance H did not undergo coupling with benzenediazonium solutions.

The nmr spectrum of compound F (in $CDCl_3$ containing DMSO-d_6) showed δ5.68, 2 H, bs; δ6.58, 2 H, d ($J = 10$ Hz); δ7.88, 2 H, d ($J = 10$ Hz). The nmr spectrum of compound H (in $CDCl_3$) showed δ2.32, 3 H, s; δ5.12, 1 H, bs; δ7.1–7.55, 4 H, m; δ7.6–7.9, 1 H, m; δ8.0–8.2, 1 H, m. The δ5.12 signal moved to lower field upon addition of more compound H to the nmr tube.

CHAPTER TEN

the literature of organic chemistry

10.1 INTRODUCTION AND ORGANIZATION

A very substantial part of the task of the organic chemist is to become familiar with the tremendous amount of literature in the field and to be able to find quickly that literature which would be useful in relation to a specific laboratory operation. This chapter is intended as an aid in carrying out this task.

We have organized this chapter by those subject areas that are likely to be of concern to organic chemists. Organic laboratory operations may involve concepts traditionally aligned with other areas of chemistry, such as inorganic and analytical chemistry. Thus these areas are listed, including, for example, biochemistry and physical chemistry textbooks.

The field of organic chemistry is, as might be expected, the one that has received the most attention. This field has been broken down into areas such as organic analysis, syntheses, mechanisms, and so on.

We have also included listings of books of a more general nature. For example, handbooks and other forms of literature that are indispensable for all types of chemists are covered here. Some aspects of the chemical literature, such as journals, computer searching, and so forth, have not been covered here. Books, such as that by Woodburn, listed in Sect. 10.8 should be consulted. Chemical Abstracts has been discussed in Sect. 2.9.

One should keep in mind that the placement of certain books into specific subject areas is somewhat arbitrary. For example, some of the books listed under "Physical Organic Chemistry" could have been listed under "Mechanisms."

In some cases comments have been included with the listed references. These are intended to give the user a more specific idea of what is contained within the book. It is hoped that this will yield more information than the title alone.

The approach that the beginner should use to extract information from this chapter is, first, to try to identify the area or areas under which a specific kind of literature could be listed. Then one should examine the list of literature in those subject areas. Finally one could consult with an instructor or a graduate student who is experienced in that subject area. This consultation could yield valuable supplemental opinions of the real usefulness of your selections.

There are a number of other places that yield useful book lists that are very helpful to the organic chemist. These include the following:

R. T. Morrisson and R. N. Boyd. *Organic Chemistry*, 3rd ed. (Allyn and Bacon, Boston, 1973), pp. 1185ff.

J. March. *Advanced Organic Chemistry*, 2nd ed. (McGraw-Hill, New York, 1977), Appendix A, pp. 1143ff. March has again provided a valuable "encyclopedia" of information for organic chemists.

W. N. le Noble. *Highlights of Organic Chemistry*, Vol. 3 in the series, *Studies in Organic Chemistry* (Marcel Dekker, New York, 1974); this is a very extensive book on organic chemistry, and the footnotes provide a large number of references to the literature of organic chemistry.

Jounal of Chemical Education. 119 West 24th St., New York, 10011. This journal, well known as a place to publish articles of pedagogical value, provides in its annual September issue a book buyer's guide. This guide lists many books available in a variety of subject areas in the general area of chemistry. The September 1976 issue listed more than 1000 books under 61 subject areas of chemistry.

10.2 ORGANIC CHEMISTRY

10.2.1 Beilstein

Beilstein's *Handbuch der Organischen Chemie* was initiated as an ambitious attempt to describe the preparation and properties of all known organic compounds. Although volumes continue to appear, it seems unlikely that the *Handbuch* will ever fulfill its original goal. Although it is in German, the *Handbuch* is a series that can be fairly easily used by chemists with no training in this language. The reader is also referred to **E. H. Huntress's** *A Brief Introduction to the use of Beilstein's Handbuch der Organischen Chemie*, 2nd ed. (Wiley, New York, 1938), and to p. 1152 of the text by March listed above. We have also discussed Beilstein in Sect. 2.9.

10.2.2 Laboratory Manuals—Introductory

R. M. Roberts, J. C. Gilbert, L. B. Rodewald, and A. S. Wingrove. *Modern Experimental Organic Chemistry*, 2nd ed. (Holt, Rinehart and Winston, New York, 1974).

J. Landgrebe. *Theory and Practice in Organic Chemistry*, 2nd ed. (D. C. Heath, Lexington, Mass., 1977). A very extensive and highly detailed laboratory manual, with many drawings of lab operations carried out by organic chemists.

J. S. Swinehart. *Organic Chemistry: An Experimental Approach* (Appleton-Century-Crafts, New York, 1969).

L. F. Fieser and K. L. Williamson. *Organic Experiments*, 4th ed. (D. C. Heath, Lexington, Mass., 1979).

10.2.3 Laboratory Manuals—Advanced

K. B. Wiberg. *Laboratory Technique in Organic Chemistry* (McGraw-Hill, New York, 1960).

N. S. Isaacs. *Experiments in Physical Organic Chemistry* (Macmillan, London, 1969).

R. B. Bates and J. P. Schaefer. *Research Techniques in Organic Chemistry* (Prentice-Hall, Englewood Cliffs, N.J., 1971); available in paperback.

M. S. Newman. *An Advanced Organic Chemistry Laboratory Course* (Macmillan, New York, 1972).

10.2.4 Mechanisms

C. K. Ingold. *Structure and Mechanism in Organic Chemistry*, 2nd ed. (Cornell University Press, Ithaca, N.Y., 1969); a very large volume and very well respected. This book, March's book, and le Noble's book (see p. 501) are the best "encyclopedias" to consult when interested in researching organic mechanisms.

E. S. Gould. *Mechanism and Structure in Organic Chemistry* (Holt, New York, 1958); a shorter text than Ingold's (just above), but a good place to learn mechanistic theory, despite its age.

L. N. Ferguson. *The Modern Structural Theory of Organic Chemistry* (Prentice-Hall, Englewood Cliffs, N.J., 1963).

G. W. Wheland. *Advanced Organic Chemistry*, 3rd ed. (Wiley, New York, 1960).

T. H. Lowry and K. S. Richardson. *Mechanism and Theory in Organic Chemistry* (Harper & Row, New York, 1976); a recent and extensive (748 pp.) development of the title subject. It includes chapters on all mechanisms (addition, eliminations, etc.) as well as chapters involving molecular orbital theory and orbital symmetry rules.

R. W. Alder, R. Baker, and J. M. Brown. *Mechanism in Organic Chemistry* (Wiley-Interscience, New York, 1971).

A. A. Frost and R. G. Pearson. *Kinetics and Mechanism*, 2nd ed. (Wiley, New York, 1961).

R. Breslow. *Organic Reaction Mechanisms,* 2nd ed. (Benjamin, New York, 1969); a brief introduction to the title topic and available in paperback.

The list of books described under "Physical Organic Chemistry" (Section 10.2.7) should also be consulted.

10.2.5 Molecular Orbital Theory

A. Liberles. *Introduction to Molecular Orbital Theory* (Holt, Rinehart and Winston, New York, 1966); a very straightforward book for chemists with very little or no background in the title subject. The math involves only algebra and very simple calculus.

M. Orchin and H. H. Jaffe. *Symmetry, Orbitals and Spectra* (Wiley-Interscience, New York, 1971).

M. Hanna. *Quantum Mechanics in Chemistry* (Benjamin, New York, 1965). A more intensive and extensive development than Liberles (see above). The book is very clearly written, and the math involved includes only algebra and simple calculus.

M. J. S. Dewar. *Molecular Orbital Theory for Organic Chemists* (Holt, Rinehart and Winston, New York, 1966); the title may be misleading, as much of this book demands substantial experience in molecular orbital theory. The chapters involving applications to organic systems and reactions are, however, easily comprehended by all organic chemists.

A. Streitwieser. *Molecular Orbital Theory for Organic Chemists* (Wiley, New York, 1961). At a level between the books by Dewar and by Liberles described just above. This book spells out procedures used in Hückel molecular orbital theory and applies these procedures to organic systems. It also involves clear and extensive applications to and correlations with organic reactions. It is unfortunate that this book came out before the development of orbital symmetry applications (Woodward-Hoffmann rules).

J. D. Roberts. *Notes on Molecular Orbital Calculations* (Benjamin, New York, 1962); an excellent, brief introduction to the title subject. It is available in paperback but contains no orbital symmetry material.

H. Zimmerman. *Quantum Mechanics for Organic Chemists* (Academic Press, New York, 1975); a textbook geared to students with both one year of organic chemistry and molecular orbital training. The book includes orbital symmetry treatments.

G. Klopman. *Chemical Reactivity and Reaction Paths* (Wiley-Interscience, New York, 1974).

W. L. Jorgensen. *The Organic Chemists Book of Orbitals* (Academic Press, New York, 1973).

10.2.6 Orbital Symmetry Rules

R. B. Woodward and R. Hoffmann. *The Conservation of Orbital Symmetry* (Academic Press, New York, 1970).

T. L. Gilchrist and R. C. Storr. *Organic Reactions and Orbital Symmetry* (Cambridge University Press, London, 1972).

(Also see the texts by Dewar and by Zimmerman cited in Section 10.2.5.)

10.2.7 Physical Organic Chemistry

L. P. Hammett. *Physical Organic Chemistry*, 2nd ed. (McGraw-Hill, New York, 1968); the author has been referred to as the "father of physical organic chemistry." This book covers all of the title subject except molecular orbital theory.

J. Hine. *Physical Organic Chemistry*, 2nd ed. (McGraw-Hill, New York, 1961); this covers reaction mechanisms, and so on, but does little with molecular orbital theory.

K. Wiberg. *Physical Organic Chemistry* (Wiley, New York, 1964).

L. N. Ferguson. *Organic Molecular Structure* (Willard Grant Press, Boston, 1977).

J. M. Harris and C. C. Wamser. *Fundamentals of Organic Reaction Mechanisms* (Wiley, New York, 1976).

J. Hirsch. *Concepts of Theoretical Organic Chemistry* (Allyn and Bacon, Boston, 1974).

A. Liberles. *Introduction to Theoretical Organic Chemistry* (Macmillan, New York, 1968); an introduction to physical organic chemistry written in a very straightforward fashion.

C. D. Ritchie. *Physical Organic Chemistry: Fundamental Concepts* (Marcel Dekker, New York, 1975).

J. E. Leffler and E. Grunwald. *Rates and Equilibria of Organic Reactions* (Wiley, New York, 1963); this book is useful for applications of kinetics and thermodynamics to organic reaction mechanisms, and it is especially valuable for extra-thermodynamic or "linear" free-energy (e.g., the Hammett σ-ρ equation) concepts.

The subject areas of "Physical Organic Chemistry" (Section 10.2.7) and "Mechanisms" (10.2.4) overlap, and thus the list of books under both subjects should be examined.

10.2.8 Qualitative Organic Analysis

N. D. Cheronis, J. B. Entrikin, and E. M. Hodnett. *Semimicro Qualitative Organic Analysis*, 3rd ed. (Wiley, New York, 1965); a fine alternative and supplemental listing of the "wet" tests described in this book and in Pasto and Johnson (see below). Earlier books by Cheronis and Entrikin are also useful.

D. J. Pasto and C. R. Johnson. *Organic Structure Determination* (Prentice-Hall, Englewood Cliffs, N.J., 1969). This was revised in 1979.

F. Feigl. *Spot Tests in Organic Analysis*, 7th ed. (Elsevier, Amsterdam–New

York, 1966); an extensive list and description of the "wet" tests of organic analysis.

F. L. Schneider. *Qualitative Organic Microanalysis* (Academic Press, New York, 1964).

10.2.9 Reactions (and Reagents)

Organic Reactions (Wiley, New York); this is a continuing series of books that started in 1942. Each chapter thoroughly discusses one reaction (e.g., the Clemmensen reduction) and with emphasis on synthesis and with extensive literature references.

R. B. Wagner and H. D. Zook. *Synthetic Organic Chemistry* (Wiley, New York, 1953; 576 reaction types, 39 chapters, methods, yields, references, physical constants (for compounds with no more than two functional groups).

L. F. Fieser and M. Fieser. *Reagents for Organic Synthesis* (Wiley, New York, 1967–1979). Volume 1 is a valuable, 1457-page listing, in alphabetical order, of reagents, catalysts, solvents, and so on, used for organic laboratory procedures. Volumes 2 through 7 are similarly organized and include new compounds and new information on compounds listed in Volume 1. In the Fieser tradition, extensive amounts of practical information such as references, commercial suppliers, drying procedures, and physical constants are supplied.

The literature areas listed as "Reactions" (Section 10.2.9) and "Synthesis" (Section 10.2.12) overlap, and thus both should be consulted in an extensive literature search.

10.2.10 Spectroscopy (Spectrometry)

R. M. Silverstein, G. C. Bassler, and T. C. Morrill. *Spectrometric Identification of Organic Compounds*, 3rd ed. (Wiley, New York, 1974); a textbook that can be used for those students who have little or no background in ir, nmr, uv, and mass spectrometry. The emphasis is on the integrated use of spectrometric information for organic structure determination.

J. R. Dyer. *Applications of Absorption Spectroscopy of Organic Compounds* (Prentice-Hall, Englewood Cliffs, N.J., 1965); a brief paperback with the same general approach as the text listed just above.

N. B. Colthup, L. H. Daly, and S. E. Wiberly. *Introduction to Infrared and Raman Spectroscopy*, 2nd ed. (Academic Press, New York, 1975).

R. T. Conley. *Infrared Spectroscopy*, 2nd ed. (Allyn and Bacon, Boston, 1972); available in paperback. This extensively outlines ir laboratory techniques and also provides many illustrations and a large number of tables of ir bands that are correlated and organized by organic functional group.

L. J. Bellamy. *The Infrared Spectra of Complex Molecules*, Vol. 1, 3rd ed. (Halsted Press-Wiley, New York, 1975); this and all of Bellamy's earlier books are helpful for analyzing the ir spectra of organic compounds.

K. Nakanishi. *Infrared Absorption Spectroscopy* (Holden-Day, San Francisco, 1962.

L. M. Jackman and S. Sternhell. *Applications of N.M.R. Spectroscopy in Organic Chemistry*, 2nd ed. (Pergamon Press, Elmsford, N.Y., 1969). An extensive expansion of the first edition. The newer edition contains essentially the same introduction to basic principles as the first edition. To this has been added extensive amounts of pmr data. No cmr data or theory is included.

H. Budzikiewicz, C. Djerassi, and D. H. Williams. *Interpretation of Mass Spectra of Organic Compounds* (Holden-Day, San Francisco, 1964).

K. Biemann. *Mass Spectra and Organic Chemical Applications* (McGraw-Hill, New York, 1962).

Introduction to Spectroscopic Methods for the Identification of Organic Compounds. Vols. 1 and 2, edited by F. Scheinmann (Pergamon Press, Elmsford, N.Y., 1970).

E. S. Stern and T. C. J. Timmons. *Electronic Absorption Spectroscopy in Organic Chemistry* (St Martin's Press, New York, 1971); effectively the 3rd edition of Gillam and Stern's earlier book on uv.

G. C. Levy and G. L. Nelson. *Carbon*-13 *Nuclear Magnetic Resonance for Organic Chemists* (Wiley-Interscience, New York, 1972); a brief introduction to the field as well as a source of cmr data.

J. B. Stothers. *Carbon-13 N.M.R. Spectroscopy* (Academic Press, New York, 1972); an extensive and high-level development of the theory and an extremely extensive and valuable source of data (especially chemical shifts).

F. W. Wehrli and T. Wirthlin. *Interpretation of Carbon-13 N.M.R. Spectra* (Heyden, New York, 1976); the title is a little misleading, as the book includes a complete introduction to the theory and measurement of cmr. The extensive development of the interpretation of cmr spectra for structure determination is supplemented by the physical chemistry of the spectral phenomena being scrutinized.

W. Kemp. *Organic Spectroscopy* (Halsted-Wiley, New York, 1974).

Sadtler Standard Spectra. Sadtler Research Laboratories, 3316 Spring Garden St., Philadelphia, Pa. 19104; large collections of prism ir, grating ir, 60 and 100-MHz pmr and cmr spectra are available.

Atlas of Spectral Data and Physical Constants for Organic Compounds. 2nd ed., Vols. 1–6 (CRC Press Inc., Boca Raton, Fla., 1973); this lists spectral data, b.p., m.p., solubility, and densities for 21,000 organic compounds. Data is tabular only, but references to the Sadtler and other collections are given.

Aldrich Library of Infrared Spectra. 2nd ed. (Aldrich Chemical Co., Milwaukee, Wisc, 1975). This provides the ir spectra of 10,000 compounds organized by chemical class. Since several spectra occur on a given page, one can survey the spectral changes associated with structural variations within a chemical class.

L. F. Johnson and W. C. Jankowski. *Carbon-13 N.M.R. Spectra* (Wiley, New York, 1972); this is a collection of the cmr spectra of 500 common organic compounds.

10.2.11 Stereochemistry

E. L. Eliel. *Stereochemistry of Carbon Compounds* (McGraw-Hill, New York, 1962); a thorough basis for the basic principles and a discussion of applications involving conformational analysis and organic reaction mechanisms.

K. Mislow. *Introduction to Stereochemistry* (Benjamin, New York, 1965); an intensive development of the title subject, including symmetry groups and reaction mechanisms. Available in paperback.

E. L. Eliel. *Elements of Stereochemistry* (Wiley, New York, 1969); a paperback book that serves as an introduction at a level that supplements undergraduate textbooks.

10.2.12 Synthesis

Organic Syntheses (Wiley, New York, 1921 to date); this is a collection of organic preparations that have been tested. This is to be contrasted with the untested procedures in the primary literature, which all too frequently cannot be reproduced. Use of this series is supplemented by R. L. Shriner and R. H. Shriner, *Organic Syntheses Collective Volumes I, II, III, IV, V Cumulative Indices* (Wiley, New York, 1976).

C. A. Buehler and D. E. Pearson. *Survey of Organic Syntheses*, 2 Vols. (Wiley, New York, 1970, 1977).

S. R. Sandler and W. Karo. *Organic Functional Group Preparations*, Vols. 1–3 (Academic Press, New York, 1968–1977).

I. T. Harrison and S. Harrison. *Compendium of Synthetic Methods* (Wiley, New York, Vol. 1, 1971; Vol. 2, 1974; Vol. 3, 1977, Ed. by L. Hegedus and L. Wade).

S. Patai (Ed.). *Chemistry of the [Functional Group]* (Wiley, New York), a series of books each outlining the chemistry of a certain functional group (hydroxyl group, ether linkage, etc.)

E. H. Rodd's Chemistry of Carbon Compounds. edited by Coffey (Elsevier, Amsterdam, 1964 to date).

Houben-Weyl. *Methoden der organischen Chemie* (Georg Thieme Verlag, Stuttgart, 1952 to date).

H. O. House. *Modern Synthetic Reactions*, 2nd ed. (Benjamin, New York, 1972).

For additional books that relate to synthesis, the titles listed under the area titled "Reactions (and Reagents)," Section 10.2.9) should be consulted.

10.3 ANALYTICAL CHEMISTRY

L. R. Snyder and J. J. Kirkland. *An Introduction to Modern Liquid Chromatography* (Wiley-Interscience, New York, 1974).

C. Mann, T. Vickers, and W. Gulick. *Instrumental Analysis* (Harper & Row, New York, 1977).

H. H. Willard, L. L. Merritt, jr, and J. A. Dean. *Instrumental Methods of Analysis*, 5th ed. (Van Nastrand. New York, 1974).

B. L. Karger, L. R. Snyder, and C. Horvath. *An Introduction to Separation Science* (Wiley, New York, 1973).

E. Stahl. *Thin Layer Chromatography* (English translation) (Springer-Verlag, New York, 1965); a thorough and well-illustrated book on this subject.

R. C. Crippen. *Gas Chromatography* (McGraw-Hill, New York, 1973); the subject is dealt with under an organization similar to this text; that is, it is organized for organic qualitative analysis.

D. A. Skoog and D. M. West. *Principles of Instrumental Analysis* (Holt, Rinehart and Winston, New York, 1971).

D. A. Skoog and D. M. West. *Fundamentals of Analytical Chemistry*, 2nd ed. (Holt, Rinehart and Winston, New York, 1969).

E. Heftmann. *Chromatography: A Laboratory Handbook of Chromatographic and Electrophoretic Methods* (Van Nostrand-Reinhold, New York, 1975).

N. Hadden et al. *Basic Liquid Chromatography* (Varian Aerograph, Walnut Creek, Calif., 1971).

H. M. McNair and E. J. Bonelli. *Basic Gas Chromatography* (Varian Aerograph, Walnut Creek, Calif., 1968).

L. Meites. *Handbook of Analytical Chemistry* (McGraw-Hill, New York, 1963).

D. G. Peters, J. M. Hayes, and G. M. Heiftje. *Chemical Separations and Measurements* (Saunders, Philadelphia, 1974); a very intensive coverage of "wet" and instrumental analytical methods.

10.4 BIOCHEMISTRY

A. L. Lehninger. *Biochemistry*, 2nd ed. (Worth Pub., New York, N.Y. 1975); a fundamental text of 1104 pages. Contains many problems that can be assigned to students.

S. P. Colowick and N. D. Kaplan. *Methods of Enzymology*, Academic Press, New York, N.Y. This is an on-going series consisting of many volumes.

D. E. Metzler. *Biochemistry* (Academic Press, New York, 1977); this text includes indexed literature references that are omitted by most texts for survey courses.

Handbook of Biochemistry. 3rd ed. by G. D. Fassman, (CRC Press, Cleveland, Ohio, 1976).

H. R. Mahler and E. H. Cordes. *Biological Chemistry*, 2nd ed. (Harper & Row, New York, 1971).

R. M. C. Dawson, D. C. Elliott, W. H. Elliot, and K. M. Jones. *Data for Biochemical Research*, 2nd ed. (Oxford University Press, New York, 1969); an extensive description of "wet" procedures and instrumental procedures used by biochemists accompanied by many literature references to additional procedures.

Tabulated below are functional group systems of importance to biological systems and a guide to references that describes analytical procedures for these systems:

Compound Class	Analysis method/comment	Reference no.[a]
Amino acids (A.A.)	A.A. analysis/automated column chromatography	11
DNA	Colorimetric/diphenyl amine reaction	8
Enzymes	Density gradient centrifugation/sedimentation behavior	3
Lipids	Method for isolation and purification/animal tissues	10
Peptide chains	SDS Gel electrophoresis/no. of chains and molecular weight	1
Proteins	Disc electrophoresis/ acrylamide gels	9
Proteins	Iodination with ^{125}I/ membrane surfaces	2
Proteins	Phenol method/modification of Folin-Ciocalteu method	4
Sugars	Colorimetric/sugars and related substances	6
Sugars	Paper chromatography	7
Sulfhydryl (thiol)	Tissue sulfhydryl groups	5

[a] Numbers correspond to the following references:

1. K. Weber and M. J. Osborn, *J. Biol. Chem.*, **244**, 4406 (1968).
2. J. J. Marchalonis, *Biochem. J.*, **113**, 299 (1969); R. O. Hymes, *Proc. Nat. Acad. Sci. (U.S.A.)*, **70**, 3170 (1973).
3. R. G. Martin and B. Ames, *J. Biol. Chem.*, **236**, 1372 (1961).
4. O. H. Lowry, N. J. Rosebrough, A. L. Farr, and R. J. Randall, *J. Biol. Chem.*, **193**, 265 (1951).
5. G. L. Ellman, *Arch. Biochem. Biophys.*, **82**, 70 (1959).
6. M. Dubois, K. A. Gilles, W. K. Hamilton, P. A. Rebers, and F. Smith, *Anal. Chem.*, **28**, 350 (1956).
7. W. E. Trevelyan, D. P. Proctor and J. S. Harrison, *Nature*, **166**, 444 (1950).
8. K. Burton, *Biochem. J.*, **62**, 315 (1956).
9. B. J. Davis, *Ann. N.Y. Acad. Sci.*, **121**, 404 (1964).
10. J. Folch, M. Lees and G. H. S. Stanley, *J. Biol. Chem.*, **226**, 497 (1967).
11. D. H. Spackman, W. H. Stein, and S. Moore, *Anal. Chem.*, **30**, 1190 (1958).

10.5 GENERAL CHEMISTRY

The Merck Index of Chemicals and Drugs (Merck and Co., Rahway, N.J). The 9th edition came out in 1976. This is an extensive listing of chemicals of biological importance that are cross-indexed by chemical name, generic name, and trade name. The formulas and physical, medical, and toxicological properties are listed for each chemical compound. A list of name organic reactions is also included.

Handbook of Tables for Organic Compounds (CRC Press, Inc., 2000 N.W. 24th St., Boca Raton, Fla.). 3rd ed., 1967. This is an extensive list of b.p. and m.p. data for compounds and m.p. data of their derivatives.

Handbook of Chemistry and Physics (CRC Press Inc., 2000 N.W. 24th St., Boca Raton, Fla.). The 55th edition appeared in 1974–1975.

Dictionary of Organic Compounds, 5 vols. (Oxford University Press, New York, 1965). Names, properties, and references for 40,000 organic compounds.

R. S. Drago. *Physical Methods in Chemistry* (Saunders, Philadelphia, 1977); a very handy book of data, charts, theory, and so on, describing various spectral and other physical-analytical methods.

D. B. Summers. *The Chemistry Handbook: Facts, Figures and Formulas* (Willard Grant Press, Boston, 1970).

10.6 INORGANIC CHEMISTRY

F. A. Cotton and G. Wilkinson. *Advanced Inorganic Chemistry; A Comprehensive Treatise*, 3rd ed. (Wiley-Interscience, New York, 1972); a near-encyclopedic coverage of descriptive and theoretical inorganic chemistry.

K. F. Purcell and J. C. Kotz. *Inorganic Chemistry* (Saunders, Philadelphia, 1977).

J. E. Huheey. *Inorganic Chemistry, Principles of Structure and Reactivity* (Harper & Row, New York, 1972).

F. A. Cotton. *Chemical Applications of Group Theory*, 2nd ed. (Wiley-Interscience, New York, 1971); an excellent introduction to the topic and to simple Hückel molecular orbital theory.

F. Feigl and V. Anger. *Spot Tests for Inorganic Analysis* (English translation by R. E. Oesper), 6th ed. (Elsevier, Amsterdam–New York, 1972).

L. Pauling. *The Nature of the Chemical Bond*, 3rd ed. (Cornell University Press, Ithaca, N.Y., 1960).

D. P. Shoemaker, C. W. Garland and J. I. Steinfeld. *Experiments in Physical Chemistry*, 3rd ed., McGraw-Hill, New York, 1974.

10.7 PHYSICAL CHEMISTRY

G. M. Barrow. *Physical Chemistry*, 3rd ed. (McGraw-Hill, New York, 1973).

E. B. Wilson, J. C. Decius, and P. C. Cross. *Molecular Vibrations: The Theory of Infrared and Raman Vibrational Spectra* (McGraw-Hill, New York, 1955).

G. W. Castellan. *Physical Chemistry*, 2nd ed. (Addison-Wesley, Reading, Mass., 1971).

F. Daniels and R. A. Alberty. *Physical Chemistry*, 4th ed. (Wiley, New York, 1975).

10.8 CHEMICAL LITERATURE

H. M. Woodburn. *Using the Chemical Literature; A Practical Guide.*, (Marcel-Dekker, N.Y., 1974).

L. F. Fieser and M. Fieser, *Style Guide for Chemists,* (Reinhold, N.Y., 1960).

APPENDIX ONE

handy tables for the organic laboratory

Composition and Properties of Common Acids and Bases

	Sp. Gr.	% by Wt.	Moles per Liter	Grams per 100 ml
Hydrochloric acid, conc.	1.19	37	12.0	44.0
Constant-boiling (252 ml conc. acid + 200 ml water, b.p. 110°)	1.10	22.2	6.1	24.4
10% (100 ml conc. acid + 321 ml water)	1.05	10	2.9	10.5
5% (50 ml conc. acid + 380.5 ml water)	1.03	5	1.4	5.2
1 *N* (41.5 ml conc. acid diluted to 500 ml)	1.02	3.6	1	3.7
Hydrobromic acid, constant-boiling (b.p. 126°)	1.49	47.5	8.8	70.8
Hydriodic acid, constant-boiling (b.p. 127°)	1.7	57	7.6	97
Sulfuric acid, conc.	1.84	96	18	177
10% (25 ml conc. acid + 398 ml water)	1.07	10	1.1	10.7
1 *N* (13.9 ml conc. acid diluted to 500 ml)	1.03	4.7	0.5	4.8
Nitric acid, conc.	1.42	71	16	101
Sodium hydroxide, 10% solution	1.11	10	2.8	11.1
Ammonium hydroxide, conc.	0.90	28.4	15	25.6
Phosphoric acid, conc. (syrupy)	1.7	85	14.7	144

Composition of Common Buffer Solutions

pH	Components
0.1	1 *N* Hydrochloric acid
1.1	0.1 *N* Hydrochloric acid
2.2	15.0 g Tartaric acid per liter (0.1 *M* solution)
3.9	40.8 g Potassium acid phthalate per liter
5.0	14.0 g KH-Phthalate + 2.7 g $NaHCO_3$ per liter (heat to expel carbon dioxide, then cool)
6.0	23.2 g KH_2PO_4 + 4.3 g Na_2HPO_4 (anhyd.) per liter
7.0	9.1 g KH_2PO_4 + 18.9 g Na_2HPO_4 per liter
8.0	11.8 g Boric acid + 9.1 g Borax ($Na_2B_4O_7 \cdot 10H_2O$) per liter
9.0	6.2 g Boric acid + 38.1 g Borax per liter
10.0	6.5 g $NaHCO_3$ + 13.2 g Na_2CO_3 per liter
11.0	11.4 g Na_2HPO_4 + 19.7g Na_3PO_4 per liter
12.0	24.6 g Na_3PO_4 per liter (0.15 *M* solution)
13.0	4.1 g Sodium hydroxide pellets per liter (0.1 *M*)
14.0	41.3 g Sodium hydroxide pellets per liter (1 *M*)

Pressure-Temperature Nomograph for Vacuum Distillations

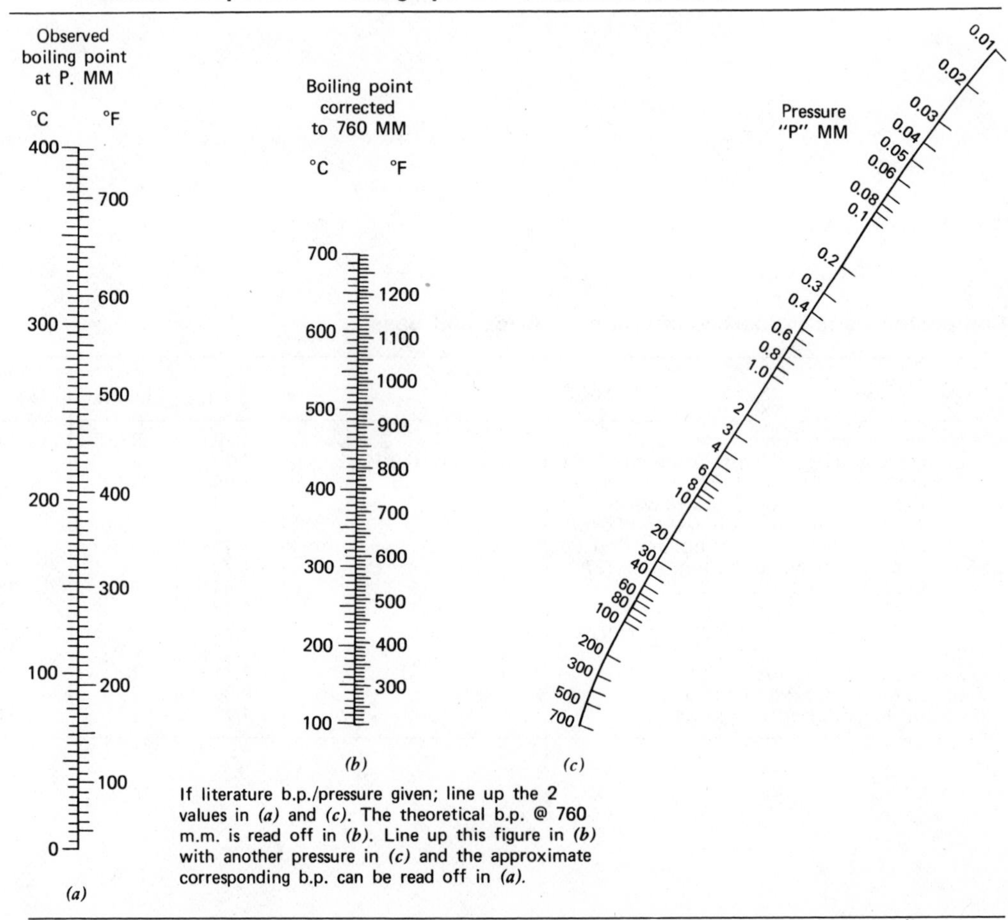

If literature b.p./pressure given; line up the 2 values in *(a)* and *(c)*. The theoretical b.p. @ 760 m.m. is read off in *(b)*. Line up this figure in *(b)* with another pressure in *(c)* and the approximate corresponding b.p. can be read off in *(a)*.

Elution Solvents for Chromatography

Alumina adsorbent		Silica adsorbent
Fluoroalkanes	Increasing polarity* ↓	Cyclohexane
Pentane		Heptane
Isooctane		Pentane
Petroleum ether (light)		Carbon tetrachloride
Hexane		Carbon disulfide
Cyclohexane		Chlorobenzene
Cyclopentane		Ethylbenzene
Carbon tetrachloride		Toluene
Carbon disulfide		
Xylene		Benzene
Di-*i*-propyl ether		2-Chloropropane
Toluene		Chloroform
1-Chloropropane		Nitrobenzene
Chlorobenzene		Di-*i*-propyl ether
		Diethyl ether
Benzene		Ethyl acetate
Ethyl bromide		2-Butanol
Diethyl ether		
Diethyl sulfide		Ethanol
Chloroform		Water
Methylene chloride		Acetone
Tetrahydrofuran		Acetic acid
1,2-Dichloroethane		Methanol
Ethyl methyl ketone		Pyruvic acid
1-Nitropropane		
(Acetone)		
1,4-Dioxane		
Ethyl acetate		
Methyl acetate		
1-Pentanol		
Dimethyl sulfoxide		
Aniline		
Diethylamine		
Nitromethane		
Acetonitrile		
Pyridine		
Butyl cellosolve		
2-Propanol		
1-Propanol		
Ethanol		
Methanol		
Ethylene glycol		
Acetic acid		

* The ability of a solvent to elute chromatographically depends on the compound being eluted as well as the adsorbent; thus, exceptions to these orders will be found. In general, however, as one goes down these lists, one finds solvents of increasing "polarity" that, in principle, more readily elute the more polar compounds.

Supports for Columns for Gas Chromatography

Support	Manufacturer[a]	Support	Manufacturer[a]
	Diatomaceous Earth and Firebrick		
Aeropak	Varian Aerograph	GC 32 (Celatom)	Eagle–Picher
Anakrom	Analabs	GC Super Support	Coast Engineering Lab.
C-22 Firebrick	Johns–Manville	Hyflo	Johns–Manville
Celite 545	Johns–Manville	S-80	R. Grutzmacher
Chromosorb	Johns–Manville	Sil-o-cel Brick	Griffin and George
Diatom	Debton Co.	Sterchamol	Sterchamol–Werke
Diatoport S	Hewlett–Packard	Supelcoport	Supelco
Embacel	May and Baker	Varaport (see Aeropak)	
Gas Chrom	Applied Science Labs.		
	Halocarbons		
Anaport[b]	Analabs	Haloport F[c]	Hewlett–Packard
Chromosorb T[c]	Johns–Manville	Kel-F[b]	3M Company
Fluoropak 80[d]	Fluorocarbon Co.	Tee Six[c]	Analabs
	Porous Polymers[e]		
Chromosorb 100 Series	Johns–Manville	Porapak	Waters Assoc.
	Glass and Silica[f]		
Anaport	Analabs	Corning Porous Glass[g]	Corning Glass Works
Cera Beads	Analabs	Porasil	Waters Assoc.
Corning GLC-Beads	Corning Glass Works	Glassport	Hewlett–Packard

[a] For company addresses, see A. J. Gordon and R. A. Ford, *The Chemist's Companion* (Wiley-Interscience, N.Y., 1972). p. 403.
[b] A chlorofluorocarbon polymer.
[c] Teflon 6 (Dupont).
[d] A fluorocarbon polymer
[e] A polymeric vinyl benzene (used for polar compounds such as acids, etc.).
[f] For high molecular weight compounds at low temperatures.
[g] Vycor 7930 glass.

Salt-Ice Mixtures for Cooling Baths[a]

Substance	Initial temperature (°C)	g salt/100 g H_2O*	Final temperature (°C)
Na_2CO_3	−1 (ice)	20	−2.0
NH_4NO_3	20	106	−4.0
$NaC_2H_3O_2$	10.7	85	−4.7
NH_4Cl	13.3	30	−5.1
$NaNO_3$	13.2	75	−5.3
$Na_2S_2O_3 \cdot 5H_2O$	10.7	110	−8.0
$CaCl_2 \cdot 6H_2O$	−1 (ice)	41	−9.0
KCl	0 (ice)	30	−10.9
KI	10.8	140	−11.7
NH_4NO_3	13.6	60	−13.6
NH_4Cl	−1 (ice)	25	−15.4
NH_4NO_3	−1 (ice)	45	−16.8
NH_4SCN	13.2	133	−18.0
NaCl	−1 (ice)	33	−21.3
$CaCl_2 \cdot 6H_2O$	0 (ice)	81	−21.5
H_2SO_4(66.2%)	0 (ice)	23	−25
NaBr	0 (ice)	66	−28
H_2SO_4(66.2%)	0 (ice)	40	−30
C_2H_5OH(4°)	0 (ice)	105	−30
$MgCl_2$	0 (ice)	85	−34
H_2SO_4(66.2%)	0 (ice)	91	−37
$CaCl_2 \cdot 6H_2O$	0 (ice)	123	−40.3
$CaCl_2 \cdot 6H_2O$	0 (ice)	143	−55

* H_2O means water (*l*), except where ice is listed parenthetically.
[a] Addition of the substance listed in the first column can, but does not always, lower the bath temperature to that listed in the last column. The final temperature may not be quite that low due to insufficiently crushed ice, and so forth.

Liquid Media for Heating Baths*

Medium	M.P. (C)	B.P. (°C)	range (°C)*	Flash point (°C)	Comments
H_2O (*l*)	0	100	0–80	None	Ideal in limited range
Ethylene glycol	−12	197	−10–180	115	Cheap; flammable; difficult to remove from apparatus
Dow Corning 330 Silicone Oil	<−148	—	30–280	290	Becomes viscous at low temperature
20% H_3PO_3, 80% H_3PO_4†	<20	—	20–250	None	Water-soluble; nonflammable; corrosive; steam evolved at high temperature
Triethylene glycol	−5	287	0–250	156	Water-soluble; stable
Dow Corning 550 Silicone Oil	−60	—	0–250	310	Noncorrosive; can be used to 400° in N_2 atmosphere
Glycerol	18	290	−20–260	160	Supercools; water-soluble; viscous
Paraffin	~50	—	60–300	199	Flammable
Dibutyl phthalate	—	340	150–320	—	Viscous at low temperature
Wood's metal (50% Bi, 25% Pb, 12.5% Sn, 12.5% Cd)†	70	—	70–350	—	Oxidizes if used at >250° for long period of time
Tetracresyl silicate	<−48	~440	20–400	—	Noncorrosive; fire-resistant; expensive

* Range of useful temperatures for a bath that is open to the atmosphere; all of these baths except water should be used only in a hood.

† Wt.%.

Solvents for Extraction of Aqueous Solutions*

Solvent	B.P. (°C)	Flammability†	Toxicity†	Comments
Benzene	80.1	3	3	Prone to emulsion;‡ good for alkaloids and phenols from buffered solutions
2-Butanol	99.5	1	3	High-boiling; good for highly polar water-soluble materials from buffered solution
Carbon Tetrachloride	76.5	0	4	Easily dried; good for nonpolar materials
Chloroform	61.7	0	4	May form emulsion; easily dried
Diethyl ether	34.5	4	2	Absorbs large amounts of water; good general solvent
Diisopropyl ether	69	3	2	May form explosive peroxides on long storage; good for acids from phosphate-buffered solutions
Ethyl acetate	77.1	3	1	Absorbs large amount of water; good for polar materials
Freon 11	24	0	1	Freons good for volatile non-polar compounds; fairly expensive
Freon 113	47.7	0	1	
Methylene chloride	40	0	1	May form emulsions; easily dried
n-Pentane	36.1	4	1	Hydrocarbons easily dried; poor solvents for polar compounds
n-Hexane	69	4	1	
n-Heptane	98.4	3	1	

* Data in this table are taken mostly from A. J. Gordon and R. A. Ford, *The Chemists Companion* (Wiley-Interscience, New York, 1972).

† 4 means most toxic or flammable, 4>3>2>1; 0 = nonflammable.

‡ *Emulsions* may form during the extraction of aqueous solutions by organic solvents, making good separation very difficult, if not impossible. Their formation is especially liable to occur if the solution is alkaline; addition of dilute sulfuric acid (if permissible) may break up such an emulsion. The following are general methods for breaking up emulsions: saturation of the aqueous phase with a salt (NaCl, Na_2SO_4, etc.); addition of several drops of alcohol or ether (especially when $CHCl_3$ is the organic layer); centrifugation of the mixture, one of the most successful techniques.

Drying Agents of Moderate Strength for Organic Solvents

Drying agent	Capacity*	Speed†	Comments
$CaSO_4$	$\frac{1}{2}H_2O$	Very fast (1)	Sold commercially as Drierite with or without a color indicator; very efficient. When dry, the indicator ($CoCl_2$) is blue, but turns pink as it takes on H_2O (capacity $CoCl_2 \cdot 6H_2O$); useful in temperature range −50 to +86°. Some organic solvents leach out, or change the color of $CoCl_2$ (acetone, alcohols, pyridine, etc.).
$CaCl_2$	$6\,H_2O$	Very fast (2)	Not very efficient; use only for hydrocarbons and alkyl halides (forms solvates, complexes, or reacts with many nitrogen and oxygen compounds).
$MgSO_4$	$7\,H_2O$	Fast (4)	Excellent general agent; very inert but may be slightly acidic (avoid with very acidic-sensitive compounds). May be soluble in some organic solvents.
Molecular sieve 4A	High	Fast (30)	Very efficient; predrying with a more common agent recommended (see below for details on molecular sieves). Sieve 3A also excellent.
Na_2SO_4	$10\,H_2O$	Slow (290)	Very mild, inefficient, slow, inexpensive, high capacity; good for gross predrying, but do not warm the solution.
K_2CO_3	$2\,H_2O$	Fast	Good for esters, nitriles, ketones and especially alcohols, do not use with acidic compounds.
NaOH, KOH	Very high	Fast	Powerful, but used only with inert solutions in which agent is insoluble; especially good for amines.
H_2SO_4	Very high	Very fast	Very efficient, but use limited to saturated or aromatic hydrocarbons or halides (will remove olefins and other "basic" compounds).
Alumina (Al_2O_3) or silica gel (SiO_2)	Very high	Very fast	Especially good for hydrocarbons. Should be finely divided; can be reactivated after use by heating (300° for SiO_2, 500° for Al_2O_3).

* Moles of water per mole of agent (maximum).

† Relative rating. For first five entries, number in parentheses is relative drying speed for benzene—low number, rapid drying; order may change for slow agents with change in solvent. [B. Pearson and J. Ollerenshaw, *Chem. Ind.*, 370 (1966).]

More Powerful Dehydrating Agents for Organic Liquids

Agent*	Products formed with H_2O	Comments
Na†	$NaOH, H_2$	Excellent for saturated hydrocarbons and ethers; do *not* use with any halogenated compounds.
CaH	$Ca(OH)_2, H_2$	One of the best agents; slower than $LiAlH_4$ but just as efficient and safer. Use for hydrocarbons, ethers, amines, esters, C_4 and higher alcohols (not C_1, C_2, C_3 alcohols). Do *not* use for aldehydes and active carbonyl compounds.
$LiAlH_4$‡	$LiOH, Al(OH)_3, H_2$	Use only with inert solvents [hydrocarbons, aryl (not alkyl) halides, ethers]; reacts with any acidic hydrogen and most functional groups (halo, carbonyl, nitro, etc.). Use caution; excess may be destroyed by slow addition of ethyl acetate.
BaO or CaO	$Ba(OH)_2$ or $Ca(OH)_2$	Slow but efficient; good mainly for alcohols and amines, but should not be used with compounds sensitive to strong base.
P_2O_5	$HPO_3, H_3PO_4, H_4P_2O_7$	Very fast and efficient; very acidic. Predrying recommended. Use only with inert compounds (especially hydrocarbons, ethers, halides, acids, and anhydrides).

* The best dehydrating agents are those that react rapidly and irreversibly with water (and not with the solvent or solutes); they are also the most dangerous and should be used only after gross predrying with a less vigorous drying agent (see below). These agents are almost always used only to dry a solvent prior to and/or during distillation. Although $MgClO_4$ is one of the most efficient drying agents, is not recommended because it can cause explosions if mishandled. See D. R. Burfield, K.-H. Lee, and R. H. Smithers, [J. *Org. Chem.*, **42,** 3060, (1977), for studies of desiccant efficiency.]

† J. T. Baker Co. sells an alloy of 10% Na, 90% Pb called Dri-Na; this dry, granular agent reacts only slowly with air, but is as efficient as Na wire for drying ethers, etc. See L. F. Fieser and M. Fieser, *Reagents*, Vol. 2 (Wiley, New York, 1969), p. 385.

‡ A less dangerous, but equally efficient alternative is $Na(CH_3OCH_2CH_2O)_2AlH_2$, known as Vitride (Realco Chemical Company, 783 Jersey Avenue, New Brunswick, N.J. 08902; available from Eastman Kodak).

Water Detection—Quantative Determination of Small Amounts

Karl Fischer titration is one of the most sensitive techniques available for the actual measurement of very small quantities of water in organic liquids. The Karl Fischer reagent is a solution of iodine, sulfur dioxide, and pyridine usually in methanol as a solvent. Reagent solutions are available commercially (Fischer Scientific; Matheson; A. H. Thomas) as well as titration unit (A. H. Thomas). For more detail, see N. D. Cheronis and T. S. Ma, *Organic Functional Group Analysis* (Wiley-Interscience, New York, 1964), pp. 472–475. An extensive study of the method can be found in J. Mitchell and D. Smith, *Aquametry* (Wiley-Interscience, New York, 1948).

A more rapid and versatile determination of water by glc has been developed which uses a Poropak column [J. Hogan, R. Engel, and H. Stevenson, *Anal. Chem.*, **42,** 249 (1970)]. For details of the technique, write for "Trace Water Analysis in Organics" from Fischer Scientific Company; detection of 1 to 10 ppm water are accomplished with high precision and accuracy.

Other methods of water analysis are reviewed in *Organic Solvents*, Vol. 2 of *Techniques of Chemistry* (Wiley-Interscience, New York, 1971).

APPENDIX TWO

equipment and chemicals for the laboratory

The following representative list of laboratory supplies suitable for a course in identification has been included in response to many inquiries. Variation from one laboratory to another in the equipment available may require much of the equipment suggested here to be modified or omitted.

APPARATUS

Individual Desk Equipment

It has been found convenient to assign each student a standard organic laboratory desk equipped with the usual apparatus required for the first year's work in organic preparations. This is supplemented by a special kit containing apparatus for carrying out classifications tests and preparation of derivatives on a small scale.

Suggested Locker Equipment

5 Beakers, 50 ml, 100 ml, 250 ml, 400 ml, 800 ml

1 Burner, Tirrill, with 3 ft of tubing

1 Burner, micro
1 Burner chimney
1 Burner lighter
1 Burner tip, wing top
1 Clamp, three-prong
2 Clamps, extension, with holder
1 Clamp, screw
1 Clamp, test tube
1 Cork ring, Suberite, 110 mm
1 Cylinder, 10 ml, graduated
1 Cylinder, 100 ml, graduated
1 Filter block
11 Flasks, Erlenmeyer. 2×25 ml, 2×50 ml, 2×125 ml, 2×250 ml, 1×500 ml
2 Flasks, filter, 125 ml, 250 ml
2 Funnels, Buchner, 55 mm, plastic
2 Funnels, Hirsch, 000, 0000
1 Funnel, separatory, 250 ml
1 Funnel, stemless, 75 mm
1 Funnel, short stem
2 Rings, 4 in. & 2 in., iron
3 ft sections of rubber tubing, pressure
6 Test tubes, 13×100 mm
6 Test tubes, 16×150 mm
1 Thermometer, −5–360°C
1 Tongs, pair crucible
2 Watch glasses, 100 mm

Suggested Supplementary Kit

1 Pipet, graduated 1 ml
1 Pipet, graduated 10 ml
5 Disposable pipets
1 Filter flask, 50 ml
1 Stainless steel spatula, small
1 Porcelain crucible and cover, no. 00
1 Desiccator, small
1 Flask, distilling, 25 ml[1]
1 Flask, distilling, 10 ml[1]
2 Side-arm test tubes, Pyrex, 15×125 mm
1 Side-arm test tubes, Pyrex, 20×150 mm

General Laboratory Equipment

Community supply of iron ware. Ring stands, all types of clamps, clamp holders.

1 Drying oven, electric, constant temperature, 20 to 250°

1 Drying rack, hot air, for glassware

1 Large desiccator, vacuum, with vacuum pump

1 Triple-beam balance, Ohaus, capacity 500 g on center beam, 0–10 g on front beam, 0–100 g on back beam

---

[1] 1 Kit ground-glass Distillation equipment.

1 Electrically heated metal block with 500° thermometer or a Mel-Temp capillary melting-point block with 500° thermometer.
Glass tubing and rods, assorted

Special Laboratory Equipment

The following equipment should be kept free from corrosive fumes.

Quantitative analytical balances, chainomatic or Mettler H5. About one balance should be available for each five students.

1 Refractometer, Bausch & Lomb, Abbe type 3L or Valentine (Industro-Scientific Co., Philadelphia, Pa.).

1 Polarimeter, Rudolph laboratory model 63 with sodium lamp and 1 dcm and 2 dcm polarimeter tubes.

1 Gravitometer, Fisher–Davidson.

1 Infrared spectrophotometer with all accessories. The Infracord 137 model of Perkin–Elmer Corp. is suitable.

Items Obtained by Temporary Loan from Instructor or Storeroom

Platinum wire, about 2 in. long, no. 20, sealed into a glass rod
Platinum foil, about 1 in. square, sealed into a glass rod
Polarimeter tubes, one 1 dcm, one 2 dcm long
Nmr Tubes, TMS and solvents (CCl_4, $CDCl_3$, acetone-d_6, DMSO-d_6, and D_2O and DSS.)
Ir cells, mulling oils, and calibration window
Small agate mortar and pestle
Cylinder of Dry Nitrogen Gas

CHEMICALS FOR LABORATORY SHELF

Organic Compounds

The following chemicals are useful for carrying out solubility and classification tests and for preparing some (but not all) derivatives. It has been found convenient to provide a set of bottles of about 100 ml capacity, using wide-mouth g.s. bottles for solids and tincture-mouth g.s. bottles for liquids. Larger bottles (about 500 ml) should be used for common solvents such as acetone, benzene, chloroform, carbon tetrachloride, dioxane, ether, ligroin (70–90°), and toluene. For a class of 20 students, about 20 to 50 g of the organic compounds may be placed in these shelf bottles, although, for economy in purchasing chemicals, they may be bought in units of 50, 100, or 500 g. The actual amounts needed per student will naturally vary with the nature of the unknowns, the intelligence with which the classification tests are selected, and the manipulative skill of the student. Everyone is reminded that many of these compounds are toxic (Appendix Four can be consulted).

Acetanilide
Acetic anhydride
Acetone
Acetonitrile
Acetophenone
Allyl alcohol
Allyl chloride
p-Aminophenol
Ammonium benzoate
Aniline
Aniline hydrochloride
Anisole
Azoxybenzene
Benzaldehyde
Benzamide
Benzene
Benzidine
Benzoic acid
Benzonitrile
Benzophenone
Benzyl alcohol
Benzylamine
Benzylchloride
S-Benzylthiuronium chloride
Biphenyl
p-Bromoaniline
Bromobenzene
p-Bromobenzenesulfonyl chloride
p-Bromobenzyl bromide
p-Bromobenzyl chloride
S-(*p*-Bromobenzyl)thiuronium chloride
p-Bromophenacyl bromide
p-Bromophenylhydrazine
p-Bromophenyl isocyanate
n-Butyl alcohol
sec-Butyl alcohol
t-Butyl alcohol
n-Butyl bromide
sec-Butyl bromide
t-Butyl bromide
n-Butyl ether
n-Butyraldehyde
Camphor
Carbon disulfide
Carbon tetrachloride
Chloroacetic acid
α-Chloroacetophenone
Chlorobenzene
p-Chlorobenzhydrazide
S-(*p*-Chlorobenzyl)thiuronium chloride
Chloroform
p-Chloronitrobenzene
p-Chlorophenacyl bromide
Cholesterol
Cinnamic acid
Cyclohexane
p,p'-Diaminodiphenylmethane
p-Dibromobenzene
Diethylamine
Diethylene glycol
p-Dimethylaminobenzaldehyde
Dimethylaniline
Dimethylsulfoxide
2,4-Dinitroaniline
m-Dinitrobenzene
o-Dinitrobenzene
3,5-Dinitrobenzoic acid
3,5-Dinitrobenzoyl chloride
2,4-Dinitrochlorobenzene
2,4-Dinitrophenylhydrazine
3,5-Dinitrophenyl isocyanate
p-Dioxane
Diphenylcarbamyl chloride
Ethyl acetate
Ethyl acetoacetate
Ethyl benzoate

Ethyl bromide
Ethylene bromide
Ethylene glycol
Formalin
Formic acid
Fructose
Galactose
Gardinol (sodium lauryl sulfate)
Glucose
Glycerol
n-Heptaldehyde
n-Heptyl alcohol
Hexane
Hydroquinone
p-Hydroxybenzoic acid
Hydroxylamine hydrochloride
Iodic acid reagent
p-Iodophenacyl bromide
Isoamyl alcohol
Isopropyl alcohol
Lactic acid (85%)
Lactose
Lauryl alcohol
Ligroin (70–90°, pet. ether)
Maleic anhydride
Maltose
Mercaptoacetic acid (thioglycolic acid)
Mesitylene
Methanesulfonyl chloride
Methone
Methyl alcohol
Methylaniline
Methyl benzoate
Methyl cellosolve (2-methoxyethanol)
Methyl iodide
Methyl *p*-toluenesulfonate
Naphthalene
α-Naphthalenesulfonyl chloride
β-Naphthol
α-Naphthoquinone
β-Naphthylhydrazine
β-Naphthyl isocyanate
α-Naphthyl isothiocyanate
m-Nitroaniline
p-Nitroaniline
p-Nitrobenzaldehyde
p-Nitrobenzazide
Nitrobenzene
m-Nitrobenzenesulfonyl chloride
m-Nitrobenzhydrazide
p-Nitrobenzoic acid
p-Nitrobenzoyl chloride
p-Nitrobenzyl chloride
Nitromethane
p-Nitrophenol
p-Nitrophenylhydrazine
p-Nitrophenyl isocyanate
3-Nitrophthalic anhydride
Oxalic acid
Pentane
2-Pentanol
2-Pentene
Petroleum ether (40–60°)
Phenol
o-Phenylenediamine
α-Phenylethylamine
Phenylhydrazine
Phenylhydrazine hydrochloride
p-Phenylphenacyl bromide
Phenylsemicarbazide
Phloroglucinol
Phthalic anhydride
Phthalimide
Picric acid
Piperazine
Piperidine
i-Propyl alcohol
n-Propyl alcohol
Pseudosaccharine chloride
Pyridine (over potassium hydroxide)
Resorcinol
Salicylaldehyde
Salicyclic acid
Semicarbazide hydrochloride
Sucrose
Sulfanilic acid
Tartaric acid
Tetrachlorophthalic anhydride
Tetrahydrofuran

Thiourea
Toluene (anhyd.)
p-Toluenesulfonic acid
p-Toluidine
p-Tolylsemicarbazide
Triethanolamine
Triethylamine
1,3,5-Trinitrobenzene
Trityl chloride
Urea
Xanthydrol
Xylene

Inorganic Compounds

The amounts suggested are suitable for a class of 20 students, although some of the bottles of the more commonly used reagents may have to be refilled several times in a semester. Many of these compounds are toxic and this subject is discussed in Appendix Four.

Compound	Amount
Ammonium carbonate	100 g
Ammonium chloride	100 g
Calcium chloride (anhyd.)	1 kg
Chromium trioxide	100 g
Copper sulfate (anhyd.)	50 g
Ferrous ammonium sulfate	100 g
Hydrazine hydrate	25 g
Hydrazine sulfate	50 g
Hydriodic acid (57%)	100 g
Iodine (resublimed)	100 g
Iron powder	100 g
Lead acetate	150 g
Lead peroxide	100 g
Magnesium sulfate (anhyd.)	1 kg
Magnesium turnings	100 g
Mercuric bromide	10 g
Mercuric iodide	10 g
Mercuric nitrate	100 g
Mercuric oxide	50 g
Mercuric sulfate	50 g
Phosphoric acid (85%)	1 kg
Potassium bromide	100 g
Potassium carbonate (anhyd.)	1 kg
Potassium cyanide	100 g
Potassium hydroxide (pellets)	1 kg
Potassium iodide	100 g
Potassium permanganate	200 g
Potassium persulfate	25 g
Potassium thiocyanate	100 g
Sodium acetate (anhyd.)	200 g
Sodium acetate (cryst.)	100 g
Sodium bicarbonate	1 kg
Sodium bisulfite (anhyd.)	100 g
Sodium bisulfite	1 kg
Sodium borohydride	100 g
Sodium carbonate (anhyd.)	1 kg
Sodium chloride	1 kg
Sodium cyanide (**toxic**)	500 g
Sodium dichromate	1 kg
Sodium hydroxide (pellets)	1 kg
Sodium metal (under toluene)	50 g
Sodium nitrate	100 g
Sodium nitrite	200 g
Sodium nitroprusside	5 g
Sodium sulfate (anhyd.)	1 kg
Sodium sulfide (cryst.)	50 g
Sodium sulfide nonahydrate	100 g
Sodium thiosulfate (cryst.)	100 g
Stannous chloride	100 g
Tin (granular)	200 g
Zinc (dust)	200 g
Zinc chloride (anhyd.)	100 g

Miscellaneous Compounds, Reagents, and Solutions

It is convenient to make up 500 to 1000 ml of the solutions and keep them on a side shelf or table in the laboratory. Directions for their preparation are given in Chapter 6 and in textbooks on qualitative and quantitative inorganic chemistry.

Aluminum/Nickel alloy (W. R. Grace)
Ammonium hydroxide solution (2%)
Benedict's reagent
Bogen's indicator solution
Boron trifluoride bis (acetic acid) complex (See Chpt. 6, Proc. 51a)
Bromine in carbon tetrachloride solution (5%)
Bromine-water (saturated)
tetra-n-Butylammonium bromide
Calcium chloride solution (saturated)
Ceric nitrate reagent [need $(NH_4)_2Ce(NO_3)_6$]
Chlorine water (fresh)
Congo red paper
Cotton
Cottonseed oil
Darco (decolorizing carbon)
2,4-Dinitrophenylhydrazine reagent
Fehling's solution no. 1
Fehling's solution no. 2
Ferric chloride solution (5%)
Ferrous ammonium sulfate solution (50 g/liter)
Fuchsin-aldehyde reagent
Glass wool
Universal indicator solution (Eastman A 4953)
Hydrochloric acid, standard solution (0.25 *N*) (5-gal carboy)
Hydrochloric acid–zinc chloride solution (Lucas' reagent)
Hydrogen peroxide (3%)
Hydroxylamine hydrochloride reagent
Iodine-potassium iodide solution
Kieselguhr
Lead acetate solution (1%)
Magnesium-potassium carbonate mixture
Mercuric nitrate
Mineral oil (Stanolind)
Nickel chloride-carbon disulfide reagent
Nickel chloride-5-nitrosalicylaldehyde reagent
Nitric acid, dilute (5%)
Norit (decolorizing carbon)
Periodic acid reagent
Phenolphthalein indicator solution (1% in 95% ethanol)
Phosphoric acid solution (20%)

Potassium fluoride solution (20%)
Potassium hydroxide, alcoholic (10%)
Potassium permanganate solution (2%)
Silver nitrate, alcoholic (2%)
Silver nitrate, aqueous (5%)
Sodium bicarbonate solution (5%)
Sodium bisulfite, aqueous-alcoholic
Sodium carbonate solution (10%)
Sodium ethoxide, alcoholic (1%)
Sodium hydroxide, standard solution (0.1 *N*)(5-gal carboy)
Sodium hydroxide solution (20%)
Sodium iodide in acetone
Sodium nitroprusside
Sodium plumbite
Starch-iodide test paper
Sulfuric acid, dilute (10%)
Thymol blue indicator solution (0.2% in 95% ethanol)
Zirconium-alizarin test paper

Chemicals Kept Under a Hood

Acetyl chloride
Aluminum chloride (anhyd.)
Ammonium polysulfide
Benzenesulfonyl chloride
Benzoyl chloride
Bromine
Chlorosulfonic acid
Hydrogen sulfide water
α-Naphthyl isocyanate
α-Naphthyl isothiocyanate
Nitric acid (fuming)
Phenacyl bromide
Phenyl isocyanate
Phenyl isothiocyanate
Phosphorus oxychloride
Phosphorus pentachloride
Phosphorus trichloride
Phthaloyl chloride
Sulfuric acid (fuming)
Thionyl chloride
p-Toluenesulfonyl chloride
p-Tolyl isocyanate

Reagent-grade chemicals in 250-ml g.s. bottles
- Acetic acid, glacial
- Ammonium hydroxide, sp gr 0.90, 28% NH_3
- Hydrochloric acid, concentrated sp gr 1.19, 37% HCl
- Hydrochloric acid solution, 10%
- Nitric acid, concentrated, sp gr 1.42, 70% HNO_3
- Sodium hydroxide solution, 10%
- Sulfuric acid, concentrated, sp gr 1.84, 95% H_2SO_4

Special Chemicals

Ethanol 95%
Ethanol, absolute
Ethyl ether, absolute
Chloroplatinic acid

Unknowns

Compounds for use as unknowns should be carefully selected for purity, and typical examples may be chosen from the tables in Appendix 3, but not necessarily limited to the compounds listed there. About 5 to 10 g of a solid and 10 to 15 ml of a liquid are sufficient. Mixtures should also contain 5 to 10 g of a solid and 10 to 15 ml of a liquid plus larger amounts of a suitable solvent.

As the experience and technique of the student improve toward the end of the semester, much smaller amounts of unknowns may be given and the solubility, classification tests, and preparation of derivatives carried out on a much smaller scale (about one-tenth the amounts specified in the experiments and procedures).

APPENDIX FOUR

toxicity of organic compounds

Recently the area of the toxicity of organic compounds has received a great deal of attention. For this reason we include the four tables that follow. Although this is hardly a comprehensive treatment, we believe that it is at least a preliminary effort in dealing with a serious problem.

The first table, entitled "Allowable Exposure to Common Compounds," is a list of a few hundred organic compounds and the time-weighted averages for them. These averages correspond to the maximum exposure to the compounds allowed by Occupational Safety and Hazard Administration (OSHA) for an 8-hour work period of a 40-hour work week. They can be calculated by using the following formula, where the exposure, E, is related to the concentration, C_i, during some time period i, of duration t_i hours;

$$E = \frac{C_a t_a + C_b t_b + \cdots C_n t_n}{8}$$

Thus, although one can be exposed to acetaldehyde of a concentration greater than 200 ppm, those periods of exposure must be counterbalanced by exposure to lower concentrations so that the 200 ppm average is maintained for the 8-hour period.

The second table, entitled "Ceiling Exposure Values for Organic Compounds," lists the compounds for which at no time shall the exposure exceed the value listed in the table. A survey of those listed will show that they are often compounds that are in structural classes on the fourth table in this series (see below).

A third table entitled "Allowable Exposure, Ceiling Exposure and Maximum Duration of Brief Exposure to Especially Toxic Organic Compounds," lists the 8-hour weighted averages and the ceiling concentrations acceptable for highly toxic compounds, such as benzene. It also lists the length of time for which a person may be exposed to one of these compounds at a somewhat higher concentration. For example, a chemist can be exposed to 2000 ppm of methylene chloride for 5 minutes in a 2-hour period, 1000 ppm for a longer period, but must not be exposed, on the average, to more than 500 ppm for an 8-hour period.

The fourth table, "Chemical Classes of Toxic Compounds," lists the names or structures or abbreviations for compounds, largely organic, that are well established to be toxic. One of the major purposes of this table is to assist chemists in developing a feeling for those classes of compounds that are very frequently found to be toxic. Thus, exceptional care should be exercised in handling any representative of these classes.

These data are taken from lists set by the Occupational Safety and Health Administration and the National Institute of Occupational Safety and Health (NIOSH), Cincinnati, and such data are subject to change; thus more recent releases by these government agencies should be consulted. A good general book for this subject is *Safety in Working with Chemicals*, by M. E. Green and A. Turk, MacMillan, New York, 1978. This book covers many physical hazards as well as chemical toxicity associated with the chemistry laboratory.

Table 1. Allowable Exposure to Common Compounds

Substance	p.p.m.[a]	mg./m^{3}[b]
Acetaldehyde	200	360
Acetic acid	10	25
Acetic anhydride	5	20
Acetone	1000	2400
Acetonitrile	40	70
Acetylene tetrabromide	1	14
Acrolein	0.1	0.25
Acrylamide–Skin		0.3
Acrylonitrile–Skin	20	45
Aldrin–Skin		0.25
Allyl alcohol–Skin	2	5
Allyl chloride	1	3
Allyl propyl disulfide	2	12
2-Aminoethanol, see Ethanolamine		
2-Aminopyridine	0.5	2
Ammonia	50	35
n-Amyl acetate	100	525
sec-Amyl acetate	125	650
Aniline–Skin	5	19
Anisidine (*o*, *p*-isomers)–Skin		0.5
p-Benzoquinone, see Quinone		
Benzoyl peroxide		5
Benzyl chloride	1	5
Biphenyl	0.2	1
Boron trifluoride	1	3
Bromoform–Skin	0.5	5
Butadiene (1, 3-butadiene)	1000	2200
2-Butanone	200	590
2-Butoxyethanol (Butyl Cellosolve)-Skin	50	240
Butyl acetate (n-butyl acetate)	150	710
sec-Butyl acetate	200	950
tert-Butyl acetate	200	950
Butyl alcohol	100	300
sec-Butyl alcohol	150	450
tert-Butyl alcohol	100	300
n-Butyl glycidyl ether (BGE)	50	270
Butyl mercaptan	10	35
p-tert-Butyltoluene	10	60
Camphor	2	....
Carbon black		3.5
Carbon monoxide	50	55
Chlordane–Skin		0.5
Chlorinated camphene–Skin		0.5
Chlorinated diphenyl oxide		0.5
Chlorine	1	3
Chlorine dioxide	0.1	0.3
α-Chloroacetophenone (phenacylchloride)	0.05	0.3
Chlorobenzene	75	350
o-Chlorobenzylidenemalononitrile (OCBM)	0.05	0.4
Chlorobromomethane	200	1050
2-Chloro-1,3-butadiene, see Chloroprene		
1-Chloro-2,3-epoxypropane, see Epichlorohydrin		
2-Chloroethanol, see Ethylene chlorohydrin		
Chloroethylene, see Vinyl chloride		
1-Chloro-1-nitropropane	20	100
Chloropicrin	0.1	0.7
Chloroprene (2-chloro-1,3-butadiene)–Skin	25	90
Coal tar pitch volatiles (benzene soluble fraction)		
anthracene, benzo[*a*]pyrene, phenanthrene, acridine, chrysene, pyrene		0.2

See footnotes at end of table.

Table 1. Allowable Exposure to Common Compounds—Continued

Substance	p.p.m.[a]	mg./m[3b]
Cresol (all isomers)–Skin	5	22
Crotonaldehyde	2	6
Cumene–Skin	50	245
Cyclohexane	300	1050
Cyclohexanol	50	200
Cyclohexanone	50	200
Cyclohexene	300	1015
Cyclopentadiene	75	200
2,4-D		10
DDT–Skin		1
Diacetone alcohol (4-hydroxy-4-methyl-2-pentanone)	50	240
1,2-Diaminoethane, see Ethylenediamine		
Diazomethane	0.2	0.4
Dibutyl phthalate		5
p-Dichlorobenzene	75	450
Dichlorodifluoromethane	1000	4950
1,1-Dichloroethane	100	400
1,2-Dichloroethylene	200	790
Dichloromethane, see Methylene chloride		
Dichlorofluoromethane	1000	4200
1,2-Dichloropropane, see Propylene dichloride		
Dichlorotetrafluoroethane	1000	7000
Dieldrin–Skin		0.25
Diethylamine	25	75
2-Diethylaminoethanol–Skin	10	50
Diethylether, see Ethyl ether		
Difluorodibromomethane	100	860
Diisobutyl ketone	50	290
Diisopropylamine–Skin	5	20
Dimethoxymethane, see Methylal		
Dimethylacetamide–Skin	10	35
Dimethylamine	10	18
Dimethylaniline (*N,N*-dimethylaniline)–Skin	5	25
Dimethylbenzene, see Xylene		
Dimethylformamide–Skin	10	30
2,6-Dimethyl-4-heptanone, see Diisobutyl ketone		
1,1-Dimethylhydrazine–Skin	0.5	1
Dimethyl phthalate		5
Dimethylsulfate–Skin	1	5
Dinitrobenzene (all isomers)–Skin		1
Dinitro-*o*-cresol–Skin		0.2
Dinitrotoluenes–Skin		1.5
Dioxane–Skin	100	360
Diphenyl		
Epichlorohydrin–Skin	5	19
Ethanolamine	3	6
2-Ethoxyethanol–Skin	200	740
Ethyl acetate	400	1400
Ethyl acrylate–Skin	25	100
Ethyl alcohol (ethanol)	1000	1900
Ethylamine	10	18
Ethyl *sec*-amyl ketone (5-methyl-3-heptanone)	25	130
Ethylbenzene	100	435
Ethyl bromide	200	890
Ethyl butyl ketone (3-Heptanone)	50	230
Ethyl chloride	1000	2600
Ethyl ether	400	1200

See footnotes at end of table.

Table 1. Allowable Exposure to Common Compounds—Continued

Substance	p.p.m.[a]	mg./m[3b]
Ethyl formate	100	300
Ethylene chlorohydrin–Skin	5	16
Ethylenediamine	10	25
Ethylenimine–Skin	0.5	1
Ethylene oxide	50	90
Fluorotrichloromethane	1000	5600
Formic acid	5	9
Furfural–Skin	5	20
Furfuryl alcohol	50	200
Heptane (*n*-heptane)	500	2000
Hexachloroethane–Skin	1	10
Hexachloronaphthalene–Skin		0.2
Hexane (*n*-hexane)	500	1800
2-Hexanone	100	410
Hexone (Methyl isobutyl ketone)	100	410
sec-Hexyl acetate	50	300
Hydroquinone		2
Isoamyl acetate	100	525
Isoamyl alcohol	100	360
Isobutyl acetate	150	700
Isobutyl alcohol	100	300
Isophorone	25	140
Isopropyl acetate	250	950
Isopropyl alcohol	400	980
Isopropylamine	5	12
Isopropyl ether	500	2100
Maleic anhydride	0.25	1
Mesityl oxide	25	100
Methyl acetate	200	610
Methylacetylene (propyne)	1000	1650
Methyl acrylate–Skin	10	35
Methylal (dimethoxymethane)	1000	3100
Methyl alcohol (methanol)	200	260
Methylamine	10	12
Methyl *n*-amyl ketone (2-Heptanone)	100	465
Methyl butyl ketone, see 2-Hexanone		
Methyl Cellosolve–Skin	25	80
Methyl Cellosolve acetate–Skin	25	120
Methylcyclohexane	500	2000
Methylcyclohexanol	100	470
Methyl ethyl ketone (MEK), see 2-Butanone		
Methyl formate	100	250
Methyl iodide–Skin	5	28
Methyl isobutyl carbinol–Skin	25	100
Methyl isocyanate–Skin	0.02	0.05
Methyl methacrylate	100	410
Methyl propyl ketone, see 2-Pentanone		
Morpholine–Skin	20	70
Naphtha (coalter)	100	400
Naphthalene	10	50
α-naphthylthiourea		0.3
Nickel carbonyl	0.001	0.007
Nicotine–Skin		0.5
p-Nitroaniline–Skin	1	6
Nitrobenzene–Skin	1	5
p-Nitrochlorobenzene–Skin		1
Nitroethane	100	310
Nitroglycerin–Skin	0.2	2
Nitromethane	100	250
1-Nitropropane	25	90
2-Nitropropane	25	90
Nitrotoluenes–Skin	5	30

See footnotes at end of table.

Table 1. Allowable Exposure to Common Compounds—Continued

Substance	p.p.m.[a]	mg./m^3[b]
Nitrotrichloromethane, see Chloropicrin		
Octachloronaphthalene–Skin		0.1
Octane	500	2350
Oxalic acid		1
Pentachlorophenol–Skin		0.5
Pentane	1000	2950
2-Pentanone	200	700
Perchloromethyl mercaptan	0.1	0.8
Petroleum distillates (naphtha)	500	2000
Phenol–Skin	5	19
p-Phenylenediamine–Skin		0.1
Phenyl ether (vapor)	1	7
Phenylhydrazine–Skin	5	22
Phosgene (carbonyl chloride)	0.1	0.4
Phthalic anhydride	2	12
Picric acid–Skin		0.1
Platinum (soluble salts) as Pt		0.002
Propargyl alcohol–Skin	1	
Propane	1000	1800
n-Propyl acetate	200	840
Propyl alcohol	200	500
n-Propyl nitrate	25	110
Propylene dichloride	75	350
Propylenimine–Skin	2	5
Propylene oxide	100	240
Pyridine	5	15
Quinone	0.1	0.4
Rhodium, soluble salts		0.001
Selenium compounds (as Se)		0.2
Silver, soluble compounds		0.01
Strychnine		0.15
2,4,5-T		10
1,1,1,2-Tetrachloro-2,2-difluoroethane	500	4170
1,1,2,2-Tetrachloro-1,2-difluoroethane	500	4170
1,1,2,2-Tetrachloroethane–Skin	5	35
Tetraethyllead (as Pb)–Skin		0.075
Tetrahydrofuran	200	590
Tetramethylsuccinonitrile–Skin	0.5	3
Tetranitromethane	1	8
Thallium (soluble compounds)–Skin as Tl		0.1
Tin (organic cmpds)		0.1
o-Toluidine–Skin	5	22
1,1,1-Trichloroethane	350	1900
1,1,2-Trichloroethane–Skin	10	45
1,2,3-Trichloropropane	50	300
1,1,2-Trichloro-1,2,2-trifluoroethane	1000	7600
Triethylamine	25	100
Trinitrotoluene–Skin		1.5
Uranium (soluble compounds)		0.05
Vinyltoluenes	100	480
Xylenes	100	435

[a] Parts of gas per million parts of contaminated air by volume at 25°C, and 760 mm Hg.

[b] Milligrams of particulates per cubic meter of air.

Source: taken from Table Z-1, OSHA 2077, General Industry Standards.

Table 2. Ceiling Exposure Values for Organic Compounds

Substance	ppm[a]	mg/m^3 [b]
Allyl glycidyl ether	10	45
Butylamine	5	15
Chloroacetaldehyde	1	3
Chloroform	50	240
o-Dichlorobenzene	50	300
Dichloroethyl ether—*skin*	15	90
1,1-Dichloro-1-nitroethane	10	60
Diglycidyl ether	0.5	2.8
Ethyl mercaptan	10	25
Methylhydrazine	0.2	0.35
Methyl mercaptan	10	20
Terphenyl	1	9
Vinyl Chloride	500	1300

[a] Parts of gas permillion parts of contaminated air by volume at 25°C, and 760 mm Hg.
[b] Milligrams of particulates per cubic meter of air Source: taken from Table Z-1, OSHA 2077, General Industry Standards.

Table 3. Allowable Exposure, Ceiling Exposure and Maximum Duration of Brief Exposure to Especially Toxic Organic Compounds

Material	8-hour time weighted average	Ceiling exposure concentration	Acceptable brief exposure above the ceiling concentration	
			Concentration	Maximum duration
Benzene	1 p.p.m.	5 p.p.m.		
Carbon disulfide	20 p.p.m.	30 p.p.m.	100 p.p.m.	30 min.
Carbon tetrachloride	10 p.p.m.	25 p.p.m.	200 p.p.m.	5 min. in any 4 hours
Ethylene dibromide	20 p.p.m.	30 p.p.m.	50 p.p.m.	5 min.
Ethylene dichloride	50 p.p.m.	100 p.p.m.	200 p.p.m.	5 min. in any 3 hours
Formaldehyde	3 p.p.m.	5 p.p.m.	10 p.p.m.	30 min.
Lead and its inorganic compounds	$0.2\ mg/m^3$			
Methyl chloride	100 p.p.m.	200 p.p.m.	300 p.p.m.	5 min. in any 3 hours
Methylene chloride	500 p.p.m.	1000 p.p.m.	2000 p.p.m.	5 min. in any 2 hours
Organo (alkyl) mercury	$0.01\ mg/m^3$	$0.04\ mg/m^3$		
Styrene	100 p.p.m.	200 p.p.m.	600 p.p.m.	5 min. in any 3 hours
Trichloroethylene	100 p.p.m.	200 p.p.m.	300 p.p.m.	5 min. in any 2 hours
Tetrachloroethylene	100 p.p.m.	200 p.p.m.	300 p.p.m.	5 min. in any 3 hours
Toluene	200 p.p.m.	300 p.p.m.	500 p.p.m.	10 min.
Hydrogen sulfide		20 p.p.m.	50 p.p.m.	10 min. only if no other measurable exposure
Mercury		$1\ mg/10\ m^3$		

Table 4. Chemical Classes of Toxic Compounds

Aliphatic Amines
Aromatic Hydrocarbons
  benzene, toluene, xylenes, naphthalene, anthracene, etc.
Aromatic Amines
  anilines, naphthylamines, benzidines, etc.
Chlorinated Aliphatic Hydrocarbons
  $CHCl_3$, CCl_4, CH_3CCl_3, $CHCl_2CH_2Cl$, $Cl_2C{=}CHCl$, $CH_2{=}CHCl$
Chlorinated Aromatic Hydrocarbons
  DDT, PCB's, 2, 4, 5-T, 2, 4-D
Diazo Compounds
Hydrazines
  NH_2NH_2, arylhydrazines, alkylhydrazines
Inorganic Compounds
  Compounds of the elements As, Be, Bi, Cd, Co, Cr,
  Halogens, Hg, Mn, Ni, P, Pb, Se, Te, Tl, Th, V
Methyl fluorosulfonate ("Magic Methyl")
4-Nitrobiphenyl
Nitrogen Mustards $[RN(CH_2CH_2Cl)_2]$
N-Nitroso Compounds

INDEX

Numbers in parentheses are boiling points. Numbers in square brackets are melting points. All other numbers refer to pages.

The nomenclature is quite varied. Both common names and systematic names are used. The student should recognize the fact that there are many variations in everyday usage and consult a textbook or the indices of *Chemical Abstracts* in order to locate structures corresponding to the names.

The main compounds in the tables of Appendix III are indexed (with melting points, boiling points, and page numbers), but the derivatives are not indexed. The procedures for making derivatives in Chapter 6 are indexed, but the numerous special derivatives and reactions are not indexed. These may be found by consulting the references cited under each class of compounds discussed in Chapter 6.